URBAN PLANNING THEORY

COMMUNITY DEVELOPMENT SERIES
Series Editor: Richard P. Dober, AIP

Volumes Published and in Preparation

TERRAIN ANALYSIS: A Guide to Site Selection Using Aerial Photographic Intrepretation / Douglas S. Way
URBAN ENVIRONMENTS AND HUMAN BEHAVIOR: An Annotated Bibliography/Edited by Gwen Bell, Edwina Randall, and Judith E.R. Roeder
APPLYING THE SYSTEMS APPROACH TO URBAN DEVELOPMENT / Jack LaPatra
DESIGNING FOR HUMAN BEHAVIOR: Architecture and the Behavioral Sciences / Edited by Jon Lang, Charles Burnette, Walter Moleski, and David Vachon
ALTERNATIVE LEARNING ENVIRONMENTS / Edited by Gary J. Coates
BEHAVIORAL RESEARCH METHODS IN ENVIRONMENTAL DESIGN / Edited by William Michelson
STRATEGY FOR NEW COMMUNITY DEVELOPMENT IN THE UNITED STATES / Edited by Gideon Golany
PLANNING URBAN ENVIRONMENT / Melville C. Branch
LANDSCAPE ASSESSMENT: Values, Perceptions, and Resources / Edited by Ervin H. Zube, Robert O. Brush, and Julius Gy. Fabos
ARCHITECTURAL PSYCHOLOGY / Edited by Rikard Küller
THE URBAN ECOSYSTEM: A Holistic Approach / Edited by Forest W. Stearns
URBAN PLANNING THEORY / Edited by Melville C. Branch
INSTRUCTIONAL MEDIA AND TECHNOLOGY: A Professional's Resource / Edited by Phillip J. Sleeman and D.M. Rockwell
NEIGHBORHOOD SPACE / Randolph T. Hester, Jr.
PLANNING FOR AN AGEING SOCIETY / Edited by M. Powell Lawton, Robert J. Newcomer, and Thomas O. Byerts

EDRA Conference Publications:
EDRA 1/Edited by Henry Sanoff and Sidney Cohn
EDRA 2/Edited by John Archea and Charles M. Eastman
ENVIRONMENTAL DESIGN RESEARCH, VOL. I: Selected Papers / Edited by Wolfgang F.E. Preiser
ENVIRONMENTAL DESIGN RESEARCH, VOL. II: Symposia and Workshops / Edited by Wolfgang F.E. Preiser
RESPONDING TO SOCIAL CHANGE / Edited by Basil Honikman

CDS/15

URBAN PLANNING THEORY

compiled,
edited, and
introduced
by

Melville C. Branch

University of Southern California

Distributed by

HALSTED PRESS A division of John Wiley & Sons, Inc.

Community Development Series, Volume 15
Library of Congress Catalog Card Number: 75-1327
ISBN: 0-470-09640-3

75 76 77 5 4 3 2 1

Manufactured in the United States of America.

LIBRARY OF CONGRESS CATALOGING IN PUBLICATION DATA

Library of Congress Cataloging in Publication Data

Branch, Melville Campbell, 1913- comp.
Urban planning theory.

(Community development series ; v. 15)
Includes bibliographical references and index.
1. Cities and towns--Planning. I. Title.
HT166.B723 309.2'62 75-1327
ISBN 0-470-09640-3

Exclusive Distributor: Halsted Press
A division of John Wiley & Sons, Inc.

SERIES EDITOR'S PREFACE

The objective behind the Community Development Series is clearcut: to provide practicing professionals, their clients, and other interested persons with informative, useful, readily applicable reference and technical books on planning, programming, designing, and constructing the built environment. However, like many a clearcut objective in an increasingly intricate world, the scope is large, the intentions universal, attainment difficult and complicated.

A case in point is urban planning, a science and art that continually develops in response to human, organizational, and institutional needs. How can one cover the essential knowledge in a field so vast and dynamic?

In this uncertain setting, Melville C. Branch's *Urban Planning Theory* is especially welcome as a unique collection, knowledgeably gathered, systematically organized, selective, yet comprehensive in its coverage.

Dr. Branch's book is also important in substantive ways for present practice embodies accepted theory, and today's theory is tomorrow's practice.

Theory is a fundamental kind of knowledge that establishes benchmarks for those looking ahead and for those concerned with matters closer at hand.

Whether as a beginning hypothesis awaiting verification or modification, or as an organized body of generalizations guiding professional actions, theory is useful in defining problems, discovering principles, reaching conclusions, and shaping intuitive thought in the search for workable solutions.

In publishing this reference book, with its synoptic introduction by Dr. Branch summarizing basic issues, we believe a diverse group of readers will be served. The book will be particularly valuable for those entering planning as a profession and for those in allied fields who wish to know about comprehensive urban planning theory.

Richard P. Dober, AIP

PREFACE

This book introduces persons in fields directly concerned with city planning to the scope, subject matter, challenge, and opportunities of comprehensive urban planning theory. At present, these fields include urban and regional planning, urban and regional studies, public administration, systems and civil engineering. Also very much involved are economics, law, business administration, political science, sociology, urban geography, architecture, landscape architecture, geology, and management science.

Only a small part of comprehensive urban planning theory can be covered in one book. To be thorough and complete on a subject that is now so extensive would require many volumes. The compiler has selected readings which he believes represent comprehensive urban planning theory today and for the intermediate future. They are not limited to those which agree with his personal views and conclusions. The intent is to introduce the reader to the scope and content of this body of theory, to interest him in its diverse subject matter, and to stimulate him to explore further.

Some of the many opportunities to contribute to urban planning theory are noted in the Introduction following this Preface. Satisfying and rewarding careers can be founded on developing theoretical competence in any one of the numerous areas of planning consideration that are not now supported by sufficient theoretical knowledge.

The most critical requirement in the selection of readings was clear and succinct coverage of the subject matter to be included. Often, this was found in articles from periodicals or other short publications, because they usually condense essential subject matter in the least space, or in sections of books. The pioneering or leading theorist may not therefore be the author, but he is almost certainly cited in the paper selected for this book. Some readings were written years ago, but are completely relevant today and maintain their uniqueness or outstanding excellence. The contents in general are the outcome of material collected by the compiler over many years and reflect the approach represented in his bibliography on comprehensive urban planning published in 1970. Most selections have been among readings required in his graduate course on comprehensive urban planning theory.

Naturally, the different divisions of the book reflect the compiler's view of comprehensive planning theory which has resulted from his experience planning for a city, a large corporation, and two of the military services. Philosophy and mathematics could have been added as separate divisions, but they are considered as underlying and threading through most of the subject matter of the readings. Mathematics is also not included separately because most readers cannot understand theories expressed in mathematical terms, although they abound today in economics, regional science, management science, operations research, and engineering. However, references are cited in many of the readings in this book which the mathematically knowledgeable person can explore. Throughout, readings are limited to those which do not require particular knowledge of any kind.

The scope of the subject matter of city planning today is confirmed by the many fields of knowledge represented in these readings: architecture, landscape architecture, engineering, urban and regional planning, economics, sociology, political science, psychology, public administration, business management, physical science, mathematics, law. Authors include academicians, people in business and the military services, practicing planners and other professionals, even parliamentary (court) reporting in the person of Ebenezer Howard. The periodicals and books in which the readings appeared originally are also varied.

No attempt has been made to edit the different readings into some uniform style, nor all citations into the same bibliographical form. References are grouped at the end of each reading to facilitate typesetting. Some readings are only those portions of the original material that deal with the subject matter covered here. The greatest shortening is made of Rexford Tugwell's seminal paper "The Fourth Power." In the original version 35 years ago, Dr. Tugwell carried forward two lines of thought as he often did at that time, one in the text proper and the other in extensive footnotes. The footnotes have been eliminated.

Acknowledgement of sources and permissions to reprint are referenced at the bottom of the first page of each selection. Appreciation is expressed to the publishers and authors cooperating in making this book possible.

Melville C. Branch
April 1975
Los Angeles, California

CONTENTS

First, *the scientific nature of the necessary theories must be* quantitative *rather than qualitative. Many theories . . . lack firm quantitative statements of . . . relationships and are to this extent not very useful for predictive purposes.*

Second, *a closely related difficulty has to do with notions of* causality . . . *definitions of causality may be completely lacking in which case the theories and their associated models become descriptive statements about associations and co-relations between events. . . .*

Third, *modern social science has experienced some difficulty in looking at* whole systems . . . *transportation, education, housing, retail trade, and many other aspects of urban activities or urban facilities are studied as "systems" without adequate recognition of the fact that they belong to a larger system which actively interacts with them.*

Fourth, *as a corollary of the preceding point, there does not now exist an adequate theory of complex social systems such as might be useful in planning at the most general level. . . .*

Fifth, *the limitations of theory and disciplinary practice in the portrayal of systems are compounded by the* lack of technical means for manipulating representations of these systems. . . .

Sixth, *a most serious difficulty with modern social science and planning analysis is its* lack of generality. . . . *Planning requires* predictions *of a duration and at a level of detail which most social scientists are not prepared to accept as feasible. . . .*

Seventh . . . *The intuitively attractive appeal of statistical uncertainty in predicting urban affairs is in my view not valid except in the case of that unusual class of events which are very small in number and very large in importance. . . .*

Britton Harris, "Generating Projects for Urban Research," Environment and Planning, *Vol. 2, 1970, pp. 4-7.*

Melville C. Branch

INTRODUCTION

Dictionary definitions of theory range from contemplation to a well worked out hypothesis or principle. For the purposes of this book, theory is the best available generalized knowledge about cities and their planning which shapes conclusions and actions with respect to them. Theory is knowledge of broad applicability that represents our comprehension of the phenomenon or subject matter in question. Of course, the less it is limited to a single or several cases, the more general the theory and the more widespread its application.

Theories are not necessarily less powerful or persistent because they are not exact or are interpreted differently by different people. Speculative formulations may be important because they meet a present need or because of particular circumstances. They may represent the most that can be generalized at the time. However, in the Western world today, theories are considered more scientific and significant if they are supported by careful observations or measurements that can be validated by qualified persons.

Naturally, ambiguity is reduced when theories are formulated precisely, logically, or mathematically. If repeated use confirms their value and validity, they become part of the continuum of basic knowledge which is the substantive stem of the intellectual field or area of endeavor. This stem evolves by the continuous formulation of better theories.

No field or endeavor can establish itself and advance without a sound theoretical base. Without theory, explicit or implicit as the case may be, sound generalizations cannot be made. Experience must be relearned repeatedly. Past mistakes are repeated. New situations must be approached without general guidelines. And the body of knowledge comprising the discipline has no continuing structure of principles, coherent thought, or confirmed conclusions to guide decisions and ensure a progressive advancement of knowledge and accomplishment.

Only bits and pieces of such a theoretical structure exist for comprehensive city planning today. This is not surprising. Although town planning has been practiced for at least 5,000 years, it has been a simple activity compared to our concept of city planning today. Communities were small in size. Their components did not include the many technological developments that have so greatly complicated cities recently. Economic and industrial activities were simple by comparison. Usually, a king or another authority arbitrarily determined the nature and extent of planning.

Under these circumstances, the physical form and features of the city were the paramount concerns. Hence the importance in early town planning of physical situation, site, size, and the enclosure and internal arrangement of cities. It is only since the development of long-range artillery and the urban effects of the Industrial Revolution that protection from outside attack no longer confines and shapes the spatial form of cities.

Since the turn of the last century, tremendous complication has occurred in cities in industrialized nations. New components that have been introduced or greatly expanded include the railroad, streetcar, subway and elevated rapid-rail transit, automobile, helicopter, radio, telephone, television, electricity, natural gas, nuclear power, water purification, sewage treatment, and solid waste disposal. All forms of production and employment have become increasingly specialized and sensitively interdependent. Health, recreational, and other services have been greatly expanded and complicated.

Populations have grown in size and become more long-lived because of decreases in death rates. Around the world, mass migration of people from countryside to metropolitan centers continues. The number of large cities and the average size or urban places have increased greatly. Most recently, the mass media of communication have opened the eyes and appetites of all people within sight of a television screen to the life styles and material possessions of the advantaged, and intensified their reactions to current events.

Thus cities are larger and the necessary scope of their planning is far more complex than was the case some 75 years ago. Geographical, geological, physical, economic, political, legal, social, psychological, cultural, and technological considerations are all involved inseparably. Recent recognition of environmental impacts as a formal concern of city planning has added another dimension to what was already an awesome intellectual task. And decisions which determine the course of city planning are the product of a prolonged process of obtaining political consensus or acceptance among many partisan and special interests.

Yet another level of analytical consideration is added by the recognition in recent years of global as well as intraurban, regional, and national interdependencies. Few urban places today are isolated from economic, socio-political, or environmental effects generated in another part of the world. Air transportation, television, various forms of pollution, and international monetary matters are tying the world together as never before. Logically, such global effects should be taken into account in the city planning of every community, but this is beyond our analytical capability at present.

As a consequence of these developments and awarenesses, city planning in the modern world represents the most ambitious intellectual and directive effort attempted by man. Although national planning is undertaken at a higher governmental level, it does not at present encompass the number and diversity of elements that seem more understandable and potentially attainable at the local level. This analytical complexity of the urban organism and of planning its development is fully recognized by few people because they do not have reason to realize that it affects every aspect of city planning and municipal management. Even many city planners do not consider the urban organism completely and deeply, but accept the simplistic view characteristic of past practice. They are divided concerning the scope of their activity, the content of the field, and the importance of theory. Yet every so-called purely practical act involves in fact numerous underlying theories.

How closely theories are involved in what many planners consider a practical matter is illustrated by the concept of the neighborhood unit, described in one of the readings in this book. Land use, transportation, spatial design, and several social theories underlie the assumptions of neighborhood theory: that pedestrian circulation to and from a neighborhood center reduces automobile traffic, is enjoyable and beneficial for those walking, and encourages social contact between inhabitants. On the other hand, pure capitalistic market theory suggests that neighborhood units in the United States promote segregation by excluding the poor, the aged, and other minority and disadvantaged groups. Neighborhood unit theory also incorporates theories of limiting population density or otherwise retaining neighborhood identification as population increases.

Neighborhoods are the basic building blocks for the theory that cities would function better and deteriorate less rapidly if they were composed entirely of these discrete physical units. Opposing theory holds that places of employment, shopping, recreation, social contact, and higher education are so dispersed throughout urbanized areas in industrialized countries that the interactions presumed within the neighborhood unit do not in fact occur.

Law provides an example of theory indirectly affecting neighborhood units. It has not been possible in the United States to separate such units spatially as required by the underlying concept. This is because legal theory and practice require compensation to property owners at "fair market value" when the "highest and best use" of their land is "prevented" for city planning or other public purposes. Under these conditions, funds to purchase open space between neighborhood units are not available.

Most of the impetus for broadening the limited kind of physical city planning which has not been effective in the United States in the past has come from outside the city planning profession, as have most advances in theory. The profession has tended to underestimate the importance of theory, partly because of the overwhelming operational difficulties city planning staffs face regularly which demand undivided attention, and partly because very little comprehensive city planning theory is well enough established to apply in practice. Also, the much-expanded scope of city planning as viewed today has multiplied the difficulties of formulating valid theory.

Another consequence of recent developments is recognition that all fields of knowledge are relevant to urban planning in one way or another, directly or indirectly. Actually, this has always been true because cities are a primary product of civilization and a microcosm of human existence and condition, and have therefore always involved or been connected in some way with all knowledge. It is for this reason that comprehensive city planning today is an amalgam of concepts and theories developed by many fields for their own purposes.

Despite this eclecticism, there is a methodological core which is predominantly if not exclusively the central concern of city planning. At present, this includes: analytical correlation of functional elements and subsystems of cities; projecting them separately and together into the future; use of master plans as a directive mechanism; and instruments of effectuation such as approval of the subdivision of land, determination of land use by type, population density limits, restriction of the height and coverage of buildings, conditional approval of designated projects, and evaluation of environmental impacts.

To these will be added before long: municipal tax policies that reinforce rather than thwart city planning objectives; effective methods of urban redevelopment; municipal capital works programs closely correlated with the master city plan; areal jurisdictions large enough for urban planning purposes; and greater independence from speculative pressures and special interests when their desires violate established planning policies and plans, or run counter to the general public interest.

If city planning as now constituted does not develop its own supportive theory, this will be done by others because the role and responsibiliteis of comprehensive city planning cannot be ignored. The many diverse elements and activities of a city must be correlated somehow to reduce conflict and waste, if not to prevent chaos. Continuity must be ensured to attain urban objectives requiring years of consistent policy and action. Some analytical mechanism or master plan must be maintained that portrays how available resources are to be allocated to carry out the desires and decisions of the body politic and elected representatives. This requires a sound body of theory to support individual decisions and actions, and promote consistent activities in the aggregate.

For the purposes of this book, comprehensive urban planning theory is divided into 11 categories.

LEGACY

There are, of course, numerous concepts employed or reflected in city planning today which originated in the past. Recognition of the spatial efficiency of the radial-concentric pattern of primary streets is an outgrowth of circular defense walls permitting the shortest enclosure of primitive communities, and of radial streets providing easy access from entrance gates to village center. Axial composition of streets for ceremonial, monumental, and aesthetic purposes is age-old.[1] Greek officials called *astynomi,* whose duty it was to prevent encroachment of abutting businesses into Athenian streets, represented ancient recognition of the relationship between land use and traffic circulation that underlies modern traffic-flow theories of limited access and is so important generally in city planning today.[2]

Basic patterns of subdivision are found in ancient cities. Formulation of the theory of building code regulation and the several purposes of building height limits began long ago.[3] Ancient Mohenjo-daro is reported in the Tamil classics to have had districts "zoned" for musicians, actors, and prostitutes.[4] The theory of neighborhoods of limited size, bordered by principal streets, with pedestrian pathways leading to a neighborhood center with an elementary school or other local facilities is a modern formulation of a spatial feature which developed naturally in many cities in the past, was adopted as the basic unit of design in British garden cities, and has been incorporated in the physical pattern of most new towns ever since.[5]

Another example of our legacy from the past are the Utopias, which serve a useful purpose although they are not intended to be practical, do not apply to the present but to the distant future, and presume such extensive socio-economic, technological, or cultural changes that they could not possibly be realized for many years. Utopias have helped us recognize key problems and critical causes as well as obvious symptoms. They involve imaginative and exploratory thinking and promote it in others, a stretching toward the possible rather than customary acceptance of the ordinary. Occasionally, entirely new concepts are proposed which take root and are eventually realized because they reflect trends and forecast advancements which most people do not recognize.[6] Some of the theoretical concepts mentioned in the section immediately following are Utopian as well as design formulations.

DESIGN

The emphasis on physical city planning in the past and the predominance of architects and engineers in its practice have produced a multitude of spatial design theories. These have ranged all the way from the Utopian conceptualizations of Plato, Aristotle, or Vitruvius to the "broadacre city" of Frank Lloyd Wright or Hilbersheimer's proposed layout of cities to reduce industrial smoke blown by prevailing winds into urban areas containing people.[7] The garden designs of LeNôtre at Versailles carried out landscape theories concerning the visual pleasure of multiple vistas along straight alleyways radiating from *rond-points,* a pattern applied by Baron von Haussman to the vehicular street system of Paris and later by L'Enfant to Washington, D.C.[8] Illustrating the impossibility of anticipating some developments, these designs for pedestrians and horse-drawn carriages turned out a century later to be one of the worst possible arrangements for automobile circulation.

Although separating automobiles and pedestrians at different levels in congested parts of cities is considered sound theory, it is rarely achieved.[9] Another concept calls for the exclusion of almost all private automobiles from the central city.[10] Circulation theories that advocate urban rapid-rail transit systems are based on their higher carrying capacity, faster average speed if station stops are sufficiently separated, feasible construction either at ground level, elevated or underground, and land-use dynamics which intensify population densities along these corridors and thereby increase the percentage of total passenger trip miles carried by the rapid-rail system.[11] A few people are considering theories for handling the much expanded local air traffic which can be expected within municipalities in the future.[12]

The neighborhood concept mentioned previously as a legacy from the past has been extended to a theory of "organic" city planning by comparing the spatial form of cities with the cellular form of biological organisms – an analogy challenged by most scientists. It proposes that all urban growth, development, and redevelopment take place by discrete physical-spatial units of limited size. In this way, according to theory, deteriorating and disordering effects or urban sprawl may be eliminated.[13] Also proposed by a few ardent advocates are megastructure cities built on land, in the air on thin stems like immense mushrooms, over water, or even underwater.[14]

Investigations of the visual awareness and reaction of city dwellers to their surroundings have given rise to theories of civic design that would incorporate certain physical features in the cityscape to provide spatial identification, aesthetic pleasure, and other subtle visual satisfactions.[15] "Design with nature," required for survival in primitive town planning and desirable throughout history for various reasons, has been extended recently into a theory and process of partial environmental planning analysis.[16]

These are but several examples of the many design theories proposed for cities. Because they are easier to conceive and specialized knowledge is not an absolute prerequisite, there are probably more theories relating to design and they are of greater diversity than is the case for any of the other divisions of comprehensive urban planning theory adopted for this book.

FORM

Cities are dynamic organisms. Throughout most of history, they were confined within defensive perimeter walls and subject to strong governmental control. Nonetheless, internal changes occurred continually as populations grew, land uses evolved, new features and facilities were built, and old structures were regularly replaced. When peripheral enclosure was no longer necessaryfor protection and new means of transportation were developed, cities could spread and sprawl into the surrounding countryside.

Urban form is also affected by situation and site, underlying geology, seismism, flood plains, and other physical factors – not to mention population and speculative pressures reflecting economic, political, and social conditions. Theories professing to explain urban form have been few and far between because of this complexity, and unsuccessful so far as predictors of spatial development. Nor is there a basic urban form demonstrably preferable for new towns as initially built that will also function well as the city ages and loses most of its new-town characteristics.

Neither the concentric-zone theory of successive outward "invasion," nor the sector, multiple-nucleii, and stage theories of urban growth have proved descriptively adequate.[17] New approaches such as studying household activity patterns or attempts to correlate urban form with local air pollution and energy conservation are still experimental or partial theories. Long-established urban forms still predominate in practice: rectangular; radial-concentric; star-shaped patterns with their built-up arms along major radial transportation routes; a few linear layouts; and irregular but carefully conceived patterns such as those often used for new communities. At smaller scale, the cul-de-sac and superblock are the most widely employed concepts of neighborhood form.[18]

An existing city can be shaped into a predetermined spatial form only by consistent and progressive action during many years. In most countries, cities have not elected to attain a predetermined form or it has not been possible or practical to do so. Only in new towns and large-sized additions to existing cities are urban forms regularly realized as planned. In the United States, the separate and individual actions of numerous private developers have in large part determined spatial arrangement, within such engineering and subdivision constraints or directives as the particular city has applied. Because of this mixture of different economic, speculative, and governmental forces and decisions by individual developers, predicting urban form in the United States today is next to impossible. Deliberately directing development and redevelopment toward a particular urbanwide form cannot now be effected.

Yet the question of the best spatial form for cities or large parts of them is vital because it relates to the kind of physical environment that can be created and the overall efficiency of the city as determined by the pattern of its major transportation routes, utility lines, and other elements of form. Lately, attention has been drawn to urban form by discussion of the desirability of limiting the size of cities, and preventing continuous urban sprawl at their periphery by requiring that this development take place in community units rather than by the tacking of one subdivision on to another without interruption, by premature and piecemeal subdivision, or by unplanned "leapfrogging" of development into the surrounding countryside – all of which have been allowed for so many years in the United States.

Location theory seeks to explain the distribution of urban places in space, and the disposition of different land uses within cities. For example, according to central place theory, under certain economic and geographical conditions communities tend to locate in a spatial pattern of hexagons according to a hierarchy of sizes. In turn, the urban form and functioning of individual cities are related to their place in this pattern. Location theory, therefore, is involved in forecasting urban development, in metropolitan and regional planning, and in the location of new towns. It reminds us of the powerful bonds between cities and their geographical-spatial surroundings.[19]

INFORMATION

Information is certainly vital to city planning in many ways. It must be sufficient for the purposes of the planning. It must be accurately indicative of what it purports to describe or represent. Its percentage significance or other determination of reliability should be known publicly as well as privately. It must be available when needed and at a cost that can be afforded. It should be in such form that correlation is facilitated with other data maintained for city planning purposes. Last and by no means least, it must be understandable, explainable, or acceptable to those using the information as a basis for their conclusions, decision, and action.

Until now, city planning in the United States seems to have operated on the assumption that the more data the better, whether or not the information could be used productively before it was outdated or before it could be incorporated meaningfully in planning analysis. It is almost routine to claim that more and more data are necessary to reach conclusions and make decisions. Too often, however, requests for more information are an excuse for inaction, non-commitment, or the perpetuation or increase of staff size or operating budget. Also, it is well known in politics and government that problems can often be avoided, delayed, defused, or obscured by claiming that more information is needed, whether or not this is so.

Probably the main problem of information for city planning has been the illusion that indiscriminantly gathering more information would somehow lead to analytical comprehension of the functioning of the urban organism. To the contrary, more and more data piled on top of limited analytical understanding hinders rather than helps analytical advancement.[20]

Most of the information used by city planning agencies in the United States today is obtained from other organizations and pertains mainly to population, land use, physical development, and circulation. To supply the broader range of city planning intended in the future will require more precise evaluation of informational needs than has been provided in the past. Responses to questions heretofore generally ignored will be necessary. What distillation of data, which a municipality can afford, should be maintained as the analytical base for the decisions and actions which are the end-product of the city planning process? How closely up-to-date should different elements of core information be maintained? Can the data be correlated to reveal the dynamic interrelationships among different components of the urban organism that are crucial for city planning? Will this body of information be used by those making city planning decisions?

Although the information needed is never exactly the same for any two communities, one would expect to find useful responses to these questions in the literature of city planning. But one will look in vain for helpful theory relating to the selection, collection, and processing of city planning information. And there is no priority list of the most important data which a newly appointed director of city planning could use as a beginning in a community undertaking comprehensive planning for the first time.[21]

Informational matters will be critical when city planning becomes an active force directing communities in the United States. Or perhaps city planning will not be accepted as a viable municipal endeavor until it identifies the minimum information necessary to direct a city constructively and demonstrates how these data can be correlated and used as the main basis for urban analysis and city planning decision. Limitation of funds and the greater comprehensiveness of city planning will force careful selection of what data to maintain, in what form, and how often renewed. The widespread use of computers for data processing and storage will also establish requirements.[22]

Information for city planning will be viewed more scientifically than it has been in the past.[23] The "bit" or another unit of measurement will be employed to evaluate information with greater precision. The reliability of data will be closely calculated and their significance as they are used will be carefully appraised. The true costs of acquiring and processing information will be determined regularly, rather than ignored or deliberately understated as is so often the case today. Information theory, scientific sampling, mathematical statistics, management science, and the biology and psychology of information retention and processing within the minds of human decision-makers will be taken into account in developing and maintaining the core of information for city planning.[24] In addition to providing the technical-analytical information on which plans and related conclusions, decisions, and actions are based, the core of information must also contain a record of actions taken, as a legal reference and as evidence of having met the requirements of due process. It also serves as a factual basis for the quantitative simulation of the city which is the main means of urban planning analysis: providing data concerning the past, disclosing trends, identifying interrelationships, making projections, and displaying the consequences of alternative plans.

Perhaps most important is how closely up-to-the-minute must planning information be? Other things being equal, the closer the planning process to "real-time," the higher the costs of collecting and processing data, the more complex and fine-tuned the entire operation, and the more sensitive the system to "noise" and its own internally generated disturbances. How "real-time" or up-to-the-minute should comprehensive city planning be to perform successfully?[25] For this and other reasons, the need for and utilization of information should be evaluated periodically. After years of repeated collection, certain data may no longer be required or may be available from another source. Even if still needed, they may now be less vital than other information or not worth the comparative cost of collection. Is reliability being maintained or have events affected accuracy and changes are in order?

Related to these core analytical data are other informational activities by various organizations in the community. If data are gathered by each unit of the municipal government so differently that separate statistical results cannot be compared or combined, creation of a core of urban analysis is impossible, made much more difficult, or is prohibitively expensive. For example, if different data are collected by geographical areas which do not coincide spatially, making the data consistent is costly and time-consuming. If changes are made in the precise way information is gathered over time, the data may no longer be sufficiently comparable to examine historical trends. Similarly, data collected at intervals of several years cannot be compared without some correlation which requires their adjustment and reduces their combined accuracy. Are there occasions when private enterprise might modify its collection of certain information for its own purposes in order to coordinate with information gathered by the municipality, if this would not impair the private use or increase the cost of collecting the information?

PROCESS

Procedures for subdividing land, zoning, and preparing and adopting master plans for an entire city, parts of a city, or special development projects have been established and legalized over a period of many years.[26] Outside these areas of procedural agreement, the process of city planning is under vigorous discussion, with various procedures being tried and further changes anticipated. The situation will not be resolved, however, until it is decided what it is that city planning in the United States can and should accomplish. Clearly, process cannot be considered apart from the purpose and substance of the endeavor it represents. Process is therefore an integral part of continuing discussion of the nature and scope of city planning. Any idea that the feasibility of planning processes and means of effectuation can be disregarded in theoretical formulations is incomprehensible to those experienced or concerned with the application of city planning theory in real life.

The planning processes best suited to levels of government other than municipalities, to business enterprises, and to the military services vary with the nature, purposes, and operational characteristics of the activity. But new or improved processes developed in one type of planning have been adopted by other applications. As one example, the concept of city planning as a staff activity, which has predominated in the United States for many years, derives from the theory and practice of military staff services.

Despite significant differences, there are common characteristics among different forms of planning, for example: the ordered selection, collection, and processing of the information that is the "raw material" for making plans and decisions; comparable logical, judgmental, mathematical, and other methods of analysis; similar ways of reaching conclusions, making decisions, formulating plans, selecting strategies, tactics, and means of implementation; and finally determining the frequency of periodic reappraisal.

Theories relating to a wide variety of urban applications are naturally more general and basic than those for more specific uses. Logically, they precede more precise formulations of the city planning process that are best suited to certain cities, particular governmental organizations, special kinds of activity, different nations and cultures, stages of economic development, and emergency situations. This is the realm of general systems and organizational theories, and of procedures claiming wide applicability such as planning, programming, and budgeting, and evaluations such as cost-benefit analysis or social and other indicators.[27] The fields of management science, operations research, and systems engineering in particular seek to develop basic knowledge of planning processes in mathematical form that applies to many organizations and activities of man.

From the variety of applied theories concerning the planning process within government, business, and the military services – and from basic theories underlying these applications – comprehensive city planning must derive its own formulation of the optimum process of planning cities, its own general principles for using information, performing analysis, making recommendations, and implementing and monitoring the results. Several concepts are being advanced to change the master city planning process as normally conducted today.[28] In part, they reaffirm original intentions for the master plan that have not yet been achieved.[29] They result from recognition that city planning is largely ineffective in the United States today and that cities must be better planned if we are to avoid even deeper difficulties than those that now exist. In large part, the new concepts derive from improvements in functional, subsystem, and comprehensive planning by business and the military.[30]

The idea of 20-year "end-state" plans produced at intervals of 10 or more years with little effectuation in between is being replaced by the concept of continuous informational inputs, analytical formulation, results, and feedback.[31] Long- and short-range plans and operating budgets are viewed as one continuum constituting the planning process. The plan is not a periodic printed document but a flexible master referent which changes as often as developments and decisions require. City planning is no longer seen as a special somewhat isolated function but as an integral and meaningful element of municipal government, the keystone of its many functional and subsystem planning activities.[32]

ANALYSIS

Although all aspects of planning are inseparably connected in one activity, analysis is the most important component of the endeavor. If a city cannot be analyzed correctly and usefully, the entire planning process is crippled because the information needed cannot be determined, plans are largely conjecture, and decisions concerning the community as a whole reflect opinion rather than comprehension. Obviously, adequate knowledge is prerequisite to sound city planning.

Yet the process of planning must continue regardless of whether analysis is weak or non-existent, for survival is as inevitable a compulsion for a city as for individual man. Modern cities cannot function without efficient management: to supply their populations with food and water; remove wastes; maintain necessary transportation and communication; and provide protective, health, and the many other services required today. Planning analysis is critical as urban components become more diverse, sensitively interdependent, and technically complex. Without analysis or structured judgment, the direction of cities is too often the product of unconstructive conflict between particular interests and partisan politics, or the belated consequence of crisis conditions. Decisions vital to the city as a whole, the total public interest, and the general welfare are based on partial analysis, sudden response, transitory opinion, random choice, or best guess.

Unfortunately, analysis is considerably weaker today than any other component of the comprehensive city planning effort. The methods now employed or advocated are borrowed from many fields, often without adjustment to the urban application or without further improvement. Many of the young people preparing for a career in city planning today are more interested in activism than analytical knowledge. Considering the present state of city planning in the United States, this is understandable. But it leaves the field exposed to challenge on the most serious of grounds: the extent and validity of its basic knowledge. A field without a firm theoretical base is in a defensive position not only in the eyes of opponents and competitors, but with respect to its own self-confidence. Whether analytic advancement can follow or is a necessary prerequisite to more effective city planning is a key question.

The development of urban analytical understanding is slow because, as already pointed out, it is one of the most difficult intellectual endeavors undertaken by man. Although this is quite obvious upon reflection, it is often forgotten or ignored. A few facts concerning the city of Los Angeles illustrate the required range of consideration and involvement. Some 3 million inhabitants occupy approximately 800,000 parcels of land in an area of 480 square miles. Enclosed within this space are several independent municipalities and the city itself is part of a larger contiguous built-up territory containing 8 million people and 78 independent municipalities within the surrounding county.

The forces that direct the development of the city – besides those exerted by international conditions and events and federal, state, county, and other local governments and political organizations – are multitudinous and diverse. There is the municipal government with its many departments, its various powers, and its politics. There are the financial institutions that underwrite private development. Industries provide an economic base. Thousands of businesses, enterprises, and institutions supply and service the people of the city and others elsewhere. Organizations galore meet the incredible variety of people's needs and wants: religious bodies, recreational and social clubs, professional associations, commercial groups, fraternities of many kinds, and a thousand others.

The municipal existence depends on maintaining a functioning environment of air, water, drainage, waste disposal, energy sources, transportation, and communication. All this and more is affected by local geography, geology, and climate, and the sociocultural context of attitudes, laws, and customs extending back into history. Also to be taken into account are aesthetics, amenity, convenience, beliefs, and other intangibles which are important although they are difficult or impossible to measure. Small wonder that comprehensive urban analysis is slow in coming.

Of course, many fields of knowledge contribute directly to increased analytical understanding of cities. Mathematics provides ways of expressing and measuring the accuracy, correlation, or dynamic interaction of quantified concepts and information, as well as statistical accuracy, inference, and probability. Physical science supplies an enormous accumulation of knowledge concerning the facilities of a city: its buildings and other structures, utilities, means of transportation and communication, and underlying and overlying environment. The biological, social, psychological, and ecological sciences generate knowledge of man as an individual and social animal, and as a species. In one way or another, every field contributes to the storehouse of knowledge that is relevant to an organism which is as indicative of almost every aspect of man's existence as is the city.

From this wide range of inputs will evolve a body of theory and principles expressly and specifically for comprehensive urban planning analysis: concerned on the one hand with the dynamics of city growth, development, and spatial form and design; and on the other hand with how the most critical elements and aspects of urban operations can be integrated, projected, alternative solutions formulated, conclusions reached, decisions made, and means of implementation selected or devised. The selections in this book under Analysis introduce several aspects of this development, which could not be covered completely in several volumes devoted solely to the subject.[33]

Establishing objectives for city planning illustrates the present paucity of analytical theory. Everyone agrees that by definition planning must be directed toward an objective. Some people believe urban inhabitants should be queried directly to determine specific objectives for city planning. Others believe that only a few persons have the time, interest, and background information to specify objectives for complex endeavors. Some maintain that establishing the objectives of government is the primary purpose and end-product of our political process. There are city planners who advocate determination of as many objectives as possible at the neighborhood level. Others maintain that the foremost desires of different groups of people are well known or readily learned.

All seem to agree that *how* to attain objectives is the critical question, which the average urban dweller cannot be expected to answer. They also agree that little money is normally available for objectives over and above those already established or committed for the future. Yet thousands of hours and dollars are often spent by city planning staffs and citizen groups in idealistic selection and specification of goals and objectives that are desirable in themselves, but totally beyond the capability of the community or requiring greatly increased taxes, new laws and regulations, or years of persistent effort. An accepted theory of establishing objectives for city planning would not only greatly improve the analytical process itself, but would also direct such supportive efforts into areas of realistic and most constructive consideration.[34]

SIMULATION

Simulation can be considered part of comprehensive urban planning analysis, but it is classified separately in this book because of its special significance as the central analytical mechanism of planning, and because it has attained procedural and legal identity in city planning in the form of master city plans. Therefore, some of the concepts of master city planning presented earlier under Process are also relevant to Simulation. Of course, as already indicated, the division of this book into sections is to facilitate absorbing and recalling the material. The contents comprise in fact an indivisible whole.

Simulation is the critical analytical mechanism of comprehensive planning. It is the main means of representing the complex urban organism so that the dynamic interactions among parts can be calculated when possible, approximated when this is all that can be done, or at least sensed sufficiently to strengthen judgments. Simulation is an analogy of the city in "miniature" or much simplified form, which makes it possible to project repercussive effects of a particular trend or assumption concerning the future on various interrelated elements of the city as represented in the simulation. In one of various forms, it facilitates gauging the effects of alternative policies, plans, or specific decisions. To the extent the urban organism is correctly represented, the simulation permits estimation of at least some of the consequences of an unexpected condition, emergency, or disaster.[35]

Conceptually, the master plan is the central simulation of city planning. It shows what is planned and how it is to be accomplished over different periods of time. Its various elements fit together and interact as much like they do in actuality as knowledge and ingenuity permit. The master plan provides at all times the representation of the city as a whole into which are fitted plans for smaller areas, functional elements, subsystems, and specific projects. It is the main analytical means of showing or judging what can be done physically, economically, politically, socially, environmentally, and legally. It incorporates a distillation of the best data available which can be used for many informational purposes besides those directly involved in its use as the master plan. It is a combination of different forms of expression: numerical, mathematical, cartographic, graphical, and written statement.

The master city plan simulation is also a legal instrument. Its content at the time of a city planning decision constitutes the analytical justification or supportive "evidence" if the decision is appealed to the courts. It is the official reference for financial institutions, land developers, builders, businessmen, individual citizens, and parties of all sorts interested in the existing city or plans for its future. The master plan simulation portrays publicly the collective intentions of the community – what the municipality proposes in comprehensive city planning.

Mathematical models are widely used today for purposes of simulation. They feature the rigor and simplification of numbers and their mathematical treatment. They are usually conceived and programmed for the computer. They can be stored, retrieved quickly in whole or in part, and repeated changes and recalculations are readily made. The quantitative effects of alternative assumptions or decisions on the simulation in model form are quickly computed. Of course, the strengths and weaknesses, validity or invalidity of this form of modeling are precisely those of the mathematical formulation which *is* the simulation.[36]

Mathematical models and computers are the key mechanisms of gaming. They permit the many interactive calculations which are the essence of the exercise. They make it possible by the rapidity of computer calculation to simulate an extended period in real life within the short time available for playing the game. They allow each participant to make his own input and receive his separate output correlated with the actions of other participants. As indicated in one of the selections in this book, gaming can be used for teaching, experimentation, entertainment, theory, diagnosis, operations, and training.[37]

Of course, urban simulations take different forms. Besides mathematical models, there are physical models, maps and photographs, graphical abstractions, statistical formulations, and written descriptions. Each is the best form of simulation for certain elements or aspects of the urban organism and its comprehensive planning.

MANAGEMENT

As mentioned previously, comprehensive city planning in the United States no longer spends all its energies attempting to formulate a grand master plan for 20-25 years hence. It is recognized that planning the resolution of a critical problem within the next several years may be more important than developing a longer-range plan, for example, when the short-range problem is severely disruptive or stands in the way of further progress. An extreme case is a community confronted with an emergency situation or natural disaster, which must apply all its resources to restoring normal operations as quickly as possible.

To be effective, city planning must spring from careful consideration of the present. For plans to be realized, they must represent extensions of the current situation and either existing or potential capabilities. Unless they meet these requirements, plans are imaginary rather than realistic. Current operations, short-range, middle- and long-range plans constitute one continuum of planning activity, with all stages significant although they are programmed for different times.

Management is therefore a more important element of comprehensive urban planning theory than it was some years ago. It is now part of analytic consideration and, in the opinion of practicing city planners, an important part of their activities on the job. When city planning staffs are assigned the responsibility, in the name of the chief municipal executive or the legislature, of coordinating departmental planning and ensuring that plans and commitments are carried out, they are necessarily concerned with management: income and application of funds, budgeting and programming, operating efficiency, span of control, and other matters associated with business management and public administration.

In their own work of collecting information, maintaining a core analysis of primary data, and preparing comprehensive master city plans, planners compound the difficulties if they do not manage these activities skillfully. In addition, at least until city planning is accepted and established as a foremost responsibility of municipal government, successful city planning depends as much on maintaining constructive relations with the managers of different municipal departments as on some superlative qualities of the plans themselves. Management skills are also needed to prevent the progressive bureaucratic inertia and inefficiency which plagues so many governments today.

Not long ago, the plans recommended to legislative decision-makers were long-range and largely imaginary, with little expectation or even intention that they would be accepted and acted on in any significant way. Now that comprehensive city planning is proposed as a significant effort, municipal decision-making becomes most important. How must city planning fit the preferences and procedures of decision-making by municipal legislators, mayors, city managers, chief administrative officers, and departmental general managers so that they will use rather than reject it?[38] If the traditional city planner considers the importance and realities of the present, and the municipal executive regards planning as his highest managerial responsibility, do these two approaches ultimately produce a common form of comprehensive urban planning?

Although the man-computer relationship is significant in all managerial activities, it is most important in government with its responsibilities for the general public interest. With computers so widely used today, human decision-makers find it increasingly difficult to comprehend computer programs and confirm the accuracy of resultant calculations. Unless the decision-maker knows the assumptions underlying the program and understands the methods of calculation, he must accept the results on faith or reject them until he acquires the necessary knowledge. This situation occurs more and more often as information and knowledge become more specialized and difficult for any one person to comprehend. No municipal or corporate chief executive can hope to be knowledgeable in all fields or areas of expertise involved in the decisions he must make: finance, taxes, statistics, mathematical operations analysis, law, engineering, various physical and social sciences, public and labor relations.

With more and more data processing and analysis for decisions produced by computer, how much decision-making is really being made by computer programmers rather than by executives who think that they are making the decisions? How can individual managers or even management committees make and accept responsibility for decisions when more and more of the substance behind them is not fully understood or cannot be checked for accuracy. No management theory has yet been advanced for solving this fundamental problem of how the chief executive can attain sufficient understanding of increasingly diverse and specialized inputs to be able to correlate them meaningfully and reach sound conclusions.[39]

Management is also, of course, an important consideration in determining how comprehensive city planning should be institutionalized, discussed briefly in the next section. For planning to be established so that it cannot be managed efficiently would be paradoxical indeed.

INSTITUTION

How comprehensive city planning is best organized and established as a legal institution within municipal government has been debated for many years.[40] There are those who believe – to paraphrase former French Premier Georges Clemenceau's remark concerning war – that city planning is too important to be left to municipal legislators and bureaucrats. Therefore, city planning commissions should be retained in some form to represent the general public and private enterprise which builds and shapes the greater part of cities in the United States. An opposing view maintains that commissions are responsible for the poor showing of city planning because commissioners are mostly incompetent, prejudiced for their own special interest, and unwilling to work hard enough to do the job. For these reasons, their recommendations are seldom heeded and rarely executed by law-makers.

A theory long discussed and familiar to those in the field regards city planning as an executive function, attached to the office of the mayor, city manager, or other chief administrative officer.[41] Another concept maintains that city planning is the primary responsibility of the municipal legislature;

therefore, city planning staff should be attached and responsive directly to the city council.[42] So far, no municipal legislature in the United States, except possibly in small or new towns, has accepted city planning as a primary obligation and activity. Still another theory advocates greater use of temporary task forces to study particular problems and propose solutions, plans, or constructive actions to the regular operating units of city government.[43] As private enterprise seeks to sell services and products that might be used in city planning and its effectuation, corporations and professional consultants suggest that many functional plans and comprehensive urban planning itself would best be done by them as contract projects or on retainer.

A quite different theory postulates that planning is more than the highest responsibility and expression of governmental management as it is conceived today. It proposes planning as a "fourth power" or additional branch of government with new constitutional responsibilities. The planning branch at each level of government would maintain the plans in force for the jurisdiction involved, supply the different units of the government with the basic information they require for their planning, and provide the executive and legislative branches with projective analyses of the effects of policies and actions they are considering. This concept institutionalizes comprehensive planning as a pivotal activity of government which accumulates society's intentions and commitments for the future that are expressed in plans and formal actions, as a continuing reference for those legislatively and operationally concerned and for the interested public.[44]

Through the years various concepts have been advanced concerning metropolitan regional planning. It is widely recognized, of course, by those involved that planning by many different municipalities for their separate jurisdictions within a much larger continuous built-up area is inefficient with respect to areawide services such as water supply, transportation, or waste disposal, and largely ineffective with respect to such environmental problems as air and water pollution. One set of proposals calls for special bodies to coordinate plans, such as regional planning commissions or associations of officials from different jurisdictions in the area. Another type of proposal suggests more basic change such as abolishing separate municipalities in favor of the county government that contains them, or creating new metropolitan governments.

Arguments are made that municipal government is much too centralized in downtown city halls, disassociated from the people it is intended to serve, and oblivious to their wishes. To correct this situation, city planning should be decentralized and more decisions made locally by districts and neighborhoods.[45] Communication between people and their municipal government should be greatly improved, with more effort by city hall to determine public desires and attitudes. At the same time, it is recognized that most urban problems require analysis, resolution, and action on a larger and larger scale – to the point some people propose that county or new metropolitan governments take over city planning for municipalities within their jurisdiction, as mentioned in the preceding paragraph.

The desirability and feasibility of such proposals depend on the answers to a number of interrelated questions. Exactly what functions and decisions should be decentralized? How will the necessary coordination or conformance between decentralized decisions and citywide policies, plans, and actions be assured? How will decisions at the neighborhood level that are against the municipal general interest be prevented? Who will make the decentralized decisions, by what process? How will this apparatus relate to existing political organization and representation? Do improved communications between people and central city planning affect the need for or type of desirable decentralization? What are the responsibilities of the city planner in the public interest?[46]

Four theoretical areas are involved in this set of closely connected questions: (1) comprehensive city planning (required level, process, and provisions for analysis and action in the total municipal interest); (2) management science (operational consequences of centralization and decentralization); (3) political science (local representation and organization); and (4) communications (interaction between people and planning). Conflicts and alternatives brought up by these several aspects must be resolved before the initial question of decentralization can be soundly evaluated. This is an excellent example of the many situations in city planning which should not be resolved on the basis of a single consideration. When related consequences are undesirable or impractical, the precipitate conclusion may be abandoned later after time, effort, and money have been wasted.

Theories concerning the organization and institutionalization of comprehensive city planning are certainly relevant today, since the federal government is continuing its efforts to promote and shape local planning, revenue sharing calls for a higher order of state planning, and national land use legislation and a national planning agency are actively discussed.

ENVIRONMENT

Intensified concern with environment adds a new group of elements to the physical-spatial and several socioeconomic components which are usually considered in city planning in the United States. There are many new effects to measure and evaluate: air and ground pollution (dust, temperature, moisture, wind, chemical and biological contamination); noise; vibration; electromagnetic disturbance; odor; light; hazards (fire, explosion, radiation, subsidence, earthquake); aesthetics; amenity; compatibility with surroundings. These are in addition to those elements already incorporated in most municipal building codes or zoning ordinances: protecting water supplies, drainage, sewerage, and soil stability. And in recent years, the interrelationships between energy production and environmental quality have been highlighted by increasing oil pollution of oceans and harbors, questions concerning nuclear safety and the disposal of atomic wastes, and proposed strip mining.

The analytical difficulties resulting from this proliferation of considerations is nowhere more evident than in the current confusion brought about by environmental impact statements. Progress has been made in measuring environmental effects separately, such as certain air pollutants, noise, or various chemical contaminants of water. But no one knows yet how to evaluate all significant environmental impacts or their combined effects. And determining the impact of a single small project on an entire air basin or watershed is possible at best for only a few effects. No uniform statements or methods of analysis have been developed and widely accepted.

Methods of charting and comparing different features of the natural terrain provide a basis for judgment in highway route selection and the location and siting of other projects.[47] Matrices are proposed which list environmental factors along one side and word descriptions of the severity of their effect along the second axis. But there is no adequate theory of how to measure, correlate, and combine these many effects, except judgmentally with full knowledge that the human mind cannot encompass or place in logical order the multitudinous interactions and comparative values involved.

These difficulties of fully considering the impacts connected with small projects are compounded with "complex sources": large multi-purpose developments such as shopping centers or "industrial parks." Their environmental effects can thread throughout an entire community or metropolitan region. They enlarge further the diversity of considerations and comprehension needed to evaluate the wider range of their environmental impact. For example, if ecology is regarded as part of environmental examination, analysis is further complicated by including all flora and fauna, and such concepts as spatial competition, societal organization, dominance, or symbiosis occurring within the city and its neighborhoods.[48]

It will be some time before the literal requirements of the national environmental policy act can be carried out. They call for: determination of the "environmental impact" of the proposed action; identification of "adverse environmental effects" which cannot be avoided if the proposal is implemented; indication of "alternatives"; the "relationship between local short-term uses of man's environment and the maintenance and enhancement of long-term productivity"; and "any irreversible and irretrievable commitments of resources which would be involved. . . ." We cannot yet successfully analyze the relatively few elements with which we have been concerned in city planning. Meeting the new environmental requirements calls for intellectual, theoretical, and analytical attainments of the highest order. When accomplished, this achievement will in itself confirm comprehensive urban planning as a distinctive and highly significant field of knowledge and action.

The question of density soon intrudes into any discussion of environment, because the number of people on the land is obviously one of the most impactful considerations. Beyond the limitations imposed through zoning, city planning practitioners have been able to evade this question until recently. Except for a few locations in the largest cities, population density has been of concern in the United States mainly as it relates to single-family and single-class neighborhoods. As people continue to concentrate in cities, density must be considered *sui generis* with respect to the many causes and effects of different degrees and kinds of crowding.[49]

Advances in the technology of transportation, the construction of highrise and megastructure buildings, municipal utility engineering, and fire and police protection allow greater concentrations of people on the land than ever before. But serious questions are being raised concerning the adverse socio-psychological effects of crowding, such as increased urban congestion, greater danger from emergencies and disasters, and impaired amenity and aesthetics.[50] Very much involved are the social effects of physical environment, discussed for years in the fields of city planning, architecture, civic design, and sociology.[51] As yet, there are no reliable theoretical or research indicators of the density limits at which the problems and costs for cities and their populations are greater than the benefits.

EMERGING CONCEPTS

Comprehensive urban planning is ripe for new theoretical concepts and formulations of all kinds.[52] Basic theory is more urgently needed than ever to provide sound principles to guide city planning in practice. Broadening of coverage and deepening the content of the field have created a vacuum of needed thought and knowledge. There are few aspects of urban planning as it is viewed today for which adequate theory exists.

The concept of social planning in the United States, although advanced some years ago, is still unresolved in theory and practice. Besides political and humanitarian pressures to consider lower-income, poor, aged, infirm, and other disadvantaged people, it is clear that city planning cannot be analytically realistic if it does not take into account the social concerns that are an inevitable part of such planning in the modern world and its primary justification. It is recognized that information, analysis, decisions, and actions concerning many social factors are required to properly plan the urban economy, housing, municipal tax policies, health, welfare and other service programs, or the nature and location of most municipal facilities. But theories of how best to incorporate such social considerations into comprehensive urban planning are lacking.

"Social indicators" are suggested to identify social needs and measure progress toward their solution. So far, however, few of these indicators are available that meaningfully measure the social condition or quality intended, or that cannot be interpreted in different ways.[53] Nor has forecasting social outcomes under different assumed conditions been carried much further than the idea itself.[54]

For various reasons, many planners believe that the full range of income, age, ethnic, educational, and other socioeconomic categories should be represented among the first inhabitants of "new towns," among their populations as they grow, and in more residential areas of older cities. But whether to formally adopt and implement this objective is not yet decided in the United States. If and when the decision is made, theories are lacking on how best to attain and maintain this diversity of population.

The designs of almost all new towns throughout the world incorporate spatially differentiated neighborhood units of from 2,000 to 5,000 people. They are also proposed as basic "growth units" for city planning in the United States.[55] Despite this widespread advocacy, there are those who argue that planned neighborhood units promote segregation.[56] As pointed out previously, this relates to the social effects of spatial design, such as any effects of the physical arrangement of dwelling units on neighborhood sociability or crime. Because of these concerns, theoretical justification or refutation of the neighborhood-unit concept as presently advocated or in modified form would certainly be worthwhile, since the concept has been so persistently accepted and applied whenever possible for many years.

Forecasting conditions and events is another theoretical area significant for planning, but as pointed out in one of the readings in this book much must be accomplished before such projections are meaningful and reliable.[57] Even if only moderately successful social and technological forecasts are achieved, the value of organized forethought in the form of projections and plans is greatly increased. Such forecasting will require greater knowledge of the basic mechanisms and characteristics underlying human behavior. There may be analogies between human and other animal behavior which will contribute to our understanding of the urban organism; or knowledge derived from ethological research might be applied directly in planning.[58] Theories concerning the attitudes and reactions of humans should help determine what kind of community will provide the most satisfying and beneficial experience of living in cities.[59]

Because of worsening congestion at the center of cities, the substitutability of communication for transportation in larger cities is being examined. Will the present concentration of activities and people "downtown" be necessary or desirable in the future? Do modern communications and trends in the production of goods and services suggest that downtown concentrations could and should be gradually decentralized throughout the metropolitan area, except for those functions which clearly need to be located at the spatial center of the city? Would such decentralization be operationally and economically feasible, reduce congestion, and improve urban environments? Obviously, this is a most important question which requires development of a theoretical concept, its formulation into indicative research, followed by pilot tests.[60]

In recent years, new technological developments have followed one another so rapidly that it is not surprising that public control or programming of their introduction into the community is being debated. The first disruptive development widely reported in the United States was the introduction about 20 years ago of hard detergents which were not biodegradable and required large municipal expenditures to restore the proper functioning of sewage treatment plants. More recently, introduction of the jumbo aircraft necessitated changes at major airports around the world, at a time when airlines were hard-pressed financially: new or greatly enlarged maintenance hangars, new passenger facilities, reinforcement of runways and taxiways, and improved ground transportation systems to relieve the traffic congestion around airports which was beginning to discourage some people from air travel and to depress surrounding neighborhoods and cities generally.

The inevitable question raised by such impactful technological developments is whether municipalities or society as represented by other governmental units should have a say in how suddenly they are saddled with major changes and costs over which they have little or no control. The lead time for most such developments is 10 years or so. To the extent possible, serious city planning must include forecasting the effects of technological advances in the community and reducing the municipal problems and costs that they produce.[61]

The likelihood of even more impactful technological developments in the future – together with greater awareness of the vulnerability of our earth to cosmic events and irreversible global changes induced by man – pose the question of how much man can and should try to plan, as well as how he should do it. The limitations attributed throughout history to supernatural forces are expressed by Shakespeare in the words of Hamlet: "There's a divinity that shapes our ends, Rough-hew them how we will." Robert Burns spoke more generally in his oft-quoted line: "The best laid schemes o' mice and men, Gang aft a-gley." Besides these divine and random effects, the directive influences of genetics are beginning to be understood. In addition, there are phenomena beyond man's consciousness and present understanding, now acknowledged in the scientific community by its formal recognition of research into extrasensory perceptions and forces.

This has led some students of city planning to conclude that complete knowledge will never be available to support the conclusions and decisions of comprehensive urban planning. Rather it must concentrate on adjusting to uncertainty, identifying trends and coming events, and applying a continuous combination of direction and adjustment to the elements that it can affect. In part, the theory of incremental planning reflects this view.[62] Some people go further, believing that the most fundamental concerns of comprehensive urban planning are beyond human capacity to control or even significantly affect, such as population growth, worldwide pollution and military expenditures, war, and social discord. Others believe that controlled biological, psycho-social, and medical developments will lead to a high degree of genetic and behavioral manipulation of man by man, in which planning will perforce be preeminent. The implications of such basic assumptions for planning theory and practice are self-evident.[63]

Most important, of course, is the question of whether there is a theory of planning in general, and more specifically for comprehensive urban planning. If the present dictionary definition is accepted, there will be such a general theory because so much is included in the word *theory:* contemplation, speculation, and their results; general or abstract principles; a plan or scheme theoretically constructed; a hypothesis; or mathematical theorems. The question is not so readily answered with the more rigorous definition that a theory is a formulation of relationships concerning a phenomenon or organism which permits predictions concerning its behavior or development to be made and subsequently tested to indicate the theory's validity. Certainly, theories meeting this more rigorous definition exist for many of the components or functional elements involved in comprehensive urban planning, and for some subsystems or combinations of elements. Whether an overall comprehensive planning or general systems theory of universal applicability can be developed remains to be seen.

Involved in this important question is the proliferation in recent years of various kinds of coordinative planning, each of which proclaims itself distinct with an incipient theory of its own. The list now includes: physical planning, social planning, systems planning, management science, policy science,[64] urban design, advocacy planning, incremental planning, environmental planning, ecological planning, and comprehensive health planning. Are these different categories of coordinative planning simply different concepts, formulations, or applications of a common process of comprehensive planning that encompasses them all? Similarities among them and the ease with which the subject matter of one can be incorporated within the logical content of another suggest that this may be the case, and that one general planning theory will apply to most variations.

A new and radically different set of considerations for planning lies on the horizon. Possible ways of controlling behavior and modifying personality now comprise an impressive and depressing list: altering the genetic code, gene selection by controlled mating, nutritional influences, drugs, surgical intervention, environmental and behavioral manipulation, monitoring, and others.[65] Prior determination of sex is already a reality and discussion of its implications for society is appearing in print.[66] If at some time in the future the full array of manipulative forces is accepted by or imposed upon society, the scope of the controls that will have to be planned and enforced is awesome and frightening indeed. It seems inevitable that more and more "human engineering" will be involved in comprehensive planning. It is by no means too early or too imaginary to begin to theorize on how man can plan to retain essential and desirable freedoms as these directive and manipulative forces are developed and applied.

REFERENCES

[1] Thomas Adams, *Outline of Town and City Planning,* New York (Russell Sage), 1935: Early Efforts in Town and City Planning, pp. 33-139.

[2] Reference 1 above, p. 56; Pierre Lavedan, *Histoire de l'urbanism,* Paris (Henri Laurens), 1926: Vol. I, Antiquité, Moyen Age, p. 145; Robert B. Mitchell and Chester Rapkin, *Urban Traffic: A Function of Land Use,* New York (Columbia University Press), 1954: The Influence of Movement on Land Use Patterns, pp. 104-132 (Reading No. 4).

[3] Pierre Lavedan, Reference 2 above, pp. 115, 146.

[4] C.P. Venkatarama Ayyar, *Town Planning in Ancient Deccan,* Madras, 1916(?), pp. VII-199; Sir John Marshall (Editor), *Mohenjo-daro and the Indus Civilization,* London (A. Probsthain) 1931: Vol. I, Chap. XVI, Architecture and Masonry, pp. 262-286; Richard Babcock, *The Zoning Game – Municipal Practices and Policies,* Madison, Wis. (University of Wisconsin), 1966: The Purpose of Zoning, pp. 115-125 (Reading No. 5).

[5] Clarence A. Perry, "The Neighborhood Unit Formula," in: William Wheaton (Editor), *Urban Housing,* New York (Free Press), 1966, pp. 94-109 (Reading No. 3); Ebenezer Howard, *Garden Cities of Tomorrow,* Cambridge, Mass. (M.I.T.), 1965:

The Town-Country Magnet, pp. 50-57 (Reading No. 2).
[6]Leonard Reissman, *The Urban Process,* Cities in Industrialized Societies, New York (Free Press), London (Collier-Macmillan), 1970: The New Visionary: Planner for Urban Utopia, pp. 39-68 (Reading No. 1); Lewis Mumford, *The Story of Utopias,* New York (Boni and Liveright), 1922, 315 pp.
[7]Melville C. Branch, "Three Graphic Reconstructions of Early City Forms," *Journal of the American Institute of Planners,* Feb. 1959, pp. 26, 34, 40; Paul D. Spreiregan, *Urban Design: The Architecture of Towns and Cities,* New York (McGraw-Hill) 1965: The Roots of Our Modern Concepts, pp. 29-48 (Reading No. 6); Ludwig Hilbersheimer, *The New City,* Chicago, Ill. (Theobold), 1944, pp. 113-122 ff; C. Peter Rydell and Gretchen Schwarz, "Air Pollution and Urban Form: A Review of Current Literature," *Journal of the American Institute of Planners,* Mar. 1968, pp. 115-120.
[8]Thomas Adams, Reference 1 above, pp. 101-106, 125-128.
[9]Paul Ritter, *Planning for Man and Motor,* New York (Macmillan), 1964, 384 pp.
[10]Colin Buchanan, *Traffic in Towns,* Middlesex, England (Penguin), 1964, 263 pp.
[11]A. Scheffer Lang and Richard M. Soberman, *Urban Rail Transit: Its Economics and Technology,* Cambridge, Mass. (M.I.T.), 1963, 139 pp.
[12]Melville C. Branch, *Urban Air Traffic and City Planning – Case Study of Los Angeles County,* New York (Praeger), 1973, 97 pp.
[13]Eliel Saarinen, *The City: Its Growth, Its Decay, Its Future,* New York (Reinhold), 1943: Principles of Town-Building, pp. 8-20 (Reading No. 8).
[14]Paolo Soleri, *Arcology: The City in the Image of Man,* Cambridge, Mass. and London (M.I.T.), 1969, 122 pp.; Kenzo Tange, *Kenzo Tange, Architecture and Urban Design,* New York (Praeger), 304 pp.; Athelstan Spilhaus, "The Experimental City," *Daedalus,* Fall 1967, pp. 1129-1141.
[15]Kevin Lynch, *The Image of the City,* Cambridge, Mass. (M.I.T.), 1960: The City Image and Its Elements, pp. 46-49, City Form, pp. 95-117 (Reading No. 7).
[16]Ian L. McHarg, *Design with Nature,* Garden City, N.Y. (Natural History Press), 1969: The Plight, pp. 19-29, A Step Forward, pp. 33-34 (Reading No. 40).
[17]Hans Blumenfeld, "On the Concentric-Circle Theory of Urban Growth," *Land Economics,* May 1949, pp. 209-212; Hans Blumenfeld, "Theory of City Form, Past and Present," *Journal of the Society of Architectural Historians,* July-Dec. 1949, pp. 7-16; Department of City Planning, *Concept Los Angeles,* The Concept for the Los Angeles General Plan, Los Angeles, Cal., Jan. 1970: Centers, pp. 13-17; David L. Birch, "Toward a Stage Theory of Urban Growth," *Journal of the American Institute of Planners,* Mar. 1971, pp. 78-87; F. Stuart Chapin, Jr., *Urban Land Use Planning,* Urbana, Ill. (University of Illinois), 1965: Toward a Theory of Urban Growth and Development, pp. 69-99 (Reading No. 10).
[18]Albert Z. Guttenberg, "Urban Structure and Urban Growth," *Journal of the American Institute of Planners,* May 1960, pp. 104-110 (Reading No. 9).
[19]Edwin von Böventer, "Spatial Organization Theory as a Basis for Regional Planning," *Journal of the American Institute of Planners,* May 1964, pp. 90-100 (Reading No. 11).
[20]Ida R. Hoos, "Information Systems and Public Planning," *Management Science,* Aug. 1967, pp. 817-831 (Reading No. 12).
[21]Melville C. Branch, *Planning Urban Environment,* Stroudsburg, Pa. (Dowden, Hutchinson, & Ross), 1974: Continuous City Planning, pp. 212-219.
[22]Melvin M. Webber, "The Roles of Intelligence Systems in Urban-Systems Planning," *Journal of the American Institute of Planners,* Nov. 1965, pp. 289-296.
[23]Andrew Vazsonyi, "Automated Information Systems in Planning, Control and Command," *Management Science,* Feb. 1965, pp. B-2 - B-40 (Reading No. 13).
[24]Herbert A. Simon, "How Big Is a Chunk?," *Science,* 8 Feb. 1974, pp. 482-488.
[25]John Dearden, "Myth of Real-Time Management Information," *Harvard Business Review,* May-June 1966, pp. 123-132 (Reading No. 14).
[26]Donald G. Hagman, *Urban Planning and Land Development Control Law,* St. Paul, Minn. (West), 1971, 559 pp.
[27]Charles J. Hitch, "Development and Salient

Features of the Programming System," in: Samuel A. Tucker (Editor), *A Modern Design for Defense Decision,* Washington, D.C. (Industrial College of the Armed Forces), 1966, pp. 64-68, 70-78, 83-85 (Reading No. 19); Planning-Programming-Budgeting System: A Symposium, *Public Administration Review,* Dec. 1966, pp. 243-310; A.R. Prest and R. Turvey, "Cost-Benefit Analysis: A Survey," *The Economic Journal,* Dec. 1965, pp. 683-735; Reference 53 below (Reading No. 48).

[28] Alan Black, "The Comprehensive Plan," in: William I. Goodman and Eric C. Freund (Editors), *Principles and Practice of Urban Planning,* Washington, D.C. (International City Managers' Association), 1968, pp. 349-350, 357-359, 371, 373 (Reading No. 15).

[29] Rexford G. Tugwell, "Implementing the General Interest," *Public Administration Review,* Autumn 1940, pp. 33-36, 39-41, 48 (Reading No. 16).

[30] Charles J. Hitch, Reference 27 above; P. Wood, "The Use of a Program Planning Budgeting System to Improve the Coordination and Implementation of Physical, Economic and Social Planning," *Papers and Proceedings,* Vol. II, Copenhagen, Denmark (International Federation of Housing and Town Planning), Sept. 1973, pp. 527-535 (Reading No. 20).

[31] E.S. Savas, "Cybernetics in City Hall," *Science,* 29 May 1970, pp. 1066-1071 (Reading No. 17).

[32] Melville C. Branch, *Continuous City Planning,* Planning Advisory Service Report No. 290, Chicago, Ill. (American Society of Planning Officials), 1973, 26 pp. (Reading No. 18).

[33] James Hughes and Lawrence Mann, "Systems and Planning Theory," *Journal of the American Institute of Planners,* Sept. 1969, pp. 330-333 (Reading No. 22); Richard A. Johnson, Fremont E. Kast, and James E. Rosenzweig, "Systems Theory and Management," *Management Science,* Jan. 1964, pp. 367-384 (Reading No. 23); Jay W. Forrester, "The Impact of Feedback Control Concepts on the Management Sciences," New York, Foundation for Instrumentation Education and Research, 1959, 24 pp. (Reading No. 24); Jay W. Forrester, "A Deeper Knowledge of Social Systems," *Technology Review,* Apr. 1969, pp. 22-31 (Reading No. 25); Jay W. Forrester, "Industrial Dynamics: A Major Breakthrough for Decision Makers," *Harvard Business Review,* July-August 1958, pp. 37-66; Jay W. Forrester, *Urban Dynamics,* Cambridge, Mass. (M.I.T.), 1968, 256 pp.; Harper Q. North and Donald L. Pyke, "Technological Forecasting to Aid R&D Planning," *Research Management,* 1969, pp. 289-296 (Reading No. 26).

[34] Melville C. Branch, "Goals and Objectives in Civil Comprehensive Planning," in: John Lawrence (Editor), *OR 69, Proceedings of the Fifth International Conference on Operational Research,* London (Tavistock), 1970, pp. 55-63 (Reading No. 21).

[35] Alan J. Rowe, "Simulation – A Decision-Aiding Tool," *Proceedings of the American Institute of Industrial Engineers,* Sept. 1963, pp. 135-144 (Reading No. 27).

[36] Richard L. Van Horn, "Validation of Simulation Results," *Management Science,* Jan. 1971, pp. 247-258 (Reading No. 28); Ira S. Lowry, "A Short Course in Model Design," *Journal of the American Institute of Planners,* May 1965, pp. 158-165 (Reading No. 29).

[37] Martin Shubik, "On the Scope of Gaming," *Management Science,* Part II, Jan. 1972, pp. P-20 - P-36 (Reading No. 30).

[38] Henry Mintzberg, "Managerial Work: Analysis from Observation," *Management Science,* Oct. 1971, pp. B-97 - B-109 (Reading No. 33).

[39] Melville C. Branch, "Simulation, Mathematical Models, and Comprehensive City Planning," *Urban Affairs Quarterly,* Mar. 1966, pp. 15-38 (Reading No. 31); David L. Johnson and Arthur L. Kobler, "The Man-Computer Relationship," *Science,* Nov. 1962, pp. 873-879 (Reading No. 32); Samuel G. Trull, "Some Factors Involved in Determining Total Decision Success," *Management Science,* Feb. 1966, pp. B-270 - B-280 (Reading No. 34).

[40] James H. Pickford, "The Local Planning Agency: Organization and Structure," in: William I. Goodman and Eric C. Freund (Editors), *Principles and Practice of Urban Planning,* Washington, D.C. (International City Managers' Association), 1968, pp. 525-531 (Reading No. 35).

[41] Robert A. Walker, "The Implementation of Planning Measures," *Journal of The American Institute of Planners,* Summer 1950, pp. 122-130 (Reading No. 36).

[42] T.J. Kent, *Urban General Plan,* San Francisco, Cal. (Chandler), 1964, 213 pp.

[43] P.H. Nash and Dennis Durden, "A Task-Force Approach to Replace the Planning Board," *Journal of the American Institute of Planners*, Feb. 1964, pp. 10-22.

[44] Rexford G. Tugwell, "The Fourth Power," *Planning and Civic Comment*, April-June 1939, pp. 1,2,4,6,11,16,17,26,30,31; Rexford G. Tugwell (Karen W. Hapgood), "The Tugwell Proposed Constitution," *Planning*, Dec. 1970, pp. 143-144. (Reading No. 37)

[45] Edmund M. Burke, "Citizen Participation Strategies," *Journal of the American Institute of Planners*, Sept. 1968, pp. 287-294 (Reading No. 39).

[46] Charles Abrams, "The City Planner and the Public Interest," *Columbia University Forum*, Fall 1965, pp. 25-28 (Reading No. 38).

[47] Ian L. McHarg, Reference 16 above. (Reading No. 40)

[48] Leonard Reissman, Reference 6 above. The Ecologists: Analysts of Urban Patterns, pp. 93-121.

[49] George Macinko, "Saturation: A Problem Evaded in Planning Land Use," *Science*, 30 July 1965, pp. 516-521 (Reading No. 41).

[50] Bernard Asbell, "The Danger Signals of Crowding," *Think*, July-August 1969, pp. 30-33 (Reading No. 43).

[51] Irving Rosow, "The Social Effects of the Physical Environment," *Journal of the American Institute of Planners*, May 1961, pp. 127-133 (Reading No. 42).

[52] Richard S. Bolan, "Emerging Views of Planning," *Journal of the American Institute of Planners*, July 1967, pp. 233-245 (Reading No. 44).

[53] Walton J. Francis, *A Report on the Measurement and Quality of Life and the Implications for Government Action of the "Limits to Growth,"* Washington, D.C. (Dept. of Health, Education, and Welfare), Jan. 1973: A Report on the Measurement and the Quality of Life, pp. 6-16 (Reading No. 48).

[54] Dennis L. Meadows, "Toward a Science of Social Forecasting," *Proceedings of the National Academy of Sciences, USA*, Dec. 1972, pp. 3828-3831 (Reading No. 51).

[55] Max O. Urbahn, "The First Report of the National Policy Task Force," *MEMO*, Special Newsletter, Washington, D.C. (American Institute of Architects), Jan. 1972: Growth Unit, pp. 4-7.

[56] Reginald R. Isaacs, "Are Urban Neighborhoods Possible?" (Part I), "The 'Neighborhood Unit' Is an Instrument of Segregation" (Part II), *Journal of Housing, NAHO*, [1966?], 9 pp.

[57] Reference 54 above.

[58] Robert M. Griffin, Jr., "Ethological Concepts for Planning," *Journal of the American Institute of Planners*, Jan. 1969, pp. 54-60 (Reading No. 49).

[59] Stanley Milgram, "The Experience of Living in Cities," *Science*, 13 Mar. 1970, pp. 1461-1467 (Reading No. 50).

[60] Paul Gray, *Prospects and Realities of the Telecommunications/Transportation Tradeoff*, Los Angeles, Cal. (Graduate School of Business Administration, University of Southern California), Sept. 1973, 32 pp. mimeo.

[61] Michael S. Baram, "Social Control of Science and Technology," *Science*, 7 May 1971, pp. 535-538; Committee on Public Engineering Policy, National Academy of Engineering, *A Study of Technology Assessment*, Washington, D.C. (Committee on Science and Astronautics, U.S. House of Representatives), July 1969, 208 pp.; Harper Q. North and Donald L. Pyke, Reference No. 33 above.

[62] Charles E. Lindblom, "The Science of 'Muddling Through.'" *Public Administration Review*, Spring 1959, pp. 79-88 (Reading No. 45).

[63] Beryl L. Crowe, "The Tragedy of the Commons Revisited," *Science*, 28 Nov. 1969, pp. 1103-1107 (Reading No. 47).

[64] Dennis A. Rondinelli, "Urban Planning as Policy Analysis: Management of Urban Change," *Journal of the American Institute of Planners*, Jan. 1973, pp. 13-22 (Reading No. 46).

[65] Gardner C. Quarton, "Deliberate Efforts to Control Human Behavior and Modify Personality," *Daedalus*, Summer 1967, pp. 837-853; Nicolas Wade, "Genetic Manipulation: Temporary Embargo Proposed on Research," *Science*, 26 July 1974, pp. 332-334.

[66] Charles F. Westoff and Ronald R. Rindfuss, "Sex Preselection in the United States: Some Implications," *Science*, 10 May 1974, pp. 633-636.

PART I — LEGACY

Looking back over the course of Western Civilization since the fifteenth century, it is fairly plain that mechanical integration and social disruption have gone on side by side. Our capacity for physical organization has enormously increased; but our ability to create a harmonious counterpoise to these external linkages by means of cooperative and civic associations . . . has not kept pace with these mechanical triumphs. By one of those mischievous turns, from which history is rarely free, it was precisely during this period of flowing physical energies, social disintegration, and bewildered political experiment that the populations of the world as a whole began mightily to increase, and the cities of the Western World began to grow at an inordinate rate. . . .

We have now reached a point where fresh accumulations of historical insight and scientific knowledge are ready to flow into social life, to mold anew the forms of cities, to assist in the transformation of both the instruments and goals of our civilization. Profound changes, which will affect the distribution and increase of population, the efficiency of industry, and the quality of Western Culture, have already become visible. To form an accurate estimate of these new potentialities and to suggest their direction into channels of human welfare, is one of the major offices of the contemporary student of cities. Ultimately, such studies, forecasts, and imaginative projects must bear directly upon the life of every human being in our civilization.

Lewis Mumford, The Culture of Cities, *New York (Harcourt, Brace) 1938, pp. 7, 10.*

1

Leonard Reissman

THE VISIONARY: PLANNER FOR URBAN UTOPIA

To the visionary the city is a problem environment, one that has developed without plan and one more sensitive to the narrow economic demands of the moment than to the lasting moral and social needs of individuals. He sees in the industrial city what the practitioner has seen: congestion, slums, blight, inefficiency, and all the other consequences of civic irresponsibility. Although both types are impressed by the same urban realities, they differ markedly in the interpretations they give to them. For the practitioner, congestion, blight and the rest of the shoddy inventory are the problems; for the visionary they are but the symptoms. The basic problem is the ethos and organization of industrial society itself. This difference separates the visionary from the practitioner. Because of it they are oriented differently in their study of and solutions for urban problems. The practitioner, as earlier described, is committed by his job to the piecemeal solutions that

---

 Selected from The Urban Process: Cities in Industrialized Societies *by Leonard Reissman, The Free Press, New York (1970), 39-68.*

can be enforced quickly, such as widening streets, zoning, partial urban renewal, or streamlining some phase of governmental organization. The visionary is a vehement opponent of these temporary measures. Rather, he insists that radical surgery is the only way to save the patient. The point is sharply made by Mumford, who establishes the visionary's perspective in this respect.

> *Much recent housing and city planning has been handicapped because those who have undertaken the work have had no clear notion of the social functions of the city. . . . And they did not, apparently, suspect that there might be gross deficiencies, misdirected efforts, mistaken expenditures, here that would not be set straight by merely building sanitary tenements or widening narrow streets.*

The visionary concentrates, then, on practical problems but is concerned with long-term qualitative considerations and social needs, and approaches these problems through a social ideology and an aesthetic philosophy. In brief, an urban problem for the visionary includes anything that violates his high standards of morality and aesthetics. His solution is usually nothing short of a massive reconstruction of metropolitan society. Standards of such a high order, after all, should not be compromised. The practitioner is committed to a job; the visionary is dedicated to an ideal.

The visionary's ideals are contained in The Plan: a blueprint, more or less detailed, for building into reality those forms, those values, and those qualities which he believes the city must contain. Sometimes, he even takes the trouble to indicate how we can reach that goal. In its fullest form, the blueprint includes not only plans for buildings, homes, and the general physical format of the city, but also definitions of what urban social institutions are to be included, and even the new psychology of the urbanite that is to emerge from all this. This is no picayune puttering with street plans or building facades or zoning regulations. It is a manifesto for an urban revolution.

The ideological roots that sustain the visionary go back to the middle of the nineteenth century and the protest, predominantly by socially conscious intellectuals, against the evils of industrialism. The protest centered on the effects of industrialism rather than on industrialism per se. It was more in opposition to the social system than to the machine. The benefits of industrial technology for human progress were more or less conceded. The argument, however, was with the social system that subordinated man to the machine and to the profit motive. The answer lay not in wrecking the machine but in controlling its social and ecological consequences. These indictments probably reached their highest pitch in the writings of Marx, in Engels' description of English factory life, and in Booth's classic studies of life and labor in London. These writings described with care the human consequences of industrialism: the effects of poverty, child labor, the erosion of the human spirit, and the senseless lives of the mass of people caught up by the factory and the city. It was in this period that dehumanization or alientation were understood in their crudest sense: depriving the individual of the barest essentials of humanity. By our own time of affluence, they are much more subtly and sophisticatedly defined.

What added fire to the intellectual's protests, aside from the real misery they saw, were the intolerable social discrepancies created by industrial society and, above all else, the differences between rich and poor. For the intellectuals the discrepancy was a betrayal of a promise. About 100 years before, with the onset of the industrial revolution, society had been dedicated to a new social philosophy that emphasized human dignity, the rule of law, the triumph of reason, and freedom. These new values were to replace the aristocratic rigidity, religious dogmatism, and monarchical absolutism condoned by the medieval philosophy of society. What the intellectuals had not seen was that the new ideology was meant to apply only to a small segment of society, not to all of it. By the middle of the last century, the intellectuals apparently were shocked into reality and they saw a social nightmare instead of the promised dream. Mandeville's *Fable of the Bees,* an allegory that argued that society as a whole benefited even as individuals suffered from the laissez-faire economic activity of its citizens, became a fairy story not even accurate for activity in the hive. The rule of law became a codified disregard of social responsibility. The new ideology became based more on the unassailable primacy of the natural order, of which the social order was conceived to be a part, than on the ability of man to shape his social order. In short, the gap between the promise of industrialism and its reality, especially in the city, became too wide to be ignored.

The disenchantment with industrial society emerged in one of three ways, depending upon the value placed on industrialism and the social change thought to be needed. (1) Reaction against it all, by which industrialism was entirely, if naively, discredited. The machine, the factory, and the city were considered to be beyond salvation in that they could not add anything worthwhile to society. In a reaction against industrialism the tightly comforting security of medievalism was sought through its image of a rediscovered rural Utopia. This philosophy has continued, in one form or another, up to the present, where it has become centered around the small community as the alternative to the metropolis. (2) Reform of some features of industrial society to keep such advantages as labor-saving machinery, release from monotonous tasks, and the comforts that machines could fashion. The reformers championed what Mannheim has called a "spatial wish," the projection of Utopia into space. By controlling industrialism for the benefit of all in a new social environment, the reformer argued, man could once again progress. Applied to the city, this view became the basis for the "Garden City" and its variations. (3) Revolt was yet a third alternative. The revolutionary accepted industrialism as a necessary historical phase; history neither could be set back to some earlier epoch nor could it be stopped. Industrial society could not be preserved as it was, nor could it be remodeled, even in part. Instead, a massive reconstitution was required, by which all existing institutions, values, and social mechanisms would be replaced by a new social order forging into reality the unfulfilled promises of industrialism. Mannheim, once again, has called these wish fulfillments "chiliasms," projections of dreams into time, the social Utopias.

The visionary at one time or another has been identified with all three of these disenchanted responses to industrialism. Ebenezer Howard set forth one such spatial Utopia in some detail. Later visionaries, such as Frank Lloyd Wright, perceived that tampering with the urban pattern necessarily involved changing economic mechanisms, political administration and social philosophy. And throughout much of the writing by visionaries, the simple desire to return to a rural civilization is obvious again and again.

PLANNING AS A SOCIAL MOVEMENT

The above description of the visionary, of his reactions to the industrial city, and of the manner he chose to enforce his values, all imply the elements of a social movement. A "social movement" is used here in its sociological meaning, to identify a concerted response by a group in support of a set of values. It need not involve organization, although the town planning movement did have organization. It does not require that the participants know they are part of a movement, although such was the case here. Social movements, in other words, are not always conscious, explicit, or fully organized. It is possible, within the meaning intended here, for individuals separately to promulgate an idea or a cause, either without steps to attain the end or without consciousness of having company in their endeavor.

I have used this conception of a social movement to avoid the implication that the visionaries were tightly organized behind a single set of beliefs. Quite the contrary, for the visionaries were too individualistic to be led, too authoritarian to be politicians in a democracy, and, in some cases, too egomaniacal to compromise their beliefs in order to move further toward their ultimate goals.

To add perspective to the analysis of the visionary's plans, therefore, let me cast him explicitly into that framework, into the role of leader of a social movement.

Planning, as understood by the visionary, should be kept clear from what is usually understood by that term today. For the visionary, planning expresses an ideology of urban reform and revolt. It goes beyond the piecemeal planning of the practitioner, for one thing. For another, it has a facade of rationality that is characteristic of most planning, but in the case of the visionary there is less reason and more emotion behind the facade.

Four criteria can be used to establish the visionaries and their plans as part of a social movement. First, the plans contained a set of propositions that could unify belief, even though these propositions were differently enunciated by different people. Most visionaries agreed that the city had deteriorated, that the industrial city was inhumane, and that the need for change was great. Although they disagreed about how to effect urban changes, there was no argument about the need for change.

Second, the movement had a history. Its point of origin, as earlier noted, was in the general protest against the effects of industrialism. The continued growth of technology, of science, of a factory society and a machine culture added more substance to the visionary's argument and new examples of how inhuman industrial cities could become. The expansion of the industrial city, coupled with a prevailing philosophy more attuned to economic demands than to social needs, assured the visionary planners of a cause and identified an enemy – landowners, industrialists, and the like. The perseverance of these problems helped give the movement relevance, over time, and time in which to develop a sense of tradition and continuity. Social movements sparked into being by a topical issue, no matter how vital, dissolve once the issue has disappeared or else try to stay alive by moving on to other causes. In the case of town planning, the movement was able to capitalize on existing protests against industrialism, and it had sufficient time to mobilize some of those sentiments in its own behalf.

Third, the movement created a following. The followers, then as now, have been intellectuals predominantly, because they were the first to recognize the discrepancies between what was promised and what was attained by industrialism. The visionaries never seemed to be wise in the ways of mass politics, or didn't want to be, so the movement never spread. Mass protest movements properly led stay close to the ground to pick up a following, but Utopian movements fly too high to be pursued by the masses. In certain cases, the visionaries were snobs by temperament, more entranced by their own ideals than dedicated to recruiting a following. The planning movement did find followers, however, although intellectuals predominated among it. What is more, the movement gained sufficient political strength to put some of its plans into effect: Howard's two garden cities, Letchworth and Welwyn, the Greenbelt towns of the New Deal era in the United States, and the British New Towns Act in 1946. The fact that there were followers, then, did give the movement some political reality. The importance of city planning and town planning today is further proof that we are considering more than the wishful dreams of a handful of politically ineffectual intellectuals.

Finally, the town planning movement had an organizing myth, which did much to make it cohesive. The function of the planning myth was to condense a complicated intellectual message into shorthand which could readily be translated into action. The myth thereby became an appealing part-truth, attractive for its simplicity, as are most myths.

The myth put forth by visionaries was based on urban characteristics that were certainly real. The evils within the industrial city at the turn of the century were plentiful and easily described. The reason for the degraded urban situation, the explanation went on, was that the city had been allowed to grow without plan, or at least without reference to human values. Economic competition had been elevated to the status of a natural law, rationality was measured primarily by profit, and self-interest was considered the primary instinct of man. Moral values and civic responsibility were twisted to serve economic values. Little wonder the industrial city developed as it did. However, argued the planners, man is rational. He can plan and thereby create a better, more harmonious, and more humane environment for himself. The economic forces and scientific knowledge that produced the industrial city could be mobilized to build anew the planned city.

The plan does not necessarily prescribe social revolution. Rather, existing social forces can simply be directed into other channels in order to realize the full potentials of an industrial society. Howard, for example, was careful to point out that his plan for the garden city was neither socialistic nor communistic. The plan, then, stated the conditions and the means for attaining a new environment. It seemed on the surface desirable, feasible, and necessary. The myth that it contained was a naive view of human motivation and of political structure: that reason is enough to change society from what it is to what we would like it to be. Above all, the visionary made the tacit assumption that his values were shared by most other people. This assumption proved to be the fatal flaw of most such plans.

Perhaps the best way to understand the visionary's contribution to an urban sociology is to look at some examples. Three have been chosen, each one because it is representative of a particular type. (1) Ebenezer Howard, the father of city planning and creator of the "garden city," a term that Osborn notes has become part of all modern languages: *Cité-Jardin, Gartenstadt, Cuidad-jardin, Tuinstad;* (2) Frank Lloyd Wright for the brash plan; (3) Lewis Mumford for an approach that is sociologically informed. Many others, such as Saarinen, Le Corbusier, Gropius, Neutra, Sitte, the Goodmans, and Gallion are not included. They are important in their own ways, but a detailed analysis would not have added greatly to the fulfillment of the main intent of this discussion, to present a range of sociologically relevant ideas. After all, the main purpose is not to assess the relative merits of lineal cities, ribbon developments, superblocks and the *ville radieuse,* but to consider the nature of urban society.

EBENEZER HOWARD: THE GARDEN CITY

Howard's book, *Tomorrow: A Peaceful Path to Real Reform,* was first published in 1898, and with slight revisions reappeared in 1902 under the title *Garden Cities of To-Morrow.* The book placed Howard among the intellectuals of the protest movement of the last century. Apparently his protest was independently arrived at, for Howard did not pore over books or steep himself in the literature, at least according to F.J. Osborn's evaluation of him.

Though not a scholar, Howard did know something of the intellectuals' protest of his time. He was a reader not only of the daily press but also of Edward Bellamy's *Looking Backward* and Henry George's *Progress and Poverty.* In any case, Howard tasted the flavor of protest that filled the period.

In a singular example of the organizing myth at work, Howard began by identifying the problem in a way calculated to win agreement from different quarters. It was

> *a question in regards to which one can scarcely find any difference of opinion. It is well nigh universally agreed by men of all parties . . . that it is deeply to be deplored that the people should continue to stream into the already overcrowded cities, and should thus further deplete the country districts.*

He took a short step to the solution.

> *All, then, are agreed on the pressing nature of this problem, . . . and though it would doubtless be quite Utopian to expect a similar agreement as to the value of any remedy that may be proposed, it is at least of immense importance that . . . we have such a consensus of opinion at the outset. . . . Yes, the key to the problem how to restore the people to the land – that beautiful land of ours, with its canopy of sky, the air that blows upon it, the sun that warms it, the rain and dew that moisten it . . . will be seen to pour a flood of light on the problems of intemperance, of excessive toil, of restless anxiety, of grinding poverty. . . .*

Howard's complaint was not only against urban congestion, but against the values of the industrial system that produced the city. It is worth quoting his remarks here in some detail, for in them Howard clearly showed the basis of his protest.

> *These crowded cities have done their work; they were the best which a society largely based on* selfishness *and* rapacity *could construct, but they are in the nature of things entirely unadapted for a society in which the* social side *of our nature is demanding a larger share of recognition. . . The large cities of today are scarcely better adapted for the expression of the fraternal spirit than would a work on astronomy which taught that the earth was the center of the universe be capable of adaptation for use in our schools. Each generation should build to suit its own needs; and it is no more in the nature of things that men should continue to live in old areas because their ancestors lived in them, than it is that they should cherish the old beliefs which a wider faith and a more*

> *enlarged understanding have outgrown. . . . The simple issue to be faced, and faced resolutely, is: Can better results be obtained by starting on a bold plan on comparatively virgin soil than by attempting to adapt our old cities to our* newer *and* higher *needs? Thus fairly faced, the question can only be answered in one way; and when that simple fact is well grasped, the* social revolution *will speedily commence. [Italics added.]*

These bold statements can be considered the garden city manifesto, and like the manifesto of Marx and Engels it identifies the evil, pinpoints the causes, and suggests the solution.

The crowded, industrial cities house the evils of our civilization. These evils have developed because the qualities of selfishness and rapacity have been valued long beyond their usefulness to society. They are now clearly out of tune with the social demands of a more developed, industrial society. In society, as in science, man must abandon outmoded methods and create new ones to meet current demands. Man is not, in Howard's view, at the mercy of uncontrollable social and natural forces. If the seventeenth and eighteenth centuries increased man's ability to control nature, then the task of the nineteenth century was to mobilize man's ability to control society. It was to this belief that Howard was dedicated, and it was to this conviction that his appeal was aimed.

To simplify his argument, Howard devised a now classic image: three magnets labeled "town," "country," and "town-country" are grouped around a rectangle labeled "The People: Where Will They Go?" The town and country each contained advantages (the positive pole) and disadvantages (the negative pole). Only the third, the town-country, was free of disadvantages, taking the best from the other two. The town magnet has the attractions of high wages, employment opportunities, amusements, edifices, and well-lit streets. However, it repels because of high rents and prices, excessive hours of toil, distance from work, foul air, slums, and gin palaces. The country magnet has the attractions of nature, sunshine, woods, and low rents; it repels because If lack of sociability, low wages, lack of amusement, deserted villages, and the lack of public spirit. It is only a clerical task to list the advantages of both as part of the town-country magnet: nature, opportunity, low rents and low prices, high wages, pure air and water, no smoke, no slums, bright homes, freedom, and cooperation. There are no disadvantages to the town-country, or garden city, life.

This was nature's way, Howard maintained in his argument, and it drew upon science, reason, and natural law. In the images and structure of his arguments and the facile way he led the argument to a determined conclusion, Howard exhibited the Utopian mentality. He was sincere, convinced, and dedicated. His main objectives, W.A. Eden has observed, went well "with the ordinary aspiration of the class to which Howard belonged, the somewhat earnest, chapel-going or chapel-emancipated lower middle class which had lately acquired political power and was destined to inaugurate a revolution by returning the Liberal Party with its high majority at the General Election of 1906."

Howard's plan was to purchase about 6,000 acres of open country and in the center, on 1,000 acres, to construct Garden City, for a population of 30,000. A greenbelt, that is, countryside, was to surround the city with a natural wall much the same as the wall of the medieval city. The purpose of the barrier for the garden city, as for the medieval city, was to protect it against invasion and encroachment, not from the barbarians but from London's overflow population. At the same time, the greenbelt would also restrict urban growth from within. Much of his discussion of the plan concerned the financing of the venture: providing balance sheets which proved the plan to be economically sound and the Garden City to be a self-sufficient proposition. The sizes of building lots, street widths and locations, the location of public parks, factories, and agriculture were specified, although Howard emphasized that his description was meant to be suggestive, not final. A hypothetical budget was drawn up for roads, bridges, schools, a library and museum, parks and ornamentation, sewerage disposal, and for a sinking fund. From a reading of this description today, Howard emerges as a colonial clerk neatly setting down the row of figures on the balance sheet. Howard must have known intimately the Victorian mentality that he wanted to persuade, and probably shared it.

Howard's brief inventory of the obstacles to his plan was hardly complete, nor did he even spend much time arguing over the few he did choose to identify. Human nature, as it was then conceived, was a major obstacle. After all, people are selfish and not given to altruism. Howard's answer, such as it was, maintained that these qualities could not be allowed to make any difference. The garden city was economically feasible and socially necessary. If it failed, we were only inviting catastrophe, for cities would continue to grow unplanned and make us their victims. On the other hand, Howard held, once the Garden City was built and its feasibility proved, similar cities could be started and, as a result, a new day would dawn for England and eventually for mankind. Man would no longer be the victim of his industrial creation, but its master. He would discover that his real wealth lay in the land rather than in the marketplace or factory.

Most Utopias remain dreams, and their authors die convinced that they held a secret to which the world unfortunately would not listen. Still other Utopias, put to the tests of reality, failed because they were poorly conceived, with insufficient attention paid to the complex reality they sought to reform. So too did Howard taste the bitter fruit of reality. He lived to see two garden cities begun, and to see both flounder. He saw enough to show him, if not to convince him, that there were more obstacles to the realization of his plan than he had imagined. Still, there were some benefits, finally proven by the permanence of the experiments and by the financial value of the bonds of the garden cities, although many years after his death in 1928. The plan failed ostensibly for financial reasons, but more important, its failure could be traced to the revolutionary character of its ideas. For better or worse, Howard advocated an ideology of government ownership or control. To Victorians raised in the philosophy of laissez faire, the truth of which seemed self-evident, Howard was advocating no less than revolution. Neither his explicit denial of sympathy with communism and socialism, then active on the Continent, nor his bookkeeper's account sheet could disguise the revolutionary consequences of his ideas. Those who followed Howard were also to learn that their leader was talking of more than architectural reforms alone. Yet, Howard's intent was not to be revolutionary, but simply to pursue his assumptions to a realistic conclusion. Any plan or urban change such as he had in mind must challenge existing social mechanisms because those very mechanisms produced the evil he complained of.

Five years after the publication of Howard's book, a company was formed to build Letchworth, the first garden city. The book became a best seller. A sizable amount of money was raised, although not enough. Of the authorized capital of £300,000, only £100,000 of shares were subscribed beyond the £40,000 pledged by the directors of the newly formed company. Heavy interest charges were incurred for the additional mortgages that were needed and a heavy financial burden was thus placed on the venture from the beginning. Money remained a problem, along with the inevitable dissension between the business and ideological leaders of the project. Nor were matters helped by the fact that Letchworth was a bad site, having few natural attractions. Public interest in the project quickly dissipated. After 43 years, all the accumulated dividends were finally repaid. "It was . . . indefatigable faith," wrote Rodwin, "which made Letchworth survive, despite all the difficulties, neglect, and derision."

With the cavalier attitude, obstinacy, and unquenchable optimism that are so characteristic of dedicated egomaniacs, Howard took an option on another piece of land in 1920, Letchworth's marked lack of success notwithstanding. Welwyn Garden City was located 20 miles northwest of London, on the same highway and railroad as Letchworth. Money, once again, was a problem, but this time the financing was more complicated by the economic depression of that period and the competition from the Letchworth venture. Tight money meant higher interest rates, and consequently the financial burden was heavier than it had been for Letchworth. The government advanced some funds, but these loans were sporadic, grudgingly given, and carried with them a right to participate in the management. Again, as in Letchworth, dissension was ever present between the lenders worried for their money and the planners worried for their ideals. After two major reorganizations, in 1931 and in 1934, a large part of the share capital was written off and restrictions on dividends were removed. Howard had died in 1928. Welwyn remained, although its development did not proceed fully according to plan.

Aside from lack of funds and dissension, the garden cities encountered less obvious but equally damaging obstacles. These were due to a naivete about sociological factors, which, by the way, even later planners did not avoid. Rodwin has noted these in his evaluation of Howard. First, Howard wanted to create a politically independent city. This requirement opened the door to a whole set of new problems concerning revenue and jurisdiction, besides further complicating the already complex administrative relationships between cities. By insisting on independence, Howard overlooked the nexus that existed, and continues to exist, between urban centers in an industrial nation.

Second, the direction of population movement did not fulfill Howard's prediction. Instead of people moving constantly into the large cities, as Howard had predicted for London, such migration leveled off and was redirected into the region around the large central city. *Urban* congestion, therefore, did not remain the key problem, *metropolitan* congestion was to complicate the problem even more. Third, the automobile and mass transportation significantly altered the urban journey from what it had been in Howard's time. He thought that people should walk to work, as one way of being close to nature. The automobile, as fact and as symbol, changed the urbanite's outlook; he would not remain a pedestrian. Urban traffic thereby became a major problem. Whether or not the garden city would have solved or would have exacerbated that problem is hard to say, even though it is a question contemporary planners take into account. Finally, Howard apparently gave little thought to human motivation; how to entice people to move into the garden city. His assumptions concerning human nature and human motivations were primitive and naive. It is an ideological disease of visionaries, Howard included, that they assume their values to be the best for all. Do people in fact, want to live close to nature? Do they want to exchange concrete for meadows? The answer obviously must be that some do and some do not. It is the human variety that the Utopian planner so frequently overlooks. He makes the unwarranted assumption either that people are alike or that they can be molded, shaped, and moved to conform to design, very much as buildings can. Rodwin concluded:

> *Unawareness of many of the pitfalls, coupled with the extraordinary loyalty to the idea of garden cities, ranks high among the factors which account for the survival of the two towns. Whatever judgment one may form of the experiment, the fact is that the leaders succeeded in their initial objective. [Only, it must be added, if that objective is very narrowly defined]. . . . In the process of development, the towns also pioneered some significant planning innovations, including use and density zoning, a form of ward or neighborhood planning, employment of an agricultural greenbelt to control urban size, and unified urban land ownership for the purpose of capturing rising land values for the benefit of the residents.*

Welwyn and Letchworth were not the end of the garden city movement nor of its importance to the present discussion. During World War II, government officials in London foresaw the need for postwar housing and, to their credit, the necessity for directing some of the population out of London itself. Several Town and Country Planning Acts were passed and a Town and Country Planning Ministry created, all before 1945. In 1946, the New Towns Act was passed and "the building of cities for the first time in contemporary Western history became a concern of long-term national policy." These developments are best shown in the story of Stevenage, one of the towns built under the national act, whose problems have been so excellently assessed by Harold Orlans. Stevenage gives us a more recent perspective on the consequences of Howard's ideas and, at the same time, conveniently displays the principal limitations of sociological understanding of the visionaries.

Stevenage in 1945, immediately prior to its selection as a site under the New Towns Act, was in its development and location somewhere between a village and a town. It was within London's metropolitan influence, being a dormitory suburb for clerks and businessmen who commuted to London. However, it still valued its local traditions and agricultural activity. The project called for building a garden city in Stevenage to house a population of 30,000, and one proposal forecasted a population double that size within 10 years. The partners in this venture were the Stevenage Urban Council and the Ministry of Town and Country Planning. Although a spirit of cooperation was present at the initial discussions about the project, it was not long before dissension, the ever-present threat in social planning, broke out between the partners. The local authorities resented the manner in which the Ministry pushed ahead with its plans without local consultation. The Ministry, apparently intent on making this project a national model, neglected its public relations in the local community.

Local residents organized themselves into the Residents' Protection Association to protest the Ministry's "tyranny of the acquisition of houses and lands, and the tyranny of control from Whitehall over homes." The residents took legal action against the Ministry, and even though they lost the case, they did succeed in stopping any development during the time, so that only 28 houses were built in the 4½ years after the project started, instead of over 300 that were to have been completed. A key to the conflict, Orlans believed, was the ideological split between the rural conservatism of Stevenage residents and the public-welfare ideals of the Ministry. What the residents of Stevenage soon came to realize was that the garden city development meant the effective end to their tradition of local independence by the shift to greater public ownership and control. Relations had improved by 1950, after the court action, but the Residents' Protection Association remained for the "exploitation of opportunities to obstruct the progress of the New Town and to secure all possible concessions for property owners and ratepayers." Such was the stormy history of the garden city plan in Stevenage.

The brief history of the Stevenage project is meant to highlight certain sociological features, principally the kinds of objections that can develop when the visionary's urban ideals are concretized. I leave out the obvious economic and political problems that are raised and look at what are presumably picayune matters. Orlans' analysis is valuable because he has indicated not only the problems of the politicians but also those of the technicians.

One explicit aim of the Stevenage plan, as it is of most town plans, was to create a *balanced* community: "We want to revive that social structure which existed in the old English villages, where the rich lived next door to the not so rich, and everyone knew everybody." The economic bases for this idea are perhaps reasonable, especially the attempt to achieve some economic stability by having more than one industry or business located in the town. As the above quotation shows, however, behind the desire for balance is frequently the unconscious wish for the norm, for what Orlans has called "the golden mean in which all the parts of a community and all citizens would work together harmoniously and without friction."

Not all town planners, of course, agree on the desirability of social balance. Even among those at Stevenage, a counter argument was advanced in support of a *homogeneous,* not a mixed community, on the grounds that a deliberate mixture of social classes would hamper the spirit of neighborliness. It is hard to say which argument is more naive. Both are based on private values more than on social wisdom. There is only the narrowest factual basis upon which to make a choice of population, and this basis is the social consequences that would follow from each alternative. The choice actually made, however, depends heavily on the values the individual planner chooses to hold. Howard, for example, wished to move the working man out of an unhealthy city into the middle-class atmosphere of the garden city, where he might spend his time in the "healthy and fascinating pursuit of gardening." Other planners envisioned a new city with all social strata transformed by a mystical, architectural osmosis into an equalitarian society. The troubles with this kind of social engineering are manifold: Whose values are to guide the plan; once enforced, who controls society and how does he make it go in the proper direction?

One illuminating instance of the planner's naïveté concerning Stevenage centered on the "Reilly Green," or common, which took its name from a plan devised by Sir Charles Reilly. The arguments supporting a common, around which small neighborhood units would be constructed, claimed that it would remedy such defects as "loneliness, juvenile delinquency, parental cruelty, poor health, declining birth-rate, late marriage, ignorance, property-possessiveness, etc." More judicious planners, undoubtedly, would not care to make so broad a claim. But the essence of most plans is to restructure social life according to some ideal, and to do it primarily through the manipulation of buildings and space and only secondarily, perhaps, through the manipulation of the people themselves. As one of the architects in the national ministry reported:

> *We do not claim that a sensible physical arrangement of houses and other buildings normal to a good residential neighborhood will automatically produce a friendly and neighborly spirit. But we do claim that it will give considerable initial advantage to the development of a healthy social life.*

This statement illustrates a common misconception among town planners: that architectural forms can alter social forms. Some planners, like Frank Lloyd Wright, go even further, contending that architectural change is the *sine qua non* of social change. It is a narrow assumption, similar to the misconception that slums cause criminal behavior and delinquency. We know now that complex behavior and motives cannot be explained merely by simple physical facts. Tenements do not make criminals any more than mansions make law-abiding citizens. Social contacts in the family, school, and church, as well as other broader social phenomena, and many individual aspects of the personality shape the human product. The visionary, nevertheless, seems to insist on his belief that the individual who is put in a magic house, in a magic setting, and surrounded with what are essentially the trappings of middle-class life will emerge a stolid, socially acceptable human product in the middle-class tradition. It is a highly doubtful assumption. Even granting its truth, one might ask: Why establish middle-class traditions as the epitome of the good life?

Another aim of the visionaries is to reconstruct social life, and this aim is frequently expressed in the "planned neighborhood." The planners of Stevenage were no exception. Arguments may develop over the optimum size of the neighborhood, whether the cul-de-sac is a way to achieve neighborhood social contacts, or where to place the community center to maximize intended neighborhood relationships. There is little disagreement, however, about the desirability of creating a neighborhood to begin with. The neighborhood seems to be the *sine qua non* of every planner's dream. It is a primary element in his ideology, and in the neighborhood, he believes, is the basis of social control to effect wanted social changes. Planners believe, as Orlans concluded the New Town planners believed,

> *that sociability [the planned neighborhood] and community activity could be organized or, at least, encouraged by a congenial physical environment and genuine social reform which would conteract the consequences of industrialism, occupational specializations, and class segregation and conflict.*

But, as Ruth Glass has correctly stated,

> *the return to the small self-contained urban [neighborhood] unit appears to be a forlorn hope. The existing trend is for a progressive division of labor and of interests. . . . This trend can be controlled but it cannot be canceled.*

The visionary, like Howard, raised in a milieu of protest against the evils of industrialism, stands fair to be disappointed. His cause may be just, his vision bright, his side that of the angels. Morally and intellectually impelled to transmogrify industrial, urban man into a middle-class, provincial, conforming, garden-city species, the visionary claims too much and knows too little. The Plan has him hypnotized into believing that this all is really feasible. But the dream cannot withstand reality, which he seems imperfectly to appreciate. He has, unfortunately, come to believe that massive social forces, with the impetus of centuries behind them, can be contained and redirected toward nirvana by an exquisitely designed community and a neatly planned neighborhood.

"Must Utopia, realized, always disappoint?" Orlans asks, and answers:

> *To be persuasive and practical (to persuade different kinds of people and to be practiced in different times and places) a Utopian idea must be relatively simple and generalized. But life is more complicated than any simple idea, and*

probably than any *idea or image, one can have of it — "the inexpressible complexity of everything that lives" is how Tolstoy, for all his genius in expressing that complexity, put it. This is the rock upon which Utopia, and reason itself, founders.*

FRANK LLOYD WRIGHT: BROADACRE CITY

Wright was more than simply the American counterpart to Ebenezer Howard and his Broadacre City more than just the American version of Garden City. Howard, as a true Victorian, after carefully adding up the economic costs and arguing for the feasibility of the garden city, had accepted much of the prevailing ideology. An overcrowded and congested London was bad business whereas a planned garden city was good business. Wright, on the contrary, never entered the marketplace to sell his plans. He much preferred to be the prophet on the mount shouting "Doom!" to the multitudes below. Wright felt the city and the industrial civilization that produced it must perish. They were the consequences of diseased values, and to achieve health, new values had to be established in a new environment. Wright was more consciously a social revolutionary than was Howard. He was prepared to recognize social mechanisms and willing to alter them. Howard's aim was to build a few garden cities to prove that they were feasible, and by this publicity to have the revolt against the city initiated by society itself. Wright was more impatient. He wanted the wholesale decentralization of cities carried on simultaneously with the creation of Broadacre City. He had no patience with businesslike arguments to support his plan. Perhaps one could not blame him for his impatience and his loftiness, since he was so convinced that human civilization would be strangled by its industrial creation unless decisive and total action was taken. Any less drastic plan would have been hypocritical.

As the citizen stands, powerful modern resources, naturally his own by uses of modern machinery, are (owning to their very nature) turning against him, although the system he lives under is one he himself helped build. Such centralizations of men and capital as he must now serve are no longer wise or humane. Long ago — having done all it could for humanity — the centralization we call the big city became a contripetal force grown beyond our control, agitated by rent to continually additional, vicarious powers.

The city, according to Wright, has perverted our values and has become the environment of false democracy, false individualism, and false capitalism. We have, by our inaction, allowed ourselves to be overwhelmed and dominated by falsity. "The citizen," Wright argued, "is now trained to see life as a cliche." He must be trained to see life as natural for "only then can the democratic spirit of man, individual, rise out of the ground. We are calling that civilization of man and ground . . . democracy." As for capitalism,

Out of American "rugged individualism" captained by rugged captains of our rugged industrial enterprises we have gradually evolved a crude, vain power: plutocratic "Capitalism." Not true capitalism. I believe this is entirely foreign to our own original idea of Democracy.

The cause of these perversions of our basic social values in the city is industrial civilization, where most of the visionaries locate the blame. Wright's contribution was to specify the causes more precisely. First, among these is *land rent,* and Henry George is resurrected as a guide to salvation. The rent for land has contributed to the "overgrowth of cities, resulting in poverty and unhappiness." Land values are artificial monsters that have taken over the destiny of the city, thereby removing us further from the natural state of mankind. Second, *money,* "A commodity for sale, so made as to come alive as something in itself — to go on continuously working in order to make all work useless. . . . The modern city is its stronghold and chief defender." Here the Puritan and Jeffersonian in Wright emerges, berating man to get back to the land and to honest labor. Third, *profit.*

By the triumph of conscienceless but "rugged individualism" the machine profits of human ingenuity or inspiration in getting the work of the world done are almost all funneled into pockets of fewer and more "rugged" captains-of-industry. Only in a small measure . . . are

> *these profits . . . where they belong; that is to say, with the man whose life is actually modified, given, or sacrificed to this new common agency for doing the work of the world. This agency we call "the machine."*

In these few words, Wright has fairly condensed the Marxian theory of surplus value. Fourth, *government* and *bureaucracy.*

> *In order to keep the peace and some show of equity between the lower passions so busily begotten in begetting, the complicated forms of super-money-increase-money-making and holding are legitimatized by government. Government, too, thus becomes monstrosity. Again enormous armies of white-collarites arise.*

These are Wright's beliefs on the state of industrial civilization. The need for revolt is clear; the means are at hand.

> *Infinite possibilities exist to make of the city a place suitable for the free man in which freedom can thrive and the soul of man grow, a City of cities that democracy would approve and so desperately needs. . . . Yes, and in that vision of decentralization and reintegration lies our natural twentieth century dawn. Of such is the nature of the democracy free men may honestly call the new freedom."*

How emphatically this point of view, so characteristic of the visionary, separates him from the mundane practicality of the practitioner. For Wright, the city in its present form cannot be saved, nor is it worth saving. A new environment must be envisioned and built. It must be one that is developed out of our technology, but one that excises the diseased growth that has infected our basic and still sound values. The plan is Broadacre City, realized by "organic architecture" or "the architecture of democracy."

Broadacre City was a more detailed Utopia than Garden City. Wright had drawn not only the ground pattern (1 acre to the individual) but also planned homes, buildings, farms, and automobiles. He also clearly specified the activities that would be permitted. Wright held definite views, to say the least, not only about architecture, but music, education, religion, and medicine as well. He was an authoritarian, some would say a messianic figure, as sure of the true and the good and the beautiful as were Christ and the early Christian prophets, along with Lao-tse and Mohammed, whom he sought to emulate.

Wright's plans for the physical setting and social order of Broadacre City were comprehensive. They contained small factories because the newer technology has made the centralized large factory obsolete, wasteful, and constricting. Office buildings housing the financial, professional, distributive, and administrative services necessary for business would be organized as a unit. Professional services would be decentralized and made readily accessible to the clients. Banks, as we know them, would be abolished and in their place there would be a "non-political, non-profit institution in charge of the medium of exchange." Money no longer would have the power it now has; therefore, the need for its "glamorization" would be removed. Markets and shopping centers would be designed as spacious pavilions to make shopping itself a pleasant and aesthetic experience. There would be apartments, motels, and community centers. Radio would carry great music to the people. "The chamber music concert would *naturally* become a common feature at home." [My italics.] Churches would be built, but the "old idea" of religion would be replaced by a more liberal and nonsectarian religion. There would be less concern with the hereafter, with superstition, with prejudice, and with deference to authority. With this new religion man, though still humble, would be made more understanding of himself and more democratic toward others.

Wright also had plans for education and the material to be taught in the schools and university of Broadacre City. He would replace the specialized, mass product of the universities of his day with a student who would obtain a deeper understanding of nature's laws governing the human spirit. Education would be a total and continuous process for the resident of Broadacre. Aside from the schools, this would be accomplished by "style centers," and by

> *television and radio, owned by the people [which would] broadcast cultural programs illustrating pertinent phases of government, of*

city life, of art work, and [would have] programs devoted to landscape study and planting or the practice of soil and timber conservation; and, as a matter of course, to town planning *for better houses.*

This plan is not Utopian, Wright argued, but rather a description of elemental changes that he saw "existing or surely coming." Either the vision is realized or society as we know it is doomed.

This long discourse . . . is a sincere attempt to take apart and show . . . the radical simplicities of fate to which our own machine skills have now laid us wide open and [to] try to show how radical eliminations are now essential to our spiritual health, and to the culture, if not the countenance, of democratic civilization itself. These are all changes valid by now if we are to have indigenous culture at all and are not to remain a bastardized civilization with no culture of our own, going all the way down the backstairs of time to the usual untimely end civilizations have hitherto met.

With Frank Lloyd Wright, the visionary's argument found its most dramatic and radical expression, and its most completely detailed one. Wright magnified Howard's plan and spelled out more specifically the visionary's discontent and rebellion against the industrial city. In Wright's words, the planned Utopia became a loud protest against the evils of industrialism. His architectural philosophy was, at the same time, a radical social ideology. Wright recognized this and did not hesitate to make the connection clear. His principal contribution to the study of the city, if one does not care to accept his dream or his philosophy, was in the repeated insistence on the relationship between the city and the society that produced it. The contemporary city, for Wright, was a product of industrial civilization. One could not understand the first without the second, which included understanding all of its institutions: the political system, social stratification and the economic order, religion and education. Wright might be excused for his authoritarianism, for his failure to consider the motivations of individuals, for his brash structuring of existing social relationships into something he wanted. For he did grasp something of the underlying complexity that sustained the city as a social environment. That he refused to consider what others wanted, or what others thought, was due to his conviction that he was absolutely right. Can the prophet, after all, have any doubts?

LEWIS MUMFORD: THE NEW URBAN ORDER

Mumford added greater social realism to the visionaries' argument. He was more aware than most of economic forces and social ideologies and their effect in shaping the city. For that reason, his analysis of the "new urban order" was sociologically informed. He showed little patience with the social engineers and architectural planners who have become so hypnotized by their own goals that they show little understanding of the existing environment from which they must begin.

I have selected Mumford for balance, to exemplify the more realistic dimensions of the visionary type. *The Culture of Cities,* both as urban history and as a sociological analysis of the contemporary city, is exceptional. Mumford, of course, is not alone in possessing realism, but, more than anyone else, he has combined his realism with an understanding of the city, and expressed it in a book that deserves its rank as a classic. Arthur B. Gallion, for example, in *The Urban Pattern,* has presented an urban history that is as informed and as sociologically realistic as Mumford's, yet in his analysis of the contemporary city Gallion seems to have forgotten the social variables he specified at the beginning, and he assumes the Utopian mentality. In somewhat similar fashion, Percival and Paul Goodman, in *Communitas,* have shown an exceptionally keen understanding of the forces of urban society, and they have incorporated that understanding into a standard by which to evaluate the plans of others. Their own plan for the future city tends, however, to fall short when measured against the same standard. This is due, I strongly suspect, to the fact that any plan for a future city must avoid many aspects of social, economic, and psychological relevance. It is impossible, it would seem, to take all these factors into account. The Goodmans try to guard against this by insisting upon the flexibility of their plan and upon its suggestiveness rather than its concreteness, as did Howard before them. Even so, it seems that The Plan must raise more problems than it solves.

It is not Mumford's history of urbanism that is my concern, but rather his analysis of what he has called the "social basis of the new urban order." Mumford began with an inventory of the architectural and sociological components that are available today for urban reconstruction. Modern architecture, he argued, has new materials to use and a wide engineering knowledge upon which to depend. Even more important, these new materials as the products of a "collective economy" are meant to be used by all persons; "one's economic position may entitle one to a greater or smaller quantity, but the quality is fixed." "Collective largesse," not niggardliness, is the hallmark of our industrial civilization, what the Goodman's have aptly labeled a "technology of surplus." In short, the materials are available for urban reconstruction and in sufficient quantity and quality. Furthermore, architectural knowledge is also at hand to make use of these materials.

Modern hygiene has given us the knowledge not only to combat disease but to prevent it. The city of the future, unlike the city of the past or of the present, could now be built as a life-supporting environment. What has been learned from the past is that health is a collective responsibility: water and waste disposal, for example, have become the accepted responsibilities of urban governments. More important for the future, however, positive attitudes toward health and hygiene have become dominant, and the life-destructive environment of the city will not be readily tolerated. "The drift to the suburbs," Mumford contended as early as 1938, "which has been one of the most conspicuous features of the growth of cities during the past half century, was one response to the more constant concern with health and education that has characterized the life of the middle classes." Perhaps so, but the wider emphasis upon health is certainly not misplaced.

The prolongation of youth is another value emphasized on the contemporary scene. In the earlier decades of industrialism, youth was cut short by child labor, and life itself was shortened by a studied avoidance of concerns with safety and the basic requirements of health. At present, however, the emphasis has changed; youth is prolonged through education, through sports, and through the medical gains that have prolonged life generally.

Another value that has undergone change has been the openhearted acceptance of the new, as if for its own sake. It is a quality that Mumford called "the capacity for renewal," a positive willingness to look to the future rather than to the past for our direction. The monument has lost its significance for the contemporary city where men no longer glorify the past or allow themselves to be chained to it. "Instead of being oriented, then, toward death and fixity, we are oriented to the cycle of life, with its never-ending process of birth and growth and renewal and death."

The rejection of the monuments of the past has been coupled with the rejection of uniformity in man himself. From an interest in caste, we have turned to an interest in the individual personality. Even though occupation, regional background, or other major social categories may direct personalities toward conformity, the individual still retains a uniqueness and a character that has come to be idealized in the present. At the same time, Mumford argued, socialization has been an equally dominant demand; that is, an appreciation for the collectivity and for what it can add to the real meaning of individual liberty. The "dogmas of private property and individual liberty" of a century or more ago overlooked the highly central role that society must play in creating and guaranteeing such liberties.

The effect of these changes and conditions has been to give us the social philosophy, the technology, and the material means to alter our environment. What is more, the city can be altered to benefit more than just the few. The mode of existence that was once thought to be the natural privilege of only the aristocracy is now available to all as their right.

This, then, is the meaning of the change that has been slowly taking place in our civilization since the third quarter of the nineteenth century. The increase of collectivism, the rising of municipal and governmental housing, the expansion of cooperative consumers' and producers' associations, the destruction of slums and the building of superior types of community for the workers – all these are signs of the new biotechnic orientation.

The change has been abetted by education, which has become as vital for the modern city as religion was for the medieval city. Through education, the masses can be transformed into intelligent individuals seeking to achieve common ends through cooperation and understanding. The transformation is, of course, still incomplete, primarily because we still treat education as a mass commodity rather than as a more individualized and private experience. The desired conditions for education become "small groups, small classes, small communities; in short, institutions adapted to the human scale."

Mumford has specifically noted that the social forces he has described as "bases of the new urban order" are not all operative. The gap between the present and the future as he foresees it is still there. This kind of recognition has made Mumford unique among the visionaries. When confronted with the fact of the contemporary city, he does not abandon the understanding he has shown in his earlier analysis of urban history. The same complex forces must be assessed in reading change into the future. Mumford has said himself that

social facts are primary, and the physical organization of a city, its industries and its markets, its lines of communication and traffic, must be subservient to its social needs. Whereas in the development of the city during the last century we expanded the physical plan recklessly and treated the essential social nucleus, the organs of government and education and social service, as mere afterthoughts, today we must treat the social nucleus as the essential element in every valid city plan.

The plan for the future city, then, must take account of social relationships and must be cognizant of the functions meant to be served by the planned urban arrangements of the future. The point that most visionaries have consistently overlooked is that their own personal values, no matter how sincere, are simply not legitimate grounds upon which to insist that the city be recreated. Instead, as Mumford was at great pains to explain, the social needs and desires of the urban community must themselves be part of the equation and must provide the grounds for rebuilding. Neither can architectural design be used as the sole, or even primary, basis for rebuilding the city. This argument is but a variation on past arguments – justly held in moral contempt – by which the factory and its demands were allowed to dictate the character of the urban environment. The emphasis by architects on, say, the functional home rather than upon the functional needs of a society – whatever they are – is misplaced and just as unfortunate as the tragic, misplaced emphasis in the last century on industrial, as opposed to social, development. Is Wright's design for urban living any more realistic, desirable, or democratic than what has emerged during the past century in a city catering to the demands of industry? Neither the nineteenth century's or Wright's solution has given fair voice to the people and their desires. The means taken to achieve a goal, as John Dewey has emphasized repeatedly, is itself part of that goal and will inevitably determine it. In the present context, then, the visionary's authoritarian insistence upon his plan cannot lead to democratic consequences. It is this kind of error that Mumford has avoided, in large measure, because he has been content to show the needs rather than to draw the plan itself.

CONCLUSIONS

The visionary has made several contributions to a sociology of the city. Behind the plans he has expounded, behind his sustained note of protest and his prophecies of urban doom – from metropolis to necropolis – there are to be found ideas that are integral to an understanding of the city. Social scientists, uncomfortable perhaps with the artistic language of the visionary and unwilling to understand the nature of his protest, have therefore omitted from consideration an important segment of information about the urban environment.

The visionary has succeeded even better than the social scientists in indicating the multiplicity of factors that, in effect, create the city as we know it.

In some cases the visionary has done this explicitly. All three visionaries discussed here, for example, have shown a lively comprehension of the city's industrial roots. That these have not been sufficiently detailed is as much an indication of the limited understanding we all possess as of shortcomings peculiar to the visionary alone. In other instances, the visionary has shown the complexity of the urban environment implicitly, by the naïveté of his plans and by his failure in trying to realize them. Human motivations and human needs have most obviously been overlooked in many of the plans of the visionaries, or else have been seriously oversimplified. That people want to live in small communities and want to get back to nature is by no means an established fact, yet the visionary frequently assumes it to be so. A more proper balance can be provided by the sociologist in this respect from his knowledge of social values and social norms. However, even the sociologist frequently has been naive in appreciating the urban complexity that he has chosen to study.

Beside the motivations of the individual there must be considered the structure of society itself. Political organizations and allegiances must inevitably become involved in any planned urban change, for the alteration sooner or later calls for a shift in the existing constellations of power. Similarly, the profit motive and the economic structure are implicated. Either the change must show a profit for those willing to support it or it must be convincingly shown that change is necessary for survival. Rarely do the visionary's plans satisfy the former condition, and as of now they do not satisfy the latter condition, except for an elite group of disciples. What the visionary's failures give the social scientist should be an appreciation for the social mechanisms that he is trying to understand. Even though the scientist's immediate interest may not be in planned change, this evaluation of the visionary should contribute a measure of understanding of the factors involved. The studies of Utopias are sociologically relevant not because of an interest in Utopias, but because of an interest in the reasons why they fail. A morbid conclusion, perhaps, but true.

One final point: the study of the city carries with it, often implicitly, a social ideology and a social philosophy. The visionary has not run from protest but, on the contrary, has been eager to make his protest evident. Sociologists, on the other hand, disciplined in the need for objectivity, often let such values creep in the back door, as they stand at the front door proclaiming their objectivity. For years, although increasingly less so now, sociologists, nostalgic for a return to a rural way of life, studied the city critically and angrily. Even urban ecologists, as will be shown, made much of the objectivity of their methods. Yet their analysis often overlooked one of their own basic value assumptions: that the city was the result of economic rationality. Objectivity in science is necessary; about that there is little dispute. However, such objectivity is achieved not by hiding one's biases and refusing to look at them, but by recognizing them openly so that they do not clutter one's conclusions.

The industrial city, as the visionaries have shown, is an historical product of values that lie at the core of our civilization. It is inconceivable that any serious study of the city can avoid recognizing those values.

2

Ebenezer Howard

THE TOWN-COUNTRY MAGNET

I will not cease from mental strife,
Nor shall my sword sleep in my hand,
Till we have built Jerusalem
In England's green and pleasant land.
BLAKE

Thorough sanitary and remedial action in the houses that we have; and then the building of more, strongly, beautifully, and in groups of limited extent, kept in proportion to their streams and walled round, so that there may be no festering and wretched suburb anywhere, but clean and busy street within and the open country without, with a belt of beautiful garden and orchard round the walls, so that from any part of the city perfectly fresh air and grass and sight of far horizon might be reachable in a few minutes' walk. This the final aim.
JOHN RUSKIN, *Sesame and Lilies*

The reader is asked to imagine an estate embracing an area of 6,000 acres, which is at present purely agricultural, and has been obtained by purchase in the open market at a cost of £40 an acre, or £240,000. The purchase money is supposed to have been raised on mortgage debentures, bearing interest at an average rate not exceeding £4 percent. The estate is legally vested in the names of four gentlemen of responsible position and of undoubted probity and honor, who hold it in trust, first, as a security for the debenture-holders, and secondly, in trust for the people of Garden City, the Town-country magnet, which it is intended to build thereon. One essential feature of the plan is that all ground rents, which are to be based upon the annual value of the land, shall be paid to the trustees, who, after providing for interest and sinking fund, will hand the balance to the Central Council for the new municipality, to be employed by such Council in the creation and maintenance of all necessary public works – roads, schools, parks, etc.

The objects of this land purchase may be stated in various ways, but it is sufficient here to say that some of the chief objects are these: To find for our industrial population work at wages of *higher purchasing power,* and to secure healthier surroundings and more regular employment. To enterprising manufacturers, cooperative societies, architects, engineers, builders, and mechanicians of all kinds, as well as to many engaged in various professions, it is intended to offer a means of securing new and better employment for their capital and talents, while to the agriculturists at present on the estate as well as to those who may migrate thither, it is designed to open a new market for their produce close to their doors. Its object is, in short, to raise the standard of health and comfort of all true workers of whatever grade – the means by which these objects are to be achieved being a healthy, natural, and economic combination of town and country life, and this on land owned by the municipality.

Garden City, which is to be built near the center of the 6,000 acres, covers an area of 1,000 acres, or a sixth part of the 6,000 acres, and might be of circular form, 1,240 yards (or nearly three-quarters of a mile) from center to circumference. (Figure 1 is a ground plan for the whole municipal area, showing the town in the center; and Figure 2, which represents one section or ward of the town, will be useful in following the description of the town itself – *a description which is, however, merely suggestive, and will probably be much departed from.)*

Six magnificent boulevards – each 120 feet wide – traverse the city from center to circumference, dividing it into six equal parts or wards. In the center is a circular space containing about 5½ acres, laid out as a beautiful and well-watered garden; and, surrounding this garden, each standing in its own ample grounds, are the larger public buildings – town hall, principal concert and lecture hall, theatre, library, museum, picture-gallery, and hospital.

The rest of the large space encircled by the "Crystal Palace" is a public park, containing 145 acres, which includes ample recreation grounds within very easy access of all of the people.

Running all round the Central Park (except where it is intersected by the boulevards) is a wide glass arcade called the "Crystal Palace," opening on to the park. This building is in wet weather one of the favorite resorts of the people, whilst the knowledge that its bright shelter is ever close at hand tempts people into Central Park, even in the most doubtful of weathers. Here manufactured goods are exposed for sale, and here most of that class of shopping which requires the joy of deliberation and selection is done. The space enclosed by the Crystal Palace is, however, a good deal larger than is required for these purposes, and a considerable part of it is used as a Winter Garden – the whole forming a permanent exhibition of a most attractive character, whilst its circular form brings it near to every dweller in the town – the furthest removed inhabitant being within 600 yards.

Passing out of the Crystal Palace on our way to the outer ring of the town, we cross Fifth Avenue – lined, as are all the roads of the town, with trees – fronting which, and looking on to the Crystal Palace, we find a ring of very excellently built houses, each standing in its own ample grounds; and, as we continue our walk, we observe that the houses are for the most part build either in concentric rings, facing the various avenues (as the circular roads are termed), or fronting the boulevards and roads which all converge to the center of the town. Asking the friend

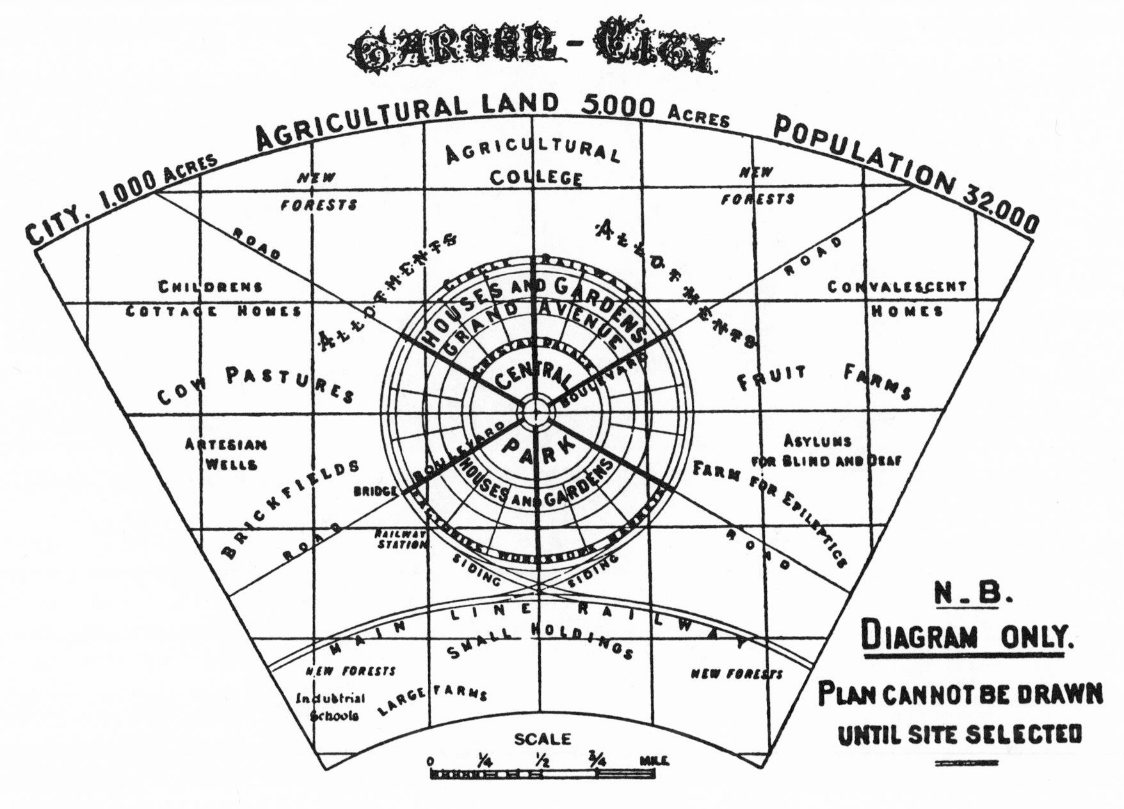

Figure 1
Garden City and Rural Belt

who accompanies us on our journey what the population of this little city may be, we are told about 30,000 in the city itself, and about 2,000 in the agricultural estate, and that there are in the town 5,500 building lots of an *average* size of 20 feet x 130 feet – the minimum space allotted for the purpose being 20 x 100. Noticing the very varied architecture and design which the houses and groups of houses display – some having common gardens and cooperative kitchens – we learn that general observance of street line or harmonious departure from it are the chief points as to house building, over which the municipal authorities exercise control, for, though proper sanitary arrangements are strictly enforced, the fullest measure of individual taste and preference is encouraged.

Walking still toward the outskirts of the town, we come upon "Grand Avenue." This avenue is fully entitled to the name it bears, for it is 420 feet wide, and, forming a belt of green upwards of 3 miles long, divides that part of town which lies outside Central Park into two belts. It really constitutes an additional park of 115 acres – a park which is within 240 yards of the furthest removed inhabitant. In this splendid avenue six sites, each of 4 acres, are occupied by public schools and their surrounding playgrounds and gardens, while other sites are reserved for churches, of such denominations as the religious beliefs of the people may determine, to be erected and maintained out of the funds of the worshippers and their friends. We observe that the houses fronting on Grand Avenue have departed (at least in one of the wards – that of which Figure 2 is a representation) – from the general plan of concentric rings, and, in order to

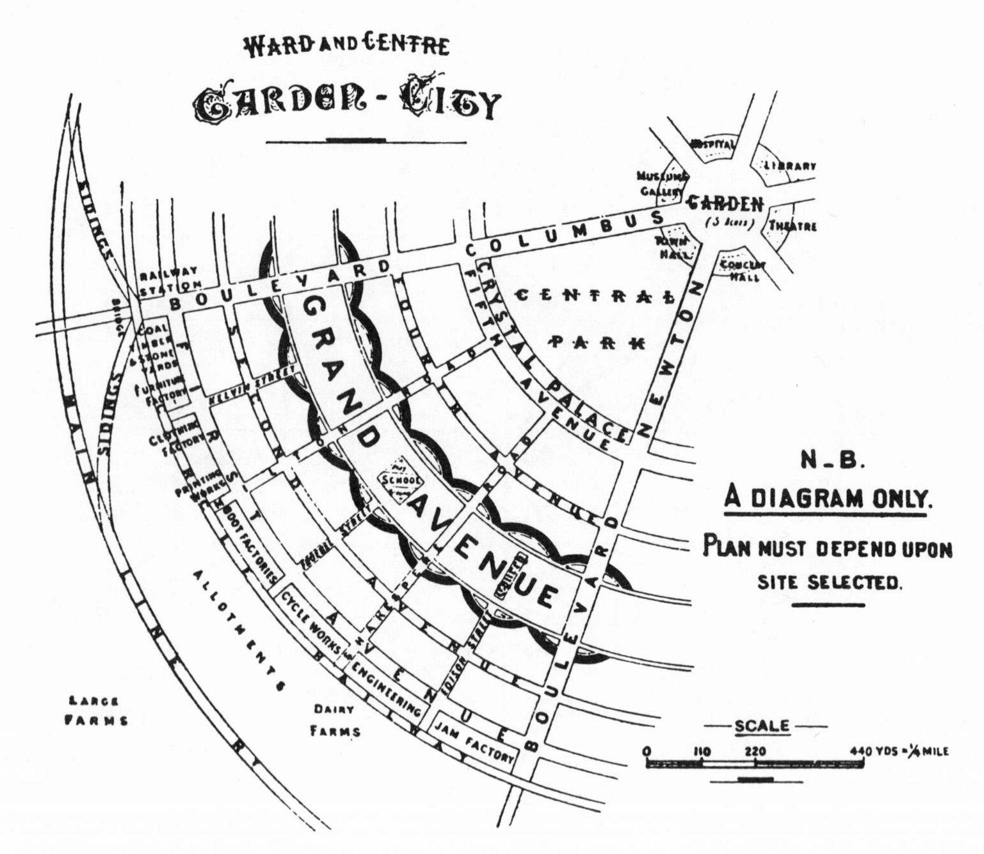

Figure 2
Ward and center of Garden City

ensure a longer line of frontage on Grand Avenue, are arranged in crescents – thus also to the eye yet further enlarging the already splendid width of Grand Avenue.

On the outer ring of the town are factories, warehouses, dairies, markets, coal yards, timber yards, etc., all fronting on the circle railway, which encompasses the whole town, and which has sidings connecting it with a main line of railway which passes through the estate. This arrangement enables goods to be loaded direct into trucks from the warehouses and workshops, and so sent by railway to distant markets, or to be taken direct from the trucks into the warehouses or factories; thus not only effecting a very great saving in regard to packing and cartage, and reducing to a minimum loss from breakage, but also, by reducing the traffic on the roads of the town, lessening to a very marked extent the cost of their maintenance. The smoke fiend is kept well within bounds in Garden City; for all machinery is driven by electric energy, with the result that the cost of electricity for lighting and other purposes is greatly reduced.

The refuse of the town is utilized on the agricultural portions of the estate, which are held by various individuals in large farms, small holdings, allotments, cow pastures, etc.; the natural competition of these various methods of agriculture, tested by the willingness of occupiers to offer the highest rent to the municipality, tending to bring about the best system of husbandry, or, what is more probable, the best *systems* adapted for various purposes. Thus it is easily conceivable that it may prove advantageous to grow wheat in very large fields, involving united action under a capitalist farmer, or by a body of cooperators; while the cultivation of vegetables, fruits, and flowers, which requires closer and more personal care, and more of the artistic and inventive faculty, may possibly be best dealt with by individuals, or by small groups of individuals having a common belief in the efficacy and value of certain dressings, methods of culture, or artificial and natural surroundings.

This plan, or, if the reader be pleased to so term it, this absence of plan, avoids the dangers of stagnation or dead level, and, though encouraging individual initiative, permits of the fullest cooperation, while the increased rents which follow from this form of competition are common or municipal property, and by far the larger part of them are expended in permanent improvements.

While the town proper, with its population engaged in various trades, callings, and professions, and with a store or depot in each ward, offers the most natural market to the people engaged on the agricultural estate, inasmuch as to the extent to which the townspeople demand their produce they escape altogether any railway rates and charges; yet the farmers and others are not by any means limited to the town as their only market, but have the fullest right to dispose of their produce to whomsoever they please. Here, as in every feature of the experiment, it will be seen that it is not the area of rights which is contracted, but the area of choice which is enlarged.

This principle of freedom holds good with regard to manufacturers and others who have established themselves in the town. These manage their affairs in their own way, subject, of course, to the general law of the land, and subject to the provision of sufficient space for workmen and reasonable sanitary conditions. Even in regard to such matters as water, lighting, and telephonic communication – which a municipality, if efficient and honest, is certainly the best and most natural body to supply – no rigid or absolute monopoly is sought; and if any private corporation or any body of individuals proved itself capable of supplying on more advantageous terms, either the whole town or a section of it, with these or any commodities the supply of which was taken up by the corporation, this would be allowed. No really sound system of *action* is in more need of artificial support than is any sound system of *thought.* The area of municipal and corporate action is probably destined to become greatly enlarged; but, if it is to be so, it will be because the people possess faith in such action, and that faith can be best shown by a wide extension of the area of freedom.

Dotted about the estate are seen various charitable and philanthropic institutions. These are not under the control of the municipality, but are supported and managed by various public-spirited people who have been invited by the municipality to establish these institutions in an open healthy district, and on land let to them at a pepper-corn rent, it occurring to the authorities that they can the better afford to be thus generous, as the spending power of these institutions greatly benefits the whole community. Besides, as those persons who migrate to the town are among its most energetic and resourceful members, it is but just and right that their more helpless brethren should be able to enjoy the benefits of an experiment which is designed for humanity at large . . .

3

Clarence A. Perry

THE NEIGHBORHOOD UNIT FORMULA

An instrument that is required to develop house building into a large-scale industry is a new form of cooperative relationship into which a municipality and a construction corporation could enter. By its terms the city would use its powers to place a large building plot within the reach of a corporation and, in return, the latter would erect upon the plot a residential development of a character yielding public benefits not attainable under existing real estate methods. Before the city could obtain from the legislature power to condemn land and turn it over to the corporation, however, the lawmakers would have to be convinced that the resulting benefits would be substantial enough to justify another extension of eminent domain powers. The only device that will meet this need is a formula defining the requirements which projects would have to meet in order to become the subject of this bargain. Such an instru-

---

 Selected from Urban Housing, *L.C. Wheaton, ed., The Free Press, New York (1966), 94-109. Originally published in* Housing for the Machine Age, *by Clarence A. Perry, Russell Sage Foundation, New York (1939), 49-76.*

mentality would enable a corporation to shape up the right kind of project and the municipality could use it as a yardstick in determining whether a submitted project would give citizens the stipulated benefits.

To be serviceable in the highest degree this formula should meet certain requirements:

1. It should not be a detailed plan of a model residential development, since there are many local conditions which a specific plan would not meet. Instead it should state principles and standards in definite, objective terms which the professional planner could apply in preparing a plan suited to the topography and other characteristics of a particular site.

2. It should be expressed in city planning terms, since it would deal with building plots, highways, recreation spaces, uses of land, location of public buildings, and those public services which require structures large enough to involve site planning.

3. To be practical, it should describe a project sufficiently self-contained so that, with the boundaries fixed, it would be possible to go forward with construction without waiting for the planning of adjacent areas. Many an excellent project has failed of realization because there was no way of detaching it from plans relating to surrounding districts and dealing with the project by itself.

As to working out the content of the formula, the method of procedure seems clear. It should cover both dwellings and their environment, the extent of the latter being – for city planning purposes – that area which embraces all the public facilities and conditions required by the average family for its comfort and proper development within the vicinity of its dwelling. In this study that area is called the family's "neighborhood." The facilities it should contain are apparent after a moment's reflection. They include at the least (1) an elementary school; (2) retail stores, and (3) public recreation facilities.

The conditions surrounding the dwelling which a family most consciously seeks come under the head of residential character. This quality depends upon many and varied features. In an apartment house district, it may rest upon location of the site, architecture of the building, or the character of its courts. In a single-family district, harmony in the style of dwellings, amount of yard devoted to lawns and planting, and comprehensiveness and excellence of the entire development plan govern residential quality. Most important in this day of swift-moving automobiles is street safety. This can best be achieved by constructing a highway system that reduces the points where pedestrians and vehicles cross paths and that keeps through traffic entirely out of a residential district.

The formula for a city neighborhood, then, must be such that when embodied in an actual development all its residents will be taken care of as respects the following points: they will all be within convenient access to an elementary school, adequate common play spaces, and retail shopping districts. Furthermore, their district will enjoy a distinctive character, because of qualities pertaining visibly to its terrain and structure, not the least of which will be a reduced risk from vehicular accidents.

NEIGHBORHOOD UNIT PRINCIPLES

A formula which, it is believed, meets all the above requirements was elaborated and published in volume 7 – *Neighborhood and Community Planning* – of the Regional Survey of New York and Its Environs. In that publication, "the neighborhood unit – a scheme of arrangement for the family-life community" is set forth in detail. Essentially, it consists of six principles:

1. *Size:* A residential unit development should provide housing for that population for which one elementary school is ordinarily required, its actual area depending upon its population density.

2. *Boundaries:* The unit should be bounded on all sides by arterial streets, sufficiently wide to facilitate its bypassing, instead of penetration, by through traffic.

3. *Open spaces:* A system of small parks and recreation spaces, planned to meet the needs of the particular neighborhood, should be provided.

4. *Institution sites:* Sites for the school and other institutions having service spheres coinciding with the limits of the unit should be suitably grouped about a central point, or common.

5. *Local shops:* One or more shopping districts, adequate for the population to be served, should be laid out in the circumference of the unit, preferably at traffic junctions and adjacent to similar districts of adjoining neighborhoods.

6. *Internal street system:* The unit should be provided with a special street system, each highway being proportioned to its probable traffic load, and the street net as a whole being designed to facilitate circulation within the unit and to discourage its use by through traffic.

The six principles enumerated do not constitute the description of a real estate development or of urban neighborhoods in general. Together they do not make a plan. They are principles which a professional planner – if so disposed – can observe in the making of a development plan. If they are complied with, there will result a neighborhood community in which the fundamental needs of family life will be met more completely, it is believed, than they are now by the usual residential sections in cities and villages.

In this scheme, the neighborhood is regarded both as a unit of a larger whole and as an entity. It is not held, however, that in an ideal city plan the whole municipality could be laid out in neighborhood units. It is recognized that a city is composed of various areas each of which is devoted to a dominant function. There are industrial districts, business districts, and large areas used as parks and cemeteries. A neighborhood unit would have local retail business areas, but besides these there would also be downtown or main business districts and subsidiary business centers serving large sections.

It is apparent that the unit scheme can be fully applied only to *new* developments. Thus it is limited to the unbuilt areas around the urban fringe and to central deteriorated sections, large enough and sufficiently blighted to warrant reconstruction. Nor is it expected that the whole of a residential section, in any practical plan, could be laid out in unit districts. There would generally be irregular areas, set off by main highways, railways, streams, quarries, or parks, of a size or location that would make them unsuited for inclusion in a unit plan.

In more detail, just what do the six principles of the neighborhood unit formula involve? Professional planners and other specialists may wish to refer to the original presentation in volume 7 of the *New York Regional Plan.* For the general reader the following condensation of that presentation will suffice.

Size Relations

> *A residential unit development should provide housing for that population for which one elementary school is ordinarily required, its actual area depending upon population density.*

What, as to the number of residents, does this requirement really indicate? According to authorities in school administration, a public school equipped with an auditorium, a gymnasium, and other accessories should have a capacity of from 1,000 to 1,600 pupils. Such a costly plant, they hold, should handle a sizable load. As a matter of fact, however, public schools are being built in small communities for 500 and even fewer pupils, while large cities, like New York, are constructing schools with 2,500 seats. The most practical procedure in determining the standard for a given city is to ascertain from the local school board the number of pupils it considers requisite for a model elementary school, and to adopt that figure.

The next question is: What proportion of the total population is represented by boys and girls in the elementary school-age group? This ratio varies greatly in American cities, but the average is about one sixth. On the basis of the standard capacities given above, an efficient urban school district may range in population from 3,000 to 9,600, or say 10,000 persons. From the standpoint of educational service it is evident that considerable latitude is allowable in fixing the population standard.

There are, however, several other requirements that must be met, in which area or distance is a factor. Suppose that we take a neighborhood of 6,000 people. Would the area of such a district be so great that children living at its border would have to walk too far in attending a school at its center? Obviously, the answer depends on how closely dwellings are placed. Let us take the density of 37.5 persons per acre that is frequently found in single-family subdivisions. The requisite plot would contain 160 acres. If it were square, it would be ½ mile, or

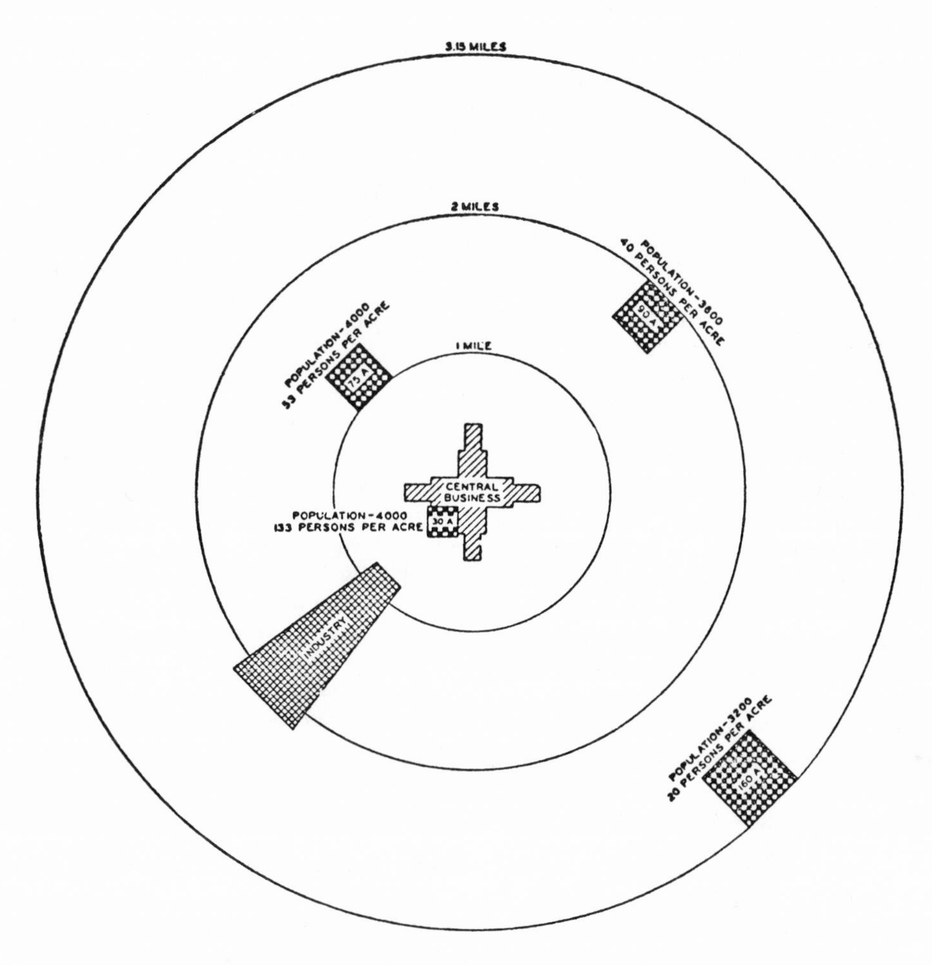

Figure 1
Neighborhood unit scheme applied to a city of 100,000 population; size as effected by density and location. Source: Regional Plan Association, Inc., New York City.

2,640 feet, on a side. A circle with a radius of a ¼ mile could be inscribed within it. Thus a school located at its center would be within a ¼ mile of all its families except those living in the corners, outside the circle, and the farthermost of those would be only a 1/3 mile from the center.

Take an area less populous – that of 5,000 people in a 200-acre plot, a common suburban density of only 25 to the acre. Here the great bulk of the residents would be within a radius of only 156 feet beyond a ¼ mile.

In these two examples we have dealt with probably the greatest travel distance that would be required by a commercial development or an efficient school administration. There might be sparser subdivisions, but there is no good reason why they should be greater than 200 acres, or indeed than 160 acres. When we move toward the center of the city, higher densities are usually encountered. Even though school districts may contain more people, the congestion is so much greater that there is a net shrinkage in the size and the distance to be traveled. Pupils in slum sections seldom have to walk as much as ¼ mile.

Let us now see what educators say about the

school radius. "Children of the elementary school grade should not be required to travel more than one-half mile to school." The Committee on School House Planning of the National Education Association made practically the same recommendation in these words: "In cities it is generally agreed that the contributing area for an elementary school may have a radius of one-half to three-quarters of a mile." From these statements it is clear that the desirable population for a school, when housed at current densities, will live well within the travel distance which our educators say should not be exceeded.

Will a population size that meets the educational requirements fit the other neighborhood service radii? Among city planners it is a rule of thumb that families should be able to find a grocery or drug store within ½ mile of the home – the same maximum travel standard as the school. According to the unit formula, shopping districts are located in the periphery of the neighborhood, instead of, as in a village, at the center. There can be as many districts as the population's buying capacity requires, and they may be located at points in the rim where convenience indicates. In the 160-acre plot of ½ mile on the side, four districts at the corners would bring shopping facilities within ¼ mile for most families, while only two districts at diagonally opposite corners would place these facilities within a 1/3 mile for a majority of the residents. Thus our school district size comes within the travel requirements for retail shopping.

As to the playground service, wide experience shows that most children will not travel more than a ¼ mile to use a playground. A large schoolyard at the center of a square 160-acre plot would be within this ¼ mile radius for the bulk of the children, and even the farthermost families would be only 1/3 mile away. Many times it will be feasible to provide more than one play space, bringing this facility still nearer its patrons. In single-family districts, where home yards take care of small tots, a public playfield is more especially needed for the baseball and other large-space games of older youngsters. Under the topic "Open Spaces," the recreational service will be discussed in detail.

Size is a factor in two other aspects of the model neighborhood community – the achievement of a distinctive residential quality, and the possession of a rich associational life. Since both these points can be more appropriately discussed in detail later, only an assurance need be given here that the unit formula does satisfy the two requirements as to area and as to desirable population.

Boundaries

> *The unit should be bounded on all sides by arterial streets, sufficiently wide to facilitate its bypassing, instead of penetration, by through traffic.*

The most important reason for wide highways as boundaries arises from their relation to street safety. The unit district is not too large, according to the opinion of an authoritative engineer, to be treated as a partly closed cell in urban street systems, without doing violence to the highway requirements for general circulation.

With adequate express channels in the circumference of a unit, through traffic will have no excuse for invading its territory, and its own internal streets can fairly and deliberately be made inconvenient and forbidding for vehicles having no destination within the neighborhood confines. If any of the original boundaries of a unit district are not suited for through traffic, they should be widened by taking land, if necessary, from the unit area. Sometimes there will be temptation to use an existing park, a stream, or a railway as a unit boundary, in place of lining that side with a wide highway. It is not a safe thing to do, since the absence of a channel for through traffic on that side will generally force such traffic into one of the neighborhood's own internal streets.

Another value of wide and conspicuous boundaries is that they enable residents and public in general to see the limits of the community and visualize it as a distinct entity. Like a fence around a private lot, they heighten the motive for local improvement by defining the area of local responsibility: "Here neighborhood maintenance ends." Residential pride may also be stimulated by the erection of ornamental arches and architectural markers of various sorts.

Open Spaces

> *A system of small parks and recreation spaces, planned to meet the needs of the particular neighborhood, should be provided.*

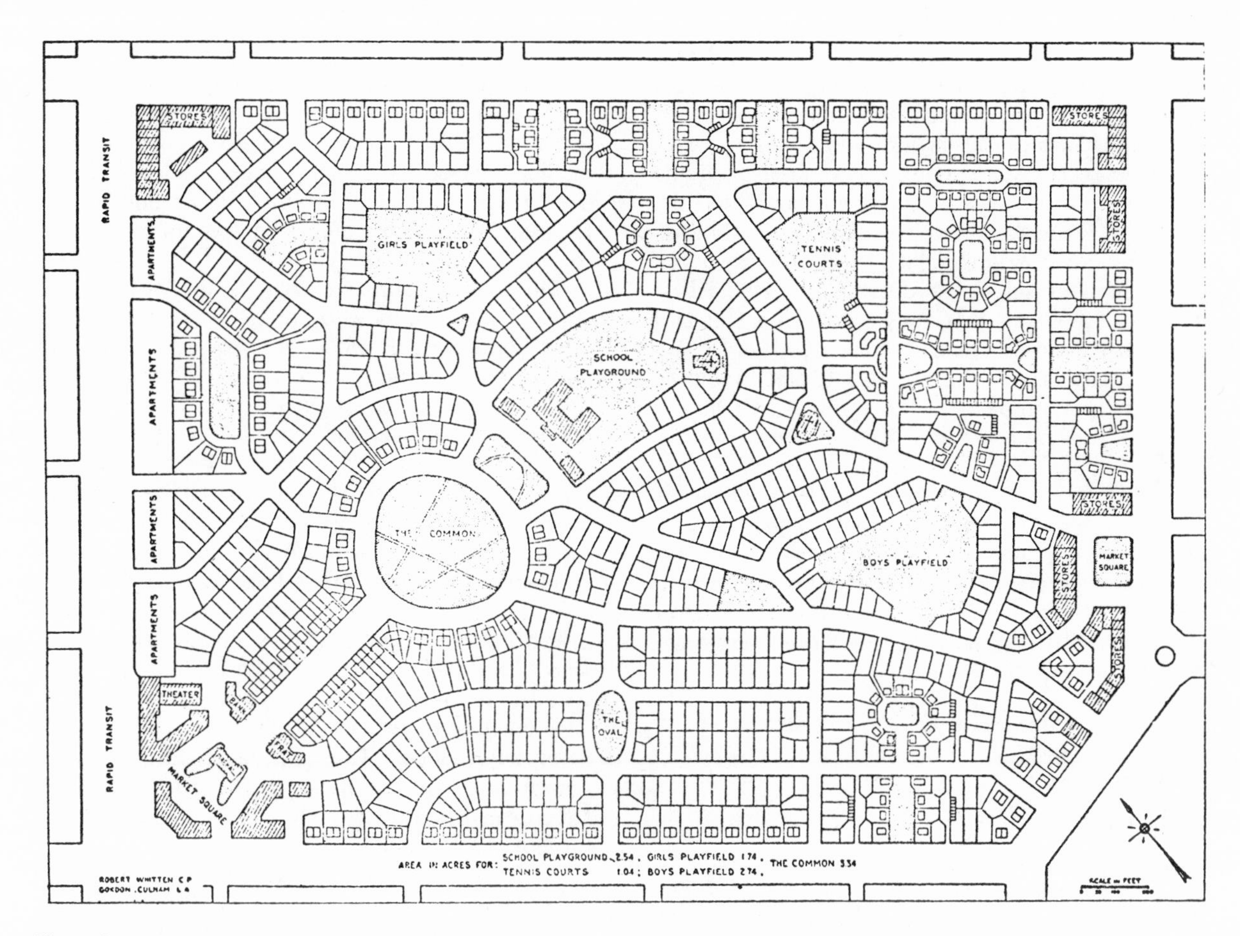

Figure 2
A 160-acre neighborhood unit subdivision; liberal recreation spaces gained through comprehensive planning. (Source: Robert Whitten, Architect, assisted by Gordon Culham, Landscape Architect.)

The large-scale advantage of unit planning shows up with special clearness in the ease with which it makes possible abundant provision for play close to the family dwelling. Having a large area to draw on, and a definite and numerous clientele to provide for, the economies and efficiency of quantity production are attainable in a high degree. A few feet taken off from the depth of a number of lots and put together in a playground that will serve all the owners produces a valuable community asset without appreciable loss to anyone.

Again, the custom-made planning – which is inherent in the unit scheme – makes it possible to avoid the wastes incidental to undifferentiated subdivision layouts. These are especially noticeable in the traditional rectangular street system, wherein two or three standard street widths are applied by a rule of thumb long before the mappers can know much about future traffic requirements. Under such circumstances it is not strange that many streets are found eventually to be of excessive width. In a unit project the function of each internal highway is determined at the time it is laid out, and it can therefore be precisely adapted to its traffic load.

The layout of a 160-acre subdivision, accommodating 6,000 residents, developed in a study of a district in the Borough of Queens, carried on by the late Robert Whitten, is shown in Figure 2. An examination of it will reveal that 10.6 (8.6, plus greens and circles 2.0) percent of its area is devoted to small parks, playgrounds, and other open spaces, all of which was gained through unit advantages over the standard gridiron system. Three quarters of the gain came from savings in street area and the remainder from shallower lots. Because of the narrower streets it was calculated that the street improvements for this neighborhood plan would cost some $400,000 less than they would in a standard layout of the same size and character.

What part of a neighborhood unit plot should be set aside for small parks and playgrounds? Obviously no hard and fast rule can be laid down. In the Whitten study – a single-family district – the recreation areas covered 17 acres, or 10.6 percent of the total area. That suggests a flat 10 percent as a good figure to aim at in open or suburban unit subdivisions. Oftentimes it should be possible to exceed it. The kinds of uses which may be made of 16 acres – 10 percent of a quarter-section – are indicated in the following table:

Recreation and Park Spaces in a Single-Family Neighborhood Unit of 160 Acres

Kind of Area	*Access*
School grounds – building site and playgrounds for the younger children	3.00
Playfields – one for boys and another for girls	5.50
Tennis courts – 12 courts	2.25
Common or civic square	2.25
Small parks – planted ovals and circles	3.00
Total	16.00

Setting up a standard recreational allotment for apartment house units is a more difficult matter. Naturally, these developments will vary widely in density and total area. The arrangement of structures and highways – governed by uncontrollable conditions surrounding the unit – will in certain instances permit larger play spaces than will be possible in others. Again it is not easy to draw a line between "space around the building" and a landscaped court which serves, or should serve, the purposes of a park.

The social objective to be kept in mind is plain. The practice of allowing city landowners to load their premises so heavily with apartment houses that no space is left on the plots for the normal physical and moral development of tenants' children should be restricted. The neighborhood unit scheme furnishes a method whereby this evil can be avoided in future large multifamily developments with a minimum cost to property owners and a maximum benefit to society.

The guiding principle then is the need of the children and youths who are to live in these apartment house unit districts. Ordinarily, the larger the population the larger the recreation allotment is the rule that should be applied. There is, however, a minimum allowance, and that has been passed when the space is so small that there is insufficient room for sports like baseball and football. Whether a neighborhood community covers 40, 80, or 160 acres, its youths will be much alike and have the same developmental needs. These large-space sports mean as much to the training of growing boys and youths in factories and offices as they do to those in high schools and colleges. Unless these facilities are accessible for the margin of work days and on holidays, large portions of the classes who most need them will never enjoy them. The large playfields in the big central and suburban parks, provided in most cities, are crowded on holidays and too far away for use at the end of a work day.

A distribution of recreation spaces is presented in the accompanying table which should be regarded as the minimum provision in an apartment house unit that is planned to meet the outdoor needs of its children and youths. It is too liberal for a downtown slum area.

It is probable, however, that even this minimum provision will be found too spacious to include in a unit designed for the reconstruction of a central slum district in the larger cities. But there is a housing program which would meet this situation.

Rebuild these costly areas in, say, 20- or 30-acre neighborhood units. There is no physical reason why such developments should not contain spacious lawns, shrubbery, pools, handball courts, and gymnasiums. Through the use of such common facilities residents would become rapidly acquainted. People who could afford these apartments would be well able to enter into cooperative schemes for the recreational life of their boys and girls. For example, such a group would have little difficulty in arranging for a weekend, all-year rural camp for their children, in which every Friday afternoon they would be transported by bus.

Recreation Spaces for an Apartment House Unit

Kind of Use	*Acres*
Small children's playground, next to school	1
Older children's playground, next to school	2
Combined baseball and football field (300 feet by 435.6 feet)	3
Hockey field for girls (200 feet by 300 feet	1-1/3
Site of school building, landscaped area and grandstand	2-2/3
Total	10

Children of the higher-income groups who enjoy a home in the country and one in the city, attend high or preparatory school, college, or university, and thus have superior advantages, do not need large recreational areas near their homes. They belong manifestly to the class for whom the downtown apartment units should be planned and constructed. For families that have less access to rural outdoor life, fuller recreational provision should exist within the neighborhood of their homes.

In apartment house units, landscaped courts may also serve certain play purposes. There may be lawns on which toddlers may romp and sand boxes where they may dig – under the eyes of nurses or mothers. Active or noisy play, however, in an enclosed court is generally objectionable. The chief enjoyment which courts can contribute is the delight to the senses that comes from lawn, shrubbery, and flowers. Is there any standard as to the amount which should be required in multifamily units?

A certain minimum requirement is set up by zoning ordinances. This usually is for the purpose of assuring to tenants adequate air and light, and belongs in a sanitary rather than an esthetic category. If an apartment house unit could not achieve more openness than municipal zoning ordinarily secures, it should be considered lacking in one of its chief virtues.

A standard of light which has been suggested by city planning authorities and which does automatically secure considerable openness requires that each structure be separated from its neighbor by a distance equal to its height. When this is observed, each ground-floor window commands a view of at least 45 degrees of sky. While this standard should never be violated, it ought generally, in a unit project, to be exceeded. Within the large frame of a school district unit, it should be easy to stagger the dwellings, or arrange them in echelon or step form, thereby securing attractive vistas in great abundance. As an example, this effect has been achieved in Plan B, known as the Mathews Plan, for the World's Fair district.

Of course, the unit plan is supposed to provide only for the strictly neighborhood needs. These do not include golf links, sea beaches, zoological museums, woodland picnicking grounds, or any of the other opportunities usually associated with the large city or suburban park.

In a word, then, the unit scheme sets up the principle that every urban neighborhood catering to families should contain within its own boundaries facilities for a normal recreational life, shaped to fit local conditions. Furthermore, if play spaces can be incorporated in the neighborhood plan while it is being formed, they will not only be much more efficient but their cost will be hardly noticeable.

Institution Sites

Sites for the school and other institutions having service spheres coinciding with the limits of the unit should be suitably grouped about a central point, or common.

We now come to the organization of the neighborhood community center. Its structural components are obvious. First is the elementary school. Next is a branch of the public library, unless as in some progressive cities this is included in the school plant. In well-to-do neighborhoods there might be a separate community building for social, club, and indoor recreational activities. There is no practical reason, however, why this structure should not be combined with the school. It would usually possess an auditorium equipped for stage productions, a gymnasium, a pool, and some smaller meeting rooms. Located adjacent to the school, the pupils could make use of the auditorium and gymnasium without interfering very much with the adult and end-of-the-day, or holiday, occasions.

The difficulties encountered in a combined use of buildings are mainly administrative. A community clubhouse, in these modern days, must permit smoking, keep open until late hours, and altogether encourage an atmosphere of gaiety and freedom that is quite foreign to the traditional school. If local residents took an active part in the management of both institutions, however, it should be possible to make a practical cooperative arrangement. This might be worked out most satisfactorily with a community building equipped and administered primarily for its social and recreational purposes, but so arranged that pupils, entering through covered passageways, could make a scheduled use of the auditorium and gymnasium. Removing the clubhouse atmosphere would be simply a janitorial task. Instructors could bring the pedagogical atmosphere with them.

A place in the community center area should be reserved for a church, provided it is known in advance that its parish is to be generally conterminous with the neighborhood. Community congregations in which several denominations have joined for carrying on worship and other religious activities do exist, and it is obvious that such bodies would find the integrated neighborhood a congenial environment. A plot reserved for a community church could be used for residences or some other neighborhood institution if circumstances later prevented the carrying out of the original plan. A church whose members will come largely from outside the neighborhood should be placed at a street junction in the periphery of the unit. The weddings, funerals, christenings, and other ceremonies which take place in churches generally crowd the vicinity with motor cars. A neighborhood should not begrudge the street space required for the occasional ceremonies of its own residents. However, in selecting the components of the neighborhood community, care can well be taken to avoid assigning central or interior locations to any institution that frequently draws together large crowds of strangers.

For the same reason, and for its own intrinsic advantage, a commercial motion picture house, a fraternal lodge, or any other institution whose supporting population is ordinarily larger than that of a school-district neighborhood, should be located, not

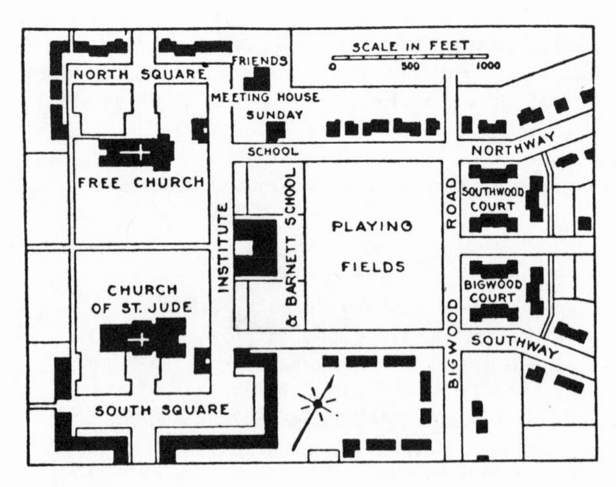

Figure 3
Central square, Hampstead garden suburb, England. (Source: New York Regional Plan, volume 7.)

within a unit, but at the point where a cluster of neighborhoods comes together, generally a business center.

If school, library, community house, and church can be grouped around a common in the center of the unit, that will be the most convenient arrangement for the residents. Such a disposition will also make possible an architectural effect of a great civic value. Fronting upon a square, the facade of each structure will be viewed from a greater perspective, and the motive for endowing it with a dignified design will be enhanced. As one of a group, there will also be reason to fit it into an attractive composition. Thus the square itself will be invested with a meaning, a symbolism, more significant than the mere sum of its parts. It will be a visible sign of unity. Obviously, this effect, if it is to be nicely wrought, will have to be worked out by an architect or planner possessed of skill and taste. When the unit is constructed under a single comprehensive management, there should be no difficulty in securing adequate talent for this purpose.

The square itself will be an appropriate location for a flagpole, a memorial monument, a bandstand, or an ornamental fountain. In the common life of the neighborhood it will function as the place of local celebrations. Here, on Independence Day, the flag will be raised, the Declaration of Independence be recited, and the citizenry urged to patriotic deeds by eloquent orators.

Local Shops

One or more shopping districts, adequate for the population to be served, should be laid out in the circumference of the unit, preferably at traffic junctions and adjacent to similar districts of adjoining neighborhoods.

Under this head we approach a municipal problem of great importance – the zoning of business and residential districts. That the present methods have not protected residential character to the degree that was expected is admitted by even the best friends of zoning. An explanation for this condition is offered by the Saint Louis City Plan Commission.

When zoning was first undertaken, there were no scientific data as to the relative amount of land needed for various types of urban land use. Lacking such data and standards, it was but natural that the early zoning was unscientific and, consequently, failed to exert a beneficial influence in stabilizing population and in molding the form and character of the city.

Residential property pays more taxes than all other classes of property combined, but zoning has failed to protect this huge investment. More than one-third of all residence property in Saint Louis is zoned for a lower classification, such as commerce or industry.

Commercial areas are 94 percent overzoned.

Further evidence of the same tendency is given by Edward M. Bassett – the "father of zoning" – in his recent authoritative work entitled Zoning:

It cannot be denied that municipalities of all sorts, especially towns, have been prone to place too much street frontage in business districts.

In some towns one hundred times as much street frontage has been placed in business districts as is likely to be used for business purposes in two generations.

In view of the fact that the protection of residential quality from the depressing effect of nearby business was one of the foremost reasons for the institution of zoning, the evidences of failure would at first sight seem highly discouraging. Such an attitude, however, would overlook the very substantial improvement in man's control of his environment which zoning has wrought. There are thousands of fine dwelling areas whose stability has been greatly strengthened by zoning. Most important of all is the fundamental principle that it has firmly established: that the uses of property are matters of public concern, so vital that the state is justified in using its police power in a reasonable regulation of them. The device is sound. It is the method of applying it wherein improvement is needed.

The difficulties that surrounded zoning when it was first launched were great. Since zoning was a public regulative measure, there was a natural necessity, and desirability, to bring the whole city within its scope. Zoners began with the central districts and worked toward the periphery. They had no fixed pattern, and had they possessed one they could not have applied it to the huge structural improvements of the downtown sections of our complicated commercial cities. An area that was dominated by dwellings they zoned as a residential district even though it contained some stores and factories. Non-residential structures could remain but no new ones could be erected. In a similar way they laid down business and "unrestricted" districts, factories or any type of use being permitted under the latter category. It was a workable method. It is difficult to see how, under the circumstances, a better one to begin with could have been devised.

In unbuilt areas, the task of the zoners was still harder. They had no existing structures to serve as a guide. It was plain that low land and land along railways or streams was suitable for industrial purposes, and that frontage on main highways was desirable for stores. Drawing the lines which determined the precise limits of those use-districts, was not easy, however, and yet they were very important, since they also defined the residential zones.

The characteristics of business and industrial plots being so much plainer, and those uses being regarded economically as so much more important, it is not strange that in the allotment of land they were taken care of first and residential use given what remained.

Furthermore, the central areas, where zoners first worked, were so largely devoted to business and industry that the experience with them yielded exaggerated notions as to the requirements of those uses. It was natural that the importance of these sections should govern the treatment of unimproved sections. May this not explain why the method has not more satisfactorily protected the character of residential districts?

When a zoner lays out use-districts in an unimproved area according to a zoning ordinance, he is allotting plots to specific functions. He is engaged in city planning. He must have not only a basis for determining what *kind* of land shall be devoted to a particular function, but also a notion as to how *much* of each kind of functioning is required per unit of population. Where now is he to seek these standards? Shall he take them from the wild, uncontrolled urban growth which zoning found when it came into existence? If not there, where or how shall he acquire the necessary standards?

It is here that the neighborhood unit scheme comes forward with a suggestion. Its offering does not directly deal with industrial districts or existing business or residential districts, except where they need rebuilding for the use of dwellings. Its field is that of relating retail business to new residential districts. Its underlying principle is simple. The logical way to effect this relation is to break the new area up into more or less uniform divisions and then apportion to each part the number and kinds of stores which will be needed by the residents for whom that part will be planned.

The basis suggested for making such divisions is the neighborhood unit, that is, a public school district of from 3,000 to 10,000 people. The area occupied by the dwellings of such a population need not exceed a ½ mile radius, which, city planners hold, is the maximum distance families should have to travel to reach a retail store. Also, within this same unit the dwellings – as has been stated – are conveniently located as respects not only the school but playgrounds and other neighborhood institutions. The unit is self-contained except as to places of work of its inhabitants and those services which ordinarily reach a city-wide clientele.

Coming now to the main detail – what is the number of retail stores which should be provided for a given population? Obviously, no city planner, however proficient, has the final answer to this question. It is a problem that does not permit exact determination. Several studies have, however, been carried on which throw light upon the matter. Here it is sufficient to say that considering their conclusions, they have led the writer to the belief that *one* per store per 100 of population to be accommodated is a fair working rule.

The truth is that we do not yet possess enough experience or data upon which to base a scientific determination. Consider then how much better off in this respect we should be if the planning and construction of neighborhood units were to become a recognized practice. In each case there would be somewhat similar conditions. The unit developments would have easily discovered characteristics, and their business data would be capable of more accurate interpretations. Furthermore, such data, continually reinforced by new experience, would be more readily available for neighborhood planning because there would then be an agency charged with the function of developing and publishing this scientific knowledge. The government would be called upon to participate in the shaping and supervision of the unit developments, and public interest would require it to be concerned – as in principle it is now – in the business zoning of the neighborhood districts.

As to location of the unit business zone, several considerations demand a new principle. Villages, towns, and cities uniformly have business at the center. They grow up around commercial and industrial activities. The neighborhood contemplated by the unit scheme, however, is primarily a residential district and its workers go daily to occupations in other parts of the city, mainly downtown districts. The reasons for not locating its business zone in the center of the unit pertain to (1) the welfare of the community, and (2) the intrinsic interests of business.

In the first place, the unit area is so compact that a collection of stores anywhere in its interior would extend their contact with dwellings and their blemish upon residential quality. Again, the supplying of goods for these stores would bring numerous trucks across the paths of boys and girls going to school and playfields, as well as occasion noise and traffic in an area where quiet and tranquility are desirable. Furthermore, the unit is too small to accommodate both a civic and a shopping group in its central region.

To understand what constitutes a good location for a store we must think about business practices. In selecting a site for a new chain store, the management sends out experts to make counts of the number of persons passing points which are under consideration. Those locations showing the highest rates are, other conditions being equal, the most desirable. Accessibility to population is a prime requisite in a store site.

It is one of the advantages of the unit scheme that it makes good business locations more definite and more easily found. Each neighborhood being a concentration of families whose workers pass daily through one, two, or possibly three main portals, the canalization of traffic is automatic. These portals are naturally located at the transit stations, or traffic junctions, in the main highways which bound the unit. Despite the telephone, the facile delivery services, and the automobile, residents frequently find it convenient to stop at a local store on their way to or back from "town." It is that convenience and the courses of the traffic streams which determine the neighborhood portals as the proper locations for local shopping districts.

Another advantage in having business in the periphery of a unit is that frequently it is only across the street from a retail district in an adjacent neighborhood. Sometimes four such districts, in the corners of their units, will be found in a cluster. Thus residents of those four neighborhoods will all enjoy a wider range of shopping opportunity. Such a collocation will not displease shop proprietors. Aggregations of similar lines of business take place spontaneously in a downtown district, each shopkeeper seeking to locate where customers interested in his line are most likely to be found. Even the disadvantage of greater competition does not counteract this commercial tendency.

The straddling of a main highway with business activity does, however, create a traffic problem. Shoppers on foot will not be so much bothered, since they will stream across the street anyway, and often with the protection of a traffic light. But motorists, not wishing to stop at that point, will complain. This difficulty will be still more vexatious in the case of a regional highway 200 feet wide. Even pedestrians would not like that condition, and a frequent stopping of the traffic would greatly reduce the value of an express channel.

The best solution of this difficulty would seem to be that of the bridge or the underpass. In a situation where the main problem would be to enable pedestrians – not vehicles – to get freely across a main highway, it might be possible to provide an underpass, accessible by stairs or a ramp, on each side of a street. The expense of this construction could be recouped, and funds for maintenance be provided, through the erection of underground stores, fronting on the underpass, thus creating a shopping arcade and extending the business facilities of the district.

Where it was desirable to provide a channel for vehicles across a main highway, in a business area, the same principle could be applied. Either a bridge or an underpass could be lined on both sides with stores. Of course, a larger layout and more extensive planning would be involved – especially in the case of an overpass – but there seems little question but that both shopping and traffic convenience could be served in this way. The new store sites thus created "out of the air" – literally *in* the air – would bring a revenue that would take care of both construction and maintenance. An overpass authority, the legal body similar to a bridge or a housing authority, could be set up to handle a number of such projects. The insertion of a 200-foot arterial highway in built-up sections does, after all, bring up problems for the neighborhoods which line them. Perhaps business opportunities will help us to cross them.

As to the shape of neighborhood business districts – that is a technical matter and planners are referred to the studies made by Clarence Stein, Catherine Bauer, and others. Several principles, however, are apparent. Stores should be bunched rather than strung along a street. Parking space and rear service entrances should be provided. Frontage and depth of stores will vary according to kinds of merchandise or service. Chain-store experience on these points will be useful in planning. Special architectural consideration should be given to the business structure at the point where stores stop and dwellings begin. A flower shop, or store of equal attractiveness, at the edge of the district will soften the harshness of this boundary.

In the unit scheme, a business district is a custom-made product. To fashion shops for various purposes and provide them with parking spaces and service lanes it is clear that the planner must not be bound by use-zones which follow the traditional street and block lines. And, finally, the value of a business district is created by the purchasing capacity of its residents. A developer in a position to build both stores and dwellings has a broader basis for meeting price competition and making profits. Comprehensive developments are promoted by the unit scheme.

Internal Street System

> *The unit should be provided with a special street system, each highway being proportioned to its probable traffic load, and the street net as a whole being designed to facilitate circulation within the unit and to discourage its use by through traffic.*

When a residential district is laid out on a gridiron street pattern, there are no points, within it or on its border, in the reaching of which some residents would not have to travel two sides of a triangle. Through the unit scheme this inconvenience can be largely avoided. Before internal streets are planned, the principal destinations of residents in their daily movements will be definitely known. These will be the portals in the circumference where the traffic junctions and business districts are located, and the civic center where the school is placed. Channels for more or less direct movement toward these main destinations can be laid out and their capacities can be approximately adjusted to the volume of the streams they will carry.

In the pattern that will result from this process, there will probably be a combination of both radials and circumferentials. There may also be culs-de-sac. Obviously, the contours of the terrain will have to be considered, as well as the other factors of vista, economical lot subdivision, and street utilities. But there should be no necessity for making these interior streets continuous with similar streets in the adjacent neighborhood, or for even requiring their openings on the boundary highway to correspond with openings across the highway. Such a requirement would not only tend to invite through traffic into the unit but also rob it of its independence. In every way the principle should be observed that internal streets are to serve exclusively the purposes of residents of a particular district, and all the planning and engineering ingenuity should be directed to that end. Since boundary streets are made extra wide to facilitate the bypassing of a unit by through traffic, a planner can feel free to shut traffic out from the interior, so far as his means permit.

The street pattern produced by this method may seem like a maze to strangers, visitors and department-store delivery trucks, but this difficulty can be met by posting maps of the neighborhood, under weather-proof frames, in police booths and at the portals of the district.

If in the process of highway specialization we adapt parkways and boulevards to the needs of vehicles, why should we not fit neighborhood streets to the special requirements of pedestrians? Every thoughtful person, taking his car out of the garage, moves cautiously while in the vicinity of his neighbors. In a residential district, he is always near somebody's neighbors. Why should he object to driving slowly for the few minutes required to reach the boundary highway where he will be able to speed? Suppose that the neighborhood streets do seem like a labyrinth, and some of them are so narrow that, with cars parked at the curb, passage for other cars is a slow process – are not the safety and tranquility thus obtained worth more than the convenience that is sacrificed? If we are going to do everything possible to reduce the present frightful casualty rate from vehicular accidents, we should apply a preventive principle of this sort to the planning of family-life residential districts.

Finally, we come to the matter of residential character. From what has been said it is plain that the unit scheme will not, of itself, confer upon a development the quality that comes from costliness of construction or wealth of landscaping. On the other hand, it enables a modest project to secure an amenity from parks and planted recreation spaces that is not ordinarily possessed by the average commercial development.

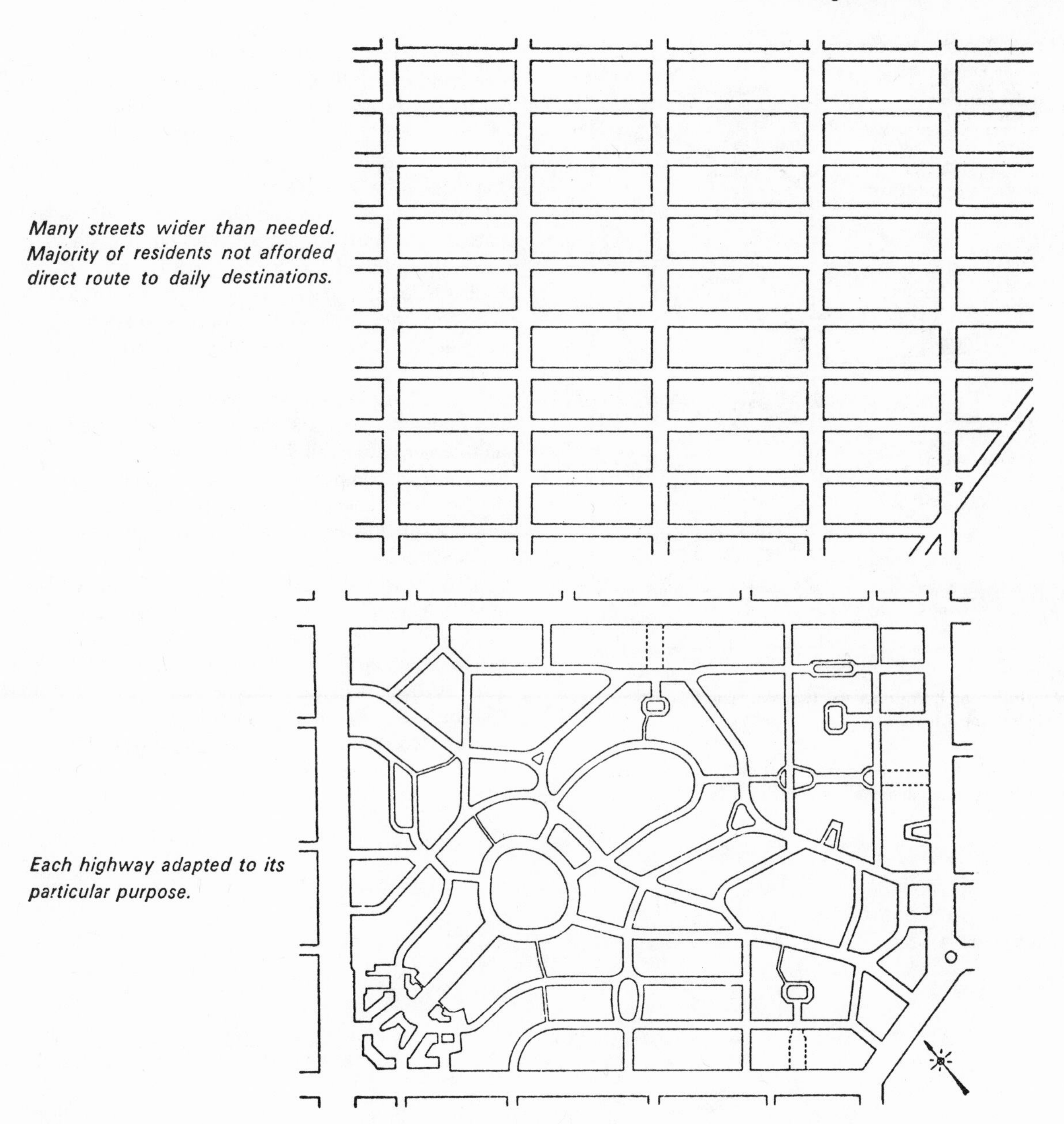

Figure 4
Gridiron and specialized street systems. (Source: New York Regional Plan, volume 7.)

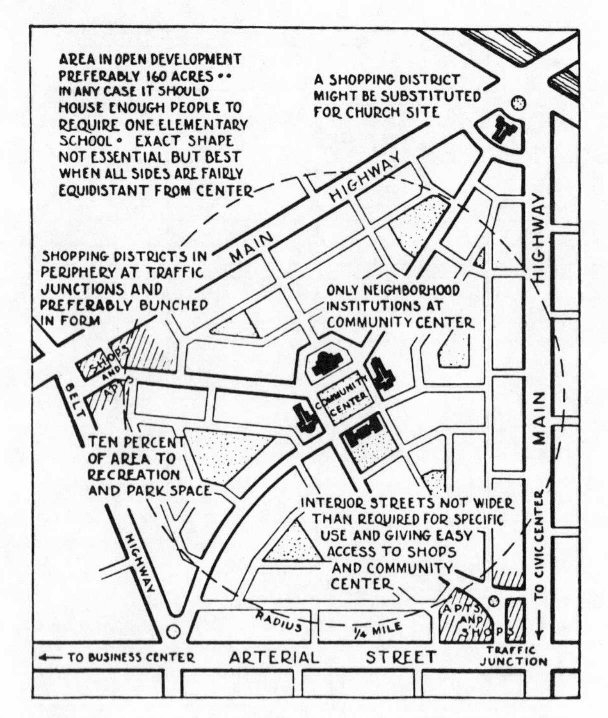

Figure 5
Neighborhood unit principles. (Source: New York Regional Plan, volume 7.)

One important merit of unit planning consists in its ability to save a residential district from the miscellaneousness that characterizes most urban neighborhoods. When a tract is subdivided into many uniform lots, sold to whomsoever will buy, and built up in accordance with the tastes and means of different owners, it is not likely to have a character that is definite or outstanding.

By reason of its wide boundaries and special street pattern, a unit is almost certain to stand out geographically as a district community. It will probably have a special name. But appearance is not the only basis of distinction. A district can be distinguished because of the affection which its residents have for it. When it has a complete equipment for the vicinity needs of its families; when the public services are nicely adapted to population requirements and all its component parts are integrated by a comprehensive plan – then you have a neighborhood community that is bound to be marked because of the esteem in which it is held by its residents.

If we will enable powerful special corporations to construct family homes, we can make certain that there will be not only a large supply of reasonably priced *new* ones, but that each house will be built into its own neighborhood – an environment that will be right because it also will be planned and constructed, along with its dwellings, to specifications based upon family needs.

4

Robert B. Mitchell and Chester Rapkin

THE INFLUENCE OF MOVEMENT ON LAND USE PATTERNS

The spatial arrangement of establishments emerges from the character of their mutual relationships and their relative positions and functions in the urban organizational structure. In a broad sense relationships (communication of one order or another) between establishments form a network over the urban scene. Every establishment must have some contact with others if it is to operate. Establishments vary in the number of contacts with others, and in the regularity and frequency with which these contacts occur. Within establishments the relative importance of some relationships will be considerably greater than that of others. If contacts through intermediate establishments are included, the network of relationships can be made to extend over the nation and across the face of the globe. For the purposes of analyzing the patternization of urban land uses, however, consideration must be limited to relationships that occur in a given urban area.

There will be an attempt on the part of a given establishment to achieve maximum accessibility to others with which it interacts, if there is the need for face-to-face contacts of persons or for the movements of goods between them. This can be done by locating proximately, or along physical channels of movement that provide ready accessibility. Since each establishment has a number of ties to others that may be scattered spatially, it will be pulled in different directions. The ties or group of ties of greatest significance to the operation of the establishment will be an important determinant in the choice of location. Thus, relating movement of persons and goods to the location of an establishment and examining movement as a factor in the spatial distribution of establishments are not two problems but two aspects of the same problem. The locational choice of an establishment is governed (among other things) by the location of *other* establishments and by the existing channels of movement.

Essentially, movement as a locational force is reflected in an attempt on the part of an establishment to minimize the amount of movement necessary between it and other establishments with which it engages in transactions. Expressed in dollar terms, minimization of movement cost is not necessarily based upon distance or time, but upon variations in the cost (e.g., classified freight) of transporting goods or persons to and from the establishment.

There are, of course, many factors and forces that prevent the optimum arrangement of land uses toward this end. Some of these relate to the type and availability of space at a given location, rent levels and land values, and cost of relocation. Others include such factors as tradition, inertia, long leases, legal limitations on use, and imperfect knowledge. An establishment may find that no space is available at its optimum location, that the available space is not suited to its use, that there are legal and social restrictions on its use, that rents or land values are too high, or that the cost of relocation exceeds the savings that may be effected. It may also be bound to its present location by a long-term lease or by inertia or tradition. Although attention . . . is focused on the manner in which the necessity for movement among establishments influences the locational distribution of land uses, this should not convey the impression that the other factors are of little, or even of lesser, importance. In fact, it is essential that allusion be made to these factors in the course of the development of the discussion. Major consideration, however, will be given to movement as a strategic factor in the determination of location and in the shaping of land use patterns . . .

THE LOCATIONAL ROLE OF MOVEMENT

The way in which the movement of persons and goods influences an establishment's choice of location is related to its functions and activities. It was previously stated that an establishment consists of a bundle of activities, each of which has its own particular movement requirements. Corresponding to these requirements are locational requirements, which conceivably may be the same for all activities of an establishment, but which will usually show more or less variation. Unless an establishment is completely subdivided by function, only one location can be chosen. The choice of this location, other things being constant, will be governed by the function (and its corresponding movement associations) that is of primary importance to the establishment.

Since the nature of the necessary movements associated with the primary function is the determining locational factor, it must be analyzed in some detail. One of the most fruitful ways of approaching the problem is to examine the movement systems that are related to the establishment in question.

It has been suggested that systems of movement are basic instruments for analyzing the relationship between movement and land use. Systems are essentially abstractions, and to be given substance they must be meshed with the location of establishments to which they refer. This process takes the form of analyzing systems by type of trip in order to discover the type of trip that is associated with the primary function and the manner in which its characteristics influence the location selected by the establishment.

Just as there are many activities conducted by an establishment, there is similarly a packet of movement systems associated with each activity and, of course, with the primary activity. An establishment engaged in the selling of goods will serve as the nonbase destination for persons on round trips or as an intermediate destination for persons on interbase trips. For persons employed at the establishment, it will be a base for home-to-work trips and for round trips made during the course of the workday or on lunch hour. As a nonbase destination, movement to the establishment may consist primarily of visitors who are members of the general public, or it may be comprised of a highly specialized type of clientele. Although each of these systems may have conflicting locational requirements, the primary activity will determine the set of movement systems that are accommodated in the final choice of location. A single location, however, may be eminently suited to more than one activity. A department store that is centrally located may be readily accessible as a destination for round trips, as a nonbased destination for home-to-work trips, or as an incidental stop in the course of a compound interbase trip. It will also provide accessibility to its employees on home-to-work trips.

In analyzing the primary function in terms of movement systems, these focal questions are to be asked: What types of trips are involved in the movements associated with the primary function? Does the establishment serve its primary function as a base or as a destination, or both? If it is a destination, is it a primary destination or an incidental stop? What is the locational distribution of establishments that are trip origins if the primary function operates as a destination, or of trip destinations if the primary function operates as a base?

If establishments which serve as origins or destinations for a given establishment are widely dispersed, accessibility cannot be in terms of proximity; it must be expressed in terms of a central location that is readily accessible to all these establishments. Here the choice of location is governed by the nature of the channels of movement. If the establishments that serve as origins or destinations are confined to a small area, accessibility can be achieved through proximity.

Department stores, other outlets for shopping goods, banks, and other types of businesses that are in large measure primary destinations for persons on round trips and that have an area of dispersion widely scattered throughout the city tend to locate at places where passenger traffic facilities converge. Similarly, types of businesses that require the services of employees with specialized training or skills also seek central locations. Conversely, less specialized branches of trade and service, which have more concentrated areas of dispersion, distribute themselves outside the main shopping centers and are thereby more accessible to buyers in one section of the city. Convenience goods establishments, which for the most part are visited incidentally in the course of a round trip or an interbase trip seek locations at important traffic intersections and along principal streets in all sections of the city.

Establishments which are devoted mainly to the handling of large quantities of goods and which are both destinations and origins for goods shipped into and out of the city tend to locate in the transport zones of urban areas – the waterfront, at sites along railway lines, and in the terminal and switching districts. The industrial zone in large cities "is neither compact nor particularly central, but stretches out along water ports, radial rail lines, and belt railways."[1] Establishments that handle goods of less bulk that can be shipped by truck, the destinations of which are either scattered about the city or concentrated in the central business district, tend to locate on the fringes of the downtown area.

THE FRICTION OF SPACE

Up to this point, movement has been discussed only in terms of distance and time. In actuality the calculation is made in money terms – the cost of shipping goods and transporting persons, plus the time cost of moving persons. In this total, goods movements will carry a heavy weight if there is a high goods-persons ratio, and persons movement will be heavily weighed if the reverse is true. But transportation cost is not the only economic consideration.

According to Haig, accessibility means "contact with relatively little friction."[2] Reducing the friction of space involves two classes of costs: the cost of transportation and the cost of site rentals. Haig maintains that these are connected phenomena: "Rent appears as a charge which the owner of a relatively accessible site can impose because of the saving in transportation costs which the use of the site makes possible."[3] Central sites afford maximum accessibility, but these are limited in number. Establishments and activities which can make the greatest savings in transportation costs will bid up the rentals for these sites,[4] and others less able to pay will be compelled to utilize less advantageous locations. "The two elements, transportation costs and site rentals," Haig has termed "costs of friction."[5]

Chamberlin, however, takes issue with Haig's definition, at least in regard to retail uses:

> *. . . rent is not paid in order to save transportation charges. It is paid in order to secure a larger volume of sales. Buyers and sellers alike are scattered over a wide area. Movement among them is so impeded that one place within the area gives advantages in securing the custom of a portion of the buyers. It affords a market which is, to a degree, distinct from the whole. The amount of product each seller can dispose of is not indefinitely large at the prevailing price. It is very definitely limited by location; if it were not, department stores would locate in outlying districts, securing the same volume of business, and increase their profits by the saving of rent. If we regard the whole area as one market it is clear that rent is paid because it contains elements of monopoly. Spatial differentiation results in demand curves which have a negative slope instead of being perfectly horizontal.*[6]

These seemingly contradictory views may be reconciled by reference to the role that the friction of space plays in the production functions of different types of establishments. Space, on the one hand, may be considered as an input factor that enters into the cost relationships in terms of rent, transportation, time (labor cost), etc.[7] On the other hand, space can also be considered in terms of output; that is, a firm not only sells goods and services but also convenience. In this sense alternative locations will have different influences on the level of the demand curve, with more accessible locations tending to raise the curve, and less accessible locations tending to lower it. Here, too, the strategic factor is movement expressed in transportation cost. The cost, however, is not incurred by the establishment but by the persons who visit it. Cost to the visitors must include such elements as time, convenience, and pleasantness as well as dollar outlay. On the whole, the relative importance of these input and output factors in the operation of specific kinds of enterprises constitutes a significant factor in the locational distribution of activities in an urban area.[8] It should be noted that in most cases in which location is related to cost, the establishment's principal activity in terms of movement is as a base of operations. In those cases where location is related to demand, the principal activity centers around the establishment as a destination of one type or another.

LINKAGE AND PROXIMITY

Factors that lead establishments to seek proximate locations serve to explain the clustering of uses that is characteristic of urban land use patterns. For certain types of movements and for certain classes of establishments, a high premium is placed on time (as an influence on cost or demand). When this is the case, interacting establishments require locations that are adjacent or in close proximity. The necessity for face-to-face contacts in the financial district and the large sums of money involved in transactions tend to keep financial institutions and their corporate customers within short walking distance of each other. In the garment center, the prevalence of subcontracting, the rapid adjustments in production that must be made in response to changes in fashion, and variation in orders also dictate tight spatial integration. When this is coupled with the need for proximity to local outlets as well as the convenience to out-of-town buyers, the present location of the garment center in New York is readily understood. Compact groupings such as these may be said to reflect the great number of linkages among the clustered establishments.

Linkage, then, may be defined as a relationship between establishments characterized by continuing or frequently recurring interaction. It is associated with the movement of persons and goods between the linked establishments and generates a tendency on the part of linked establishments to seek proximate locations.[9] The examples that follow present illustrations of the various types of linkage which may be delineated.[10]

Establishments are linked if they participate in transactions involving the movement of persons and goods. A machine repair shop and a factory that it serves are linked establishments. The smaller concern would tend to move toward the larger, particularly if the smaller is the supplier and the larger is the customer. In cases of this sort the factory may be considered as a *dominant* use and the machine shop may be characterized as a *subordinate* use. A dominant land use may also consist of a group of similar establishments, no one of which is dominant in itself (garment center in New York). The group of establishments taken together, however, constitute a dominant use. The movement between dominant and subordinate uses consists of interbase trips in which the individuals operate in their roles as members of their respective establishments.

Linkage exists when an establishment serves the members of another establishment. Thus a restaurant or a barbershop near a large industrial plant draws its customers predominantly from the workers in the plant. Despite the fact that there may be no direct transactional relationships between the restaurant and the plant as establishments, nevertheless consistent and routinized persons movement exists between them. The plant, in this case, may be considered as a dominant use, while the smaller establishments may be termed *ancillary* uses. Ancillary uses may be nonbase destinations for round trips or incidental stops on interbase trips. In these cases, the individual moves in a personal role, and not as a member of an establishment.

Establishments are linked if they serve common customers. Macy's and a small specialty shop located on Thirty-fourth Street in New York City have customers in common. It is evident that a larger proportion of the customers of the specialty shop will also be customers of Macy's than vice versa. Macy's, by virtue of its great variety of merchandise, is able to attract large numbers of people, and the specialty shop capitalizes on the traffic. In this case Macy's may be considered as the dominant use and the specialty shop may be termed a *satellite* use. Macy's and Gimbels, however, by virtue of the fact that each can attract large masses of people in its own right, cannot be considered to be in dominant-satellite relation to each other. Macy's and Gimbels, having equal importance, can be termed *codominant.*

Satellite uses are, as a rule, incidental stops for round trips destined to the dominant use. Codominant establishments may both be of equal importance as a destination for a particular trip, but in many instances one becomes an incidental stop for trips that have the other as their primary destination.

Proximity among establishments frequently exists for reasons other than linkage. Establishments which are proximate but not linked may be considered as *noncomplementary* land uses. In fact, the side-by-side existence of nonlinked establishments may militate against each other's interests. For example, an establishment requiring the regular movement of a large volume of goods will impede the operations of a next-door establishment that requires ready access to a large number of persons.

Noncomplementary land uses which are not linked but which at the same time do not interfere with each other's movements may be termed *compatible.* Noncomplementary land uses which interfere with each other's operation may be considered *incompatible.* In actuality, however, few noncomplementary uses will be entirely compatible or incompatible. Consequently, these cases should be classified according to their degree of compatibility, i.e., the extent to which they interfere with each other. In the following illustrations, characterizations of compatability are to be considered as indications of the section of a scale in which the cases are likely to fall, rather than as absolute designations.

Establishments may be attracted to the same address or locality because of prestige reasons. Insurance companies and the Curtis Publishing Company occupy space in the vicinity of Independence Square in Philadelphia because of a desire to be associated with an important historical symbol. Before Curtis moved its printing and distributions operations to Sharon Hill, these uses were incompatible. Now that only some of the editorial offices are maintained at Independence Square, the establishments may be considered compatible. Curtis Publishing Company on the masthead of *The Saturday Evening Post* gave its address as Independence Square, Philadelphia.

Establishments may tend to proximity because they require space with the same or similar physical characteristics or rent levels. An accountant and a wholesaler without stocks may locate in the same building despite the fact that they may never deal with each other. Establishments in this category will tend to have similar types of operations. These are compatible uses.

Establishments may locate close to each other because they require the same general location. *The New York Times* and the Paramount Theatre are at Times Square for reasons of accessibility to a widely dispersed population. The *Times* wishes to be able to send its reporters and distribute each edition to all sections of town as expeditiously as possible. The Paramount wishes to be accessible to as many customers as possible. Because it shows first-run films and features prominent entertainers, its market is the entire city, not just one neighborhood. During the late evening hours when the newspaper trucks are loaded and dispatched at the same time that crowds are pouring out of the theaters, these uses are incompatible. (At 5 a.m., when there is no conflict of schedules, can these establishments be called compatible?)

Establishments may be proximate because of historical accident. The Yankee Stadium was placed in the lower Bronx to be as close as possible to the residents of other boroughs. The Bronx County Court House was located two blocks away because its site – on a wide street in a pleasant residential area, adjacent to two parks – was one of the few remaining in the built-up sections of the borough that provided a suitable setting for a public building.

It may be argued that the concept of linkage should encompass all establishments that have movement between them, and that proximity is merely a secondary concomitant of the necessity for transporting persons and goods. If proximity, however, is not stipulated as part of the definition, there is no clear distinction between linkage and some aspects of movement systems. Systems of movement are said to be functionally related to organized systems of business or social action. One major order of persons movement is composed of systems related to kind of establishment. All major systems of goods movement are also related to establishments as bases or destinations – systems based on process of action, systems based on kind of trip, and systems based on kind of commodity.

Linkage, however, may be considered a subsystem of movement in which the destination-and-origin establishments are specified and proximity between them is stipulated. This distinguishes linkage from movement systems which are defined as kinds of movement . . . without regard to concrete time or location.

Linked establishments will be characterized by movements among each other, while the nonlinked but proximate establishments will not. The analysis of linkages provides a clue to the relation between land uses and traffic (via systems of movement), and proximate establishments, both linked and nonlinked, provide a basis for analyzing the structure of movement, that is, the totality of traffic at a particular place and time.

AN ILLUSTRATION OF LAND USE PATTERN

In a recent study of the central district of Philadelphia an attempt was made to reveal the manner in which land uses arranged themselves in the downtown area of that city. Primary emphasis was placed upon the way in which the activities and their associated movement requirements influenced the locational choice of various types of establishments. For descriptive and analytical purposes all nonresidential establishments in the area were classified into six basic land use types:

1. *Retailing:* Every type of establishment selling goods primarily to the customer, and including department stores and specialty shops with citywide appeal and convenience stores serving the resident or daytime populations of the central city.
2. *Manufacturing:* All establishments engaged in the production of fabricated goods and ranging all the way from loft manufacturing to large factories.
3. *Wholesaling with stocks:* All lines of wholesaling maintaining stocks on the premises and selling to the trade out of that stock. Display rooms and warehouses are the facilities generally required.
4. *Wholesaling without stocks:* All intermediaries handling sales transactions without maintaining stocks on the premises. While this catagory includes brokers and commission men, it is much broader. In a number of lines of wholesaling such as coal and lumber it is customary for the wholesaler to take

ownership but not physical possession of the merchandise. The usual requirement is for office space.

5. *Business services:* This category covers many activities in which the customers are other establishments and in which services are sold rather than goods. Law firms, advertising agencies, real estate, banking, insurance, and engineering are among the subclasses. Business services account for a large percentage of the demand for office space.

6. *Consumer services:* This is the broadest of all the six categories as to type of facility although it covers only establishments providing services to the consumer. It ranges from repair shops and barbershops to museums and churches, with a wide range of public, personal, and professional service in between. It will be seen that this category, broad as it is, has a definite place in the pattern of relationships among establishments.[11]

In analyzing the distribution of establishments, an attempt was made to arrive at the delineation of functional areas by designating every block in terms of its principal activity. Functional type was defined in terms of the combinations of activities and their relative dominance block by block. If two types of establishments together accounted for 50 percent or more of the nonresidential floorspace in any block, the block was taken to be characterized by that combination. The six types of establishments can be arranged into 15 different groups of two each, but the calculation revealed that only three of the combinations were sufficient to account for over 90 percent of the blocks in the central city. The combinations were (1) manufacturing and wholesaling with stocks, (2) business services and wholesaling without stocks, and (3) retailing and consumer services.

These patterns, gross though they are, reveal the tendency for certain types of establishments to sort themselves into cohesive areas within the central district in accordance with the functions and activities that they perform in common. In this sorting process many locational factors come into play and the exigencies of the movement and wholesaling with stocks are classes of activity that require large amounts of space for the storage and handling of goods. They seek low-priced space generally found in the older buildings, which tend to be concentrated in the older sections of the area. Within this group there is a major subdivision. The establishments that deal with the downtown department stores and other types of retail shops are found on the periphery of the concentrated portions of the retail shopping area, and those that supply markets outside the area tend to locate at truck and rail terminals and along the riverfronts.

There are not only positive factors that draw these two types of establishments together; there are also forces that repel other types of establishments. Noise, odors, and other unpleasant characteristics of establishments fabricating, storing, or assembling large quantities of goods are, of course, among them. But perhaps the most repellent is the fact that the movements of large quantities of goods interferes with other types of movements. Large trucks block the streets, and the movement of goods from the buildings to the trucks as well as the temporary storage of crates and cartons on the streets impedes the free movement of pedestrians along the sidewalk.

The second type of functional area consists of establishments devoted to business services and wholesaling without stocks. Both of these types of establishments deal with business rather than with consumers and neither handles goods on the premises. Both typically require and can afford accessibility and prestige in location. Their space needs are primarily for office facilities. As a consequence, the concentration of these establishments is to be found in the vicinity of Broad and Market Streets (in the newer buildings) and around Independence Square, where the insurance companies seek to associate themselves with the historical symbol of the area.

Retailing and consumer services are the two groups of establishments that constitute the third type of functional area. These two types of activity are held together primarily by consumer traffic. While they usually are found together, they are not usually in balanced proportions. Large retailing establishments, by drawing consumer traffic, create opportunity nearby for specialty shops and small service establishments. Large service establishments, such as public or semipublic institutions, create opportunity in the immediate neighborhood for small retail stores.

Thus the locations of establishments and the patterns of location are influenced by the functions and activities of establishments (the principal activity in particular) and the movements associated with the conduct of these activities. Viewing these movements in terms of the systems of which they are a part facilitates the analysis of establishments in terms of bases and destinations, their relationships to the spatial distribution of establishments with which they interact, and the tendency for certain types of establishments to seek proximate locations. Other factors such as the quality of space or rent levels have also been noted, and in the analysis of a real situation these cannot be omitted. In fact, in many individual instances these may be of equal or greater importance.

In an integrated study of locational influences, however, it is possible to deal with seemingly disparate forces within a modification of the framework presented by Haig's concept of the friction of space. In the original statement of the concept, transportation costs and rent are presented as balancing forces. If the desire to raise the demand schedule is placed alongside the incentive to minimize transport costs, many of the other locational influences can be treated within this intellectual structure. Prestige, for example, can be considered as a demand factor in which the establishment seeking a prestige location does so in an attempt to maintain or enlarge its clientele.

DYNAMIC RELATIONSHIPS

Up to this point the discussion has centered around the manner in which the need for movement influences the present location and spatial distribution of establishments — the static situation. But since land use patterns are constantly changing, a comprehension of the forces which bring about changes is of importance. In fact, one principal reason for examining a static situation is to derive some insight into its dynamics. The question now is this: How are changes in movement requirements related to changes in the pattern of land uses?

For the pattern of land uses to change, there must be a change in the spatial distribution of activiteis due to a rearrangement of establishments or to an alteration in their activities. Since the vacant tracts, parks, and streets are integral elements in the land use pattern, changes in these will also, by definition, vary the pattern of uses. Thus, changes in the distribution of land uses may come about through processes that alter the activities of establishments at a given location, the character of improvements on the land, or the street system and other channels of movement.

The forces that lead to changes in land use can be segregated into two broad groups. The distinction between these sets of forces is to be found in the terms "accommodation" and "accessibility." The first includes such factors as changing space requirements or rent paying ability or other reasons that relate to the physical setting — the building space — in which an establishment is housed. Technological, administrative, or organizational change or economic vicissitudes may so alter the activities of an establishment that it is impelled to seek a different kind, amount, or quality of space from that which it presently occupies. This category also includes fortuitous factors, such as the necessity to seek new quarters because the building in which an establishment occupies space is slated for demolition or because a lease cannot be renewed. Factors related to accommodations as a force bringing about land use change, as well as questions centering around the ability to pay for more satisfactory quarters or locations, are, of course, crucial considerations. But here they are accepted as given, since it is not within the scope of the present study to probe into questions regarding the nature of change in space requirements of establishments, or the character of activities that can desire the highest economic return from the use of a site.

The second set of factors, those related to accessibility, is concerned with the necessity for moving persons or goods. These relate to changes in land use through alterations in the nature of an establishment's relationship to others or in the spatial distribution of establishments with which it regularly interacts. The desire to secure accessibility to a different set or distribution of establishments is the underlying force in this set of factors.

CHANGING MOVEMENT REQUIREMENTS

Since the functions and activities of an establishment give rise to certain types and patterns of movement, changes in the internal structure of establishments can be expected to alter its movement requirements. Changes in internal structure come about as a result of variations in technology and in business organization. A striking example of this is to be found in the recent changes that have occurred in some branches of wholesaling business. With standardization of items and purchase from catalogue or through specifications, a large section of the wholesaling business is conducted without stocks and no longer requires locations that are readily accessible to water, rail, or truck terminals. Instead business is conducted by telephone or personal contact from offices in the center of town, a location most convenient to customers and sources of business contacts.

In retailing, there has been a marked trend from corner grocer to large supermarket. This has served to increase the volume and type of goods shipped to retail food establishments of the latter type, and has extended the destination area for and number of trips to each of these shops.

Sometimes sheer growth in size will bring about a change in associated movement by enabling an establishment to subdivide its constituent function and to move one of the constituents to a new and more propitious location. Many manufacturing plants have maintained a head office and salesroom in the downtown district but have moved their fabricating functions to the peripheral areas. Department stores have separated sales and display from warehousing and storage, locating the latter in areas suited for goods movements. In each case the separation of functions or activity has altered the nature of the movement associated with the downtown establishment by changing the goods-persons ratio, the pattern of home-to-work trips, and the nature of necessary contacts with other establishments.

The second way in which change in movement relates to changes in land use is to be found in external relationships among establishments. This includes the response to shifts in the location of establishments with which a given firm, agency, or household interacts, as well as the effect of changes in social organization on the number and type devoted to specific types of activities.

The various types of linkage and the compatibility of nonlinked but proximate establishments provides an important tool in the analysis of the external factors in land use dynamics. Any change in the location of a dominant establishment will tend to be accompanied by changes in the location of the subordinate, ancillary, or satellite land uses. Nonlinked uses that compete for space but have mutually harmonious relationships regarding goods and persons movements can be expected to remain in proximity, other things being equal. Incompatible land uses, on the other hand, interfere with the operations of both establishments and provide incentive for one or the other to move to a more suitable location.

Once a change has occurred, even if the incentive for a shift in land use comes from forces not related to movement, movement relationships and requirements will be altered. Through the interplay of forces, a change in land use caused by a factor unrelated to movement may create a situation in which movement relationships are released as dynamic factors. An example may best illustrate how such a chain reaction may be brought about. The telephone and other long-distance methods of communication and control no longer make it necessary for the management of manufacturing firms to be located in the factory. The production units can be placed in an outlying area and the administrative offices located in office buildings in the central city at places convenient for the assembly of persons. Hence the tendency for offices to pile up in big buildings and concomitant generation of persons traffic. This in turn will drive out uses like wholesaling with stocks that are unable to operate satisfactorily in heavy mixed traffic. On the other hand, uses like small job printing, telephone, stationery, and some types of consumer services will be attracted to the area.

In this illustration the interrelationship between changes in land use and changes in movement is indicated by a chain of events. The nature of this interrelationship will become clearer when the major processes through which land use change comes about are analyzed.

THE RELOCATION OF FIRMS OR HOUSEHOLDS

Perhaps the first process that comes to mind when land use change is discussed is shifting of firms or households from one location to another. A bank or insurance company will move its headquarters from an old to a new center, as will other types of nonresidential establishments. Perhaps the most mobile type of establishment, however, is the American household. Estimates in recent years indicate that, on the average, over one in five changes place of residence each year. Unfortunately, similar estimates are not available for nonresidential establishments, but one may venture that the rate is considerably lower.

The principal reasons why a firm or household may be attracted to a new location have already been indicated in large part (changes in movement requirements due to internal shifts in activities and external shifts in the location of establishments); but there are additional reasons for the relocation of establishments. For example, the existence of uses in the immediate area that impede the ready movement of goods or persons will operate as a factor to repel a given use from its present location. Changes in the street system or in other channels of movement may cause an establishment to change location. This factor is treated in some detail under a subsequent heading.

CHANGES IN THE ACTIVITIES OF ESTABLISHMENTS AT A GIVEN LOCATION

The second process by which land use change comes about is through changes that occur internally in an establishment that remains located at a given site. Retail stores, for example, have responded to a shift in the composition of traffic passing the establishments: men's clothing stores located along the paths of people traveling to and from work have taken in lines of women's wear in recent years to capitalize on the increase in the number and proportion of women on home-to-work trips. Department stores have introduced grocery and other types of food departments on their street floors.

Another type of internal change that results from a change in movement requirements is the subdivision of functions and the removal of one or more to other locations more suited to their movement requirements. The relocation of a constituent function or activity may come about because uses are incompatible or because the pattern of linkage changes. Thus what might have been a manufacturing area at one time may now be an administrative district occupied by the main offices of manufacturing firms.

DIFFERENTIAL GROWTH RATES AND CHANGES IN AGGREGATE ACTIVITIES

A change in the land use composition of an area may come about because of differential rates of growth among establishments of various types either in an urban area as a whole or in certain subareas within a city. This results in changes in the relative proportion of establishments devoted to various activities. It is considered by some to be a more significant long-run process of change than the shifting of firms or households from one location to another.[12]

The trend toward services and distribution activities has been observed repeatedly in studies of the entire national economy.[13] In the United States, as in other advanced industrial countries, the proportion of the labor force employed in agriculture has declined, the percentage in trade, finance, government, and service occupations increased, and the remaining groups have changed little in their total share of the labor force. The same tendencies are seen in the distribution of national income and aggregate payments by industry. At the same time, it has been universally recognized that the population of the United States has steadily become more urbanized. But what may not have been so evident is that the balance of activities within the range of urban life has been shifting steadily away from production toward distribution and services, and that this shift has had formative influence on the land use patterns in urban areas.

The structural change in the activities of the central district of Philadelphia, for example, is revealed by an index of the rate of net growth of the number of establishments of various types.[14] For all types of establishments combined, the ratio between the total number of establishments in 1906 and the number in 1949 is equivalent to an index of 1.35. Variations from this average indicate the extent to which various types of activities have shifted their relative importance in the intervening 43 years. The following table indicates the indexes for five major types of establishments:

Construction	.83	Wholesaling	1.24
Manufacturing	1.12	Business services	1.45
Retailing	1.16	Consumer services	1.58

Because of lack of detail in the early records, it was impossible to segregate wholesaling establishments handling stock from those which held title but not physical possession of goods. Data for a later period indicate the varying tendencies between these two types of establishments. Between 1934 and 1949 the goods-handling wholesaling establishments increased from 2,040 to 2,120, or 4 percent, while the nongoods-handling wholesaling establishments increased from 1,351 to 2,570, or 90 percent, during the same period of time.

The change in these proportions is, of course, reflected in a shift in the composition of land use, and therefore is a change in pattern simply on the basis of a changed amount of land or building space devoted to each kind of establishment in aggregate. But in addition to the aggregative effects, the different types of establishments that lie behind the shift in proportions give additional impetus to changes in land use patterns. The demise of an establishment at one location and the birth of another at a different location may make for change in the pattern of land uses even if the new and deceased establishments were of the same types. But if these two are of different types, the effect of the changing establishment mix on the spatial distribution of land uses is accentuated.

An illustration of the manner in which land use change comes about as a result of growth and attrition in the number of establishments is provided by the Philadelphia Central District Study.[15] A historical study of the survival rates among nonresidential establishments over the period 1906-1949 was made for a sample of 39 blocks located in the central district. Data were derived by listing according to type of activity all establishments that were located in the sample blocks in the year 1930 and then tracing the change backward to 1906 and forward to 1949. In 1930 there were 5,038 establishments in the 39 blocks studied, representing a large proportion of all establishments in the central business district. Of these, only 1,143, or 23 percent, were located anywhere in the central district in 1906, and only 1,542, or 31 percent, still existed in the central district in 1949. These data provide a striking indication of the very large turnover of business establishments in the core area of a large city.

The influence of the turnover of establishments on a small area is revealed by the land use composition of standard Philadelphia Planning Commission Block #1S-10E for benchmark dates of 1906, 1930, and 1949. The block is characterized by two principal functions, business services (predominantly insurance) and wholesaling. For the years of record the proportion of business service establishments to total establishments consistently declined from 68 to 58 to 32 percent, while wholesaling rose from 14 to 24 to 33 percent of total. The records further reveal that of 376 establishments in the block in 1930, 236 did not exist in 1906, and of these, 129 (55 percent) were business services and 64 (27 percent) were wholesaling. The distribution of the new firms tended to shift the balance in favor of wholesaling. Between 1930 and 1949, it was attrition rather than growth that accentuated the change in the established direction. In this 19-year period the number of establishments located in the survey block declined from 376 to 121. But of the 121 establishments in the block in 1949, only 11 had been in existence in 1930 and 110 were new establishments. The composition of the new establishments again was such as to shift the proportion in favor of wholesaling.

It is interesting to observe the general spatial distribution of establishments over this period. Most of the enterprises which had moved to the survey block between 1906 and 1930 (and were still there in 1930) had come from a two-block radius, indicating little geographic movement of the focus of business services and wholesaling. By 1949 a shift in the location of the functional area was evident. Of the 108 enterprises which were in the survey block in 1930 and had moved to other locations in the central business district, 61 (56 percent) remained within a two-block radius of their original location while the remainder (44 percent) tended to cluster in a new and established concentration 12 to 14 blocks away.

It is thus suggested that a major process of land use change is to be found in the location of new establishments which reshape existing concentrations. The new establishments usually perform new types of functions or variants of the functions of existing establishments. Moreover, a new establishment has the opportunity to choose a location most suitable for its purposes and, unlike an existing firm, is unfettered by the cost of moving, disruption of business, inertia, or the loss of goodwill associated with an established location. The effect of the birth and death of firms will also be magnified by the sharply varying space requirements of the existing and remaining firms.

In the land use shifts resulting from the birth and death of establishments the role of movement is readily seen. Business firms may die for any of a host of reasons, one of which may be wrong location. For goods-handling firms, transportation costs may have been too heavy because channels of movement were inadequate or proximate uses were antagonistic. Retail shops or other types of establishments dependent upon visits from members of the general public may find that their location is inconvenient to the clientele they expected to attract or that persons who were "customers" in the past have moved to other areas.

A new establishment has the advantage of being able to choose a location most propitious to its movement requirements, within the limitations of the rent level and the availability of existing space. But by the same token, since it has not as yet commenced operation, it cannot be certain that the interactions it expects with a group of predesignated establishments will materialize in the form anticipated. The original location is thus based upon estimates that contain a good measure of guesswork. The degree of uncertainty is diminished if an establishment is not the first of its type to locate in an area, for then it can make judgment on the basis of the experience of others.

The shifts in the aggregate composition of land uses resulting from differential growth rates can be assessed only in part in terms of movement of goods and persons. The drift from the production of goods to the production of services is a national phenomenon that accompanies a rising standard of living. Within urban areas, the high cost of transporting goods through the congested city streets encourages the trend from goods-handling to persons-handling establishments, as it did in the central district of Philadelphia. If the changing movement pattern has not been a cause, it has surely been a consequence of the altered character of urban activities.

Variations over time in the kinds of activities that are conducted in cities are extremely significant factors in the relationship between movement and land use dynamics. Changes in activities and in the establishments performing them alter systems of action and their corresponding movement systems. A major consequence is manifest in the realignment of establishments that constitute the bases and destinations of various movement sytsems. Another concomitant is a change in the daily rhythms of movement.

The drift from goods- to persons-handling establishments brings about shifts from goods- to persons-movement systems, and influences the pattern of daily variations in the volume of persons movement. Workers employed in goods-handling firms, for example, are usually required to report earlier or leave later than the 9 a.m. to 5 p.m. employees of establishments dealing with persons. Goods-handling establishments frequently work more than one shift. Thus the trend from goods to persons activities will cause more home-to-work trips to be concentrated during the peak hours of travel or to coincide with trips for other purposes. Even without an increase in the total daily volume of traffic, congestion may be aggravated at crucial periods of the day.

CHANGES IN THE AVAILABLE SUPPLY OF SPACE

One of the most dramatic and readily observable factors in land use change is to be found in the construction of new buildings. As a rule, the new structure is put to a different use than the structures that previously occupied the site. Occasionally, the general category of use is the same, but there are differences in quality; for instance, a modern apartment building replaces several ancient single-family structures. If the building has been constructed on vacant land, the nature of the change is obvious.

Although new construction is the most evident means of increasing the total amount of building space,[16] the space devoted to a particular type of use may be increased within the limits of the existing supply. Many an old private residence now houses offices or stores, as a walk through the central district of any city will reveal.

The construction of new buildings, however, is a conditon rather than a process of change. Additional space either at existing concentrations or at new locations serves to facilitate the working out of other forces in process. It does this by making available additional physical accommodations both for enterprises that wish to move from their present locations and for new establishments. A new structure may provide a more strategic location for existing firms and for the estimated movement requirements of new establishments. If the structure has been built for the express occupancy of a stipulated establishment, its choice of location was in large measure governed by its movement requirements, expressed in terms discussed previously.

When establishments are attracted to new buildings because of the quality of space or the prestige associated with the occupancy of modern structures, they may form new foci of activities previously located elsewhere. Establishments that are either directly linked or linked in ancillary or satellite fashion will be attracted to the new concentrations. In fact, buildings frequently make provisions for this by constructing suitable facilities for the linked uses – witness the shops at any large rail or bus terminal or the doctors' offices on the ground floors of large apartment houses.

Change in the available amount and type of building space is also brought about by direct governmental action in the fields of slum clearance and urban redevelopment. The motives for these programs are only indirectly related to movement. Their primary purpose is to make decent, safe, and sanitary dwellings available at low rent to families in the lower ranges of the income distribution or to make strategically located land available for private reuse at a value in consonance with the new use. However, before any local redevelopment proposal became eligible for federal grant or loan under Title I of the Housing Act of 1949, the local public agency had to certify that it was in accordance with a master plan, which presumably had a section dealing with problems of movement, circulation, and transportation.

Although the bulk of local redevelopment programs focus on reuse of land by establishments, a few center their consideration on the street system and other movement channels. In Mobile, Alabama, for example, a proposed redevelopment site is directly in the path of the shortest connection between two major thoroughfares leading from a residential district to the industrial and dock sections.[17] As the site is now laid out, traffic from the residential areas must weave through a tangle of narrow streets to reach the industrial areas. The Broad-Beauregard Streets project has as its major purpose the provision of a wide connecting street, which is expected to remove the present traffic bottleneck. Many other redevelopment proposals also provide for revision in the street system, but for the most part these are incidental to other objectives.

CHANGES IN THE STREET SYSTEM AND OTHER MOVEMENT CHANNELS

The street system is, of course, a basic element in the land use pattern of a city. In many urban areas, the proportion of land occupied by streets exceeds any other use. On the average, one third of the developed area of a city is devoted to streets. Street area amounts to 72 percent as much as the combined area occupied by private residential, commercial, and industrial building.[18]

The importance of the street system, however, is not due to the volume of land that it occupies. The street system constitutes the framework of a city. The size, shape, and orientation of blocks, lots, and, in some cases, of buildings are determined to a considerable degree by the street pattern. It exerts an influence on the pattern of land uses by establishing the basic avenues of accessibility. Streets not only provide for the movement of persons and the transportation of commodities, they also furnish light, air, and means of access to abutting property. Below the surface of streets are found most of the city's utility installations – telephone and telegraph cables, the distribution systems for electricity, gas, water, and heat, and facilities for the disposition of drainage water and sewage.

The streets in the central business district of many American cities show little alteration from, and in fact may be identical with, those that served the area during a less populous and more leisurely time. Many of the early street systems were platted, but others just developed in random and casual fashion. This situation is aggravated by the fact that many cities have developed their entire street pattern by the mechanical extension of the original plat. Even the designed street systems, perhaps efficient in their day, have outlived their original purpose. The great number of crosstown streets on Manhattan Island at one time served the river to river traffic, while persons and goods were carried to the upper sections of the island by boat. The few north-south streets now carry the burden of traffic, and the narrow cross streets are inadequate as feeders, difficult for passage, and inefficient for access.

The streets of a city appear to be one of the most permanent elements in the entire man-made physical structure. Once utilities are installed and the abutting lots improved, street widening or relocation is exceedingly difficult and very costly. The rebuilding of an entire street system, even for a small city, is virtually unthinkable.

Changes in streets and the street system occur nevertheless, for the most part within the framework of the existing street pattern. Streets are widened or extended, or their use regulated. Less frequently some are closed or others opened. Only rarely are there drastic changes. These usually take the form of the provision of a new type of street – a circumferential highway or a parkway. Most of the important changes occur in other channels of movement – the addition of a subway system or an elevated route, the institution of a trolley or bus line.

How do changes in channels of movement come about and how are the antecedents of change related to movement? It is, of course, quite obvious that changing movement requirements are the direct long-run cause for changes in the street system and in some of the other channels of movement. Changing movement requirements can be viewed in two ways, first in terms of variations in the volume of movement, and second in terms of variations in the kind of movement.

Change in the volume of movement may result from city growth, even in the absence of change in basic urban structure. Thus the more intensive use of land brought about by the construction of a greater amount of building space per unit of land, or by more intensive occupancy of a given amount of space, will in and of itself generate an increase in the volume of movement. Changes of this order increase the total burden of traffic on the channels that serve the areas of intensive use and bring pressure to increase their capacity either by traffic management, street widening, or the provision of additional surface, subsurface, or overhead channels.

Changes in the kind of movement on a particular channel occur as a result of shifts in the land use pattern. Goods-handling establishments in an area may give way to establishments dealing primarily with persons, or the reverse may happen, making it necessary to adapt the streets or other channels to different traffic requirements. City growth may be extensive as well as intensive, with new urban developments forming on the outskirts of the built-up areas. In the transformation of vacant to occupied land or rural to urban uses, it is usually necessary to tie the new developments to existing concentrations with roads, bus lines, commuting trains.

Perhaps the most significant influence on movement channels is the change in movement requirements brought about by technological advances in modes of transportation. The introduction and widespread use of the private automobile has been the greatest single factor in altering the functional requirements of the street system. The use of motor truck for commodities transport has liberated large goods-handling establishments from the rail siding and the dock, but it has created the need in other areas for special streets adapted to heavy truck traffic.

Just as changes in kind or intensity of land use by establishments bring pressure for changes in the channels of movement, changes in channels tend to affect the distribution of establishments by altering existing paths of movement and avenues of accessibility. Even minor changes, such as traffic regulations which establish one-way streets or shift bus stops, will have some influence on land use, particularly in regard to retail shops. Changes in channels can reduce accessibility as well as increase it. Congestion in a channel may render a street that was previously satisfactory inadequate for current movement requirements. These changing patterns of accessibility generate attempts on the part of establishments to locate more conveniently in relation to others with which they interact. Growth patterns are also influenced. The construction of a new road or the extension of a bus or subway line will lead to further development of outlying areas, particularly those directly accessible to the new channel.

The interpenetration of influences is of particular importance to the traffic engineer and the city planner. The existing distribution of establishments of itself is not a sufficient basis for developing a program of streets or highways or subway construction. Nor can a projected distribution of establishments based upon the extension of existing trends suffice. The design of any new facility must take into account the land use changes that the additional or altered channel can be expected to bring about (as well as the additional traffic that the new facility will generate). Thus the design of a new channel or the redesigning of an existing channel of movement should be a process of iteration, in which the proposed facility, the existing land use pattern, and its anticipated changes are brought into consonance.

REFERENCES

[1] Edgar M. Hoover, *The Location of Economic Activity* (New York, 1948), p. 128.

[2] Robert Murray Haig, *Regional Survey of New York and Its Environs* (10 vols.; New York, 1927), I, 38.

[3] *Ibid.*

[4] In the Loop district of Chicago the aggregate value of land in 1926 was one-fifth of the total for the city, an area 1,000 times as great as the central business district. Homer Hoyt, *One Hundred Years of Land Values in Chicago* (Chicago; 1933), pp. 336-37. Peak values in New York City in 1927 were to be found on Broadway, and on State and Madison in Chicago in 1930. R.D. McKenzie, *The Metropolitan Community* (New York; 1933), p. 234. In St. Paul maximum land values were at the principal points of traffic convergence. Calvin F. Schmidt, "Land Values as an Ecological Index," *Research Studies of the State College of Washington*, March 1941, pp. 31-36. Similarly, at the turn of the century land values in the business sections of Council Bluffs, Salt Lake City, Duluth, Seattle, Atlanta, Toledo, Columbus, Richmond, Kansas City (Missouri), and Minneapolis were highest at the main traffic intersections. Hurd, *op. cit.*, pp. 134-43. According to Spengler, the effect of the construction of a new transit facility is neither direct or uniform. In his study of the influence of subway construction on land-value trends in New York, he concludes that "In some cases . . . land value changes . . . along or near traffic arteries . . . have been found to exceed 1,000 percent. In others, actual losses have been experienced. . . . Moreover, in a large number of areas there has been a kind of indifference to the presence of transit facilities, the value of land along such routes behaving in no different manner than that for an entire section of which it is a part. . . . A transit facility . . . permits . . . an emergence of land values, the values being determined largely by other factors." Edwin H. Spengler, *Land Values in New York in Relation to Transit Facilities* (New York, 1930), pp. 130, 133.

[5] Haig, *op. cit.*, I, 39.

[6] E.H. Chamberlin, *The Theory of Monopolistic Competition* (Cambridge, Mass., 1942), p. 215.

[7] In this regard Walter Isard suggests the incorporation of a new factor in the production function. The new factor is "distance inputs," a mode of indirect labor, the price of which is measured in terms of transportation rate ("Distance Inputs and the Space Economy," *Quarterly Journal of Economics*, LXV (May and August 1951), pp. 181-98, 373-99.

[8] It is assumed in this study, however, that establishments have already made their compromises between cost or demand factors and rent.

[9] The concept of linkage presented here is similar to that suggested by Professor Florence. He says that "owing to intransportability of semi-manufactured materials or products or to the need of contacts for the interchange of knowledge, there may be *linkage* with the makers of a particular product or a variety of products." He further points out that, if one of two linked industries is dispersed, the other will also tend to be dispersed (i.e., tend to have the same coefficient of localization). Thus, although he does not state that establishments of linked industries will be proxi-

mate, it is reasonable to draw such an inference on the basis of the agreement between their localization coefficients. P. Sargant Florence and W. Baldamus, *Investment, Location, and Size of Plant* (Cambridge, England, 1948), p. 180.

[10] For suggested indexes of linkage, see pp. 150-54.

[11] *Philadelphia Central District Study* (Philadelphia, 1950), pp. 5-6.

[12] The same observation has been made in regard to the shift in the locational distribution of industrial plants for the nation. From the experience of the years 1928 to 1933, it was concluded that "a very small fraction of industry has moved its location during this period." Carter Goodrich et al., *Migration and Economic Opportunity* (Philadelphia, 1936), p. 340. See also E.M. Hoover, *Location of Economic Activity* chap. 9, "The Process of Locational Change," pp. 145-65. Hoover similarly maintains (p. 150) that "most shifts in the location of any specific industry are essentially geographic differentials in the rate of growth of the industry with the actual relocation of firms or plants playing a minor role."

[13] See, for example, John D. Durant, *The Labor Force in the United States, 1890-1960* (New York, 1948); and Simon Kuznets, *National Income, A Summary of Findings* (New York, 1946), Tables 11 and 12.

[14] *Philadelphia Central District Study*, pp. 60, 101.

[15] Data for this section were drawn from *Philadelphia Central District Study*, pp. 58-61; *First Phase of the Philadelphia Central District Study*, pp. 7-43 to 7-45 and from the two maps of CPC Block 18-102 (no page number); and *Number of Central District Establishments in Each Block by Kind of Business, 1949;* Supplementary Tables, Vol. 3.

[16] In housing, for example, there are no years in which demolitions exceeded the volume of new dwelling units. Although there are no similar records for nonresidential structures, the same appears to be true.

[17] *A Preliminary Report on Urban Redevelopment for Mobile, Alabama* (Mobile, n.d.), p. 7.

[18] Harland Bartholomew, *Urban Land Uses* (Cambridge, Mass., 1932), Table 40, p. 105.

5

Richard F. Babcock

THE PURPOSE OF ZONING

In the perfect market, natural zoning would result. RATCLIFF

The witness, I have inferred, takes the position that good zoning requires that this property-holder . . . should be protected against himself.
Record on Appeal. *Corthouts v. Town of Newington,* 140 Conn. 284 (1953)

The most fun as well as the safest path for the amateur commentator is to keep things anecdotal. But even the practicing attorney, faced with the duty to discharge immediate assignments, is not free from the impulse to point the way. Hence my intention . . . to consider a few of the issues underlying most disputes involving public regulation of private land.

Why do we have zoning anyway?

It is indicative of the chaotic nature of the subject that there is no generally accepted answer to this

---

 Selected from The Zoning Game – Municipal Practices and Policies, *by Richard F. Babcock, The University of Wisconsin Press, Madison, Wisconsin (1966), 115-125.*

question. At the start I suggested that zoning caught on as an effective technique to further an eminently conservative purpose: the protection of the single-family-house neighborhood. In spite of all the subsequent embellishments, that objective remains paramount. As might be expected, such a motive is rarely articulated as a rationale for this popular device, either by the supporters or critics of zoning.

There are, however, some plausible theories offered in support of zoning. I am not concerned here with the deeper psychological motivations which drive many of the backers of zoning. These may vary from a fear of Negro infiltration to a vague identification of zoning with "good government." While they cannot be ignored by anyone practicing in this field, I am concerned here with more rational purposes of zoning.

We can dismiss the early legal fictions which were created to validate zoning under the jurisprudence of the 1920s. The early proponents of zoning claimed that the single-family district was insulated to prevent the spread of fires. Minimum house size requirements were supposedly related to public health. Billboards were said to endanger public morals because of the promiscuous activities which took place behind them. No one really believed these fictions in 1920 and no one believes them today. And today our courts have progressed beyond the need for this type of shibboleth.

I have found in circulation two relatively rational theories of the purpose of zoning, which I refer to as the "property value" theory and the "planning" theory. For the purpose of clarifying the issues, let me state these theories in a generalized fashion.

THE PROPERTY VALUE THEORY

To most real estate brokers and promoters, and to some land economists, lawyers, and judges, zoning is a means of maximizing the value of property. The use of property, under this theory, is basically determined by the dynamics of the market. Denver attorney George Creamer speaks for this view:

> *The dynamics of a community, so long as that community remains economically free, dictate the uses to which land will inevitably gravitate, whatever expedient of zoning be employed. Zoning otherwise employed than as a braking mechanism is probably misapplied, and, historically, is probably futile.*

Although the exponents of this theory purport to believe in the dominance of free market forces, they are strong supporters of zoning. This paradox can be understood only by the realization that under this theory the "proper" zoning of property is determined by market forces. Zoning is merely an adjunct to the market mechanism.

The basic axiom of this theory is that each piece of property should be used in the manner that will ensure that the sum of all pieces of property will have maximum value, as determined by market forces. In other words, every piece of property should be used in the manner that will give it the greatest value (i.e., its "highest and best use") without causing a corresponding decrease in the value of other property. The zoning ordinance can achieve this goal by prohibiting the construction of "nuisances," provided the common-law concept of nuisance is extended to include any use which detracts from the value of other property to a degree significantly greater than it adds to the value of the property on which it is located.

In the property value theory, for every piece of land there is a "proper" zoning classification. Above every town there exists a Platonic ideal zoning map, waiting to be dropped into place. This map shows for each piece of property the use or uses which will give to the sum of all property the greatest total value. This theory is what enables many judges to determine the proper zoning classification for property based solely on the estimates of appraisers of the value of the property and surrounding property under various zoning classifications. The property value theory requires only ordinary arithmetic, and has the appeal of all simple solutions to complex problems.

A corollary of the property value theory is that the planners tend to be meddlers who, by their tinkering, upset the natural market forces. Zoning, properly concerned, does no more than protect the market from "imperfections" in the natural operation of supply and demand. Professor Richard Ratcliff, economist at the University of Wisconsin, is an outspoken proponent of this view:

We start with the premise that the arrangement of community land uses should be the product of social preferences; and that, but for the imperfections of the real estate market, the market interactions of demand and supply would create a city so organized. Thus we view city planning as a device for releasing the basic forces of demand rather than inhibiting them.

If Professor Ratcliff's outlook had too much the appearance of economic determinism, it remained for Allison Dunham, University of Chicago law professor, to soften Ratcliff's dictum and to provide a rationale by which all local control over private development could be justified: public control in all events is justified, but in some cases there must be compensation to the landowner.

Dunham, the lawyer, sensed the social flaw in economist Ratcliff's theory. "With respect to private land use decisions," he said, "considerations of economy and efficiency are reflected in the market price, but no beneficial or detrimental impact of a land use upon other lands is reflected in the market for a particular land use." The public may take (regulate) says Dunham but "[t]he public need not compensate an owner when it takes (restricts) his privileges of ownership in order to prevent him from imposing a cost upon others; but when the state takes (uses or restricts) his property rights in order to obtain a public benefit it must compensate him."

As for the role of the planner, Dunham decrees: "The city planner may interfere with and supervise the land use decisions of a private developer only because of the interaction of one land use upon another and only then where the private developer's land use adversely affects others."

I am not sure I understand what he means by "interaction" but I suspect Professor Dunham is saying that any control by a municipality over private land use is justified by that municipality's goals, but that development cannot be forbidden or regulated without compensation unless it has a direct and demonstrably adverse impact on neighboring land. The community may without compensation stop X from developing a subdivision of half-acre lots if the consequence would be overflowing septic tanks ("a cost upon others"); but the community cannot, without compensation, require Y to build only on 3-acre lots if 3-acre lots are required not for reasons of public health but simply because a majority happen to prefer that kind of living.

There are two difficulties with this market-oriented theory, even with Dunham's cash sweetener.

In the first place, under this theory any land use which unduly lowers the value of neighboring property has to be a "nuisance." Whether the municipality, in order to regulate, should pay or need not pay depends on the degree of adverse impact. But the blind concentration on property values can hide less savory values that may not be entitled to protection, whether by uncompensated regulation or by cash payment. The property value theory does not ask why a particular development has an adverse impact on values of neighboring property.

New Jersey lawyer-planner Norman Williams illustrates:

When the argument is made that property values will be affected what is meant is simply that some factor is present which some people may dislike, and which may therefore tend to result in a net reduction in the number of people interested in buying property in the area affected – thus tending to push values down. The real question is always a simple one – what is the factor which is involved? Some factors which affect property values (or which are thought to do so) are legitimate subjects for public regulation, by zoning or otherwise; others are not. For example, the invasion of factories and the movement of Negroes into a residential neighborhood both may be thought to affect property values. Yet one is obviously a proper subject for zoning protection, while the other is not. The fact that property values may be affected gives reason to look into the situation, but by itself tells nothing about whether governmental protection is appropriate.

To use zoning as a tool solely for protecting the values of neighboring property is an extreme form of parochialism our society cannot afford in the twentieth century. As Finley Peter Dunne pointed out, it is possible to cheer too loudly for the rights of property:

But I'm with th' rights iv property, d'ye mind. Th' sacred rights an' th' divine rights. A man is lucky to have five dollars; if it is ten, it is his dooty to keep it if he can; if it's a hundred, his right to it is th' right iv silf-dayfinse; if it's a millyon, it's a sacred right; if it's twinty millyon, it's a divine right; if it's more thin that, it becomes ridickilous. In anny case, it mus' be proticted. Nobody mus' intherfere with it or down comes th' constichoochion, th' army, a letther fr'm Baer an' th' wrath iv Hivin.

I am disturbed for another reason by the Ratcliff-Dunham apology for zoning. Neither spokesman defines the scope of the "public" whose interest justifies some public limitation on the free market forces. There appears in their thinking a view that private development can be limited by regulation (or compensation) only where there is a direct and adverse impact upon a neighbor's land, or upon the municipality in which the land happens to be located. Both views assume a parochial definition of the "public" that will be affected by land development. Ratcliff sees the "imperfections of the real estate market" as equivalent to essentially local forces, and Dunham obviously is concerned with impact upon land uses in the immediate neighborhood. In the former case, I suspect the emphasis is explained by a lack of interest in regional consequences of local land use development; the Dunham amendment suggests not indifference but a firm rejection of the idea that metropolitan or regional interests have any place in the municipal regulation or, indeed, the taking of private land.

I believe that in many cases the reasonableness of zoning should be determined by reference to factors far more complex than a simple balancing of values of neighboring property, whether or not cash boot is tossed on to the scales. This doubt leads to my disenchantment with the other doctrinal justification for zoning.

THE PLANNING THEORY

I suppose that every city planning student is required to write on the blackboard a hundred times the Planner's Oath: "Zoning is merely a tool of planning." Walter Blucher asked some years ago if the zoning tail was wagging the planning dog. The question points up the view of the planner that zoning is only a minor appendage to the essential body, city planning.

Standard planning dogma requires that a planner invited to prepare a municipal zoning ordinance go through the following ritual: first, a very junior planner makes a survey of the municipality and prepares a map showing the land uses. On the basis of this and reams of other data and of consultations with community leaders, a very senior planner prepares a "comprehensive plan" for the community, which indicates the community's idea of what it wishes its future to be. The planner then sets forth a number of means for "implementing" the plan, including, typically, a capital improvements program, a subdivision control law, and a zoning ordinance. In the planner's view, understandably, the zoning ordinance is merely one of a number of methods of effectuating an overall municipal plan.

I do not suggest that this exercise should be abandoned. It is a laudable if not legally essential exercise for a community to analyze and articulate its communal goals and objectives before it enacts controls over the use of private land. This discipline, required by law if not in fact in England, has not achieved the responsible status it should have in this country. This is Harvard Professor Haar's "impermanent constitution," the comprehensive municipal plan. It is said that only when the community has, in its plan, set forth and exposed to public scrutiny its goals and desires can the arbiter, required to settle land use disputes, measure the reasonableness of the implementing ordinances. Professor Haar explains:

To the professional planner, the dependence of zoning upon planning is relatively simple and clear. The city master plan is a long-term general outline of projected development; zoning is but one of the many tools which may be used to implement the plan. Warnings have constantly emanated from the planners that the two must not be confused. . . .

The legal implications of this theory seem manifest. A city undertaking to exercise the land regulatory powers granted to it by state enabling legislation should be required initially to formulate a master plan, upon which regulatory ordinances, of which the zoning ordinance is but one, would then be based. Such ordinances could be judicially tested not only by constitutional standards of due process and equal protection, but also by their fidelity to the specific criteria of the master plan.

Thereby is born the "principle" that a zoning ordinance must (should) be based upon a plan.

What, then, is a plan? Hugh Pomeroy defined it this way:

Well – what do we mean by comprehensive plan? The nearest I can come to defining it is this: it is a plan that makes provision for all the uses that the legislative body of that municipality decides are appropriate for location somewhere in that municipality. That's Number 1. Number 2, it makes provision for them at the intensities of use that the legislative body deems to be appropriate. In Number 3 – the locations that the legislative body deems to be appropriate. That is the mechanical concept. Beyond that the plan should consistently represent developmental objectives for the community. And if you can have a good enough statement of developmental objectives, then I don't think that a deviation from a particular mechanical device such as regulations by districts – the departure from that – violates the attribute of comprehensiveness that consists in endeavoring to carry out these objectives.

Perhaps what troubles me about this definition is not what it says but what it implies. The corollary to this precept, accepted by many lawyers (though not Professor Haar), and by most planners and laymen, is that the validity of local land use laws should be measured *only* by their consistency with the municipal plan. This is no more of a valid purpose for zoning than is the concept of the use district. The public disclosure of municipal objectives may be a necessary first step by which equal treatment of similarly situated individuals within the municipality can be determined. To this extent the municipal plan serves as a useful intramural yardstick for the municipal regulations. The local plan in this sense is imperative as a device to bring some consistency and impartiality in local administrative decisions among residents of the same municipality.

It is an error, however, to dignify the municipal plan with more authority than this limited function. But to measure the validity of zoning by the degree to which it is consistent with a municipal plan does just that. The municipal plan may be just as arbitrary and irresponsible as the municipal zoning ordinance if that plan reflects no more than the municipality's arbitrary desires. If the plan ignores the responsibility of the municipality to its municipal neighbors and to landowners and taxpayers who happen to reside outside the municipal boundaries, and if that irresponsibility results in added burdens to other public agencies and to outsiders, whether residents or landowners, then a zoning ordinance bottomed on such a plan should be as vulnerable to attack as a zoning ordinance based upon no municipal plan.

The trouble then, with the planning theory of zoning is that by deifying the municipal plan it enshrines the municipality at a moment in our history when every social and economic consideration demands that past emphasis on the municipality as the respository of the "general welfare" be rejected.

It has to be conceded, however, that zoning has been a huge success in most of our suburbs if the planning theory of zoning means doing with land what the municipality alone wants done, provided it announces its intentions in advance. If planning is designed to provide that environment which a majority of voters within the boundaries of a particular municipality believe they want, zoning has been remarkably successful and I predict it will prosper. Indeed, if planning is intended to achieve not only physical amenities but also to accomplish some unstated or whispered social and political objectives, zoning has been far more effective than its originators dared expect. In this sense, far from being a "negative tool," zoning has been a positive force shaping the character of the municipality to fit its frequently vague but nevertheless powerful preconceptions.

If, when we speak of planning, we postulate objective standards for physical environment and let the social chips fall where they may, then zoning as an implement of planning has not merely failed but has been instrumental in the failure of planning. This failure is pernicious. Like another noble experiment with about the same birthdate as zoning, it erodes the civic conscience by permitting us to wrap our selfish antidemocratic aims in a garment of public interest.

I suppose what really disturbs me is that, because zoning is the most universal of all the legal tools for shaping the character of the municipality, any unwise use of the process has a far greater impact upon our national character than does the abuse of a less widely employed device for control of land use. The zoning power is so fragmented that its abuse does not have a dramatic impact. Dollar venality in the execution of one urban redevelopment project will receive strident and outraged attention from the metropolitan press, while daily evidence of intellectual dishonesty and moral corruption in the application of zoning in our suburban areas is accepted as a civic norm. If you are of the school which has as its premise that each of the hundreds of municipal units in a politically fragmented system of local government may regulate as it pleases and exclude whom and what it chooses to exclude, you should embrace the present state of affairs and the existence of a municipal plan is sufficient. If, however, you suspect, as I do, that the current practice impinges not merely upon property rights but upon some less tangible values which are important in a democratic society, it is time to redefine our goals and to restate the planning theory of zoning in the hope that this exercise may lead us to reshape our implements for land use control.

In my opinion there can never be any single foreordained purpose of zoning. Both the planning theory and the property value theory of zoning set forth valid goals for some people in some situations. Their proponents err only when they set up their hypothesis as the one valid purpose of zoning. They err when they try to turn zoning into a tool to implement only their own local purposes.

Zoning needs no purposes of its own. Zoning is no longer a "movement" like the single tax or prohibition; zoning is a process. It is that part of the political technique through which the use of private land is regulated. When zoning is thought of as a part of the governmental process, it is obvious that it can have no inherent principles separate from the goals which each person chooses to ascribe to the political process as a whole.

While we should not insist that zoning have "purposes," we can insist that the zoning process be exercised in accord with certain principles, that the "means" if not the "ends" of zoning be governed by neutral principles. This necessitates an inquiry into (1) whether zoning has spawned its own indigenous set of principles, or whether it is subject merely to the principles applicable to other forms of governmental action; and (2) whether it makes sense to restrict the zoning debate to only two parties: the landowner and the local municipality.

PART II DESIGN

Urban design is not just an academic discipline, or a pastime for visionary planners and architects. Neither is it coldly oriented to physical things rather than to people and their experiences. It has to do above all, with the visual and other sensory relationships between people and their environment, with their feeling of time and place and their sense of well-being.

Application of good urban design produces a logic and cohesion in the physical form of the city and its districts. It is concerned with both preservation and development, and not with one to the exclusion of the other. It teaches that man can do great things in cities, but it also teaches him that he must have the humility to live with the environment rather than attempt to master it. . . .

The Urban Design Plan, For the Comprehensive Plan of San Francisco, May 1971, p. 3.

6

Paul D. Spreiregan

THE ROOTS OF OUR MODERN CONCEPTS

The Greeks had created an urban form made for people. The Romans built upon their theories but not without discarding some of Greece's most important urban ideals. In medieval times ideas parallel to those of ancient Greece found a new expression. Again these more humane concepts were discarded by the Renaissance, whose aims were loftier and of considerable artistic quality. The design of cities in the Renaissance had been an instrument of state control. In the eighteenth and nineteenth centuries it became a technique for greedy speculation. At the same time, however, a new breed of design theorists entered the scene. Sometimes their thinking was quite practical, like Valentine Knight's. Sometimes it was Utopian. Sometimes it relied on extravagant mechanical concoctions. Sometimes it rejected all semblances of engineering technology as unnecessary, proposing instead a return to nature. But all these ideas strove

---

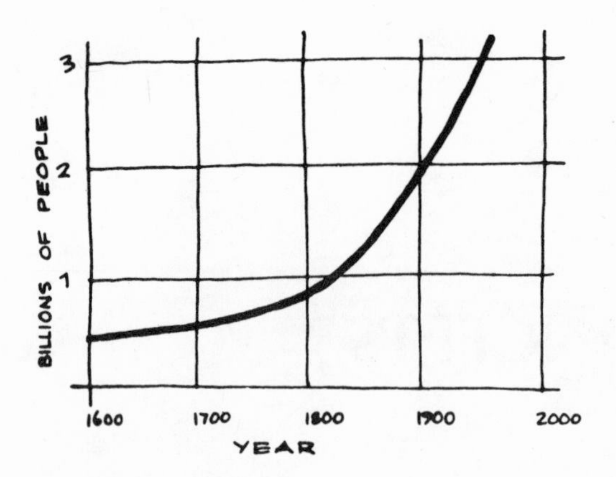

World population: 1600-1950.

John Wood's design for a duplex worker's cottage.

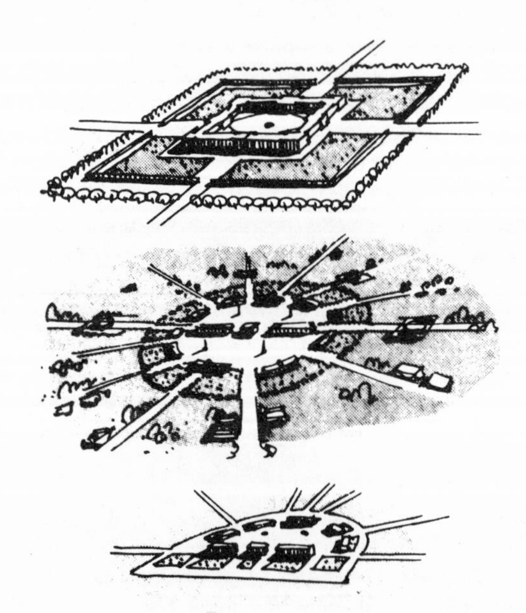

Ledoux's three plans for Chaux.

toward one objective: the design of cities as a place to live for all, with particular emphasis on the needs of the working classes.

There is no distinct line of separation between the architects of the grand baroque cities and those oriented toward the well-being of the general population. By way of illustration, John Wood the younger, architect of the Royal Crescent at Bath, also produced a book of designs for workers' houses in the healthful open countryside. But the French architect Claude Nicolas Ledoux is best remembered as one of several late eighteenth- and early nineteenth-century theorists who brought intense analysis and rationale to the design process. He also launched a new era of urban design, one that gave as much attention to workers as it did to society's ruling classes.

Ideal Towns and Worker Towns

The principal urban work of Ledoux was the design of Chaux, a town for salt workers in France. Its construction dates from 1776. In all, Ledoux made three plans for Chaux. The site was in open countryside between two villages, with a pair of overland roads intersecting at right angles at the center. In the first version of his plan, a quadrangle of buildings – workers' homes, common buildings, and allotment gardens – was arranged as a 1,000-foot square formed by tree-lined avenues. In later revisions of his plan, Ledoux changed the square to an ellipse and, finally, to a semiellipse with roads radiating into the surrounding countryside. He may have preferred the semiellipse for better sunlight orientation. Along the roads Ledoux envisioned informal groupings of houses. This was one of the first plans to advocate informal grouping as part of an overall design concept. Ledoux thought of his design as an ideal plan "wherein everything is motivated by necessity." The workers would even grow their own food. He published all his plans in an influential book called *Architecture* in 1804.

Implicit in Ledoux's thinking was the self-sufficiency of the worker towns. Ledoux was an architectural visionary, soon to be followed by men of similar inclinations. In 1799 the English social reformer Robert Owen started construction of an

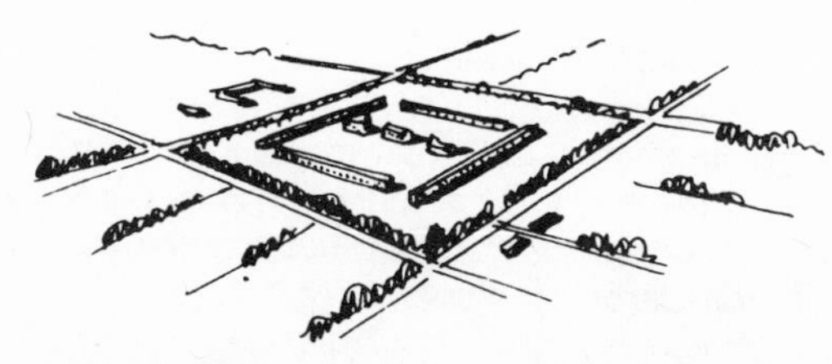

Robert Owen's "Village of Unity and Mutual Cooperation."

New Lanark, Scotland. Started in 1785 and, in 1799, operated by Robert Owen as an enlightened social and manufacturing community. Owen left in 1825. The mills are still in use.

François Fourier's "Phalanstery."

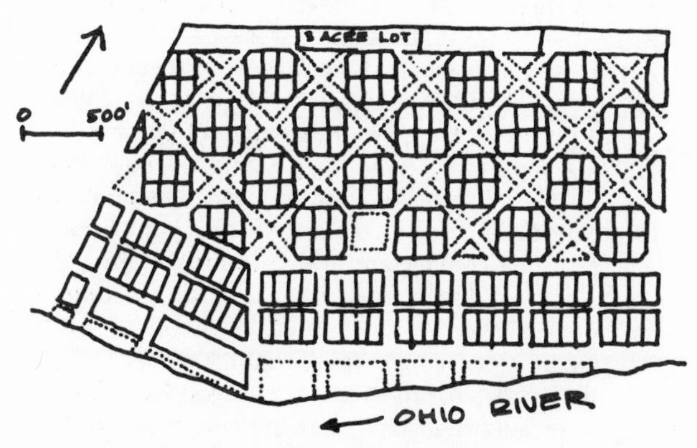

1802 plan of Jeffersonville, Ind. The upper portion "laid out after the plan suggested by President Jefferson."

industrial village at New Lanark mills near Manchester. Owen pondered deeply the question of the worker community. His thoughts led him to theories which he expounded and which proved highly influential. Owen's ideal was a community of 800 to 1,200 people on at least 600 to 1,800 acres. Each community would be self-sufficient and there would be no child labor. The community would have recreational and educational facilities. Several Owenite communities were set up in England and the United States. Owen's son started one in Indiana, called New Harmony. A group of New England transcendentalists created Brook Farm in Massachusetts. A Frenchman named Cabet was one of the more adventuresome Owenites. He proposed a Utopian settlement to be called "Icarus" on the Red River in Texas. When it failed, he joined with the Mormons in their search for their promised land and helped lay out Salt Lake City.

In 1829 the French social reformer François Fourier published *The New World of Industry and Society.* His visions were far more rigid than Owen's, for Fourier would put 1,620 people, 400 families, into one large palace-like building which he called a "Phalanstery." The building he pictured resembled a Renaissance palace. A succession of "ideal" proposals followed. In 1849 James Silk Buckingham published *National Evils and Practical Remedies* and proposed "Victoria," a glass-roofed town like the Crystal Palace. Robert Pemberton planned "Happy Colony" for New Zealand, a series of ten circular town districts laid out along the lines of Chaux.

In the United States, Dr. Benjamin Richardson proposed "Hygeia," a town spaciously laid out for fresh air and health. Thomas Jefferson had advocated a simple expedient to achieve this. Jefferson's favorite city was a simple grid in which every other block was a park — a grid city built as a checkerboard. Jefferson felt that this grid variation could easily be adapted and several were actually built. By leaving alternate blocks empty, the city would be more spacious, would have more light and air, the danger of the spread of fire would be lessened, and the town would be more handsome. Unfortunately, the towns laid out in this way sold their public squares to developers when municipal finances were depleted.

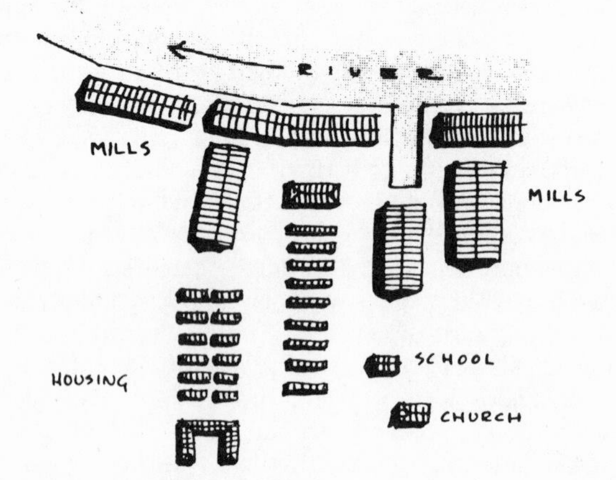

Schematic plan of Lowell, Mass.

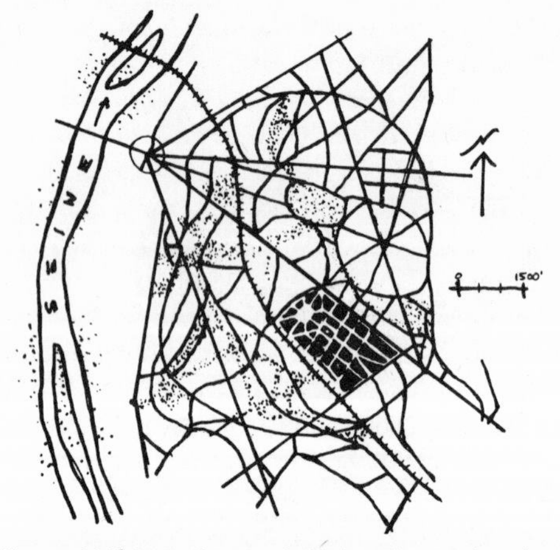

The village of Vésinet, France, built in a former hunting forest. The curved paths were 19th-century alterations.

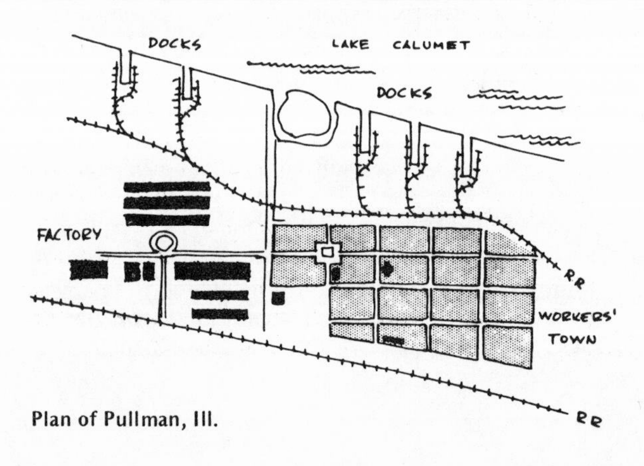

Plan of Pullman, Ill.

Planned Industrial Towns

Several planned industrial towns were built in the nineteenth century. The scale of organization of large manufacturing operations necessitated a well-organized corps of workers. The English were the main developers of these large-scale operations in their textile plants, and the Americans were quick to learn their lessons. A small mill village of workers' houses was erected as early as 1812 in Georgiaville, Rhode Island. Francis Cabot Lowell perfected mill operations in Waltham, Massachusetts, and in 1816 built a mill town in Harrisville, New Hampshire. In 1822 he built another one, Lowell, Massachusetts, in which he tapped a hitherto unused source of labor: young New England farm girls who came to work for a few years in order to earn a dowry.

In 1859 the town of Vésinet was started in France. Designed by an architect named Olive, its plan was a remarkable combination of classical French landscape architecture and English parks. Olive laid out an artful network of axial streets for the houses and interspersed the whole town with meandering swaths of green space. This design anticipated, by half a century, the design of the twentieth-century garden city.

Around the Krupp factories of Essen, Germany, a number of communities were built, starting in 1863. Called *Siedlungen,* or worker colonies, by 1925 they amounted to 25,000 houses in about a dozen communities. In 1879, Pullman, Illinois, was started as a town for factory and workers. The overly paternalistic management held the employees in all but feudal bondage, which resulted in a famous strike. By 1890 the town numbered 11,000 people. In 1887 the W.H. Lever Soap Company built Port Sunlight, a worker community near Liverpool. In 1889 the Cadbury Chocolate Company built Bournville, a garden community near Birmingham. In 1906 Gary, Indiana, was laid out by a steel corporation, a "made to order" city. Kohler, Wisconsin, is a descendant of these nineteenth-century towns.

While many towns were rather good in design, they were almost insignificant in comparison to the enormous workers' agglomerations being built in the world's expanding industrial cities. Only large and wealthy companies could afford to build them, and even then they were considered rather risky.

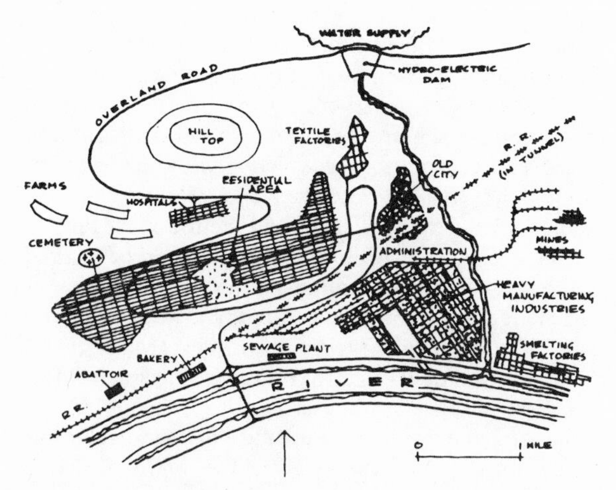

Tony Garnier's Cité Industrielle.

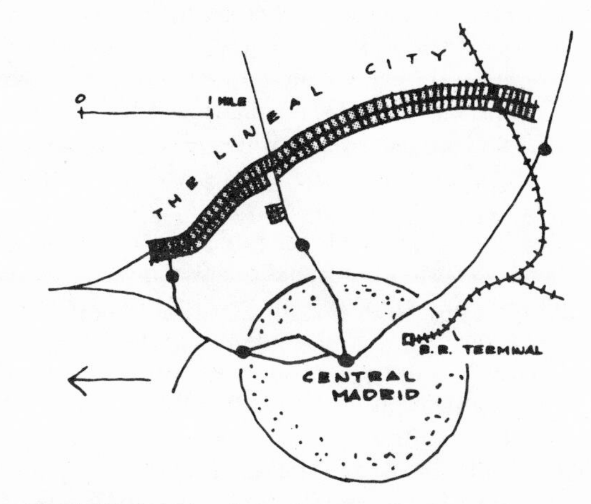

Soria y Mata's Ciudad Lineal.

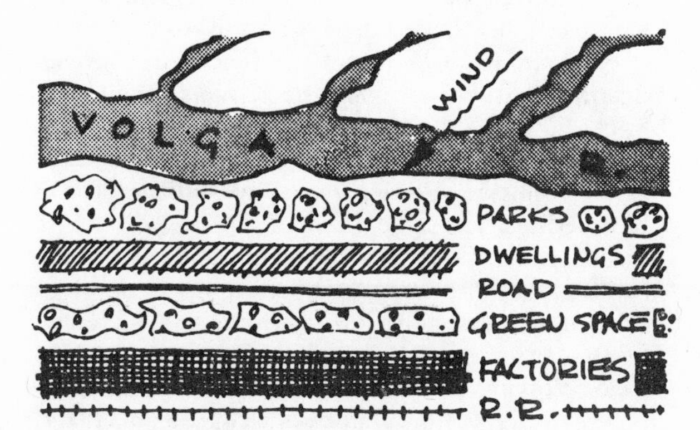

Stalingrad, Russia. An example of an actual linear city based on linear communication.

The idealization of the industrial city was expressed by a young French architect, Tony Garnier, in designs he made between 1901-1904 while on a scholarship at the French Academy in Rome. Garnier designed a hypothetical industrial town, which he called *Une Cité Industrielle.* He created an imaginary site consisting of high plateau and level valley, all alongside a river. The plateau would be used as the residential portion of the city; the valley for the factories. The total population of the city, including an imaginary existing old town incorporated in the plan, was to be about 32,000. A dam would furnish hydroelectric power. Garnier's separation of the city's parts anticipated modern zoning but with considerably more wisdom, for he first decided where things should be most suitably located and then laid out areas according to those decisions. He also located complementary urban facilities together where they belonged — a relationship sometimes denied by improper zoning ordinances.

In this respect Garnier's plan is incredibly detailed. A hospital is on a high hill. Cemeteries have fine natural vistas. Smelting factories and mines are at respectful distances. The plan included detailed locations for sewage plant, abattoir, bakery, and civic center. Garnier showed testing grounds for cars and even airplanes! He used a grid plan for the residential area with 100- by 500-foot blocks; the short cross streets would accommodate major circulation, thus diverting through traffic from the long residential streets. He drew working drawings for many buildings, utilizing the newly developing technique of reinforced concrete. Garnier returned from Rome to his native Lyons, where he spent his career trying to carry out his ideas on individual buildings.

Not many years later the Dutch architect J.J.P. Oud built a small worker colony near Rotterdam which is reminiscent of Garnier's plans. As a matter of fact it comes closer to embodying Garnier's ideals than any other town plan built. In the world of letters George Bernard Shaw suggested that such planned industrial towns had their merits, as in his play *Major Barbara.*

Urban Design and Machines

Practical inventions have always fascinated designers. For example, the complexities of Renaissance fortification design eventually found their way into the design of villas. In the nineteenth century, mechanical inventions grasped the imagination of many designers, as they still do to a large extent. The Spanish businessman and engineer Don Arturo Soria y Mata was of such a mind. He had created Madrid's first streetcar and telephone system. In 1882 he suggested the idea of *La Ciudad Lineal,* or the *Linear City,* in which he proposed that the logic of linear utility lines should be the basis of all city layout. Houses and buildings could be set alongside linear utility systems supplying water, communications, and electricity. Soria y Mata thought that linear cities could crisscross the entire globe and actually built a linear city on the outskirts of Madrid. Stalingrad is the outstanding example of a planned linear city.

View of Sant'Elia's Città Nuova.

Concept of a floating city from a sketch by Kiyonori Kikutake.

In 1899 the Russian-born geographer, author, and revolutionary Peter Kropotkin published a book called *Fields, Factories and Workshops: or Industry Combined with Agriculture, and Brainwork with Manual Work.* In it he suggested the use of electricity to allow towns to be built anywhere. He advocated minimal government and maximum individual self-sufficiency, for he saw only tyranny in large organization. The idea of freeing town development and locale by the use of electricity was important to town designers of this era, like Garnier.

Among the inventions of the nineteenth century which influenced urban form, the railroad stands foremost. Railroads for long distances or short commuter runs were highly profitable. They opened up large land areas for speculation. The possibilities of transportation technology extended well beyond the railroad, and were fascinating. The 1893 World's Fair in Chicago had a moving sidewalk which carried footsore tourists along a long pier projecting into Lake Michigan. A world's fair in Paris had a moving sidewalk to connect its exhibits. The world's first monorail system was built in Wuppertal, Germany, around the turn of the century and is still in operation.

After the turn of the century, Edgar Chambless, an American, proposed a city with all vehicles running on the rooftops of a continuous building. Undoubtedly he was influenced by Soria y Mata. The same idea keeps popping up in different guises. It was recently proposed in England under the name "Motopia," this time with the rooftop roads laid out as a grid. In 1910 an inventive Frenchman, Eugene Henard, published *Les Villes de l'Avenir* (The Cities of the Future), in which he proposed buildings on stilts, traffic circles, underpasses, and airplanes landing on rooftops. Henard may have influenced Le Corbusier's early urban concepts.

The Italian futurist architect Antonio Sant'Elia provided a new, perhaps frightening vision of what might come: an enormous metropolis – *La Città Nuova* – based on motion, with every element of its design implying either horizontal or vertical circulation. Sant'Elia was inspired by the complex plans for the Grand Central area of New York City, a horizontal skyscraper with buildings connected above ground by pedestrian walks and vehicular roads. Many contemporary urban scenes are living examples of Sant'Elia's vision. Such an example is the approach road to the George Washington Bridge in New York City with its air-rights apartment towers above.

Just recently a French book called *Où Vivrons-Nous Demain?* (Where Will We Live Tomorrow?) was published with many illustrations of visionary urban design. Most of the illustrations would be at home in science fiction magazines. Some Japanese architects who call themselves the "Metabolism Group" have produced underwater cities, biological cities, cities that change their own forms, and cities that are built as pyramids.

The possibilities of modern science and engineering in the improvement of the city are vast, but there are dangers also. This potential, and its concomitant pitfalls, has been the theme of much literature, most notably *Looking Backward, 2000-1887* by Edward Bellamy, published in 1887, and several books by H.G. Wells that were published between 1902 and 1911.

Visionary thinking is to be expected in a society which depends so much on science and machines. Most of the ideas cited here are fanciful. Many of them are still frequently proposed, and some of them have actually been put into highly useful operation. Visionary thinking is inevitable in any growing society. In a society that depends heavily on machines, visionary thinking naturally takes a mechanistic approach to solving problems. Some of the mechanical devices proposed for modern cities have been highly useful. Others have simply allowed undesirable conditions to worsen. For example, a rail rapid-transit system can cause further crowding in an already crowded city. It can also be the principal transit means for organizing satellite cities around a central metropolitan core. Any judgment of a visionary idea with mechanical overtones is twofold: we must evaluate its possibilities for improving the quality of urban life and we must be sure that we have the means to achieve proper usage of a new device. Otherwise, it can simply make things worse in the name of expedience or in the guise of progress.

A New Attitude Toward Nature

The technological advances of the nineteenth century were not all greeted as signs of progress. The nineteenth-century industrial city was all too horrible evidence of technology at work. While some designers saw possibilities for innovation that could harness technology to human purposes – the designers of industrial towns and self-sufficient communities – there were others who saw in it only disaster. The chief spokesmen of this point of view were the Frenchman Viollet-le-Duc, the Englishman John Ruskin, and the American Henry David Thoreau.

Viollet-le-Duc and John Ruskin popularized a return to the simpler Christian virtues of the Gothic period. In England their sentiments found expression in a crafts movement led by William Morris. Between 1875-1881, with architect Norman Shaw, he succeeded in creating Bedford Park, a picturesque residential area on the outskirts of London, linked to the city by rail – which, incidentally, heralded the commuter suburb. The Gothic revival of the nineteenth century in churches, in houses, and later in American college campuses was evidence of the popularity of Ruskin's and Viollet-le-Duc's outlook. Thoreau, meanwhile, sought complete refuge in nature, where he conducted a personal experiment in independent existence. The ideals of these men still have considerable appeal. Frank Lloyd Wright, for example, once commented that the Gothic period was the last original architectural era, and Le Corbusier wrote a book nostalgically titled *When the Cathedrals Were White.*

The Conservationists and the Park Movement

Of all nineteenth-century thinkers, perhaps the most profound is the one least known, at least by the general public: the American George Perkins Marsh. Marsh was dismayed at the wasteful land practices around his native New England. A man of brilliant mind – he mastered 20 languages by the age of 30 – he was a naturalist, humanist, historian, geographer, and practical politician. In overgrazing and overcutting he saw major causes of land erosion and river flooding, the foolish despoilment of the land. He, too, had his predecessors, including such men as conservation-minded John Quincy Adams, ornithologist James Audubon, naturalists William Bartram and Henry David Thoreau. In 1849 Marsh was appointed minister to Turkey, and in 1861 Lincoln appointed him minister to Italy. These travels gave him an opportunity to see the geography of Europe and the Middle East, which deepened his thinking and confirmed his practical ideas for using the land's resources in ways that would not destroy its bounties. In 1862 his thoughts were collected in a book called *Man and Nature.*

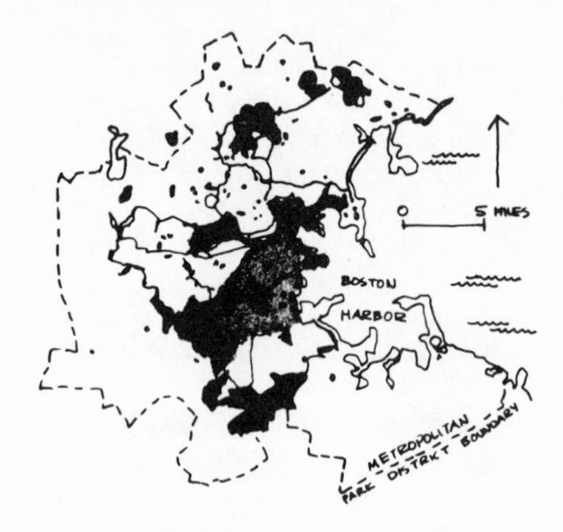

The Boston Metropolitan Park System. Originally planned by Frederick Law Olmsted, it was enlarged by Charles Eliot. The city of Boston is shown shaded.

The Public Gardens, Boston, Mass. A central feature of the Boston park system.

Marsh's book was widely read. It was an introduction to ecology, an encyclopedia of land facts, depicting the deterioration of land as a result of man's ignorant disregard of the laws of nature. Marsh also explained the interrelationship between plant and animal life. He assailed the myth of superabundance and described how despoiled land could be restored. Marsh, in short, was the founder of modern conservation, putting man in the position of cooperation with nature. He had a considerable influence on America's great conservationists, including Carl Schurz, Theodore Roosevelt, John Wesley Powell, Gifford Pinchot, George Norris, John Muir, Rachel Carson – people associated with our national parks and preservation measures, exercised or advocated.

Marsh's ideas applied to land at regional and continental scale. It is one of the principal contributions to the art and knowledge of the use of the earth. However, it was not the only major American contribution. In the more specific area of urban design stands the American park system. The foremost name in this enterprise is that of Frederick Law Olmsted.

Olmsted came to the public's attention prior to the Civil War as a social reformer through articles on slavery published in the *New York Times.* He was aware of the increasingly rapid urbanization of the United States, noting that urban population doubled between 1840 and 1860, mostly because of immigration. He felt that the improper use of both land and labor was damaging to democracy. He was concerned as well with the moral disintegration that large formless cities engender – formless in physique and social community. Olmsted, also a farmer, saw in landscape design a solution to these ills. Disdaining aristocracy, he nevertheless admired the landscaping of English estates. The urban park could be an aid to social reform, giving the downtrodden city dweller uplifting communion with nature.

Olmsted's first opportunity came with the design of Central Park in New York City. A plan had been drafted by an Army officer for a large city park, but it was little more than a military drill field and drew stinging criticism. In the face of public clamor, a competition was held, and won by Olmsted in 1859. His notes on the design of Central Park, and the thoughts that ran through his mind as the plan proceeded, are brilliant insights into the creative mind of an artist at work in a city.

Olmsted designed several other parks in the New York City area and throughout the United States. The original park plans for San Francisco, Buffalo, Detroit, Chicago, Montreal, and Boston are his work. He also had a hand in the Yosemite Park bill during Lincoln's presidency. He believed that cities should plan for two generations ahead, maintain sufficient breathing space, be constantly renewed; and that urban design should embrace the whole city. The city should exist to serve its inhabitants. His attitudes toward comprehensive park planning were summed up in 1870 in his book *Public Parks and the Enlargement of Towns.*

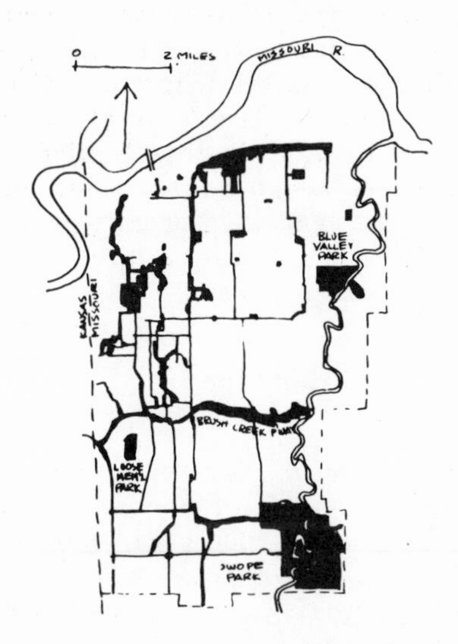

The Kansas City, Mo., park system. Parks, parkways, and boulevards. It was originally planned by George Kessler.

Olmsted was followed by other great urban park designers – notably, Charles Eliot, who completed Olmsted's Boston park system; George Kessler, who laid out the Kansas City park system; and Jens Jensen, who designed Chicago's original park system. This gave many American cities fine urban parks, but the practice, regrettably, has not been maintained as Olmsted insisted it must. To be successful a park and open space program must be continuous. The present misuse of metropolitan open space might not exist had we followed Olmsted's advice. Olmsted had many colleagues here and abroad. Haussmann's landscape architect, Alphand, has been called "the French Olmsted." In Germany an interesting and widely adopted idea was advanced by Daniel Schreber, a physician and educator. Schreber proposed small gardens for children, for healthful play and exercise; later *Schrebergärten* were also used by the elderly, to raise vegetables. In effect, he popularized the idea of the urban playground in Europe.

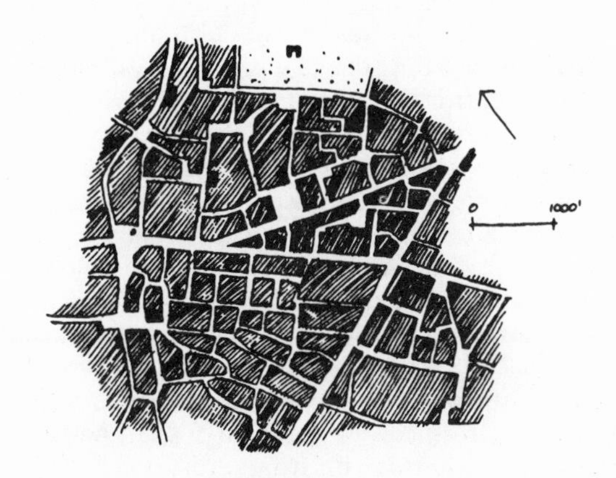

Informal plan of Camillo Sitte.

Reviewing the Past

The nineteenth century was also an age of exploration into the past. Archeology was becoming a science. The architectural academicians kept an eager eye on the discoveries, for they were busily engaged in classifying and dispensing the sum of the world's architectural knowledge. Much of this knowledge was improperly applied.

In 1889 a Viennese architect, Camillo Sitte, published *An Architect's Notes and Reflections upon Artistic City Planning.* His book provided deeper understanding of the various modes of urban design of the past, and so became an invaluable guide to contemporary design. Sitte may have been influenced by a German colleague, J. Stubben, who in 1880 published a book called *Städtebau,* in which he advocated the careful preservation of old cities as

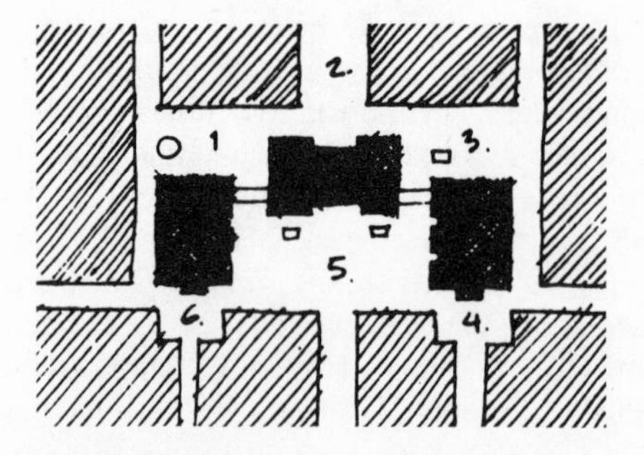

Formal plan of Camillo Sitte. Three buildings on a small site forming six plazas.

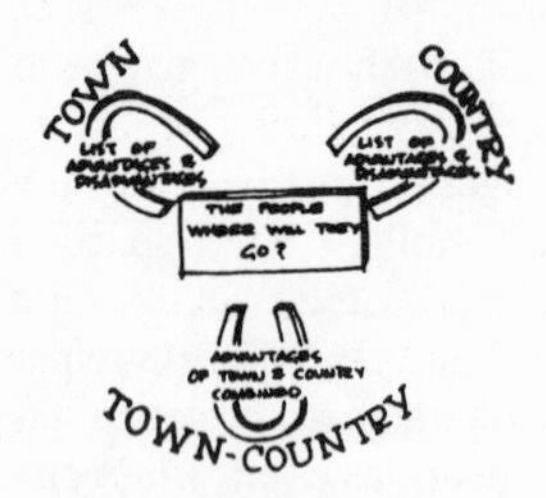

Ebenezer Howard's diagram of the "three magnets."

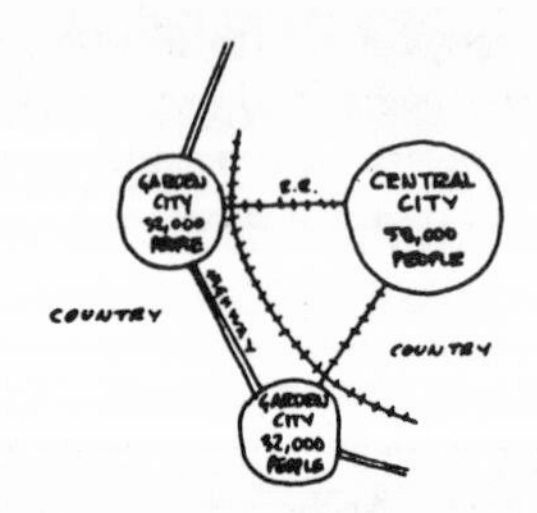

Howard's central and garden cities.

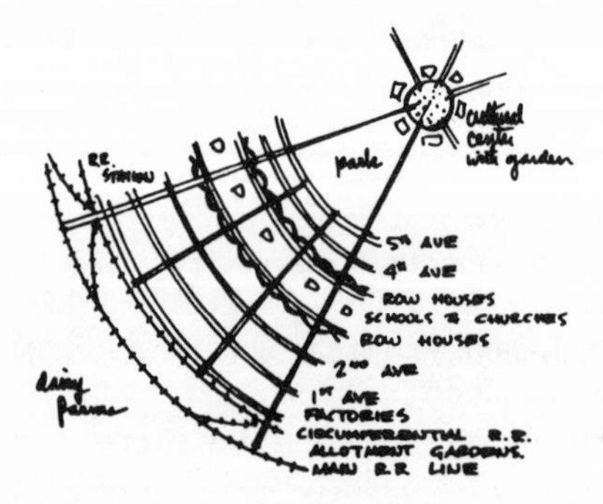

Ebenezer Howard's schematic diagram of a garden city.

they are enlarged – possibly in reaction to Haussmann. Sitte described the design of medieval and Renaissance cities, delving into the principles of arrangement, proportion, scale, and purpose with clarity and objectivity. His book has sometimes been regarded as a defense of the contrived, the irregular, and the picturesque – qualities of the Gothic and medieval towns which Ruskin had popularized. At a deeper level, however, Sitte's work was an argument, not for superficial style, but for underlying principle. In his city designs, Sitte applied these principles with sensitive appropriateness. His designs for civic centers were classically formal in layout. His plan for a small village, on the other hand, is rustically informal, with winding streets following the terrain. Sitte regarded no design element as sacred and inveighed against "paper architecture," formal or informal, which failed to fulfill its design promise.

The Garden City Movement and a Scientific Approach

In his influential book *Tomorrow: A Peaceful Path to Social Reform* (1898), Ebenezer Howard, an English parliamentary stenographer, showed how workable and livable towns could be formed within the capitalist framework. He started with discussions of the optimum size for towns – a subject frequently debated by theorists – and concluded with a cluster concept: a central city of 58,000 people surrounded by smaller "garden cities" of 30,000 people each. Permanent green space would separate the city and towns, serving as a horizontal fence of farmland. Rails and roads would link the towns, which would have their own industries, the nearby farms supplying fresh foods. All increases in land values would accrue to the town and its "stockholders," the townspeople.

Howard's proposal was accompanied by diagrams showing the attractions of the town, the country, and then of both, when ideally combined. The functional relations between the central city and its surrounding garden cities were also depicted, as well as the overall concept of a garden city and its internal layout. Howard's detailed thinking was not limited to physical design or to studies of optimum population sizes. He also made a precise financial analysis of what it would cost to build a garden city and how its

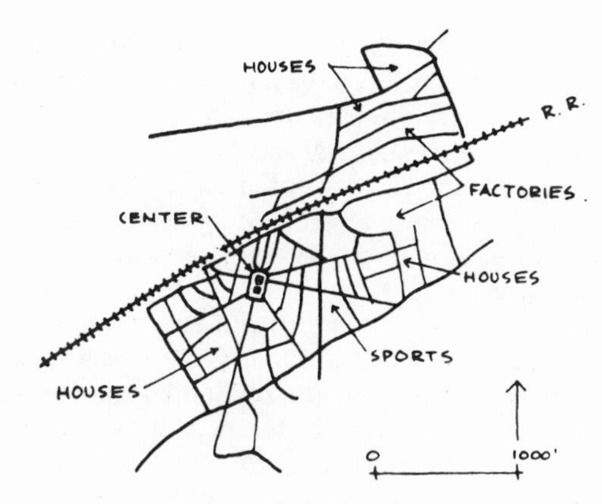

Letchworth, the first garden city.

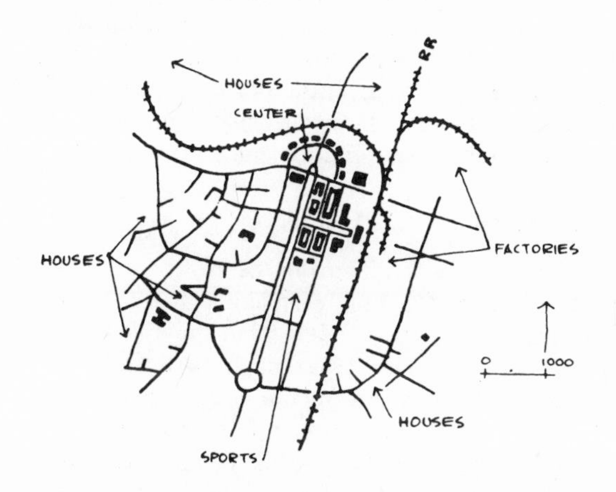

Welwyn, the second garden city.

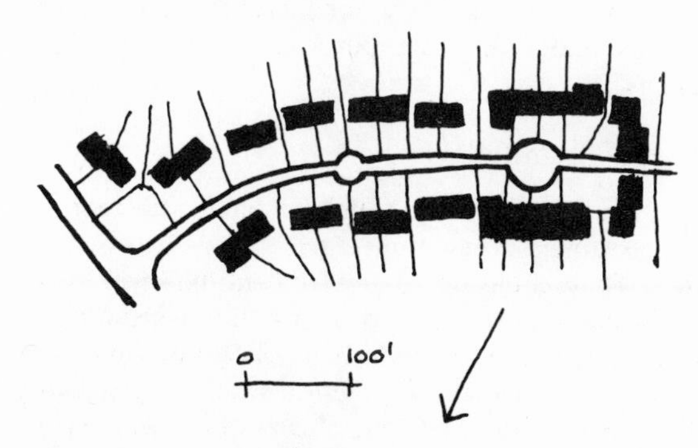

Grouping of houses in Welwyn.

operating costs would be met. Therein lay the strength of his proposal. He showed how it could be accomplished.

The idea received great acclaim, and in 1902 a garden city was started, Letchworth, about 35 miles from London. It was planned by architects Barry Parker and Raymond Unwin. Unwin, an exponent of low-density planning, had demonstrated his design principles in the Hampstead Gardens suburb, also near London. The plan of Letchworth was a combination of landscaping, informal street layout to suit topography, and a main axis focusing on a town center. Sports fields, a train station, houses, and factories were all included.

The factories failed to materialize at the outset and Letchworth became a satellite for London, thus revealing the principal difficulty of Howard's ideas. It was a risky venture depending on coordination between homebuilders and industry – which was not easy to bring about. Nevertheless the idea prevailed and in 1920 a second garden city was started, Welwyn. It was designed by architect Louis de Soissons and was more successful than Letchworth in terms of Howard's original concept. Today it is a center of the British film industry.

Howard's analytical approach was an indication of the almost scientific study that modern city building requires. The city is so large and its operations so complex that its proper understanding can only be gained by the full application of precise analysis. The Scottish city planner Patrick Geddes was the man who established this tool. His approach was not only analytical and comprehensive but also stressed the social basis of the city. The analytical survey was the principal groundwork from which any plan would be conceived.

Geddes was a prolific writer, lecturer, and planner. His writings include reports on park development, cultural institutions, applied sociology, the function of a civic museum, and the actual techniques of an urban survey. His most widely read book, *Cities in Evolution,* was published in 1915. In this work he coined the term "connurbation" to describe the waves of population inflow to large cities, followed by overcrowding and slum formation, and then the wave of backflow – the whole process resulting in amorphous sprawl, waste, and unnecessary obsoles-

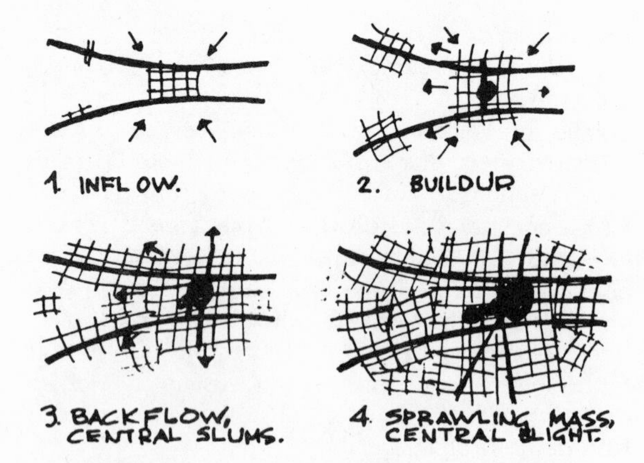

Geddes explanation of "connurbation."

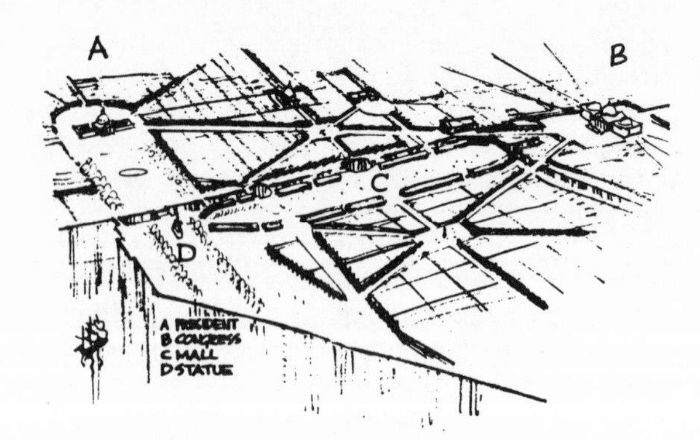

L'Enfant's concept for central Washington, 1791. (After Elbert Peets.)

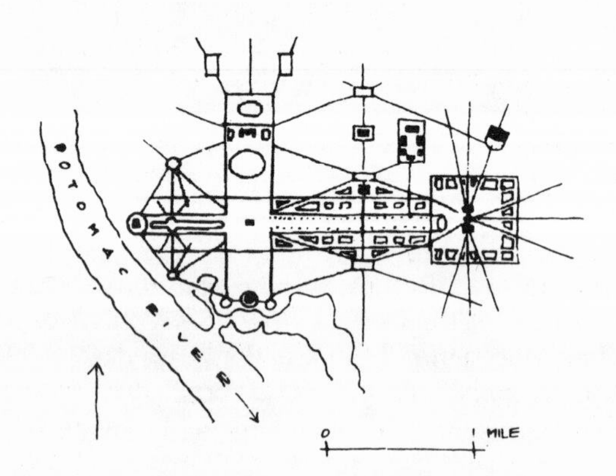

McMillan Commission plan for central Washington.

cence. As a planner he laid out some 50 cities in India and Palestine. Geddes' contributions to the understanding of the city stand alongside Marsh's contributions to the understanding of land. Geddes was concerned with the relationship between people and cities and how they affect one another. Marsh's concern had been with the interrelationship between man and nature.

Thus, just after the dawn of the twentieth century, the foundation of ideas for building the modern city was in place. From Ledoux to Geddes extends a course of ideas 100 years in the making. From the days when the plow first broke the soil and the herdsman built a fence there has been an unbroken tradition of designing cities, a tradition as old as civilization itself. Beset with mighty thoughts, mighty experiments, and mighty deeds, an encyclopedia of urban design experience was the heritage given to our era. It remains only to know of this experience, to put it to use, to adapt it to particular conditions, and, where necessary, to build upon and expand it.

THE AMERICAN EXPERIENCE

Two world wars, a major depression, waves of immigration, and a host of major and minor crises have erased from the public's mind the fact that this country once enjoyed what might well be called a golden age of urban design: the years between about 1890 to the Great Depression. Often narrowly termed the "City Beautiful Era," this period drew upon almost every idea in the history of designing cities. In many cases the ideas were enlarged upon significantly. Emphasis on formal design distinguishes the period somewhat unfairly, for there were broad social motives underlying the purposes of the ambitious designs.

The City Beautiful Movement

In our world's fairs of the late nineteenth century, we proclaimed a new hope and a fresh image for our cities. They could be far nobler than the small towns of our many farming regions, and the ugliness of our large industrial cities could be displaced by handsome works of civic art. The world's fairs were living examples of civic art to match any of the wonders of modern Europe or ancient Rome. Their design took advantage of our latest technology, from moving sidewalks to the spanning of large interior spaces with iron trusses. Although facade architecture was decried by Louis Sullivan, who saw in it only sham and a denunciation of his own search for an original American architectural expression, the buildings were, in fact, harmoniously designed as groups. There were ample pedestrian places with fountains, trees, and places to sit and relax. Here, in elegant, albeit, foreign clothes was the promise of America come to life.

The fairs were sometimes designed as renewal operations. The land upon which they were built became a new quarter of the city after the fair was over. For example, Jackson Park in Chicago was the site of its world's fair of 1893, and San Francisco's Marina district and Treasure Island were originally fairgrounds.

In 1901 the AIA held a national conference on city beautification in Washington, D.C. The McMillan Commission was then formed to prepare a plan for the improvement of central Washington. Some of the country's foremost artists constituted the group, including Daniel Burnham, Augustus St. Gaudens, and Frederick Law Olmsted. They toured Europe for inspiration and returned to propose a grand classical concept of landscape architecture with axes, mall, focal points, and pools – in effect reviving the original L'Enfant plan for the city. This, together with the example of the world's fairs, initiated a countrywide program of civic improvement efforts: the City Beautiful Era.

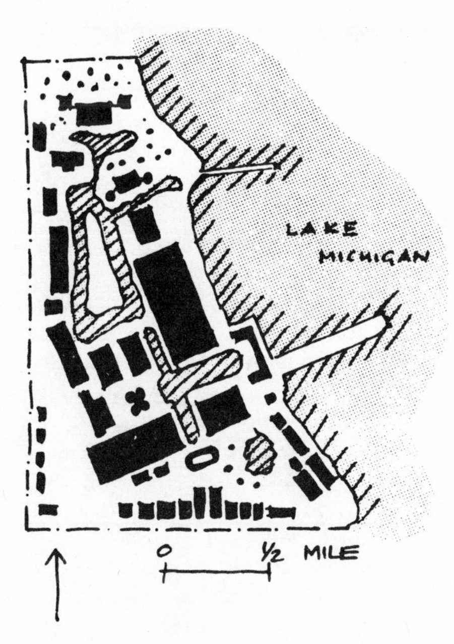

The Chicago World's Fair, 1893.

A city hall, a county court house, a library, an opera house, a museum, and a plaza were employed as the building blocks of civic centers the country over. A dome for the city hall was an essential article of pride and architectural accent. The movement spread to embrace public works of all sorts. Bridges were designed by architects as pieces of sculpture. River embankments were made into classical garden terraces. The vision of the classical world, neatly arranged, became a chief architectural concept for many a college and university: Columbia, M.I.T., the University of California at Berkeley, and the University of Washington in Seattle. The American railroads, at their zenith, built Roman basilicas and baths as grand portals to Chicago, New York, Washington, and many other cities. In 1917 the AIA published a book of all the projects, proposed or accomplished, throughout the land.

The City Beautiful Era was by no means limited to civic centers or fine public buildings. Daniel Burnham made plans for the whole of Chicago, San Francisco, Manila, and several other cities. Burnham is regarded as the father of American city planning, and his remark, "Make no little plans . . . they have no power to stir men's blood . . . ," still rings. Burnham's concept of a city was a totally designed system of main circulation arteries, a network of parks, and clusters of focal buildings. In his plans are to be seen the last use of French Renaissance principles applied at the largest scale possible.

There was considerable activity in the creation of planned residential communities. Roland Park in Baltimore was started in 1892 as a garden suburb for commuters. Its lessons of finance and design were copied in many similar developments throughout the country. All the large cities with flourishing economies and well-to-do families were only too ready for such commuter suburbs. The Country Club area in

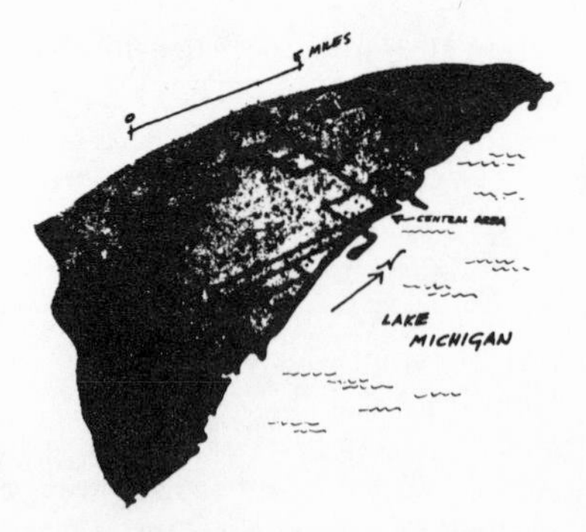

Concept of the parkways, boulevards, and Lake Shore parks of Burnham's plan for Chicago.

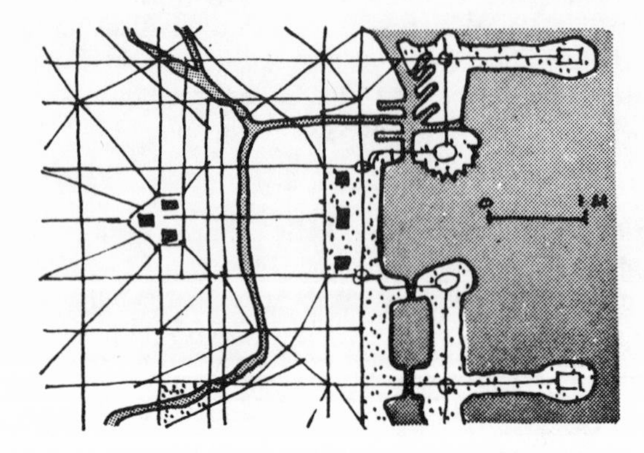

Concept of central Chicago, from Burnham's plan.

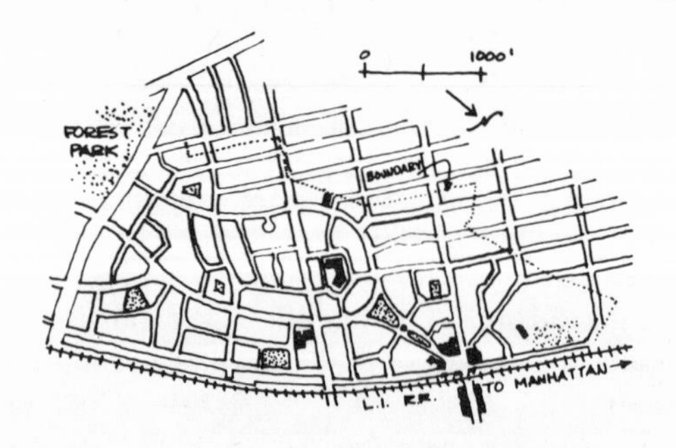

Forest Hills Gardens, L.I. Curvilinear streets, parks, public buildings, and a commuter railroad. Note the difference in street layout within the project boundary from the typical layout of blocks.

Kansas City is one of the more famous. Indeed, the image of the green American suburb became the ideal of all urban American families. Forest Hills Gardens in Long Island was built as a commuter suburb for Manhattan in 1911, financed by the Russell Sage Foundation.

American architects promptly rose to the problems of burgeoning cities. From these beginnings, the American city planning profession came into being. In the early 1920s city planning was very popular and plans were made for almost all our cities. Zoning was introduced in 1916 and generally adopted within a decade or two over the entire country as a means for enforcing city plans. From abroad came many lessons, too, particularly from England and the garden city movement. English architect-planners lectured in the United States and the early English books on city planning had an avid audience here. Low-density planning and grouped housing designs were advocated and acclaimed. Tempering this activity was an atmosphere of social reform. The civic center and commuter suburbs were handsome indeed, but what of the residential needs of the average man? Among those who felt that we could produce not only better homes but better communities was a group that included several deep-thinking architects.

The New Communities Movement

In the early 1920s, discussions were held on community problems in meetings in New York City. Among the participants were Clarence Stein, Frederick L. Ackerman, Lewis Mumford, Henry Churchill, Henry Wright, and Alexander Bing. Some of these names are still remembered; some, unfortunately, are not. They all felt that the piecemeal development of the residential communities on endless gridiron tracts was wasteful and unnecessary – worst still, it did not produce the kind of housing and communities we were capable of creating. The common practice of laying out block-pattern streets long before the builder arrived on the scene prevented clustered community design and the interspersal of open and built-up spaces, as the English planners had been advocating and as we ourselves had done. Other built-in restrictions, too, prevented us from building well-designed communities. These men set out to

Typical block development in Long Island in the 1920s. Duplex houses; narrow side yards; poorly lit and poorly ventilated side rooms; no common play space.

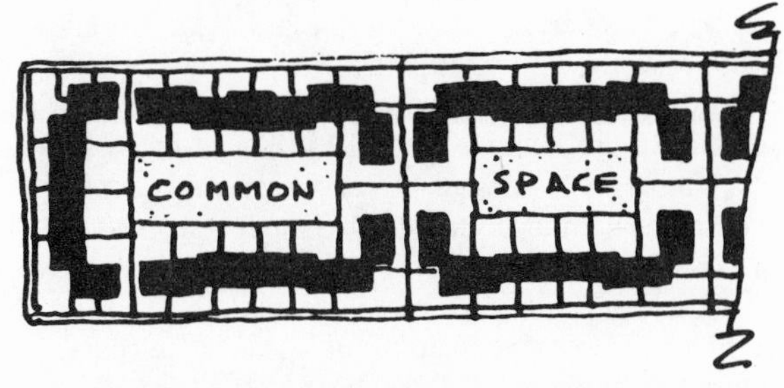

The Sunnyside idea. Row houses eliminating useless side yards; well-lit and well-illuminated rooms; useable private yards, plus ample common play space.

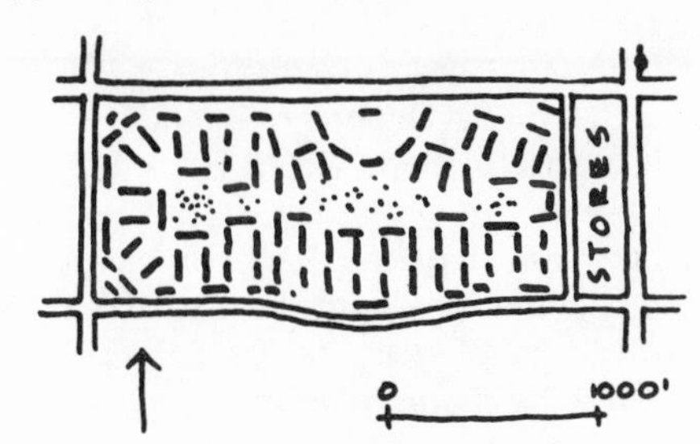

Baldwin Hills Village, Los Angeles.

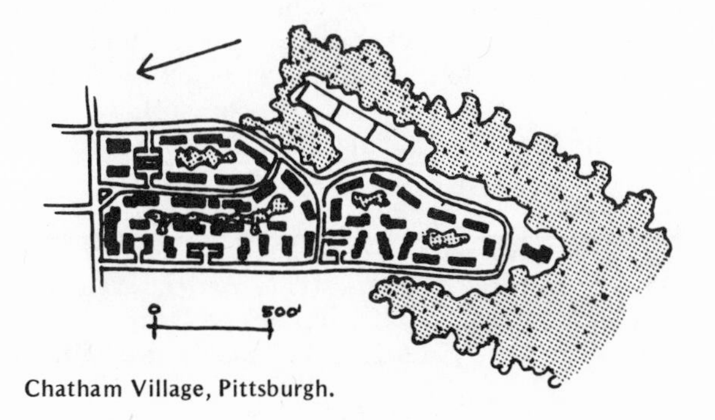

Chatham Village, Pittsburgh.

show how this situation could be corrected. By forming home building corporations, financed by prosperous companies seeking long-term investments, well-designed communities could be built, like Roland Park, the Country Club district, and Forest Hills Gardens. The English experience furnished ideas, and Sunnyside Gardens, Long Island, heads the list of accomplishments that resulted.

The designers of Sunnyside had to contend with an inefficient and unchangeable block pattern. They could overcome this difficulty only partially by siting the houses as row-house quadrangles with common garden space inside. The basic problem was that the blocks were too long and narrow and that there were too many through streets. The block pattern had been platted by an engineer who had spent his life laying it out. He refused to allow any alterations. This was typical of platting practice throughout the country – municipalities laid out plats and sometimes entire utility systems as an inducement to speculators. Wright and Stein decried this practice and proposed carefully considered alternatives. Henry Wright's *Rehousing Urban America* (1934) and Clarence Stein's *Towards New Towns for America* (1951) tell the full story of this group's work.

The early residential communities, such as Roland Park, had curvilinear road layouts which suited their generally rolling topography. However, all houses fronted on a through street. By the 1920s, automobile traffic was recognized as a formidable fact of urban life, for no landscaped street, however pastoral and winding, was free of the annoying intrusions of through traffic. The solution was the creation of traffic-free groupings of houses. Auto traffic should be made to serve the houses without despoiling the whole neighborhood. The answer to this problem was the "superblock," an island of green, bordered by houses and carefully skirted by peripheral automobile roads. Parking areas were conveniently located along the peripheral roads in carefully sited clusters. Chatham Village in Pittsburgh and Baldwin Hills Village in Los Angeles are among the best examples of the idea. However, a superblock was not a whole community or a whole town. What of the larger community? Indeed, what of a whole town?

The opportunity to demonstrate a design concept to answer this larger problem came at Radburn, New

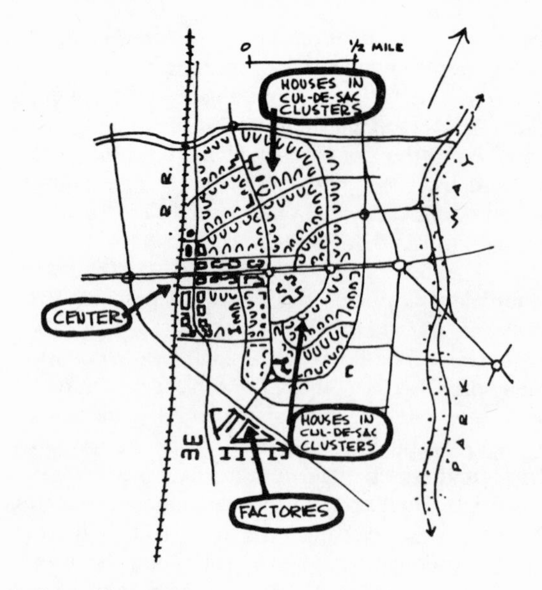

Radburn, New Jersey.

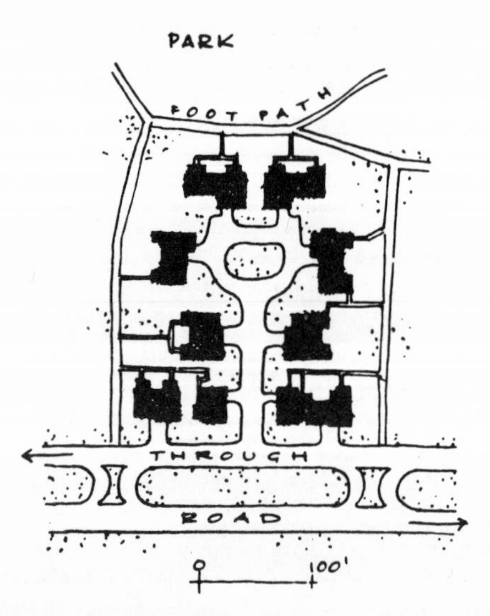

Radburn, New Jersey. Site planning of houses around a cul-de-sac.

Jersey, less than an hour's travel time from Manhattan. The Radburn idea was to create a series of superblocks, each around an open green with the greens themselves interconnected. Within the greens, pathways led to schools, shopping, and other centers. The greenways were pedestrian ways. Where they crossed a street they bridged over it or passed under it. The automobile circulation did not interfere with, or endanger, the pedestrian. Auto access to houses was by means of a short dead-end road. Hence the houses were arranged as cul-de-sac clusters around a stub service drive. The main circulation streets were also kept generally free of parked cars to allow unhindered flow of through traffic.

Unfortunately, Radburn was never entirely completed because of the Depression. It is one of the most important designs ever conceived for the modern residential community. Essential to the Radburn idea was the scheduling and coordination of its construction. No town can be built as a whole overnight – unless it is built for a special purpose or by one large company. In Radburn, industries were to be established to create jobs for the inhabitants – as in Howard's garden cities. As new industries furnished new jobs, new houses would be built, of course in carefully designed cluster communities. Through coordinated town development – industries and houses – investors in both would be assuring each other's investment.

An essential aspect of the Radburn idea was the organization of the town into cohesive neighborhoods, described by Clarence A. Perry, in 1929, in a book called *The Neighborhood Unit.* The idea would be applicable not only to new towns but to large city areas. One objective of replanning old cities became the creation of neighborhood centers and the physical delineation of neighborhhod groups. The "Radburn idea" was a new town idea. Of course, all American towns had been new towns originally. The nineteenth-century commuter suburbs were a form of "new town" in their time. Gary and Pullman were new industrial or satellite towns, as the American G.R. Taylor observed in two books: *Satellite Cities, A Study of Industrial Suburbs* (1915) and *The Building of Satellite Towns* (1925). Taylor's reasoning reinforced Ebenezer Howard's belief that metropolitan

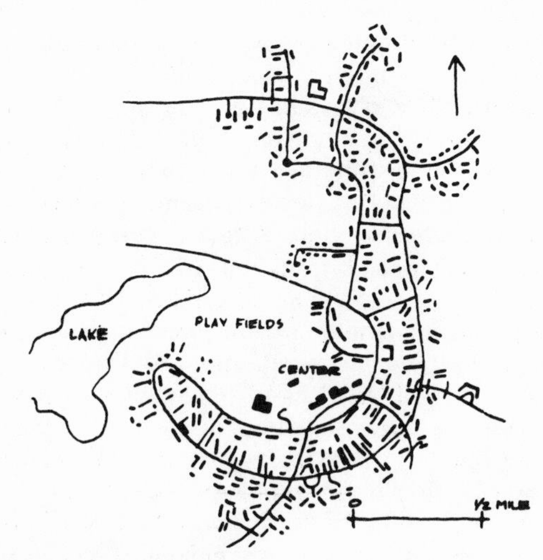

Greenbelt, Maryland.

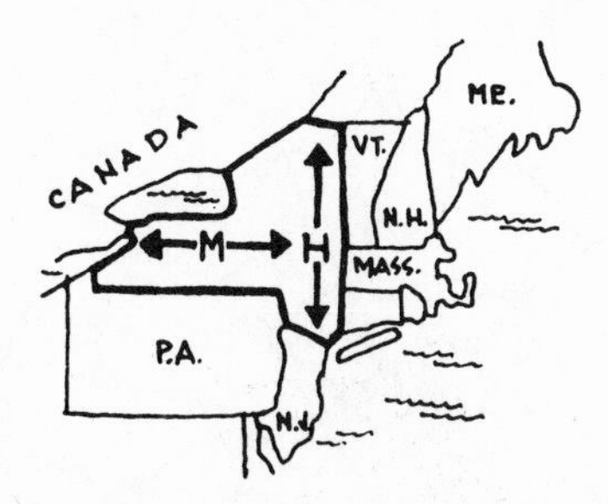

New York State. "M" is the Mohawk Valley; "H" is the Hudson River Valley.

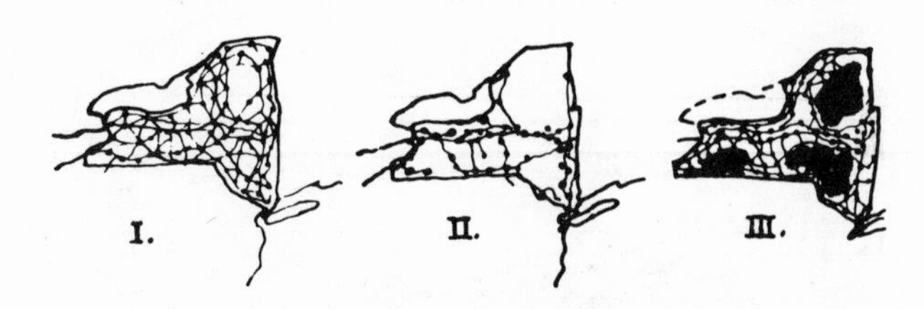

I. New York State, 1840-1880. Rapid development of natural resources; small towns economically independent; industry served by water power and canals. Near the end of this period the drift toward the new rail lines had started.

II. New York State, 1880-1920. Development of central rail routes; concentration along main-line transportation; use of steam power; agricultural competition with the West. Town growth is in the central valley belts and the towns off the main line decline.

III. New York State, the future. New forces of influence are the automobile, good roads, and electric power transmission. Rather than a return to the past dispersion and distribution pattern, it is possible to make a more effective use of all economic resources, to divert growth from New York City, and to develop areas in accord with their most favorable use – industry, agriculture, recreation, water supply, and forest reserves.

growth could and should be directed into a colonization movement.

Regional Planning

However, these concepts of community planning were part of a far larger picture, namely, the city or town in its total physical and economic region. A town thrives because its region is healthy; it dies as its region declines. America had, by the 1920s, more than one impoverished region and more than one ghost town. Bad soil practices in farming or in lumbering could wipe out a whole area's economy. The decline of small towns and the growth of large cities were both outcomes of the same regional picture. In many cases, however, growth and decline were caused by conditions that could be altered. Small-town demise and large-city expansion could be checked within significant limits. Howard and Taylor had shown that satellite colonization was an alternative. The real answer to the larger questions could be found only in a regional outlook. That outlook had its roots in the thoughts of Marsh and Geddes and its champions in Henry Wright and Benton MacKaye.

Working under a commission chaired by Clarence Stein, Henry Wright produced the *Report of the Commission on Housing and Regional Planning for the State of New York.* Wright, with a simple group of drawings, showed how the towns of New York State had first been small trade centers for an agricultural society. Western expansion and the development of Midwestern farms, producing food more cheaply, caused New York's farms to dwindle. Simultaneously, however, industrialization took hold, favoring towns that lay along streams and roads; streams were the early sources of power and roads

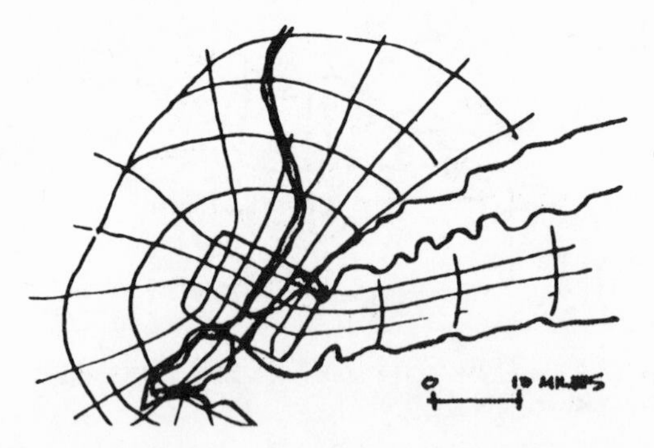

Concept of highways from the regional plan of New York.

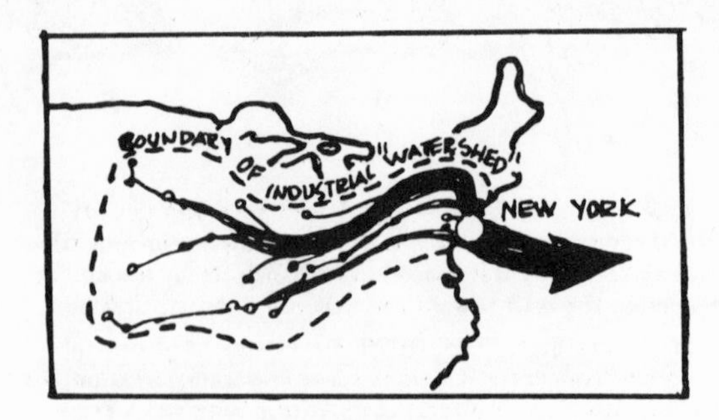

MacKaye's diagram of the U.S. stream of industrial productive flow through the Port of New York. He compared this to the flow of streams and rivers in a watershed area. (From "The New Exploration.")

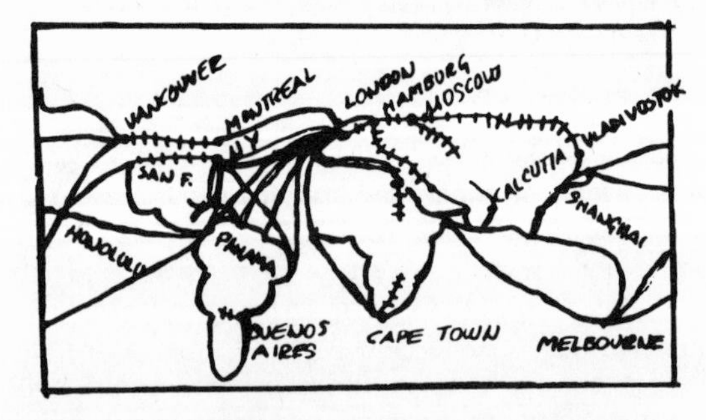

MacKaye's diagram of the world's major shipping and rail routes – the main arteries of flow between the modern world's metropolitan areas. (From "The New Exploration.")

facilitated distribution of industrial products. Thus the small farm towns declined, and the Hudson River and Mohawk River valleys became the spine of New York's industrial economy. New York City became a world port for distributing the products of the American industrial belt. It also became its financial and managerial heart. Henry Wright's plan explained these phenomena and proposed channeling these forces into a better development pattern. The former farm hinterlands could become recreational and dairying lands. The industrial spine would improve transportation, and New York City would be a strong central core for a constellation of smaller communities like Radburn. Wright's plan is still one of the finest models of regional planning ever produced. The evidence of its validity is that while it was not officially adopted, its key recommendations were eventually realized, in some cases at the hands of people who had never known of the plan. The regional approach to planning led to the formation of the Regional Planning Association of America (RPAA).

In the late 1920s another significant regional plan was made for the New York City metropolitan area by the Regional Planning Association of New York. RPAA and RPA of New York are sometimes confused. The latter was a study which embraced 22 counties, 500 municipal districts, 10 million people and parts of New York State, New Jersey, and Connecticut. This plan, however, was far smaller in scope than Wright's. In 1928 an eight-volume survey was produced, two volumes of which were plan proposals. All of this was done under the direction of the Scottish planner Thomas Adams. This was the most complete plan study ever done for an American metropolitan area, but it was criticized by the RPAA, which felt it was too limited in its physical scope.

Planning, if it is to be successful, must start at the beginning. The beginning, according to Wright's approach, is the total situation of a village or a metropolis in its statewide and multistate region. Benton MacKaye was to go even further.

MacKaye was originally a forester, a man in the tradition of the great American conservationists. He saw the nation as a series of component parts whose natural divisions were river drainage areas, not state lines. For him the Connecticut and Colorado River

MacKaye's diagram of the relative strength of the world's civilizations in terms of population. (From "The New Exploration.")

MacKaye's diagram of the relative strength of the world's civilizations in terms of resources. (From "The New Exploration.")

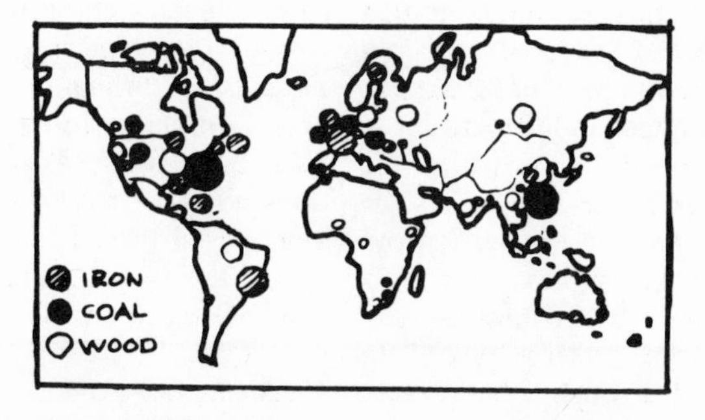

MacKaye's diagram of the location of the prime natural resources of modern industrial civilizations: iron, coal, and wood. (From "The New Exploration.")

basins were truer pictures of our country's component parts than its political lines. The country, to MacKaye, was really a number of natural "flow systems." MacKaye also saw towns and cities as part of these flow systems – determined, to be sure, by natural resources but built by men. If the two systems did not operate in harmony, either the land would be ruined or the towns fail. MacKaye's greatest accomplishment was the Appalachian park and trail system. MacKaye realized that the entire Eastern seaboard's population needed a system of open spaces commensurate with its continuing growth. Indeed, he titled his plan *An Open-space System for Appalachian America.* Incidentally, MacKaye proposed this idea in an article in the *AIA Journal* in 1921. In 1928 MacKaye published *The New Exploration, A Philosophy of Regional Planning.* Along with Geddes's and Mumford's writing, it is one of the best statements of the regional outlook. A second edition was published in 1962. MacKaye also realized the necessity of separating through traffic from residential areas. He spoke of the "townless highway" (today's interstate system which bypasses cities) and the "highwayless town" (like Radburn) as mutually interdependent.

MacKaye, like his colleague Wright, also used simple sketches to explain these ideas. But where Wright sketched New York State, MacKaye sketched maps of the entire world. He discussed the world's resources, climates, history, and major routes of trade and ports. He showed how all were inseparably related. In a page or two he summed up all the economic theories of the past. He showed New York City as the entry and exit portal for the entire American industrial empire; how Boston, Philadelphia, and Baltimore were portals on this main stream of flow with New York City; and how Pittsburgh, Detroit, and Akron were ancillary parts of the industrial stream. In short, he showed how the country worked as a combination of man's cities on nature's land.

In the 1920s, the battle of conservation was far from won, but it had gained a strong foothold. The campaign for conservation had been initiated by George Perkins Marsh and those men, like MacKaye, who followed him. MacKaye, however, started a new campaign. The exploration of the wilderness and conservation practices had to be expanded to include cities. That was the "new exploration" that MacKaye advocated so eloquently in his book.

The Depression furnished us with the manpower and opportunities to expand on the urban design ideas which had been proposed so far. Through WPA projects many urban parks, playgrounds, pools, and bathing beaches were created. Their design often derived from the old City Beautiful concepts, and the "greenbelt" towns built during this time were patterned after the Radburn idea.

Regional studies were also undertaken by the National Resources Planning Board, according to the outlook of Henry Wright and Benton MacKaye. The Tennessee Valley Authority put these ideas into practice and remains the greatest accomplishment ever achieved on this scale. MacKaye conceived the "multi-use programs" of TVA, and he proposed numerous programs which could be put into operation to the benefit of the TVA region, ranging from fish breeding to rural electrification. These were the side effects of a regional enterprise.

The American story of urban design of the last 50 years is little known on a popular level and often misunderstood where it is known. By far, it is one of the richest stories in the annals of urban design — even though its accomplishments were not too numerous. The men who authored its deeds were part of a worldwide fraternity. Throughout continental Europe, similar stories were unfolding.

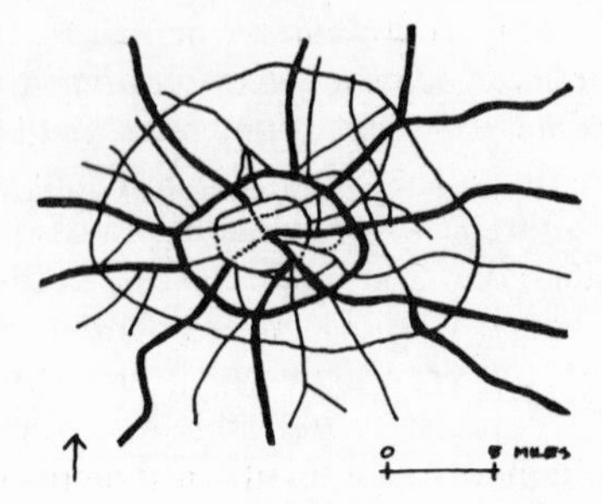

Abercrombie's plan for London. Radiocentric road system.

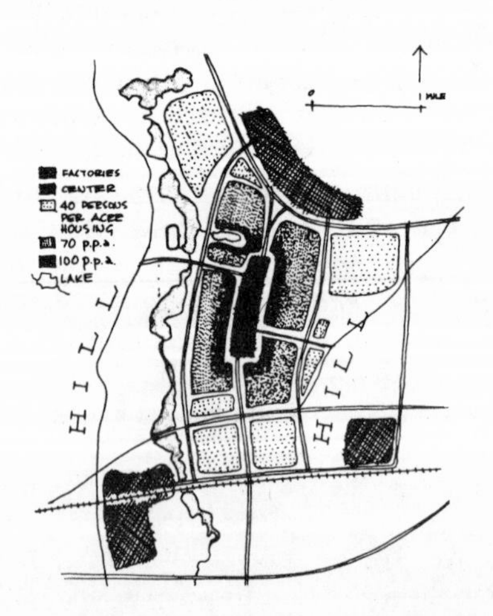

Plan of Hook. A proposed "new town" for 100,000 set in a valley. Though not built, it was a radical design departure which paved the way for Cumbernauld.

ACHIEVEMENTS ABROAD

Several garden cities were built in France — not long after Howard had proposed the original idea in England. The first, Dourges, was built in 1919 by a railroad. It was followed by Longueau, Tergnier, and Lille-la-Déliverance.

In England, Raymond Unwin had written *Nothing Gained by Overcrowding* (1903) and *Town Planning in Practice* (1909). F.J. Osborne's *New Towns After the War* (1918) advocated garden cities balanced with a central metropolis. The titles of these books indicate the kind of discussion held in England since Howard. In 1940 Sir Anthony M. Barlow headed a commission which studied urban problems and produced *The Report of the Royal Commission of Distribution of Industrial Population,* which encouraged industry to locate away from large cities in planned communities. In 1943 Sir Patrick Abercrombie and J.H. Forshaw published *The County of London Plan*. The Barlow report paved the way for the British New Towns policy, and the Abercrombie plan showed how London could be improved while retaining a large bulk of the population.

The English New Town movement has produced about [30] new towns to date. The early ones were rather spread out with large open spaces, indeed sometimes far too large. Some towns lacked cohesive-

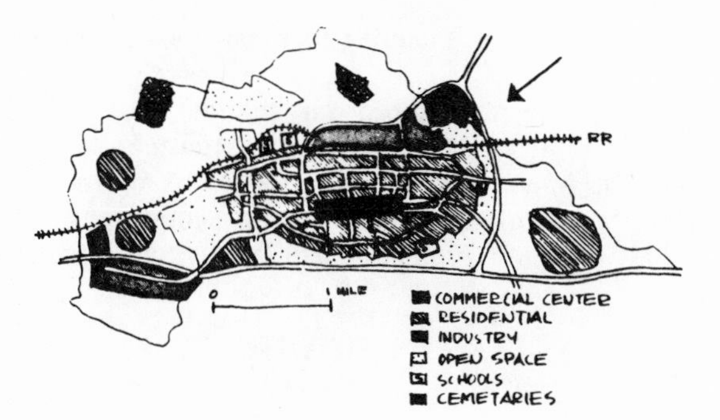

Land use plan of Cumbernauld.

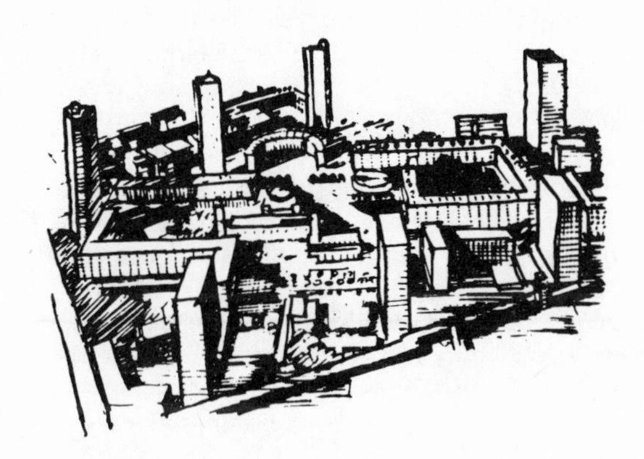

The Barbican Development, London.

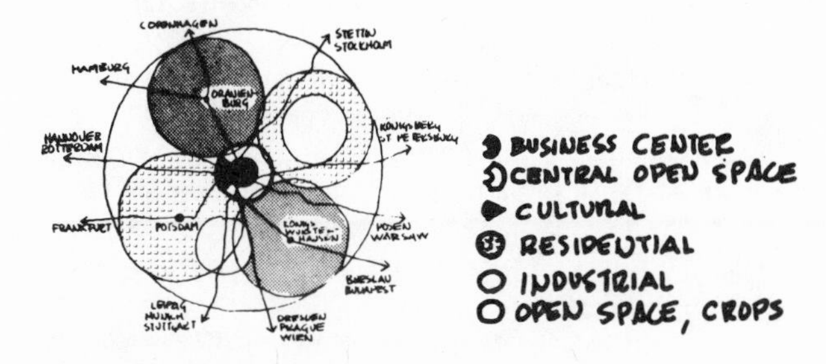

Martin Mächler's plan for Greater Berlin, 1920.

ness and convenience, the spaces being distances to overcome rather than amenities. A radical change in New Town design was introduced in the 1950s in the plan for Hook, wherein houses were very closely grouped along a multilevel linear spine. This decreased distances and all communal facilities could be reached by a short walk. The open countryside was right at the edge of town, just as accessible to the pedestrian as was the town center. The circulation concept of Hook, in fact, was that of an ocean liner. Because of a political controversy it was not built, but a second one, Cumbernauld, was planned and is proceeding toward completion at this writing. Recently a vast new plan for Southwest England was drafted as a further attempt to check London's growth.

An attempt to incorporate the demands of the automobile in modern town building is exemplified in London's Barbican area. Actually, there are two Barbican areas, or projects. One is a 63-acre area cleared by wartime bombings. The second is a 28-acre zone along the Thames just east of London Bridge. Both are high-density projects with tall towers. Their main design significance is their multilevel circulation system. Elevated walkways link various parts of the projects. The Barbican scheme attracted the attention of the world's planners not because it was a new idea – da Vinci had proposed it, the Adam brothers had used it, and Rockefeller Center was based on it – but because it was being built in response to the postwar pressures of the automobile, pressures which every modern city was feeling.

Traffic congestion became a main concern of the English planners. The automobile was making intolerable intrusions into cities and towns. A report published in 1963, *Traffic in Towns*, deals with the long-term problems of traffic in urban areas. It designates traffic as a servant, not a wanton force to wreak havoc in cities. The town is foremost a place to live, with traffic playing a service role. Many old towns have a limited capacity for traffic, which should not be exceeded. Other towns can be reshaped to advantage with increased traffic, but only up to a point. New towns, like Cumbernauld, can be designed with totally new circulation concepts. This kind of investigation is sorely needed for the United States.

The story of Germany would repeat many ideas so

far outlined, as would those of Italy, Holland, Switzerland, the Scandinavian countries, the Soviet Union, and Israel. In the early 1920s, the Berlin architect Martin Mächler produced a typical diagram of basic metropolitan planning, showing the fundamental relationships between the central core, its cultural facilities, outlying residential and industrial areas, open space, and circulation. Sweden's story traces a history of planning that goes back centuries. Its new towns are based on Howard's original concepts, much advanced and refined. In Finland, Tapiola is a new satellite town of Helsinki, and in Holland, Amsterdam South is a satellite of Amsterdam. Many of their lessons and innovations are pertinent to our problems here in the United States

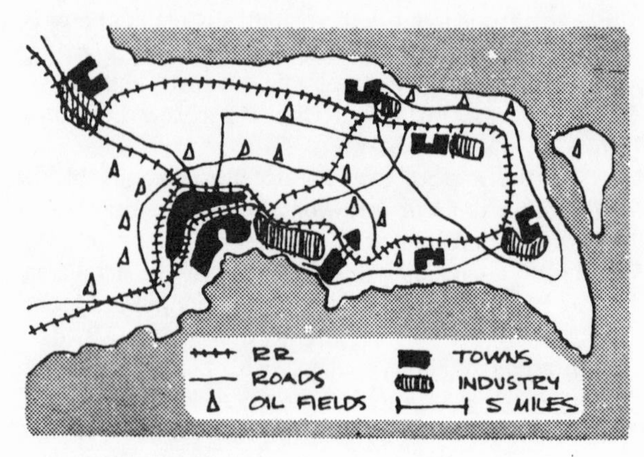

Baku. Plan for a satellite-city group in Russia, 60,000-80,000 people in each town.

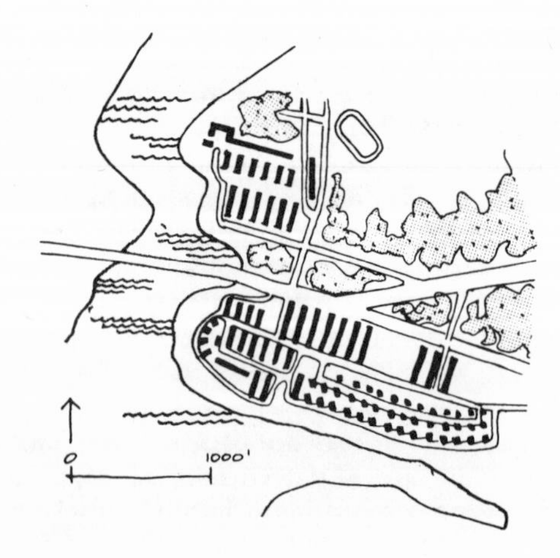

Part of West Kungsholmen, Stockholm.

MEN OF MODERN ARCHITECTURE AND PLANNING

Foremost in these years of search have been the ideas and writings of modern architects. Eliel Saarinen, for example, produced a prize-winning plan for Helsinki in 1911. He proposed orderly and distinct subcommunities, open space, and an overall circulation pattern to suit Helsinki's island and bay topography. At Cranbrook, where Saarinen taught for many years, the teaching of architecture and urban planning were closely allied. Walter Gropius came to the United States from his Bauhaus school in Germany via England and took the same approach in his teaching at Harvard. In 1943 Saarinen wrote *The City,* proposing the decentralization of large cities. E.A. Gutkind discussed this concept in detail in *The Twilight of Cities,* published in 1962. The idea of at least articulating urban segments is of great importance. Ludwig Hilbersheimer proposed that cities be laid out in relation to prevailing winds so as to avoid the smoke of factories. He later modified his ideas to include avoidance of atomic fallout particles – before the destructive effects of larger bombs invalidated any such defensive possibilities. Richard Neutra wrote *Rush City Reformed,* in which he showed how a modern city could use modern transportation technology to avoid congestion. Le Corbusier fused the ideas of modern architecture and city form with modern technology. In 1922 he unveiled *Une Ville Contemporaine,* a hypothetical plan for a city of 3 million people – not an abandonment of the congested industrial city, but, rather, a rearrangement of its form exploiting the new technology. His ideas are traceable to Eugène Hénard's *Les Villes de L'Avenir* and Tony Garnier's *Une Cité Industrielle.*

Le Corbusier's proposed rearrangement would create three distinct areas: a central business city with

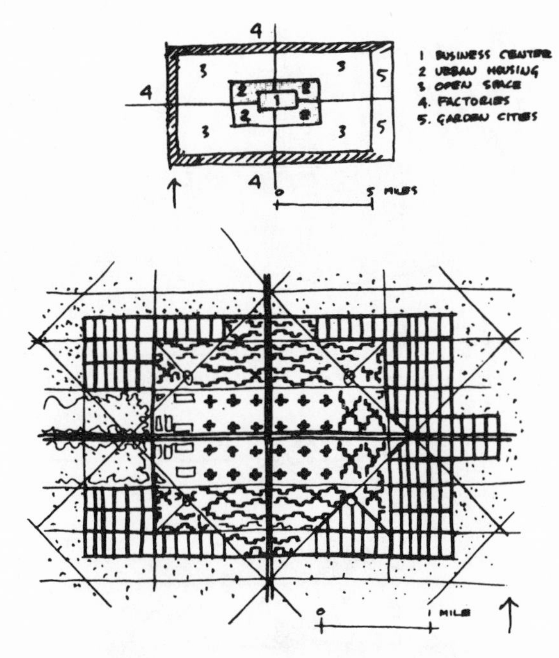

Le Corbusier's plan for a city of 3 million people (whole city above, central portion below).

Le Corbusier's Plan Voisin for Paris.

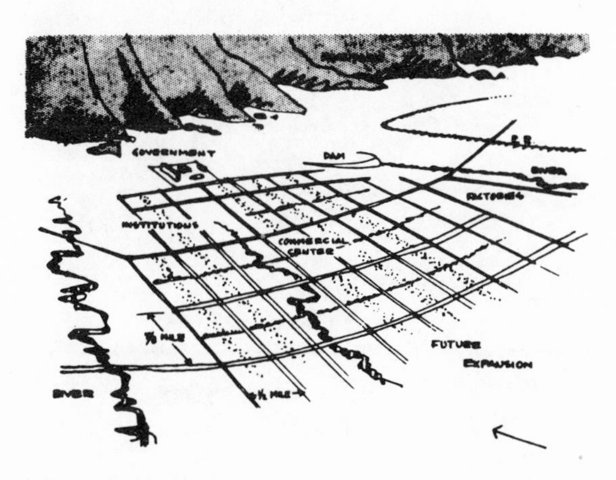

Chandigarh, India.

400,000 inhabitants in 24 tall skyscrapers; an encircling residential zone of 600,000 occupying multistory continuous slabs; then garden houses for 2 million. The plan had a crisp geometric form, with roads creating large rectangles interwoven with major diagonals. Le Corbusier had four major objectives: to decongest the center city; increase density; improve circulation; and, all the while, provide more natural verdure, light, and air.

Three years later came his *Plan Voisin* (Neighborhood Plan), embodying the same goals but applied to a large section of Paris. Eighteen 60-story towers replaced the crammed-in houses of central Paris, freeing the ground for high-speed circulation, parks, cafes, shops – and people. The vitality of the streets of Paris was to be released from its traditional corridor setting to extend freely in all directions.

In 1935 he proposed *La Ville Radieuse,* refining many of his earlier concepts. Long rectilinear buildings meandered in zigzag fashion to cover only 12 percent of the ground surface. And in 1937 came *Le Plan de Paris,* an even more advanced development, showing how all Paris could be rebuilt without destroying its magnificent old architectural monuments. Le Corbusier felt that the old role of the street as an artery for pedestrians and vehicles was no longer possible. The two had to be separated. If, as he proposed in his designs, the road were elevated and connected directly to buildings, the ground would then be free for omnidirectional pedestrian movement. It would also be free for recreational use. High urban density, essential to the modern city, would be realized through the use of multistory skyscrapers and multistory buildings, meandering as slabs in a fret pattern across the landscape. From the ground level they would barely be visible despite their size, for they would be screened by dense groves of trees. Through the years, Le Corbusier had been applying his ideas – on paper – to many cities of the world, punctuated in a handful of places by a few accomplishments. As his sketches increased, his fame multiplied. He became the leading spokesman and fountainhead of the "International Movement."

Not until the last decade did he get the opportunity to design an entire city: Chandigarh, India. His plan, done in collaboration, followed the earlier and brilliant concepts of Matthew Nowicki, himself strongly influenced by Le Corbusier. Its great significance is its regional flavor – its embrace of local Indian culture in decidedly modern terms. It is a series of neighborhood enclaves, arranged in a grid pattern and interconnected by a carefully articulated circulation system.

Le Corbusier also showed how massive design problems could be handled by large groups of high and low buildings; in effect, he brought cubism to large-scale architectural compositions. Coupling these concepts of architectural mass to modern construction, his ideas have been followed the world over. Le Corbusier's influence and popularity stem from his use of modern forms and designs in his architecture and the convenience of his large block-like compositions for planning large-scale developments. His architecture and town planning were in tune with modern technology and administrative organization.

Le Corbusier's writing on architecture includes much discussion of city design in particular. *When the Cathedrals Were White* (1947) is a brilliant essay on modern technology as well as on American life generally, and on New York City particularly. *Concerning Town Planning* (1948) emphasizes urban housing. Lewis Mumford has been critical of Le Corbusier, feeling that his designs are spectacular rationalizations of gigantism of the modern metropolis. Mumford would prevent the giant from getting so big and, where it has, cut it down to size. Mumford would start with a regional approach. But so, too, would Le Corbusier. His urban studies include sketches and discussions of countries as a whole, and show their "flow systems" and urban hubs in a style almost identical to MacKaye's. Le Corbusier recognizes large cities as a fact of life, but has never denied the village or the region in which both operate.

Le Corbusier was instrumental in organizing the Congres International d'Architecture Moderne (CIAM), an international group of architects and planners who discussed the urbanization problems that were appearing in all the world's cities. In their Athens Charter of 1931 they proclaimed their dedication to the service of urban planning. The English CIAM organization, the MARS group, proposed a plan for rebuilding London. The whole population would be redistributed in 16 finger corridors, all connected by a major circulation spine and encircling

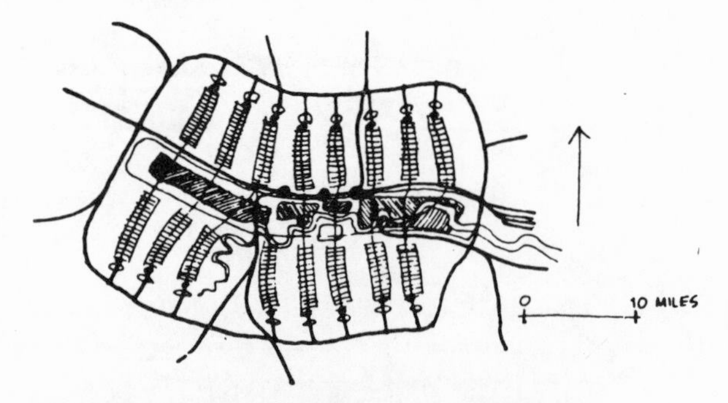

1938 MARS plan for London. An east-west spine of commerce and industry with sixteen residential corridors.

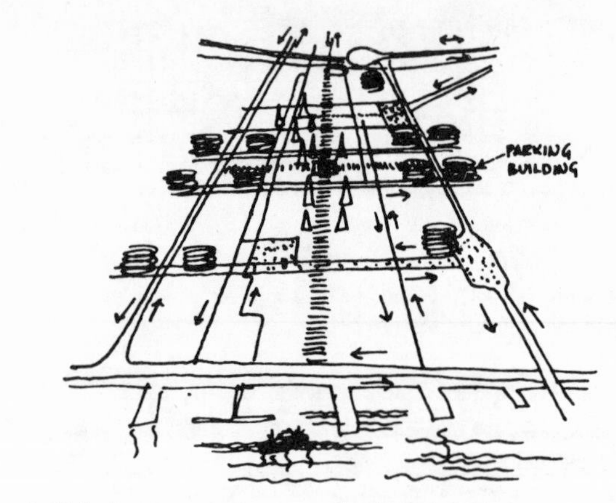

Louis Kahn's movement pattern for Philadelphia.

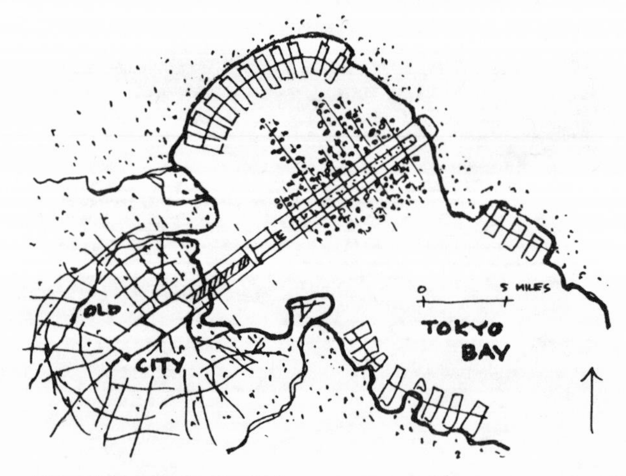

Kenzo Tange's plan for Tokyo.

circulation loop. Le Corbusier conceived the CIAM grid — a graphic file system for recording pertinent information in an urban study and for explaining a plan. The grid had four component sections: work, residence, circulation, and leisure.

Great study has been concentrated on the problems of circulation and urban design. For example, Louis Kahn has made important designs for central Philadelphia showing circulation — main arteries, stop-and-go streets, and parking towers — as key determinants of urban form, all organized into a symphony of circulation.

One of the most profound designs to appear for the modern city is Kenzo Tange's plan for Tokyo. Like Kahn, Tange puts great emphasis on the role of circulation as a determinant of urban form. Kenzo Tange's proposal is based on penetrating analysis which casts light on our own problems. Setting out to make a plan for Tokyo, he concluded that the primary problem of large cities (Tokyo has 10 million people) is circulation. The very life of a city of 10 million depends on the ability of its inhabitants to communicate with each other face to face. Through careful study, Tange saw that once a city grew beyond 2 or 3 million inhabitants this vital communication became very inefficient in a radiocentric city form created by spontaneous growth. He examined every possible alternative. Tower buildings in large open spaces were no solution, he felt, because they still caused congestion in circulation systems. Overall densities are not altered by a tower pattern. His proposed solution envisioned the construction of a new Tokyo over Tokyo Bay, hung on a series of suspension bridges. Vehicular circulation would be segregated according to speed, and dwelling and work areas stacked in several levels. Tange developed detailed cost estimates and concluded that they would not exceed the construction outlays contemplated for Tokyo in the next several decades.

Frank Lloyd Wright, in contrast to Le Corbusier or Tange, showed the way for abolishing the city. He followed Howard, Geddes, and the social reformers in their distrust of the modern monster city and, especially in his early years, echoed the views of his great teacher Louis Sullivan, who built his most important works in the city but had no affection for it. In 1932 Wright published *The Disappearing City* and later *Broadacres* — proposing that every family live on an acre of land. Present-day American suburbs are crude microcosms of what Wright would have made into art.

But at the end of his career, perhaps in recognition of the difficulties of land supply and logistics in applying the Broadacres plan to an America grown immense in population, Wright (possibly with tongue in cheek) unveiled a scheme for a superskyscraper a full mile high. Ten or so of these could replace all Manhattan's buildings and free the land for greenery. Wright's site design for the Marin County Civic Center north of San Francisco is a slice out of his old Broadacres plan.

Among the world's foremost urban theorists today are Constantine Doxiadis, Charles Abrams, and Buckminster Fuller. Their thinking could be categorized along the lines of MacKaye, Le Corbusier, and the other regionalists. Doxiadis has addressed himself to the problem of urbanization on a worldwide scale, and his major designs have been made for countries

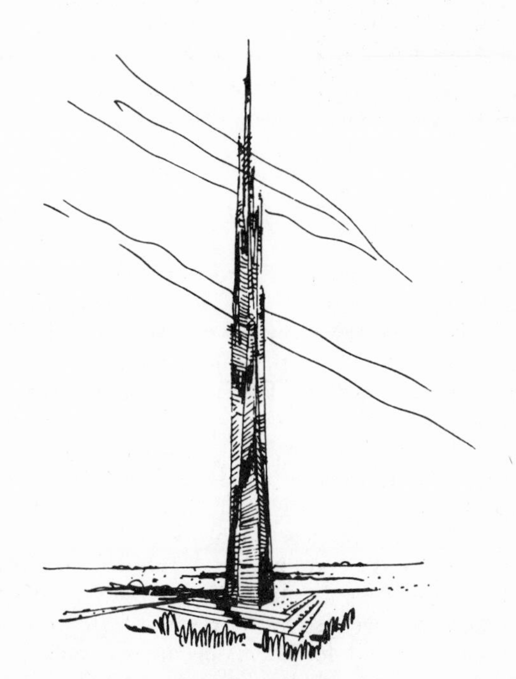

Frank Lloyd Wright's mile-high skyscraper.

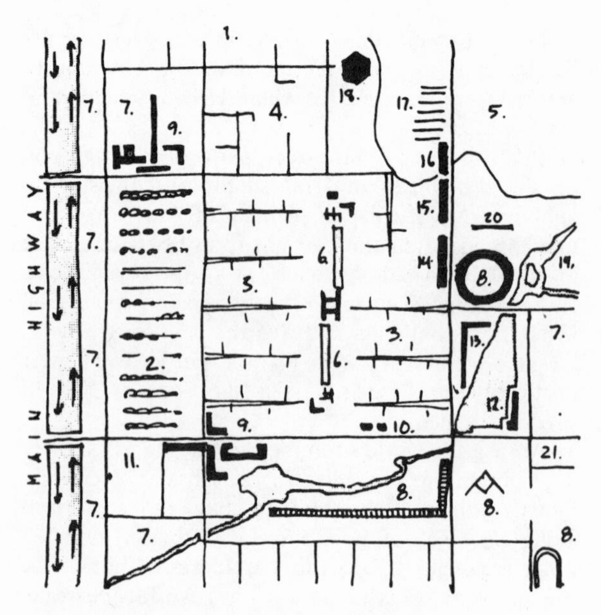

Frank Lloyd Wright's Broadacres:

1. Small farms
2. Orchards
3. Small houses
4. Medium houses
5. Large houses
6. Schools
7. Small industries
8. Sports and recreation
9. Markets
10. Clinics
11. Tourists' hotel
12. County seat
13. Arts
14. Arboretum
15. Aquarium
16. Zoo
17. University
18. Community church
19. Sanitarium
20. Hotel
21. Airport

where the economy and productive system can be coordinated by policy and decree. Doxiadis' best work is in the newly developing nations of Africa and the Middle East, where this is possible. In *Architecture in Transition* (1963), he explains his total view. Case studies in his magazine *Ekistics* show many of his plans – also including detailed programs and schedules. He recently published his Ekistics grid, a system for recording planning data and ordering the planning process. It is reminiscent of the old CIAM grid. Doxiadis's contribution – and it is not a small one – has been to state the problems of modern urbanization with scientific clarity and to propose a rational method for addressing those problems. He approaches town planning as a science, which includes planning and design as we traditionally know it, as well as the contribution of the sociologist, geographer, economist, politician, social anthropologist, ecologist, demographer, etc. All this he assembles into a total rational and human approach which he calls "ekistics" – the science of human settlements.

Charles Abrams sees housing as one of the prime fields of endeavor for solving urban problems. In *Man's Struggle for Shelter in an Urbanizing World* (1964) he discusses the prime role of housing in urban development and in national policy. Abram's experiences and thought extend from highly developed to completely underdeveloped nations. Buckminster Fuller continues the internationalist train of thought best of all. His *Inventory of World Resources – Human Trends and Needs* (1963) is a work par excellence in assessing the current state of the world's production and productive energy – and in suggesting how it can be turned to man's complete advantage.

No modern thinker, however, has surpassed the scope and depth of Lewis Mumford, who has seen fit to devote his energies to writing. His all-embracing views have been consistently confirmed over the last 40 years, in which time he has authored 20 books and innumerable articles. Mumford has been urging that the fundamental needs of society be the basis for the judicious use of our technological power, a power whose wonders obscure the better ends they might bring. These ends are the harmonious life of civilized social groups in ecological balance with the particular place occupied – a notion too readily, and incorrectly, dismissed as latter-day romanticism. Inherent in Mumford's thinking is the need for recognizing the physical limitations of human settlements. One of his most eloquent and, at the same time, concise expressions of his philosophy was the speech he delivered in June 1965 in Washington, D.C., at the AIA national convention. The great summary of his thought is contained in his largest volume, *The City in History*, published in 1961.

CONCLUSION

The vastness of current urban problems confounds the average mind. It occupies a large fraternity of theorists who may have produced in the last five years more books on cities than in the whole history of civilization. Indeed the complexity of the problem has spawned a complexity of language which raises some doubt as to usefulness – but no doubt as to expenditure of energy in behalf of a vital human problem.

The city is as much a physical object in three dimensions as it is anything else. As a physical object it can be designed – perhaps as artfully as the gardens of Versailles, as practically as the town of Ferrara, and as humanely as the towns of the ancient Greeks. Indeed, for those who do not care to look in those directions we can draw encouragement from our own history of accomplishments. After all, it is the physical city which is the result of all planning efforts, whether dealing with economy, sociology, or transportation. It is the physical city we have to live in. It is the physical city we must design.

With this sketch of the history of urban design, we can examine with some understanding the current state of the art as it applies to us. It is an extremely rich state at present, one which asks now, as it has asked throughout history, to be put to use. If our problems seem vast, we must remember that our means are greater than those of any society in the past. Our problems are compounded by the multiplicity of actions needed to yield a decision to act. Here the history of urban design is helpful because it gives us ideas as to what we can act on, the things we can decide to do.

For today's city dweller, a history of urban design is an insight into the current state of things as well as a suggestion of better alternatives. For the politician, it is a demand for administrative techniques and actions that will realize the vast store of urban design possibilities. For every urban administrator and engineer, the history of urban design insists on responsible and considered decisions. For the designer, the merits of studying the history of urban design may go further. A knowledge of its store of ideas is not so much a matter of correct interpretation, for interpretations and even acknowledgments of historical events will vary from person to person and from era to era. The value of this knowledge is that it stimulates one's thoughts toward a deeper understanding of his own period. It enables him to act with far greater depth of understanding and scope of vision. And, not the least, it sharpens his taste for excellence as nothing else can.

POSTSCRIPT

January 1975

The design of our habitat (cities, farms, wilderness) is more than any other thing that we do the consequence of all our social enterprises. Our cities and their settings are us. More than that, they are an accumulation of the sum of the endeavors of a society, and of the interruptions to these endeavors. In comprehending them we understand ourselves and recognize that we constitute the society which in fact designs our cities.

Town planning or urban design requires knowledge of the past, an understanding of the present, and a sense of future needs and possibilities.

With respect to the past, Daniel Boorstin's work in American history effectively treats the context of our urban design. John Rep's work in the history of American city planning is unsurpassed; it begins with the European settlement of North and South America and ends approximately one quarter way into this century. Well worth knowing is the work of Roy Lubove which starts with the social reformers of the end of the nineteenth century and extends through the 1930s. There is no better explanation of the relation between American patterns of settlement and the landscape than an unedited little pamphlet entitled *Ten Lectures on the Historical Geography of the U.S. as Given in 1933,* written by the University of Chicago geographer Harlan Barrows. For general scope and insight there are Henry Churchill and Benton MacKaye. Lewis Mumford's many works grow stronger with the years.

To learn from observation, there is much urban design to be seen directly in America. We are aided in our understanding of it by foreign observers who are often clearer-eyed than those who live close by.

One must go abroad to see the potential of urban design because the nature of American government limits it to acting as mediator between contending interests. Our government is not in a position to initiate or take anticipatory action. It waits until needs are overwhelmingly apparent and until most of the population are so persuaded.

An unparalleled source of scientifically based study, however, is the U.S. National Resources Planning Board which existed for about a decade first as an independent federal agency, later within the Executive Offices of President Franklin D. Roosevelt. Many of these reports are relevant today. Some relate to urban design. Others incorporate an approach to land use which is a direct predecessor of the ecological work of Philip H. Lewis and Ian McHarg [Reading No. 40].

Some day, the world's knowledge of settlement or community design should be collected and organized. The late E.A. Gutkind made an effort in this encyclopedic direction. The computer and the satellite make such a project possible: the first by comparing and simulating urban forms, the second by periodically recording the patterns of communities around the world.

. . . Urban design is perhaps especially subjected to the forces of change since it deals with the harmonious relationship between structures and the human beings that use these structures. The dilemma caused by the relative permanence of man-made structures in a dynamic society poses questions that have urban design implications. Traditionally, civic design has meant an imposed formalism, strong enough to resist changes and at times large enough to accommodate it, i.e., baroque city planning. The imposition of design through the force of dominant will is, however, no longer possible in a pluralistic, democratic society unless done under the guise of function, i.e., freeway design. . . . This general lack of design will, has brought about a vacuum of indecision and resignation in the design of cities as a series of accidental and uninformed decisions geared to immediate problem solving rather than toward well-established goals. The need exists to look at the city as a system wherein there can be a forum for idea and response as to the quality that the city wants and is able to achieve.

Department of Community Development, Seattle Urban Design Report, Determinants of City Form, *Seattle, Washington, 1971, p. 90.*

7

Kevin Lynch

THE CITY IMAGE AND ITS ELEMENTS

There seems to be a public image of any given city which is the overlap of many individual images. Or perhaps there is a series of public images, each held by some significant number of citizens. Such group images are necessary if an individual is to operate successfully within his environment and to cooperate with his fellows. Each individual picture is unique, with some content that is rarely or never communicated, yet it approximates the public image, which, in different environments, is more or less compelling, more or less embracing.

This analysis limits itself to the effects of physical, perceptible objects. There are other influences on imageability, such as the social meaning of an area, its function, its history, or even its name. These will be glossed over, since the objective here is to uncover the role of form itself. It is taken for granted that in actual design form should be used to reinforce meaning, and not to negate it.

The contents of the city images so far studied,

---

which are referable to physical forms, can conveniently be classified into five types of elements: paths, edges, districts, nodes, and landmarks. Indeed, these elements may be of more general application, since they seem to reappear in many types of environmental images. These elements may be defined as follows:

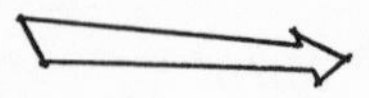

1. *Paths:* paths are the channels along which the observer customarily, occasionally, or potentially moves. They may be streets, walkways, transit lines, canals, railroads. For many people, these are the predominant elements in their image. People observe the city while moving through it, and along these paths the other environmental elements are arranged and related.

2. *Edges:* edges are the linear elements not used or considered as paths by the observer. They are the boundaries between two phases, linear breaks in continuity: shores, railroad cuts, edges of developments, walls. They are lateral references rather than coordinate axes. Such edges may be barriers, more or less penetrable, which close one region off from another; or they may be seams, lines along which two regions are related and joined together. These edge elements, although probably not as dominant as paths, are for many people important organizing features, particularly in the role of holding together generalized areas, as in the outline of a city by water or wall.

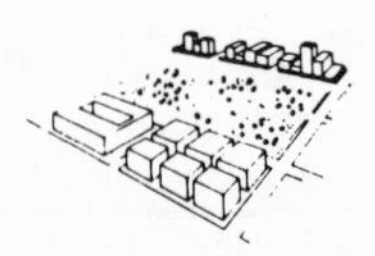

3. *Districts:* districts are the medium-to-large sections of the city, conceived of as having two-dimensional extent, which the observer mentally enters "inside of," and which are recognizable as having some common, identifying character. Always identifiable from the inside, they are also used for exterior reference if visible from the outside. Most people structure their city to some extent in this way, with individual differences as to whether paths or districts are the dominant elements. It seems to depend not only upon the individual but also upon the given city.

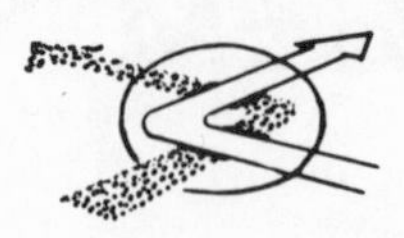

4. *Nodes:* nodes are points, the strategic spots in a city into which an observer can enter, and which are the intensive foci to and from which he is traveling. They may be primarily junctions, places of a break in transportation, a crossing or convergence of paths, moments of shift from one structure to another. Or the nodes may be simply concentrations, which gain their importance from being the condensation of some use or physical character, as a street-corner hangout or an enclosed square. Some of these concentration nodes are the focus and epitome of a district, over which their influence radiates and of which they stand as a symbol. They may be called cores. Many nodes, of course, partake of the nature of both junctions and concentrations. The concept of node is related to the concept of path, since junctions are typically the convergence of paths, events on the journey. It is similarly related to the concept of district, since cores are typically the intensive foci of districts, their polarizing center. In any event, some nodal points are to be found in almost every image, and in certain cases they may be the dominant feature.

5. *Landmarks:* landmarks are another type of point reference, but in this case the observer does not

enter within them; they are external. They are usually a rather simply defined physical object: building, sign, store, or mountain. Their use involves the singling out of one element from a host of possibilities. Some landmarks are distant ones, typically seen from many angles and distances, over the tops of smaller elements, and used as radial references. They may be within the city or at such a distance that for all practical purposes they symbolize a constant direction. Such are isolated towers, golden domes, great hills. Even a mobile point, like the sun, whose motion is sufficiently slow and regular, may be employed. Other landmarks are primarily local, being visible only in restricted localities and from certain approaches. These are the innumerable signs, store fronts, trees, doorknobs, and other urban detail, which fill in the image of most observers. They are frequently used clues of identity and even of structure, and seem to be increasingly relied upon as a journey becomes more and more familiar.

The image of a given physical reality may occasionally shift its type with different circumstances of viewing. Thus an expressway may be a path for the driver and edge for the pedestrian. Or a central area may be a district when a city is organized on a medium scale and a node when the entire metropolitan area is considered. But the categories seem to have stability for a given observer when he is operating at a given level.

None of the element types isolated above exist in isolation in the real case. Districts are structured with nodes, defined by edges, penetrated by paths, and sprinkled with landmarks. Elements regularly overlap and pierce one another. . . .

DESIGNING THE PATHS

To heighten the imageability of the urban environment is to facilitate its visual identification and structuring. The elements isolated — the paths, edges, landmarks, nodes, and regions — are the building blocks in the process of making firm, differentiated structures at the urban scale. What hints can we draw from the preceding material as to the characteristics such elements might have in a truly imageable environment?

The paths, the network of habitual or potential lines of movement through the urban complex, are the most potent means by which the whole can be ordered. The key lines should have some singular quality which marks them off from the surrounding channels: a concentration of some special use or activity along their margins, a characteristic spatial quality, a special texture of floor or facade, a particular lighting pattern, a unique set of smells or sounds, a typical detail or mode of planning. Washington Street may be known by its intensive commerce and slot-like space, Commonwealth Avenue by its tree-lined center.

These characters should be so applied as to give continuity to the path. If one or more of these qualities is employed consistently along the line, then the path may be imaged as a continuous, unified element. It may be a boulevard planting of trees, a singular color or texture of pavement, or the classical continuity of bordering facades. The regularity may be a rhythmic one, a repetition of space openings, monuments, or corner drugstores. The very concentration of habitual travel along a path, as by a transit line, will reinforce this familiar, continuous image.

This leads to what might be called a visual hierarchy of the streets and ways, analogous to the familiar recommendation of a functional hierarchy: a sensuous singling out of the key channels and their unification as continuous perceptual elements. This is the skeleton of the city image.

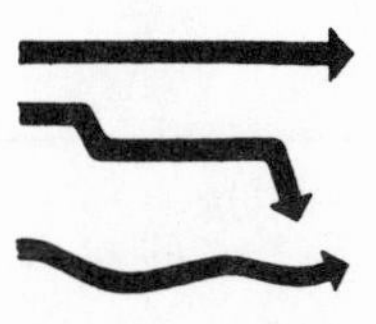

The line of motion should have clarity of direction. The human computer is disturbed by long successions of turnings, or by gradual, ambiguous curves which in the end produce major directional shifts. The continuous twistings of Venetian calli or of the streets in one of Olmsted's romantic plans, or the gradual turning of Boston's Atlantic Avenue, soon confuse all but the most highly adapted observers. A straight path has clear direction, of course, but so does one with a few well-defined turns close to 90 degrees, or another of many slight turns which yet never loses its basic direction.

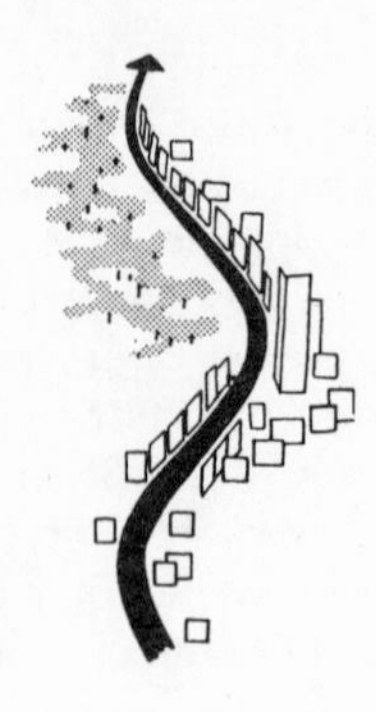

Observers seem to endow a path with a sense of pointing or irreversible direction, and to identify a street with the destination toward which it goes. A street is perceived, in fact, as a thing which goes toward something. The path should support this perceptually by strong termini, and by a gradient or a directional differentiation, so that it is given a sense of progression, and the opposite directions are unlike. A common gradient is that of ground slope, and one is regularly instructed to go "up" or "down" a street, but there are many others. A progressive thickening of signs, stores, or people may mark the approach to a shopping node; there can be a gradient of color or texture of planting as well; a shortening of block length or a funneling of space may signal the nearness of the city center. Asymmetries may also be used. Perhaps one can proceed by "keeping the park on the left" or by moving "toward the golden dome." Arrows can be used, or all projecting surfaces facing one direction might have a coded color. All these means make the path an oriented element to which other things can be referred. There is no danger of making a "wrong-way" mistake.

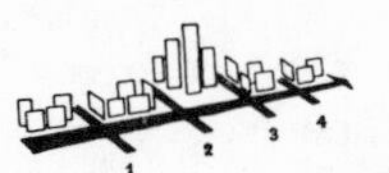

If positions along the line can be differentiated in some measurable way, the line is not only oriented, but scaled as well. Ordinary house numbering is such a technique. A less abstract means is the marking of an identifiable point on the line, so that other places may be thought of as "before" or "after." Several checkpoints improve the definition. Or a quality (such as the space of the corridor) may have a modulation of gradient at a changing rate, so that the change itself has a recognizable form. Thus one could say that a certain place is "just before the street narrows down very rapidly," or "on the shoulder of the hill before the final ascent." The mover can feel not only "I am going in the right direction," but "I am almost there" as well. Where the journey contains such a series of distinct events, a reaching and passing of one subgoal after another, the trip itself takes on meaning and becomes an experience in its own right.

Observers are impressed, even in memory, by the apparent "kinesthetic" quality of a path, the sense of motion along it: turning, rising, falling. Particularly is this true where the path is traversed at high speed. A great descending curve which approaches a city center can produce an unforgettable image. Tactile and inertial senses enter into this perception of motion, but vision seems to be dominant. Objects along the path can be arranged to sharpen the effect of motion parallax or perspective, or the course of the path ahead may be made visible. The dynamic shaping of the movement line will give it identity and will produce a continuous experience over time.

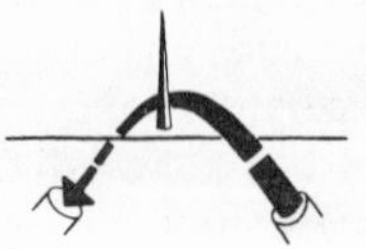

Any visual exposure of the path, or its goal, heightens its image. A great bridge may do this, an axial avenue, a concave profile, or the distant silhouette of the final destination. The presence of the path may be made evident by high landmarks along it or other hints. The vital line of circulation becomes palpable before our eyes, and can become the symbol of a fundamental urban function. Conversely, the experience is heightened if the path reveals the presence of other city elements to the traveler: if it penetrates or strikes them tangentially, if it offers hints and symbols of what is passed by. A subway, for example, instead of being buried alive, might suddenly pass through the shopping zone itself, or its station might recall by its form the nature of the city above it. The path might be so shaped that the flow itself becomes sensuously evident: split lanes, ramps, and spirals would allow the traffic to indulge in self-contemplation. All these are techniques of increasing the visual scope of the traveler.

Normally, a city is structured by an organized set of paths. The strategic point in such a set is the intersection, the point of connection and decision for the man in motion. If this can be visualized clearly, if the intersection itself makes a vivid image and if the lie of the two paths with respect to each other is clearly expressed, the observer can build a satisfactory structure. Boston's Park Square is an ambiguous joining of major surface streets; the junction of Arlington Street and Commonwealth Avenue is clear and sharp. Universally, subway stations fail to make such clear visual joints. Special care must be taken to explain the intricate intersections of modern path systems.

A joint of more than two paths is normally quite difficult to conceptualize. A structure of paths must have a certain simplicity of form to make a clear image. Simplicity in a topological rather than a geometrical sense is required, so that an irregular but approximately right-angled crossing is preferable to a precise trisection. Examples of such simple structures are parallel sets or spindle forms; one-, two-, or three-barred crosses; rectangles; or a few axes linked together.

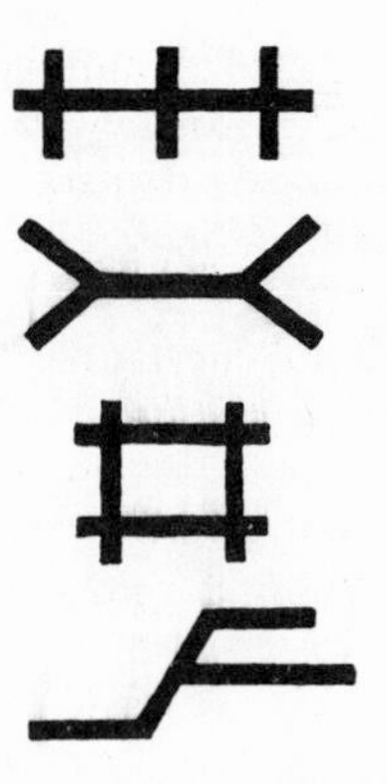

Paths may also be imaged, not as a specific pattern of certain individual elements, but rather as a network which explains the typical relations between all paths in the set without identifying any particular path. This condition implies a grid which has some consistency, whether of direction, topological interrelation, or interspacing. A pure gridiron combines all three, but directional or topological invariance may by themselves be quite effective. The image sharpens if all paths running in one topological sense, or compass direction, are visually differentiated from the other paths. Thus the spatial distinction between Manhattan's streets and avenues is effective. Color, planting, or detail might serve equally well. Naming and numbering, gradients of space, topography, or detail, differentiation within the net may all give the grid a progressive or even a scaled sense.

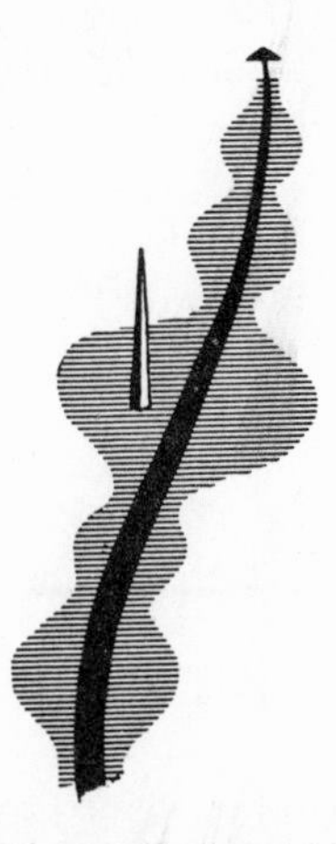

There is a final way of organizing a path or a set of paths, which will become of increasing importance in a world of great distances and high speeds. It might be called "melodic" in analogy to music. The events and characteristics along the path – landmarks, space changes, dynamic sensations – might be organized as a melodic line, perceived and imaged as a form which is experienced over a substantial time interval. Since the image would be of a total melody rather than a series of separate points, the image could presumably be more inclusive, and yet less demanding. The form might be the classical introduction-development-climax-conclusion sequence, or it might take more subtle shapes, such as those which avoid final conclusions. The approach to San Francisco across the bay hints at a type of this melodic organization. The technique offers a rich field for design development and experiment.

DESIGN OF OTHER ELEMENTS

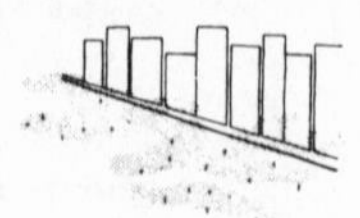

Edges as well as paths call for a certain continuity of form throughout their length. The edge of a business district, for example, may be an important concept, but be difficult to discover in the field because it has no recognizable continuity of form. The edge also gains strength if it is laterally visible for some distance, marks a sharp gradient of area character, and clearly joins the two bounded regions. Thus the abrupt cessation of a medieval city at its wall, the fronting of skyscraper apartments on Central Park, the clear transition from water to land at a seafront, all are powerful visual impressions. When two strongly contrasting regions are set in close juxtaposition and their meeting edge is laid open to view, visual attention is easily concentrated.

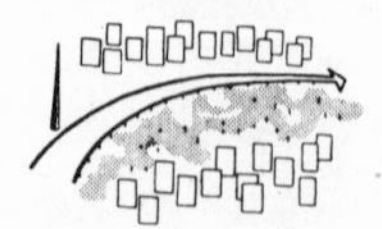

Particularly where the regions bounded are not of contrasting nature, it is useful to differentiate the two sides of an edge, to orient the observer in the "inside-outside" sense. It may be accomplished by contrasting materials, by a consistent concavity of line, or by planting. Or the edge may be shaped to give orientation along its length, by a gradient, by identifiable points at intervals, or by individualizing one end with respect to the other. When the edge is not continuous and self-closing, it is important that its ends have definite termini, recognizable anchors which complete and locate the line. The image of the Boston waterfront, which is usually not mentally continuous with the Charles River line, lacks a perceptual anchor at either end, and is in consequence an indecisive and fuzzy element in the total Boston image.

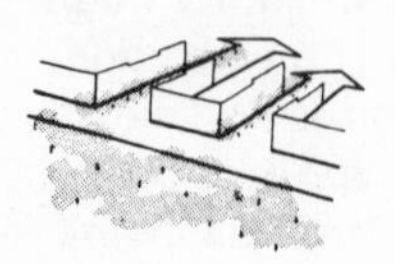

An edge may be more than simply a dominant barrier if some visual or motion penetration is allowed through it – if it is, as it were, structured to some depth with the regions on either side. It then becomes a seam rather than a barrier, a line of exchange along which two areas are sewn together.

If an important edge is provided with many visual and circulation connections to the rest of the city structure, then it becomes a feature to which everything else is easily aligned. One way of increasing the visibility of an edge is by increasing its accessibility or use, as when opening a waterfront to traffic or recreation. Another might be to construct high overhead edges, visible for long distances.

The essential characteristic of a viable landmark, on the other hand, is its singularity, its contrast with its context or background. It may be a tower silhouetted over low roofs, flowers against a stone wall, a bright surface in a drab street, a church among stores, a projection in a continuous facade. Spatial prominence is particularly compelling of attention. Control of the landmark and its context may be needed: the restriction of signs to specified surfaces, height limits which apply to all but one building. The object is also more remarkable if it has a clarity of general form, as does a column or a sphere. If in addition it has some richness of detail or texture, it will surely invite the eye.

A landmark is not necessarily a large object; it may be a doorknob as well as a dome. Its location is crucial: if large or tall, the spatial setting must allow it to be seen; if small, there are certain zones that receive more perceptual attention than others: floor surfaces, or nearby facades at, or slightly below, eye level. Any breaks in transportation — nodes, decision points — are places of intensified perception. Interviews show that ordinary buildings at route decision points are remembered clearly, while distinctive structures along a continuous route may have slipped into obscurity. A landmark is yet stronger if visible over an extended range of time or distance, more useful if the direction of view can be distinguished. If identifiable from near and far, while moving rapidly or slowly, by night or day, it then becomes a stable anchor for the perception of the complex and shifting urban world.

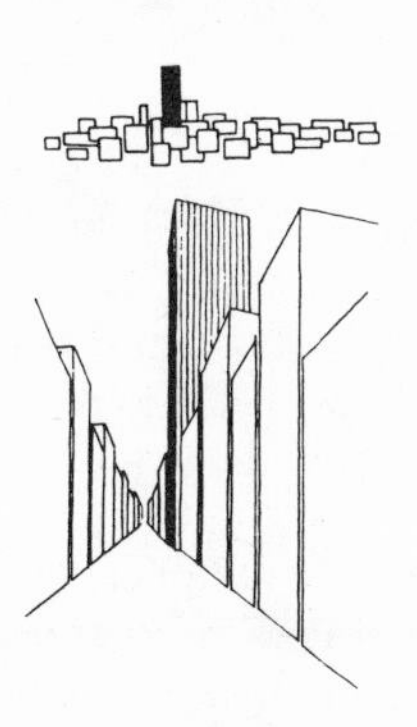

Image strength rises when the landmark coincides with a concentration of association. If the distinctive building is the scene of an historic event, or if the bright-colored door is your own, it becomes a landmark indeed. Even the bestowal of a name has power, once that name is generally known and accepted. Indeed, if we are to make our environment meaningful, such a coincidence of association and imageability is necessary.

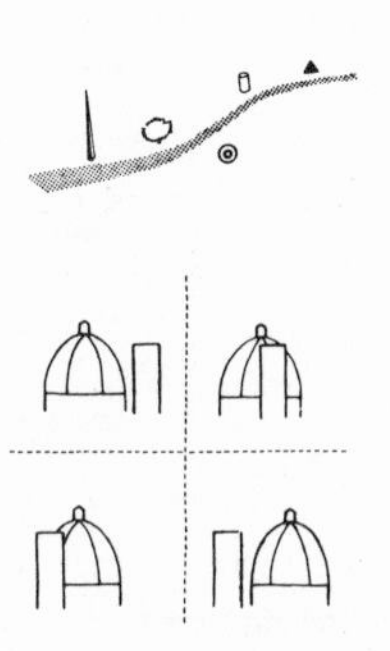

Single landmarks, unless they are dominant ones, are likely to be weak references by themselves. Their recognition requires sustained attention. If they are clustered, however, they reinforce each other in a more than additive way. Familiar observers develop landmark clusters out of the most unpromising material, and depend upon an integrated set of signs, of which each member may be too weak to register. The marks may also be arranged in a continuous sequence, so that a whole journey is identified and made comfortable by a familiar succession of detail. The confusing streets of Venice become traversible after one or two experiences, since they are rich in distinctive details, which are soon sequentially organized. Less usually, landmarks may be grouped together in patterns, which in themselves have form, and may indicate by their appearance the direction from which they are viewed. The Florentine landmark pair of dome and campanile dance about each other in this way.

The nodes are the conceptual anchor points in our cities. Rarely in the United States, however, do they have a form adequate to support this attention, other than a certain concentration of activity.

The first prerequisite for such perceptual support is the achievement of identity by the singular and continuous quality of the walls, floor, detail, lighting, vegetation, topography, or skyline of the node. The essence of this type of element is that it be a distinct, unforgettable *place,* not to be confused with any other. Intensity of use strengthens this identity, of course, and sometimes the very intensity of use creates visual shapes which are distinctive, as in Times Square. But our shopping centers and transport breaks which lack this visual character are legion.

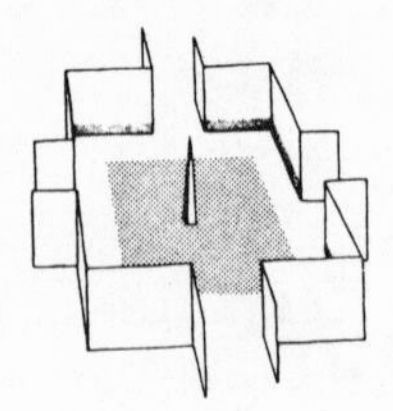

The node is more defined if it has a sharp, closed boundary, and does not trail off uncertainly on every side; more remarkable, if provided with one or two objects which are foci of attention. But if it can have coherent spatial form, it will be irresistible. This is the classic concept of forming static outdoor spaces, and there are many techniques for the expression and definition of such a space, transparencies, overlappings, light modulation, perspective, surface gradients, closure, articulation, patterns of motion and sound.

If a break in transportation or a decision point on a path can be made to coincide with the node, the node will receive even more attention. The joint between path and node must be visible and expressive, as it is in the case of intersecting paths. The traveler must see how he enters the node, where the break occurs, and how he goes outward.

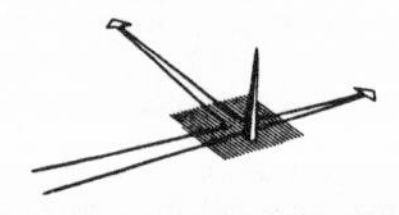

These condensation points can, by radiation, organize large districts around themselves if their presence is somehow signalized in the surroundings. A gradient of use or other characteristic may lead up to the node, or its space may occasionally be visible from outside; or it may contain high landmarks. The city of Florence focuses in this manner around its Duomo and Palazzo Vecchio, both standing in major nodes. The node may emit characteristic light or sound, or its presence be hinted at by symbolic detail in the hinterland, which echoes some quality of the node itself. Sycamores in a district might reveal the proximity of a square noted for a heavy plantation of these trees, or cobblestone pavements lead up to a cobbled enclosure.

If the node has a local orientation within itself – an "up" or "down," a "left" or a "right," a "front" or a "back" – then it can be related to the larger orientation system. When recognized paths enter in a clear joint, the tie can also be made. In either case, the observer feels the presence of the city structure around him. He knows in what direction to move outward to reach a goal, and the particularity of the place itself is enhanced by the felt contrast with the total image.

It is possible to arrange a series of nodes to form a related structure. They can be linked together by close juxtaposition or by allowing them to be intervisible, as are the Piazze S. Marco and SS. Annunziata in Florence. They may be put in some common relation to a path or edge, joined by a short linking element, or related by an echo of some characteristic from one to the other. Such linkages can structure substantial city regions.

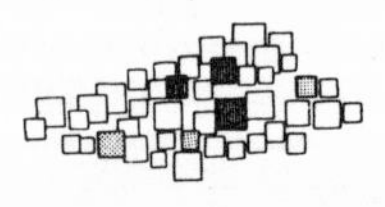

A city district in its simplest sense is an area of homogeneous character, recognized by clues which are continuous throughout the district and discontinuous elsewhere. The homogeneity may be of spatial characteristics, like the narrow sloping streets of Beacon Hill; of building type, like the swell-front row houses of the South End; of style or of topography. It may be a typical building feature, like the white stoops of Baltimore. It may be a continuity of color, texture, or material, of floor surface, scale or facade detail, lighting, planting, or silhouette. The more these characters overlap, the stronger the impression of a unified region. It appears that a "thematic unit" of three or four such characters is particularly useful in delimiting an area. Persons interviewed usually held together in their minds a small cluster of such characters: such as the narrow sloping streets, brick pavements, small-scale row houses, and recessed doorways of Beacon Hill. Several such characters can be held fixed in a district, while other factors are varied as desired.

Where physical homogeneity coincides with use and status, the effect is unmistakable. The visual character of Beacon Hill is directly reinforced by its status as upper-class residence. The more usual American case is the reverse: use character receives little support from visual character.

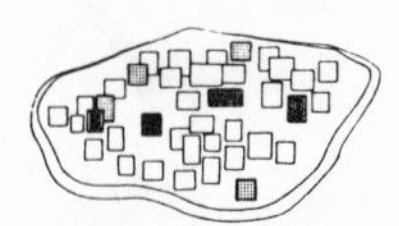

A district is further sharpened by the definiteness and closure of its boundary. A Boston housing project on Columbia Point has an island-like character which may be undesirable socially but is perceptually quite clear. Any small island, in fact, has a charming particularity for this reason. And if the region is easily visible as a whole, as by high or panoramic views, or by the convexity or concavity of its site, then its separateness is sealed.

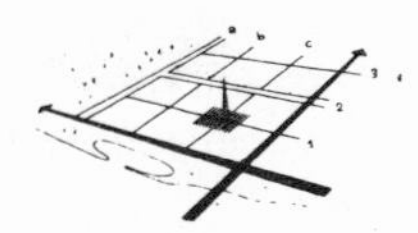

The district may be structured within itself as well. There may be subdistricts, internally differentiated while conforming to the whole; nodes which radiate structure by gradients or other hints; patterns of internal paths. The Back Bay is structured by its network of alphabetized paths, and usually appears clearly, unmistakably, and somewhat enlarged on most sketch maps. A structured region is likely to be a more vivid image. Furthermore, it tells its inhabitants not simply "you are somewhere in X," but "you are in X, near Y."

When suitably differentiated within, a district can express connections with other city features. The

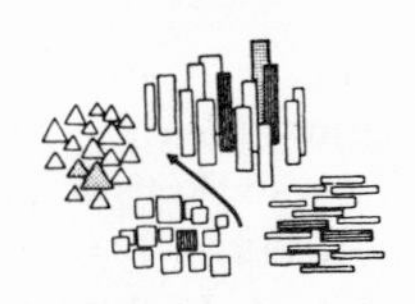

boundary must now be penetrable: a seam, not a barrier. District may join to district, by juxtaposition, intervisibility, relation to a line, or by some link such as a mediating node, path, or small district. Beacon Hill is linked to the metropolitan core by the spatial region of the Common, and therein lies much of its attraction. Such links heighten the character of each district, and bring together great urban areas.

It is conceivable that we might have a region which is not simply characterized by homogeneous spatial quality, but is in fact a true spatial region, a structured continuum of spatial form. In a primitive sense, such large urban spaces as river openings are of this nature. A spatial region would be distinguished from a spatial node (a square) because it could not be scanned quickly. It could only be experienced, as a patterned play of spatial changes, by a rather protracted journey through it. Perhaps the processional courts of Peking or the canal spaces of Amsterdam have this quality. Presumably they evoke an image of great power.

FORM QUALITIES

These clues for urban design can be summarized in another way, since there are common themes that run through the whole set: the repeated references to certain general physical characteristics. These are the categories of direct interest in design, since they describe qualities that a designer may operate upon. They might be summarized as follows:

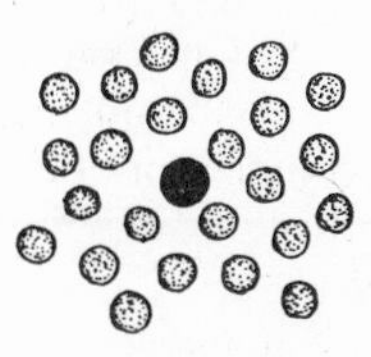

1. *Singularity* or figure-background clarity: sharpness of boundary (as an abrupt cessation of city

development); closure (as an enclosed square); contrast of surface, form, intensity, complexity, size, use, spatial location (as a single tower, a rich decoration, a glaring sign). The contrast may be to the immediate visible surroundings, or to the observer's experience. These are the qualities that identify an element, make it remarkable, noticeable, vivid, recognizable. Observers, as their familiarity increases, seem to depend less and less on gross physical continuities to organize the whole, and to delight more and more in contrast and uniqueness which vivify the scene.

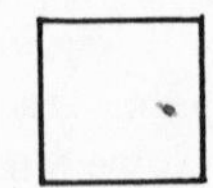

2. *Form simplicity:* clarity and simplicity of visible form in the geometrical sense, limitation of parts (as the clarity of a grid system, a rectangle, a dome). Forms of this nature are much more easily incorporated in the image, and there is evidence that observers will distort complex facts to simple forms, even at some perceptual and practical cost. When an element is not simultaneously visible as a whole, its shape may be a topological distortion of a simple form and yet be quite understandable.

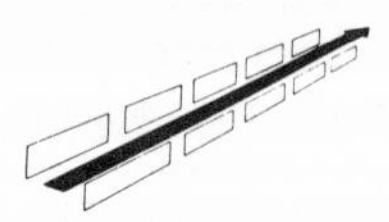

3. *Continuity:* continuance of edge or surface (as in a street channel, skyline, or setback); nearness of parts (as a cluster of buildings); repetition of rhythmic interval (as a street-corner pattern); similarity, analogy, or harmony of surface, form, or use (as in a common building material, repetitive pattern of bay windows, similarity of market activity, use of common signs). These are the qualities that facilitate the perception of a complex physical reality as one or as interrelated, the qualities which suggest the bestowing of single identity.

4. *Dominance:* dominance of one part over others by means of size, intensity, or interest, resulting in the reading of the whole as a principal feature with an associated cluster (as in the "Harvard Square area"). This quality, like continuity, allows the necessary simplification of the image by omission and sub-

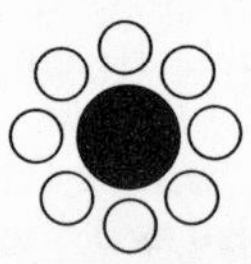

sumption. Physical characteristics, to the extent that they are over the threshold of attention at all, seem to radiate their image conceptually to some degree, spreading out from a center.

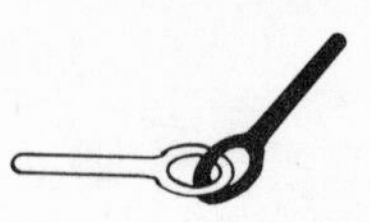

5. *Clarity of joint:* high visibility of joints and seams (as at a major intersection, or on a seafront); clear relation and interconnection (as of a building to its site, or of a subway station to the street above). These joints are the strategic moments of structure and should be highly perceptible.

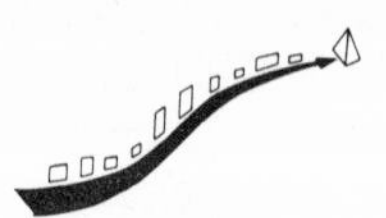

6. *Directional differentiation:* asymmetries, gradients, and radial references which differentiate one end from another (as on a path going uphill, away from the sea, and toward the center); or one side from another (as with buildings fronting a park); or one compass direction from another (as by the sunlight, or by the width of north-south avenues). These qualities are heavily used in structuring on the larger scale.

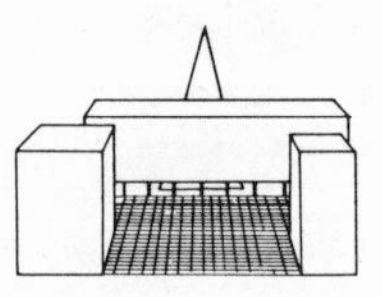

7. *Visual scope:* qualities which increase the range and penetration of vision, either actually or sym-

bolically. These include transparencies (as with glass or buildings on stilts); overlaps (as when structures appear behind others); vistas and panoramas which increase the depth of vision (as on axial streets, broad open spaces, high views); articulating elements (foci, measuring rods, penetrating objects) which visually explain a space; concavity (as of a background hill or curving street) which exposes farther objects to view; clues which speak of an element otherwise invisible (as the sight of activity which is characteristic of a region to come, or the use of characteristic detail to hint at the proximity of another element). All these related qualities facilitate the grasping of a vast and complex whole by increasing, as it were, the efficiency of vision: its range, penetration, and resolving power.

8. *Motion awareness:* the qualities which make sensible to the observer, through both the visual and the kinesthetic senses, his own actual or potential motion. Such are the devices which improve the clarity of slopes, curves, and interpenetrations; give the experience of motion parallax and perspective; maintain the consistency of direction or direction change; or make visible the distance interval. Since a city is sensed in motion, these qualities are fundamental, and they are used to structure and even to identify, wherever they are coherent enough to make it possible (as "go left, then right," "at the sharp bend," or "three blocks along this street"). These qualities reinforce and develop what an observer can do to interpret direction or distance, and to sense form in motion itself. With increasing speed, these techniques will need further development in the modern city.

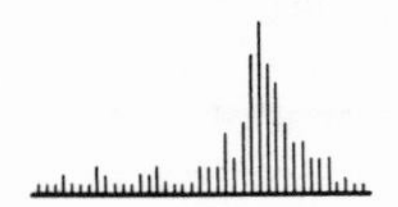

9. *Time series:* series which are sensed over time, including both simple item-by-item linkages, where one element is simply knitted to the two elements before and behind it (as in a casual sequence of detailed landmarks), and also series which are truly structured in time and thus melodic in nature (as if the landmarks would increase in intensity of form until a climax point were reached). The former (simple sequence) is very commonly used, particularly along familiar paths. Its melodic counterpart is more rarely seen, but may be most important to develop in the large, dynamic, modern metropolis. Here what would be imaged would be the developing pattern of elements, rather than the elements themselves — just as we remember melodies, not notes. In a complex environment, it might even be possible to use contrapuntal techniques: moving patterns of opposing melodies or rhythms. These are sophisticated methods and must be consciously developed. We need fresh thought on the theory of forms which are perceived as a continuity over time, as well as on design archetypes which exhibit a melodic sequence of image elements or a formed succession of space, texture, motion, light, or silhouette.

10. *Names and meanings:* nonphysical characteristics which may enhance the imageability of an element. Names, for example, are important in crystallizing identity. They occasionally give locational clues (North Station). Naming systems (as in the alphabetizing of a street series) will also facilitate the structuring of elements. Meanings and associations, whether social, historical, functional, economic, or individual, constitute an entire realm lying beyond the physical qualities we deal with here. They strongly reinforce such suggestions toward identity or structure as may be latent in the physical form itself.

All the above-mentioned qualities do not work in isolation. Where one quality is present along (as a continuity of building material with no other common feature), or the qualities are in conflict (as in two areas of common building type but of different function), the total effect may be weak or require effort to identify and structure. A certain amount of repetition, redundancy, and reinforcement seems to be necessary. Thus a region would be unmistakable which had a simple form, a continuity of building type and use, which was singular in the city, sharply bounded, clearly jointed to a neighboring region, and visually concave.

THE SENSE OF THE WHOLE

In discussing design by element types, there is a tendency to skim over the interrelation of the parts into a whole. In such a whole, paths would expose and prepare for the district, and link together the various nodes. The nodes would joint and mark off the paths, while the edges would bound off the districts, and the landmarks would indicate their cores. It is the total orchestration of these units which would knit together a dense and vivid image, and sustain it over areas of metropolitan scale.

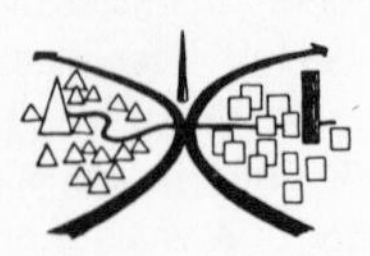

The five elements – path, edge, district, node, and landmark – must be considered simply as convenient empirical categories, within and around which it has been possible to group a mass of information. To the extent that they are useful, they will act as building blocks for the designer. Having mastered their characteristics, he will have the task of organizing a whole which will be sensed sequentially, whose parts will be perceived only in context. Were he to arrange a sequence of ten landmarks along a path, one of these marks would have an utterly different image quality than if it were placed singly and prominently at the city core.

Forms should be manipulated so that there is a strand of continuity between the multiple images of a big city: day and night, winter and summer, near and far, static and moving, attentive and absent-minded. Major landmarks, regions, nodes, or paths should be recognizable under diverse conditions, and yet in a concrete, rather than an abstract way. This is not to say that the image should be the same in each case. But if Louisburg Square in the snow has a shape that matches Louisburg Square in midsummer, or if the State House dome by night shines in a way that recalls that dome seen in the day, then the contrasting quality of each image becomes even more sharply savored because of the common tie. One is now able to hold together two quite different city views, and thus to encompass the scale of the city in a way otherwise impossible: to approach the ideal of an image which is a total field.

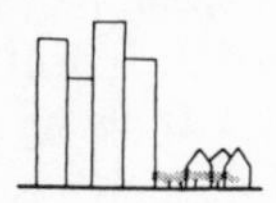

While the complexity of the modern city calls for continuity, it also furnishes a great delight: the contrast and specialization of individual character. Our study hints at an increasing attention to detail and to uniqueness of character, as familiarity develops. Vividness of elemeents and their precise tuning to functional and symbolic differences will help to provide this character. Contrast will be heightened if sharply differentiated elements are brought into close and imageable relation. Each element then takes on an intensified character of its own.

Indeed, the function of a good visual environment may not be simply to facilitate routine trips, nor to support meanings and feelings already possessed. Quite as important may be its role as a guide and a stimulus for new exploration. In a complex society, there are many interrelations to be mastered. In a democracy, we deplore isolation, extol individual development, hope for ever-widening communication between groups. If an environment has a strong visible framework and highly characteristic parts, exploration of new sectors is both easier and more inviting. If strategic links in communication (such as museums or libraries or meeting places) are clearly set forth, those who might otherwise neglect them may be tempted to enter.

The underlying topography, the pre-existing natural setting, is perhaps not quite as important a factor in imageability as it once used to be. The density, and particularly the extent and elaborate technology of the modern metropolis, all tend to obscure it. The contemporary urban area has man-made characteristics and problems that often override the specificity of site. Or rather, it would be more accurate to say that the specific character of a site is now perhaps as much the result of human action and desires as of the original geological structure. In addition, as the city expands, the significant "natural" factors become the larger, more fundamental ones, rather than the smaller accidents. The basic climate, the general flora and surface of a large region, the mountains and major river systems, take precedence over local features. Nevertheless, topography is still an important element in reinforcing the strength of urban elements: sharp hills can define regions, rivers and strands make strong edges, nodes can be confirmed by location at key points of terrain. The modern high-speed path is an excellent viewpoint from which to grasp topographic structure at an extensive scale.

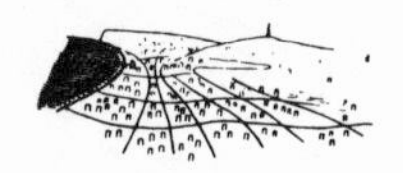

The city is not built for one person, but for great numbers of people of widely varying backgrounds, temperaments, occupations, and class. Our analyses indicate a substantial variation in the way different people organize their city, in what elements they most depend on, or in what form qualities are most congenial to them. The designer must therefore create a city which is as richly provided with paths, edges, landmarks, nodes, and districts as possible, a city which makes use of not just one or two form qualities, but of all of them. If so, different observers will all find perceptual material which is congenial to their own particular way of looking at the world. While one man may recognize a street by its brick pavement, another will remember its sweeping curve, and a third will have located the minor landmarks along its length.

There are, moreover, dangers in a highly specialized visible form; there is a need for a certain plasticity in the perceptual environment. If there is only one dominant path to a destination, a few sacred focal points, or an ironclad set of rigidly separated regions, there is only one way to image the city without considerable strain. This one may suit neither the needs of all people, nor even the needs of one person as they vary from time to time. An unusual trip becomes awkward or dangerous; interpersonal relations may tend to compartment themselves; the scene becomes monotonous or restrictive.

We have taken as signs of good organization those parts of Boston in which the paths chosen by interviewees seemed to spread out rather freely. Here, presumably, the citizen is presented with a rich choice of routes to his destination, all of them well structured and identified. There is a similar value in an overlapping net of identifiable edges, so that regions big or small can be formed according to taste and need. Nodal organization gains its identity from the central focus and can fluctuate at the rim. Thus it has an advantage of flexibility over boundary organization, which breaks down if the shape of regions must change. It is important to maintain some great common forms: strong nodes, key paths, or widespread regional homogeneities. But within this large framework, there should be a certain plasticity, a richness of possible structures and clues, so that the individual observer can construct his own image: communicable, safe, and sufficient, but also supple and integrated with his own needs.

The citizen shifts his place of residence more frequently today than ever before, from area to area, from city to city. Good imageability in his environment would allow him to feel quickly at home in new surroundings. Gradual organization through long experience can less and less be relied upon. The city environment is itself changing rapidly, as techniques and functions shift. These changes are often disturbing to the citizen emotionally, and tend to disorganize his perceptual image. The techniques of design discussed here may prove useful in maintaining a visible structure and a sense of continuity even while massive changes are occurring. Certain landmarks or nodes might be retained, thematic units of district character carried over into new construction, paths salvaged or temporarily conserved.

METROPOLITAN FORM

The increasing size of our metropolitan areas and the speed with which we traverse them raise many new problems for perception. The metropolitan region is now the functional unit of our environment, and it is desirable that this functional unit should be identified and structured by its inhabitants. The new means of communication which allow us to live and work in such a large interdependent region could also allow us to make our images commensurate with our experiences. Such jumps to new levels of attention have occurred in the past, as jumps were made in the functional organization of life.

Total imageability of an extensive area such as a metropolitan region would not mean an equal intensity of image at every point. There would be dominant figures and more extensive backgrounds, focal points, and connective tissue. But, whether intense or neutral, each part would presumably be clear and clearly linked to the whole. We can speculate that metropolitan images could be formed of such elements as high-speed highways, transit lines or airways; large regions with coarse edges of water or open space; major shopping nodes; basic topographic features; perhaps massive, distant landmarks.

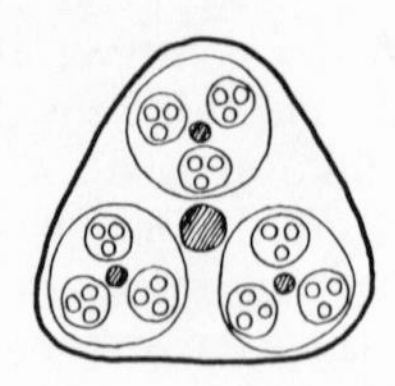

The problem is none the less difficult, however, when it comes to composing a pattern for such an entire area. There are two techniques with which we are familiar. First, the entire region may be composed as a static hierarchy. For example, it might be organized as a major district containing three subdistricts, which each contain three sub-subdistricts, and so on. Or as another example of hierarchy, any given part of the region might focus on a minor node, these minor nodes being satellite to a major node, while all the major nodes are arranged to culminate in a single primary node for the region.

The second technique is the use of one or two very large dominant elements, to which many smaller things may be related: the siting of settlement along a seacoast, for example or the design of a linear town depending on a basic communication spine. A large environment might even be radially related to a very powerful landmark, such as a central hill.

Both these techniques seem somewhat inadequate to the metropolitan problem. The hierarchical system, while congenial to some of our habits of abstract thinking, would seem to be a denial of the freedom and complexity of linkages in a metropolis. Every connection must be made in a roundabout, conceptual fashion, up to a generality and back to a particular, even though the bridging generality may have little to do with the real connection. It is the unity of a library, and libraries require the constant use of a bulky cross-referencing system.

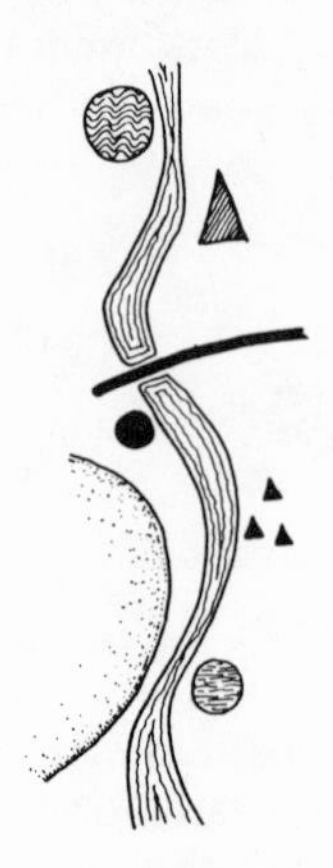

Dependence on a strong dominant element, while giving a much more immediate sense of relation and continuity, becomes more difficult as the environment increases in size, since a dominant must be found that is big enough to be in scale with its task, and has enough "surface area" so that all the minor elements can have some reasonably close relation to it. Thus one needs a big river, for example, that winds enough to allow all settlement to be fairly near its course.

Nevertheless, these are two possible methods, and it would be useful to investigate their success in unifying large environments. Air travel may simplify the problem again, since it is (in perceptual terms) a static rather than a dynamic experience, an opportunity to see a metropolitan area almost at a glance.

Considering our present way of experiencing a large urban area, however, one is drawn toward another kind of organization: that of sequence or temporal pattern. This is a familiar idea in music, drama, literature, or dance. Therefore, it is relatively easy to conceive of and study the form of a sequence of events along a line, such as the succession of elements that might greet a traveler on an urban highway. With some attention and proper tools this experience could be made meaningful and well shaped.

It is also possible to handle the question of reversibility, that is, the fact that most paths are traversed in both directions. The series of elements must have sequential form taken in either order, which might be accomplished by symmetry about the midpoint or in more sophisticated ways. But the city problem continues to raise difficulties. Sequences are not only reversible, but are broken in upon at many points. A carefully constructed sequence, leading from introduction, first statement, and development to climax and conclusion, may fail utterly if a driver enters it directly at the climax point. Therefore, it may be necessary to look for sequences which are interruptible as well as reversible, that is, sequences which still have sufficient imageability even when broken in upon at various points, much like a magazine serial. This might lead us from the classic start-climax-finish form to others which are more like the essentially endless, and yet continuous and variegated, patterns of jazz.

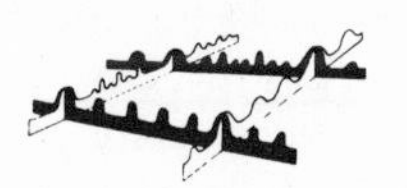

These considerations refer to organization along a single line of movement. An urban region might then be organized by a network of such organized sequences, any proposed form being tested to see if each major path, in each direction and from each entry point, was possessed of a formed sequence of elements. This is conceivable when the paths have some simple pattern such as radial convergence. It becomes more difficult to image where the network is a diffuse and intersecting one, as in a gridiron. Here the sequences work in four different directions throughout the map. Although on a much more sophisticated scale, this is akin to the problem of timing a progressive traffic-light system over a network.

It is even conceivable that one might compose in counterpoint along these lines, or from one line to another. One sequence of elements, or "melody," might be played against a countersequence. Perhaps, however, such techniques would wait upon a time when there is a more attentive and critical audience.

Even this dynamic method, the organization of a network of formed sequences, does not yet seem ideal. The environment is still not being treated as a whole but rather as a collection of parts (the sequences) arranged so as not to interfere with each other. Intuitively, one could imagine that there might be a way of creating a *whole* pattern, a pattern that would only gradually be sensed and developed by sequential experiences, reversed and interrupted as they might be. Although felt as a whole, it would not need to be a highly unified pattern with a single center or an isolating boundary. The principal quality would be sequential continuity in which each part flows from the next -- a sense of interconnectedness at any level or in any direction. There would be particular zones that for any one individual might be more intensely felt or organized, but the region would be continuous, mentally traversable in any order. This possibility is a highly speculative one: no satisfactory concrete examples come to mind.

Perhaps this pattern of a whole cannot exist. In that case, the previously mentioned techniques remain as possibilities in the organization of large regions: the hierarchy, the dominant element, or the network of sequences. Hopefully, these techniques would require no more than the metropolitan planning controls now sought for other reasons, but this also remains to be seen.

THE PROCESS OF DESIGN

Any existing, functioning urban area has structure and identity, even if only in weak measure. Jersey City is a long step upward from pure chaos. If it were not, it would be uninhabitable. Almost always, a potentially powerful image is hidden in the situation itself, as in the Palisades of Jersey City, its peninsular shape, and its relation to Manhattan. A frequent problem is the sensitive reshaping of an already existing environment: discovering and preserving its strong images, solving its perceptual difficulties, and, above all, drawing out the structure and identity latent in the confusion.

At other times, the designer faces the creation of a

new image, as when extensive redevelopment is underway. This problem is particularly significant in the suburban extensions of our metropolitan regions, where vast stretches of what is essentially a new landscape must be perceptually organized. The natural features are no longer a sufficient guide to structure, because of the intensity and scale of the development being applied to them. At the present tempo of building, there is no time for the slow adjustment of form to small, individualized forces. Therefore, we must depend far more than formerly on conscious design: the deliberate manipulation of the world for sensuous ends. Although possessed of a rich background of former examples of urban design, the operation must now proceed at an entirely different scale of space and time.

These shapings or reshapings should be guided by what might be called a "visual plan" for the city or metropolitan region: a set of recommendations and controls which would be concerned with visual form on the urban scale. The preparation of such a plan might begin with an analysis of the existing form and public image of the area, using the techniques rising out of this study.... This analysis would conclude with a series of diagrams and reports illustrating the significant public images, the basic visual problems and opportunities, and the critical image elements and element interrelations, with their detailed qualities and possibilities for change.

Using this analytical background, but not limited thereby, the designer could proceed to develop a visual plan at the city scale, whose object would be to strengthen the public image. It might prescribe the location or preservation of landmarks, the development of a visual hierarchy of paths, the establishment of thematic units for districts, or the creation or clarification of nodal points. Above all, it would deal with the interrelations of elements, with their perception in motion, and with the conception of the city as a total visible form.

Substantial physical change may not be justified on this aesthetic score alone, except at strategic points. But the visual plan could influence the form of physical changes which occur for other reasons. Such a plan should be fitted into all the other aspects of planning for the region, to become a normal and integral part of the comprehensive plan. Like all the other parts of this plan, it would be in a continuous state of revision and development.

The controls employed to achieve visual form at the city scale could range from general zoning provisions, advisory review, and persuasive influence over private design, to strict controls at critical points and to the positive design of public facilities such as highways or civic buildings. Such techniques are not in principle very different from controls used in the pursuit of other planning objectives. It will probably be more difficult to gain an understanding of the problem and to develop the necessary design skill than it will be to obtain the necessary powers, once the objective is clear. There is much to be done before far-reaching controls are justified.

The final objective of such a plan is not the physical shape itself but the quality of an image in the mind. Thus it will be equally useful to improve this image by training the observer, by teaching him to *look* at his city, to observe its manifold forms and how they mesh with one another. Citizens could be taken into the street, classes could be held in the schools and universities, the city could be made an animated museum of our society and its hopes. Such education might be used not only to develop the city image, but to reorient after some disturbing change. An art of city design will wait upon an informed and critical audience. Education and physical form are parts of a continuous process.

Heightening the observer's attention, enriching his experience, is one of the values that the mere effort to give form can offer. To some degree, the very process of reshaping a city to improve its imageability may itself sharpen the image, regardless of how unskillful the resulting physical form may be. Thus the amateur painter begins to see the world around him; the novice decorator begins to take pride in her living room and to judge others. Although such a process can become sterile if not accompanied by increasing control and judgment, even awkward "beautification" of a city may in itself be an intensifier of civic energy and cohesion.

8

Eliel Saarinen

PRINCIPLES OF TOWN-BUILDING

In the search for principles, we should not be satisfied with such principles as have only local and limited bearing, or are mere man-made doctrines and theories. Our endeavor must be to discern those principles which are inherent in the nature of things from time immemorial and are valid in any circumstance. Therefore, to approach our problem from the right angle, it is important to go down to the mother of things, to nature, so that we may find there such processes as can be considered analogous to the process of townbuilding. Now, the process of townbuilding – by means of town design – must be to bring organic order into the urban communities, and to keep this organic order continuously vital during the growth of these communities. Fundamentally, also, this process is analogous to the growth of any living organism in nature, and, inasmuch as there is no difference of underlying principles between one living organism and any other, we would do well to study these principles in organic life in general. Analogously, we must learn to understand these general

---

Selected from The City: Its Growth, Its Decay, Its Future *by Eliel Saarinen, Reinhold Publishing Corporation, New York (1943), 8-20.*

organic processes so that we may be able to maintain order in the city's physical organism. For sure, we must learn a lesson of basic consequence.

Therefore, let's for a moment dwell in the realm of nature and see what the microscope is able to reveal. Not much manipulation, however, is needed with the microscope to discern in organic life two phenomena; the existence of individual cells, and the correlation of these cells into cellular tissue. In itself, this relevation might seem an insignificant matter, yet it is amazing to learn that the whole universe, from the most microscopic to the utmost macroscopic, is constituted along this dual thought of individuals as such and of the correlation of these individuals into the whole. Furthermore, one learns that vitality in all life manifestation depends, first, on the quality of the individual and, second, on the quality of correlation. Consequently, there must exist two fundamental principles according to which these two mentioned qualities are constituted so as to foster and maintain vitality in the course of things. In fact, by a closer study of natural processes we will perceive two fundamental principles, *Expression* and *Correlation*, of which the former principle brings individual formshapings into true expression of the meaning behind these forms, and of which the latter brings the individual forms into organic correlation.

THE PRINCIPLE OF EXPRESSION

Supposing one could follow the seed's gradual growth into a tree, and that this could be done every moment in every part of the organism. One then would observe billions and again billions of ever-new patterns of cell tissue, all emerging from one another according to inherent potentiality, all tuned in a key that is characteristic of the species concerned, and all organized into symbolic design representing that very species. Furthermore, supposing one were to follow the processes along two different lines – say, along the growths of an elm and an oak, respectively – one then would discern two different worlds of form pattern, respectively characterizing the two tree species, and running throughout the two modes of growth, in the inner cell structures as well as in the design of trunks, branches, and foliages. In other words, one would discover two different rhythm worlds; one of the elm and only of the elm, the other of the oak and only of the oak. And so evident is the difference of these two rhythm worlds that already at a distance one is able to differentiate between elms and oaks, because of expressive rhythm characters.

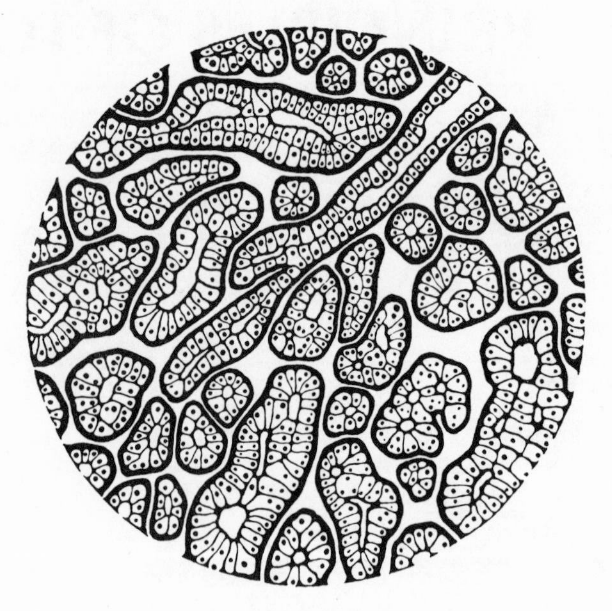

Figure 1
Healthy cell tissue: microscopic "community planning."

It is fully obvious that these expressive rhythm characters are innate potentialities already in the respective seeds of elms and oaks, and that from these latter the said potentialities are conveyed to every single cell throughout the respective organisms. In this manner, every single cell throughout these organisms becomes an expressive exponent of the species it belongs to. Speaking in general terms, this is a statement which does not concern only elms and oaks or any other species of organic life, but nature's form manifestations at large. That is, any form manifestation in nature is a true expression of the meaning behind this form-manifestation; and this is a rule from which there is no exception in all the universe. It is a principle: the said principle of expression.

Man belongs to the realm of creation and is subject

to the principle of expression. Therefore, whatever forms man brings forth through his endeavor and work, if honest, must be true expressions of his life, emotions, thoughts, and aspirations. Man's art, at best, is a significant testimony in this respect, for, by studying the various form worlds of the great cultural epochs of the past, we do find that each of these great epochs had its own form characteristics through which the best characteristics of its time and people came into expression. This expressiveness, however, did not concern only these great epochs as such, but it concerned even each individual form detail that was created within the framework of these epochs. Accordingly, even the most inconspicuous objects – the "cells" so to speak – can tell their true story through the expressiveness of their shapes. Take for instance the story of that humble chair. During thousands of years there have been made millions and more millions of chairs, yet you may select one of these quite at random – except for those that stupid and pernicious imitation has brought forth – and you can trace its origin to time and people. The truer this chair expresses the best of its time and people, the more it possesses those qualities that could further the growth of that cultural tree of its epoch. And the more there were of those forms which possessed similar expressive qualities, the stronger the expressive formation of that cultural tree was able to grow.

What is true about chairs and other minor objects must be so much the more true about more important features such as, for example, individual buildings. Now, individual buildings are just those "cells" which constitute the major material through which towns and cities are built. Therefore, bringing now our original question as to the quality of the individual – in this case, the individual building – to an issue, the answer is close at hand: in urban communities the quality of the individual building, no matter how humble, must be a true expression of the best of its people and time. This is an answer which is based on the principle of expression. The demand of this principle is a directive from which there can be no exemption if the aim is to achieve positive results. For, in case the principle of expression is disregarded – as it so often is by those who erect all kinds of trashy buildings – the consequences in the city's case are bound to be as devitalizing as they would be in the case of the tree, if false cells were brought into its cellular tissue.

THE PRINCIPLE OF CORRELATION

Turning now our attention to the matter of correlation as it appears in nature's manifestations, we will find an enlightening example in the landscape, when looking through microanalytical lenses.

There are myriads of cells which capriciously, though in reciprocal action, shape the tree into a manifest species. Thousands of these species, because of an enigmatic leaning to coherence, are formed into the concerted unity of the forest. There are myriads of molecules which, although of millions of varieties, constitute, in mutual cooperation, mountains, cliffs, valleys and lakes. There are myriads of particles which independent of one another, yet in instinctive unison, form congruous mass effects of clouds. There are myriads of molecules which individually, yet in constant adherence, break the rays of the dawning sun into light and shadow and into the brilliancy of a rich variety of color. All these myriads of molecular particles in trees, mountains, cliffs, lakes, and skies – and in countless other things – are brought into a single picture of rhythmic order: into the landscape.

Now, is there anything supernatural in all this?

Certainly not. It is just as natural a thing as it could be, because we are used to such manifestations, and because we do not seem to harbor for a single moment the thought that, if nature in her actions were not governed by the principle of correlation, cosmos would dissolve into chaos. This kind of harmonious condition in nature has come to us as a heavenly gift in such a direct manner that we scarcely have felt our own obligation to keep our environment in order in a like harmonious manner. But we must learn to understand our obligations in this respect. Particularly in the building of cities, we must learn to understand that it would prove just as disastrous to the city, if the principle of correlation did not exist, as it would prove to the landscape – to the "city of nature" – if the same principle should cease to function.

Many are scarcely conscious of the significant meaning of correlation, although subconsciously they might act in accordance with its command. Take, for example, just such a little detail from everyday life as the arranging of a few flowers in one's room. One might pick a number of these flowers in order to have a greater effect, yet automatically one assorts the flowers into a rhythmic bouquet and into a pleasing correlation with the miscellaneous features of the room. How well this is done depends on one's subconscious sensitiveness to the principle of correlation.

In cases where sensitiveness to the principle of correlation is alert, the correlative effects are carried throughout the whole field of action, from the room to streets and plazas and to the towns and cities at large. Such was the case in olden times when the creative impetus was still strong and principles were in command. Whoever, if he were sensitive, has traveled in those countries where towns were built in ancient times has no doubt felt that correlation is not an empty esthetic theory. He has felt that any group of buildings, in the country or in the town, was conceived as a rhythmic ensemble of forms of man and forms of nature. He has felt that any street or plaza was made a correlative product of many artists in accord. He has felt that any town, with all its various and varying units, was made an organism of masses and proportions where the rhythmic characteristics of building groups and skyline sprang from the characteristics of the time and the people. He has felt that the proper correlation of forms was consistently carried through the whole organism, beginning with the minor things in the rooms and residences and ending with the highest pinnacles of towers and turrets. He has felt that all this happened "once upon a time."

All this happened until man thought he was intelligent enough to get along with his practical reasoning only. Proud of his expedient practicality, he accumulated forms upon forms without proper correlation of forms. And his towns and cities became often a heterogeneous accumulation of forms. The principle of correlation was entirely forgotten.

THE PRINCIPLE OF ORGANIC ORDER

When speaking about expressive and correlative tendencies in nature's form shaping, it is obvious that these tendencies are not independent trends, but two phases of the same process: the process toward *organic order.* In other words, the principles of expression and correlation are not independently functioning principles, but rather daughter principles of that all-governing mother principle of organic order — which, in fact, is the very principle of architecture, nature's manifestations are carried out and kept in function. As long as this is the case, and

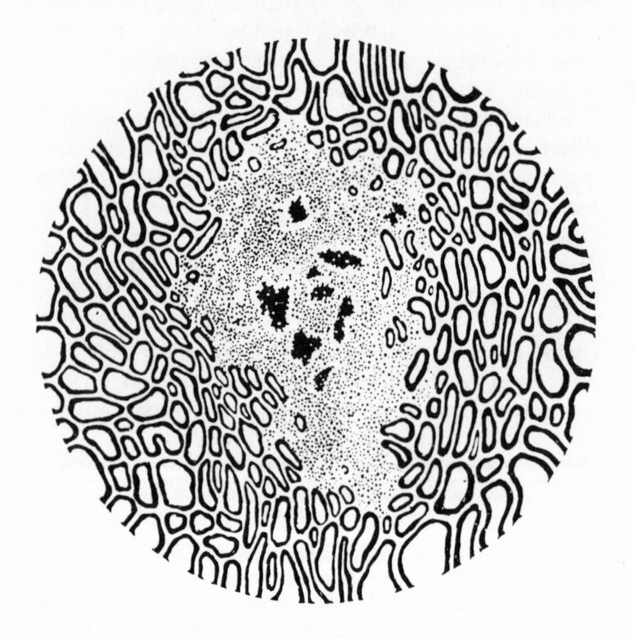

Figure 2
Disintegrating cell tissue: microscopic "slum growth."

the expressive and correlative faculties are potent enough to maintain organic order, there is life and progress of life. Again, as soon as this ceases to be the case and the expressive and correlative faculties are impotent to prevent disintegration of organic order, decline and death enters. This is true, no matter whether it happens in the microscopic tissues of cell structure where cancer causes disintegration, or in the hearts of the large cities of today where compactness and confusion cause slums to spread.

These facts are open to all eyes and minds to see and to comprehend. Only a short glance through the microscope is enough to convince that organic cell tissue means life and health, whereas disintegration of cell tissue means sickness and death. Really, our normal senses, if not actual reasoning, have accustomed us to this simple truth. We understand organic life in nature as a manifestation of rhythmic order, and whenever we find exceptions from that order – in plants, in trees, in animal bodies, and in man – we realize that sickness has crept into the organism. And we are conscious of the fact that this sickness is contagious and tends to spread itself, if preventive provisions have not been made.

In spite of this simple truth – open to all eyes and comprehensible to all minds – its consequences, insofar as civic development is concerned, are astonishingly often disregarded. Not only are slums spreading themselves because of previous thoughtless planning, but even new cities, towns, suburbs, and satellite communities are planned in the same slum-breeding manner. This kind of procedure brings conditions from bad to worse, for, instead of decreasing the evils of poor planning, continuous poor planning increases the evils. It is not too early, therefore, to take these matters seriously into consideration in order to learn what to do, how to do, and how not to do. With regard to this, the understanding of fundamental principles as they have been described is of supreme importance.

In the light of these fundamental principles one can follow the development of towns and cities during the past, and one is able to understand the basic reasons why these developments were successful or unsuccessful. Successful results sprang from an instinctive awareness of these principles, whereas unsuccessful outcomes were due to a lack of this sense. For this reason, when the question arises as to why the towns and cities of today are so frequently heterogeneous and discordant, the answer can be found in the fact that these principles have been forgotten or unperceived. Again, when it is asked how the growth of towns and cities can be conducted toward greater expressiveness and better unity, the answer is, and must be, that this can be achieved only when the town designers – and all the cooperating designers and architects for that matter – have absorbed the meaning of these principles into the very blood of their veins.

THE ARCHITECTURAL MOMENT

Because the principle of organic order is the underlying law of nature's architecture, the same principle must be recognized as the underlying law of man's architecture as well. Such is the case when the art of building is correctly understood. Architecture is not – as many have believed during the long period of its gradual decline – a stylistic decoration which can be arbitrarily pasted on the surface of a structure. Architecture must be definitely understood as an organic and social art form with the mission of creating about man a culturally healthy atmosphere by means of proportion, rhythm, material, and color. As such, architecture embraces the whole form world of man's physical accommodations, from the intimacy of his room to the comprehensive labyrinth of the large metropolis. Within this broad field of creative activities, the architect's ambition must be to develop a form language expressing the best aims of his time – and of no other time – and to cement the various features of his expressive forms into a good interrelation, and ultimately into the rhythmic coherence of the multiformed organism of the city.

When the architects are corporately sincere in their profession and are acting in accordance with the thoughts outlined, then – and only then – can town and city be made a healthy place in which to live. Just as any living organism can be healthy only when that orgnaism is a product of nature's art in accordance with the basic principles of nature's architecture, exactly for the same reason town or city can be healthy – physically, spiritually, and culturally – only when it is developed into a product of man's art in accordance with the basic principles of man's architecture. Surely, this must be accepted as the fundamental secret of all town building.

Although the architectural nature of all town building is fully logical and scarcely disputable, there is, as things stand at present, much reason for stressing this fact. In earlier days when senses were alert and principles were in the blood, town building matters were automatically solved along architectural lines. Later on when a divorce took place between architecture and town planning . . . the architectural profession became involved in stylistic escapades, whereas the development of towns was left to the mercy of the surveyor. Since then, generally speaking, town building has been regarded as a practical matter-of-course in terms of two-dimensional planning – and so long has this mode of procedure already lasted that it has become a rigid tradition. It is against this rigid tradition – because of its dissatisfactory results – that we must fight. . . .

SYNOPSIS

[It is our intent] to stress those points that are essential in the solution of town building problems. It is highly important to have these points clear in our minds, before we attempt anything constructive for civic improvement and development. Let us therefore recapitulate these points, and by doing so, still more emphasize their significance.

1. We stress the importance of adequate town building from the point of view of the whole nation's welfare, material and cultural.

2. We make it clear that social research is a necessary background for civic improvement and development. This social research, it is said, must constitute a permanent institution so as to fertilize the city's continuous development in a socially constructive spirit. The cultural phase of this social problem is particularly stressed. . . . Whatever is considered the best for man, from the point of view of inner cultural growth, must be established as the governing principle in the shaping of a healthy environment.

3. We emphasize that any civic improvement and development must happen in accordance with appropriate means and methods. With regard to this, superficial town planning is found lacking, because neither has it the right approach nor can it show satisfactory results: more thoroughgoing and dynamic town building – by means of town design – had best take its place.

4. We find it mandatory that fundamental principles should be followed in all town building, for only this guarantees satisfactory solutions and lasting results. These fundamental principles, we find, are the principles of expression, correlation, and organic order, of which the last is the all-governing mother principle and the fundamental principle of architecture in all creation.

5. The architectural nature of town building is emphasized, particularly stressing the fact that only when town or city is a product of human art can it be physically, spiritually, and culturally healthy. In other words, the architectural aspect of town or city inevitably mirrors the social status within that town or city as well as the cultural ambitions of the population. Consequently, any investigation of town-building matters, to be accurate and significant, must be essentially an investigation or architectural standards.

PART III FORM

In analyzing the determinants that have shaped and are shaping the city's form, it appears that the dynamic technology of our time has become increasingly important and perhaps the dominant factor. The ecological factors including the natural setting of the city, the city's past history, and early platting pattern have all been subjected to rapid technological and subsequent social change, the effect of which is still not clear. However, it is becoming apparent that although man is very adaptable to change in his environment, the fundamental needs of men have been subjugated to the demands of technology. The result is a vast change in scale which serves the need of the technological innovations that ironically enough were originally meant to serve the citizens' need. This reversal in roles whereby decisions as to the quality of the city has been primarily based on the technique rather than human values, has reduced the citizen from a decision maker for his own life to a consumer of ready-made and limited alternatives. In order to return the citizen to his role as a participant in decision making he must be made aware of the various factors that constitute the morphology of the city.

. . . the new developing physical form of the city must be conceived on a regional scale. . . . There also must be a general realization that the new evolving urban pattern include such factors as the electric grid, telephone, radio, television, and other means of information storage and transmission that have replaced some of the need for face-to-face contact. These means of communication are additional important determinants to the [more familiar morphological] phenomena. . . .

James Braman, Determinants of City Form, *Seattle, Washington (Department of Community Development) 1971, p. 17.*

9

Albert A. Guttenberg

URBAN STRUCTURE AND URBAN GROWTH

The effects of the automobile on urban life and on cities which achieved their present structure in horse-and-buggy days are well known. The private automobile makes possible a wider spatial range of personal activity. Corresponding to this wider range is a community larger in area, and its precise size must be found. With its speed and flexibility, the automobile is also changing the time-distance relationships among established activity centers with the result that a new pattern of urban centers is emerging. As growth takes place, old links are broken, new ones are sought, and problems of future structure arise which involve the fortunes not only of individuals but of whole communities.

The present article was written with these problems in mind. It offers no solutions. Instead, the purpose here is to identify some of the critical elements and relationships of urban structure so that the probable consequences of different solutions can be better investigated. I hope it will be received as it

---

was written – not as a mere exercise in the theory of urban structure, but as an analytic aid in city and metropolitan planning.

The discussion is presented in three parts. The first explains how structure and form result from an effort to relate people and facilities over metropolitan distances. The second part discusses briefly the mutual influences of structure and form on the one hand, and community size and growth on the other. The third part is an attempt to derive the major features of the future metropolis from the present by analyzing the effects of a change in transportation efficiency. No doubt, the results are questionable, inasmuch as the analysis manipulates but a few elements whose relationships are more surmised than known. Even so, I hope this material will be of some interest as an attempt to grasp the dynamic interdependency of urban phenomena and systematically to explore the consequences of planned change.

Also of some interest, I believe, will be the method used here. Too frequently, one gets the impression that planning concepts include two kinds of statements of necessarily independent origin, that is, theoretical statements about the nature of the city and practical statements or objectives. I propose to show here what an intimate relationship can exist between these two kinds of statements. For example, I begin with an objective – that of overcoming distance between people and facilities – and use it to produce a definition of urban structure. By pursuing this method, a continuum of planning concepts is produced, some of them practical and some of them theoretical, but all of them in touch with each other.

URBAN STRUCTURE

The Elements of Structure

Assume the spatial separation of all people and of all the kinds of facilities[1] that they need in order to live in both the biological and cultural senses of that word. The sum of all distances between each person and each kind of facility is total distance. The objective is to overcome total distance.[2] In working toward this objective there are only two means available. People can be transported to facilities or facilities can be distributed to people. Each method applied in the extreme produces a distinctive kind of city.

In Figure 1[3] the function of transportation is to overcome total distance between people and facilities. In Figure 2 transportation has no function, total distance being overcome by means of distribution.

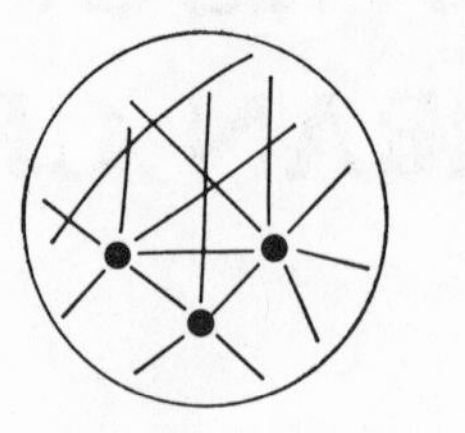

Figure 1

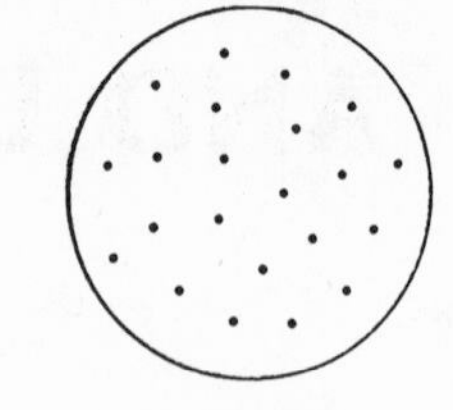

Figure 2

Actually, neither transportation nor distribution can do the whole job, because not all people are mobile and not all facilities can be distributed.

People may be place-bound because of their age (the young and the old) or because of their social role (as women with young children). Some facilities can't be distributed because they are underpinned by natural resources or because, to exist at all, they must exist at a physical or economic size which prevents indefinite multiplication.

These constraints require the use of both transportation and distribution to overcome total distance between people and facilities. Part of the distance must be overcome by means of local facilities. The remaining or *residual* distance must be overcome by means of the transportation system. Accordingly, we combine Figures 1 and 2 to give a truer picture of urban structure.

In this brief analysis we have identified the major functional parts of urban structure and given them a meaning in relation to a single objective. The major parts are the distributed facility, the undistributed facility, and the transportation element. In overcoming distance the first and the third have complementary roles: more distributed facilities mean less residual distance and less need for transportation capacity; fewer distributed facilities mean more residual distance and greater need for transportation capacity. Practical limits to both concentration and distribution are set by place-bound people and place-bound facilities. Figure 3 shows the anatomy of the human settlement conceived as a system of relating facilities to people. It also represents the fundamental elements in any plans for the settlement. Certain facilities can be distributed throughout the area in close physical proximity to their users.[4] Other facilities can be distributed only in the sense of being made accessible through the transportation system.[5]

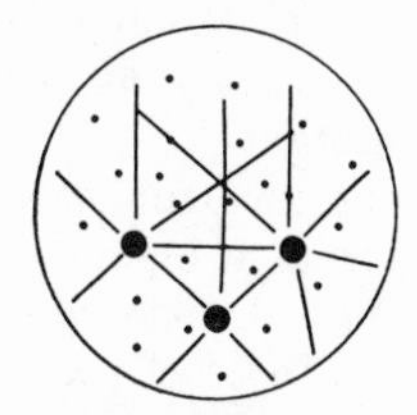

Figure 3

Urban Structure and Urban Form

Our first objective was to overcome total distance between people and facilities. A second objective, related to the first, is to reduce residual distance, that is, the distance remaining between people and facilities after some facilities have been localized.

Residual distance can be reduced by adjustments in the spatial form of the community. Form is used here to mean the way the structural components just described are arranged on the ground – the pattern they make. The effects of this arrangement on residual distance are depicted in Figures 3, 4, and 5.

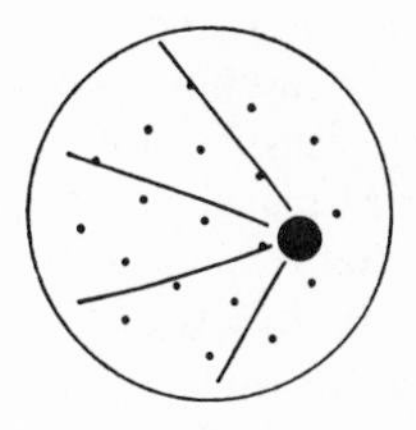

Figure 4

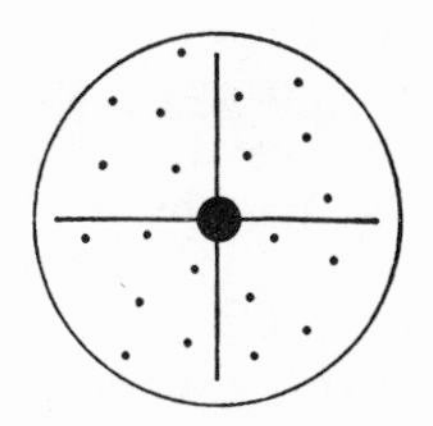

Figure 5

The obvious effects of the arrangement in Figure 3 is much movement back and forth between people and undistributed facilities. Consolidate these facilities at one point and the distance to be covered will decrease (Figure 4). It will decrease even further if the consolidated facilities are put at the center of population (Figure 5). The pursuit of the objective, to reduce residual distance, requires that one activity center in the urban field be privileged with respect to size and location. This fact gives functional meaning to the core of the city or metropolis and to the radial shape of the transportation system serving it.[6]

The Hierarchical Aspect of Structure

Limits to both concentration and dispersion are set by immobile persons and by facilities which cannot be distributed. But not all persons are equally immobile, and not all facilities are fixed or concentrated in the same degree. Mobility differs according to age, sex, and income, whereas distribution is a matter of economic plant size or of the spatial scatter of natural resources. These facts explain the hierarchical tendency of urban structure: people must travel different distances to facilities and the facilities themselves have different service radii. Also, they account in part for the existence of central places of several sizes and the corresponding differentiation of the highways which serve them into major and minor arterials and local streets (Figure 6).

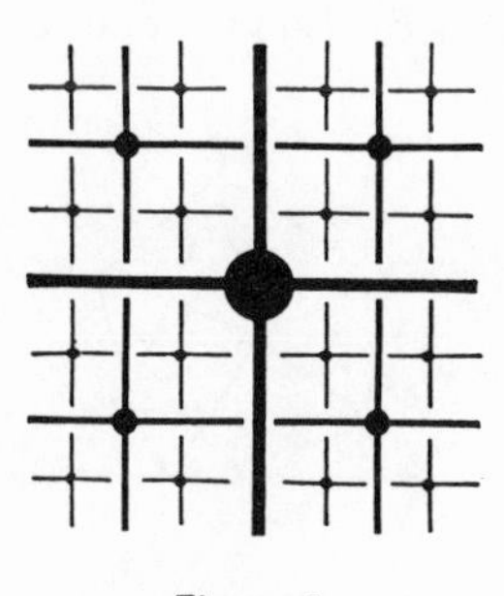

Figure 6

As there is more than one level of community, community planning must be carried on at more than one level.

The Whole and the Part

Bringing people and facilities together creates three urban structural elements. Of these, one, the distributed facility, becomes the basis for a local organization of human activities. The undistributed facility and the transportation element become the basis for regional organization. The existence of these two levels of community is responsible for the development of a regional as opposed to a local interest in local land. For the local community the local area is "home." For the regional community the local area is a unique geographical resource to be used for its own purposes. Where these different viewpoints produce conflicting demands on the same land, a technical solution is required in the form of a set of defined land use relationships which permit the simultaneous expression of both local and regional functions in the same area with as little friction between them as possible. Of course, the possibilities are different in different parts of the field. Structural accommodation is more difficult toward the center where the great regional paths converge and where regional facilities come to predominate.

Density

We consider next the factors which give to any location in the urban field its value. These factors bear upon another dimension of urban structure – density. By the value of a location we mean here its desirability as a place of residence or business.

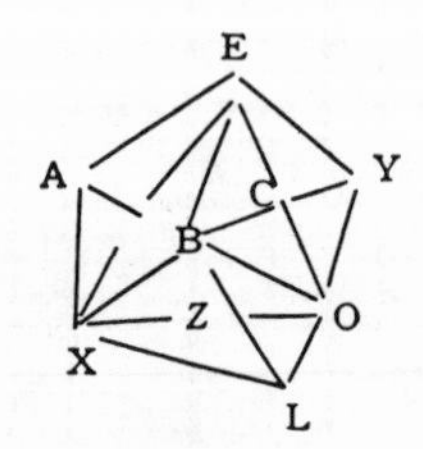

Figure 7

In Figure 7 let *X* represent the locus of one kind of facility and *Y* another. If a person (or establishment) needed access only to *X* (or *Y*), he would locate there. What about the person whose needs or preferences require him to be close to both *X* and *Y*? This is a question as to whether the several possible locational requirements of an individual are practically compatible, and if so, how so.

As a matter of fact, the needs of our hypothetical individual are not beyond compromise. He can reconcile them spatially to some extent by locating at a point where access to both opportunities is adequate but where access to neither is optimal. If *X* is one point of opportunity and *Y* is another, then there is likely to be a point *Z* from which both are accessible in proportion to the needs or interests of the person who seeks them. In this simple example the relationship of *Z* to *X* and *Y* constitutes a good part of its value as a location to the individual or establishment in question.[7]

If all persons had the same needs or preferences, all would try to locate at one point, for example at point *Z*. But all persons don't have the same needs and preferences, and this fact accounts for the value of points *A*, *B*, *C*, . . . , *Z*, and for the population of the whole field.

But the whole field is not evenly populated. For, although every point is valuable to some people, not all are valuable to the same number of people. There are two reasons for this. In the first place, social forces operate so as to create in most persons like needs and interests. Second, certain points in the field, because of their locations, offer better access than others. Such points are able to accommodate diverse locational interests.

Where, in terms of access, one place is supreme, as in a metropolitan region (Figure 5), every other place acquires a value which is some fraction of the value of the chief place, the amount depending on how well it is able to substitute for the chief place. The better it can substitute, the more desirable it becomes as a place of residence or business. Hence, it tends to be more intensively occupied with households, other establishments, or structures. As places differ in their ability to substitute for the chief place, or center, they differ in their crowdedness, that is, their density.

As a rule, as one leaves the center of a region, ease of access declines, and, with access, substitute or referred value also declines (Figure 8, line *BP*). A region ends where it is no longer possible to substitute for the center in any respect or any amount. The rise or fall of referred value with distance from the center is the basic *economic* density gradient which underlies all forms of the physical density gradient.[8]

The general slope of the economic density gradient

depends mainly on transport efficiency in substituting more peripheral for more central location. Irregularities in the slope are of two kinds. Positive

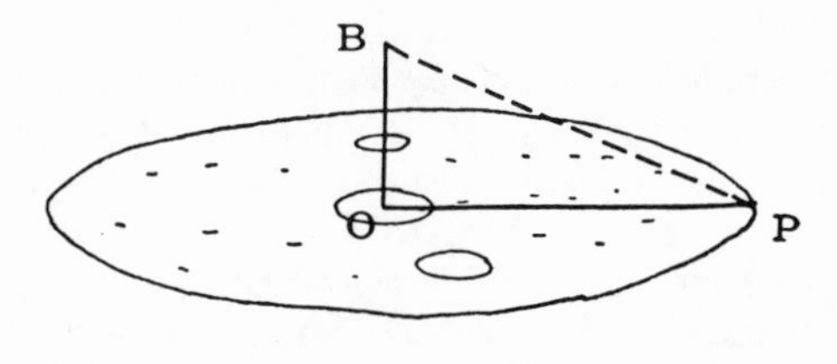

Figure 8

deviations are caused by transport advantages (e.g., an expressway ramp which gives quicker access to the center from a point *M*, Figure 9, than from other points at about the same distance from the center); negative deviations, conversely, are caused by transport disadvantages (e.g., a bad stretch of road).

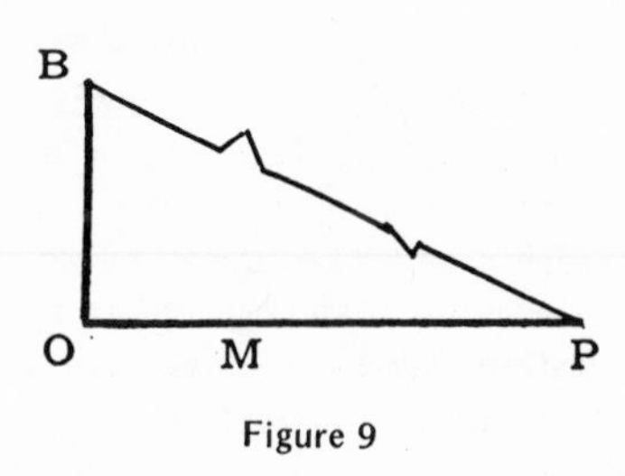

Figure 9

The relation between economic density and physical density is not necessarily direct. Whether or not high economic density at a given point is translated into high physical density depends on local site conditions, on whether or not prospective users need a lot of space, and on their ability to pay for space. High economic density may go with low physical density, and low economic density with high physical density, but these combinations are exceptional.

URBAN STRUCTURE AND URBAN GROWTH

The Spatial Extent of a Community in Relation to Its Structure

Place-bound persons and place-bound facilities cause the community to have a stable spatial structure. In turn, this structure affects the community's spatial extent.

Figures 10 and 11 are similar to Figures 1 and 2 except that symbols have been used to represent loci of different classes of major activities. A characteristic of Region 1 is the separation of place of work, play, and residence at a regional scale. In Region 2 separation of activities is local.[9]

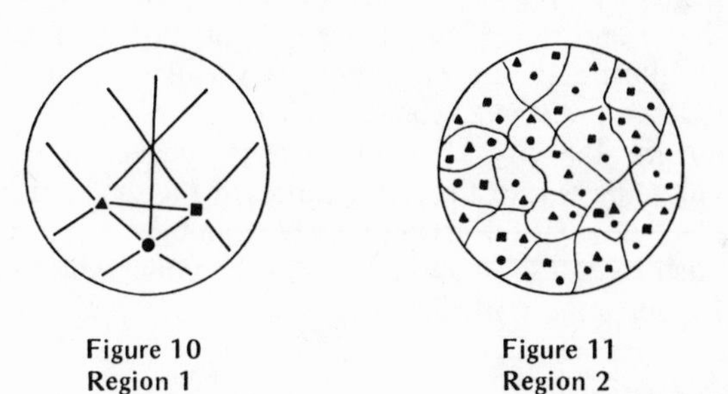

Figure 10
Region 1

Figure 11
Region 2

Corresponding to these structural differences there are implied differences in the spatial range of personal activity. Whereas in the second region a person would hardly have to travel at all, daily routine might require him to traverse Region 1 several times over. One can imagine that the size of the second region, no matter how great it might be, would impose no strain on the time or energy of any of its inhabitants. On the other hand, the size of Region 1 is subject to a practical limit – a limit determined by the ability of people to reach unique activity centers, and represented in Figure 10 by the circumference of the circle. Growth beyond this limit requires structural changes in the form either of faster or more far-reaching transportation, or of new regional centers. These considerations permit us to give to the concept growth a definite meaning.

Growth and Its Consequences

Growth involves an increase in size and an adjustment to size. In biological forms these two processes occur at the same time or so close together that they can hardly be distinguished. Cities, however, increase in size first and adjust to size only gradually. As a result transitional relationships often confuse the features of the old and new equilibriums. But always emerging in the growth of a city is the perduring structure of the human settlement at a larger scale – that is, distributed facilities, nondistributed facilities, and highways, usually with more space between them.

The emerging scale involves the welfare of individuals as well as of whole communities. The political community is affected when, as a result of adjustments to scale, its boundaries cease to coincide with the boundaries of the functional community. Figure 12 illustrates the possibility: a vacuum caused by excessive distance comes to be served by centers outside the city. The influence of the new centers ranges deep into the service areas of the old, causing the reorientation of considerable numbers of people. Such a situation poses problems of conflict which are beyond a merely technical solution.

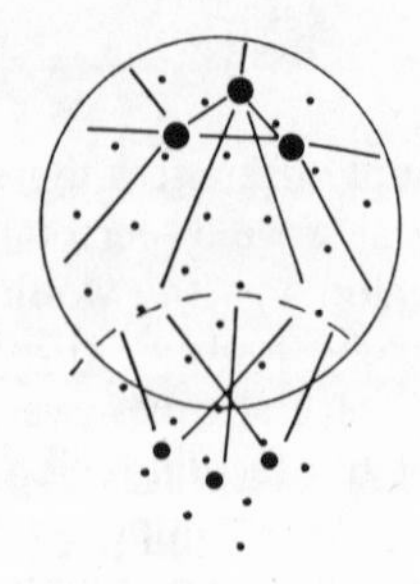

Figure 12

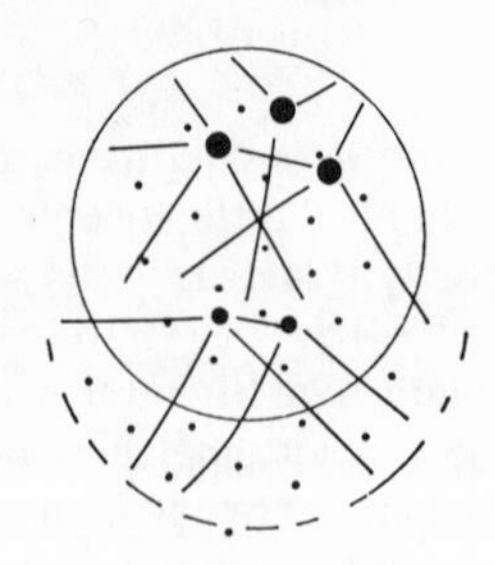

Figure 13

Or the new centers may rise within the political community. In this case, also, old established centers may lose their supremacy, but this loss must be weighed against the advantage to the political community of retaining its custom and winning new customers beyond the boundaries (Figure 13).

Older centers may keep their importance by means of better transportation – transportation being, as we have shown, the functional equivalent of distributed facilities.

Whether they occur in the form of new centers or as improved transportation facilities, structural adjustments to greater scale affect individuals by changing the positions of their homes or places of business relative to the locations of other homes and other business places and to the great activity centers of the community. Any point in the city derives its personal, economic, or social value in part from its position in the total structure. When the structure changes, all points are thereby dislocated. This fact implies loss for some individuals and gain for others, depending on their interests and on the relationships of the points they occupy to the emerging structure.

When growth occurs, population movements ensue, caused by many people seeking points in the new structure corresponding to points in the old. As a result, some areas develop and others decay, both processes giving rise to social and economic problems to be met by means of development and redevelopment.

TRANSLOCATION

The Meaning of Translocation

Growth and the adjusting movements of populations mean a new region – new in the sense that the absolute locations of people and activities will be different even if the relative locations are the same or similar. If the community is to proceed from the present to the future without wasting effort or resources, then the corresponding parts of the present and future region must be identified ahead of time. But how can places in different regions be compared? A prior question must be: how can we compare different places in the same region?

Suppose we say that places can be compared in terms of their access to other places, especially to central places. Then places in the same region are comparable if they have equal access to the regional center. Places in different regions are comparable

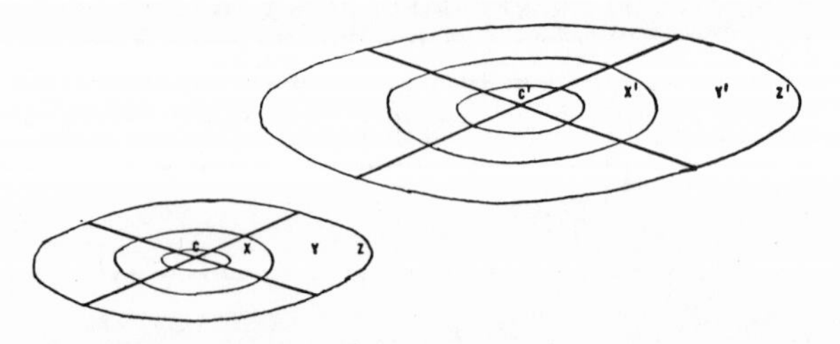

Figure 14

(correspond) if they have equal access to their respective regional centers. In Figure 14, *C* can be reached from X as easily as C' can be reached from X'. Therefore X and X' are corresponding locations. In like manner, Y and Y' and Z and Z' are corresponding locations.

If A is the present region and B the region of the future, is it reasonable to expect that the function now performed at X will one day be carried on at X'? Perhaps. Units of different kinds of activity, as well as different units of the same kind, compete for a site. The site goes to that activity which can best use it, in the sense of realizing from its use the highest material return. As activities differ in their ability to use different sites, that is, in their ability to survive at different time-distances from the center, they become spatially sorted. In this way originate those broad areas over which one kind of activity prevails, or one quality, or one density, or a typical mix of different activities, or, perhaps, no activity at all. *Insofar* as time-distance is the sorter, we can expect locators in A to take up corresponding positions in B unless they have changed their time-distance requirements in the meantime.

Of course, in the interval between A and B it may happen that the economic composition of the region changes, in which case B will be not just a larger version of A but will be a somewhat different animal. If such a metamorphosis does occur, it will not be enough to identify corresponding parts in terms of time-distance from the center. In fact, in this event, the idea of translocation is open to question, since it presumes the identity of the translocator.

But suppose that the economic composition of the region doesn't change and that within it all locators keep their relative time-distance requirements. Then we can expect that, relative to the center, the time-shapes[10] of the two regions will be alike. Does this mean that the space-shapes must also be alike, as we have made them in Figure 14? Here, too, the answer must be conditional. It depends mainly on the transportation system in the two regions. The transportation system, to repeat, is a means of substituting one place for another. As transportation systems differ in their design and these designs differ in their substitution effect, the spatial pattern of activities differ.

Where there is one big center of overwhelming importance and transportation efficiency is equal in all directions, the rate of substitution in all directions is also equal, and the prevailing form of the city will, theoretically, be circular (Figure 15).

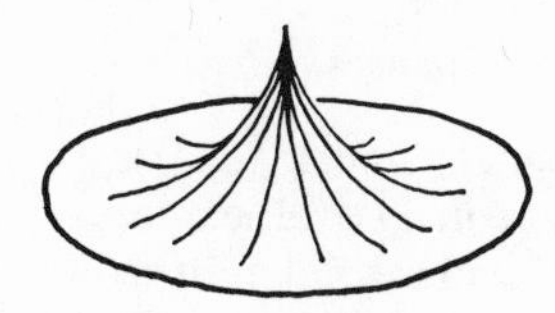

Figure 15

Simple radial form appears where there is one big center and a few of the routes serving it offer superior transportation (access). Displacement from the center is more intensive along these routes, and there is less demand for the area between them (Figure 16).

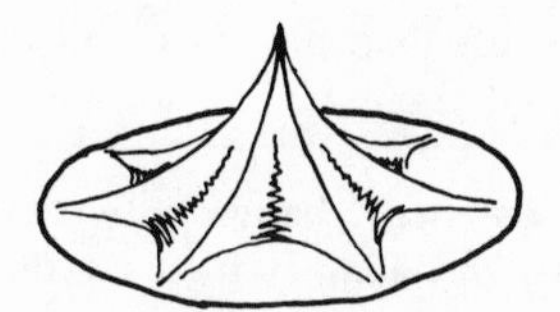

Figure 16

Displacement of activities from the center along major radial routes will be continuous if access to these routes is continuous; otherwise, it will concentrate at points of superior access (e.g., at interchanges with major circumferential routes). In this case, the prevailing spatial pattern of the activities will be not merely radial, but also nodal. This is the kind of form which a radial freeway system is likely to induce (Figure 17).

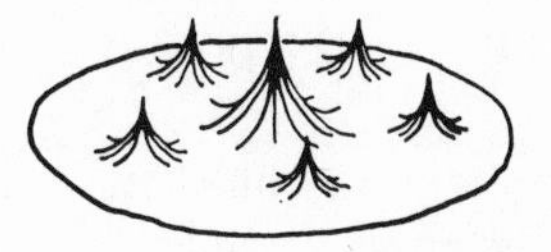

Figure 17

Insofar, then, as the transportation system in B is similar in design to the system in A, the spatial pattern of activities in the two regions will be similar, *unless* the pattern comes to be distorted by other factors such as accidents of terrain, public or private intervention, or by the influence of other regions and their centers; for as regions expand in area, the possibility of their bumping into each other grows.

Size and Density Changes

If *B* does remain similar to *A* in its space-shape, it will be different in two other important respects – first in its size, and second in its density distribution.

As with location, changes in the densities at which all activities are carried on follow as a matter of course from changed time-distances. The general slope of the density gradient depends upon transport efficiency in substituting more peripheral for more central locations (Figure 18, line *BP*). A new level of efficiency means a flatter slope, a territorially larger region, a more spread-out center, and a lower regional density, assuming, of course, that the total population increases less than proportionally to the increase in area and that there are no unusual restrictions on space supply.

New time-distances change potential in every part of the region, throwing it into conflict with actual densities which are past values realized. The choice is to put things where they belong once more (i.e., in the right amount), or else, in the public interest, deliberately to bear the cost of not doing so in the form of unrealized value, inflated value, or subsidy.

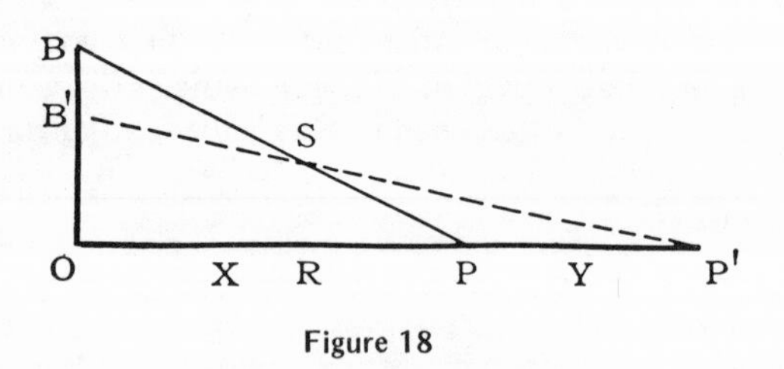

Figure 18

The curves in Figure 18 illustrate some of the basic relationships assumed to hold here between the regional center, the size of the region, transport efficiency, and density distribution. The line segment *RP′* represents that range of the region where new conditions of accessibility will cause future density (*SP′*) to exceed present density (*SP*) by a definite amount. For the range represented by the segment *OR* the reverse is true – there future density will be lower than present density. By this criterion, then, the range *OR* is overdeveloped, *RP′* is underdeveloped, and at point *R* where the two curves cross, present and future development are identical.

For every point (*X*) on the line *OP′* it would be possible to specify a difference between present and probable future density representing the relative demand for redevelopment. Thus, of all points in the overdeveloped range of the region depicted in Figure 18, the demand for change is greatest at *O*, the regional center. The demand for change in the underdeveloped range is greatest at *P*, the edge of the present region. At *R* very little change, if any, would be required.

Figure 18 suggests that theoretically the question of redevelopment priorities is not a valid one. The reason for this is that redevelopment means redistribution. A reduction of development at point *X* requires a corresponding change at other points. All would have to be undertaken at the same time.

On the other hand, Figure 18 does give us some clue as to where scarce public redevelopment monies should be concentrated. For example, at *O*, and again at *P*, the forces for change would appear to be greatest, and we would expect the marginal effect of every public dollar spent there to be correspondingly large.

In this section we have tried to find an intelligible connection between the present and future metropolis by analyzing the effect on the former of an assumed change in transportation efficiency. The shortcomings of this analysis are obvious – above all, its rudimentary character. Even the few relationships presumed to hold here are, as yet, hypothetical and require further investigation. We have no doubt made too much depend on a single factor, access, while other factors are omitted from consideration altogether.

A different kind of criticism is possible on the grounds that no account is taken of the financial, social, cultural, and political consequences of structure and of their reactive effect on the total system.[11] Surely, Figures 15, 16, and 17 imply great differences in cost, in quality of social and cultural life, and, if the metropolitan area is composed of independent communities, differences in the balance of economic and political power among these communities.

These considerations are beyond the scope of the present article. Here we hoped only to stress the connectedness of urban phenomena and to spur its further investigation – a purpose of more than academic importance. For only when it is accomplished will it be possible for a community to foresee the results of its own actions and to make an intelligent choice among proposed structural alternatives.

NOTES

[1] The term facility is used here to mean any fixed physical instrument of artificial *or* natural origin (e.g., a beach) which people use to meet a need or fulfill a purpose. So defined, the term includes anything from the corner mailbox to a factory, but excludes vehicles or anything normally portable.

[2] A necessary objective since in order for people to live at all, or to participate in their culture, the distance between them and facilities must be bridged.

[3] In Figures 1 and 2, dots represent facilities or clusters of facilities, and lines represent transportation routes. People are assumed to be scattered over the whole field and are not represented symbolically.

[4] For example, a small park or playground, an elementary school, a neighborhood convenience store, or movie house. A dwelling unit would be the "most distributed" facility.

[5] For example, a beach, a department store, a regional park; many major work places also fall into this class.

[6] Residual distance is physical distance. Both physical distance and time-distance are reduced by adjustments in spatial form. Time-distance may be further reduced by improvements in transportation efficiency; e.g., by new or improved highways, or operational changes which speed traffic flow.

[7] The other component in the value of a site is the intrinsic properties of place – such as soil, slope, or elevation.

[8] Economic density is the value of a place as a vantage point for access to all opportunities in the region as a group.

[9] The meandering lines in Figure 11 represent boundaries, not highways.

[10] In this context, "time-shape" means the time-distance relationships among all activities, not a sequence of events.

[11] However, see section on Urban Structure and Urban Growth.

10

F. Stuart Chapin, Jr.

TOWARD A THEORY OF URBAN GROWTH AND DEVELOPMENT

The strategic role that theory plays in man's quest for understanding and knowledge is so basic, so well documented, and so widely accepted that it seems unnecessary to stress its essential importance for the planning field. Yet until recently there has been very little effort within the procession going into theoretical research. A relatively young field such as planning normally faces a long period of "finding itself" before it begins to develop schools of thought and acquire theoretical traditions. Under the most favorable conditions theory building is a slow process, particularly so if it is to be systematic and rigorous by contemporary standards of scientific inquiry. The time involved from the first statement of theory, through the intermediate stages of empirical tests of its critical components, the successive reformulations and retests, to the stage when it becomes operational is painfully long. It is so long that components of theory often pass into operational use long before the full system can be made to work in practice.

The pressures on the profession to meet new emerging situations in creative and effective ways have been heavy in the post-World War II period, and the field has been pushed into maturing and assuming responsibility more rapidly than was generally anticipated. There is a growing sense of urgency that tactical efforts at solving urban growth problems require much stronger strategic support, and there is generally a new interest and sense of mission in advancing theoretical research. In the 1950s there were several important efforts of this kind beginning to develop, and during the 1960s some of this work can be expected to become operational. While there is this developing sense of need and expectancy from the profession, there is another reason for stressing the importance of theoretical research. Not only is it needed to advanced knowledge and understanding in the field, but it supplies the guidelines, the rationale, and the *raison d'être* for planning studies undertaken in practice. Without these guidelines, the field remains in an ad hoc status. This is not to say that the field has been developing without theoretical foundations. Because of its close ties with such disciplines as economics, public administration, and sociology, and a conversancy with areas of theory in these fields relevant to planning, the profession has shown an alertness to and has put to use new developments from these related fields. Indeed, the results of research from these fields have often been received with more enthusiasm in planning than in the field from which they originated. But as the planning field expands further, it cannot remain dependent solely on the work of related fields, with "a little of this and a little of that" as a substitute for its own research and development effort. Indeed, the future growth of the profession is no less dependent on a basic research effort than is the case for the natural sciences, which have been emphasized so strongly in recent time.

At the present stage in the field's development, theoretical research is developing along two related lines. One major emphasis is on the *process of planning* and its relationship to decision-making theory, and the other focuses on the *subject matter of planning:* the urban complex, the larger region, and sometimes entire nations. In the sense that they constitute the means and ends of planning, they are related areas of theory specialization. For purposes of land use planning we shall be concerned mainly with the second area of specialization, and particularly with urban spatial structure and urban growth. . . .

CRITERIA FOR AN ADEQUATE THEORY

As used here, "theory" refers to a system of thought which through logical verbal or mathematical constructs supplies an explanation of urban areas – why they exist, how their growth and change occur, and what the basic structure and form components are in the urban scene. In addition to supplying an explanatory rationale, theory frequently provides a rationale for prediction.[1] By observation of the similarities or regularities in the behavior of phenomena in reality, it seeks to establish likely future states of these phenomena under differing assumptions of behavior. At still a third level of application, theory may supply guidelines for exercising choice among alternatives in the course of deciding on a plan. Thus, we see that there are different cutoff points in the development of theory which fall to the discretion of the researcher. It should be noted that Utopian concepts of the city are not included here. While constituting an important area of study in the planning field, this kind of system of thought does not constitute theory in the usage of this chapter.

Normative and Explanatory Theory

At the outset it is useful to differentiate between theory which concerns itself with explanation (the what is and the why) and theory which is normative (the what ought to be). In emulating the dispassionate view of the physical sciences, social sciences have a predisposition to keep the normative element out of theory, making anything to do with the "what ought to be" a part of the application of theory and the choice among alternatives which comes into play when a theory is applied. This is the more detached approach. Occasionally the scientist will ask, "What will be the effect if this is assumed or that change introduced?" This takes him into a predictive type of theory and involves a more clinical approach. Of course, when these questions are asked, it is only one further step to pose questions in alternatives, and then to move into normative emphases. While the traditional social science disciplines lean toward the explanatory type of theory, it is the very nature of a profession which has a prescribing function in practice to favor theory with a normative emphasis. Actually, the kinds of theory which seem to be developing in the field today follow both emphases. Some work focuses on the development of systems for explaining and predicting growth – the "what is," the "why," and the "what consequences" emphasis. This work seeks to understand more fully how the city grows and develops and how plans may alter these "normal" or "natural" processes. The normative theory may address itself to these same questions, but it also seeks to supply guidelines for discovering and analyzing various mixes of qualities, and how these might be put in optimal combinations to fulfill some predetermined requirements (goals). However, there is generally no intent to supply answers as to what the optimal pattern of development should actually be, only guides as to how to arrive at "what ought to be," once the community's development goals have been identified. To put it in operational terms, under normative theory on urban spatial structure, planners do not set goals; rather, they would employ this theory to determine the structure and form choices open to an urban area, given a certain combination of goals.

Formulation of Theory

Theory has been defined as consisting of a system of thought used in explaining a phenomenon. In a formal sense, a system of thought is composed of subsystems made up of concepts, some of which are from known states of knowledge with known relationships, some of which consist of propositions of fact or relationships still to be proved. Since functional relationships are the connective links between components of theory and are at the root of explaining a phenomenon, the form of relationship becomes important. Ackoff has identified three forms of functional relationship which theory may put forward. One is the cause and effect relationship which involves *deterministic causality* – a relationship which has one and only one result. Another is what he calls a "producer-product" relationship which involves *probabilistic causality* – a relationship which due to chance factors yields variable results, but nevertheless results which tend to cluster around one theoretical end product. The third form of relationship may not involve causality at all; it is *correlation* and deals with the tendency for variables to behave in a similar manner together, that is, to change or not to change their values together. A finding of correlation does not establish causality, but it may be useful in inferring causality.[2]

As our state of knowledge in a field improves, many of the truths once taken to possess deterministic causality are found to involve probabilistic causality, sometimes only correlation. Even in the natural sciences, where we have long expected absolutes, there is now a sense of caution about deterministic relationships and a disposition to express relationships in terms of probabilities. In planning, where we are concerned with fixed improvements of long life, it would simplify our task immeasurably if we could construct all our theory in terms of deterministic causality. However, we know that land development and urban growth are so deeply involved with human behavior, and human behavior is so complexly affected by chance considerations, that we anticipate, at least at our present state of knowledge, that probabilistic causality is likely to be more prominent in our theoretical research when at length our directions of development in theory settle down and theories of land development and urban growth become widely accepted.

Causality has occupied logicians and scientists for a long time, and we cannot digress here to go into the long-standing controversy over the determinism and indeterminism of causality. For our purposes we shall note the sense of caution that goes with claiming cause-and-effect relations. But as Simon demonstrates in his series of essays on this subject, there are good and sufficient situations in empirical work where causation can be discussed with a clear conscience.[3] In any event, in a field which is just beginning to develop conceptual systems, we can anticipate that theory building will take advantage of the license of a "pick-and-shovel" approach and hypothesize deterministic causality until proved otherwise. When theory begins to reach more mature stages in the field, and certainly when in more advanced work on theory interest becomes directly centered on the methodological aspects of theory testing, the differentiation between causation and association can then become quite important.

It will be helpful at this point to distinguish between levels of attainment in theory formulation. These levels fall into a descending order of scope, from a "general theory" to a "middle-range theory" to a "small theory." Omitted from the list at the upper end is the notion of a "universal theory" and at the lower end of a "field theory." A universal theory is not included simply because it is doubtful that human behavior of so many diverse forms as that which is involved in the growth and development of urban areas can soon be reduced to one all-inclusive system. What is taken up in this chapter are the early stages of work of a middle-range character. Most of the conceptual systems aim toward becoming general theories, but to become full-blown theories of this level involves a rigorousness and systematicness that is not yet in sight. The middle-range theory is an intermediate level en route to a general theory. Usually limiting assumptions are necessary because of the state of the research arts. Small theories are generally associated with empirical studies of very specific dimensions and limited scope, but where replication strongly supports hypotheses made in the theory. Perhaps the most primitive level of theory formulation is found in field theory, which is essentially a contrived rationale borne out in one particular situation but not replicated.

As suggested above, the statement of theory may consist of both proved and unproved concepts. Research concerned with urban spatial structure is likely to deal with such mixtures of concepts. An unproved concept consists of hypotheses, that is, statements of what is expected to be found when the hypotheses are eventually tested. In classical social science research, the test of hypotheses follows set rules in formulating the hypothesis, making assumptions, conducting statistical tests of the probability that a particular result is due to errors of different types, and reaching conclusions. When confirmed, a hypothesis may be stated as an empirical generalization and, under certain circumstances where theory reaches a formal stage of formulation, it may become a law. Formal statements of theory often involve preconditioning statements in the form of axioms or postulates. These are to be distinguished from assumptions which are preconditioning statements of a more technical character employed in the test of a theory or particular hypotheses composing parts of a theory.

In the formal traditions of theory building, scientific inquiry may explore statements of functional relationship by deductive or inductive means. It is beyond the scope of this discussion to go into the conventions involved in the formal formulation of deductive and inductive propositions. For our purposes, we will simply observe that a theory may be constructed so as to reason from a known general proposition to an unknown specific proposition, or it may be built around an approach which infers truth from empirical observation, with the validity of the theory being developed on the basis of consistency in findings from one case to another. The first is a deductive form of theory, and the second is an inductive form. While noting these conventions, we must hasten to point out that the state of theory formulation in the planning field has not yet reached a stage where these distinctions can be applied with any great meaning. They are briefly touched upon at this juncture to suggest the general kind of formal standards that the planning field must eventually seek to achieve.

Criteria of Adequacy

Where the objective of theory building is not simply a quest for truth for knowledge's own sake, but includes the eventuality of applying the theory in improving the lot of man (an objective which is characteristic of a profession), we are concerned not only with the conventions of formulating theory, but also with specifying performance criteria. We wish to ensure that theory meets the requirements which we foresee will be important in its application in the field.

There are at least four criteria of adequacy which will be useful to bear in mind in reviewing the work summarized below.[4] One criterion holds that a theory must have a dynamic aspect if it is to have utility in representing the process by which cities are structured and by which they grow. A second requirement is that the theory be susceptible of empirical verification, that it be capable of being tested. A third requirement is that the theory have an internal logic and consistency. The rigor of the logic and consistency may vary from a very general and somewhat summary form where concepts and relationships are very broadly stated to a very formalized form where all propositions and relationships are spelled out in detail. Finally, the theory must not be so abstract as to have no relation to reality. Indeed, it should seek to represent the phenomena under study as they actually occur or appear to function in reality.

As will become abundantly clear, by these standards the present stage in theory building relating to urban spatial structure is spotty, and in general the field has a long way to go to measure up to requirements of this kind. The field appears to be still in a primitive stage of conceptualization of urban phenomena, still seeking to establish what the proper content of urban spatial structure theory should be. This simply underscores the observation made earlier about the urgency and the great need there is for fundamental research in planning.

EMERGING THEORETICAL ORIENTATIONS

It has been indicated that theoretical work in urban spatial structure is the type of theory which has special relevance for urban land use planning. In the usage here, urban spatial structure is concerned with the order and relationship among key physical elements of urban areas as they evolve and pass through transformations in time and space. But in all the theoretical work so far undertaken, either implicit or as an integral part of the conceptual framework, is a second antecedent area of theoretical significance concerned with activities of people and their institutional entities and the interaction these activities create. Implied is a causal relationship between two pairs of concepts. One pair focuses on human behavior: (1) place-related patterns of interaction (activities), and (2) patterns of interaction among different place activities (movements or, more broadly, communications). The other pair focuses on physical structure and form: (1) space adapted for activity use, and (2) channels developed for movements and other forms of communication. Depending on the scope intended, theoretical work may be concerned not only with spatial relationships between user activities and between space uses at a particular moment in time, but it may also extend to relationships in a dynamic framework, focusing on interaction patterns and on space use patterns and their interrelationships in an evolutionary sense over time.

An extensive amount of work of theoretical significance has been accumulating over the years, but it is widely dispersed through the social science fields and, for land use planning needs, is somewhat fragmentary and limited in its direct applicability. In a stricter usage, where emphasis is placed on linked systems of thought as we have been discussing them (for example, linkage between activity systems and land use patterns), the task of summarizing theoretical work is not so overwhelming. The problem is not so much one of identifying work from widely dispersed sources or of making selections from a great proliferation of work, but rather one of how far to go in summarizing conceptual systems that are so recent and, in some respects, still so tentative in the originator's thinking as to be still in a working stage of formulation. It is because of this tentativeness that we shall refer to them as conceptual frameworks or theoretical orientations rather than as theories per se.

It is premature to identify schools of thought in theory building, and there may be some question as to the meaningfulness of classifying tendencies in the work emerging today. At this stage only the most general observations about these tendencies can be made. We now turn to the conceptual systems which have been advanced in recent time.

A Communications Theory Approach to Urban Growth

Meier approaches the task of a theory of urban spatial structure by asking what, after all, is the quintessence of the city?[5] Is there a common pattern that holds through time, one perspective in the behavioral sciences that provides a logical basis for building a theory of urban development? After examining human settlements as they emerged from the beginnings of civilization, following the changes as they might be seen by an archaeologist, anthropologist, historian, and natural scientist, and after considering man's behavior in cities, looking at human activities as the economist, social psychologist, human ecologist, and political behaviorist might, Meier concludes that the one common element in all these perspectives is *human communication.* Whether viewed in very concrete terms of marketplace transactions or in the more abstract notion of the transmission of culture, he is saying that the human communications process possesses all the requisite requirements of an organizing concept for a theory.

Meier conceptualizes the city in terms of systems of interaction prompted by man's urge to maintain communications (in the general sense) with his fellow man. At the present stage in man's state of development, transportation and communications technology supply the principal media of interaction. While noting that cities have always exerted a strong attraction for growth because of the opportunities for face-to-face transactions that they offer, Meier holds that technological developments are reducing the necessity for face-to-face interaction, and transportation overloads are imposing limiting conditions on opportunities for interaction through transportation systems. With the substitution of communications for transportation, communications becomes increasingly important as a focus for studying the city. In noting that overload crises in communications systems are in prospect, Meier anticipates control mechanisms being invoked to correct for these overload conditions. Thus the communications system (in the narrow sense) offers what he considers to be the basis for understanding human communications (in the general sense) and the activity systems that arise out of the human relations involved.

Having satisfied himself on the validity of using a communications system as a basis for building a theory of urban growth, Meier develops a set of requirements for the communications process. Specifying that there must be (1) a sender, (2) a message, (3) a channel, (4) a receiver, (5) an attention span on the part of the receiver, (6) a common language, (7) time for the process to take place, and (8) one or more purposes to be served, he proposes to construct a representation of the city from the information content of communications flows. Information would be measured and recorded in a double-entry accounting system in much the same manner as origin and destination traffic studies record traffic flows today. The unit of measurement for information transmitted would be the "hubit," which Meier defines as "a bit of meaningful information received by a single human being" – a per capita concept of units of information received.[6] According to Meier, by obtaining a sample of communications flows in a metropolitan area, information theory can be used to construct a set of social accounts which can then become the basis for explaining activity systems.

Meier does not indicate fully the manner in which the framework would be used by a planner in a predictive application, say for the year 2000. However, he suggests that his concept of the "urban time budget," which estimates the proportions of a day's time a person would spend in various forms of public communication (as opposed to private or personal communications), would provide a means of making projections.[7] Given an estimate of the population for the year 2000 for a particular metropolitan area, one surmises he would construct a set of sender-receiver accounts of information flows or transactions broken down into "activity sectors," for example, leisure activity patterns of various forms, or wholesale-to-producer or wholesale-to-retailer activity patterns of various types. On the basis of assumed states of technology, presumably he would trace out spatial loci of activities, and once location relationships are established, he would assign space to activities and modes of movement according to standards developed to correspond to the technology assumed. This aspect of his framework will undoubtedly be made clearer as he extends his work further.

In order to study some of the variables which he hypothesizes would affect the flows of information, he has undertaken two pilot-type, exploratory investigations of communications flows – one, of the January 1959 speculative activity that swamped the American Stock Exchange, and the other, of the functioning of a library system. In both studies he was able to investigate relationships among concepts of error, stress, and capacity in a communications system and demonstrate that these factors affect the dynamics of a communications system, indicating that the construction of activity patterns must ultimately take into account behavioral variables of the most complex order.

On the basis of this progress report on Meier's

work, it is clear that it possesses a distinct behavioral emphasis on the study of the city, and tends more toward the explanatory than the normative emphasis. His work surely reflects a very strong feeling for the dynamic, not only in the usual time sense but also in his concern for constructing the evolutionary sequence in human behavior patterns. Clearly, he is intent on empirical verification of this conceptual framework and builds it with this end in view. Surely there is a compelling logic in this system of thought concerning spatial organization and growth of cities. There is every indication that a formal and rigorous statement of theory will emerge. Perhaps because it is in a working stage of formulation, the internal continuity of the way in which the framework is eventually to be put to use is not entirely clear. How the analysis would proceed from sampling information flows, to identifying transactions, to constructing activity patterns, to defining space use patterns is not yet clearly established. As is perhaps true of any work which is still in progress, there are parts of the schema which are more developed and therefore more easily understood than other parts. Until the work nears a more formal stage of formulation, the exposition of concepts and analytical sequence is likely to remain uneven.

A Framework Emphasizing Human Interaction

Webber also utilizes interaction as the basic organizing concept of his theoretical system. He views urban communities in two related perspectives – one in which human interaction occurs in a particular metropolitan community, and one in which it extends to widely scattered places over the face of the earth.[8] He calls the first a "place community" and the second a "nonplace community." With modern transportation and communications developments having the effect of stretching distances, he notes that individuals, firms, organizations, and institutions more and more have contacts, conduct transactions, and maintain communications on a global basis. Thus their ties may extend to a variety of nonplace communities as well as exist within a particular urban place. To distinguish them from the urban place, he calls these nonplace communities "urban realms."

It is this total concept of the urban community which is the distinctive flavor of Webber's approach. He calls for an understanding of the interaction systems which extend into larger urban realms as well as those which fall within a particular metropolitan area. Thus he holds that the study of systems of interaction within the urban region is no longer a complete and sufficient scope for metropolitan planning. According to Webber's concept, what goes on within the spatial confines of an urban place must be interpreted in the framework of all the ties that the community may have with the world at large. He notes that individuals may or may not engage in all their activities in a place community and, according to whether they are scientists, manufacturers, or writers, their interest communities extend to differing realms. These same individuals may participate in several non-work-related interest communities – in the arts, in recreation, in public service, and so on. In contrast, some persons, such as the butcher, the factory worker, or the clerk, may have interest communities which at present are completely contained within the place community. So today metropolitan planning requires a view which considers how the place population may also be a part of various realm populations, each with what Webber refers to as its own "space field" for interaction, some global, some national, and some in various regional contexts.

In both the place and nonplace view of the urban community, Webber emphasizes the importance of viewing the city as a "dynamic system in action." This dynamic feature is traced through "linkages," which he defines as "dependency ties" relating individuals, groups, firms, and other entities to one another. He terms these "the invisible relations that bring various interdependent business establishments, households, voluntary groups, and personal friends into working associations with each other – into operating systems." His spatial counterpart of this aspatial view of linkages involves three related perspectives. First is a view of the city in terms of spatial patterns of *human interactions* – the flow of communications, people, goods, and so on; second is a view of the *physical form* of the city – the space adapted for various human activities and the pattern to networks of communications and channels of transportation; and third is a view of the city as a configuration of *activity locations* – the spatial distribution of various types of activities by economic functions, social roles, or other ways of classifying activities.

Using these three perspectives of the city, Webber develops a six-way cross-classification system for describing urban spatial structure. Under his schema, he would measure spatial linkages, that is, the flows of information, money, people, and goods; he would study the channels used and space forms adapted for human interaction; and he would examine locations of activities. These observations would be classified as follows:

Dimension	*Interaction component*	*Physical component*	*Activity component*
Size of phenomenon	Amplitude	Capacity	Volume
Degree to which phenomenon piles up in major concentric forms around a point	Focality	Nucleation	Centralization
Propensity for phenomenon to pile up at points of lesser concentration	Subfocality	Subnucleation	Subcentralization
Degree of pile-up per unit (e.g., pile-up per 100 contacts between people, per square mile of area)	Intensity	Density	Concentration
Relative togetherness of like phenomena	Affinity	Clustering	Localization
Relative degrees of mixture	Insularity	Separation	Segregation

With the elements of the metropolitan community thus identified, measured, and classified, Webber recognizes that there is still a step beyond, one of using this framework in the investigation of the directions that growth and development might take. While this step is still to be made, he indicates it would involve an analysis of interaction in terms of the locational behavior of various types of establishments.

In Webber's work we have a conceptual system which is extremely broad in scope but still in an early stage of development. Presently, it consists of a framework for describing the city, one that is explanatory rather than normative. Webber emphasizes the importance of the dynamic aspect of a theory, but how the classification system will be used and how he will use it in the behavioral approach he favors for the analysis of interaction systems is not yet entirely clear. We know that he places a high premium on making these systems continuous and dynamic, and so we anticipate that some kind of interaction model to represent these as operating systems will become a central concern in the next stages of his theory building. It is premature to examine the empirical content of his work in its present stage, but his concern for the problem of measurement in both the place and nonplace aspects of his schema indicates that empirical tests are very much in his thinking. Certainly, Webber places heavy emphasis on achieving an internal logical consistency to his conceptual framework, and his concern for the study of locational behavior of the principal agents of interaction would indicate that as his work progresses he will be seeking a close representation of reality.

A Conceptual System Focusing on Urban Form

Lynch and Rodwin view the city as being made up of what they call "adapted space" for the accommodation of human activities and "flow systems" for handling flows of people and goods.[9] Although they differentiate between activities and flows on the one hand and adapted space and flow systems on the other, so far they have devoted their main effort to the latter level of analysis, which they equate with the study of "urban form." The distinctive feature of their conceptual system is the emphasis they place on the formulation of goals as an integral part of their framework. Their work begins with the study of urban form; it then focuses on the specification of goals; and finally it draws upon the goal-form analysis to indicate the nature of the planner's task in efforts aimed at shaping urban form in line with the goals that have been identified.

In their conceptual framework they are concerned first with a system for analyzing urban form. (Were they starting with activities and flows, they might well focus initially on a system for analyzing *interaction.*) Lynch and Rodwin propose evaluating urban form by six analytical categories: element types, quantity, density, grain, focal organization, and generalized spatial distribution. "Element types" is a category for differentiating qualitatively between basic types of spaces and flow systems; and, as might be expected, "quantity" has to do with amounts – a measure of the size of particular types of adapted spaces and flow systems. "Density," expressed either as a single measure or as a range of measures, has to do with compaction (of people, facilities, vehicles) per unit of space or capacity of channel. "Grain" is their term to indicate how various elements of urban form are differentiated and separated. Adapted spaces and flow systems may be fine-grained or coarse-grained according to the extent of compaction or separation in their internal components (houses, skyscrapers, streets) and how sharp or blurred these form elements are at the edges where transition occurs from one element to another. "Focal organization" is concerned with the spatial disposition and interrelations among key points in the city (density peaks, dominant building types, major breaks between forms of transportation). "Generalized spatial distribution" is the patterned organization of space as it might be seen from the air at a high altitude. This six-part classification system is the basic analytical tool they propose for classifying urban form.

The second major conceptual problem which the Lynch-Rodwin framework seeks to deal with is the formulation of goals utilizing this analytical tool. They point out that the problem is not alone one of identifying out of the multitude of possible goals those that have significance for urban form, but it is also one of specifying the goals in concrete terms which leave no doubt as to how they are to be realized. The identification of goals is one aspect of the problem, and the specification of content is a second aspect. With respect to the first, Lynch and Rodwin point out that goals must to some extent be determined in the normal democratic processes, with community-held goals being carefully differentiated from the planner's personal goals (which would tend to emphasize the goals of only one segment of society). But at the same time they would give careful attention in the choice of goals to the planner's goals as a professional and urban designer, where, they point out, he has a proper role to play in seeing that more advanced values take their place in the community value system beside the familiar ones of long standing. They suggest that the choice of goals have first a human and then an economic basis. Thus goals relating to urban form are fundamentally concerned on the one hand with relationships between man and his environment and between man and man, and on the other hand with the efficiency of these relationships – maximizing the return and minimizing the cost in both a social and an economic sense. The specification of goal content derives from the analytical framework they devised in the first instance. Thus the goals would be specified in terms of type of adapted space and flow system, quantity, density, grain, focal organization, and the spatial distribution pattern. Some would have quantitative emphases, some would deal more in qualitative concepts, and all would be subject to continuing checks as to relevance and reasonableness.

The final aspect of the Lynch-Rodwin framework is concerned with the application of the goal-form statements in the study of the city and in establishing what emphases will be needed in the plan that eventually is to emerge. Through the use of simple cross classifications of the six components of their system of analysis applied to both adapted spaces and flow systems, they demonstrate how these two elements of urban form interact under different goal emphases. In sum, Lynch and Rodwin view the framework as a means for analyzing urban form in a systematic and logical manner. They think of it as a means of posing the problems for planning, but they leave the solution to the planning task which follows from their framework.

This conceptual system has been developed to deal with a particular aspect of what we look for in a total system of theory. Although they acknowledge the important role which the interaction level of study has in theoretical formulations, so far Lynch and Rodwin have limited themselves to urban form, directing their attention to the physical implications of human interaction. Their work is concerned with the rationale of planning for the city rather than a framework for analyzing the structure of the city and explaining how growth occurs. In this sense they are providing a framework which has special significance for plan-making. In focusing on goal formulation, they have injected an essentially normative emphasis into their scheme. However, they do not specify "what should be"; rather, they indicate how goal combinations can be analyzed in deriving "what should be" in a particular locality and how these in turn may be integrated systematically into the planning process. The importance of a theory being a dynamic one is recognized by Lynch and Rodwin, and, in the sense that the sequence from goal formulation to form analysis in their conceptual system has a continuous and dynamic interrelationship, it is dynamic in conception. But in the sense we have been using "dynamic" to signify the organizing aspect of theory which takes account of the evolutionary process of urban development, their framework is as yet incomplete. Their work reflects sensitivity to the importance of empirical verification. This has since become particularly evident in Lynch's studies of the perceptual form of the city.[10] They have given careful attention to the logical continuity of their conceptual system, and there is clearly a great sense of responsibility for tying their work closely to reality.

Accessibility Concepts and Urban Structure

Although all conceptual systems of the kind we have been discussing sooner or later become concerned with accessibility as an element inherent in the physical organization of space and movement systems, some work gives this concept a more central role in building theory. Much of the recent work on accessibility concepts have been primarily focused on transportation.[11] Although this work has had a very considerable impact on research in urban spatial structure, we do not attempt to include in this summary any report on work which is primarily transportation oriented.

Guttenberg develops a theoretical approach to urban structure and city growth which utilizes accessibility as an organizing concept – what he calls "a community effort to overcome distance."[12] In the sense that human interaction is the underlying reason for minimizing distance, he is implicitly viewing interaction as the basic determinant of urban spatial structure. However, his work focuses primarily on the physical facility aspect of a total system of theory. In place of the simple two-part view (space use for activities and interconnecting systems of transport and communications), he identifies three components. He subdivides the first into "distributed facilities" and "undistributed facilities," with these being a function of the third component, "transportation." The rationale states that if transportation is poor, the workplaces, trade centers, and community services will tend to assume a pattern of distributed facilities; if it is good, these activities will assume more concentrated patterns in the form of undistributed facilities. Thus Guttenberg maintains that urban spatial structure is intimately tied up with the aggregate effort in the community to overcome distance.

In his framework, he sees the spatial gradation of density outward from distributed and undistributed facilities as a function of access. He points out that his distributed centers of activity acquire a value in accordance with the substitutability of that place for the chief place, with the physical density gradient outward from these centers corresponding closely with, but not necessarily directly coincident with, the economic density gradient. In the context of his framework, therefore, the slope of the economic density gradient is closely related to transport efficiency as it enables outlying locations to substitute for more central locations.[13]

In examining the implications of growth for his concept of urban structure, he points out that the transportation system holds the key to the way in which growth proceeds. The transportation decisions made from one year to another will result in a constantly changing urban structure, with the emphasis shifting along the continuum between the situation with highly distributed centers to the situation with one major undistributed facility. He implies that there is some limit in the ability of the undistributed facility continuing indefinitely to function as the only major center as compared to the capacity that distributed centers have for absorbing growth. As growth occurs, structural adjustments to overcome distance can take the form of either new centers or improved transportation facilities. Commonly both occur. With the enlarged scale and resulting changed relationships among home, work, and various activity centers, population movements ensue. With these shifts the areas which do offer the accessibility that people seek develop, and those which do not, decline in a social and an economic sense. With growth, the enlarged scale alters the density gradient. If transport efficiency is improved, favoring the substitution of outlying for central locations (as has been the case in Los Angeles), the slope of the density gradient is flattened, the region spreads out, and, depending upon the amount of population influx in relation to the area added, the density may go down.

How may such a view of urban structure and urban growth be used in anticipating urban form in the future? Guttenberg acknowledges that transport efficiency is not the sole variable. He notes that activities may choose a location in relation to a central place for reasons other than time-distance. For example, a change of economic composition in the region may produce new location patterns. However, assuming such other things are constant (similar economic conditions, terrain, tastes, and so on), he maintains that accessibility in terms of time distance serves to sort activities spatially. If the additional assumption is introduced that the transportation system remains similar over time, he points out that there will be comparability in accessibility and therefore we may anticipate similar patterns in the distribution of activities in the region. He does not discuss the complexities of prediction involved when constraints are relaxed and one by one the elements held constant are allowed to vary, but it is clear that by introducing differing combinations of assumptions the interplay of these elements quickly becomes exceedingly complex.

In the present stage of its development, this conceptual framework centers mainly on the physical aspects of a theoretical system of urban structure and growth. Its distinctive feature is the emphasis it gives to the interplay between the location of urban activities and transport efficiency. In the sense that activity concentrations and transportation are continuously interacting and that accessibility provides an organizing rationale for urban structure and a regulating concept for urban growth, the framework is a dynamic one, supplying an evolutionary basis for explaining urban form. While it has a well-developed logical context and a direct relation to reality, Guttenberg's statement gives no indication as to how this framework is to be translated into an analytical system and given empirical form.

Other work which should be cited here includes Hansen's use of the accessibility concept in the analysis of the growth of residential areas and Voorhees's use of the concept in the analysis of other use activities.[14] However, both are primarily dealing with the pragmatic aspects of prediction rather than the formulation of a more general system of thought governing urban spatial structure. Hansen defines accessibility as "a measurement of the spatial distribution of activities about a point, adjusted for the ability and desire of people or firms to overcome spatial separation." His concept of accessibility is very similar to Guttenberg's, and he has formulated a model which has useful immediate applications.

Economic Models of Spatial Structure

In some respects quite different from the approaches discussed so far, the economic models approach the conceptualization of urban spatial structure in the traditions of economic theory. The roots of this work go back to an agricultural land development concept advanced by von Thünen in the early 1800s; work in more recent time has been done by Alfred Weber, Lösch, Isard, and others.[15] The economic approaches discussed here make use of what is known in economics as equilibrium theory. Since forms of notation used in equilibrium theory are specialized and some of the technical aspects of the concepts involved in this theory are outside the scope of this discussion, this work is presented in somewhat abridged form. The student versed in economic theory will want to pursue this line of theoretical development in the original sources of work cited here.

Essentially, the view taken in these approaches sees urban development processes as economic phenomena. The organizing concept is the market mechanism and the sorting process it provides in the allocation of space to activities. In the work on urban spatial structure, this involves allocation of space in both quantitative and locational aspects to various users according to supply-and-demand relationships and a least-cost concept in an equilibrium system.

Wingo's work provides perhaps the most systematic and rigorous statement of urban spatial structure in the framework of equilibrium theory.[16] Traditionally, economists have dealt with location as a constant, and there has been a disinterest or an unwillingness to examine location as a variable. In his work, Wingo lifts this constraint. He seeks to give explicit recognition to the way in which policy affects the market and how in turn these effects are reflected in urban spatial structure. In this sense, he is seeking to relate theory to real-world situations. However, in addition he seeks to bring developments in spatial models into closer harmony with general economic theory and to relate theoretical work on location to the broader concepts of the urban economy.

Directing his attention mainly to residential development, Wingo develops first a concept of transportation demand, considering the spatial relationship between home and work. With the journey to work viewed as "the technological link between the labor force and the production process," he defines demand for movement as the total employment of an urban area multiplied by the frequency of work – in other words, the number of trips required to support the production process. As Meier has done, Wingo recognizes the propensity for urban society to substitute communications for transportation and stresses the necessity for taking into account technological developments in this respect. The supply aspect is expressed in terms of the capacity of a movement system – a measure of its ability to accommodate movements between home and work. Drawing on a somewhat similar concept of accessibility as that discussed above, he uses as a unit of measurement the cost of transportation based on the time spent in movement between points and the out-of-pocket costs for these movements, expressed in money equivalents for distance and number of trips.

The central problem of this kind of economic model is to achieve an equilibrium distribution of households of particular rent-paying abilities to sites with a particular structure of rents. Wingo achieves this location equilibrium by substituting transportation costs for space costs. Thus, on the supply side, he utilizes transport costs to establish the distribution of household sites at varying position rents. He defines position rent as "the annual savings in transportation costs compared to the highest cost location in use." On the demand side, if prices for other goods competing for the household dollar are held constant, the rents households are willing to pay are based on the class utility concept, which holds that the greater the unit rent, the fewer the units of space consumed. Clearly, this view of space use immediately involves density, and the smaller the quantity of space consumed in the more accessible locations, the higher the density. The spatial distribution of these densities in the urban area involves the density gradient concept noted earlier, with the slope falling off from the center of the city to the outskirts. To get at the characteristics of demand in the spatial context, Wingo constructs a demand schedule and utilizes appropriate position rents from this schedule to determine the point at which prices and densities are in equilibrium.

The economic model that Wingo advances functions under the usual behavioral axiom that those who control residential space and households who seek space will each behave to maximize their returns. He specifies as givens: the locations of employment centers, a particular transportation technology, a set of urban households, the marginal value the worker places on leisure, and the marginal value households place on residential space. Wingo then uses his model to determine the spatial distribution of densities and rents, and the spatial distribution, value, and extent of land required for residential use. For the derivation of the elements to the model as well as the mathematical form, and for his discussion of the empirical advantages and limitations of the model, the reader is referred to the source.

Although it is beyond the scope of this discussion to go further into this work as it relates to economic theory, we may note that as a theoretical system it is the most developed one considered so far. The market mechanism furnishes the dynamic aspect and organizing concept for the theoretical framework. The conceptual system is rigorously stated, and logical consistency is carefully observed throughout. Wingo has sought to maintain a close contact with the real world, and although empirical tests of this work are still to be made, we do have indications from experience cited below that this kind of economic model can be made operational.

Alonso uses the market mechanism in a somewhat different manner to distribute space users to urban land.[17] Instead of developing a demand function, he uses "bid price curves," which in interaction with the price structure of land are used as a basis for distributing agricultural, business, and residential users to sites (the space users that he has selected for attention). Beginning at the center of the city, land is "put up for bid," and on the basis of these curves the bid for the most central site is compared to the next preferred alternative, with this preferred alternative being the marginal combination of price and location for that particular use. On the basis of the steepest bid price curve, the highest bidder takes the most central site; the next highest bidder corresponding to the second steepest curve takes the next most central site still available; and so on. For the first space user the price paid at the bid location is determined by the price at the marginal location, but for the next user what was a marginal location for the first user becomes the equilibrium location for him, with his bid price determined by the price at the marginal location for this site, and so on down the chain.

There are several models with more of an empirical content which should be noted. One is Herbert and Stevens's linear programming approach to distributing households by maximizing their rent-paying ability.[18] Harris integrates a refined version of this model into a system of growth models he has developed for the Penn-Jersey Transportation Study.[19] Another example comes from Artle's work on the Stockholm economy in which he suggests the use of a simple regression model and an income-potential model to get at the clustering of establishments as they become distributed to retail sites.[20] Still another piece of work of particular promise for planning is found in Lowry's linked series of models to determine the distribution of various use activities in the Pittsburgh metropolitan area based on pre-established assumed locations of "basic" employment and on indices of trip distribution developed from an area-wide transportation study.[21] As another example, Garrison proposes a general simulation model of "urban systems" which would involve an analysis of urban structure and growth by means of a whole series of models.[22] Finally, the work of the RAND Corporation in the use of linked mathematical analyses should also be noted.[23] These are indicative of a whole series of developments in mathematical and economic models, but because they are presented primarily in methodological terms, those now in operational form are more appropriately taken up elsewhere.

Decision Analysis and the Structure of Cities

As still another approach to urban spatial structure, the author proposes that a conceptual system based on the *values-behavior patterns-consequences* framework . . . offers an organizing concept for theory development. In its most basic form and viewing the components in reverse order, this framework seeks explanations for any particular man-induced phenomenon being studied (in this instance, urban growth and development) in terms of human behavior (interaction patterns), with such behavior patterns being a function in turn of people's values (or the attitudes held concerning interaction).[24]

In common with the view taken by several of the foregoing conceptual systems, under this framework it is a behavioral axiom that human beings tend to concentrate at various places on the earth's surface in the satisfaction of needs and desires for interaction. Interaction is a form of behavior, then, growing out of complex interrelations among men and their various institutions – for example, the interrelations between individuals, or households, or firms, or the interrelations among individuals and households, households and firms, and so on. These interrelations may take rational or irrational forms. The term "behavior patterns" is used to refer to rational and overt forms of interaction, both the forms which cluster spatially into duplicating or near-duplicating patterns and remain relatively unchanged for extended periods of time, and the forms which take variable spatial patterns and appear in different locations from one day to the next. For example, common duplicating behavior patterns are involved in transactions between wholesale and retail firms or between customer households and a supermarket. A very well known behavior which tends to appear in cluster-like patterns is visiting behavior. Nonclustering behavior patterns, those that are more scattered, less pronounced, and variable in spatial location from one day or week to the next, are illustrated in the circuits of salesmen visiting customer firms, or family drives on Sunday afternoons, or outings in the country. . . . These patterns of interaction can be classified into *activities* ("within-place" interaction) and *communications* ("between-place" interaction), but for the present we shall think of the two as a related combination under the generic term "behavior pattern." Although as planners we are concerned with both clustering and nonclustering patterns of behavior, to get at the elements of urban spatial structure we are particularly interested in the way duplicating or clustered patterns build up in space. Further, we are concerned with clustered patterns in a time sense, that is, the repetitive aspects of these patterns in daily, weekly, or seasonal cycles.

Out of the universe of behavior patterns which are constantly evolving, then, certain ones have a spatial importance, show a tendency to duplicate in clusters, and occur in rhythmical and repetitive forms in the course of a day, a week, or a season. These behavior patterns are constantly undergoing adjustment and reformation in response to value orientations, but at the same time some of these behavior patterns have sufficient importance in their spatial, duplicating, repetitive characteristics to produce outcomes in physical structure and form. As in most of the other conceptual systems reviewed here, these physical forms are composed of (1) *spaces* adapted to the various forms of place-related interaction, and (2) *interconnecting channels* for the various forms of movement-related or communications-related interaction – one growing out of the need for *activities* and the other for *communications* (in its narrower meaning). Yet to understand fully urban concentrations as outcomes of human actions requires an understanding of relevant antecedent behavior patterns and value orientations. Thus there is an evolutionary basis for the study of ordered arrangements of urban form that we see on the ground, and our ability to predict how these arrangements may occur in the future is dependent upon our ability to anticipate the full evolutionary sequence. At the same time, it is necessary to recognize that this evolutionary sequence from values to interaction to outcomes is modified by a feedback aspect – by the effect that these outcomes may have on subsequent interaction and on value orientations, and all the secondary, tertiary, and other successive effects.

The means by which interaction patterns become translated into structure-form outcomes is found in the *location behavior* of households, firms, government, and institutional entities. Location behavior is viewed as a sequence of action in the same three-part framework, growing out of the needs and desires of day-to-day interaction. While occurring on a day-in-and-day-out basis, for any particular household or firm, location actions occur infrequently. The location action, then, constitutes a different type of behavior pattern from the daily activity patterns that we have been discussing. But at the same time, the location action is the instrumentality by which the activity patterns of the first type are accommodated in physical form. For this strategic reason, under the present conceptual system location behavior is given special attention.

In our society, the marketplace and the council chambers of government (local and frequently nonlocal) provide the means by which location-specific activities and movements (or, more broadly, communications) are translated into place-fixed use areas and interconnecting circulation systems in the metropolitan area. These two mechanisms mediate location decisions. Neither can function without the other; each must be responsive to the other. In part because one involves many widely diffused and independent decisions while the other involves relatively few and concentrated decisions, and in part because government must represent the overriding public interest, in matters of space use and transportation it falls to government to assume leadership in seeking harmony between the two systems. Insofar as metropolitan planning is responsive to both systems, it becomes the means of securing this harmony.

Now clearly the manner in which the market and government interact in mediating location behavior is extremely complex. One way to follow this process is to focus on the *decision* as the critical point to the behavioral sequence in a location action. Under a decision analysis approach, of the many kinds of decisions by which space is adapted and put to use and movement systems established, two groups can be differentiated. One group involves what we may call "priming decisions" in the sense that they are seen to trigger the other group, what we may call "secondary decisions," with the two together accounting for the development as a whole. Priming decisions are made in both the public sector (for example, those involved in major highway locations or utility locations) and in the private sector (for example, decisions on large-scale investment in land, on the location of industries with large employment or the location of major shopping centers or Levittown-type developments). They set the stage for secondary decisions, for example, park acquisition or street-widening decisions in the public sector, or small-scale subdivision, mortgage-financing, lot-purchasing, or home-building decisions in the private sector.

Priming actions tend to develop from single decisions or a mix of discrete decisions of some strategic importance in setting off a chain reaction of development, and secondary actions usually consist of clusters of decisions (for example, clusters of household decisions) stimulated by, but following from, the strategic actions. Because it would be impractical to unravel and deal with the separate effects of all kinds of decisions, the emphasis in this framework is on priming decisions – discovering what mix of these actions tends to trigger other actions and thus influence the course of events which accounts for the pattern of development that subsequently emerges. Therefore, under this conceptual system, land development is viewed as the consequence first of certain strategic decisions which structure the pattern of growth and development and then of the myriad household, business, and governmental decisions which follow from the first key decisions. This view of the land development process is a greatly oversimplified representation of what actually occurs. Initial investigations of the priming-secondary decision aspect of the rationale suggest that the "queuing order" or sequence of priming decisions bears an important relationship to the generating power which a mix of these decisions exerts on secondary decisions.[25] To take account of decision sequence and the differential lag times in the impact of different decisions may call for more detailed subclassifications of the present priming-secondary basis of classifying decisions. How much more elaborate the system can feasibly be made is a problem for future study.

Meanwhile, an experimental model has been developed for testing this conceptual approach.[26] The initial emphasis of this work is on experimentation with household location. The model is designed to test planning proposals which seek to shape growth and development into prescriptive forms that will fulfill interaction needs. As preconditions, it assumes the locations of major employment centers (the only "given" in the private sector), and it assumes certain location-specific, government-prescribed public works decisions and policy positions. It then seeks to simulate market-mediated location decisions of households, firms, and institutions. It looks to the basic needs and desires of interaction and how these become translated into key decision factors, such as space availability and the value man ascribes to land and its particular mix of access, amenity, and conveniences. In short, "given" certain postures of government plus an assumed pattern of basic employment, the model is addressed to the question: How can these location decisions shape those of the private sector to produce the pattern of development set forth in a plan? In the priming-secondary decision nomenclature, priming decisions are "givens," and the task of the model is to simulate secondary location decisions – a distribution of households to available and usable land pursuant to a predetermined evolutionary sequence of priming actions. The basic approach used in this model can be extended to other land use decisions, and it can be made to simulate the renewal of outworn areas.

Thus, if we consider priming decisions as structuring elements of urban growth and therefore of special concern in a metropolitan area plan, such a model offers a tool for planning. Indeed, a distinctive aspect of this approach is its utility as an aid to decision making by governing bodies, planning agencies, developers, and many other groups. If the influence on land development of a particular mix of priming decisions with a particular timing sequence can be simulated, this means that it is possible for governing bodies, for example, to take this knowledge into account in choosing among the decisions. A governing body might use the model to study the effects of one particular decision, say a decision on the location of an expressway, and, holding all other factors constant, determine what the probable pattern of household settlement would be for each of several alternative routes and what each of these settlement patterns would mean in terms of other public costs. As suggested, another application uses the model in an opposite way. Given several alternative schemes (for example, a compact single-centered emphasis in the pattern of growth, a diffused pattern of growth, and a cluster pattern of growth), each predicated on achieving one defined set of goals but following different policy emphases, the model might be used to establish what mix of priming decisions would tend to produce each form of development. With this kind of output from the model, relative long-range costs and benefits implied in each policy choice can be evaluated in reaching a decision on what plan and policy to pursue.

In effect, this model therefore is a simplified statement of the conceptual view of urban structure and growth set forth at the outset. The location decisions simulated by such a model are seen as satisfying interaction needs and desires of urban residents, firms, and other entities. Indeed, priming decisions are viewed as the result of pressures which arise from these daily interaction needs and desires. If this is the case, there is a premium on studying not only prevailing *activity patterns* but also *attitudes* of different population groups for insights they supply into the likely future stability of these patterns. It may be anticipated that future work on this conceptual system will emphasize two lines of research development: (1) the analysis of activity patterns from data obtained by home interviews, firm interviews, and so on, and (2) the analysis of attitudes from a paralleling line of inquiry in the course of these interviews.[27] These areas of inquiry are seen as supplying parameters for the location model. . . .

The conceptual system we have been reviewing is an explanatory one. It seeks an explanation of urban spatial structure through the study of location behavior and more particularly through the analysis of selected classes of location behavior in terms of daily activity patterns and related value systems. In this kind of approach norms may be introduced in the form of alternative choices in the decision process, but the normative aspect is not an integral part of the conceptual system. Particular stress is placed on the use models in making empirical tests of aspects of the system and in making the approach dynamic, with a direct relation to reality. In its present early stage of development, the framework is still loosely formulated. However, as experience in the use of models in testing the conceptual system accumulates, we can look toward a more rigorous statement of the conceptual system.

Comparative Aspects of Theoretical Approaches

There are some important similarities in these systems of thought we have been reviewing and some significant differences. Explicitly or implicitly each system is directly concerned with the development of a framework which identifies and describes regularities in patterns of human interaction in space and explains their origins and transformations in time wherever population aggregates in urban areas. Along with these rather fundamental relationships in space and time is a common concern for accessibility, a concept of great importance in spatial relationships. In describing interaction patterns, most of these conceptual systems make a distinction between patterns of intraplace and interplace interaction, the former having importance for the adaptation of space and the latter involving communications between spaces. "Space adaptation" and "communications," of course, are counterparts for "land use" and "circulation". . . .

While there is this fundamental common base, other similarities may be found when comparisons are narrowed to two or three approaches. Some similarities are based on acknowledged cross-ties, as in Meier's acceptance of Webber's urban realm concept, or in Webber's adoption of Meier's communications emphasis. Some appear to be simply parallelisms among the conceptual systems. Thus, while Wingo sees man's economic behavior in the marketplace as the medium which regulates location decisions, with adjustments introduced to take account of noneconomic factors influencing these decisions, the author's work in effect sees the urban social system setting the context for locations decisions, with government and the market serving as the medium which regulates location decisions.

Yet with these similarities and parallelisms, there are distinctive differences in approach. The differences are in part a function of the background, specialization, and research biases of the person advancing the approach. Beyond this, one senses differences in conception of what would constitute a proper set of criteria for theory building whether or not they are consciously considered in any particular approach. Meier's work reflects a unique perspective across the expanse of both physical and social sciences, and, drawing on this broad scope of interest, he directs attention to the opportunities for synthesizing elements from different fields in a theory of urban spatial structure. Deeply concerned with the implications of technological and social changes for patterns of human interaction, Webber emphasizes the placelessness of many new forms of activities and the importance of the dynamic aspect of theory building to take these changes into account. Lynch and Rodwin have focused their attention on one part of the broad area we have been discussing, dealing with this part in some depth and laying particular stress on a normative approach. Wingo's work reflects a strong inclination toward a rigorously classical and systematic approach to theory building. Guttenberg has a sensitivity to the policy implications of theory and the alternatives open to metropolitan areas in making transportation choices. The author's work with his associates at North Carolina places a strong emphasis on relating theory on urban structure and growth to theoretical work in public policy and decision making.

It should be clear by this time that there is a significant effort going into theoretical research, a healthy variety to approaches being explored, and a developing emphasis on an experimental view. Some question the present-day emphasis in the social sciences on the use of experimental designs. This is in part a reaction against the heavy emphasis in these times on the natural sciences. The advent of symbolic logic and mathematical models is seen by some as a hocus-pocus and a dream world of faddism seeking to mystify with esoteric language and mathematical learning. There may be some excesses of this kind, but the real difficulties are more likely to come from the blind hope of formula hunters who use models without looking into the qualifying conditions which apply in any particular situation. It should be noted, too, that the distrust and occasional eruption of impatience with the use of mathematics in theory building is not peculiar to the planning field. Old-line theorists in economics, political science, and sociology have frequently spoken out on this matter.[28] The planning field has come to this controversy more recently, but the arguments are not too different. Harris offers a cogent brief for the place of models in city planning, and the reader would do well to look up his comments on this subject if he has lingering doubts about the utility of mathematics in planning theory.[29]

SOME GUIDES FOR LAND USE PLANNING

What does this glimpse into examples of work in progress tell us about the scope and approach to be used in land use planning? Remembering the tentative nature of this work and the fact that it is still in the process of formulation, what can be said? What guides can be safely set down at the outset for the land use planning task? The answers to these questions are to be found in elements common to all the conceptual systems we have been examining. In this respect, this work provides at least two leads for the land use planning task we are about to take up – one concerned with a point of view about land use, and the other, with a process of study. . . .

What can be said about a land use point of view can be set down fairly simply in the form of two guiding considerations. The first is the notion that land use is concerned with human activity in a very broad sense. It is concerned with living patterns of households, productive patterns of industries, selling patterns of retail and personal service establishments, and the many other classes of activity patterns that exist and interact as elements in the urban social system. Thus, from this viewpoint, "land use" means a great deal more than existing or proposed improvements visible on the ground. Indeed, it may be said that space use has no intrinsic meaning separate and apart from human activity – the reasons for which the space was put to use in the first place. Land use maps of the conventional type are useful in answering the question "where" at a particular moment in time, but they tell us very little about "why," nor do they say very much about "where" *over time.*

This brings us to the second requisite of an adequately inclusive point of view toward land use. This is the view of land use as a constantly evolving and continuously changing phenomenon. An evolutionary view has great importance in one's outlook toward the land use plan. For example, if we take what is sometimes called the "unitary approach" in which emphasis is placed on the single, detailed scheme, we are essentially ignoring change, or we are taking the view that planning intercepts change, neutralizes it, and thus obviates the necessity of providing for it. More consistent with the trend of this discussion is the "adaptive approach" – an evolutionary scheme which through the medium of development policies is progressively adjusted in the flow of time to take account of unpredictable elements of technological and social change.[30]

Indeed, a proper recognition of this dynamic or evolutionary aspect of land development involves the very process of study followed in developing a land use plan. So, finally, we may ask what guides grow out of the work we have been examining that offer assistance in organizing for the task of land use planning. Ultimately, we may expect some very direct and tangible guides from theory itself. Indeed, when theory flows more directly into practice, and as we substitute more advanced systems of analysis for "hand" methods, the guidelines singled out here as leads may in fact be built into the analytical framework itself. Four emphases tend to crop up again and again in the work above:

1. The necessity of a system of analysis which is continuous rather than discontinuous.
2. The importance of the system taking account of activity linkages where changes in one element has the effect of altering the climate for change in another element. Recognition of the feedback influence which gives rise to the need for modifying change initially instituted.
3. Significance of the random aspect of human interaction and the importance of a probabilistic view toward development rather than a deterministic view.
4. The interrelation of policy, proposal, and action – phased to lead successively from one to the next.

These are complicated ideas, and we do not deal with them here.

REFERENCES

[1] Prediction is used here in a time sense. Of course, in a formal hypothesis-testing usage, prediction concerns explanatory theory too. In this sense relationships are predicted, and tests are then made to establish the existence, direction, degree, and nature of these predicted relationships. In using prediction in a time sense, theory is dealing with probabilities, not absolute determinism nor the mysticism of prophecy.

[2] Russell L. Ackoff, *Scientific Method: Optimizing Applied Research Decisions,* New York: John Wiley & Sons, Inc., 1962, p. 16.

[3] Herbert A. Simon, *Models of Man,* New York: John Wiley & Sons, Inc., 1957.

[4] Adapted from Rosalyn B. Post, *Criteria for Theories of Urban Spatial Structure: An Evaluation of Current Research,* unpublished manuscript, Chapel Hill, N.C.: Department of City and Regional Planning, University of North Carolina, 1964.

[5] Richard L. Meier, *A Communications Theory of Urban Growth,* Cambridge, Mass.: MIT Press, 1962.

[6] *Ibid.,* p. 131. A "bit" refers to the unit of information which can be handled in a binary system of digits, the system used in modern digital computers. By making use of the off-on combinations of an electrical circuit, the computer is able to carry coded messages in the binary system.

[7] *Ibid.,* pp. 48-54, 129-132. He excludes private communications as having no cultural significance and thus irrelevant in the study of the cultural processes of the city which shape growth.

[8] Melvin M. Webber, "The Urban Place and the Nonplace Urban Realm," in Webber (ed.), *Explorations into Urban Structure,* Philadelphia: University of Pennsylvania Press, 1964.

[9] Kevin Lynch and Lloyd Rodwin, "A Theory of Urban Form," *Journal of the American Institute of Planners*, Nov., 1958.

[10] Kevin Lynch, *The Image of the City*, Cambridge, Mass.: Harvard University and Technology Presses, 1960.

[11] See work of J. Douglas Carroll and his associates in Detroit, Chicago, and Pittsburgh on metropolitan area transportation studies (for example, J.R. Hamburg, "Land Use Projection and Predicting Future Traffic," *Trip Characteristics and Traffic Assignment*, Highway Research Board Bulletin 224, National Academy of Sciences, 1959); see also the work of Alan M. Voorhees in the Baltimore, Hartford, and Los Angeles transportation studies (for example, C.F. Barnes, "Integrating Land Use and Traffic Forecasting," *Forecasting Highway Trips*, Highway Research Board Bulletin 297, National Academy of Sciences, 1961).

[12] Albert Z. Guttenberg, "Urban Structure and Urban Growth," *Journal of the American Institute of Planners*, May 1960. [Reading No. 9]

[13] Robert M. Haig made a similar point 35 years ago: "Since there is insufficient space at the center to accommodate all activities which would derive advantages from location there, the most central sites are assigned, for a rental, to those activities which can best utilize the advantages, and the others take the less accessible locations. Site rents and transportation costs are vitally connected through their relationship to the friction of space. Transportation is the means of reducing that friction, at the cost of time and money." See Robert M. Haig and Roswell C. McCrae, "Major Economic Factors in Metropolitan Growth and Arrangement," *Regional Survey of New York and Its Environs*, I, New York: Regional Plan of New York and Its Environs, 1927, p. 39.

[14] Walter G. Hansen, "How Accessibility Shapes Land Use," *Journal of the American Institute of Planners*, May 1959; Alan M. Voorhees, "Development Patterns in American Cities," *Urban Transportation Planning: Concepts and Application*, Highway Research Board Bulletin 293, National Academy of Sciences, 1961.

[15] J.H. von Thünen, *Der isolierte Staat in Beziehung auf Landwirtschaft und Nationalökonomie*, Hamburg, 1826; C.J. Friedrich, *Alfred Weber's Theory of Location of Industries*, Chicago: University of Chicago Press, 1928; August Lösch, *The Economics of Location*, New Haven, Conn.: Yale University Press, 1954; Walter Isard, *Location and Space-Economy*, New York and Cambridge, Mass.: John Wiley & Sons, Inc., and Technology Press, 1956.

[16] Lowdon Wingo, Jr., *Transportation and Urban Land*, Washington, D.C.: Resources for the Future, Inc., 1961.

[17] William Alonso, "A Theory of the Urban Land Market," *Papers and Proceedings of the Regional Science Association*, VI, Philadelphia: University of Pennsylvania, 1960. For a fuller statement, see Alonso, *Location and Land Use: Toward a General Theory of Land Rent*, Cambridge, Mass.: Harvard University Press, 1964.

[18] John D. Herbert and Benjamin H. Stevens, "A Model for the Distribution of Residential Activity in Urban Areas," *Journal of Regional Science*, Fall 1960.

[19] Britton Harris, "Experiments in Projection of Transportation and Land Use," *Traffic Quarterly*, Apr. 1962.

[20] Roland Artle, *Studies in the Structure of the Stockholm Economy*, Stockholm: The Business Research Institute at the Stockholm School of Economics, 1959.

[21] Ira S. Lowry, *A Model of Metropolis*, Santa Monica, Calif.: The RAND Corporation, Aug. 1964.

[22] William L. Garrison, "Toward a Simulation Model of Urban Growth and Development," Proceedings of the Symposium in Urban Geography, Lund, 1960, *Lund Studies in Geography*, Lund, Sweden: C.W.K. Gleerup, 1962.

[23] Issued from time to time since 1960, a series of memoranda contain reports of these investigations. (Processed material, Santa Monica, Calif.: The RAND Corporation.)

[24] Based on the behavioral science schema developed by the Urban Studies Program, Institute for Research in Social Science, University of North Carolina. See Chapin, Introduction, in F. Stuart Chapin, Jr., and Shirley F. Weiss (eds.), *Urban Growth Dynamics*, New York: John Wiley & Sons, Inc., 1962.

[25] F. Stuart Chapin, Jr., and Shirley F. Weiss,

Factors Influencing Land Development, Chapel Hill, N.C.: Institute for Research in Social Science, University of North Carolina, in cooperation with the U.S. Bureau of Public Roads, Aug. 1962.

[26] Thomas G. Donnelly, F. Stuart Chapin, Jr., and Shirley F. Weiss, *A Probabilistic Model for Residential Growth,* Chapel Hill, N.C.: Institute for Research in Social Science, University of North Carolina, in cooperation with the U.S. Bureau of Public Roads, May 1964.

[27] For examples of attitude studies of urban residents, see John Gulick, Charles E. Bowerman, and Kurt Back, Chapter 10, and Robert L. Wilson, Chapter 11, in Chapin and Weiss (eds.), *Urban Growth Dynamics.* (See Ref. 24.)

[28] For arguments pro and con, see James C. Charlesworth (ed.), *Mathematics and the Social Sciences,* a Symposium, Philadelphia: American Academy of Political and Social Science, June 1963.

[29] Britton Harris, "Plan or Projection: An Examination of the Use of Models in Planning," *Journal of the American Institute of Planners,* Nov. 1960.

[30] For a discussion of the "unitary" and "adaptive" views toward a plan, see Donald L. Foley, "An Approach to Metropolinan Spatial Structure," in Webber (ed.), *Explorations into Urban Structure,* pp. 63-75. (See Ref. 8.)

. . . there is a design *orientation, which regards urban form as a phenomenon subject to control and direction for a variety of purposes such as defense, efficiency, beauty, and healthfulness. Historically, this stream of thought has had a limited but powerful influence, expressed mainly through a few master-planners and master-architects. It seems not unlikely that the most successful and influential of these possessed not only a creativity of design, but an implicit understanding of the fundamental sources and scope of urban function and city building resources — thus, some measure of analytical and scientific understanding. The relative importance of this aspect of social thought about cities for our present purposes has been established in the past century by two major changes in real life. First, as a result of technological advance and social change, the size and complexity of our urban concentrations has grown enormously. Second, the relative expansion of the private market economy in urban land, and the growth of a pluralistic society, have greatly complicated the processes of decision-making and control in urban development.*

Britton Harris, "Some Problems in the Theory of Intra-Urban Location," Operations Research, *Vol. 9, No. 5, September-October, 1961, p. 698, 699.*

11

Edwin von Böventer

SPATIAL ORGANIZATION THEORY AS A BASIS FOR REGIONAL PLANNING

In regional planning, as in all economic decision making, existing and expected potentialities have to be evaluated in the light of certain welfare ends. In this article, I shall be concerned with the *potentialities* that the economy offers, defining the goal as the maximization of real income per head. Thus I shall consider regional planning in its role of *adaptive planning* (to use John Friedmann's terminology), with the national rate of savings, the rate of technological progress, and changes in population being taken as given. My main concern will be the question: What insights can be derived from the theory of spatial organization of the economy about the *optimal spatial distribution* of economic activities and hence of new investment in a growing economy, from a long-term point of view?

In analyzing this question, the real-income concept will be modified: the costs of commuting, including the monetary equivalent of the psychic

---

costs of commuting, should be subtracted from each family's income. For both theoretical and practical reasons no exact figures can be derived for this negative item; but it is obvious that with the expansion of cities and the growth of suburbs commuting costs have been increasing rapidly in almost all modern economies, and therefore some allowance for these costs should be made in the evaluation of alternative regional plans. Administrative problems of decision making and of policy implementation will not be discussed here.

At the beginning, two extreme situations should be mentioned which would rule out most kinds of planning. The first would be one which takes all prices, individual input and output quantities, *and all locations* as variables and thus aims at a spatial general-equilibrium model. On this basis, equilibrium *conditions* may be stated, but no quantitative generalizations can be derived, and regional planning would be virtually impossible. If, in addition, it were assumed that all the decisions of individuals always lead to the optimum solution from society's point of view, no regional planning would be necessary.

The other extreme situation would occur if we were to establish certain regularities about the actual spatial distribution of economic activities and settlements (such as city rank-size rules or observations about the economic gravitational forces of existing centers) *and to consider the observed relationships as largely unchanging and immutable.* In this case, most regional planning would be useless, because it could not change anything anyway.

In establishing a basis for regional planning, one has to take a position somewhere between these extremes, while remaining aware that there is *some* validity in both positions – fortunately, one may add, because it is important to have, on the one hand, a great deal of freedom so that planning can change the conditions for individual decisions in the direction that is thought beneficial for society, and, on the other hand, some rigidities, which are desirable because they can serve as guideposts to action. There are, in addition, some basic economic interrelationships which do not reduce the flexibility of the system but make possible certain generalizations which are of help in devising regional plans. These include interregional or intermetropolitan relationships which can be derived from general economic theory and foreign trade theory.

In my approach to planning, therefore, I would use the results of location theories, foreign trade theory, and input-output studies (with industrial complex analysis as one method of integrating them[1]), and combine these with the results of theories of spatial structure. The theoretical basis of both the following remarks and of regional planning as such will be a combination of certain rigidities (such as input-output relations), certain spatial interdependencies, and individual flexibilities of locational choice, with additional consideration of the regularities and systematic shifts in technology and in consumer preferences.

THE BASIS OF REGIONAL PLANNING

The starting point of regional planning is the total of the available *resources,* the *production possibilities* as determined by such restraints as the state of technology and the institutional structure of the economy, and the *production possibilities actually realized* on the basis of individual choices and governmental decisions at the various administrative levels – with due regard to their distribution in space. As a second step, the expected future changes in technology and behavior (including institutional changes) have to be appraised, and projections about changes in the available resources have to be made. Third, on the basis of these projections or expectations, estimates about future production possibilities and future product mix have to be derived, subject to the various restraints and shifts in demand arising from changes in the preferences of the society or region. Fourth, there must be an appraisal of the spatial distribution of the projected activities that would follow in the absence of any *new* policy measures that might be prescribed.

It is here, then, that spatial analysis and regional planning have their contribution to make. On the basis of spatial analysis it has to be determined whether the expected spatial distribution is optimal within the given restraints and on the basis of society's preferences. If the resulting pattern is not optimal, regional planning has to devise measures for improving upon the expected result, or for effecting changes in the spatial structure necessary to stimulate long-term growth in underdeveloped regions richly endowed with natural resources, for example. It is obvious, however, that regional planning would be expected to influence, in varying degree, all the factors mentioned under the first, second, and third points. The attainable optimum is therefore itself a variable which is dependent on the results of planning.

The most important economic factors and developments which bear upon the spatial aspects of the economy and which have to be considered here may be summarized as follows: internal and external economies, transport costs, and the demand for land.[2] Space plays a twofold role: on the one hand, it is an input to production and a commodity demanded by private households for residential and recreational purposes; on the other, it is an obstacle to economic exchange, especially over long distances, since it gives rise to transport costs.

By increasing *internal economies* (within firms) and *external economies* (as between firms), technical and organizational progress leads to an increasing division of labor, not only between production processes but also within and between geographic centers of production and regions. Other things being equal, this process always fosters the tendency toward increased concentration of production in larger firms and in urban centers.

The restraining effects of transportation costs on economic exchange, in particular over long distances, have been declining. Partly as a result of modern highway construction and the cheapened cost of air travel, partly by reflecting the relative rise in value added per unit weight, the ratio of transportation costs to the price of the finished commodity has gone down. This means that more goods than ever before can be transported economically over long distances. At the same time, the mobility of both labor and capital has increased. Thus *distance has lost part of its effect as a restraint on production.*

At the same time, space in its role as an input to production has remained important. While it is possible to economize on land through the use of more capital, population growth tends to increase the demand for space as an input to food production, at least on a worldwide scale. Finally, as a result of population and real-income growth, the private demand for space for consumptive (residential and recreational) purposes has been gaining in importance.

The following have been the outstanding consequences of these developments. Rising internal and external economies, as well as the declining role of transportation costs, have promoted an increased spatial concentration of production and the growth of metropolitan areas, while increased mobility of workers and rising demand for space, both in an absolute sense and on a household basis, have resulted in the phenomenon of suburbanization within metropolitan regions. In general, the spatial separation of the various stages of production, as well as of workplace and residence and the sites where vacations are spent, has become easier. Consequently, the distances traveled by both labor and commodities have grown rapidly and may be expected to do so in the future. This is particularly true for commuting.

In discussing the relative decline of transport costs and enhanced overall mobility, it is essential to distinguish between the effects on *commodity movements, movements of labor,* and *commuting.* All three types of movement are, to a significant extent, substitutes for each other but have quite different effects on the spatial distribution of economic activities. High mobility of labor, if combined with small commodity mobility, leads to agglomeration; while growing commodity mobility (and small labor mobility) leads to a dispersion of economic activities. On the other hand, commuting and suburbanization allow a combination of the advantages of spatially concentrated production (by way of internal and external economies) and spatially dispersed consumption (large individual residential lots) – at the considerable cost of the expenses and personal impositions of long daily work trips. It is noteworthy that the two kinds of transportation costs involved appear in radically different form in national income statistics: commodity transport costs are subtracted in national income calculations, while commuting costs and the costs of population movement are not. Besides, the personal efforts involved in commuting represent additional inputs which are necessary to procure the commuter's income.

In planning decisions, these additional costs should

not be ignored. Considerations of welfare suggest that commuting costs should be subtracted if the incomes generated within alternative spatial distributions are compared. This alternative between commodity and labor movements constitutes one of the most important choices within regional planning. In the following section, locational choices will be discussed, first from the individual's point of view, and subsequently from the standpoint of economic interdependencies within the nation.

DETERMINANTS OF SPATIAL STRUCTURE

In working out its optimum arrangements, the individual producing unit has to make decisions about the scale of production, the relevant input coefficients, and the production site. These decisions are interdependent. Location theory tells us that, for an optimal choice of site, total transport costs must be minimized – for each scale of production and for each set of input coefficients and prices. At the same time, the scale of production will be heavily influenced by internal and external economies. The greater these economies are, the greater will be the scale of production and, by implication, the size of the centers of agglomeration. As long as transport costs remain high, exchange over increasing distances becomes less and less profitable; transport costs therefore tend to restrict the scale of output of a single firm, while internal economies work in the opposite direction. The optimal size of the firm is obtained as a compromise between these two forces, assuming demand, technology, and input prices as given.

Certain regularities in spatial structure emerge as a result of this optimizing process, if we assume (1) a perfectly homogeneous plain with supply conditions everywhere identical and demand distributed evenly in space, and (2) *absence of any external economies.* In this event, the supply firms for each homogeneous good will have equal size and will locate in the *centers of regular hexagons,* simultaneously minimizing transport costs for their products and maximizing distances from their nearest competitors. Since the optimal size of the firm and the optimal size of the market area will vary for different goods, a complex system of networks of markets and a high degree of specialization is obtained. Certain goods will be offered in every second settlement, others in every seventh, still others in every ninth, and so forth. This is the essence of Lösch's model of spatial structure in his path-breaking study, *The Economics of Location.*[3] It can thus be shown that production centers or cities of different size will develop and that they will tend to do so even on the specialized assumption of a homogeneous plain.[4]

Before Lösch's study appeared, the hexagonal layout of the market nets for individual commodities had already been used to supply the basis of central place theory, which was first developed in a systematic model by W. Christaller in his *Die zentralen Orte in Süddeutschland.*[5] Here, the most "national" commodity, which has the largest minimum market area per firm, is considered first. This commodity is offered only at the geographical center of the plain; additional suppliers would not be able to stay in the market without incurring losses. Less "national" commodities, with smaller minimum market areas, would be supplied from the largest center as well as by producers in smaller central places. These smaller centers will develop in the interstices left by the bigger centers which have come into existence first. This kind of derivation can be thought of as describing an historical development where new settlements are established on an initially thinly populated plain, and where, as Adna F. Weber remarked as early as 1899, "the boundary stone becomes the predecessor of the market cross" (that is, of the boundary between different market areas).[6] In this kind of derivation, there is no *specialization* as between different cities, but a *definite hierarchy of centers* develops, with every larger center offering at least as many goods as every smaller one. As to the spatial layout of *production* sites, this pattern does not correspond to reality. The finding is more realistic and certainly more significant, however, if only the *distribution* of goods and services is considered, that is, if the hierarchic pattern is applied to the tertiary sector of the economy. Moreover, a more or less regularly distributed city hierarchy would also evolve as long as the population densities did not differ too much among geographical areas (this will be discussed below).

Even on a homogeneous plain, both the hierarchic

Christaller model and the Löschian specialization model have to be modified as soon as product differentiation and personal preferences are considered. In this case, the individual market areas would overlap – but this is a relatively minor objection. Really significant reservations to an uncritical application of these models are, first, that, as a result of the development of city hierarchies, demand will vary for different geographical points and, second, that many important economic interdependencies modify the Löschian results quite markedly even for a homogeneous plain.

In the real world, with the uneven surface of the earth, the optimal site of an industrial enterprise can, *within a short-run partial analysis,* be determined as a function of such factors as the locations of the relevant raw materials, natural traffic routes, rail and highway nets, transshipment points, and major consumption centers. In addition, spatial differences in the prices and availabilities of labor, capital, and commodities must be taken into account. For regional analysis, it is important to know how these factors become effective, how they are affected in turn by the mechanism of economic adjustment, how their influences may be modified through regional planning, and what kinds of generalization are possible under more realistic conditions than the assumption of a homogeneous plain.

The answers to these questions are important because the location factors mentioned become effective through the given level of transportation costs and the external economies generated by these factors, and all of them are subject to change in the process of economic development. This is true even for raw material sites, whose pull either weakens with the decline in transportation costs and the development of cheaper substitutes, or grows if demand for the produce should rise.

As conditions of equilibrium, with perfect information, (1) each enterprise and household must have found the site where it will maximize its profits or utility function; (2) in each market, prices have to be adjusted until supply equals demand and, for each region, payments equal receipts; and (3) spatial price differentials between any two points must never be greater than the unit transport costs between the two points. This holds for spatial differences in the price of labor as much as for commodities. Even if an equilibrium is never attained, there is always a tendency for additional goods to be shipped to places where prices net of the respective transport costs are particularly high, and for labor to move to places where it can earn more (in real terms) or otherwise increase its general level of satisfaction.

It has to be stressed that, for analyses of spatial structures, the most difficult problems arise in connection with external economies. As long as external economies, in the form of agglomeration economies, are among the most important factors (but not *the* most important factor) determining spatial structure, some significant generalizations are possible. These will be described in the next section; in the following section the theoretical implications and policy consequences of the existence of agglomeration economies will be discussed.

SPATIAL STRUCTURE: POSSIBLE GENERALIZATIONS

Within a partial analytical framework, from the point of view of the individual enterprise or household, it is not possible to derive generalizations with material content upon which an overall theory of spatial organization could be built. A more comprehensive framework is required to describe the influences and interdependencies of location factors and to make possible useful generalizations. These generalizations pertain to systematic variations in space of production techniques (or input coefficients) and the size of industrial plants, shopping centers, and communities, as well as to regularities in the spatial distribution of settlements. I shall not be concerned with production techniques of individual firms, with plant sizes, or the establishment of shopping centers. The relevant points within this more comprehensive framework are regional specialization and city size. In this framework, attention will be focused not on the effects, but on the causes of price differentials. Whenever, as a consequence of growing demand, there is a shortage of goods or production factors, the adjustment mechanism works through price differentials. Thus, in low-productivity regions, nominal wages must be lower so that entrepreneurs are offered at least *one* advantage over other regions and, similarly, prices must be lower, so that goods can be shipped to other regions and compete there successfully.

In deriving generalizations about spatial structure, these interdependencies among relative shortages of goods or production factors, price differentials, and locational choices are paramount. They can best be illustrated under certain restrictive assumptions about population distribution, but the results carry over for more general situations.

If the initial spatial distribution of the population within the region to be studied is rather even and if, for sociological reasons, population mobility is small, the following general statements are possible even if the spatial distribution of the other factors of production is not in any way regular[7]:

1. The locations of primary and secondary activities are determined by raw material sites and traffic routes, as well as by "historical accidents" (as seen from the economic point of view – personal entrepreneurial decisions or political acts), all of which have led to the growth of certain centers of demand and thus attracted new industries and service activities. For the latter, it may be assumed that they are spatially distributed according to some hierarchic Christaller pattern, adapting themselves to the locations within the agricultural and manufacturing sectors and to the traffic routes of the region's economy, but also influencing these sectors in turn.

2. As a result of the partial concentration of activities in cities of various sizes, the land values, housing rents, and a number of other prices and expenditures, such as for transportation, are higher in urban than in rural districts. For this reason, and in order to attract additional workers into the cities, *nominal wages* must be higher in the cities; and the greater the rate of growth – that is, the more additional employees an urban district intends to attract – the greater the wage differential has to be.

3. As a consequence, urban centers have the advantage of greater local markets and of certain agglomeration economies of production while, as was mentioned, rural areas can offer lower wage costs. The optimal site of a new enterprise will then depend on the relative importance of the agglomeration economies, the existing wage differentials, and the level of transportation costs. With low labor mobility (and hence great wage differentials) as well as low transport costs, areas with few industries will enjoy a considerable comparative advantage for the production of labor-intensive goods which also depend on local raw materials. As long as agglomeration economies are not overwhelming and rural areas are everywhere densely populated, the economic adjustment mechanism favors the establishment of *some industries* in all parts of the region and thus works toward the establishment of *hierarchies of settlements or central places.* Which particular industries are established in the individual subregions may be determined either by the special raw material deposits available, by purely personal entrepreneurial decisions, or by other historical accidents. In this way, a certain degree of *specialization* among areas or subregions may evolve and central places of similar sizes may develop. Thus, through the pull of wage and price differentials, a certain spatial regularity in the distribution of the central places may be obtained for the region as a whole. This may be true even though no spatial regularity obtains for the *production* of all individual goods. On the other hand, as far as the *distribution* of the commodities and the services is concerned, some degree of regularity in the spatial layout may be expected.[8]

4. However, as the agglomeration economies of production and marketing grow and the labor mobility rises, and as the model comes closer to describing the real world in a highly industrialized or postindustrial society, large metropolitan areas develop, the spatial regularities diminish, and there is also a certain specialization within the services sector among cities. These tendencies will be discussed in the next section; the focus will be on the agglomeration centers and the growing demand for land.

GROWTH POLES AND AGGLOMERATION ECONOMICS

External economies may arise for a particular firm whenever *other* firms exist within the *same* branch and/or within the same location: in the latter case, the external economies take the form of agglomeration economies. In economic development at new producing centers, the successful operation of one or several existing firms offers certain external economies or diseconomies to all firms that may follow suit. These obtain either within the same industry for the country or region as a whole (where they may arise in connection with procuring raw materials and other inputs, with the establishment of retailing facilities, and so forth), or within the firm of a particular industry at the same location as *localization economies.* For firms in *other* branches which operate at the same location these economies arise in the form of *urbanization economies.* Agglomeration economies may thus be divided into localization and urbanization economies. They consist of such factors as economies in transportation (unit transport costs declining as the number of firms and their transport volumes increase), access to labor pools, growth of local demand for the products of all suppliers and resultant sales economies, changes in the administrative costs of the city as calculated per firm, and power supply economies.[9]

Growth is initiated by particular entrepreneurs who take advantage of potential markets or of available inventions and thus introduce innovations, or by passive adaptation to growing demand. A process of growth is set in motion by business firms that follow the innovators either at the same location or at some other locations. Thus growth rarely proceeds evenly: it may be viewed as starting within particular firms (by Schumpeterian *entrepreneurs*) or as being promoted within, and from, particular points in space (Perroux's *pôles de croissance*).[10] Thus the Schumpeterian concept of *entrepreneur,* developed without reference to the spatial dimension of economic development, is complemented by the concept of *growth poles* as the agents of economic development in space.

If a second business firm follows an innovating enterprise, there are always two tendencies at work which, in a dimensionless economic model, work in opposite directions for the first firm: a reduction of profits and the (positive) effects of certain external economies. The first firm producing a new commodity or introducing new production techniques at an existing center may have been able to reap windfall profits as a temporary monopolist, which are subsequently reduced by the competition of the second firm. This may also hold for the first entrepreneur at a new location. But the first firms of *other* branches may be at a great disadvantage in the beginning, particularly if they depend heavily on local demand, which may be forthcoming only after a great number of other firms have located at the same place and have attracted employees who then become customers of the formerly established firms.

For these reasons, locational patterns can be changed only slowly: regardless of whether a certain place represents a good site in the beginning, its competitive position for new firms in general improves through the very fact that a significant number of firms have chosen to settle there, and that all firms derive agglomeration economies from their combined existence and existing functional interdependencies. In general, such external economies are insignificant at very small centers. They will grow, however, as new firms appear and population increases, although the *marginal* external economies for a center may eventually decline and even become negative. The point at which such economies start to decline is not the same for all industries. The same economic conditions that give rise to positive agglomeration benefits for one branch may lead to *negative* agglomeration economies for others.

Investigations have shown that the power of an agglomeration center to attract new business is a positive function of the size of the center; in some studies, population movements from a given region to a certain center have also been found to vary in proportion to the size of the center. As long as such gravity forces work toward a social optimum, or if a socially desirable moving equilibrium is established, nothing needs to be done about them. This tendency does not appear to be true for all growth processes in urban-metropolitan centers, however, many of which appear to be too large from the standpoint of an optimal spatial distribution of economic activites. At the same time, other centers appear to be too small to be able to reap agglomeration economies that somewhat larger centers enjoy. It should be stressed at this point that there are definite theoretical reasons why one should expect this to happen whenever there are agglomeration economies which rise at first and later on decline.

If a settlement is below its optimum size from the standpoint of public services, additional firms or households could be accommodated at lower unit costs for road construction, road maintenance, and other services. Thus *marginal costs* of supplying public services to additional firms are declining and will be lower than average costs. *If the community now tries to cover its expenses through taxation in a balanced budget,* and if it does not discriminate among firms, *all firms* will have to pay taxes equal to the average of the costs of *supplying the public services.* Taxes cannot be levied on the basis of the marginal costs, because the community would incur losses. Thus each additional firm pays higher taxes than the additional community outlays it causes. On the other hand, there is no justificiation for having new or expanding firms pay *less* than already existing firms in the community.

The problem is the same for a private firm producing at declining marginal costs: this firm cannot set its prices equal to its marginal costs because it would incur losses and would be unable to survive.

The solutions suggested by welfare considerations are the same for the firm and for the community that offers its services at declining costs: prices or tax payments should be set equal to marginal cost, and the difference between marginal and average costs should be covered by subsidies. The community would run a deficit on this account, and this should be covered by transfer payments from the central or state government. From the point of view of sub-optimal settlement patterns, this approach sounds plausible: the relatively low tax rates (together with improved community services) should induce additional firms to locate at this place, until the optimum has been reached. The opposite holds for centers where *marginal* social costs are higher than their average; these centers would run a surplus, which is to be used to subsidize the suboptimal centers.

In any case, these transfer payments would aim at equalizing marginal *private* costs (including tax payments) and marginal *social* costs. If high social costs are more than outweighed by market advantages and if, for this reason, additional firms decide to locate at the big centers, there is no theoretical economic reason for interfering with this decision.

This concept of an optimal city size is a purely economic one. It is not the only one relevant in the present context. Before other considerations are discussed, however, it has to be pointed out that marginal and average cost data depend very much on what is included in the public service sector and that many more empirical investigations are necessary before such a concept can be applied generally, quite aside from the political problems of implementing such a policy.

The optimal size of cities can be determined only by taking all the relevant economic, sociological, political, and cultural factors into account, and depends on the policymaker's values with regard to the functions of urban centers.

One factor that does not enter national income statistics directly but is directly relevant for the evaluation of optimal city sizes is the rising amount of time spent in commuting. If individual commuting costs were subtracted from all incomes, if all road construction and maintenance for commuters were treated as *intermediate goods,* and if the psychic costs of commuting were taken into account, the optimal size of cities would probably appear to be much smaller than if these costs are not allowed for.

The factors mentioned so far would tend to reduce the optimal sizes of urban centers and their suburbs and would point toward a subsidization of the smaller agglomerations. There is another important fact which supports this conclusion: the long-term population decline of predominantly agricultural regions. These regions suffer in two ways. In the first instance, they lose important agglomeration and market economies to urban centers and more densely populated areas and consequently become less and less attractive as locations for manufacturing production. Second, the adjustment processes which accompany economic development are less costly and much easier to obtain if the overall rate of economic growth is high and if the adjustment can be effected by way of differential *positive* rates of growth than if most activities within a region have to be reduced absolutely. Since, in the long run, more space will undoubtedly be needed, it would be wise to constrain this temporary depopulation process now and to encourage *growth factors* in all regions by reducing incentives for population movement out of the region. It is imperative for planning to create new growth poles in such regions through a combination of general subsidies, such as tax relief, specific subsidies to industries that are suited for the region in question, and the construction of the necessary structure of public services and facilities.

PAST EXPERIENCES AND FUTURE GROWTH: THE TASK OF PLANNING

It was said in the beginning that the optimal spatial distribution of economic activities cannot be derived within a general model where almost everything is variable. The bases and yardsticks for planning have to be the existing interindustry and interregional relations, and all that economics can do within this context is to comprehend their relations, to reveal important implicit relationships and to show where the structures deviate most obviously from their optimal points.

The process of arriving at an optimal plan for a region may be divided into four closely related parts:

1. The present and expected future comparative advantages of the region have to be determined.
2. The expected movements of capital and labor into and out of the region have to be estimated and must be compared with the desired rate of migration.
3. The expected and the desired industry or activity mix of the region have to be specified.
4. The optimal spatial distribution of these activities has to be decided.

The first task consists of the determination of the activities in which the region should specialize. This task has to be solved by means of all the available tools of location theory, the theory of international or interregional specialization, input-output analysis, linear programming, and other techniques as described by W. Isard.[11] The solution depends on the factor endowments of all regions, the transportation and traditional trade connections of the region in question, and its historical development. No generalizations about the result are possible here; this depends on the specific situation.

To solve the second task, it is necessary to determine first how movements of labor and capital are related to price and interest differentials and differential regional growth rates, to job availabilities, and to investment possibilities. As the next step, projections have to be made for the critical variables, and the magnitude of the expected movements of labor and capital has to be assessed. As a third step, *desired* population and capital movements have to be derived on the basis of studies of agglomeration economies and population movements and a determination of optimal production patterns. Finally, incentives for migration have to be created so as to gear actual migration to the desired levels.

Third, the future activity mix of the region should be derived from input-output studies based on the region's factor endowments, the current and expected input-output structure of the region and of other regions whose structures are similar to the present or, better still, to the expected structure of the region. Thus the industrial structure of regions that are already further advanced economically may serve as an important yardstick under appropriate safeguards. On the basis of such input-output studies, it may be possible to estimate what the region can produce, particularly in the critical export industry sector, and to deduce how far all other secondary and tertiary activities may be expected to expand in output and employment.

The fourth task consists essentially of distributing in space the activities that have been determined as optimal for the region. It is necessary for planning considerations to be based on an empirical observation of the patterns of existing spatial structure, with their hierarchies of centers of different sizes. The spatial structure of a region depends, apart from historical or accidental factors, on such variables as population density, natural resources, and the stage of economic development. The most satisfactory way of obtaining an optimal pattern for the region would be to study the spatial structure of an economically further advanced region, and to modify this structure in the light of the welfare considerations mentioned, while allowing for historical and noneconomic factors that have shaped its spatial structure in the past. The normative spatial pattern derived in this way must of course be compared with what would be expected to evolve in the absence of planning measures.

The most important task would be to determine

where additional growth poles are needed, in which centers economic growth should be stimulated, and in which parts of the region the pull of the agglomeration forces should be reduced.

CONCLUSIONS

The concept of planning as developed in this paper recognizes the importance of historical structures. Since neither input-output relations nor spatial configurations can be derived from the thin air of pure economic theory, one has to use historical structures as a starting point. It is then that economic theory may be introduced as a tool for modifying the historically given and as a means of rendering the whole system more flexible. Its task lies in determining the optimal degree of specialization among and within regions, and by specifying the desired degree of agglomeration activities and residences within a given region. At this point, value judgments about the respective roles of the city, the suburb, and the outer belt become important considerations.

Historical observations are also needed in supplying the basis for studies of the relationships between past structural changes and the factors that induced them. These relationships must be the foundation of all policy measures designed to influence individual location and migration decisions. This principle holds for deliberate changes both in regional wage, price, and profit differentials and in infrastructure, inasmuch as these changes are meant to influence individual decisions.

As long as the economy is not expected to reach a stationary state, the above-mentioned "standards" for the region must not be interpreted as describing a fixed goal. The goal is rather to move further ahead every year, so that the main function of the standards is to indicate a desirable *direction of change.*

In this context it may be asked whether this direction of change would be expected to follow an equilibrium path or should be planned to do so. The *general* answer to this question is negative. The plan would certainly not aim at maintaining or establishing a dynamic equilibrium in the short run. It is aimed not at sustaining short-run market forces, but at interfering with them, at reducing the free market rates of growth at some places, and at creating growth poles where otherwise economic development would be insufficient, at the same time raising the overall rate of growth of the economy.

REFERENCES

[1] See Walter Isard, *Methods of Regional Analysis* (Cambridge, Mass., and New York: MIT Press and John Wiley & Sons, Inc., 1960).

[2] Edwin von Böventer, "Towards a United Theory of Spatial Structure," *Regional Science Association, Proceedings of the Zurich Meeting,* 1963.

[3] August Lösch, *Die räumliche Ordnung der Wirtschaft* (Jena, 1944). The English translation of this book is *The Economics of Location* (New Haven, Conn.: Yale University Press, 1954).

[4] For details, in particular the full catalogue of assumptions on which these results depend, see Lösch, *op. cit.;* Walter Isard, *Location and Space-Economy* (Cambridge, Mass., and New York: MIT Press and John Wiley & Sons, Inc., 1956); Edwin von Böventer, "Die Struktur der Landschaft, Versuch einer Synthese und Weiterentwicklung der Modelle J.H. von Thünen, W. Christaller und A. Lösch," in R. Henn, G. Bombach, und E. von Böventer, *Optimales Wachstum und Optimale Standortverteilung* (Schriften des Vereins für Socialpolitik, N.F. Bd. 27, 1962).

[5] Walter Christaller, *Die zentralen Orte in Süddeutschland* (Jena, 1933). For a detailed discussion, see Brian J.L. Berry and Allen Pred, *Central Place Studies: A Bibliography of Theory and Applications* (Philadelphia: Regional Science Research Institute, 1961); and von Böventer, "Die Struktur der Landschaft," *op. cit.* The derivation in the text coincides with the spirit, but (for the sake of brevity) not with the letter of Christaller's presentation: Christaller's starting point is a commodity with a sales radius of 21 km.

[6] Adna F. Weber, *Growth of Cities in the Nineteenth Century* (New York, London, 1899), p. 171, as quoted by W. Christaller, *op. cit.*

[7] See also von Böventer, "Die Struktur der Landschaft," *op. cit.*, and "Towards a United Theory of

Spatial Structure," *op. cit.* for further details, in particular with regard to (3) below.

[8] It has to be stressed that in the entire literature there exists no model within which a definite *quantitative* statement about city-size relations has been derived on the basis of all economic and sociological interdependencies, for the simple reason that these are too complex to be caught within a single model, no matter how intricate.

[9] Isard, *Location and Space-Economy, op. cit.*

[10] See François Perroux *L'Économie du XXème Siècle.* (Paris: Presses Universitaires de France, 1961).

[11] Isard, *Methods of Regional Analysis, op. cit.*

The Growth Unit is first of all a concept -- a general way of saying that America's growth and renewal should be designed and executed not as individual buildings and projects, but as human communities with the full range of physical facilities and human services that ensure an urban life of quality.

The Growth Unit does not have fixed dimensions. Its size in residential terms normally would range from 500 to 3,000 [dwelling] units – enough in any case to require an elementary school, day care, community center, convenience shopping, open space, and recreation. Enough, too, to aggregate a market for housing that will encourage the use of new technology and building systems. Also enough to stimulate innovations in building maintenance, health care, cable TV, data processing, security systems, and new methods of waste collection and disposal. Large enough, finally, to realize the economies of unified planning, land purchase and preparation, and the coordinated design of public spaces, facilities, and transportation.

National Policy Task Force, "The First Report of the National Policy Task Force," Memo, *Special Issue, Washington, D.C. (American Institute of Architects) January 1972, p. 4.*

PART IV INFORMATION

There are broadly two schools . . . the first, corresponding to orthodox management practice, declares that a study of [sufficient] information will reveal patterns and trends in the data, which will enable experienced managers to feed instructions back to the situation through its input loop — and thus modify its behavior. The second school of thought, corresponding roughly to the position of operational research, is more realistically aware of the magnitude of the problem. It says that the human brain cannot cope with all this information, and that the thing to do is to create an analytic model of what is going on.

Stafford Beer, Decision and Control, *The Meaning of Operational Research and Management Cybernetics, New York (Wiley), p. 277.*

Many are thus coming to visualize an idealized urban intelligence center, that is, an effective city planning agency, as being equally oriented to improving theory and action, for we know that the dualism that would distinguish them is false. In contrast to a mere "information center," such an intelligence center would seek to describe and explain what is going on, to report on stocks and flows, and to identify cause-and-effect behavior. Using simulation-type techniques, it would try to predict what would happen if *one course of action were taken rather than some other, and to trace the repercussions of those actions through as many of the subsystems as our theory and data permit. By thus feeding forward predictions and likely outcomes, the center would inevitably become an agent of change, affecting actions and subsequent outcomes. . . .*

Melvin M. Webber, "The Roles of Intelligence Systems in Urban-Systems Planning, "Journal of the American Institute of Planners, *November 1965, p. 294.*

12

Ida R. Hoos

INFORMATION SYSTEMS AND PUBLIC PLANNING

There is an unmistakable tendency on the part of public planners from county to federal levels to assume, since their mandate is to "plan rationally," that their first and primary need, in order to discharge this obligation, is a management information system. Implicit in this persuasion is a set of apriorities: (1) if, as public planners, they had more information, they would make better plans, and perhaps arrive at better decisions; (2) more and faster-moving information would improve the "efficiency" of governmental operations; (3) greater "efficiency" would better serve the needs of the community in particular and society at large; and (4) the design of information systems is a technical matter and best assigned to an "information expert," whose movable talent is almost universally applicable. Thanks to the cult surrounding information systems, critical inquiry into these assumptions amounts to a kind of heresy, but it is important that we examine them and review them as practiced lest the new mythology so dominate the social planning scene that only the voice of the devotees will be heard.

So as to provide the perspective for our scrutiny of these four apriorisms, we first analyze the principal elements with which they are concerned: (1) information, (2) system, (3) the information system. Clarification is necessary and proper because, while there is no gainsaying the fact that a body of organized information is essential to any systematic analytic process, confusion prevails both as to its proper constitution and to the qualifications of the "experts." There are no distinctions between quantity and quality, between the necessary and the busy.

INFORMATION

Information, data, and, especially in military parlance, *intelligence,* are terms often used interchangeably and frequently equated with *facts* and *knowledge.* As such, they enjoy acceptance in the public mind. Whether because of a historically derived reverence for knowledge or cultural heritage, we in the computerized age show an enormous respect for data. We feel comfortable with "hard facts," and the more, the better. The very concept "data bank" is permeated with virtue. Associated with the values of the Protestant Ethic, the notion not only conjures up the bright, lively, and good things associated with banking generally, – saving, interest, etc. – but it replaces the dreary and dusty archive, the dead record office.

The allure of a bankful of data, available on command, is practically irresistible to the public administrator. The data base has come to be regarded as the keystone of the art of planning and the arch of learning as well. The current generation of graduate students in almost every academic discipline are card carriers of the new genre. They can be seen on every college campus, the huge stack of IBM cards their project, the computer their hope for making sense out of and finding a hypothesis in the morass of material. No matter what their field or their topic, they first sally forth to gather data. In much the same fashion, the professional planner, whether in the employ of the CIA, the NEA, or the BSA,[a] whether dealing with pacification in Vietnam, education in the ghetto, or crime in the streets, whatever else he accomplishes, energetically collects data.

At this point, it might be well to underscore an interesting etymological anomaly. *Datum,* by origin, is something *given. Data,* the term now so familiar, is the plural form; but as conceived at present, it is something *gotten.* When recognized as such, *data* are divested of the qualities of accuracy and objectivity. In fact, the very opposite may be closer to the truth. The aggregation, selection, and organization of data are all part of a value-laden, mission-oriented process that renders absurd the notion that any information is "neutral." If this were so, it would probably be so vacuous as to be worthless, anyway. Separated from derivation, the context in which used, and the conclusions derived from manipulation, *data* is an empty concept. In operation, we shall see later, it is often fallacious and dangerous besides. C. West Churchman, discussing the social significance of computer technology, suggests that, in the context of social policy, there may be no such thing as accurate or objective information. "Instead, so-called 'information' is simply one kind of incentive, which can be used by one person or group to influence the behavior of another person or group. It is, in fact, a commodity with its own price, a commodity that serves the purpose of shaping social action."[5]

SYSTEM

The second element requiring analysis in our disquisition on the *information system* is that of *system.* This term is a coverall, and, not surprisingly, generous in scope, loose in dimensions, and imprecise in meaning. The Webster *International Dictionary* offers a range of definitions from which to choose: meaning number one is "an aggregation or assemblage of objects united by some form of regular interaction or independence; a group of diverse units so combined by nature or art as to form an integral whole, to function, operate or move in unison, and, often, in obedience to some form of control; an organic or organized whole." The second meaning is brief and to the point: "the universe; the entire known world." Number three, a bit less comprehensive, shifts attention to the nonmaterial: "an organized or methodically arranged set of ideas; a complete exhibition of essential principles or facts, arranged in a rational dependence or connection" hence number four: "a hypothesis; a formulated theory." Number five suggests structure: "a formal scheme or method governing organization, arrangement, etc., of objects or material, or a move of procedure; a definite or set plan of ordering, operating, or proceeding; a method of classification, codification, etc." Number six develops the same notion further, into "regular method or order; formal arrangement, orderliness." Meanings numbered eight through fifteen are specialized and run from *anatomical* through *legal* to *zoological.* Number seven is exceptional and worthy of sober contemplation: "the combination of a political machine with big financial or industrial interests for the purpose of corruptly influencing a government."[b] Purveyors of the systems approach, for all their claim to precision, have so far failed to reveal which definition they accept. Judging by their unanimous predilection for the plural form, i.e., the *systems* approach, we can only infer that they mean to embrace *all* the meanings with the possible exception of number seven!

Persons engaged in the analysis, design, and engineering of systems display a remarkable tendency toward solipsism. The system is what they *say* it is. This they study, this they manipulate. And by so doing, they define and delimit *other* systems, for these can only "interface with" the first system and cannot, therefore, be part of it. Paradoxically, absence of clear articulation about the system allows, at one and the same time, for both arbitrary eclecticism and broad inclusiveness. Already demonstrated as any one man's conception, a system, in the broad view, is "a set of parts coordinated to accomplish a set of goals."[6] Thus the term *system* is used freely in matters animal, vegetable, and mineral, in the inner city and in outer space. The semantic impoverishment that allows reliance on the same terminology for, say, nuclear weaponry and elementary education then leads to the assumption that systems design, engineering, and analysis as practiced in the first can be meaningfully and appropriately applied in the second. Since any system fits the description, they are alike and, therefore, amenable to the same treatment. The next step in this fallacious logic is that the person who is expert in one system is expert in them all. In practice, we find that there is just about as much justification for commiting society's malfunctioning systems to the care of a "systems expert" as to call upon a hydraulic engineer to cure an ailing heart merely because that organ is essentially a pumping system!

THE INFORMATION SYSTEM

Information means less than it says and *system* is an amorphous term, but a remarkable metamorphosis occurs when the two are joined together. The *information system* emerges as a tidy and finite entity, a commodity for sale by hardware and software merchants, the *sine qua non* of planners, business executives, and public administrators. Representing a fusing of computer technology and management science, the information system has gained rapid acceptance and enormous prestige. It begins to become evident, however, that the latter has exceeded its accomplishments in the business world where the technology was spawned. Despite aggressive sales campaigns on the part of hardware and software merchants, the rosy dreams of the 1950s have faded. Clerical costs still soar, computers handle many of the pedestrian routines but at a price. Most of the paper problems plaguing management still persist and they will proliferate. Disenchantment is being voiced in such hitherto enthusiastic columns as those of *Fortune,* where an author describes the "misguided euphoria" about computer installations and underscores the confusion as to just what constitutes and what is the purpose of an information system[2]. A survey reported in *Dun's Review* provides details on specific shortcomings. Systems reviewed are shown to inundate managers with useless information, the plethora of which obscures what might have been important. Managers cannot specify nor can the information systems supply just what is needed. Thus the $1 billion spent by U.S. industry on management information systems seems able neither to equip managers to make better decisions nor to establish justification in salubrious effects on profits[10]. For all their touted "efficiency" as an adjunct to record keeping, information systems have not provided a tool to estimate or justify their own cost effectiveness. There seems as yet to be no accounting of the very items crucial not only to the organization paying for the sophisticated technology but also and especially to the computer industry and purveyors of software in substantiation for claims made for their products. Measuring the cost effectiveness of computerized information has been the subject of many high-level conferences and the object of much professional concern[1]. But the present situation is best summed in the statement of W. Holst (Norwegian Industries Development Association): "the information problem is not solved by hurriedly spending a million pounds"[20].

Of the eudaemonia[c] of public planners who have discovered the information system and, therefore, think they can now proceed "rationally," we shall speak later. For now, it is important to report the extent to which imprecision about what an information system *is* and what it is supposed to *do* prevails as much in the social arena as in that of business affairs. Unfortunately, the confusion is only compounded when one examines the information systems proposed and designed for public use. There are, on the one hand, information systems that are supposed to help managers manage information. Such, for example, is the California Statewide Information System: "The Statewide Information System has the basic objectives of promoting maximum utilization of acquired information"[13]. There are, on the other hand, information systems that are apparently supposed to help managers manage the enterprise or organization. In the real-life situation, what starts out as the management *of* information becomes managemenet *by* information.

An example of this conception of an information center is to be seen in a proposed system for the Nassau County (New York) Department of Welfare. The project was specifically intended to "(1) establish Welfare Department goals and objectives; (2) define information requirements and managerial techniques; (3) establish information acquisition requirements; (4) establish information distribution requirements; (5) develop information feedback techniques; (6) develop decision-making techniques; and (7) develop computerized information systems." The ultimate objective was stated as "to aid the Welfare Department in optimizing programs, services, and resources to satisfy community needs." Implicit here is the notion that a computerized information flow is what is needed to improve the functioning of the welfare system and, presumably, deliver better service. In actual fact, Nassau County's poor people are suffering not because their records do not move but because they cannot. Recipients of public assistance, with annual family incomes under $5,000, have been found to be handicapped by poor health, education, and vocational skills, trapped by inadequate transportation facilities, and in pockets remote from jobs.[17]

Information Systems in Public Welfare

In California, public welfare has been the target of a number of systems studies, intended to reduce rising costs of aid to the needy. The underlying assumption is that more efficient management of information or paperflow will affect the trend. Management-minded administrators at every level of government recognize reorganization of the information system as a politically palatable device, whereas overt acknowledgment of the handicapping social and economic conditions contributing to burgeoning welfare loads could have unfavorable repercussions. Preoccupation with records, forms, and data processing is frequently a substitute for rather than an adjunct to intelligent social planning.

Aid to Families with Dependent Children was the program selected as the focus for a $225,000 study in California because, the consulting team averred, "it offers some hope of reduction using the techniques of systems analysis." This orientation ignored both the factors which determine dependency and the statutes regulating eligibility for and amount of aid; information-handling procedures assumed a key role in what was loosely defined as "the welfare problem."

Juxtaposed against a table of particular deficiencies of the current data-handling system, as though the proposed information system would correct the failures and shortcomings of the entire system of welfare and reduce dependency, was the following set of "design goals" for the information system study:

(1) to increase the flow of information in order to promote better service and management control at all levels; (2) to minimize administrative cost and improve efficiency; (3) to provide research and statistical data for State planning and program evaluation purposes; (4) to provide inquiry service for questions which cannot now be anticipated; (5) to provide fiscal data for State plan-

ning and evaluation purposes; (6) to provide a system sufficiently flexible to accommodate changes in needs, volume, policy, and/or data demands; and (7) to reduce the cost of operations below that of the present information system[23].

Review of numerous systems studies, covering a variety of subjects and executed by "experts" from a diversity of disciplines, reveals a predilection for what may be called "flip-chart analysis." Generalized objectives like the seven listed appear both in the original proposal and in the final report with attainment no closer in the latter than in the first place. In other words, the approach often turns out to be only superficially and semantically analytical, the expertness lying less with operations knowledge and more with jargon juggling. "Design goals" such as these are certainly commendable, but the information systems provide a rather doubtful vehicle for their implementation or realization. When one probes the items in the flip charts for substance or content, one finds them vague, generally applicable, and lacking in specific usefulness.

While there is no denying that current practices have shortcomings, we find little evidence that the analysts' criticisms stem from "technical analysis." They more often than not reflect a hasty review of operations and opinions gleaned from interviews with welfare personnel and others. There are no new insights here; the traditional bureaucratic complaints are merely being used as a springboard for their campaign to sell a new system. That it will overcome present deficiencies is highly problematical; in fact, it could create more trouble than it eliminates.

Notwithstanding the enticement of electronic technology and speed-of-light transmission of data, there still remains the fundamental question as to the appropriateness and relevance as well as the uses to which the information will be put. Information systems have gained ready acceptance in the innocent cloak of being the first and necessary step in the direction of rational planning. But herein lies one of the most serious dangers of information systems. Just because they may, indeed, become the basis for planning, now and in the future, the way in which they are conceived, for what purpose, and by whom remain crucial matters, unsatisfied, and usually ignored by technically oriented designers. Insensitivity to or lack of knowledge about the substantive issues are often washed out of sight in the deluge of detail enthusiastically captured.

This is clearly illustrated in a proposed welfare information system which would yield routine facts about age, sex, address, etc., and then respond to "special inquiries." For example, it could tabulate the number of cases in which the mother (unwed) was of a particular ethnic minority, with four children, under the age of six, known to have a mental history, with a police record. And, like the sorcerer's apprentice, it could keep on pouring out information – that the area in which the family lives has x number of substandard dwellings, y number of known drug addicts, and is z miles from the nearest police station. What is never made clear is how this cornucopia would "reduce the cost of operation below that of the present information system" (item 7 above).

If, in fact, reduction of cost of operations is the main concern, attention should be given to many matters besides the mere pushing of paper. It is to be recalled that little substantitation for cost reduction has been found in the experience of the business world. Far from cutting down costs or "reducing dependency," sophisticated information systems in public welfare might actually increase caseloads and costs by uncovering and bringing into the system eligible persons now outside public relief rolls. Speeding up investigative and certifying procedures might not be a clear-cut benefit to the system of welfare as we know it. If the claim of reducing cost of operations pertains to record keeping only, it has not been confirmed by actual experience. The sole reduction would have to be calculated in terms of unit cost of processing; actually, the free-flowing information would be no bargain. Without lowering administrative costs appreciably, the system would at best shift them, with effects on efficiency, however construed, speculative, conjectural, and nebulous.

Criminal Information Systems

Fraught with great significance not so much because of poor economics but rather because of bad social ethics are the many information systems being developed as a weapon in the current war on crime. A key item in the system of criminal justice proposed for the state of California, for example, was "the development of an information system linking together various agencies of criminal justice and being capable of evaluating program and system effectiveness through collection, storage, and processing of appropriate data." By this point in our exposition we should not have to pause to analyze and refute the shaky foundations for the implicit promise that the information system will yield measures of program effectiveness, however conceived, and that collection, storage, and processing of appropriate data, however defined, will improve the quality of justice. In this instance, the analysts equated "criminal justice" with law enforcement, and accepted as their data base crime statistics for the preceding five years; their assumptions and conclusions about crime present and future were built on offenders convicted in the past. Actually, the statistics reflected merely concentration of law enforcement.

Reliance on arbitrarily selected figures yielded a biased picture, encouraged preoccupation with crime-prone individuals, and diverted attention from crime-making conditions and circumstances. Not the least, although little recognized, among factors to be considered were the prevailing public attitudes toward law and order, deterntion and bail procedures, state of the court calendar, philosophy dominating administration of penal institutions, and, especially, the local political climate. Moreover, the system left out of account organized crime in its various manifestations, including police corruption. In other words, it concentrated on the hapless and helpless, those least able to defend themselves.

Convicted offender records provide a poor clue to criminality; reported crime rates do little better. For example, the Crime Analysis Unit, New York City Police Department, reported decreases of 2.7, 4.2, and 6.8 percent in index crimes for July, August, and September respectively. At the very same time, however, a separate Police Department report revealed that arrests during the first nine months of 1969 showed a rise of 17.8 percent over the corresponding period in 1968 and that arrests on narcotics charges had increased 39.4 percent. The apparent contradition was due to the fact that the Crime Analysis Unit used the seven specific categories chosen by the Federal Bureau of Investigation to represent a general level of crime activity: murder, rape, robbery, aggravated assault, burglary, larceny ($50 and over), and motor vehicle theft.

Crime, it becomes evident, is a matter of definition, institutional, cultural, legal, political, and social. If, however, one were willing to accept the simple-minded premise that crime is that which gets punished, one could more comfortably accept the next "logical" step in the development of criminal information systems. This is the determined effort on the part of almost every police department to set up an automated file to implement the capture of criminals. Police information networks, intended to aid in "law enforcement," are operating in many parts of the country, having received an enormous impetus in money and public support through the Safe Streets legislation. In California, where the recommendations of an earlier crime study were used as leverage, the Department of Justice paid Lockheed Missiles & Space Company $350,000 (from a U.S. Department of Justice Office of Law Enforcement Grant) to design a system which would meet the "total information needs" of law enforcement agencies. Its main objective was to "aid in direction and apprehension of criminals." Upon reviewing the proposed system design, even some of the dedicated law enforcement officials were a bit discomforted to learn that the proposed network called for the same items of intelligence about potential *jurors* as *criminals.* What this amounted to, therefore, was institutionalized Big Brotherhood of serious proportions, especially in view of the linkages with other information systems elsewhere in the country.

The main drawbacks of criminal information systems as currently conceived deserve brief review in anticipation of the next section of this paper, which deals with social consequences of the data bank and its implications for society. Our review has shown that crime information is based on crime as measured by law enforcement activity and definition. Police "crackdowns" on prostitution or lewd movies demonstrate the first; the level of community tolerance to certain kinds of behavior governs the second. The proposed systems would provide only for the mass gathering of baseline data. With all offenders included, persons involved in brushes with the law through civil rights marches and peace demonstrations would be counted like the burglars and rapists.

Planners concerned with improving public policy vis-à-vis crime must ask a number of questions: what objectives will be served by the criminal justice information system? Will it (1) maintain order? (2) protect society? (3) get individuals to conform? (4) increase respect or the fear of the law? (5) improve administration of the law? Then, having satisfied themselves that the system offers some socially healthy promise for a reduction in crime, planners still must face the more fundamental question about proper allocation of resources: should community money be spent on reduction of crime (assuming after the above dissertation that we had a workable definition) or on eliminating poverty and other known and long-run determinants of many forms of crime and delinquency? Although funds for general social improvement might be more effective in stemming certain kinds of criminal activity, it is an inescapable fact of political life that public attention and support are much more readily gained for the computerized law enforcement networks.

Land Use Information Systems

Among the happiest of all hunting grounds for proponents of information systems are those dealing with land use. Attracted by large federal grants and descending upon anxious planners persuaded that a data bank is a prime necessity for their and the community's good, "information experts" of all stripe busily vend their wares. And it may be noted that they meet little sales resistance. Quite the contrary. Uncertain as to goals and defensive as to bailiwick, naive about computer technology and oversold on space age management methods, public officials invite feasibility studies by persons who claim expertness in such matters. To ask an aspiring contractor for a feasibility study is, of course, tantamount to inviting a fox into one's henhouse. Not only does his review disclose feasibility but downright indispensability. The bureaucratic overlaps, the jurisdictional duplication, the antediluvian procedures — all are set forth as though newly discovered. And in neat juxtaposition is the land use information system, which, presumably, will "facilitate effective sharing of land use data between departments within a jurisdiction and between jurisdictions"[22].

The planners of one such project, for which a $200,000 contract was awarded TRW Systems Group in 1966, thought this objective could be accomplished "by first obtaining a consensus among users as to the range and type of information required, then establishing policy and standards for data exchange." The final report[4], looking and sounding more like a sales brochure than the result of professional analytic effort, was an agglomeration of platitudes. For example, "Information about land is collected and used by many different organizations at many different levels, i.e.: major agencies of the federal government within the state; major agencies of the state government; counties; cities; industrial and commercial businesses; special intergovernmental organizations and districts."

The final report, a document of 23 unnumbered pages, presented at most 11 pages of text. Sample displays and printouts accounted for considerable space. Three and a half pages were simply photographs, neither particularly illustrative nor enlightening. The equivalent of more than a page was given over to decorative but not especially relevant drawings and a full page was devoted to a gallimaufry of items — a clock face, a field, a freeway, a female fiddling with a dial, a fisherman in a canoe, a family picknicking at the seashore, a stylized cow sculpture, an elongated raccoon, and assorted skyscrapers, all pictured on a globe. The numerals *1973* accompanied this fanciful display.

Along with many vague generalizations, there was a display of what might best be called artificially hardened facts. For example, a page of tables showed "basic characteristics of the land data environment" in percentages:

Unfulfilled data needs	%
Federal	5
State	20
County	15
City	8
District	1
Private	25

Another surprisingly precise display provided a summary of "tangible cost savings," e.g., $803,000 in fiscal 1970. Such nicety would impress only persons totally unfamiliar with bookkeeping practices in the public or private sector. Even the contingencies were made to sound as though exactly computed: "If participation and services rendered exceed the estimates used in this analysis, the operation costs will be correspondingly higher; however, the benefits will increase with stronger participation." Further to demonstrate the exactness of the systems team's operations and to allay any notion that the work of information gathering is not busily done, the final report devoted a full page to a questionnaire used and half of the facing page to the following "survey facts":

> *Each questionnaire contained 412 data elements — with ten questions about each element. A total of 844 questionnaires were sent to agencies in the state. A total of 554 questionnaires were completed and returned. The resulting information amounted to 35,000 records and about ten million characters.*

The rest of the half-page was left blank.

In most land use information systems, compatibility of classification is vital to computerization. But the requirement that the data fit into fixed categories obscures important differences and nuances which may be more crucial for planners than their similarities. Selected because they are known and machine processable, the items passing for a data base are homogenized into isomorphous condition. As adjuncts to the planning process, information systems leave to be desired and yet to be realized most of the rosy promises of (1) better resource allocation and (2) improved efficiency in land usage. As to the former, an experienced government official has observed that most pertinent decisions take place at the ballot box anyway. Regarding the latter, fundamental choice issues enter into the very conception of the term *efficiency,* and, according to one RAND expert, preoccupation with the analytical or managerial tools distracts attention from fundamental issues and policies which deserve study before we even concern ourselves with "efficiency." The problem, as he sees it, "is not absence of knowledge; it is rather that appropriate actions are constrained by political factors reflecting the anticipated reactions of various interest groups"[21].

INFORMATION SYSTEMS AND THE INVASION OF PRIVACY

Alameda County, California, has PIN, a Police Information Network; the state of California has CJIS, the California Criminal Justice Information System; the United Planning Organization, an antipoverty agency in Washington, D.C., is developing the UPO bank, with about 81,000 individual records from local police, education, and welfare files. The New York State Identification and Intelligence System stores data in a centralized computerized facility on persons who have entered the law enforcement files of the 3,600 police, prosecutive, judicial, prison, probation, and parole agencies of New York State. Kansas City, Missouri, has a "municipal regulatory system." New Haven, Connecticut, is having designed for it by the International Business Machines Company a system to consolidate all the city files on individuals into a single data pool[25]. "The U.S. Secret Service Liaison Guidelines," issued to all federal and local law enforcement agencies, could, if literally interpreted, yield vast amounts of "negative information" of potentially great harm to individuals. Every military and civilian agency, every official bureau, every religious, social, and fraternal organization throughout the land is busily gathering information about people. So also are commercial organizations of diverse kinds. The Credit Data Corporation, for example, maintains personal credit files of millions of persons, possibly some 70 percent of the U.S. population[15].

The publisher of 1,400 different city directories advertises that for almost 100 years it has been "in the business of keeping track of people — who and how many they are, where and how they live, where they work and what they do." Gathered in the course of citywide, door-to-door canvasses conducted each year in about 7,000 American communities, the materials become the source record both for printing the directories and for preparing what is called "the Urban Information System." And this is for sale, eligible for federal funding, and available on tape for local processing or ready for merging or cross reference with other data stored in the company's files[16].

The information systems under consideration and in operation are capable of providing a full dossier on any individual, with complete details on birth (place, legitimacy, etc.), color, religious and political affiliations, organization memberships, school grades, military record, criminal career, financial status, and medical history. A person may have been arrested on a minor juvenile offense, may have carried a protesting placard in a parade, may have had a nervous breakdown. He might have put his name to a politically unpopular petition; he might have displayed a bumper sticker on a controversial matter. Any of these occurrences could cause him to be tabbed as a potential member of some designated "risk" group and made subject to unfavorable discrimination, if not outright harassment.

The threat of cradle-to-grave surveillance was called to public attention by the congressional hearings on the establishment of a National Data Center. An understandable desire to take advantage of computer technology in government record keeping had prompted the Bureau of the Budget, in 1961, to commission a special study for the centralization and computerization of the numerous personal records now scattered throughout various federal agencies. The Task Force, made up of highly respected specialists in economics, statistics, and similar fields, strongly recommended creation of this national data bank. Stressed were the advantages of the proposed facility: viz., centralization and integration of and ready access to information. Neglected was the inherent threat to individual freedom and privacy. The potential dangers were, however, so grave as to arouse serious public debate.

Senator Long and his committee conducted intensive inquiry into all aspects of the proposed data center. Congressman Cornelius E. Gallagher, heading a special subcommittee on invasion of privacy, assembled a vast array of documents and brought together the testimony of many authorities[18]. Although congress ultimately ruled against creating a federal data center, the geature was a sort of tilting at windmills. The linking together of hundreds of data banks at the various levels is already taking place; the result will be both statistical and regulatory federal data centers, whatever they are called. The congressional hearings were, however, not without effect: (1) they opened to a bemused public many hitherto unknown or neglected facets of the problems generated by information in a computerized age; (2) they disenchanted a beguiled citizenry on the matter of technological locks and legal safeguards; and (3) they created a climate of intelligent concern.

As to the first and third points, which are related, the Senate Subcommittee on Administrative Practice and Procedure acknowledged the virtual existence of a dossier, the chairman's portentous introduction stating, "More than two years of hearings have shown us that perhaps one of the most subtle invasions of privacy is that which is accomplished through the use of the information which the Government maintains on American citizens"[14].

A United Nations report expressed concern about possible erosion of human rights through technological developments. Emphasizing as potential dangers the very features regarded as advantages by proponents of a National Data Center, the commission issued a solemn commentary, worth pondering whenever any information system, large or small, is under consideration.

> *One of the important features of a democratic government is the doctrine of the separation of powers which makes it difficult for any branch of the government to jeopardize the fundamental rights of the individuals. Certainly, at present, the multiplicity of agencies and procedures and the resulting red tape protect the individual against undue invasion of his privacy by making it more difficult for various government officials to know enough to cause real trouble. But if all the available data are integrated and stored in a computer in a way permitting instantaneous access to the record of each person, a sword of Damocles is going to hang all the time over the head of everybody. Even the best of us have done something which can be easily blown up out of proportion, or have offended somebody who would be*

glad to deposit a little misinformation in our file. In addition, there is always the possibility of misfiling, of mistaken identities, or pure spite and vindictiveness of casual acquaintances with warped personalities. On the other hand, it seems quite impossible to envisage a process which would purify the data in the computer through properly protected legal proceedings. Considering the effort required to check the incomplete data which are now available to various agencies, when they have to decide on the employment of persons in positions which are sensitive from the point of view of national security, one can easily see that there are not enough investigators, funds, and, in case of dispute, judges to deal even with one-hundredth of the problem. It is, therefore, doubly important to consider the advisability of the whole scheme and, in case of its execution, to provide sufficient safeguards with respect to the maximum accuracy of data, their confidentiality, access to them, and the permissibility of their use in situations involving an invasion of individual privacy.[7]

For testimony about protection of privacy through technological barriers to access and legal redress, the congressional committee heard computer specialists and professors of law. Laymen soon learned that the primary objective of conventional computer hardware and software is fast and inexpensive retrieval of information. To design and construct a system with built-in guards against misuse would add greatly to the cost and thus defeat the very purpose of the system. Moreover, even though additional expenditure for safeguards might discourage some improper access, no system was judged to be impenetrable by powerful organizations for whom the particular mission at hand seemed worthwhile[3].

As for legal protection, there is none. The law is a notorious laggard with respect to technology and no redress is available until *after* damage is claimed and proved. We cannot look to the legislative system for help with respect to technology and its effects, for the legislative process needs a great deal of lead time, while technological development moves at a rapid pace. When the technology, such as the computer, the data bank, or such, is in use, vested interests influence usage. Besides the economic, there are strong political factors which affect and even impede the framing of protective statutes[26]. And privacy still remains a nonlegal concept. "Much of the history of privacy in the law is still ahead of us," observes the editor of *Law and Contemporary Problems*[9].

Many social scientists, who are as prone as any other professionals to perceive the mote in their brother's eye while ignoring the beam in their own,[d] interpreted the failure of the task force advocating the federal data center as an instance of trained incapacity. Insensitivity to social consequences, in the name of operational efficiency, might have been expected from computer technologists, but for distinguished economists and statisticians to have allowed the eclipse of considerations as vital as the right to privacy served as a chastening warning: the temptations of technology may be as irresistible to the "soft" scientist as his "hard" brother. The apologia which later came from the task force in the form of a confession of "gigantic oversight" carried an attempt at explanation for it[8]. The chairman of the task force wrote the *post mortem* of the federal data bank. He acknowledged that public fears were founded and, at the same time, suggested a list of additional abuses. However, he dismissed the idea of governmental intrusion as the "stuff of right-wing ideology." Without decisively choosing one over the other of these ideological stances, and with full recognition that a government too feeble for the welfare of its citizens in some matters may be too strong for their comfort or even their liberty in others, it is possible to believe, as I do, that the present balance of forces in our political machinery tends to the side of healthy restraint in such matters as these"[11]. The chairman's sanguine confidence in the "healthy restraint" governing the "present balance of forces on our political machinery" is not widely shared, nor can it derive much credence from recent events in the United States and other countries.

CONCLUSION

Having reviewed the conceptions, preconceptions, and misconceptions involved in information systems, with real-life experience as background and future implications as foreground, we can now reassess the assumptions listed at the beginning of this paper. The first was that if public planners had more information they would make better plans and, perhaps, arrive at better decisions. We now know that they cannot look to information systems as designed and merchandised at present to help them much. In fact, there is more likelihood that they will be inundated by an overabundance of data that will impede their efforts to understand problems in their true and dynamic dimensions. Data selected because they are machine-processable provide a shaky foundation, indeed, for community planning.

The second assumption was that more and faster-moving information would improve the efficiency of governmental operations. We now know that this, like a Sunday band in the park, sounds better than it is. On the technical side, there still remain great difficulties with storage uniformity, cross-availability of data reference items, and retrieval. If overall efficiency of agency operation encompasses dollar costs, there is no evidence that the promised economies will be realized. In fact, it could well be that, saddled with elaborate and expensive systems, government agencies will find themselves serving their information systems instead of deriving service from them. Even if there were clear-cut technical and financial advantages, the social benefits are nebulous.

The third assumption was that greater efficiency would better serve the needs of the community in particular and society at large. We now know that "efficiency" is a loaded term. Efficiency of operation could carry very high social costs if it were an instrument for centralization of control and for circumvention or stifling of democratic processes and procedures. The terms of the Faustian bargain defraud the citizen: he receives his tax bill faster, although, despite all the claims about operating economies and efficiency, it is higher every time. But his privacy is eroded with every technological advance that is adopted, presumably, to save his money.

The final assumption is that the design of information systems is a highly technical matter and best assigned to an "information expert." We now know that information is not an entity separate and apart from a context. The selection, aggregation, and manipulation of data are matters where knowledge, not mere know-how, must be applied. Insensitivity to the special problems involved, preoccupation with the mechanistic formal model, and ignorance of the stuff and substance of the real-life situation can result, if taken seriously, in designs for a neatly programmed future fraught with social disaster.

With all the technologically contrived information systems that could ever be crafted, wise and humanitarian planners will have to be aware of and take into account the economic balances of power, the sources of pressure, the political and jurisdictional realities, and the likelihood of rapid change and swift reaction as communities become more alert to their rights and responsibilities. Herein reflected are the human and social values of the society and they defy technical handling. They are incalculable, immeasurable, but all-important considerations in plans for the present and patterns for the future.

---

[a] Central Intelligence Agency, National Education Association, Boy Scouts of America.

[b] Webster's *New International Dictionary,* Second Edition Unabridged, 1935.

[c] This term is Aristotle's conception of human felicity, a life of activity in accordance with reason.

[d] Matthew, vii: 3,4,5.

REFERENCES

1 Ackoff, Russell L., "Management Misinformation Systems," *Management Science,* Vol. 14, No. 4 (Dec. 1967), pp. B147-157.

2 Alexander, Tom, "Computers Can't Solve Everything," *Fortune* (Oct. 1969), pp. 126-129, 168, 171.

3 Baran, Paul, *Communications, Computers and People,* The RAND Corporation, Santa Monica, Calif. P-3235 (Nov. 1965), p. 11.

4 *California Regional Land Use Information System,* TRW Systems Group, no date.

5 Churchman, C. West, "Real Time Systems and Public Information," Fall Joint Computer Conference, 1968, p. 1467.

6 ——, *The Systems Approach,* Delacorte Press, New York, 1968, p. 29.

[7]Commission to Study the Organization of Peace, *The United Nations and Human Rights,* Eighteenth Report (Aug. 1967), pp. 42-43.

[8]Dunn, Edgar S., Jr., "The Idea of a National Data Center and the Issue of Personal Privacy," *The American Statistician,* Vol. 21, No. 1 (Feb. 1967), p. 22.

[9]Havighurst, Clark C., Foreword, *Law and Contemporary Problems,* Vol. 31, No. 2 (Spring 1966), p. 251.

[10]Hershman, Arlene, "A Mess in MIS?," *Dun's Review,* Vol. 91, No. 1 (Jan. 1968), pp. 26-27, 85-87.

[11]Kaysen, Carl, "Data Banks and Dossiers," *The Public Interest,* No. 7 (Spring 1967), p. 60.

[12]Lockheed Missiles & Space Company, *California Criminal Justice Information System,* Preliminary System Recommendations, T-29-68-8 (Apr. 29, 1968), p. 6-1.

[13]____, *California Statewide Information System Study,* Final Report, Y-82-65-5 (July 30, 1965), p. 53.

[14]Long, Senator Edward V., Foreword, Government Dossier *(Survey of Information Contained in Government Files),* Submitted by the Subcommittee on Administrative Practice and Procedure to the Committee on the Judiciary of the United States Senate, Ninetieth Congress, First Session, Nov., 1967.

[15]Lyons, Richard D., "Blacklist Study Started by Finch," *The New York Times* (Oct. 10, 1969), and "Information Drive by Secret Service Could Affect Many," *The New York Times* (Nov. 8, 1969).

[16]Polk, R.L. & Co., *Computerized Urban Information System,* A Presentation of the Urban Statistical Division (Jan. 15, 1968), p. 25.

[17]"Poverty in Spread City – Study of Constraints on the Poor of Nassau County," A study released on Nov. 10, 1969, and conducted for the Nassau County Planning Commission under a grant from the U.S. Office of Economic Opportunity.

[18]*Privacy and the National Data Bank Concept,* Thirty-Fifth Report by the Committee on Government Operations, Ninetieth Congress, Second Session, Aug. 2, 1968, Union Calendar No. 746, House Report No. 1842, Washington, D.C., U.S. Government Printing Office, 1968.

[19]"Reported Crimes in City Continue Dip from Year Ago," *The New York Times* (Oct. 25, 1969).

[20]Report on Second International Conference on Mechanized Information, Storage and Retrieval Systems, "Easing the Search," *Nature,* Vol. 223 (Sept. 20, 1969), p. 1205.

[21]Schlesinger, James R., *Systems Analysis and the Political Process,* The RAND Corporation, Santa Monica, California, P-3464 (June 1967), p. 26.

[22]*Scope,* Office of Planning, Department of Finance, State of California (Third Quarter, 1966), p. 1.

[23]Space-General Corporation, *Systems Management Analysis of the California Welfare System,* SGC 1048R 9 (March 15, 1967), p. 1.

[24]Sperry Gyroscope Company, *A Proposed Demonstration Project for a Nassau County Welfare Information Center,* Sperry Publication No. GJ-2232-1116 (May 1966), p.v.

[25]Westin, Alan F., Hearings on *Computer Privacy,* Subcommittee on Administrative Practice and Procedure of the Committee on the Judiciary, U.S. Senate Ninetieth Congress, Second Session, Part 2, Feb. 6, 1968, Washington, D.C., U.S. Government Printing Office, 1968, pp. 279-280.

[26]____, *Privacy and Freedom,* Atheneum, New York, 1967, Miller, Arthur R., "The National Data Center and Personal Privacy," *Atlantic* (Nov. 1967), pp. 53-57.

13

Andrew Vazsonyi

AUTOMATED INFORMATION SYSTEMS IN PLANNING, CONTROL, AND COMMAND

Man controls his environment by manipulating matter and energy through information handling systems. Information is generated by recognizing and sensing events; then information is transmitted, processed, stored, and displayed so that decisions can be made to bring about desired events. The complexities of modern society created the need for a type of information management which appears to be beyond the capabilities of man's unaided intellect. New electronic computers created a promise to assist man in these tasks, but so far systems have not emerged that provide man with the degree of control he needs. During the last few years, man has been fighting a losing battle and has been losing control of his physical environment. It is the thesis of this paper that man will regain control by a man-machine symbiosis, which will reinstate man to a dominant position in the hierarchy of information networks.

The fabric of today's typical information system is a curious patchwork of automatic and human ac-

---

tivities. The *sensing* of events is being automated in commercial activities like banking, retailing, airline seat control, and in military systems like SAGE (Semi-Automatic Ground Environment). Transmission of information is performed by voice and TV through cables, radio, and microwave links. Information is *stored* by mechanical, magnetic, photographic, and optical means. Information is *displayed* by electric typewriters, high-speed printers, cathode-ray tubes, graphical, photographic, and other means. On the other hand, means to communicate *to* a computer is essentially limited to primitive keyboard devices like typewriters and keypunch machines.

Realization of efficient man-machine symbiosis and the required automation of information systems will emerge from research work performed in many fields. The most direct attack on the problem is the development of on-line real-time management planning and control, and command and control systems. Significant progress is being made on the software side by the development of more effective user-oriented computer languages and general-purpose simulation techniques. On the hardware side, more effective display techniques, computer systems for time sharing and remote control, better communication links, and new man-machine communication techniques and consoles are being developed. The problems of information retrieval are attacked both by the development of new hardware, new classification systems, and languages. Realization of problems associated with the use of complex equipment lead to increased emphasis on training, education, and the development of teaching machines.

On the more basic side, the mathematical theory of the transmission of information, signal processing, and detection has been developed. The human and animal nervous system and brain are studied from the point of view of computers. Machines with artificial intelligence and self-organizing and learning capability are being invented. Problem solving and human thinking are being studied and automated. The effect of information systems on the behavior of task-oriented groups is being studied and possible adaption of human organizations to the techniques of information processing is being explored.

Information technology plays a significant role in man's current way of life and radical changes in the future are predicted by many. The purpose of this paper is to present to potential users a semi-technical introduction to the status of automated information systems and to make a prognosis of changes that are likely to occur. Close rapport between users and inventors is a necessary ingredient to the progress and application of information technology. . . .

ON-LINE REAL-TIME INFORMATION SYSTEMS

Let us speculate now on how problem solving will be carried out in the future. One thing is predictable: humans will have to behave in a different way from the accepted practices of today. We shall have to deal with a set of computers, probably of different sizes and types, and we shall have to control remotely these computers with the aid of specially designed input-output devices. We shall have teams of people and teams of computers geographically dispersed but working together in intimate collaboration. Today, computer availability tends to control the schedules of our problem-solving activities; in the future man will dominate the computer. Computers will be designed in such a way that they will precisely match human problem-solving needs. Man and machine working jointly must form a new sort of organism with higher intellectual capability than either the man or the machine. Computers will talk to each other, people will continue to talk to each other, people will talk to computers, and computers will talk to people. Here, it is not meant necessarily, that computers will be literally controlled by voice communication, but it is implied that there will be convenient ways to generate two-way communication between men and machines.

People will solve problems at their desks, in libraries, and at home; they will not be forced to travel to computers or to send their problems to computers. In fact, probably our future place of work will look different. Our telephone system will be augmented by television and by special input-output devices with which we will be able to communicate to computers. When people in this future era will look back to 1964, they will have similar feelings as we have when we look back to those days when we did not have telephones. Granted, there is nothing basically magic about telephones and one could do everything by mail that can be done by telephone. Still, the time lag inherent in mail communications would make our current way of dealing with problems impossible. Possibly, in the future a similar condition will develop with respect to computers and we will solve many problems with computers that today we cannot solve at all.

Admittedly, many developments will have to come before we reach the state of problem solving described here. Better hardware, better programming techniques and many new concepts will have to be developed. However, the important thing to realize is that many of the components and subsystems are already in existence today, though there is no complete system yet exhibiting all the features described. A review of some of the most advanced information systems of today will show to what extent we are using automation and man-machine interaction in problem solving. It will also be seen that today's systems are quite expensive and, therefore, can be justified only in those cases where either it is impossible to solve the problem without on-line real-time processing or where the economic payoff is high. . . .

WHERE COMPUTERS EXCEL

Some tasks we can automate well, others poorly or not at all. . . . We are today sensing data in a worldwide manner, transmitting these data for processing and displaying. It is expensive to do this, but current equipment and techniques are capable of performing these functions. The airplane reservation problem discussed later shows that data can be collected, transmitted, and processed at a reasonable cost within the continent of the United States. Worldwide data processing will become more practical when costs decrease, as the problem here is primarily of improving current techniques and equipment.

As far as the display of processed information is concerned, we are doing reasonably well, but not quite as well as we would like to. The display of information by cathode-ray tubes is expensive and limited in size and content. Large-scale displays are important in certain types of problems, and as of today there is no efficient way to generate automatically large displays. When it comes to color, we are even worse off. Automatic photographic processes do not seem to fill the requirements.

As far as storing data, we are doing pretty well. Admittedly, random access memories are expensive and sequential memories are not quite as efficient to work with. We can expect that random access memories will get larger and cheaper.

The area where we are doing best is the performance of computations. When it comes to carrying out very extensive computations and logical operations, electronic computers have a most astounding performance. As a consequence of this capability, we are very well set in solving problems that are well structured and for which there are algorithms available. By an algorithm is meant a sequence of mathematical or logical steps which spell out in complete detail how the solution to the problem is to be obtained. Electronic computers have been designed to carry out precisely these kinds of functions and they are eminently suited to such tasks.

WHERE MEN EXCEL

Let us turn our attention now to problems where computers are not doing so well. One of the most serious shortcomings that we are faced with could be grouped under the heading of "recognizing patterns." This broad term covers a multitude of problems. As a relatively simple example, when messages arrive at a decision center, often these messages are not stated in a clear fashion. A human operator can efficiently decipher these messages, and correct and rearrange the data so that the computer can process the information. Unfortunately, computer techniques are inadequate to recognize these irregularities. Here is then a field where computers are doing poorly.

Going now to more complicated types of patterns, computers are indeed poor in recognizing writing or speech. Geometrical forms, reading of maps, and a number of similar ill-structured tasks are too difficult today for computers. We see then that where problems of pattern recognition occur, at least for the time being, partnership of man and machine is indicated. A difficult problem is how to divide the task between man and machine in such a way that both men and machines can contribute their best.

Another area where we are doing poorly is in the

development of computer programs. Here computers are already used, but not too well, and men still must spend a great deal of their time. A particularly vexing aspect of computerized solutions is the extreme difficulty of making modifications. There is quite a difference here between a man and a machine. When a man is taught to solve a problem and a small change is required, he can easily adapt himself to the modified problem. Unfortunately, we do not know yet how to do this with computers.

In the field of formulating goals or establishing policies the role of computers is minimal. This is a field where the kind of judgment that men have is indispensable. However, to develop policies or goals, many details are required and many routine steps and calculations must be carried out. Here then we have again an illustration where a partnership between man and machine could become beneficial.

This, however, does not mean that computers are totally incapable of solving unstructured problems. Under the name of heuristics, a great deal of work is being carried out to uncover how computers can solve problems for which algorithms are not available. We shall return to this subject shortly.

Again, we find that little automation has been accomplished in handling exceptional cases. It is much easier to teach people to handle exceptions that arise say only once a month or one out of a thousand cases than to program computers for such contingencies. The reason for this is that the computer has to check in each routine case whether an exception exists or not. Men can do this more rapidly than computers.

Another field, again, where computers are doing poorly is in the selection of alternative courses of action when there are no precise rules for selecting these alternatives. This is, again, an area where people can do much better than computers.

In the field of information retrieval, computers are useful, but not as useful as we would like to have them. When we are faced with large files where the retrieval cannot be carried out by an explicitly formulated rule, there is a need for associated memory systems and here computers are not doing well. For instance, if we wanted to make a literature survey and to find a paper we might be interested in, it would be difficult to assign the search to computers. It is believed that in these problems man-machine partnership will lead to better solutions.

However, today the realization of such man-machine systems is prevented by the lack of the possibility to communicate easily between men and machines. When we look over today's practices, we see that most computer information comes in the form of large amounts of printed sheets, which are difficult for people to assimilate. On the other hand, when man wants to communicate to the computer, this is done primarily by punch cards or to a limited extent by keyboard devices. This is an area where today's hardware fails us.

We can see, in summary, that machines are very good in certain problems, while in a number of other problems they are quite poor. Fortunately, the human mind operates quite well in some of those problems where the machines fail. To determine how machines could really extend man's intellectual ability, we should examine not only how machines work, but also how people solve problems.

HOW MAN WORKS

Whether we work in an office or in a library or whether we are just solving problems, we are surrounded by documents, reports, letters, blueprints, and graphs. We search for information, but we do not do this in a continuous fashion, like when we collect information, we usually try to get some written record for our own use: we search and take notes. Also, we do not receive all the information through our eyes; we talk to people, listen to words that may contain information. From time to time, when we have enough notes or we want to have more permanent records of our thoughts, we write a memorandum or dictate to our secretary or a dictating machine. Later we edit our drafts and prepare final memoranda, reports, books.

As a part of a process of collecting information, we attend lectures or may give lectures to others; we attend briefings and assimilate information.

We do not go around blindly collecting information, we usually have some reason to collect it: we are trying to solve some sort of a problem. Part of our mental activity is the formulation and statement of problems. This is not a one-shot exercise either, as often we state a problem in a preliminary way and then, as we go on solving it, we redefine it and state it more accurately. Often a problem and its solution are found simultaneously. As we search for information and solutions, we unfortunately must do a lot of "housekeeping" and often not much time is left for complicated tasks. We make computations on a slide rule or on a desk computer, and we obtain trial solutions to be examined later for evaluation. Often we subdivide a problem and assign parts to other people. Sometimes we might even assign a part of the problem to a computer, though today we have no convenient way to do this.

There is, however, one feature of problem solving that frustrates most of us: we really spent only a relatively small fraction of our time on what we would like to call thinking and problem solving. Most of our time is spent carrying out some routine functions. A hard problem we need to solve is how to increase our time available for problem solving.

It is also important to recognize that we solve problems not in a systematic or step-by-step way, but in an unstructured fashion. The various functions that are needed in problem solving are mixed in an unscheduled and haphazard manner. Therefore, a problem solver needs absolute mastery in ordering the various functions required to solve his problem. He wants paper and pencil on his desk. He wants a slide rule in his drawer. He wants his own telephone, dictating machine, secretary. He also needs his own computer; he wants this computer to be at his desk and be suited to his problem. However, this can only be achieved if he understands what part of the problem he can delegate to the computer. But how can he develop this understanding if he is not familiar with computer capabilities?

Our problem is then to break down the problem-solving activity into two kinds of tasks. For one set of tasks, which we would call routine tasks, algorithms are available. These tasks can be assigned to the computer and hopefully these are the tasks that computers can carry out with speed and accuracy. On the other hand, the other kinds of tasks for which we find that computers are wanting, but for which humans are suited, are to be reserved for human beings. By the intimate association of men and machines we can create a problem solving system which will allow men to spend the bulk of their time on the higher intellectual functions of thinking. However, to do this we need a better understanding of the nature of problem solving.

HEURISTIC PROBLEM SOLVING

Unfortunately we know very little about how man solves (or fails to solve) his problems.[18,39] There have been many speculative and illuminative descriptions of insight, creativity, solution finding and so on, but we find these investigations of limited use in developing man-machine partnership

On the other hand, recently there has been considerable effort in attempting to solve by machines problems for which people do not have explicit rules for finding a solution.[11,48,49] These researchers examined how problem solving actually takes place. Whereas our primary interest in this paper is to solve problems by man-machine partnership, still we find that research in the field of heuristic problem solving and artificial intelligence provides good guidelines for problem solving in general. We will try then to describe here briefly some of these new concepts and show how the search for solutions of problems can be carried out by joint man-machine effort.

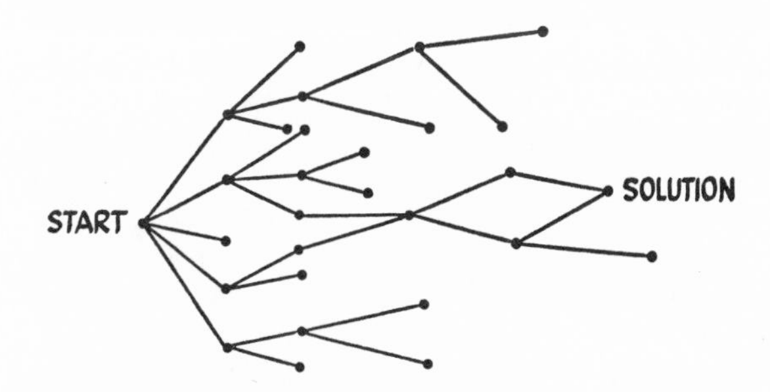

Figure 1
The search for solutions.

The graphical description of the search for solutions (Figure 1) is useful in describing the process of how to divide up problem solving between men and machines. According to this diagram, finding the solution of a problem is something like going through a maze. One starts at a certain point, one has many paths and many branch points, one could go many ways, but only a few lead to the solution. To find this solution, as one reaches each vertex of the maze, one tries to evaluate his position with respect to the solution. After the position is evaluated, one needs to make a decision, what the next move should be. This can be done presumably on the basis of evaluating the expectation that after the move one will be in a better position to find the solution than one was before. After this second move is made, one has to evaluate again the position. Note that if one had time to examine all the possible moves, one would necessarily find the solution. However, this is of little avail, as in interesting problems the number of possibilities is so great that an exhaustive study of all possible moves is impractical. In fact, this is one of the reasons for resorting to heuristic problem solving: a random search for the solution is impractical.

This description of searching for solutions can be further illuminated by a few illustrations. The problem of playing chess can be easily explained in terms of Figure 1. Every time we want to make a move there are only certain permissible moves. After we make our move, our adversary will respond with his move. Due to the enormous number of possibilities, it is impractical to examine all the possible moves. Consequently, men and machines that play chess use heuristic methods employing the type of techniques we are talking about.

Machines have been designed employing heuristic techniques that can prove theorems in logic and in geometry. In game playing the adversaries alternatively make moves; in the case of theorem proving, we are making every move, as we are playing, so to speak, against nature. The problem of machine scheduling in a factory, or more broadly speaking management planning and control, can be treated again from this same point of view.[60,61] The problem of maximizing functions with the method of steepest ascent is a search problem of this sort. Information retrieval and document classification are further fields to illustrate the potential of heuristics.

The difficulty of employing computers in entrepreneurial and military decisions is complicated by the fact that often what constitutes an acceptable solution is not defined. In fact, it is often impossible to state what the "problem" is. However, one can start by obtaining from the decision maker a tentative statement of the problem. This initial statement might be so vague that the problem can be interpreted in many ways. We may choose a particular interpretation and a particular solution. When these are brought to the attention of the decision maker, he may say that we interpreted the problem the wrong way and that we obtained an unacceptable solution. His very statement of rejecting our problem formulation and solution can serve as further clarification of his true problem. So now we can start again and iterate the process. We see then that, in these cases, statement of the problem and finding a solution go hand in hand. The need for man-machine partnership is even more compelling. The process of finding the "problem" and the "solution" can be illuminated by techniques of heuristic problem solving.

In summary, characteristics of heuristic problem solving are as follows. We do not have an algorithm, so we cannot be assured that we find a solution. On the other hand a random search is impractical. Consequently, a selected search technique is employed by recognizing patterns of improvements in solutions. As we go along solving the problem, we learn and we get more efficient in solving the problem. In addition, as we go along solving the problem, we plan how we would want to get a solution, and as we proceed, we replan and improve our plan.

The techniques of heuristics (and artificial intelligence) are important in understanding man-machine problem solving. Basically, all man-machine problem solving should be heuristic, as problems for which algorithms are available should be solved by machines alone.

Now that we have some insight into the nature of the problem of solving problems, let us turn our attention to equipment requirements for two-way communication between men and machines.

MAN-MACHINE COMMUNICATION CONSOLES

In Figure 2, an elaborate equipment is shown which could serve as a communication device between man and machine. This console is not a computer, but it is a machine with which people can talk to computers and computers can talk to people.

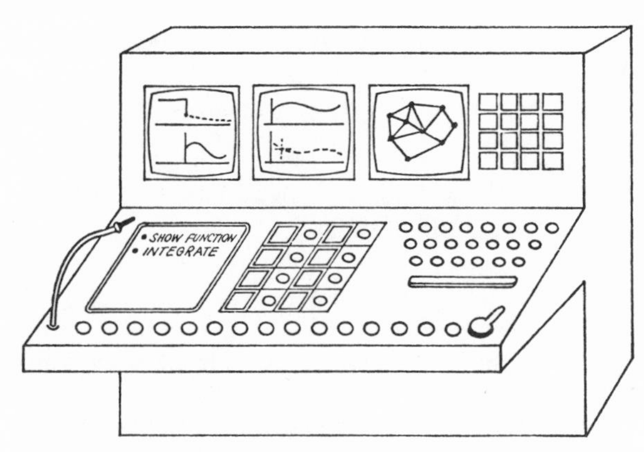

Figure 2
Man-machine consoles.

Figure 2 shows a complex console with four cathode-ray tubes, many buttons, and other devices. We do not mean to imply that every communication console between man and machine should be this complicated. In fact, sometimes a very simple keyboard or some other component of the console shown in Figure 2 would suffice. Let us go now into some of the details of this piece of equipment.

On the right side, a typewriter keyboard is shown with the aid of which man can directly type alphabetic and numeric messages to the computer. In the center, there is another set of keys which are somewhat different. Left of each key there is a blank space which can be labeled. This gives the possibility to man to label his own keyboard the way he desires. In addition, there are other special keys which direct the computer to perform certain specific routines.

The cathode-ray tubes serve for the computer to communicate to man. Alphabetic, numeric, special messages and geometrical figures can be generated by the computer. Furthermore, the tubes serve for man to communicate with the computer. With the aid of a light gun, shown on the extreme left, man can designate points or messages on the cathode-ray tubes. In addition, a cross hair is available that can be maneuvered across the face of the tube with the aid of the joy stick shown on the right side. The cross hair, like the light gun, serves the purpose of designating points on the cathode-ray tubes.

However, we do not propose that in the future every man be equipped with a console of this complexity. We imply, however, that the office of the future will take on a new look. Just as we have telephones today, the time will come when many of us will have communication devices to talk to computers.

Let us examine now somewhat in detail how such man-machine communication devices would fit into the computer system of the future and, furthermore, what kind of software will be required to run such man-machine computer systems.

First, it is unlikely that we will exclusively rely on supergiant computers with which many people can communicate. More likely, we are going to deal also with modular-type hardware. This means that there will be many arithmetic, control, and storage units, not all the same type or size. Many of these units will be connected to each other so that they can intercommunicate. There will be many input and output devices and many different types of consoles with which men can talk to computers, and vice versa. All these units will be connected together and will be capable of time sharing and interrupting so that there will be a high utilization of equipment. But more important, computers will serve people when and where there is a need to solve problems. An important innovation will be the new method of operation and the programming languages employed in these computer systems.

Today, good progress is being made on problem-oriented languages and further important improvements are coming. This does not mean that users will not have to learn more about computers. In fact, functional understanding of computers will be necessary for most problem solvers. Furthermore, programming is not going to be so sharply separated from problem solving as it is today. In fact, recent work on programming techniques indicates that programming itself and debugging of programs will be accomplished by on-line computing. The programming task itself is to be considered as one of the problems ot be solved by man-machine partnership.

Now that we have described somewhat in detail the status of current on-line real-time information systems and examined the problems of automation, we will proceed to describe a few examples of how problems can or could be solved with this kind of a system.

SOLUTION OF DIFFERENTIAL EQUATIONS

The problem of solving systems of ordinary differential equations has been the subject of extensive investigations on the part of mathematicians for hundreds of years. The theory of linear differential equations forms one of the classical branches of mathematics. The theory of nonlinear equations forms an impressive body of knowledge, but still the results obtained so far do not meet all the needs of engineering. Techniques of numerical analysis permit the solution of any system of differential equations whether linear or nonlinear. The practical need for solving systems of ordinary differential equations arose only relatively recently in connection with the design of electronic circuits and dynamic systems. As a result of this recent need, new mathematical theories have been developed. However, the decisive step in the practical solution of these equations came about not so much by theory as by the emergence of analog computers.

To illustrate our point, let us consider an extremely simple engineering problem shown in Figure 3. The problem here is to design the electrical system of a dc motor so that the performance of the system will meet certain engineering requirements. The voltage drop can be determined with the aid of Figure 3 and is characterized by the following equation:

$$V = IR + LdI/dt + Kd\theta/dt$$

In this equation, V denotes the voltage drop, I the current, R the resistance of the circuit, L the inductance, K the proportionality factor related to back electromotive force, θ the angular position of the motor, and t time.

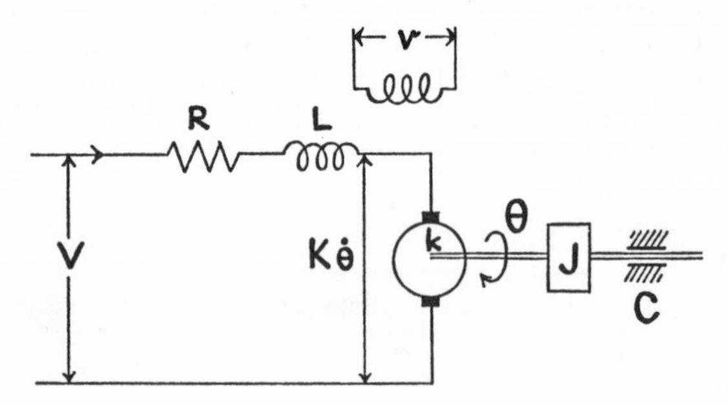

Figure 3
Direct-current motor.

The motion of the rotor can be described by Newton's equation:

$$kI = Jd^2\theta/dt^2 + CF$$

where J is the moment of inertia of the rotor, k is the constant related to the magnetic torque, and CF is the Coulomb friction torque:

$$CF = C\frac{d\theta/dt}{|d\theta/dt|}$$

(The fraction on the right-hand side takes the value +1 or -1.)

The solutions of these two equations describe the behavior of the dc motor. The equation is nonlinear owing to the Coulomb term in the second equation. As the equation is nonlinear, the classical theory of linear equations does not apply. Theory of nonlinear equations does give some insight to the problem, but does not answer the engineer's question of how this dc motor behaves. With techniques of numerical integration, the equations can be solved for any particular set of constants R, L, K, k, J, and C. However, the problem of the engineer is to determine these constants so that a "satisfactory" design results. We put the term satisfactory into quotation marks because the engineer generally does not know in quantitative terms what is a satisfactory solution. His real need is to examine a large number of possible solutions for various combinations of the constants and select by judgment the solution he considers satisfactory.

The illustration given here is a very simple one and has only six constants. In a complex system there would be many equations and many constants, and the number of solutions that the engineer may want to examine would be large. This makes it impractical to compute by hand the solutions. What the engineer needs is to start with a given set of constants, get the solution, study it and change the constants to get a new solution, examine that one, and continue this step-by-step modification process until he finds a solution that he considers satisfactory. This, however, requires that solutions be obtained rapidly and that a technique be available so that he can examine the solution and change the constants at will.

Analog computers offer precisely this opportunity. With the aid of components, such as integrators and amplifiers, one can assemble an electronic circuit which behaves in a parallel fashion to the equations to be solved. This means that the engineer can arrange for an input disturbance (or for a particular type of a load) and almost instantaneously examine the dynamic behavior of the system under investigation. An analog computer has the capability of solving linear or nonlinear differential equations in milliseconds. Furthermore, the engineer can turn knobs on the panel, which then changes the constants of the circuits in the analog computer and, thereby, he can make the analog computer solve the system of equations with desired constants.

We see then that the engineer solves his problem in a step-by-step heuristic fashion. He examines the solution of the equations for a certain set of constants and then by his intuition he changes the constants; he examines again the solution and, again, he changes the constants. By a systematic trial-and-error method, he finally reaches a set of constants for which the system behaves in a fashion that he considers acceptable.

Let us note two important aspects of the method of solution. As he examines hundreds or possibly thousands of different solutions, it is mandatory that (1) the solution for a given set of constants be obtained rapidly; (2) that he be enabled by an immediate and direct way to change the constants in the equation. There is a need for immediate two-way communication between man and the machine.

We have an example of the extension of the human intellect by man-machine partnership. The problem could not be solved by a machine alone, as there is no known sequence of steps to obtain a solution. Man could not solve the problem alone, as the solution of each step would require hours or possibly weeks of computation.

In the example in the discussion, we have referred only to analog computers. However, analog computers are limited in dealing with nonlinear problems. For this reason, currently, hybrid computer systems using analog and digital elements are being introduced. Very recently techniques have been introduced to use solely digital computers. In spite of the fact that analog computers are in general more efficient for linear equations, for complex nonlinear systems digital computers may be superior. There is today, however, the difficulty of establishing a direct two-way, man-digital computer relationship. It is safe to predict, however, that in the solution of complex systems of ordinary differential equations we will see an increased utilization of digital computers.

COMPUTER-AIDED DESIGN

We have already talked about the engineering design problem where it is necessary to interweave steps of creative thinking and relatively routine tasks. Now we want to describe somewhat in detail current research aimed to assist designers in carrying out their tasks. In particular, we want to show how, with the aid of man-machine interaction, a designer can be aided in carrying out engineering drawing functions.[8, 22, 52, 53]

The first thing one has to do is to eliminate pencil and paper, as a computer cannot interpret drawings made on the drawing board. The console shown in Figure 2 is equipped with four cathode-ray tubes and now we show how a designer can "draw" on these cathode-ray tubes.

The designer takes hold of the light gun (left side of Figure 2) and denotes points on the cathode-ray tube. As he moves the light gun across the face of the tube, the computer can be directed to draw curves passing through these points.

In Figure 4, insert 1 indicates the technique of drawing straight lines. The designer places his light gun at point *A* and moves the light gun along the dotted line. The computer is programmed in such a way that as the gun moves to points *B,C,D,* and so

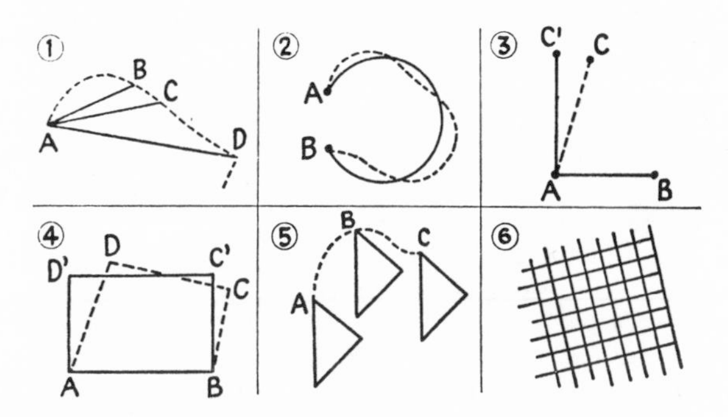

Figure 4
Automated drawing.

on, the straight lines AB, AC, and AD are automatically drawn and shown on the cathode-ray tube. To the observer it would appear as if there were a rubber band stretched between the point A and each of the points B, C, and D. When the designer stops at, say, point D, only the line AD is shown and in the computer memory the fact is stored that there is a straight line between points A and D.

Insert 2 shows how one can draw a circle. The designer starts at point A and, as shown by the dotted line, draws an approximate circular arc to point B. Then by pushing appropriate keys he instructs the computer to draw an exact circular arc starting from A, ending at B and approximating the dotted line.

In insert 3 we show how two perpendicular lines can be drawn. First, the designer marks the line AB and then he draws an approximate perpendicular AC. Then again through the keyboard, he instructs the computer to make the angle exactly 90 degrees. The computer erases the line AC and draws line AC' perpendicular to the base line AB. With a similar technique it is also possible to draw parallel lines.

In insert 4 we show how a rectangle can be drawn on the cathode-ray tube. First, the designer draws the quadrangle $ABCD$; then he instructs the computer to change this quadrangle into an exact rectangle $ABC'D'$. Note that the computer can change the quadrangle in many ways into a rectangle. If the line $C'D'$ is at the wrong height, the designer can move the line, with the aid of the light gun, up and down so that he gets a rectangle with desired height. The computer must be programmed to move line segments in parallel and so that the end points are constrained to straight lines.

If the designer desires to draw a square, he instructs the computer to change an approximate square into an exact square. Similarly, he can draw an approximate equilateral triangle and instruct the computer to change it into an exact equilateral triangle.

It often happens that a diagram is drawn but is placed at a wrong relative position. In insert 5, Figure 4, we show how a diagram that has been drawn on the cathode-ray tube can be shifted with the aid of the light gun. The designer designates one point, say point A; then he instructs the computer to move the triangle. As the light gun is moved across the face of the tube, the triangle is "pulled" along with the light gun. In this process the sides of the triangles stay parallel and the scale does not change. However, it is possible to program the computer so that a diagram can be rotated. Also, it is possible to change the scale of a diagram. We can also store any diagram in the memory of the computer and then later, when we want to use the diagram again, we can recall it from the memory. It presents no particular difficulty to erase a line that we do not wish to have and for instance, to make a dotted line out of a solid line.

In insert 6 we show that one can draw a rectangular net on the cathode-ray tube. One can change the scale of the net, rotate the net, and erase it.

In insert 7, Figure 5, we show a somewhat more complicated exercise. Suppose that we have a triangle ABC, and we want to place this triangle into the circle drawn. We put our light gun on point A and move the light gun through point A' to A'', a point on the circle. As we move the light gun, the base of the triangle BC remains unchanged, but the upper two sides are pulled along with the light gun. Now we move point B to point B'' on the circle, and point C to point C''. Now we have $A''B''C''$ as the triangle inscribed into the circle. We may want to instruct the computer to change the triangle into an equilateral triangle inscribed into the circle.

In insert 8 we show a three-bar linkage system. The original position of the three links is $PABQ$. We instruct the computer to move the linkage system,

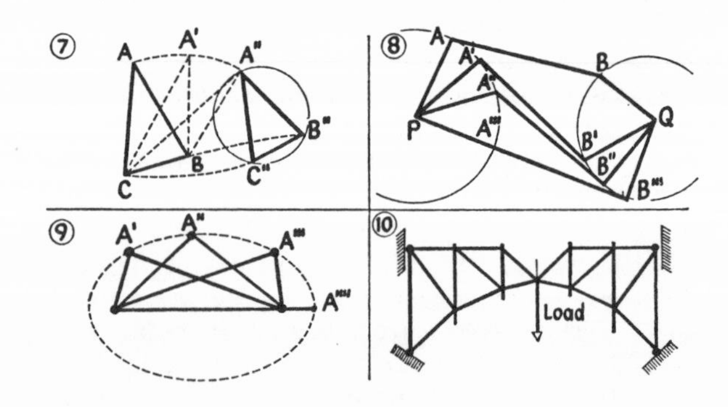

Figure 5
Automated drawing (continued).

and the lines as shown in the figure would appear on the cathode-ray tube.

The linkage we have described could be physically built; however, with the technique we are describing here, one could study more complicated kinds of linkages. In insert 9 we show a linkage for which the base is fixed and it is desired to move the apex along an ellipse. It is known that for an ellipse the sum of the distances from the foci is a constant. We can program this on our computer, and then as we move point *A*, it would stay on the ellipse.

With more complex programming techniques, it is possible to represent solid bodies in three-dimensional space. It is possible to make perspective drawings and it is possible to show the various views of the body. It is possible to specify an axis of rotation and instruct the computer to turn the three-dimensional body around this specified axis. A drawing of a three-dimensional body may be prepared, and then the designer may wish to look at it from a particular angle.

In insert 10 we show the simplified structure of a bridge. We assume that this structure was drawn by a designer on a cathode-ray tube. After he made his drawing, he can specify the loading of the bridge and direct the computer to show the stresses and strains resulting from the applied load.

There is no reason why the technique could not be further extended to the study of vibrating bodies, heat flow problems, aerodynamics, or hydrodynamics.

We see now how progress is being made in developing man-machine techniques to aid designers. It is again safe to predict that we will see interesting developments in this field.

SOLUTION OF PARTIAL DIFFERENTIAL EQUATIONS AND INTEGRAL EQUATIONS

We have already shown how ordinary differential equations are solved by integrated man-machine techniques; now let us turn our attention to other fields of mathematics. There are many mathematical problems where on-line real-time techniques will be important[5 6], but in this paper we will concentrate on one particular type of problem. In Figure 6, on the left side, the problem of the flow of a liquid from

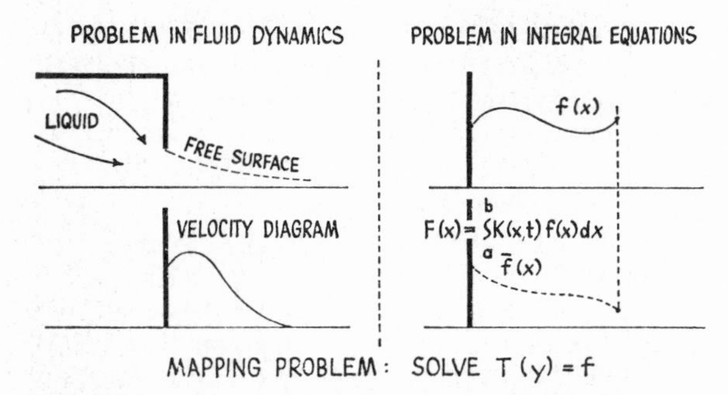

Figure 6
Solution of differential and integral equations.

an open vessel is shown. The problem here is to determine the free surface of the liquid as it leaves the opening of the vessel. This problem can be solved with classical techniques of mathematics, but the solution described here is general and applies to more complex problems than the one shown.

The reason that this problem is difficult stems from the fact that finding the location of the free surface represents a mathematically complex problem. Determination of the flow of liquids in vessels with known boundaries is relatively simple. Let us see now how, with the aid of on-line computers, one can transform the free surface problem into a simpler fixed boundary problem and how by a heuristic procedure one can solve free surface problems.

Suppose that we draw by intuition a free surface line as shown in the upper part of the figure by the dotted line. There are algorithms available to compute the flow of fluid within this trial free surface. In the lower part of the diagram we show the variation of the velocity of the fluid along the trial free surface. If we select the free surface in the wrong fashion, the velocity diagram may look as shown in the figure. The proper selection of the unknown boundary can be defined by the simple criterion that the velocity diagram must be a constant. (This follows from Bernoulli's law, as the pressure must be atmospheric along the surface and, consequently, the velocity must be constant too.) We can now state our problem of selecting the appropriate free surface of the liquid as follows: change the dotted line on the upper diagram until the velocity line in the lower diagram becomes a constant. Suppose that, by some technique, we can manipulate the dotted lines and that we can inspect the resulting velocity diagrams. By following our intuition of how velocity changes in channels, we could modify and change the free surface until the velocity diagram became a constant. This is precisely the procedure we can follow with the aid of on-line computing.

In the right-hand side of Figure 6, we show graphically a problem from the theory of integral equations. In the upper diagram we show an unknown function. The integral equation shows how the solid line in the upper diagram is transformed into the dotted line in the lower diagram. The solution of the integral equation can be defined by the criterion that the solid line is to be transformed into a dotted line, identical with the original solid line. Now, we will show how with the aid of on-line computing it is possible to draw in the solid line and observe the dotted line. With the aid of a heuristic method of altering the solid curve, it is possible to find a curve which transforms into the same curve. This is then a heuristic method of solving an integral equation with the aid of on-line computing.

Actually, both of these problems are particular cases of the problem of inverse mapping or of the problem of finding fixed points. In general, the problem is to find a curve which, by a given transformation, transforms into a desired other curve. (Find a point that transforms into itself.) We will outline now how a search for the solution can be carried out with the aid of the techniques we have been considering.

Let us follow step by step the operations that are to be performed. We already described how to draw simple curves. More complex curves such as parabolas or higher-order polynomials can be drawn by designating points on a cathode-ray tube either with a light gun or with a cross hair and by instructing the computer to put appropriate curves through these points. We need to have certain programs stored in the computer and we need appropriate buttons that can be pushed so that the computer carries out curve fitting.

In addition, one might want to draw certain functions such as sine, cosine, or logarithm. Again, it is relatively easy to store such programs in a computer and have these programs executed. Provision must be made so that parameters (such as the coefficients in polynomials) can be designated by man.

The next problem is to instruct the computer to carry out the transformations involved. For the fluid flow problem, we need to store the program for solving Laplace equation so we can compute the flow of the fluid within the boundaries designated and the associated velocity functions. For the integral equation, it is necessary to carry out the evaluation of the integrals indicated. With the very fast speed of electronic computers, the transformations can be carried out, even for complictaed problems, in fractions of seconds.

Now we are in a position to "draw" curves and to "transform" curves. The next step is to compare the transform with the desired curve. The computer can show the differences between the vertical scales. If these differences are too small to be seen on the cathode-ray tube, we instruct the computer to magnify the vertical scale, or we can obtain high accuracy by showing numerically or graphically the ratio of these values.

Now we need a way to alter our original curves as indicated by our intuition. This can be done either by drawing new curves or by altering parts of the old curve. For instance, with the aid of the light gun or the cross hair, one could designate two points on the curve, change the curve between those two points, and leave the rest of the curve unchanged. By this technique we can alter the function and instruct the computer to compute and show new transforms.

In addition, we need the capability to recall previous curves, as we might need an earlier trial solution.

A certain amount of elaboration on the keyboard to be used[9,13] might be in order at this point. For instance, we might have a key that is permanently marked "interpolate," as we may have many problems needing interpolation. Or again we might take a blind key and mark it in a particular exercise with the word "interpolate." During this particular exercise, this key will interpolate. Again, we might have certain keys designated by the name of certain functions and every time we push this particular key the appropriate function would be available for mathematical operations.

In addition, we might use the cathode-ray tube

itself as a keyboard. For instance, we might flash on the cathode-ray tube several instructions, one of them being "interpolate." With the aid of the light gun, we point to the word "interpolate" and then the computer would carry out an interpolation. It is seen then that the cathode-ray tube may represent an unlimited keyboard.

There is, however, one serious limitation of the technique so far described. Suppose we want to carry out a step in the analysis that has not been programmed in advance. How could we program the computer, so to speak, "on the way"? We need a capability to carry out programming and problem solving intermittently.

As an illustration, suppose that we are dealing with the problem of an integral equation, and in the course of the solution we find that we need to compute the sine of a function and that the sine function has not been previously programmed. Assuming that an accuracy of two terms in the power series is sufficient, we need to generate the function.

$$\sin f = f - 1/6f^3$$

Let us assume that our computer is programmed so that, when the key bearing the name of a function is depressed, the data pertaining to this function become available for mathematical manipulations. We can simply say that the key marked f contains the function f. Assume that the computer has a working register R and mathematical operations are performed by placing an entire function into R. We assume now that we have a key marked LOAD. If we push the LOAD key and after that we push the f key, the f function which was stored in the f key is put into the R register. Proceed to push the STORE key and after that the F key. Now we have the function f stored in the key F. In an abbreviated form we can say that the sequence of key pushing

(LOAD) (f) (STORE) (F)

transfers the function from the key f into the key F. Proceed to push the multiplication key and the F key again, then the multiplication key again and the F key the third time. By this sequence of key pushing, we succeed in storing f^3 in the R register. We have now described the following sequence of key pushing:

(LOAD) (f) (STORE) (F) (·) (F) (·) (F)

Proceed to push the STORE key and then the G key. This means that we have transferred the f^3 function into the G key. Proceed to push the LOAD key and then, with the numeric keyboard, type in the number -.1667 (the decimal form of -1/6). Now we have in our R register the proper factor. Proceed to push again the multiplication key and the G key. Now we have the function $-1/6f^3$ in the R register. Proceed to push the + key and then the F key and now we finally have the approximation for sine f in the R register. This second half of the program consists of the sequence of key pushing:

(STORE) (G) (LOAD) (-.1667) (·) (G) (+) (F)

By this technique we can carry out programming while we are operating the console.

However, it would be a time-consuming process everytime we want to compute this function to push this sequence of keys. By appropriate programming and use of the keys, we can assign the computation of this sine function to a new key. First push the key designated PROGRAM, then push one of the blind keys, which is to be designated in the future by sine. Proceed to push the keys in the sequence stated before. At the end, again push the PROGRAM key. Our computer can be programmed so that from now on when we push the new sine key the computer automatically "pushes" the indicated sequence of keys, and so we can obtain the approximation of sine.

The important point to recognize from this simple illustration is that it is practical to generate programs on-line real-time while we are carrying out computations and problem solving. The example shown is a simple one, and is suited to a particular problem and computer. Undoubtedly, in the future many different ways will be found for "on-the-way" programming.

Let us now summarize the essential features of solving mathematical problems with the aid of an on-line real-time computer system. We begin by establishing certain basic computer programs that are important for the solution of the problem and we store these in the memory of the computer. Then we proceed to carry out our mathematical manipulations in a heuristic fashion. During the course of the solution, we recognize that certain additional types of computer programs are required to complete the solution of the problem. These can be developed either by conventional "paper" programming techniques or by "on-the-way" programming. The solution of the problem is obtained by a heuristic procedure: routine tasks requiring large amounts of computations are carried out by the computer; steps requiring judgment are under the direct and immediate control of the problem solver.

AUTOMATED PROGRAM EVALUATION AND REVIEW TECHNIQUE

Let us now return to the problem of planning research and development of large-scale programs – for instance, those that occur in connection with space travel. One basic technique used in these programs is PERT, a network type of analysis. The basic concepts involved are shown schematically in Figure 7. The upper part of the diagram shows a network where each vertex corresponds to a particular event to be accomplished and the horizontal scale is time. The line segments represent activities to be completed so that the events can occur. For instance, event *Q* will occur after the four particular activities shown are completed. For illustration, let us focus our attention on the activity that leads from event *P* to *Q*.

As far as the diagram is concerned, *PQ* is a single activity. However, in reality, *PQ* consists of many subactivities and so we need another PERT diagram for describing activity *PQ*. In other words, for each activity shown in the overall diagram, there is a sub PERT diagram. For top management control, a

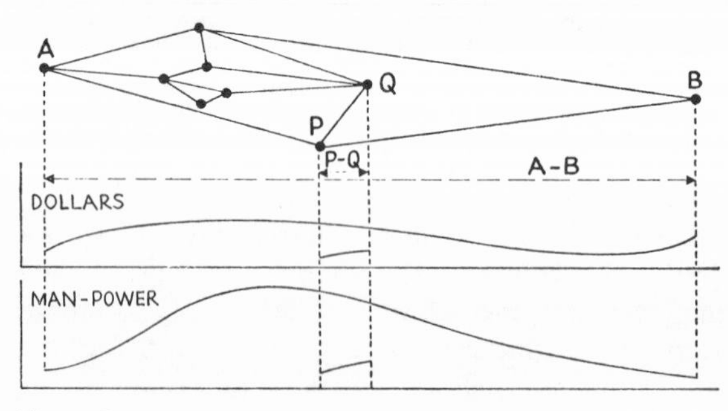

Figure 7
Program evaluation and review technique (PERT).

relatively simple overall PERT diagram would be used. For lower-level management control, each major activity is decomposed into subactivities. Each manager has his own PERT diagram pertaining to his activities. As there is a hierarchy of managers, there is a hierarchy of PERT diagrams.

Let us turn our attention now to the lower diagrams of Figure 7. Here, resource requirements associated with the program are shown. The middle diagram, for instance, shows dollars required to carry out the activities. (The short segment in the middle shows the dollars associated with activity *PQ*.) The bottom diagram shows the scientific manpower requirement associated with the program. (Again, the short curve on the bottom shows the scientific manpower associated with activity *PQ*.) There would be an additional curve corresponding to each type of resource requirement of the program.

Let us consider now how a manager deals with his control problem.[58,60] He receives overall plans from a higher echelon and his problem is to try to adjust his schedules and resources to meet this overall plan. If he thinks this can be done, he will issue directives to lower-level decision makers, who in their turn will go through their planning exercise. On the other hand, if the manager finds that he cannot meet the new overall plan, or that he needs additional resources, he will appeal to the higher echelon.

To determine whether he can meet the overall plan, he needs to consider alternative possibilities for the different activities under his control. On the basis of alternatives, he can develop various combinations and determine what he has to do to meet the required overall plan. Let us consider now how these alternative possibilities can be developed.[57]

First, consider a simple activity under the control of a project engineer. Let us ask him to prepare a time-cost relationship, as shown in Figure 8. This curve shows that a particular activity could be carried out in three different fashions. If it is done on a crash basis, it can be done in a short time but at a high cost (point *A*). There is also a minimum cost plan (point *B*), which is the most efficient way to carry out the activity. A third alternative (point *C*) has a longer time schedule requiring more total dollars, but lower rate of expenditure. The dotted lines show how dollars would be expended for each of the alternatives.

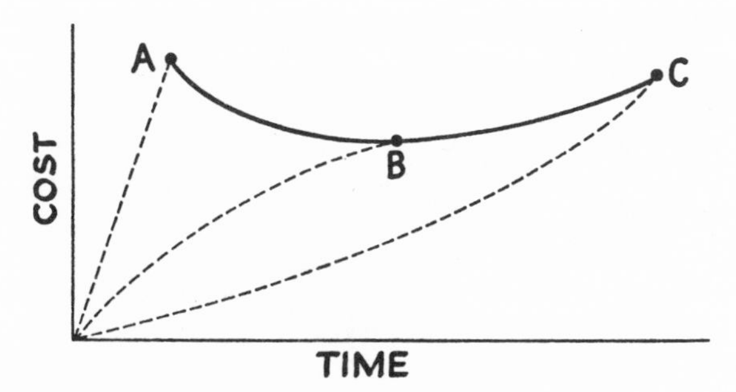

Figure 8
Development of plans and alternative plans.

Let us prepare for each simple activity a time-cost relationship. Consider the control problem of the manager who deals with the type of activity which is made up exclusively of these "simple" activities. We could say that this manager deals with a composite PERT diagram on the bottom of the hierarchy.

As he already has for each simple activity a time-cost curve of the type shown in Figure 8, he can combine these time-cost relationships in many various ways and can select a "best" time-cost relationship for the total activity under his control. In a similar fashion, we can move to higher levels of PERT charts and develop "best" time-cost relationships. For each manager we can provide a set of time-cost relationships pertaining to the activities under his control. Consequently, he in his turn can develop alternatives by combining his time-cost relationships and select the "best" possible course of action.

So far, we have talked only about dollar resources. Similar diagrams could be developed for manpower or other resources and then these diagrams could be combined. It is seen then that by this technique managers could obtain good control of large-scale programs, provided the manipulations required could be carried out in a rapid and effective manner. Now we will show how this manipulation can be done on an on-line real-time computer system.

Let us examine in detail the data processing functions involved in this proposed system[59]. First, it is of great importance that status information on the progress of projects, activities, and events be recorded as they occur. Consequently, a set of input devices (or consoles) is to be located where the work is done. We assume that there is a central computer or that there are various computers interconnected, so that as the events transpire appropriate inputs are made to the common data base. Suppose, for instance, that a certain event is completed. The man in charge would go to the nearest console and report that the expected completion date is to be replaced by the actual completion date. He would do this by depressing appropriate keys and by inspecting the PERT network pertinent to his work. The language to be used in the course of his analysis would contain sentences such as "display designated network," "display events summary," "display activity," "display critical path."

With the aid of such language, he could retrieve the information he wants and also he could modify the data. Let us proceed now to examine how a manager would prepare his new plan.

As up-to-date information is available on changes in resources, in completion dates, etc., he could generate new time-cost relationships and could enter these into the computer. Then the computer would compute new networks and display them. The manager would also direct the computer to compute the associated dollar, manpower, engineering, and other associated resource requirements. Then the manager would pose problems to the computer. He would ask what would happen if a particular event slipped by a certain length of time. The computer would recompute the whole network and report back on the consequences of such slippages. By this technique the manager could study a large number of trial solutions and select the one that looks the best. When such a "best" solution is reached, the computer could be directed to carry out all the planning computations required and store the results in the memory.

It is seen then that in the system visualized here all data would be updated as events transpire and plans would be determined in light of the most recent data. As managers and planners are "on-line," every time they consider new events and new possibilities, they will work with the most up-to-date data.

Admittedly, the establishment of such a system could not be implemented from one day to another. Perhaps we are still quite far from the realization of such an extensively automated PERT system. However, the need exists and one may predict that as the years go by we will realize a substantial part of the system described here. . . .

APPROACH FOR AUTOMATION

A review of the history of science shows that major discoveries often cause fright, distrust, and uneasiness. The discovery that the earth is not the center of the universe or the discovery by Darwin of the theory of evolution was interpreted by some as contrary to Christian religion. Thermodynamics was interpreted by some as a "dismal" science, as it seemed to imply that the universe was running down. It is quite possible that the discovery of digital computers will have a major impact on our way of life and some of today's apprehensions are a natural reaction to a major change.

We have had computers now for about [30] years and the spread of their use has been spectacular. Today, we begin to understand better the role computers should take and how they can serve mankind. We have probably, in many ways, underestimated the difficulties of computerizing solutions and perhaps we underestimated human capabilities. In some ways we have overestimated, but in some other ways underestimated, the capabilities of computers.

There are many people who feel that computers have failed us, as many predictions about computers have not come true. On the other hand, we need to recognize that we have designed computers for certain tasks, and it is not surprising that tasks for which the machines have not been designed cannot be carried out by the machines. It is becoming increasingly clear that we need not only new machines, but new techniques and new attitudes in capitalizing on the potentialities of digital computers. It is also possible that we perhaps overestimated the capability of man to adapt to new ways of doing things.

We are witnessing today the emergence of a new point of view. We are trying to make our overall plans less ambitious, but more specific. We are beginning to realize that a step-by-step approach in the introduction of computers is in order. A balanced approach among analysis, experimentation, and simulation seems to be the road to progress in digital computers.

We would like to conclude this paper with the following three points:

1. Mankind is faced with many important problems that can be solved only if we discover better ways to manage information.
2. Electronic digital data processors offer a promise to resolve the current crisis in information handling.
3. Solutions to the problems of information management will not come from purely automated systems, but from a balanced man-machine partnership.

REFERENCES

[1] "Aerospace in Perspective: The Program," *Space Aeronautics*, p. 71, Jan. 1964.

[2] Ashby, Ross. *Design for a Brain*, John Wiley & Sons, Inc., New York, 1960.

[3] Ashby, Ross. "Design for an Intelligence-Amplifier." See Ref. 47, pp. 215-234.

[4] Barnett, M.P., and Kelley, K.L. "Computer Editing of Verbal Texts. Part I. The ESI System," *American Documentation*, pp. 99-108, April 1963.

[5] Benington, Herbert. "Military Information – Recently and Presently." See Ref. 45.

[6] Bush, V. "As We May Think," *The Atlantic Monthly*, July 1945.

[7] Casciato, L. "The Control of Traffic Signals with an Electronic Computer – A New Application of Real-Time Data Processing." See Ref. 44, pp. 231-234.

[8] "Computers That Feed Many Mouths," *Business Week*, pp. 54-55, Feb. 1, 1964.

[9] Culler, G.J., and Huff, R.W. "Solution of Non-Linear Integral Equations Using On-Line Computer Control." See Ref. 40, pp. 129-133.

[10] Engelbart, D.C. "Augmenting Human Intellect: A Conceptual Framework," Stanford Research Institute, Menlo Park, Calif., Oct. 1962.

[11] Feigenbaum, A.E., and Feldman, J. *Computers and Thought*, McGraw-Hill Book Co., Inc., New York, 1963.

[12] Forrester, J.W. "Managerial Decision Making." See Ref. 17, pp. 36-91.

[13] Fried, B., and Culler, G. Proceedings of the Pacific Computer Conference, "On-Line Computations and Faster Information Processing," pp. 221-242, California Institute of Technology, Mar. 1963.

[14] Gass, S.I., et al. "Project Mercury Real-Time

Computational and Data-Flow System." See Ref. 41, pp. 33-78.

[15] Geisler, M.A., Haythorne, W.W., and Steger, W.A. "Simulation and the Logistics Systems Laboratory," RAND Memorandum, RM-3281-PR, Sept. 1962.

[16] Gozinto, Z. "How I Work," Memoirs of a Genius, Chapter 3 (still under preparation).

[17] Greenberger, M. (editor). *Management and the Computer of the Future,* John Wiley & Sons, Inc., New York, 1962.

[18] Hadamard, J. *The Psychology of Invention in the Mathematical Field,* Dover Publications, Inc., New York.

[19] Haythorne, W.W. "Information Systems Simulation and Modeling." See Ref. 42, Session 7, pp. 89-100.

[20] Hoover, W.R., Ancand, A., and Miller, T.B. "A Real Time Multi-Computer System for Lunar and Planetary Space Flight Data Processing." See Ref. 43, pp. 127-150.

[21] "How Do You Organize to Shoot the Moon?" *Business Week,* p. 37, Nov. 17, 1962.

[22] Johnson, T.E., "Sketchpad III: A Computer Program for Drawing in Three Dimensions." See Ref. 43, pp. 347-353.

[23] Kahn, Arthur B. "Skeletal Structure of PERT and CPA Programs," *Communications of the ACM,* Vol. 6, pp. 473-479, 1963.

[24] Katz, Jesse H. "Simulation of Traffic Network," *Communications of the ACM,* Vol. 6, pp. 480-486, 1963.

[25] Lessing, Lawrence. "The Transistorized M.D.," *Fortune,* Sept. 1963.

[26] Licklider, J.C.R. "Interactions between Artificial Intelligence, Military Intelligence and Command and Control." See Ref. 42, Session 8, pp. 31-56.

[27] Licklider, J.C.R. "Man-Computer Symbiosis," *IRE Transactions on Human Factors in Electronics,* pp. 4-11, Mar. 1960.

[28] Licklider, J.C.R., and Clark, W.E. "On-Line Man-Computer Communication." See Ref. 40, pp. 113-128.

[29] Malcolm, D.G. "Exploring the Military Analogy — Real-Time Management Control." See Ref. 31, pp. 187-207 (Description of SAGE).

[30] Malcolm, D.G., Rosenbloom, J.H., Clark, C.E., and Frazar, W. "Application of a Technique for Research and Development Program Evaluation," *Operations Research,* Vol. 7, pp. 646-669, Sept.-Oct. 1959.

[31] Malcolm, D.G., and Rowe, A.J. (editors). *Management Control Systems,* John Wiley & Sons, Inc., New York, 1960.

[32] "Manned Venus, Jupiter Satellites Planned," *Aviation Week and Space Technology,* p. 26, Nov. 12, 1962.

[33] McCarthy, J. "Time Sharing Computer Systems." See Ref. 17, pp. 220-248.

[34] Minsky, M. "Steps Toward Artificial Intelligence," *Proceedings of the IRE,* pp. 8-30, Jan. 1961. Also see Ref. 42, Session 8, pp. 1-23, and Ref. 11, pp. 406-450.

[35] Morgenstern, O. "How to Plan to Beat Hell," *Fortune,* Jan. 1963.

[36] Moshman, J., Johnson, J., and Larsey, M. "RAMPS — A Technique for Resource Allocation and Multi-Project Scheduling." See Ref. 43, pp. 17-27.

[37] Newell, A., Shaw, J.C., and Simon, H.A. "The Process of Creative Thinking." Presented at a Symposium on Creative Thinking, University of Colorado, Boulder, Colo., May 1958. Also RAND Publication P-1320, Jan. 1959.

[38] Perlis, A.J. "The Computer in the University." See Ref. 17, pp. 180-217.

[39] Polya, G. *How to Solve It,* Princeton University Press, Princeton, N.J., 1945. Also, *Induction and Analogy in Mathematics and Patterns of Plausible Inference,* 2 Vols., Princeton University Press, Princeton, N.J., 1954. (Also available in paperback.)

[40] *Proceedings 1962 of the Spring Joint Computer Conference,* San Francisco, Calif. AFIPS, Vol. 21, The National Press, Palo Alto, Calif., May 1962.

[41] *Proceedings of the Eastern Joint Computer Conference,* Washington, D.C., Dec. 12-14, 1961, An American Federation of Information Processing Societies Publication, Vol. 20, Macmillan Publishing Co., New York.

[42] *Proceedings of the First Congress on the Information System Sciences, Air Force Electronic Systems Division and MITRE Corporation,* Nov. 1962. (To appear as a book *Information System Science and Engineering,* the McGraw-Hill Book Co.)

[43] *Proceedings of the 1963 Spring Joint Computer*

Conference, Vol. 23, Spartan Books, Inc., New York, May 1, 1963.

[44] *Information Processing 1962,* Proceedings of IFIP Congress 62, North-Holland Publishing Co., Amsterdam, 1963.

[45] *Proceedings of the Western Joint Computer Conference,* pp. 17-37, May 1962, Los Angeles, Calif.

[46] Rome, R.K., and Rome, S.C. "Leviathan: An Experimental Study of Large Organizations with the Aid of Computers." See Ref. 42, Session 7, pp. 1-87.

[47] Shannon, C.E. and McCarthy, J. (Editors). *Automata Studies,* Princeton University Press, Princeton, N.J., 1956.

[48] Simon, H.A., "Simulation of Human Thinking and Problem Solving." See Ref. 17, pp. 94-131.

[49] Simon, H.A., and Newell, A. "Heuristic Problem Solving, The Next Advance in Operations Research," *Operations Research,* Jan.-Feb., 1958.

[50] Sprague, R.E., *Electronic Business Systems,* The Ronald Press Co., New York, 1962.

[51] Stauffer, R.B., and Lewis, T.H. "Met-Watch: A Technique for Processing and Scanning Meteorological Data with a Digital Computer." See Ref. 44, pp. 242-244.

[52] Stotz, R. "Man-Machine Console Facilities for Computer-Aider Design," See Ref. 43, pp. 323-328.

[53] Sutherland, I.E. "Sketchpad: A Man-Machine Graphical Communication System," See Ref. 43, pp. 329-346.

[54] Swanson, D.R. "Interrogating a Computer in Natural Language," See Ref. 44, pp. 288-293.

[55] Tongue, F.M. "The Use of Heuristic Programming in Management Science," *Management Science,* Vol. 7, pp. 231-237, April 1961.

[56] Ulam, S.M. "A Collection of Mathematical Problems," Chapter VIII – *Computing Machines as a Heuristic Aid,* pp. 115-144, Interscience Publishers, Inc., New York, 1960.

[57] Vazsonyi, A. "An On-Line Management System Using English Language," See Ref. 45.

[58] Vazsonyi, A. "Demonstration Models of Intellectronic Management Planning and Control Techniques," Ramo-Wooldridge, Canoga Park, California, 1962.

[59] Vazsonyi, A. "Extending Management Capability by Electronic Computers," See Ref. 44.

[60] Vazsonyi, A. "Gaming Techniques for Management Planning and Control," Foundation for Instrumentation Education and Research, Inc., New York, New York, Oct. 1960.

[61] Vazsonyi, A. "Program Evaluation and Review Techniques," Ramo-Wooldridge, Canoga Park, California, 1962.

[62] Burck, G. "On Line in Real Time," *Fortune,* p. 141, April, 1964.

[63] Langefors, Börji. "Automated Design," International Science and Technology, pp. 90-97, February, 1964.

14

John Dearden

MYTH OF REAL-TIME MANAGEMENT INFORMATION

The latest vogue in computer information systems is the so-called real-time management information system. The general idea is to have in each executive's office a remote computer terminal which is connected to a large-scale computer with a data bank containing all the relevant information in the company. The data bank, updated continuously, can be "interrogated" by the manager at any time. Answers to questions are immediately flashed on a screen in his office. Allegedly, a real-time management information system enables the manager to obtain complete and up-to-the-minute information about everything that is happening within the company.

The purpose of this article – aimed at a time span of the next five to seven years – is to raise some serious questions concerning the utility of a real-time information system for top management. I will try to show that it would not be practicable to operate a real-time *management control* system and, moreover, that such a system would not help to solve any of the

---

critical problems even if it could be implemented. I will also try to show that in other areas of top management concern a real-time system is, at best, of marginal value. It is my personal opinion that of all the ridiculous things that have been foisted on the long-suffering executive in the name of science and progress the real-time management information system is the silliest.

MEANING OF REAL TIME

One problem in any new field of endeavor is that there is frequently no universally accepted definition for many of the terms. It therefore becomes nearly impossible to question the validity of the concepts underlying the terms because their meanings are different to different people. The term "real time" is no exception. In fact, in a single issue of one computer magazine, back-to-back articles defined real time differently; and one example, cited in the first article as an illustration of what real time is *not,* appeared in the second article as an illustration of what a real-time system *is.*

Semantic Confusion

One concept of real time is demonstrated by these two quotations:

> *A real-time management information system — i.e., one that delivers information in time to do something about it.*[1]
>
> *A real-time computer system may be defined as one that controls an environment by receiving data, processing them and returning results sufficiently quickly to affect the functioning of the environment at that time.*[2]

The problem with both of these definitions is that they are too broad. *All* management control systems must be real-time systems under this concept. It would be a little silly to plan to provide management with budget performance reports, for instance, if they were received too late for management to take any action.

The following is a description of real time that comes closer to the concept of real-time as it is used by most systems and computer people:

> *The delays involved in batch processing are often natural delays, and little advantage can be obtained by reducing them. But elimination of the* necessity *for such delays opens new and relatively unexplored possibilities for changing the entire nature of the data processing system — from a passive recorder of history (which, of course, is valuable for many decisions) to an active participant in the minute-to-minute operations of the organization. It becomes possible to process data in* real time *— so that the output may be fed back immediately to control current operations. Thus the computer can interact with people on a dynamic basis, obtaining and providing information, recording the decisions of humans, or even making some of these decisions.*[3]

System Characteristics

To expand somewhat on this description, the term "real-time system" as used in this article will mean a computer system with the following characteristics:

1. *Data will be maintained "on line."* In other words, all data used in the system will be directly available to the computer — they will be stored in the computer memory or in random-access files attached to the computer. (This is in contrast to data maintained on magnetic tapes, which must be mounted and searched before information is available to the computer.)
2. *Data will be updated as events occur.* (In contrast to the "batch" process, where changes are accumulated and periodically updated.)
3. *The computer can be interrogated from remote terminals.* This means that the information stored in the computer can be obtained on request from a number of locations at a distance from the place where the data are processed and stored.

Perhaps the most widely known example of a real-time system currently in operation is the American Airlines SABRE system for making plane reservations.

POTENTIAL APPLICATIONS

With the new generation of computers, random access memories have become much less expensive than has been true until now. This fact, coupled with the advances made in data-transmission equipment and techniques, will make many real-time applications economically feasible.

Real-time methods will improve those systems where the lack of up-to-the-minute information has in the past resulted in increased costs or loss of revenue. I believe that many companies will employ real-time methods to control all or part of their logistics (the flow of goods through the company) systems. For example,

> *A manufacturer of major household appliances might have raw material and work-in-process inventories in his manufacturing plants, and finished goods inventories both in company and distributor warehouses and in dealer showrooms. There is a more or less continuous logistics flow all along the route from raw material to retail customer. If all of the data on inventory levels and flows could be maintained centrally and updated and analyzed continuously, this would not only solve many of the problems now faced by such a manufacturer, but would make it possible to provide better all-around service with lower inventory levels and lower costs (particularly in transportation and obsolescence).*

There are, of course, many other potential applications for real-time management information systems, and I believe that they will be used extensively in the next few years. However, these applications will take place almost exclusively in logistics, and, as I shall explain later on, techniques that may improve a logistics system will not necessarily improve a management control system. I want to make it clear at this point that I am not opposed to real-time systems per se. I believe they have valuable applications in operating situations. I am only opposed to using real-time information systems where they do not apply. The balance of this article will consider top management's use of real-time systems.

MANAGEMENT FUNCTIONS

As used here, the term "top management" will apply to the president and executive vice-president in centralized companies, plus divisional managers in decentralized companies. In other words, I am considering as top management those people responsible for the full range of a business activity – marketing, production, research, and so forth. I am also assuming that the company or division is sufficiently large and complex so that the executive makes only a limited number of operating decisions, if any. I believe that this is a reasonable assumption in considering real-time management information systems. A company where the president makes most of the operating decisions could scarcely be considering a sophisticated and expensive computer installation.

Six Categories

This part of the discussion considers, in general terms, the functions of top management. The purpose is to establish how a typical executive might spend his time so that we may later evaluate the extent to which his decision making can or cannot be helped by real-time computer systems. I have divided top management's functions into six general categories – management control, strategic planning, personnel planning, coordination, operating control, and personal appearances. Each is discussed below.

1. *Management control:* one of the principal tasks of a manager is to exercise control over the people to whom he has delegated responsibility. Ideally, this control consists of coordinating, directing, and motivating subordinates by reviewing and approving an operating plan, by comparing periodically the actual performance against this plan, by evaluating the performance of subordinates, and by taking action with respect to subordinates where and when it becomes necessary.

The formal management control system will, of course, vary with the type and size of business as well as with the type and amount of responsibility delegated to the subordinate. Nevertheless, all effective formal management control systems need three things:

(a) A good plan, objective, or standard. The manager and the subordinate must agree as to what will constitute satisfactory performance.

(b) A system for evaluating actual performance periodically against the plan. This would include both a clear explanation of why variances have occurred and a forecast of future performance.

(c) An "early warning" system to notify management in the event that conditions warrant attention between reporting periods.

2. *Strategic planning:* this consists of determining long-range objectives and making the necessary decisions to implement these objectives. Much of top management's strategic planning activity involves reviewing studies made by staff groups. Capital expenditure programs, acquisition proposals, and new product programs are examples of studies that fall into this area.

Another phase of strategic planning consists of developing ideas for subordinates to study – that is, instead of waiting for staff or line groups to recommend courses of action, the executive develops ideas of his own as to what the company should be doing.

3. *Personnel planning:* this important function of management deals with making decisions on hiring, discharging, promoting, demoting, compensating, or changing key personnel. In the broadest sense, this consists of organizational planning. Personnel planning is, of course, related both to management control and strategic planning. Nevertheless, there are so many unique problems associated with personnel planning that I believe it is reasonable to consider it as a separate function.

4. *Coordination:* here management's function is to harmonize the activities of subordinates, especially where it is necessary to solve a problem that cuts across organizational lines. For example, a quality control problem might affect several operating executives, and the solution to this problem might require top management's active participation. In general, this activity tends to be more important at the lower organization levels. The president of a large, decentralized company would perform less of this coordination function than his divisional managers because interdepartmental problems are more common at the divisional level.

5. *Operating control:* almost all top executives perform some operating functions. For example, I know a company president who buys certain major raw materials used by his company. Usually, the operating decisions made by top management are those which are so important to the welfare of the company that the executive believes the responsibility for making them cannot be properly delegated.

6. *Personal appearances:* many top executives spend much time in performing functions that require their making a personal appearance. This can vary from entertaining visiting dignitaries to giving out 25-year watches. (I shall assume the activities involving such personal appearances will not be affected by a real-time management information system.)

REAL-TIME PRACTICALITY?

The purpose of this part of the article is to examine, in turn, each of the management functions described above (except 6) to see whether or not it can be improved by a real-time information system.

Management Control

I do not see how a real-time system can be *used* in management control. In fact, I believe that any attempt to use real time will considerably weaken even a good management control system. (In setting objectives or budgets, it may be useful to have a computer available at the time of the budget review to calculate the effects of various alternatives suggested by management. This, however, is not a real-time system, since a computer console need be installed only for the review sessions.)

Calculating Performance. In the area of performance evaluation, real-time management information systems are particularly ridiculous. When a division manager agrees to earn, say, $360,000 in 1966, he does not agree to earn $1,000 a day or $1,000/24 per hour. The only way actual performance can be compared with a budget is to break down the budget into the time periods against which performance is to be measured. If the smallest period is a month (as it usually is), nothing short of a month's actual performance is significant (with the exception of the events picked up by the early warning system to be described). Why, then, have a computer system that allows the manager to interrogate a memory bank to show him the hour-to-hour or even day-to-day status of performance against plan?

Even assuming objectives could logically be calendarized by day or hour, we run into worse problems in calculating actual performance, and worse still in making the comparison of actual to standard meaningful. If the performance measures involve accounting data (and they most frequently do), the data will never be up-to-date until they are normalized (adjusted) at the end of the accounting period. I will not bore you with the details. Suffice it to say only that a real-time accounting system which yields meaningful results on even a daily basis would be a horrendous and expensive undertaking.

Let us go one step further. Performance reports, to be meaningful, must include an explanation of the variances. This frequently involves considerable effort and often requires the analyst to spend time at the source of the variance in order to determine the cause. Would this be done every day or oftener? Ridiculous! There is one more thing about performance reports. The important message in many reports is the action being taken and the estimated effect of this action. In other words, the projection of future events is the important top management consideration. Will this be built into the real-time system? Since this involves the considered judgment of the subordinate and his staff, I do not see how this could possibly be done even on a daily basis.

Early Warning. How about real-time for providing an early warning? Here, also, I do not see how it could be of help. Early warning has not been a problem in any top management control system with which I have been acquainted. In most instances, when situations deteriorate to the point where immediate action is required, top management knows about it. As the manager of a division ($100 million a year in sales) said to me, when I asked him how he knew when things might be out of hand in one of his plants, "That's what the telephone is for."

In any case, it is possible to prescribe the situations which management should be apprised of immediately, without even relying on a computer. Furthermore, the important thing is to bring the situation to top management's attention *before* something happens. For example, it is important to inform management of a threatened strike. Yet a real-time management information system would pick it up only *after* the strike had occurred.

In summary, then, early warning systems have been put into operation and have worked satisfactorily without a real-time system. I see nothing in a real-time management information system that would improve the means of early warning, and such a system would certainly be more expensive. (Note that here I am talking about management control systems. The early warning techniques of many logistical control systems, in contrast, could be greatly improved by real-time systems.)

My conclusion on management control is that real-time information cannot be made meaningful – even at an extremely high cost – and that any attempt to do so cannot help but result in a waste of money and management time. Improvements in most mangement control systems must come from sources other than real-time information systems.

Strategic Planning

Since strategic planning largely involves predicting the long-run future, I fail to see how a real-time management information system will be of appreciable use here. It *is* true that past data are required to forecast future events, but these need hardly be continuously updated and immediately available. Furthermore, much of the preparation of detailed strategic plans is done by staff groups. While these groups may on occasion work with computer models, the models would certainly be stored away, not maintained on line between uses.

Perhaps the most persistent concept of a real-time management information system is the picture of the manager sitting down at his console and interacting with the computer. For example, as a strategic planning idea comes to him, he calls in a simulation model to test it out, or a regression analysis to help him forecast some event; or, again, he asks for all the information about a certain subject on which he is required to make a decision.

It seems to me that the typical manager would have neither the time nor the inclination to interact with the computer on a day-to-day basis about strategic planning. Problems requiring computer models are likely to be extremely complex. In most instances, the formulation of these problems can be turned over to staff specialists. Furthermore, I think it would be quite expensive to build a series of models to anticipate the manager's needs.

Under any conditions, strategic planning either by the manager alone or by staff groups does not appear to be improved by a real-time system. Models can be fed into the computer and coefficients can be updated as they are used. Between uses, it seems to me, these models would be most economically stored on magnetic tape.

Personnel Planning

A real-time management information system does not help the top manager to solve his problems of personnel planning, although the computer can be useful in certain types of personnel data analysis. About the only advantage to the manager is that information becomes available somewhat more quickly. Instead of calling for the history of a particular individual and waiting for personnel to deliver it, the manager can request this information directly from the computer. Therefore, while a remote console device with a visual display unit *could* be used for retrieving personnel information, the question of whether it *should* be used is one of simple economics. Is the additional cost of storing and maintaining the information, plus the cost of the retrieval devices, worth the convenience?

Coordination

The coordination function is very similar to the management control function with respect to potential real-time applicability. A manager wants to know right away when there is an interdepartmental problem that will require his attention. As is the case with early warning systems developed for management control, a real-time system is not necessary (or even useful, in most cases) to convey this information. Further, I cannot see how a real-time management information system could be used in the solution of these coordination problems, except in unusual cases.

Operating Control

There is no question that real-time methods are useful in certain types of operating systems, particularly in logistics systems.[4] To the extent that a top executive retains certain operating control functions, there is a possibility that he may be able to use a real-time information system. Because of the necessity of doing other things, however, most executives will be able to spend only a limited amount of time on operating functions. This means generally that they must work on the "exception" principle. Under most conditions, therefore, it would seem much more economical for a subordinate to monitor the real-time information and inform the top executives when a decision has to be made.

It is very difficult to generalize about this situation. Here, again, it appears to be one of simple economics. How much is a real-time system worth to the manager in relation to what it is costing? I cannot believe that there would be many instances where a manager would be concerned with operating problems to the extent that a real-time information system operating from his office would be justified.

REPORTING BY COMPUTER

In recent months, there have been experiments to replace traditional published reports by utilizing consoles and display devices to report information directly to management. Although these techniques, strictly speaking, are not real time, they bear such a close relationship to real-time systems that it will be useful to consider them here.

Modus Operandi

The general idea is that the information contained in the management reports would be stored in the computer memory so that the manager could ask for only the information he needed. This request would be made from the computer console, and the information would be flashed on a screen in his office. For example, a manager could ask for a report on how sales compared with quota. After looking at this, he could then ask for data on the sales of the particular regions that were below quote and, subsequently, for detail of the districts that were out of line.

The benefits claimed for this type of reporting are as follows:

1. The manager will receive only the information he wants.

2. Each manager can obtain the information in the format in which he wants it. In other words, each manager can design his own reports. One manager may use graphs almost exclusively, while another may use tabulation.

3. The information can be assembled in whatever way the manager wants it – that is, one manager may want sales by areas and another may want it by product line. Furthermore, the manager can have the data processed in any way that he wants.

4. The information will be received more quickly.

Important Considerations

Before installing such a system, it seems to me a number of things should be taken into account.

First, what advantage, if any, does this sytem have over a well-designed reporting system? Since the storage and retrieval of data in a computer do not add anything that could not be obtained in a traditional reporting system, the benefits must be related to convenience. Is there enough additional convenience to justify the additional cost?

Second, is it possible that for many executives such a system will be more of a nuisance than a convenience? It may be much easier for them to open a notebook and read the information needed, since in a well-designed system the information is reported in levels of details so that only data of interest need be examined.

Finally, will the saving in time be of any value?

It seems to me that the two main considerations in installing such a system are the economics and the desires of the particular executive. There is one further possibility, however, that should be carefully considered. What will be the impact on the lower level executives? If these people do not know the kind of information their superiors are using to measure their performance, will this not create human relations problems?

Without going into the details, I can see many problems being created if this is not handled correctly. With a regular reporting system, the subordinate knows exactly what information his superior is receiving – and when he receives it – concerning his performance. Furthermore, the subordinate receives the information *first.* Any deviations in this relationship can cause problems, and the use of a computer to retrieve varying kinds of information from a data base is a deviation from this relationship.

THREE FALLACIES

If management information on a real-time basis is so impractical and uneconomic, why are so many people evidently enamored with this concept? I believe that the alleged benefits of real-time management information systems are based on three major fallacies.

Improved Control

Just about every manager feels, at some time, that he does not really have control of his company. Many managers feel this way frequently. This is natural, since complete control is just about impossible even with the best management control system. Since most companies have management control systems that are far from optimum, there is little wonder that a feeling of insecurity exists. In the face of this feeling of insecurity, the promise of "knowing everything that is happening as soon as it happens" has an overpowering appeal.

As explained previously, real time will not improve management control and, consequently, will not help to eliminate the insecurity that exists. What is usually needed is a combination of improved management control systems and better selection and training of personnel. Even at best, however, the executive will have to accept responsibility for what other people do, without having full control over their actions.

Scientific Management

There appears to be considerable sentiment to the effect that the scientific way to manage is to use a computer. This fallacy implies that the executive with a computer console in his office is a scientific manager who uses man-machine communication to extend his ability into new, heretofore unavailable, realms of decision making.

I believe that it is nonsense to expect most managers to communicate directly with a computer. Every manager and every business is different. If a manager has the necessary training and wishes to do so, it may be helpful for him to use a computer to test out some of his ideas. To say, however, that *all* managers should do this, or that this is "scientific management," is ridiculous. A manager has to allocate his time so that he spends it on those areas where his contribution is greatest. If a computer is useful for testing out his ideas in a given situation, there is no reason why he should have to do it personally. The assignment can just as easily be turned over to a staff group. In other words, where a computer is helpful in solving some management problems, there is no reason for the manager to have any direct contact with the machine.

In most instances, the computer is of best use where there are complex problems to be solved. The formulation of a solution to these complex problems can generally be done best by a staff group. Not only are staff personnel better qualified (they are selected for these qualifications), but they have the uninterrupted time to do it. It seems to me that there is nothing wrong with a manager spending his time managing and letting others play "Liberace at the console."

Logistics Similarity

This fallacy is the belief that management control systems are merely higher manifestations of logistics sytems.

The fact is that the typical real-time system, either in operation or being planned, is a *logistics* system. In such a system, for example, a production plan is developed and the degree of allowable variances established in a centralized computer installation. The actual production is constantly compared to plan; and when a deviation exceeds the established norm, this fact is communicated to the appropriate source. On receiving this information, action is always taken. Either the schedules are changed or the deficiency is somehow made up.

Notice that speed in handling and transmitting vast amounts of information is essential. This is the critical problem that limits many manual logistics systems; and the computer, particularly with real-time applications, goes a long way toward solving the speed problem.

In contrast, speed in processing and transmitting large amounts of data is *not* a critical problem in *management control* systems. Consequently, the improvements that real-time techniques may effect in logicstics sytems cannot be extrapolated into management control systems.

The critical problems in management control are (1) determining the level of objectives, (2) determining when a deviation from the objective requires action, and (3) deciding what particular action should be taken. The higher in the organizational hierarchy the manager is positioned, the more critical these three problems tend to become. For example, they are usually much more difficult in planning divisional profit budgets than plant expense budgets. In some instances the computer can help the manager with these problems, but I do not see how it can solve them for him. Furthermore, the use of computers in solving these problems has nothing to do with real-time.

SHORT-TERM VIEW

While real-time management information systems may be very useful in improving certain kinds of operating systems, particularly complex logistics systems, they will be of little use in improving management control. This is particularly true in the short-range time span of the next five to seven years.

The following is a checklist of questions that I believe the manager should have answers to before letting anyone install a remote computer terminal and a visual display screen in his office:

1. What will the total incremental cost of the equipment and programming be? (Be sure to consider the cost of continuing systems and programming work that the real-time systems will involve.)
2. Exactly how will this equipment be used? Be sure to obtain a complete description of the proposed uses and the date when each application will become operational.)
3. Exactly how will each of these uses improve the ability to make decisions? In particular, how will the management control system be improved?

With precise answers to these three questions, it seems to me that a manager can decide whether or not a remote terminal and visual display device should be installed. Do not be surprised, however, if the answer is negative.

LONG-RANGE OUTLOOK

What are the prospects of real-time systems, say, 15 or 20 years from now? Some experts believe that, by that time, staff assistance to top management will have largely disappeared. Not only will the staff have disappeared, but so will most of the paper that flows through present organizations. A manager in the year 1985 or so will sit in his paperless, peopleless office with his computer terminal and make decisions based on information and analyses displayed on a screen in his office.

Caution Urged

It seems to me that, at the present time, the long-term potential of real-time management information systems is completely unknown. No one can say with any degree of certainty that the prediction cited above is incorrect. After all, 15 or 20 years is a long time away, and the concept of a manager using a computer to replace his staff is not beyond the realm of theoretical possibility. On the other hand, this concept could be a complete pipedream.

Under any circumstances, many significant changes in technology, organization, and managerial personnel will be required before this prediction could be a reality for business in general. As a result, if such changes do occur, they will come slowly, and there will be ample opportunity for business executives to adjust to them. For example, I believe there is little danger of a company president waking up some morning to find his chief competitor has installed a computer-based, decision-making system so effective that it will run him out of business.

I do believe all executives should be open-minded to suggestions for any improvements in management information systems, but they should require evidence that any proposed real-time management information system will actually increase their effectiveness. Above all, no one should rush into this now because of its future potential.

The present state of real-time management information systems has been compared to that of the transportation field at the beginning of the Model-T era. At that time, only visionaries had any idea of how transportation would be revolutionized by the automobile. It would have been foolish, however, for a businessman to get rid of his horse-drawn vehicles just because some visionaries said that trucks would take over completely in 20 years.

It seems to me that this is the identical situation now. Even if the most revolutionary changes will eventually take place in management information systems 20 years hence, it would be silly for business executives to scrap present methods until they are positive the new methods are better.

REFERENCES

[1] Gilbert Burck and the Editors of *Fortune, The Computer Age* (New York, Harper & Row, Publishers, 1965), p. 106.

[2] James Martin, *Programming Real-Time Computer Systems* (Englewood Cliffs, N.J., Prentice-Hall, Inc., 1965), p. 378.

[3] E. Wainright Martin, Jr., *Electronic Data Processing* (Homewood, Ill., Richard D. Irwin, Inc., 1965), p.381.

[4] See Robert E. McGarrah, "Logistics for the International Manufacturer," *Harvard Business Review*, Mar.-Apr. 1966, p. 157.

PART V PROCESS

. . . Good comprehensive planning deals with system problems which cannot be better treated by other methods of systems management, and especially auto-controls. Good comprehensive planning limits the degree of comprehensiveness within the boundaries of manageability. Good comprehensive planning is itself subjected to cost-effectiveness analysis with special attention to the opportunity-costs of highly qualified manpower and time. Good comprehensive planning regards both balance and imbalance as often useful stages of system development and limits balance-aiming comprehensive planning to appropriate situations identified after careful scrutiny. Good comprehensive planning aims at maximizing the objective expectation of desirable impact on future reality, using whatever nominal tools are most useful for that purpose. And good comprehensive planning is continuous and iterative, regarding all subphasing as instrumental. . . .

. . . eight additional preferred features which characterize good comprehensive planning warrant more detailed attention: (1) intermediate position between policy making and strategic planning on one side and operational planning on the other side; (2) multidimensional, but manageable system as subject for planning; (3) an interdisciplinary approach; (4) high development of rationality components and high development of extra-rationality components; (5) high sensitivity to value-judgments and value assumptions; (6) high political sophistication; (7) orientation of "idealistic-realism"; (8) self-consciousness, self-evaluation and continuous self-development.

Yehezkel Dror, "Comprehensive Planning: Common Fallacies Versus Preferred Features," Essays in Honour of Professor Jac. P. Thijsse, *The Hague (Mouton) 1967, pp. 90-91.*

15

Alan Black

THE COMPREHENSIVE PLAN

. . . A comprehensive plan is an official public document adopted by a local government as a policy guide to decisions about the physical development of the community. It indicates in a general way how the leaders of the government want the community to develop in the next 20 to 30 years. Because it is general and agencies devote more of their time to charting approximately, it is not a piece of legislation. T.J. Kent, Jr., one of the leading proponents of the comprehensive plan concept, has given this definition: "The general plan is the official statement of a municipal legislative body which sets forth its major policies concerning desirable future physical development."[1]

Notice that Kent speaks of the "general plan"; this term is used interchangeably with "comprehensive plan." Another synonym, "master plan," is probably the most familiar to the ear. This phrase has fallen into disrespect among planners because of its misuse in the past to describe plans which were not general and comprehensive (such as "master street plan" or "master park plan"). The term "city plan" is also used.

It is often said that the essential characteristics of the plan are that it is comprehensive, general, and long range. "Comprehensive" means that the plan encompasses all geographical parts of the community and all functional elements which bear on physical development. "General" means that the plan summarizes policies and proposals and does not indicate specific locations or detailed regulations. "Long range" means that the plan looks beyond the foreground of pressing current issues to the perspective of problems and possibilities 20 to 30 years in the future.

Although there is some variation in the content of comprehensive plans, three technical elements are commonly included: the private uses of land, community facilities, and circulation. The first of the three is sometimes called the "land use plan," but this is a misnomer because community facilities and streets are also uses of land. Kent labels this part the "working and living areas section." Comprehensive plans may cover other subjects, such as utilities, civic design, and special uses of land unique to the locality. Usually there is background information on the population, economy, existing land use, assumptions, and community goals. Every plan includes a drawing of the community on which the major design proposals are brought together to show their interrelationships.

Among most city planners, the preparation, adoption, and use of a comprehensive plan are considered to be primary objectives of the planning program. Most of the other plans and procedures applied in the course of local planning are theoretically based upon the comprehensive plan. Many planners have chafed under the pressure of day-to-day activities which denied them the time to take a more thoughtful look at the long-range development of the community. In the past dozen years, though, the federal government has increasingly conditioned financial assistance upon conformance to a local comprehensive plan, a spur which has caused hundreds of local governments to prepare plans.

RELATIONSHIP OF THE PLAN TO OTHER DOCUMENTS

Several other documents used in local planning are often confused with the comprehensive plan – in particular, the zoning ordinance, official map, and subdivision regulations. These are specific and detailed pieces of legislation which are intended to carry out the general proposals of the comprehensive plan. The confusion is understandable because these documents are often adopted prior to a comprehensive plan, and many communities which do not have a plan do have one or more of these. Such a sequence is contrary to good planning practice, and in some states the existence of these tools in the absence of a plan may cast doubt upon the legality of this legislation.

Particularly troublesome has been confusion between the zoning ordinance and the section of the comprehensive plan dealing with the private uses of land. Both deal with the ways in which privately owned land will be used, but the plan indicates only broad categories for general areas of the city, whereas the zoning ordinance delineates the exact boundaries of districts and specifies the detailed regulations which shall apply within them. Furthermore, the plan has a long-range perspective, while the zoning ordinance is generally meant to provide for a time span of only 5 to 10 years.

Other tools of the trade which are meant to effectuate the comprehensive plan include the capital improvements program and its accompanying budget and special-purpose regulations, such as a sign ordinance. A different level of plan, sometimes called a "middle-range development plan,"[2] is supposed to implement the comprehensive plan by concentrating on a particular area of the city or a particular functional element. Such plans are more specific and have a lesser time perspective, say 5 to 10 years.

The growth of urban renewal programs since 1949 has created some confusion with the comprehensive plan, particularly when these activities are conducted by an agency distinct from the regular planning staff. More than 100 cities have had community renewal programs prepared. To some professionals this work has seemed to overlap the preparation of a comprehensive plan. The relationships among these planning efforts have not really been clarified, but they probably will evolve gradually. Urban renewal tends to emphasize residential land and the older parts of the city; geographically and functionally, it is not truly comprehensive. Community renewal programs, while considering long-range policies, tend to recommend specific improvements to be made in the near future. It seems logical to number urban renewal and community renewal programs among the activities designed to implement the comprehensive plan. . . .

THE PRINCIPAL CLIENT OF THE PLAN

The comprehensive plan must be useful to many clients, but since their needs will differ, it is necessary to determine the principal client of the plan, the one whose needs must be met first. Different views of the principal client follow from the different concepts of the role of planning in local government.

Those who believe in the independent planning commission think the plan should be designed primarily for the use of the commission and should be adopted only by the commission. This was the prevailing practice during the early years of city planning. Under this arrangement, the legislative body must refer all physical development matters to the commission for its recommendations, and can overrule the commission only by a greater-than-majority vote.

Those who see planning as a staff aid to the chief executive think the plan should be shaped to serve him. This idea has not developed very fully, for experience has been that the typical chief executive does not want a plan to follow; he wants recommendations from the planner on specific problems.

Some planners believe the principal client of the plan should be the planning staff. This idea does not really fit any of the concepts of the role of city planning. Those who hold this view consider planning to be too technical to be understood by laymen. Frequently, these planners prepare plans for themselves to use privately as a basis for making recommendations to the planning commission or chief executive.

Those who regard planning as primarily a policy-making activity of the legislative body believe the principal client of the plan is the legislature. This concept is the one adopted by the author of this chapter, since it appears to him to offer the most promise for making city planning more effective. It is, after all, the legislative body that ultimately makes the decisions which either carry out or defeat the plan. Under this concept, the plan is primarily a legislative policy instrument, rather than a complex technical document. The planners must make their technical findings and professional judgments understandable and convincing to the legislators.

THE SCOPE OF THE PLAN

Probably the greatest controversy in the planning profession today concerns the substantive scope of city planning. Should it be broad and open-ended, encompassing anything with which local government might be concerned, or narrow and limited to subjects which directly pertain to the physical development of the community? The dispute is reflected in different views of the substantive scope of the comprehensive plan.

One faction believes that truly comprehensive planning must include the planning of social, economic, administrative, and fiscal matters, many of which are obviously interrelated with physical planning. Planning is regarded as a method or approach which can be applied to any subject matter. The city planner would become a central planner. The comprehensive plan might include plans for social and economic development and other nonphysical proposals. So far there have been few examples of these types of plans. Henry Fagin has suggested that a physical plan and all the other sorts of plans should be unified in an ultracomprehensive "policies plan."[3]

The other faction would confine the scope of city planning to physical development (i.e., matters of location, size, and spatial relationships). Planning is thought of as being concerned with physical things, rather than as a particular body of techniques. This has been the traditional scope of city planning, and this view is customarily held by planners with physical design backgrounds.

This group recognizes the interdependence of physical, social, and economic factors in community development, and it concedes that a physical plan must take into account objectives, analyses, and

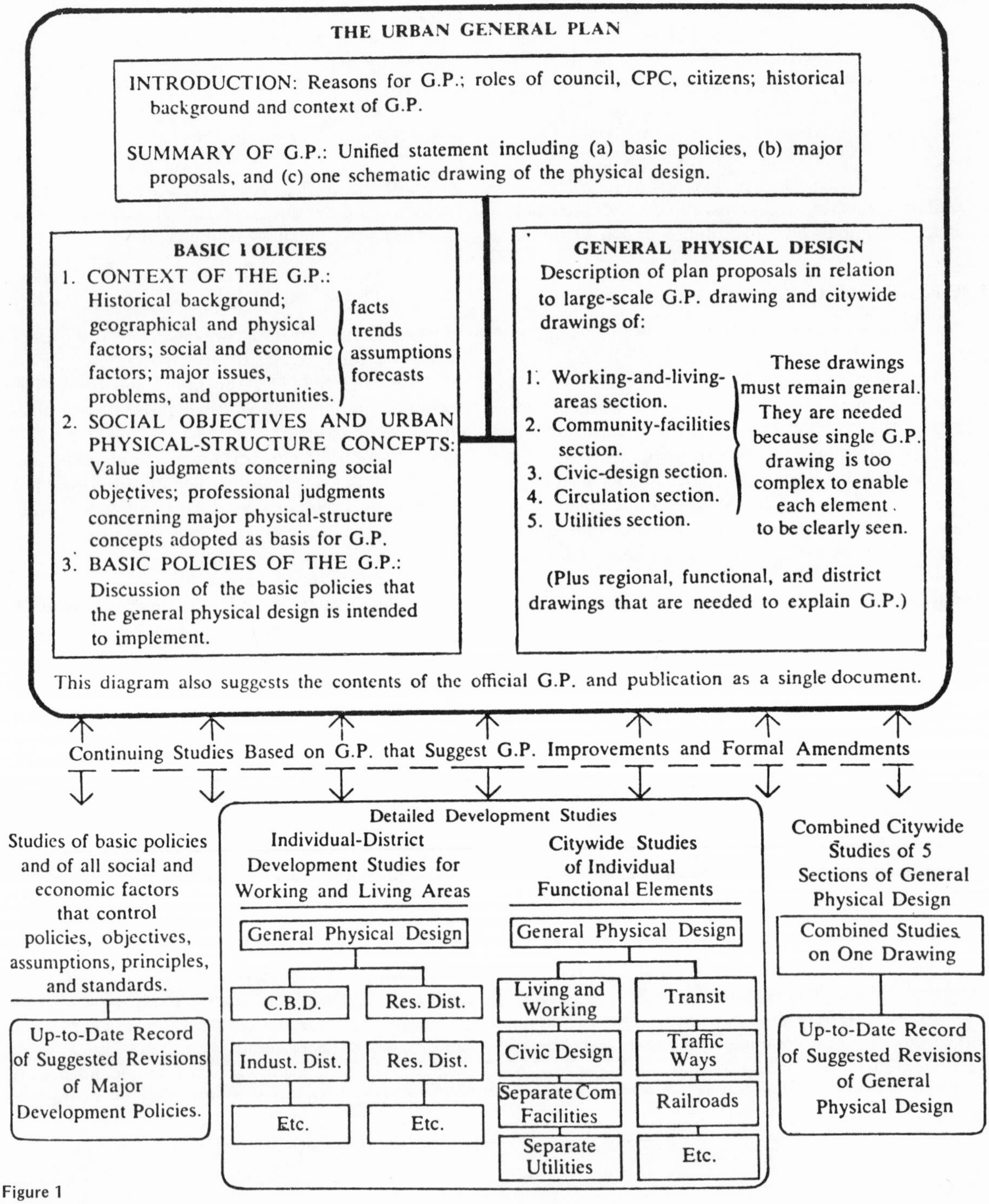

Figure 1
The urban general plan.

forecasts from the nonphysical realm. The distinction is sometimes hard to pin down, but, in general, a plan with a physical development scope will not emphasize economic and social development.

This faction acknowledges the need for nonphysical planning, but would assign it to other than city planners. Thus their argument depends upon who is considered to be a member of the planning profession.

The trend seems to be that more non-designers are entering the planning field, and the abilities available in the profession are becoming more varied and specialized. A broader scope for planning seems inevitable, although there is a stiff rearguard action by the limited-scope faction. In time, no doubt some type of policies plan will emerge.

For the time being, however, it seems advisable for any community undertaking a comprehensive plan to focus upon physical development. There is ample need for coordination in this area, and involvement in the polemics of the planning profession would probably be a needless distraction. There has been so little experience in social and economic planning that anyone entering this area must adopt the posture of an experimenter. There is little danger that a physical plan will ever become superfluous; it would certainly be an important element of any more extensive policies plan. . . .

THE COMPREHENSIVE PLAN DOCUMENT

Basic Requirements as to Subject Matter

The discussion now devolves to a concrete level, focusing on what the published plan document looks like and contains. It is not intended to prescribe exactly the composition of the plan document; the comments here should be regarded as suggestions. They do not have the cogency of the conceptual propositions advanced earlier. While there is some latitude for experimentation, there are six basic requirements which the plan document should fulfill. These have been discussed previously, and are brought together for a brief mention here.

1. The plan should be comprehensive.
2. The plan should be long-range.
3. The plan should be general.
4. The plan should focus on physical development.
5. The plan should relate physical design proposals to community goals and social and economic policies.
6. The plan should be first a policy instrument, and only second a technical instrument.

Overall Form

The comprehensive plan should be completely contained in a single, published document, which should include a large drawing showing the general physical design proposed for the entire community, written text, and whatever maps, illustrations, and tables are needed to support the text. Often the large drawing is folded and put in a pocket inside the back cover. The plan should be easy to read and use, and inexpensive enough so it can be widely distributed (without charge, if possible). The document should be designed so that it is attractive and written so that it is interesting. It should not look forbidding or ponderous.

It helps if the plan is of convenient size, perhaps 8½ x 11 inches. Some plans tend to be large and bulky, which makes them hard to carry around and use. Some provision for entering amendments should be made in the document. One solution is to use a loose-leaf binder, so that new pages can be inserted. Another is to provide a pocket where amendments may be kept.

The plan document should be self-contained, so that it will stand alone. It should not be necessary to consult other publications in order to grasp the essential ideas of the plan. Of course, some people may want greater detail, and it is appropriate to include references and a bibliography, but it should not be necessary for the average reader to undertake such research.

A final question is whether the document should be prepared as a slick or plain publication. Should the design talents of the planning staff be used to produce an eye-catcher? Often this has been done, and some very striking plan documents have resulted. However, there is an argument for conveying the image that the plan is a straightforward working document. While the legislators may be impressed with a flashy publication, they are not so likely to accept it as theirs. . . .

REFERENCES

[1] T.J. Kent, Jr., *The Urban General Plan* (San Francisco: Chandler Publishing Co., 1964), p. 18.

[2] See Martin Meyerson, "Building the Middle-Range Bridge for Comprehensive Planning," *Journal of the American Institute of Planners,* XXII (Spring 1956), pp. 58-64.

[3] Henry Fagin, "Organizing and Carrying out Planning Activities Within Urban Government," *Journal of the American Institute of Planners,* XXV (Aug. 1959), pp. 109-114. See Chapter 12 for an extended discussion of this approach.

Long-range comprehensive plans commonly reveal a desired state of affairs. They rarely specify the detailed courses of action needed to achieve that desired state. By their long-range nature they cannot do so. The development plan, in contrast, will indicate the specific changes in land use programmed for each year, the rate of new growth, the public facilities to be built, the structures to be removed, the private investment required, the extent and sources of public funds to be raised, the tax and other local incentives to encourage private behavior requisite to the plan. The development plan — which incidentally in a more limited form is required by law in England — would have to be acted on each year and made an official act for the subsequent year, much as a capital budget is put into law. Revised yearly it would become the central guide to land use control, to public budgeting and to appropriate private actions to achieve directed community improvement.

Martin Meyerson, "Building the Middle-Range Bridge for Comprehensive Planning," Journal of the American Institute of Planners, *Vol. XXII, Spring 1956, p. 62.*

16

Rexford Guy Tugwell

IMPLEMENTING THE GENERAL INTEREST

None of the classic branches of government has proved amenable to the need for a program based on public rather than on private objectives and which is also a commitment to positive and persistent action. It would be too much to say that the New York City Planning Commission achieves that result. But that is its business; and it moves in that direction. It is apparent, both from the good words which were said for it, and from the distrust which was expressed prior to its establishment, that this was expected to result. And that expectation has been, in a measure, met. Most of the borough presidents opposed it then, and they still oppose it: they represent a local rather than a general interest and the Commission is by its nature committed to guidance by citywide criteria. In the sense in which its sponsors meant it to be regarded as advisory, it has been something else; at least its advice has seldom seemed welcome in the area where its powers were brought to bear. Its usefulness in this way has come from its persistent

---

opposition to the encroachment of private interests, which local politicians often feel it desirable to champion.

This implementing, for once, of a general interest rather than of special interests is the most significant feature of the Commission's activity. Its accomplishment was provided for by furnishing it with several instruments, none of which was new, but each of which was, for the first time, brought together with the others under one agency: the master plan, the capital budget, zoning regulation, the making and custody of the city map, and control over realty subdivisions.[1]

The master plan was generally defined not as anything physical or visual like a map, but as a scheme which included the forecast of adequate facilities for "the improvement of the city and its future growth and development: . . . housing, transportation, distribution, comfort, convenience, health and welfare." This was as wide a frame of reference as could well be imagined.

It was expected that this plan would be a changing and developing intimation of the shape of things to come, yet one which would have an influence on day-to-day affairs which would have to be respected. For the Commission was to make this master plan and to act as its custodian; and no project anywhere in the city which touched any of these interests could be initiated without an inspection by the Commission and a report concerning the plan. Furthermore, the Board of Estimate could reject the Commission's view of this relationship only by a three fourths vote[2] and such issues were not to be referred to the Common Council at all.

The task of making this master plan was, to take the charter literally, one of unprecedented magnitude. It obviously required the services of experts in a dozen different technical fields, together with a large service staff to carry out the duty of comparing current projects with the plan. There was also the fact that the plan would represent serious modifications – so serious, perhaps, as to amount to complete recasting – of the programs of the various departments. For up to that time what planning had been done in the city had been done departmentally. These departmental plans were remarkable in their comprehensiveness, but they often had a very tenuous relationship with other city facilities and none at all with financial possibilities or with population expectancy. With respect to all the hundreds of services and facilities the modern city provides for its citizens, a whole series of master plans existed in the minds of departmental heads and in the files of their subordinates. Their accomplishment was determined by such factors as the political astuteness, public appeal, or administrative ability of the department heads, by the activities of pressure groups which they might build up, or by similar irrelevant uses of power.

The city might in this way be committed to capital expenditures which would prevent more necessary facilities from being built and to burdens on the expense budget which would exclude other expenditures. For, in New York State at least, city budgets have a more delicate balance than is usually appreciated. Income is not expansible beyond a point allowed in the constitution, and unbalance is not available even as a temporary resource.

It was evidently conceived, perhaps only vaguely, that a city, no less than a person, has a *Gestalt,* a behavior pattern which is distinct from (and not to be understood by any study of) its parts. This configuration, this city-in-being, is more than a system of streets, sewers, water pipes, schools, hospitals, and the like; it is more even than these taken together with houses, transit lines, and places of work and recreation. It is, in one sense, the life, the spirit, the whole which is quite apart from all the elements which analysis would reveal: it is something like an individual's character which is not to be understood by description of his bodily members. Indeed, the analytical method is one which yields few necessary results for a planner. He has to use an entirely different one which attempts to grasp the whole before it considers the parts, since these have only a derived, a contributory significance and none taken alone.

There is nothing esoteric about such an approach. It is merely more realistic, though in an age devoted to the analytical method it may not at first seem so. Beginning with the axiom that all social arrangements are man-made, it follows that they may be made, within the resources at disposal, in conformance with an objective. If this objective is broadly defined as an ideal urban life, then that in turn may be broken down into suitable allocations for homes, sanitary and health facilities, recreative and work arrangements, means for transportation, and the like. And each falls into a relationship with the others which is not too difficult to determine by careful judgment. But, by beginning at the other end and describing each part separately, and even finally adding all together, something lifeless and strange would result. In the language of natural science this difference is recognized in two terms: "additive" and "emergent." The additive is a mechanical concept in which the whole is merely the sum of parts; but the emergent is something more, as water, to use a possibly oversimple illustration, is more than hydrogen and oxygen, or as a child is more than "saltness mixed with dust." The city is an emergent, to be understood only in terms of its behavior – more than brick, steel, lights, and steam, more even than its homes and commerce. Yet those things are necessary to it in suitable proportion – a proportion determined, however, by their relative necessity to the whole. It is this whole, this emergent, this relative and conjunctural interest which, it seems, the Planning Commission was intended to represent.

The master plan contemplated under New York's new charter would be constructed with all this in mind, but also with a view to keeping its cost within the city's ability to pay. It was not to be merely a fanciful picture, but rather the design for a city which could actually be brought into being. The limitations were not imagined ones; nor were they even largely physical, such as engineers could surmount; they were mostly economic. It was necessary to know what the population would be in the future, how it could best be distributed, what kind of activities it would be carrying on, how much recreation – and of what sorts – it would need, what program for health protection would be required, how people would prefer to live, how they would move about, and so on. There were numerous technical questions whose answers were precedent even to preliminary work.

In the plans which existed in the departmental offices these questions had either been ignored or had been guessed at with bias. To substitute its own work for that of the departments, the Commission would need a public trust and confidence which could only come from belief that its work was expert – more so than that of any department. And the risk was that once the administrators smelled the limitations implicit in central planning they would begin a campaign of belittlement, even, if possible, of ridicule, which might end the Commission's existence before it was well begun. It was vital, for this reason, on the one hand, to engage well-known, even famous planners and experts, and, on the other, to keep natural enemies in check by appeasement and by mayoral discipline.

For neither of these necessary measures was there any provision in the charter. Beyond the salaries of the Commission there was no fiscal protection; and it was apparently overlooked that in setting up central planning in the City of New York a challenge was being issued to enemies who had repeatedly served notice of hostility, and who had the disposal of funds for personnel. Why the charter writers thought that the expert service they had made imperative would be consented to by those who were – and would remain – actively opposed to the Commission and its work is not apparent. There was the further point in this connection, also, that the mayor would be unlikely to sympathize with the setting up of any such staff as would be required for a first-rate job. He had to recognize influences which would lead in the other direction: persistent requests for enlargement of departmental budgets which were supported by pressure groups, and demands for economy from highly vocal taxpayers' groups. It was likely from the first that the Planning Commission would be expected to perform a miracle in producing its master plan; that it would get little sympathy from any source for delays, defects, or enmities; that, indeed, there would be many who had been instrumental in reducing its competence who would be the first to point out its shortcomings, with loud demands for its suppression.

The situation with respect to zoning was different, but in effect came to the same thing. For, although New York's was the oldest comprehensive zoning ordinance in the country, it had been amended so little since its first adoption that it was almost worse than none at all. In putting it into the hands of the Commission, the charter makers were imposing a serious task of modernization rather than furnishing an instrument of control. In the first place, the compromises necessary to secure its passage in 1916 had been so serious that it had exercised none of the controls necessary to securing reduction in density or segregation of industrial and business areas. Actually, the multiple dwellings law which was intended to reduce the worst abuses of slum housing was more restrictive than the zoning in most areas within which characteristic New York apartments had been built.

The results were to be seen not only in the areas of low-rent housing, but elsewhere as well. Such limited dividend projects as Knickerbocker Village on the lower East Side actually had a much higher density than the slums they replaced. The only superiority was their newness. Also, some of the new projects of the Housing Authority were being planned at densities averaging two and a half times those which had prevailed in the areas for which they were being substituted. These occurrences could hardly be objected to as long as it could be said that they met the zoning requirements of the city, and, of course, such crowding contributed to a reduction in costs. A good showing of economy by the Authority was thus gained at the expense of space, light, and air for tenants – a way of putting it which is perhaps unfair, since the primary fault lay in overgenerous zoning allowances. That this is so is indicated by the fact that the expensive Park Avenue district was even worse.

There was, on the whole, no dissent from the necessity for zoning reforms, such as the reduction of density, the contraction of business areas, and the like, as long as these proposals were made in general terms. No one of experience, however, had the least doubt that when the proposals were particularized and brought into focus as limitations on the freedom of certain property owners there would be strenuous opposition. The property owners who objected to such limitations were always in a strategic position. A great part of municipal revenues came from property taxes, a fact which led to the formation of taxpayers' associations with salaried secretaries whose business it was to guard realty interests from all real or imagined dangers – and even to invent them when real ones were lacking. They not only cultivated close contact with officials, but, through their advertising, disposed the avenues of opinion in their favor. The whole arrangement was one in which an injured private interest was likely to prevail over a public one, especially since the private interest was immediate and real and the public one could be made to seem merely theoretical and its advocates somewhat fanciful.

Then, too, for what needed to be done if zoning was to be an effective part of a planning program for the city, an ordinance which went back more than two decades was bound to be technically inadequate. The New York statute had too few zone differentials and was too rigid in its application; it had no provision for extinguishing nonconforming uses. It was clear that some means would need to be found for providing easier variances and for controlling certain obnoxious uses. Another need was for loosening the restrictions on parking lots and garages, even making them compulsory, so that vehicles might be got off the streets. Regulations formerly necessary to guard against explosion, fire, and the like, were now obsolete and required revision.

No one familiar with New York's situation needed to be told that all these steps were necessary. The fact was elementary. And this modernization weighed upon the Commission as a duty, since in the charter it was specified that zoning regulations should be written by the Commission and transmitted to the Board of Estimate for consideration, where they might lie upon the table no more than 30 days; they would at the end of that time become law unless modified or rejected by a three fourths vote.[2]

It was obviously felt by the framers of the charter that this kind of weighting for the Commission's advice would result in most of its actions becoming law without much change, and at the same time would relieve the Board of Estimate of having to pass upon items of a routine character. It was anticipated that the Commission would have an expert staff, that its hearings would be public and complete, and that its actions would command such public confidence that most of them would pass into statute without Board of Estimate action. Such a forecast may seem strange in view of the makeup of the Board and its susceptibility not only to local and short-run but to private pressures, while the Commission was by nature devoted to long-run and wholly general interests; but no other reasoning can account for the establishment of such a relationship as was set up. . . .

The master plan might easily degenerate into a

mere map — or a series of them — setting out a scheme for the flow of traffic, and the placing of public buildings, sewer and water systems, and the like. Most planning efforts . . . have amounted to little more than this. One means of ensuring attention to expertness and detachment was the reference of projects for comparison with the master plan. To strengthen this influence the programming of improvements was entrusted to the same agency which was to create the plan. Planning was thus given a new dimension — time was added to space. The same group which laid out sewers, transit lines, schools, and playgrounds would have to consider not only their desirability, but their possibility, that is, when and how the changes indicated might be expected to come about.

If competence was required for working out a plan, the requirement was increased when the budget and program (as it was officially known) was taken from the budget bureau and given to the planners. It appears from the *Hearings*, which record the proceedings precedent to charter change, that what was being done here was well understood by some, at least, of the framers. There was a good deal of discussion concerning the time period during which the program should be laid out in some detail; as it appeared finally in the charter the period was set at 6 years. The conception was that the budget should cover the prospective year and the program the succeeding 5 years. The budget was to be made rigid and difficult to amend, but the program, though it showed the probabilities for the future, has to be recast every year with the benefit of that year's experience. The first year of the program did not automatically become a budget for its period; it was gone over again in detail and reissued with a new succeeding program.

Budget theorists, have, for a long time, been longing for a fiscal layout which went beyond the current period. There was a special reason for it in public budgeting; it was thought of as going some way toward reducing "pork barrel" appropriations, but also it had the possibility of approaching a more normal fiscal period (if such a thing can be spoken of in an unplanned economy) than the arbitrary *annum*. Cities particularly — because they might not unbalance their budgets as the federal government could — suffered from the alternations of prosperity and depression when tax receipts rose or fell unpredictably and when demands for expenditure had a reverse relationship to income. The 6-year period chosen was doubtless a kind of trial-and-error compromise. It was not long enough to span most business cycles, but it was a gain over the annual budget. . . .

Still another city function was entrusted to the Commission: the making of the city map and its custodianship. It will perhaps have been noted that the duties already described affected more than strictly municipal functions. The capital budget and program, of course, was confined to the extension and replacement of city facilities and affected privately owned property only indirectly. But zoning was intended to be a limitation on the form and use of private property; and, as has been said, the very plots and buildings which would be most affected would inevitably be owned by those with the most influence in preventing any serious limitation of their activities. This last was no fault of the charter makers. Every city of any account has zoning regulations. They have always been approved by the theoreticians of city planning. That they have never succeeded in doing more than limiting the worst absurdities of building heights, or the most flagrant incursions of business into residential areas, did not argue that they ought to be dispensed with. The very fact that they were regarded as reformist instruments, yet could be managed in the interest of those who ought to have been regulated, had contributed first to passive acceptance and then to enthusiastic defense by realty interests, holders of mortgages, and others with similar interests. All these factors account for the prominent position of zoning matters in the Commission's work: it was a presumed limitation on private operations which was really very difficult to enforce at all.

As for the master plan, the charter laid out a frame of reference, as has been described, which would include not only municipal operations but all life and activity within the city. Actually, the master plan was thought of as coming into being slowly through the years by conformance to it of slowly replaced or extended facilities, a conformance shaped partly by zoning and partly by capital budget appropriations, the one influencing private improvement and the other giving shape to new public equipment. These were theoretically the instruments *par excellence* for creating the future. That neither was perfect nor even, perhaps, actually very effective the experts may have known well enough, and yet not have been able to think of anything acceptable to use in their places. In fact, each was evidently thought of as supplementary, since neither was completely implemented for the only purpose it could rationally be conceived to serve.

But by the apparently innocuous transfer to the Commission of the custody of the city map and the corollary power of making and revising it, both these other instruments were considerably strengthened. The master plan represented the desirable future; the capital budget and program showed it coming into being as far as city facilities were concerned; zoning indicated broadly what private interests could and could not do; but the provisions concerning the city map held the possibility that when a change was requested a considerable bargaining advantage would be available – an advantage which affected both public and private enterprises.

If a brewery wanted to expand and to close a city street, the Commission had the opportunity to examine the project in relation to the master plan, even though no reference to the plan for private changes of this sort was legally required. If an insurance company wanted to invest in a large housing development – no matter of what rental class – it would be certain to need some revision of grade or street alignment, some modification of the sewer or water system, even if it needed no change of zone. These requests gave the Commission opportunity for extending its examination of the project. Density could be inquired into, since such developments in the past had tended to increase crowding enormously no matter how much they improved construction standards. And the relation of the project to existing municipal services could be examined, since the desire for cheap land usually determined that these projects should be placed in situations where facilities were inadequate; these would then have to be provided by the city; and what at first looked like a desirable housing enterprise often turned out to involve the city in costs greater than those incurred by the builders. These are examples of the utility to the city of close guardianship for its map, and, indeed, of locating that guardianship within the same agency which administers the master plan. . . .

. . . The public's attitude toward the Commission could not be expected to be an active or informed one. The duty of drawing things together and of central planning would not be spectacular in the sense that the maps and architectural designs sometimes are. The nature of its duties would require it to check in a serious way the ambitions of able administrators, sometimes of long experience. But nothing was provided in the way of defense.

Another way in which the Commission was to control private activities was by passing on all subdivision plans, subject to the power of the Board of Estimate to reject or to modify its decisions by a three fourths vote. The Commission soon discovered a need to use this tool in the solution of a problem more important than it seemed on the surface. It had to do with checking that movement toward the periphery which was an accompaniment of decaying central areas. Neither phenomenon – developing outer areas or degenerating inner ones – could be said to cause the other. Both were the result of more fundamental causes. Checking subdivisions was perhaps an attempt to treat symptom rather than disease, and so bound to have little success; but it has to be recorded as part of the Commission's duties. . . .

The *Gestalt*, the configuration, of a city is not only something which, as has been insisted here, must be taken account of; it is also something which is created by men. It may be done badly or well; that is, the resulting social organism may function efficiently or wastefully. These results are determined more fatally than is usually admitted by the motive which is dominant in the process. Thinking and caring about the city leads to generally good results; thinking and caring about making money out of land, out of utility privileges, or even out of mortgages on realty may – and often does – lead to bad results. Every city government is tormented by the necessity for turning these latter activities in directions which will have at least a minimum of acceptability by general standards. There is an immemorial struggle among the politicians in every city to attain credit for giving advantages to those who can make private use of them. It is necessary to do this and still maintain a reputation for public service, except when utter cynicism will go unpunished. And it is sometimes, though not often, possible. In this struggle a favorite weapon is the representation of an opponent as the enemy of one after another of these private individuals and groups until the total disaffection is sufficient to discredit him. The position of the planning commission will, in the natural course of things, be an exposed one. It will often offend. It will seldom be able to do a favor. It will never be, in the usual sense, popular.

NOTES

[1] To which might be added the review of assessable improvements for their accord with the master plan, though this to be effective would need to be supplemented by a budgetary control which is now lacking.

[2] The Board of Estimate, without being designated as the city's "legislative body," nevertheless possesses most of the legislative powers. It is made up of the mayor (3 votes), the comptroller (3 votes), the president of the council (3 votes), and the presidents of the five boroughs (1 vote each, except Manhattan and Brooklyn, 2 votes each). It is thus a mixed legislative and executive body. The council, now elected by proportional representation, has far fewer powers and duties than the usual "lower house."

. . . No plan, for the middle or the distant future, at least, is expected by the agency which makes it, to be realized as drawn. That is not its utility. Rather it has the utility . . . of any . . . theoretical construct by which significant likenesses and differences are established. Only in this case, there is a further usefulness, which is perhaps the most important one flowing from the planning process; it establishes an operating harmony, a coordinative smoothness, a successfully functioning Gestalt in the present. This it does by picturing an objectively arrived at, meticulously examined, and generally agreed, view of the future. As new emergents are taken into account the view of the future shifts, and the new view is transmitted back as an influence on the present. The real utility of the Development Plan is thus seen to be that of reducing friction, avoiding waste, and gaining concurrence; it seeks to channel all efforts into one increasing movement in an agreed direction.

Rexford G. Tugwell, "The Study of Planning as a Scientific Endeavor," Fiftieth Annual Report of The Michigan Academy of Science, Arts, and Letters, *1948, p. 41.*

17

E.S. Savas

CYBERNETICS IN CITY HALL

The science which is parent to much of modern technology is cybernetics – the science of communication and control in organized systems. The word itself, introduced and popularized by Norbert Wiener[1], is transliterated from the Greek word Κυβερνήτης. That same Greek word has also entered our language in a slightly different transliteration, and translation, as the word "governor." Etymologically, therefore, there is an equivalence between a governor and a cyberneticist, between government and cybernetics.

This equivalence is worth exploring. What happens when a cyberneticist, perhaps naively, assumes that big-city government is "an organized system" and casts his practiced eye on it? What does he see, and what can he tell us about applying the principles of cybernetics to cities?

The cyberneticist brings to his task the view that an organized, adaptive system is a goal-seeking ensemble which can sense its relation to its objective and modify its behavior in order to approach the objective more closely. The simple feedback control diagram of Figure 1 is the basic tool of the cyberneticist, and it suffices to illustrate the elements of such a system. The desired condition of the system is selected by some goal-setting process, entered into a comparator, and then tested against the actual condition, which is observed and reported by some process of information feedback. Any discrepancy between the desired and the observed conditions causes the actuator to act upon the system to reduce the discrepancy. The continuing, dynamic nature of this entire process results from the disturbances – that is, causative factors outside the system which upset the system and make it necessary to apply control action to counteract their effects.

The discerning cyberneticist can identify corresponding elements of this feedback control system in city government, even though the latter is far from being a simple system with a single goal. The goal-setting mechanism, which establishes objectives and priorities, is the mayor's decision-making process. A comparison of the desired condition with the observed state of the city results in action to reduce the disparity; municipal administration (that is, the bureaucratic processes of city government) constitutes this action element of the system – a provocative thought indeed! The system being acted upon is the city and its people. It is subject to external upsets that may be classed as social, economic, political, and natural. The output, or observed condition of this living system, is the state of the city. Feedback concerning the condition is transmitted to the mayoral decision center by way of an information system.

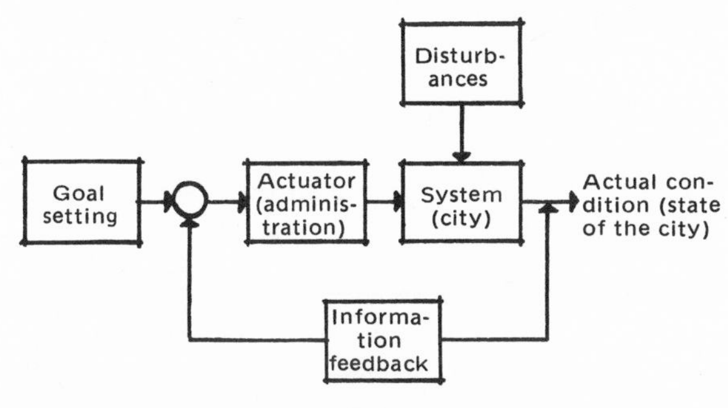

Figure 1
A basic "feedback control" diagram.

What happens when a cyberneticist, impelled by his students and his conscience to seek relevance, examines New York City's government? How would he interpret recent developments there in terms of the five basic attributes and elements of this cybernetic loop: (1) the overall dynamic characteristics of the process, (2) the information system, (3) administration, (4) goal setting, and (5) disturbances?

DYNAMICS OF URBAN GOVERNMENT

The first disturbing realization is that the natural time constants of urban systems are unrelated to the term of elected office. Thus is is impossible for an official to show visible accomplishments in 4 years on a problem which may require at least a full generation – 25 years – to solve. If Moynihan's thesis[2] is correct, for instance, then it will be necessary to a black male infant to grow to maturity, with his father as breadwinner and resident head of the house, before his family realizes its potential and acquires a life style which puts it in harmony with the community.

Similarly, if it takes a year to determine the state of the system (that is, identify a major problem in a way which suggests approaches to its solution), another year to define objectives, to plan, and to allocate resources to attack the problem (in the annual budgeting ritual), and a third year to construct, staff, and test the appropriate administrative structure for implementing the plan, this leaves precious little time, mostly the fourth (or election) year, to look for significant, tangible results. Awareness of these process dynamics may impel a political leader to settle for smaller goals, which are surer of attainment, or for highly visible acts which, initially, may be more symbolic than effective. Such acts can be further justified on the grounds that they will contribute to reelection, thereby making it possible to address the real problems systematically and fundamentally during the next term of office.

Forrester[3] has analyzed the feedback control loops implicit in the everyday business activities of industrial firms, and has showed that their complex, time-varying interactions have profound consequences. Depending upon the amounts of time required to process orders, produce goods, bring the goods to market, and report sales, it is possible to generate wild instabilities and feast-or-famine conditions in the manufacturing plant, the warehouse, or the sales office. In other words, differing time constants for related processes can cause poor performance.

For a municipal analogy, take the embarrassing example of the housing administrator who approves an application to rehabilitate a decaying building, only to discover that the building has already been demolished by his agency. The time necessary to examine, evaluate, and approve a rehabilitation application exceeds the sum of the time required for the building to deteriorate past the point of no return, the time required to detect that deterioration (information feedback), and the administrative time required to make the decision and actually to effect demolition.

The large time constants and the incredibly involved multivariate nature of the city as a system require that we employ a very large, very sophisticated, very complicated governor – that is, a control device we call government. But, as any control engineer knows, it is difficult to keep a large, sophisticated controller tuned up and functioning well, for there are always component failures, gear slippages, time lags, loose connections, nonlinear effects, and other problems. In government, as in other large organizations, the analogous shortcomings are incompetent individuals in key posts, poor coordination, bureaucratic delays, bad communications, and conventional responses to unconventional situations.

The common cybernetic approach to this kind of problem is to apply minor loop control – that is, to divide the control function among several smaller, simpler controllers. In doing this, one recognizes that *complete* decoupling of variables cannot be accomplished and that, by relegating such variables to different controllers, one is sacrificing the optimum that *theoretically* could be attained by a more integrated comprehensive "total approach."

Decentralization is an example of minor-loop control. From the standpoint both of good government and of cybernetic theory, decentralized control of some government activities makes sense. Getting decision making down into the community offers hope of getting more rapid response and more effective performance of the system. Thus the concept of school decentralization, and of participatory democracy in general, is in accordance with cybernetic principles, although getting a new controller "on line" is always troublesome, as was demonstrated during the school turmoil in New York.

As mentioned, minor-loop control results in some sacrifice. In the case of school decentralization, one might have to forego the economies of scale available in purchasing, in administration, and in handling special students. Furthermore, a decentralized school system might lead to such diversity in curriculum and student achievement that students transferring, or proceeding upon graduation, to another school would be poorly served; in other words, the close coupling among some of the variables cannot be totally disregarded.

The control engineer has a solution to this problem of coordinating independent minor loop controllers: cascade control. A high-level controller establishes objectives for the subordinate controllers, but allows the latter to achieve those objectives through local action. This is equivalent to having a central education authority set standards for curriculum and pupil reading achievement, for example, while leaving the details of implementation to local authorities.

In *Urban Dynamics*, Forrester[4] concludes that programs for job training, job creation, and low-cost housing have the perverse long-term effects of increasing the relative numbers of underemployed and accelerating the deterioration of housing. These outcomes, he theorizes, result from the complex systems; that is, the ultimate result of an action turns out to be quite the opposite of that which is intended and which intuitively appears obvious.

As an example of counterintuitive behavior, one has only to look at the elaborate, time-consuming, and costly bureaucratic system of checks and balances that is designed into municipal government to assure that the city gets fair value in its purchases. The procedures were constructed to protect against graft and corruption in contracting for supplies, equipment, and services. However, the consequence for the city is an inordinately long delay in securing bids and paying bills; the result of these delays is that many potential vendors refuse to sell to the city, and that those who do sell to it have to charge higher prices in order to make up for the additional expense of dealing with it. Thus a strategy designed to increase competition and reduce the cost of goods has the perverse effect of reducing competition and increasing the cost of goods.

INFORMATION SYSTEM

Turning now to the second of the five areas, the information system, the cyberneticist begins by examining a mayor's sources of information and the characteristics of those sources.

One of these sources is direct personal observation: the mayor sees and hears things as he goes about the city. This is a good information channel, but obviously it is exposed only to signals within a restricted portion of the spectrum and has a limited capacity.

Information input to the mayor is also supplied by his subordinates, who constitute a well-structured feedback source, but whose signals are selectively amplified. Therefore, a mayor, like any chief executive anywhere, must maintain a subliminal defense and wonder, "Why is this particular individual giving me this information at this time?" Perhaps it is an effort to get a larger piece of a limited budget, or it could be an attempt to whitewash an incipient problem in the subordinate's area of responsibility.

The press and other public media represent yet another information source for a mayor – a good source, but only for carefully filtered signals, those that portray dramatic events.

A mayor also receives information from the "establishment" – the leaders of political religious, business, labor, and academic groups, for instance. The cyberneticist recognizes, however, that these are high-impedance channels; that is, the signals transmitted by them may generate resistance and, therefore, careful matching throughout the entire communication circuit is necessary in order to conduct the signal with high fidelity from the individual members of the group to the mayor without excessive distortion.

Other channels of information connect the mayor to public officials at other levels and in other branches of government. These channels may also exhibit impedance problems.

The public constitutes a major source of information for the mayor. Four particular subgroups within this source are worth considering.

1. There are the highly vociferous individuals. The problem with their signals, of course, is that they have a rather low signal-to-noise ratio and, therefore, require long-term smoothing if meaningful information is to be extracted from them. In other words, one should look for persistent patterns in such complaints, and not overreact to isolated incidents.

2. On the other hand, special-interest groups emit signals which have very high signal-to-noise ratios, but these are biased signals. The bias must be detected and compensated for.

3. The cyberneticist also views civil disorders as information-bearing signals from the public. Unfortunately, these signals are rather powerful; they saturate the system, and this tends to set off the circuit protectors of the society, with the result that the only information received is the information that the system has failed – at least with respect to the groups immediately responsible for the disorders.

4. Finally, elections are the classical democratic institution for channeling feedback from the voters to the mayor. The problem with this channel, however, is that it is characterized by a very low sampling rate (one sampling per 4 years) and produces only one bit of information: yes or no! Polls are a means of increasing the sampling rate, and also of providing more bits of information. Elections, for example, fail to register the intensity of a voter's feelings. Why should the vote of a deeply committed individual have the same weight as that of a casual citizen whose vote is determined by a chance remark overheard at the supermarket, or by the most recent piece of campaign literature pressed upon him on election day? One might speculate about a hypothetical election where the voter has a choice of pulling one of four levers marked, respectively, "Have strong preference for candidate A," "Have mild preference for candidate A," "Have mild preference for candidate B," "Have strong preference for candidate B," where the strong preference" votes have greater weight than the "mild preference" votes. This might be worth trying for referenda, at least, if this well-known concept from the market-research field is still considered too revolutionary to apply to a choice among political candidates.

Given the characteristics of the information feedback process discussed, the cyberneticist can immediately identify ways to improve the quality, quantity, and flow of usable information to the mayor: increase the sampling rate, open more feedback channels, increase the bandwidth, enhance weak signals, match impedances, suppress noise, and correct biased signals. Recent innovations in New York City, although modest in scale, attempt to do just this.

One of these innovations was an effort to bring government from City Hall into the community by opening up several Neighborhood City Halls throughout the city and sending Mobile City Halls into various neighborhoods to find out what the local problems are and what the people are thinking – that is, to increase the sampling rate and open more channels. (One can speculate that, with computer consoles in Neighborhood City Halls, the computer may someday become the electronic equivalent of the old ward heeler, making possible convenient, decentralized data acquisition and delivery of services, but with centralized coordination and control.)

A second example of improved information feedback is the Mayor's Urban Action Task Force, now being emulated in other cities. In essence, the objective of the task force is to bridge the communication gap between the people and the executive branch of city government. High-ranking appointed officials are assigned liaison roles in specific neighborhoods, in addition to their normal duties. They tune in on the same wavelength as the community and keep in touch with local leaders, youth groups, neighborhood civic associations, and community corporations, and coordinate the delivery of municipal services in 44 areas, ranging in character from ghetto slum to upper-middle-class area.

The Action Center and the Night Mayor program, which together constitute a round-the-clock complaint bureau, serve the cybernetic role of enhancing weak signals and suppressing noise.

ADMINISTRATION

Administration is the weak link in the cybernetic loops of the cities. Goals may be set, but it is up to administrators to activate and guide the process by which broad goals are translated into specific objectives and by which resources are allocated and expended to convert objectives into achievements. This requires appropriate personnel, structure, and theory. All three are problem areas.

Salaries account for the largest part, by far, of local government costs. Yet the decades of neglect since the Depression years – when cities had their pick of employees – have converted state and local government into a refuge for mediocrity. This shows up all over the country in the form of weak managers, uncivil servants, and employees going through preprogrammed motions while awaiting their pensions. The result is mindless bureaucracies which appear at times to function solely for the convenience of their staffs, rather than for the public whom they ostensibly serve. When Servan-Schreiber alerted Europe to the challenge posed by America's managerial prowess[5], he wasn't thinking of our civic ineptness.

One step taken in New York to start changing this dismal picture was to increase executive salaries, a reform which preceded the Kappel Commission's recommendations[6] for analogous action at the federal level. In the past, a greater proportion of senior city officials came from political party circles, and city salaries (together with the fringe benefits of potential judicial appointment or other political rewards) were ample for purposes of recruiting from this limited labor pool, although not competitive in the free market for experienced executives. The increase in salary levels has enhanced the city's ability to compete for talented people from various relevant fields and to reward capable civil servants.

Another salutary change results from the "people power" marshaled by the Mayor's Volunteer Coordinating Council. Ten thousand volunteers, including corporate executives, computer experts, lawyers, and housewives, work in 61 city agencies. More important than their contributed labor is the fact that, with their independence, they can challenge the status quo. The bring new ideas and fresh approaches to ventilate administrative bureaucracies which have sealed themselves off from the outside world with the bricks and mortar of inbreeding and complacency.

It is interesting to speculate on the effect on personnel of the 4- to 8-year turnover in government administrations. The new personnel usually enter office on a platform of promises to clean up the mess, make a clean sweep, reform the system, and otherwise tighten up and improve upon the flaccid performance of their predecessors. This would seem to lead to a permanently autocratic style of government management, a style which looks upon inherited employees as an alien force of sluggards who must be brought under control and made to toe the mark. This approach is the antithesis of the democratic, participatory style of management in which people are considered to be an organization's greatest asset. The latter style has been found consistently more effective than the former in achieving high productivity, in a variety of organizations throughout the world. Perhaps this explains why government bureaucracies tend to have low productivity: the political process produces a management style which brings this about.

It is also worth remarking on the observed American tradition that lawyers be represented in upper echelons of government in disproportionately large numbers. Young lawyers in private practice, to a much greater degree than other professionals, have control over their time, meet clients with means, and deal with the power structure. This gives them an advantage in political work, and leads to their election or appointment to government posts. However, their professional training is valuable only in the judiciary (and perhaps marginally in the legislative branch, where drafting bills is a minor aspect of the legislator's work). Legal experience is of no discernible value to a line commissioner or administrator in the executive branch of city government. The best experience for running large, complex government bureaus and departments is management experience in large, complex corporations, unions, universities, or other behemoths.

STRUCTURE OF GOVERNMENT

Even the best of personnel will be ineffective if poorly organized. Incredible as it may seem, until recently the New York City government consisted of about 50 separate departments reporting directly to the mayor! This improbable structure was thoroughly overhauled and reorganized by consolidation into 10 superagencies, in 1966. This has reduced the span of control to a more manageable number of subordinates, and offers the promise of improved administration. . . .

Politics has been defined as the business of who gets what, when, and where – that is, the business of resource allocation. A principal responsibility of the administrative structure is the allocation of resources to government programs. The resources consist of matter, space, and time. Through allocation of the right kinds and amounts of matter (personnel, supplies, equipment, facilities, and money), work space, and land to well-planned programs over a considerable period, the objectives of those programs are likely to be attained.

The "business" of resource allocation is conducted within every organization and is salutary, for it is the source of alternatives for organizational actions. Recently, this business has acquired some theories to support it, and rational guides to resource allocation in government have become available.

First, there has been a determined drive to develop and apply planning-programming-budgeting systems throughout the city government. This has required arduous explication of departmental objectives, conscious relation of alternative programs to those objectives, evaluation of the alternatives in terms of cost and effectiveness, and budgatary allocations to drive the programs selected.

New York is the first city to establish a Management Science Unit, which utilizes the tools of systems analysis and operations research to identify and quantify objectives and to select suitable means for attaining them. The city also retained the Rand Corporation on a large scale to assist it in developing planning-programming-budgeting systems and to conduct systems analysis studies, primarily in the areas of police, fire, and housing. This relationship was subsequently expanded with the formation of the joint New York City-Rand Institute for applied development work on urban problems.

GOAL SETTING

Goal setting remains the most intuitive element in the city's cybernetic system, a wise choice of goals being dependent upon the acumen, sensitivity, and (apparently) extrasensory perception of the chief executive. However, he is aided in this task by information from the very same channels which report on the performance of his government, inputs which clearly affect his choice of goals and priorities.

It is important to recognize that citizens have multiple roles in the cybernetic process. Someone may be part of the system being acted upon by government, but he may also be a vocal element in the information feedback subsystem, he may be employed by the city in the administration subsystem, he may be a member of a politically powerful organization and add his strength to influence the setting of goals, and, if he takes advantage of federal mortgage subsidies and moves out of the city into the suburbs, his movement, although Brownian in scale, is a sociodemographic disturbance to the urban system.

The recent activation of community planning boards is a move to formalize participatory goal setting. Neighborhoods which are historical and topographical entities have been recognized as community planning districts. They receive staff support, a modest budget to use for determining community sentiment and goals, and a mechanism for expression that goes beyond the traditional opportunity to write letters to their elected officials.

DISTURBANCES

Let us look now at the last element of the cybernetic loop, the disturbances which affect the system. Disturbances are the independent variables which act upon the system from the outside and over which local government has no direct control.

Primarily, the cyberneticist recognizes that municipal government is only one of the agents – and a minor one at that – acting upon the system of city and people. Corporations, banks, construction unions, and medical societies, for example, all have a more significant impact upon certain of the performance variables – certain conditions of the system – than local government has. In other words, with respect to many factors in city life, the "disturbances" are more important than the explicit actions of local government.

Disturbances can be classified as social, economic, political, or natural. As an example of a social disturbance, one can cite the revolution of rising expectations that is affecting not only our cities but the entire world. Recessions, wars, inflation, and high interest rates, which have a cataclysmic effect on the economy of the city, are economic disturbances which are beyond the control of a mayor. Political disturbances affect the city when changes in administration at the state or national level have a profound impact on urban programs. And, of course, there are natural disturbances, such as the droughts, floods, tornadoes, hurricanes, and earthquakes that afflict some cities.

For examples of government-induced disturbances originating at other levels of government, one has only to consider the highway construction and mortgage policies of the federal government, which "developed" the countryside surrounding the cities and peopled it with the cities' middle class. Also, it is evident that the nation's welfare policies, particularly as implemented in certain states, have influenced the rate of migration from southern rural shacks to northern urban slums.

For a more specific, local example, consider the change in federal law which led to a change in state law that forced New York City to drop 43 percent of the participants who had been enrolled in the Medicaid program. Over 1 million people were dropped – a disturbance indeed! This large and sudden fluctuation in load caused administrative havoc and made it appear that the city was incapable of handling routine administrative matters. The legislators failed to consider the response time of the administrative mechanism when they designated the effective date of the new law. To make matters worse, the load reappeared as an equivalent disturbance in the municipal hospitals, as patients who lost their eligibility for Medicaid no longer sought private medical care and crowded the municipal hospitals.

The cyberneticist deals with disturbanced in one of two ways: through feedback control or through "feedforward" control (anticipatory control). Under feedback control, when a disturbance acts upon the system, the performance deteriorates and information feedback causes control action to be taken to counteract the disturbance and thereby restore the system to the desired performance level. In contrast, feedforward control *anticipates* the effect of the disturbance on the system, and causes action to be taken to counteract the disturbance *before* the latter can affect the performance of the system. In other words, feedforward control involves planning to accommodate predictable, externally caused changes that would otherwise impact the system.

Clearly, feedforward control (planning) would appear to be the preferred mode of handling predictable disturbances, feedback control being used only to cope with unexpected upsets. Unfortunately, two characteristics of government conspire to limit the applicability of feedforward control.

1. Problems not perceived as problems by the mass public are problems not acted upon. In other words, if an intellectual or scientific elite points to a large problem (like the population explosion, for example), this is not sufficient to cause government to mobilize large resources to counteract the expected disturbance. Only when a sufficiently large body of opinion is aroused can government begin to plan and take anticipatory control action (such as fertility research or various birth control programs).

2. Feedforward control does not work too well, because the predictive models for social phenomena are poor. We have difficulty in forecasting the magnitude of the effect that a disturbance will have on system performance, and also difficulty in calculating the kind and quantity of anticipatory corrective action that should be taken in order to cancel out the disturbance. For example, it was predicted that the New York State Medicaid legislation (an external variable) would have a certain (budgetary) effect on the financial state of the city (and state). This prediction was wrong, and the cost to the taxpayer turned out to be much greater than had been expected. The reason: state planners assumed a low enrollment rate, due to ignorance and apathy, but welfare rights groups, neighborhoods groups, and legal clinics in poverty areas were effective in making contact with, and educating, eligible patients and helping them enroll in the Medicaid program. . . .

REFERENCES

[1] N. Wiener, *Cybernetics* (MIT Press, Cambridge, Mass., 1948).

[2] *The Negro Family: The Case for National Action* (Office of Policy Planning and Research, U.S. Department of Labor, 1965).

[3] J.W. Forrester, *Industrial Dynamics* (MIT Press, Cambridge, Mass., 1961).

[4] ——, *Urban Dynamics* (MIT Press, Cambridge, Mass., 1969).

[5] J.J. Servan-Schreiber, *The American Challenge* (Atheneum, New York, 1968).

[6] Report of the Commission on Executive, Legislative and Judicial Salaries, U.S. Civil Service Commission (1968).

18

Melville C. Branch

CONTINUOUS CITY PLANNING

For the past 50 years or so, city planning in the United States has been characterized by both conceptual and operational inadequacies which have greatly limited its effectiveness:

1. It has long been assumed that a master plan should be drawn for a city 20 or more years in the future, adopted by the local legislature, and implemented by enactment of whatever laws, regulations, and monies are needed for its fulfillment.
2. The master plan is conceived as a printed publication, subject to occasional modification or complete reformulation at long intervals.
3. Very few planning departments have therefore chosen or been able to maintain anywhere near up to date either their basic information or their plans.
4. At least until now, city planning in the modern era has tended to function independently and separately from the political and municipal administrative process.

---

5. City planning has been viewed as exclusively long-range, general, and all-encompassing, and therefore distinct from and not critically related to shorter-range operations and events.

6. Because city planning has concerned itself with a future so distant and suppositional that present problems can be disregarded or projected into relative insignificance, city planners have been able to avoid the really difficult key problems and turn to those of lesser importance, those they might mitigate before they become critical, or those which will arise after the long-range plan has been published without their anticipation and discussion.

7. As a consequence of these facts – or contributing to them – professional city planners have tended to be idealistic rather than realistic, passive rather than constructively active and persistent.

8. Until recent years, there was greater concern with design and physical-spatial factors than with quantification, management, behavioral sciences, and scientific method.

Certainly, there have been exceptions to these generalizations so succinctly stated, but they are generally descriptive of the current state of city planning. While they do not apply to new towns during their construction under private ownership, they become relevant when the new town becomes a self-governing municipality.

Few master city plans in the United States have significantly shaped the development of cities. For example, one staff estimates that their large city planning department is responsible for less than ten percent of either the fact or form of municipal growth and development. A year after master plan publications are printed, they are usually outmoded in important respects and significantly less relevant. Several years later, they are largely forgotten because a new or revised version of master city plans as they have been so grandly conceived in the past usually takes about 10 years to prepare. Also, few communities can afford their high cost more often than once every decade, all the more because it has been a wishful rather than productive enterprise.

Despite this history, the delusion persists among many practicing city planners that the concept of the master city plan has been right and the record wrong. To make matters worse, past master plans defining a city as it should be two or three decades in the future incorporated only a limited number of physical-spatial elements, projected separately. Now with the emphasis on comprehensive and systems planning, it is presumed that what could not be done in the past for a few elements can now be done for all economic, social, political, and physical features and aspects of the city combined. The idea or intention is that the master city plan is presented and discussed, adopted in some legislative form, and effectuated over the time period of its projection by the necessary implementing laws and allocation of monies. Presumably, a new master city plan will be ready when the old one is realized at its completion date.

Implicit in the traditional concept of the master city plan are a number of questionable conclusions. First and foremost, it presumes that the purposes, needs, and situation of an organism as complicated as a sizable city can be conceived, projected, foreseen, and predetermined 20 or more years in the future. The projective time span of some city plans is no less than 50 years. This end-state concept assumes that people's desires, objectives, and priorities are sufficiently known and fixed to permit drawing conclusions today with certainty concerning what they will want a half-century hence – or that city planners are wise enough and will be allowed to do this for people.

Furthermore, the concept implies that people want to or will commit themselves a quarter- to a half-century in advance. It presupposes that no unexpected events or developments will occur which make it imperative or desirable to alter the plan repeatedly. It assumes that the municipality is sufficiently independent to project and act with finality on its own long-range future, either apart from regional, state, federal, or international influences and events, or according to reliable forecasts of the future behavior of these entities. Finally, it presumes that someone, the city planner or a group, has the knowledge and technical capability to analyze all important elements and aspects of a city, project them in concert, identify and quantify their numerous interactions, and explain the results and significant alternatives to the body politic and legislature clearly enough for intelligent reaction and collective decision. This complete analysis will be accomplished promptly so that the data underlying the master city plan are still valid at the time of decision. . . .

Very few democratic societies will commit themselves to a completely planned end state to be attained 20 or more years hence. This would require resolving now many basic issues and questions which are nowhere near settled, for example:

The form of the city: compact and dense, or dispersed and low density?

Shall growth or size be limited? If so, how?

What is to be the relative expenditure of money between limited-access highways and mass rail transit during the 20-year plan period – to achieve what combination of the two in the end-state condition?

Since there is never enough money to meet all needs and desires, what are the cumulative expenditures and division of ultimate facilities to be between, for example, correctional and preventive law-enforcement and fire-protection facilities; hospitals and health maintenance clinics; private and commercial flying within the urban area; or parks, playgrounds, and open space and taxable income-producing use of the land?

What changes, if any, are to be provided for or encouraged in the level and methods of education?

Are the steadily increasing per capita demands for water, electric and gas energy, solid waste disposal, and teleintercommunication to be extrapolated and met?

What standards are to be attained by the end of the 20-year period with respect to environmental pollution of the air, waters, and ground by chemicals, biologicals, radiation, noise, and too many people?

How many dwelling units are to be provided in highrise apartment structures, row and cluster housing, and detached single-family homes?

What changes, if any, are to be assumed in present land-use controls? Will land ownership and development opportunities remain essentially the same as today?

Should today's taxes, their type and percentage of family income, be projected to determine future municipal revenues?

Will there be changes in the type and extent of governmental subsidies or incentives?

Is the spatial dispersal of minority groups and low-income housing to be effected?

Are any provisions for nuclear defense to be incorporated in cities?

Such questions are not just of speculative or theoretical interest; answers are required before the end state of a city can be conceived 20 or more years in the future. For each of them, either assumptions must be made or the body politic or its legislators must decide and define now – with sufficient clarity to permit a program of progressive attainment – the characteristics and features of the city wanted at the end of the planning period and the means and methods of achieving them. To be sure, every single question and relevant detail need not be decided today. But logically as well as empirically, a future end state for a city or any other organism cannot be formulated without assumptions or conclusions concerning the characteristics of the primary elements of which it is composed, their interrelationships, underlying forces, and attitudes which determine what is likely to occur and what can and will be done, and specific mechanisms of realization. Without this prior determination, the end-state concept of a master city plan is a fantasy. . . .

Planning objectives are developed by projecting from the present into the future. Their achievement must be possible. Realism rather than idealism is the fundamental motivation, although imagination is definitely part of background thinking. It is this requirement of realism which multiplies both the intellectual and practical tasks of city planning. Fashioning an imaginary future is not difficult. It can be highly subjective and individualistic. But to determine planning objectives which are demonstrably desirable in the public or majority interest, which represent the maximum attainment the organism involved can realistically consider, and for which a program of effectuation can be formulated – this is probably society's most formidable task. . . .

THE ISOLATED FUTURE

A natural corollary of the characteristics of traditional city planning as described has been strong emphasis on the very general and the long-range future and minimization of the specific and immediate. It was presumed that after commitment to an end-state master plan, city planning or municipal operating departments could develop the strategies, tactics, mechanisms, programs, and schedules required for its attainment. This has not worked out because this follow-up or implementation, rather than being secondary, is the vital essence of real planning and the most difficult part of the process. To prepare plans without reference to reality or effectuation is relatively easy. Witness the end-state plans commonly produced in the United States which for years have presumed resources that are not available, assumed instruments or effectuation that will be a long time coming, or incorporated positive answers to socioeconomic, political, and procedural questions still strongly debated and far from consensus.

Furthermore, exclusive focus on the long-range future has discouraged city planning staffs from acquiring and maintaining the knowledge of ongoing operations, of real time and lead time, and of sequential attainment necessary to formulate a realizable master plan. The philosophies of end-state and progressive planning are different. Successful city planning requires coincident consideration of the general and the specific, strategy and tactics, long range and short range, operational and projectional, present as well as end state. . . .

BASIC CONCEPT

What must gradually replace the grand master plan delusion and the after-the-fact type of city plan is continuous master planning, in which certain elements of the city are projected far into the future, others for the middle range, some short range, and a few not at all. The master or general city plan is not a document which emerges from some costly program of mammoth effort, sporadically altered until another such effort can be made, or more usually soon discarded because it is no longer relevant. A continuous plan is always as current as need be. No pretense is made that it is an intellectually or analytically complete statement at any given moment. Rather, it is the continuously changing representation of what the municipality can and intends to carry out with respect to its future.

Specific plans and programs are formulated for those functions, elements, or projects which are of such nature of are sufficiently circumscribed that they can be analyzed with the necessary precision and reliability. Different municipal utilities illustrate functional systems which can be planned separately with precision, but less exactly as they are combined into larger subsystems of the total system constituting the city as a whole. Distribution systems for electricity, gas, and oil are easier to plan separately than together as a coordinated subsystem supplying energy. Similarly, integrating sewage, household trash, construction debris, junked automobiles, and infectious hospital waste into one subsystem of waste disposal is much more difficult than dealing with each system separately. Nonetheless, the more systems which can be planned together as larger municipal subsystems, the sounder the plan for the organism as a whole.

There is a wide variation in how much is known about different systems comprising a city, the extent to which they can be quantified, and their likely development. For example, new technologies cannot be forecast and planned as reliably as the physical systems or facilities they may vitally affect. Thus a municipal water system may require revision or replacement as the recycling of waste water is made technologically feasible and psychologically acceptable, or an urban mass transit system may be changed as new types of vehicles become practical.

Some systems are particularly difficult to correlate because normally they are associated with, belong to, or are institutionalized within different urban subsystems. Tax policies, for example, belong normally to an economic-financial subsystem of the city, but through the local property tax they vitally affect public education, the amount of open space, and the percentage of single and multiple dwelling units (and thereby urban density). It is clear that many different municipal systems cannot be treated as uniformly as required for an end-state plan.

City planning must incorporate analytically current information, conditions, and decisions to a much greater extent than is now the case. Longer-range plans must represent the outcome of a succession of actions and anticipations rooted in the past but beginning or continuing in the present. To formulate longer-range plans without working out their derivation from the present through the intermediate future is analytically invalid as well as indefensible in practice. On the other hand, city planning certainly cannot assume the operating functions of municipal departments such as finance, building regulation, traffic, public works, or social welfare. Day-to-day management and long-range planning for each such function constitute a full-time task. However, city planning is the central mechanism for synthesizing the operations, budgets, and plans of separate municipal departments with relation to the total city system and its projected future. In this particular respect, the master city plan and the corporate business plan serve the same purpose of coordinating the plans of different groups, divisons, or departments in the organization.

Continuous master planning must encompass the spectrum from the past, to the present, toward the distant future: obligations and commitments from the past, immediate needs and long-range objectives, tactics and strategy, certainties and uncertainties. Some elements, such as freeway and primary water-supply lines, are projected 50 or more years into the future. Some, such as land use in certain areas, may not be planned more than a few years ahead. Others, such as changes in restrictions on outdoor advertising, in the policies of private lending institutions investing in urban land and real property, or in the forms of government subsidy and incentive, may at times be difficult or impossible to forecast.

The city plan must incorporate information and projections for each principal urban element separately and portray their synthesis into a combined pattern of actions and objectives over time for the best benefit of the city as a whole. It must be maintained and displayed in such form that it can be revised regularly and completely, and quickly changed when need be. Above all, it must always be sufficiently up to date to serve as the basic analytical simulation of the municipality and the official reference for discussing and deciding many different matters. It is ahead of the game rather than running to catch up. To function in these ways, the master city plan cannot be so restricted by legislation requiring bureaucratic formalities that the necessary combination of fixed objectives and flexibility is impossible. It will probably include budgets for one or two years, operating plans for two to three years, and longer-range projections, policies, and plans for various periods in the future.

Master plans must be formulated in a far more flexible and dynamic way than the booklet or brochure which has been the accepted statement for master plans in the past. The first step in this process requires a mechanism of analysis and exposition in a place similar to the planning rooms of many large corporations and to the NASA spaceflight center familiar to millions of television viewers (on a larger scale and more as an operational than advance planning mechanism). Eventually, simulation for city planning purposes may be handled entirely on output display devices connected with electronic computers and videotape storage, but this will be some time coming.

ANALYSIS

Primary Elements

The analytical approach of continuous city planning is opposite to that for end-state plans. Rather than attempting or pretending to cover everything and project it accurately into the distant future, continuous analysis begins with deliberate modesty. First, the most important elements for the functioning, form, and future of the municipality are identified. . . .

While various persons might select different primary elements to be employed for continuous city planning, and the most important elements vary from community to community, a priority must be established. Clearly, everything cannot be handled at once, nor can all information be collected, much less analyzed successfully. In every form of planning, primary elements must be selected, although they may shift with time, events, different values, changes of objective, or other developments.

Information

There must be information available with which to evaluate and study these elements, from local, state, or federal governments, from private enterprise, or gathered by the planning agency itself. It must be available at a cost which is feasible and worthwhile, sufficiently reliable, and available soon enough for the intended purposes. Continued availability should be reasonably certain, and data should be sufficiently comparable over time to reveal trends. To the extent possible, the quantification of information should facilitate comparison, correlation, and projection of components.

Unless the data for a particular element meet most of these requirements, inclusion of the element in planning analysis must be postponed. Careful and controlled use of data for basic analysis is particularly important for city planning, which has been ineffective and wasteful with data, often collecting masses of information indiscriminately with little regard for its relative importance and correlation with other data. In fact, frequently, city planning staffs ask for more and more information, not because it is specifically needed, but in order to prolong the process, avoid tackling critical but difficult problems, or in a vain attempt to collect the enormous amount of information which would be needed theoretically for the end-state type of plan described earlier. . . .

Analytical Core

An analogy from corporate planning illustrates the kind of analytical core lacking in present-day city planning. For years, businesses have quantified their activities, based many of their decisions, and appraised their success on three analyses, operating statement, profit-and-loss statement, and balance sheet. Although it has long been known that these statements cover only the primary aspects of the business, they serve nonetheless to run businesses successfully. Additional information can be incorporated into this closely correlated core of information and analysis, or the factual and comparative accuracy of each element can be progressively improved. . . .

While multinational corporations now employ as many as several hundred thousand people working in several dozen different enterprises, they are not as difficult to manage and plan as a city. Ownership and control are far more centralized and absolute; motivations and objectives are simpler; and factors to be considered are far fewer. Nevertheless, the principle of a core of information and analysis applies equally to municipal management and city planning. Without it, public management is inefficient if not chaotic, and city planning is ineffective if not a delusion. Every planning activity needs such a core to provide the factual basis for rational conclusions, decisions, and planning ahead. It may be formally specified and regularly reported or exist only in the mind. The information and analysis can be neither complete nor certain because total knowledge is impossible, but it is no less necessary for sound planning.

Correlation

Because each element of a city is directly or indirectly interrelated, correlation is a fundamental part of planning and meaningful analysis. By adding another essential dimension of understanding, correlation makes it possible to determine how available resources are best allocated among the various components of a municipality to meet functional requirements for its continued operation and to achieve certain objectives through progressive change. This allocation of resources is, of course, a primary purpose of planning; correlation is required for the underlying analysis of the organism necessary to make this allocation. Theoretically, there is almost an infinity of interrelationships among a relatively small number of elements. Those few which it is feasible to quantify and study regularly must be selected. . . .

These correlations are made in various ways: by individual judgmental comparison; by collective judgments obtained in an organized way; visual comparison of data side by side; by overlay of information; by ratios established for comparative purposes and to monitor change; by mathematical correlation with or without the aid of computers. As correlations are used and their validity tested by experience, they are retained, improved, or discarded in a constant process of development.

Because in practice correlations are particularly difficult to formulate for comprehensive city planning, quantitatively meaningful interactions are formulated much less frequently than presumed or pretended. Most such correlations today are largely personal and judgmental. After one or several functions have been interrelated, further quantitative correlation is obscured or most often not attempted at all. Unfortunately, the impression is usually given that city planning presentations and recommendations are supported by reliable and conclusive correlations, when in fact they are not.

Projection

By definition, projections into the future are an inherent part of planning. Some are a matter of choice and therefore quite certain, such as a schedule of construction or program of accomplishment for the future involving no likely constraints. Some are the result of previous commitment: repayment of bonded debt or minimum levels of government service. Normally, these two can be reliably projected. Another category of elements is extrapolated from past trends, forecast from established correlations, or projected from an accumulation of knowledge concerning the community accurate enough for the purpose. This category is exemplified by population forecasts, average demand for water and energy, crime rates, or the location of private development. The reliability of these projections varies widely. They require a closer watch, more frequent review, and quick response if successive short-range changes indicate a trend.

Each element in the core of analysis is projected separately into the future in one way or another, and to the extent possible all of them are projected together so that they maintain their proper interrelationships at each interval of future time. The latter is by far the most difficult task in planning, possible at present only in part and within an expected range of error. Thus certain projections are expressed within upper and lower limits, rather than as a single quantity or trend line.

In continuous city planning, there is no mistaken attempt to formulate a complete end-state master plan by projecting all elements and significant aspects of the municipality in concert some 20 to 25 years into the future. Some elements of the city, such as its main highway or drainage systems, will remain essentially as they are now for 50 to 75 years. Others can be forecast reliably for 5 to 10 years. Some must be left open because situations or conditions make projection no better than guess. Therefore, continual information gathering, observation, and analysis are more important than attempts at positive long-range prediction.

Rather than focusing on the distant future and the end-state type of master plan, continuous city planning starts with today and works toward the future. Of immediate importance are the next several years. Not only is the next fiscal year specifically budgeted, but in addition tax rates and assessments, borrowing capacities, state and federal assistance, and other financial factors limit the funds available under any circumstances. What takes place in the next several years may have a profound effect on what is possible thereafter. It could further the achievement of city planning or postpone certain planning purposes, even make them impossible.

Therefore, continuous city planning first projects the next budget year in detail, the subsequent year in somewhat less detail and certainty, and the following 5 years as thoroughly as practicable. Every year, the 7-year span of projection is maintained by adding another year into the future. The first 2 years constitute an operating plan, combining the budgeted and immediately subsequent intentions of municipal departments with planning for the city as a whole, thereby ensuring that longer-range plans relate to reality. The projections for the next 5 years cover the "middle-range bridge," the period next most predictable and critical for the future. For it is during this time that progressive improvements and changes can be effected toward planning objectives. Some projections for this period are firm, others subject to continual scrutiny and frequent adjustment.

Projections and plans beyond the 7-year period of formal plans are made when they are reliable enough to justify the time and cost of the projective study they require and when the means for their realization can probably be attained. For example, municipal utility systems involve projections and plans which often must be started more than 7 years before realization because of the long lead time required for system design, engineering, acquisition of land, funding, approvals, construction, and a test period of operation. The important requirement is that these longer-range plans represent firm intentions to produce and the availability or strong probability of funding and other requirements for effectuation.

Other projections are expressed as policies. Thus continuous city planning might recommend formal adoption of a policy that more low-income housing be programmed if and when this is possible, because funds become available or community attitudes change, or for another reason. Or with reference to the longer-range future extending beyond the 7-year period of specificity, there might be a policy of locating this housing more widely throughout the community, closer to more centers of employment and public transportation. Such policies should be more than pious platitudes. They are intended to underlie and shape current decisions and should be abandoned when no longer relevant. They are maintained as conspicuously and up to date as is quantified material; they are treated as "facts" in different form.

Progressive Development

There is a limit, of course, to the core information and analysis which can be processed, maintained, and conceptualized with regular reference, study, and decision making. The information readily available, its reliability, the time required to revise the analysis or develop alternatives, personnel initially to establish and maintain the core analysis, its cost, and incomplete knowledge are all involved.

What is attempted and how this is organized and carried out is crucial to successful city planning, for decisions and actions are only as sound as the information and analysis on which they are founded. The information maintained and the exact ways of handling it will vary among cities, but a core of basic data can be formulated which must be incorporated in all informational-analytical mechanisms for continuous city planning.

Only those data are incorporated which have the requisite reliability, can be updated as frequently as needed, are critical to the most important decisions to be made, and are understood and absorbed by those directly involved. Above all, additional information and further analysis are introduced into the core reference only as fast as they can be interrelated with the data and analysis currently considered. At all times, the total construct of material must be meaningfully interrelated analytically and presented clearly enough for decision makers to use it as their main means of evaluation and conclusion. More complete information and conclusive analysis support better planning only insofar as they meet the fundamental requirement of consistent internal correlation. A very simple construct of data properly correlated is preferable to more material which cannot be interrelated quantitatively or judgmentally. "Model simple, but think complex" is far superior to the opposite. Any method of collection, storage, processing-manipulation, analysis, or display can be employed which contributes to meeting these basic requirements.

Form

How analytical material is formulated is important. Delineating the most influential elements of a city and quantifying their interrelationships can be a difficult task. Every possible clarification and simplification is needed by the staff to study and analyze the functioning of the city. It is also needed for legislative decision makers, who are not willing to act on the basis of staff analysis which they do not understand.

No responsible decision maker wants to act mainly on someone else's say-so. He cannot do so without in effect abrogating his decision-making role to a subordinate. When he is informationally or analytically dependent on another person, he is vulnerable to exaggeration, an unrevealed bias, or even deliberate deception. Increasingly, this is a serious problem confronting executive management generally, as knowledge becomes more compartmentalized and specialized and mathematical, often requiring more and more electronic computation understood only by the programmers. Frequently, the underlying assumptions of staff work are unstated. This basic difficulty has not yet been resolved, and there is as yet no indication of how it will be. Until then, analysis must not be too complicated and incomprehensible for decision makers to understand.

In general, graphical forms are more readily comprehended than numbers. . . . Superposition of data or their placement side by side facilitates comparison. Any spatial separation of data requires the user to remember each figure as he refers to another. Memoranda and other written materials usually involve remembering one or several figures as one turns to related data as close by as the previous or following page. An awkward thumbing back and forth is often necessary. When such separated data number three or more entities, the average user will have difficulty maintaining clear recall and comparison.

Of course, forms should be chosen which do not introduce interpretive distortions, such as improperly selected scales for ordinate and abscissa or presentations which imply that small numerical differences are meaningful when in fact they are not statistically significant. The key criterion of form is whether it makes comprehension and mental manipulation easier for the planning staff and – more importantly – for the decision makers whom they support. The core of information and analysis for continuous city planning must be directed specifically to the minds and understanding of those who make the final decisions. Otherwise, there can be serious abrogation of responsibility and grave consequences for democratic society; a computational elite could become the deciders in fact. Therefore, the core of information and analysis is formulated to meet the needs of the ultimate decision makers, providing as clearly as possible that which can be comprehended, absorbed, and used.

Flexibility

Earlier in this discussion, the necessity of up-to-date city planning was emphasized. It is repeatedly observed by planners that city planning in the United States has been ineffective largely because the few components of end-state master plans which could be acted on become quickly outmoded, since redoing this type of fixed grand master plan can be afforded only every 10 years or so. Being closely current is essential to successful continuous city planning. . . .

Another kind of flexibility required is the capability of revising the core of information and analysis quickly. A natural catastrophe, an emergency situation, or an unexpected decision of major consequence may require immediate revision. Core information, correlations, and analyses must be recalculated in as short a time as a week or as long as a month, depending on the community and circumstances. Conscientious decision makers will not tolerate more time lage than is necessary or reasonable. Both ongoing and emergency types of revision should be borne in mind when formulating the process, procedures, and mechanisms of continuous city planning. These capacities to change are crucial.

DISPLAY

Besides gathering, correlating, and selecting the best form for core data with the necessary flexibility, how they are displayed is important. First, they must be readily available to those who make city planning decisions. Those with this responsibility cannot delay for hours, much less the days or even weeks they often wait with present-day city planning, until the background information they need is developed or found in some bureaucratic labyrinth of files. City planning commissions have been known to wait literally for years before information they wanted regularly before them when making certain decisions was made available. All parts of the core of information and analysis must be available in minutes for immediate decision, for contemplation, for staff study, or to inform an individual concerned, an interested group, or the general public.

Display units should be as uniform and consistent as possible to reduce the time required for their preparation and revision and also to facilitate comparison. A "standard module" of display provides the kind of advantage obtained from other applications of uniformity and regularity. The size of the module is limited by the need for it to be transportable not only to city council chambers, but to the offices of individual lawmakers, branch administrative centers in large cities, and elsewhere in the metropolitan region for meetings and display. City councilmen, other government officials, and private groups and persons may not wish or be able to come to the planning center but would be willing to receive material if it is brought to them. . . .

The most important requirement is that displays not be typed, not be prepared in a drafting room, and not be printed. The delay inherent in typing and drafting is not fully recognized; it takes much longer than is assumed without time checks. The preparation of displays in a central drafting room requires an inordinate time because of laborious handwork and scheduling delays. Printing obviously takes considerable time under the best of circumstances.

None of these delays is acceptable in continuous city planning, which must permit as close to immediate revision as feasible. It is highly desirable that alternatives requested by decision makers be displayed within several hours, rather than days or weeks later when relevant material has faded in their memories and other matters have come to the fore. To meet these requirements, the preparation and revision of displays is done within or next door to the place where they are normally used. They are not sent out for preparation or processing. They are kept up to date and revised, alternatives are developed, and new displays are created in situ. . . .

PICTURIZATION

Since the core of information and analysis provides the representation or simulation necessary for continuous city planning, the municipality must be portrayed or pictured in ways which are sufficiently representative of its reality to permit meaningful study and valid conclusions. A map shows ground features of the city at uniform scale and correct spatial interrelationships. Statistical tabulations depict population characteristics, municipal costs, retail sales, or other facts or trends numerically descriptive of the city. Instruments measure air, water, and ground pollution. Descriptive writing may be the best way of expressing certain urban features and developments. In each case, an aspect of the city is being simulated by the most appropriate form of representation or statement.

The three-dimensional, physical, visual reality of a city is, of course, an essential manifestation of its existence and a root consideration in its planning. It is the physical body within which the multitude of component parts of the municipality function. In turn, it represents in its anatomy and form the forces exerted by these many constituent parts and those exerted on the city from outside its borders by region, state, nation, and international situations. The interrelationship between urban form and pattern and land use, transportation, environment, services, municipal costs, urban aesthetics, and every other aspect of city life is well documented in the literature of city planning.

What is not well known is the dependence of city planning staffs, commissions, and legislative decision makers on their personal visual memory to picture this physical city. . . . However, except in small places, a person cannot recall to mind a memory image of every part and prospect of the community; there is too much to be seen and remembered. Certainly, in larger cities and those changing rapidly, only a fraction of the visual scene can be recalled to mind. Maps assist in this recall, but they supply only additional two-dimensional cartographic information. Even if a city is examined in perspective from a tall building or nearby hilltop, most of the street pattern and many other features are hidden from view by intervening structures and terrain. Of course, much less can be seen standing at one point or several within a community. City planning staff, commissioners, and legislators cannot visit every site involved in their many decisions, nor is it practical to provide them with sufficiently descriptive photographs of every location. . . .

Aerial photography can provide a visualization of the three-dimensional reality of the municipality which can be acquired in no other way and which is essential for urban analysis and many city planning decisions. Although its view of the city from above is an unusual and partial one, it is the closest approximation to a complete picturization available. There are features which are not shown from above because they are beneath an intervening structure or vegetation, too small to be detected at the photo scale, or otherwise obscured or unseen, but every form of visual and photographic observation is incomplete. Those unfamiliar with aerial photographs must spend a while becoming familiar with the overhead viewpoint and learn to read and interpret with ease the large amount of information they contain. Besides this information, the photographs act as triggers to recall to view within the conscious mind various scenes in different parts of the city stored unremembered in the individual's memory.

If new aerial photographs are taken every 6 months or every year, this most important visual simulation is kept closely current for continuous city planning, and a record is accumulated which may reveal trends and support urban studies. . . .

CITY PLANNING CENTER

Information, analysis, and displays are brought together in a primary place and mechanism specifically designed for collection, preparation, use, and decision. . . . In the main part of the installation, discussants, participants, and audience are surrounded on three sides by walls with sliding displays containing core information and analysis. The recommending or decision-making body (city planning commission, committee of city council, or city council if not too large) and several members of the planning staff are on a slightly raised dais at the end of the room. . . .

Next to this auditorium and connected by the door at the rear of the dais is space for the maintenance of the core of information and analysis and official master plan, staff study and meetings, development of new displays, temporary storage of panels, and computer terminal. Here, staff groups large or small can meet while the main room is in use, and people from other levels of government or visitors from abroad can gather for working and explanatory sessions of many kinds. It is here that studies are made which are analytically tentative or exploratory only and should not be seen and construed as a serious consideration, intention, or proposed alternative. To this extent, the staff room can serve appropriately as restricted staff study space.

Connected with this staff study space is a third smaller room specifically for maintenance of the master plan. Here, desks and files are arranged to facilitate work directly on analysis-display panels as they stand in the wall space built to contain them. By making and maintaining the displays directly on the panels in this upright position, transporting the panels to and from a drafting room – a source of inflexibility, inevitable delays, and higher costs – is avoided. It reinforces the fact that continuing analysis is conducted here, utilizing information, departmental plans, and other materials flowing to and from the planning center. In this maintenance office, panels are photographed and filed as needed for the official, legal, master plan record. . . .

LEGAL INSTRUMENT

Any legal difference between city planning as practiced today and continuous city planning as proposed in this discussion is a matter of degree and not of kind. The main difference is that the legal reference in continuous planning is not a printed master plan but the recorded content of the city planning center. This includes information on analysis-display panels in use or in storage, photographic records of this information and analysis previously existing and formulated, and related legislation. The official legal master plan is that part of this material which has been adopted by the proper authority for this purpose.

Identification of the master plan at the time of a particular request, action, or decision is made by a well-organized system of reference to photographic reproductions of analysis-display panels on microfilm and to other records representing the master plan at a particular time. Essentially this is what happens today, except that continuous city planning involves keeping track of more frequent changes.

If a document showing the master plan is needed, it is produced by photographing the analysis-display panels and other records which depict it. These materials can be published as a printed report, but this expense should certainly be justified by some worthwhile purpose served only in this way. . . .

Legal requirements for city planning are met as successfully by continuous planning as by traditional planning. It is important, however, to substantiate that the more frequent changes involved in continuous planning are not arbitrary with respect to any particular matters, but reflect new information, further analysis, or conditions or events calling for revision. This substantiation is best accomplished by carefully following planning procedures which provide or support proper notification, hearings, reasonableness, consistency, due process, and public purpose.

STRATEGY OF USE

. . . The core of information and analysis must be promoted and proved as a useful and relatively uncontroversial reference which serves the self-interest of many people in the municipal government. For example, municipal departmental managers can use the data to learn what other departments are doing which affects their operations or intentions. If it is municipal policy to locate facilities together whenever feasible, a police chief will want to know where the fire chief is planning new facilities in the future. Those concerned with land acquisition will want to know about the intentions of both departments, as will the mayor, city manager, councilmen, and others responsible for viewing the municipality as a whole.

For the activities of different departments to be

coordinated, a core of information and analysis covering the current operations and future plans of each one must be available to the others and to those who must decide among competing demands for available funds. While this coordination is often resisted at first, it can be shown over time that duplication or conflict of effort can no longer be afforded, that mutual support is to everyone's advantage over time, and that coordination is mandatory in today's world of interdependencies among activities which could once upon a time operate more independently.

The strategic requirement for coordination is that each department or agency of local government present its functional budget, operating plans, and projections covering 7 years in the future. If there are units which cannot or will not cooperate at first, the city planning staff develops this information for these units as best it can. The disadvantages to one's own self-interest of not cooperating are soon apparent.

Since the information, analysis, decision, and actions recorded in the planning center constitute the legal master city plan, it becomes a necessary reference for many people. Municipal departments, the mayor, the city manager, council members, the city attorney, and others use it to determine, support legally, or explain decisions and actions. Individual applicants for city planning approvals, special interests, citizens' groups, and litigants refer to it for the information each needs. Various private enterprises not directly involved in city planning find that following continuous city planning provides information concerning the operations and intentions of local government which enables them to improve the conduct of their own affairs. The main reason these continuing contacts have not developed in the past is that city planning is rarely up to date, more often after the fact than before the fact.

Another important use of the city planning center is to consider alternatives. When the time comes periodically to propose the allocation of available resources among competing demands, many different resolutions are possible. But the total array must always be mutually consistent and must meet aggregate limitations and requirements. This is difficult to do without the planning information, analysis, and flexible means of comparison incorporated in the planning center.

Municipal departments can be encouraged to develop their own proposed solutions to matters about which they are deeply concerned but which involve other governmental units, activities, or considerations. With this kind of exercise possible in the planning center, the final resolution will be more explicable and probably more acceptable. Certainly a feeling of participation is encouraged. When alternatives have been studied by all directly concerned, the likelihood of irreconcilable disagreement is greatly reduced. Since the information shown in the city planning center indicates the parameters which workable plans must meet, citizen groups can prepare advocacy plans realistic enough for careful consideration.

Progressive involvement is the key tactic in getting continuous city planning accepted first by local government and next by the community generally. Item by item, information and analysis are made useful to the powers that be. Once this information and analysis are employed and a procedure for their regular use is established either formally or informally, it becomes progressively more difficult to abandon the now familiar process of consideration. From a small beginning of assisting municipal decision makers and others in their own self-interests, continuous city planning becomes an ingrained part of local government. . . .

19

Charles J. Hitch

DEVELOPMENT AND SALIENT FEATURES OF THE PROGRAMMING SYSTEM

It was not until 1961 that the full powers of the secretary of defense to run the department on a unified basis – spelled out by various amendments in the intervening years – were actually used. And I suggested that this situation existed principally because earlier secretaries of defense lacked the necessary tools to do so.

. . . From a modest beginning, limited to the protection of our land frontiers against the Indians and our neighbors in Florida and Canada, our national security objectives have expanded to involve us in an interlocking system of free-world military alliances with over 40 sovereign nations. We now maintain for this purpose by far the largest peacetime establishment in our history. We have, today, a force of almost 2.7 million military personnel on active duty,

---

 Selected from A Modern Design for Defense Decision, *Samuel A. Tucker, ed., Industrial College of the Armed Forces, Washington, D.C. (1966), 64-68, 70-78, 83-85. Originally published in* Decision Making for Defense, *by Charles J. Hitch, University of California Press, Los Angeles, California (1965).*

supported by about 900,000 civilians in the United States and about 250,000 overseas, the latter mostly citizens of other countries. In addition, we have almost 1 million reserve personnel and about 500,000 retired military personnel on our payrolls.

Pay alone accounts for more than $20 billion out of a total defense budget of $50 billion. The remaining $30 billion is used to procure a staggeringly large variety of goods and services from the private sector of the economy – aircraft, missiles, tanks, food, clothing, research and development, construction, and utilities. We draw from virtually every segment of the American economy and utilize a very large share of the nation's total research and development capacities. Because of the vast scope of our activities – on the land, on and under the seas, in the air and in space – and the great demands we make on our weapons and equipment, and Defense Department is vitally interested in virtually every field of scientific and technological knowledge. The value of our current inventory of equipment, weapons, and supplies is conservatively estimated at $135 billion. Our principal installations and facilities number in the thousands, and we control nearly 15,000 square miles of land. Our military operations extend around the world, and we spend almost $3 billion a year in other countries.

How, one might ask, can any one man or group of men ever hope to manage such a vast aggregation of men, equipment, installations, and activities spread all over the globe? And yet, . . . the defense effort, to be fully effective, must be managed on a unified basis, not only in the conduct of combat operations but also in the planning and execution of the programs. And, as President Eisenhower stressed in 1958: "It is . . . mandatory that the *initiative* for this planning and direction rest not with the separate services but directly with the secretary of defense and his operational advisors. . . ."[1]

The revolution in military technology since the end of World War II, alone, would make necessary the central planning and direction of the military program. The great technical complexity to modern-day weapons, their lengthy period of development, their tremendous combat power and enormous cost have placed an extraordinary premium on sound choices of major weapon systems. These choices have become, for the top management of the Defense Department, the key decisions around which much else of the defense program revolves. They cannot be made properly by any subordinate echelon of the defense establishment. They must be directly related to our *national* security objectives, rather than simply to the tasks of just one of the military services.

The revolution in military technology has not only changed the character of our military program, it has also, to a significant degree, blurred the lines of demarcation among the various services. Is the missile an unmanned aircraft, as the Air Force likes to think, or extended-range artillery, which is the Army view? Most of our major military missions today require the participation of more than one of the military services. Therefore, our principal concern now must be centered on what is required by the defense establishment as a whole to perform a particular military mission – not on what is required of a particular service to perform its part of that mission. This is not only true with regard to the planning of our military forces and programs, but also with respect to the development of new major weapon systems.

BUDGETING BEFORE 1961

. . . Prior to 1961, the defense secretaries lacked the tools to manage the overall effort on a truly unified basis, they had to resort (except in times of emergency, like Korea) to what might be described, generically, as the "budget ceiling" approach. The President would indicate the general level of defense expenditures which he felt was appropriate to the international situation and his overall economic and fiscal policies.[2] The secretary of defense, by one means or another, would allocate this figure among the three military departments. Each military department would in turn prepare its basic budget submission, allocating its ceiling among its own functions, units, and activities, and present additional requests, which could not be accommodated within the ceiling, in what was variously called an "addendum" budget, "B" list, etc. Then all the budget submissions were reviewed together by the Office of the Secretary of Defense in an attempt to achieve balance.

Let me make quite clear the fact that this was

indeed the traditional way of preparing the defense budget. Frank Pace, then director of the Bureau of the Budget, in testifying before a congressional committee in 1949 on how the defense budget was prepared in the Truman administration, described the process as follows:

> *We [the Bureau of the Budget] would provide him [the President] with certain factual information as to where certain policies would lead. From that the President sets a ceiling on the armed services, which was last year, I think, generally known as $15 billion.*
>
> *However, I think it should be explained that under the ceiling process – and this is not solely for the armed forces but exists for every department of the Government – . . .*
>
> *There is also the proviso that if within that limitation it is impossible to include certain programs which the Secretary of Defense considers of imperative importance to the national defense, they shall be included in . . . what is termed the "B" list. . . . The "B" list is what cannot be included under the ceiling.*[3]

It was recognized long ago that this was a rather inefficient way to go about preparing the defense budget. Its consequences were precisely what could have been predicted. Each service tended to exercise its own priorities, favoring its own unique missions to the detriment of joint missions, striving to lay the groundwork for an increased share of the budget in future years by concentrating on alluring new weapon systems, and protecting the overall size of its own forces even at the cost of readiness. These decisions were made by patriotic generals and admirals, and by dedicated civilian leaders a well, who were convinced that they were acting in the best interests of the nation as well as their own service – but the end result was not balanced, effective military forces.

The Air Force, for example, gave overriding priority to the strategic retaliatory bombers and missiles, starving the tactical air units needed to support the Army ground operations and the airlift units needed to move our limited war forces quickly to far-off trouble spots. The Navy gave overriding priority to its own nuclear attack forces – notably the aircraft carriers – while its antisubmarine warfare capability was relatively neglected and its escort capability atrophied. The Army used its limited resources to preserve the number of its divisions, although this meant that they lacked equipment and supplies to fight for more than a few weeks.

Moreover, because attention was focused on only the next fiscal year, the services had every incentive to propose large numbers of "new starts," the full cost dimensions of which would only become apparent in subsequent years. This is the "foot in the door" or "thin edge of the wedge" technique which the one-year-at-a-time approach to defense budgeting greatly encouraged.

Another unsatisfactory aspect of this method of attempting to exercise control and direction of the defense effort through the annual budget was the almost complete separation between budgeting and military planning (I speak here of medium- and long-range planning, including weapon systems planning – not the contingency planning for the use of existing forces).

1. These critically important functions were performed by two different groups of people – the planning by the military planners and the budgeting by the civilian secretaries and the comptroller organizations.

2. Budget control was exercised by the secretary of defense but planning remained essentially in the services. It was not until 1955-1956 that the first Joint Strategic Objectives Plan (JSOP), projecting the requirements for major forces some 4 to 5 years into the future, was prepared by the Joint Chiefs of Staff organization, but the early JSOP was essentially a "pasting together" of unilaterally developed service plans.

3. Whereas the planning horizon extended 4 or more years into the future, the budget was projected only 1 year ahead, although it was clear to all involved that the lead time from the start of a weapon development to the equipping of the forces ranged from 5 to 10 years, depending on the character of the particular development effort.

4. Planning was performed in terms of missions, weapon systems, and military units or forces – the "outputs" of the Defense Department; budgeting, on the other hand, was done in terms of such "inputs"

or intermediate products as personnel, operation and maintenance, procurement, construction, etc.; and there was little or no machinery for translating one into the other.

5. Budgeting, however crudely, faced up to fiscal realities. The planning was fiscally unrealistic and, therefore, of little help to the decision maker. The total implicit budget costs of the unilateral service plans or of the JSOP always far exceeded any budget that any secretary of defense or administration was willing to request of the Congress.

6. Military requirements tended to be stated in absolute terms, without reference to their costs. But the military effectiveness or military worth of any given weapon system cannot logically be considered in isolation. It must be considered in relation to its cost – and in a world in which resources are limited, to the alternative uses to which the resources can be put. Military requirements are meaningful only in terms of benefits to be gained in relation to their cost. Accordingly, resource costs and military worth have to be scrutinized together.

As a consequence, the secretary each year found himself in a position where he had, at least implicity, to make major decisions on forces and programs without adequate information and all within the few weeks allocated to his budget review. Moreover, every year the plans and programs of each of the services had to be cut back severely to fit the budget ceiling, by program cancellations, stretch-outs, or postponements – but only for that year. Beyond the budget year, unrealistic plans continued to burgeon – perhaps next year the ceiling would be higher.

These deficiencies did not go unnoticed in the Congress. Representative George Mahon, then chairman of the House Defense Appropriations Subcommittee and later also chairman of the full committee, addressed two letters to the secretary of defense in the summer of 1959 and fall of 1960. In his first letter he stressed the importance of looking at the defense program and budget in terms of major military missions, by grouping programs and their cost by mission.[4] In his second letter, he called "for more useful information and for a practical means of relating costs to missions."[5]

Many other students of the defense management problem had reached the same conclusion, including the group with which I had the honor to be associated at the RAND Corporation. Many of these conclusions found their way into a book, *The Economics of Defense in the Nuclear Age,*[6] which the RAND Corporation first published in March 1960, some 10 months before I was called upon as assistant secretary of defense (comptroller) to help introduce them into the defense department.

BUDGETING AFTER 1961

The new secretary of defense, Robert S. McNamara, made it clear from the beginning that he intended to be the kind of secretary that President Eisenhower had in mind in 1958 and take the initiative in the planning and direction of the defense program. In a television interview, after having been in office less than 1 month, Secretary McNamara defined his managerial philosophy as follows:

> *I think that the role of public manager is very similar to the role of a private manager; in each case he has the option of following one of two major alternative courses of action. He can either act as a judge or a leader. In the former case, he sits and waits until subordinates bring to him problems for solution, or alternatives for choice. In the latter case, he immerses himself in the operations of the business or the governmental activity, examines the problems, the objectives, the alternative courses of action, chooses among them, and leads the organization to their accomplishment. In the one case it's a passive role; in the other case, an active role. . . . I have always believed in and endeavored to follow the active leadership role as opposed to the passive judicial role.*[7]

Furthermore, Secretary McNamara made it known that he wanted to manage the defense effort in terms of meaningful program entities – of "outputs" like the B-52 force, the POLARIS force, the Army Airborne Division force, etc., associating with each output all the inputs of equipment, personnel, supplies, facilities, and funds, regardless of the appropriation account in which financed. He wanted to know and, indeed, would have to know, in order to optimize the allocation of resources, the cost of, for example, a B-52 wing – not only the cost of equipping the wing but also the cost of manning and operating it for its lifetime or at least for a reasonable period of years in the future. Only then would he be in a position to assess the cost and effectiveness of a B-52 wing as compared with other systems designed to perform the same or similar tasks.

Moreover, he wanted to know the total costs of the forces assigned to each of the major missions – the costs of the strategic offensive forces, the continental defense forces, the general purpose forces, etc. As General Maxwell Taylor had pointed out to a congressional committee in 1960:

> *... If we are called upon to fight, we will not be interested in the services as such. We will be interested rather in task forces, those combinations of Army, Navy, and Air Force which are functional in nature, such as the atomic retaliatory forces, overseas deployments, continental air defense forces, limited war expeditionary forces, and the like. But the point is that we do not keep our budget in these terms. Hence it is not an exaggeration to say that we do not know what kind and how much defense we are buying with any specific budget.*[8]

These views closely coincided with my own. The secretary and I both realized that the financial management system of the Defense Department must serve many purposes. It must produce a budget in a form acceptable to the Congress. It must account for the funds in the same manner in which they were appropriated. It must provide the managers at all levels in the defense establishment the financial information they need to do their particular jobs in an effective and economical manner. It must produce the financial information required by other agencies of the government – the Bureau of the Budget, the Treasury, and the General Accounting Office.

But we both were convinced that the financial management system must also provide the data needed by top defense management to make the really crucial decisions, particularly on the major forces and weapons systems needed to carry out the principal missions of the defense establishment. And we were well aware that the financial management system, as it had evolved over the years, could not directly produce the required data in the form desired. It was clear that a new function, which we call programming, would have to be incorporated in the financial management system. I had hoped that I would have at least a year to smooth the way for the introduction of this new function. I recall outlining the proposed programming system to Secretary McNamara in the spring of 1961 and recommending that we spend 18 months developing and installing it, beginning in the first year with a limited number of trial programs, with a view to expanding the system to include all programs during 1962. The secretary approved the proposed system but shortened my timetable from 18 months to 6. Somehow we developed and installed it, department-wide, in time to use it as a basis for the fiscal year 1963 defense budget. Submitted to Congress in January 1962, this was, of course, the first budget to be prepared wholly under the new administration.

Since the military planning function and the budget function were already well established, the role of programming was to provide a bridge between the two. It was, of course, theoretically possible to recast both the planning and budget structures in terms of major programs related to missions. In fact, the military planning operation was later adapted to the program structure, and I once thought that the budget structure should be similarly realigned. You will find on page 56 of *The Economics of Defense in the Nuclear Age* a format for such a program budget.

But the existing budget structure serves some very useful purposes. It is organized, essentially, in terms of resource categories: (1) military personnel, (2) operation and maintenance, (3) procurement, (4) research, development, test, and evaluation, and (5) military construction.

This type of structure lends itself ideally to the manner in which the Defense Department actually manages its resources. While military planning and the formulation of programs should logically be done in terms of missions and forces, the department must be managed not only in those terms but also in terms of resources. For example, we have to manage the acquisition, training, and careers of military personnel; the operation of bases and facilities; the procurement of aircraft, missiles, ships, and tanks; the research and development program; and the construction of airfields, missile sites, quarters, and other additions to our existing physical plant. The present budget structure facilitates the estimation of resource costs as well as the execution of the resource programs.

This division of the budget by broad input or resource categories also provides needed flexibility for the adjustments in the program that are inevitably required in the course of the budget year. Program priorities and requirements always change in unanticipated ways even in the course of a single year as a result of international developments, technological breakthroughs (or disappointments), and all sorts of other events. It is important not to freeze programs in appropriation bills.

Finally, the Congress, and particularly the appropriations committees, prefer the existing arrangement of the defense budget.[9] They have been working with it for more than a decade and have established an historical basis for forming judgments on the validity of the budget requests. It is much easier for an appropriations committee, for example, to review a budget request of $4.3 billion for pay and allowances for 960,000 active duty Army personnel than, say, a request of $18 or $19 billion for the major program "General Purpose Forces," or even a request of $700 million for program element "Army Infantry Divisions." Although the President, under the Budget and Accounting Act of 1921, can propose his budget in any form he pleases, it is the Congress that determines how the funds will be appropriated and this, in turn, determines how the funds will be accounted for. I now feel that the advantages of the existing budget structure far outweigh the disadvantages, which are principally mechanical, namely, the need to translate program categories into budget categories, and vice versa. This is the sort of disadvantage that modern high-speed computers are well designed to overcome.

Accordingly, we decided to leave the budget structure undisturbed and to span the gap between planning and budgeting with the new programming function. This resulted in a three-phase operation: planning-programming-budgeting.

PLANNING PHASE

The first phase — military planning and requirements determination — we envisioned as a continuing year around operation involving the participation of all appropriate elements of the Defense Department in their respective areas of responsibility. We anticipated that the Joint Chiefs of Staff organization and the planners in the military departments would play a particularly important role in this phase. What we were looking for here were not just required studies in the traditional sense, but military-economic studies which compared alternative ways of accomplishing national security objectives and which tried to determine the one that contributes the most for a given cost or achieves a given objective at the least cost. These are what we call "cost-effectiveness studies" or systems analyses.

I had originally thought that once an approved 5-year program had been developed, the Joint Chiefs of Staff organization and the military planners in the departments would concentrate their attention on specific segments of the program which might require change, and that they would propose such changes whenever the need became apparent any time during the year. When these change proposals were approved, the 5-year program would be modified accordingly, thereby providing an up-to-date, long-range plan at all times. But I had given too little weight to the need to review and analyze, at least once a year, the entire long-range program in all its interrelated parts, rather than in bits and pieces during the course of the entire year. I must confess that the Joint Chiefs of Staff saw this need more clearly than I did. They wanted to make a comprehensive program review each year to take account of the latest changes in military technology and in the international situation; so did each of the military departments; and so did the secretary.

Accordingly, the planning-programming-budgeting process now starts with the Joint Strategic Objectives Plan prepared by the Joint Chiefs of Staff organization with the help of the military planners in the services. As I noted earlier, the format of this plan has been modified to bring it into harmony with the new program structure. Thus the Joint Chiefs of Staff have the opportunity each year to recommend to the secretary of defense on a comprehensive basis the military forces and major programs which they believe should be supported over the next 5 to 8 years. The secretary of defense in the spring of each year reviews these recommended forces and programs, makes preliminary decisions, and provides to the military departments what is called "tentative force guidance" to serve as a basis for the preparation of their formal change proposals to the official 5-year program. The principal "cost-effectiveness" studies are scheduled for completion at about the same time in order to provide the secretary and his principal advisors with information in depth on the most critical and difficult requirement problems.

I recall that the first list of these requirements projects was developed by Secretary McNamara and was known at the time as "McNamara's 100 trombones." These projects were assigned to the Joint Chiefs of Staff, the military departments, and various elements of the Office of the Secretary of Defense. One, for example, dealt with the question of how many strategic bombers and missiles we would need during the next decade to destroy priority targets. Another involved an examination of requirements for airlift and sealift to meet various contingency war plans and the most economical means of providing that lift. Still another dealt with the comparative advantages and cost of (1) refurbishing existing items of ground equipment, (2) replacing them with new equipment off the assembly lines, and (3) expediting the development of still better equipment. The secretary of defense still originates many of these requirements studies. Others are originated by the Joint Chiefs of Staff, the military departments, and various elements of the secretary's staff.

PROGRAMMING PHASE

The initial development of the programming system, the second phase, was an enormous undertaking, considering the short time allowed and the fact that we had to handle simultaneously three amendments to the fiscal year 1962 budget originally prepared by the preceding administration. The problem here was to sort out all the myriad programs and activities of the defense establishment and regroup them into meaningful program elements, i.e., integrated combinations of men, equipment, and installations whose effectiveness could be related to our national security objectives. These are the basic building blocks as well as the decision-making levels of the programming process. As I noted earlier, the B-52 bomber force, together with all the supplies, weapons, and manpower needed to make it militarily effective, is one such program element. Other examples are attack carriers, F-4 fighter wings, the manned orbiting laboratory development project, and recruit training. Wherever possible, program elements are measured in physical terms such as numbers of aircraft per wing, numbers of operational missiles on launchers, numbers of active ships, and so forth, as well as in financial terms, thus including both "input" and "output" – costs and benefits. Of course, such program elements as research projects can only be measured in terms of inputs.

Costs are measured in terms of what we call "total obligational authority" – the amount required to finance the program element in a given year, regardless of when the funds are appropriated by the Congress, obligated, placed on contract, or spent. Now, admittedly, this is something of a compromise.[10] It would be preferable to cost the program in terms of expenditures, or ideally in terms of resources consumed. However, the accounting difficulties appeared so great that we did not attempt that approach. Moreover, as long as the budget is in terms of obligational authority, the program must be, for the program has to be firmly anchored to the budget. We do not even find it necessary to cost individual program elements in terms of cash expenditures. We have a much better idea of the full cost of 100 Minuteman missiles, for example, than of the phasing of the actual expenditures year by year. And from the point of view of planning and decision making we are far more interested in the full cost of a program – in "cost to complete" – than in the precise phasing of the costs.

To tie in with the "branch points" at which critical decisions must be made, we subdivide program costs into three categories: research and development, investment, and operation (Figure 1). Because of the great expense involved in just developing a new weapon system to the point where it could be produced and deployed, a determination to go ahead with full-scale development is, in itself, a major decision. There are few major weapon system developments today that can be accomplished for less than $1 billion. For example, we will have spent $1.5 billion making two prototypes of the B-70. We have already spent $1.5 billion developing the Nike-Zeus antiballistic missile system and are now spending a comparable amount on the Nike X. We spent $2.3

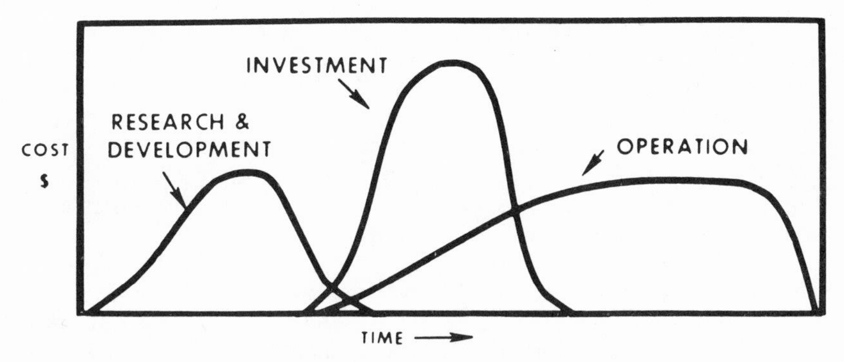

R&D: Development of a new capability to the point of introduction into operational use

INVESTMENT: Costs required beyond the development phase to equip forces with new capability

OPERATION: Recurring costs required to operate and maintain the capability

Figure 1
Cost Categories

billion developing the Atlas ICBM, $2.6 billion on Titan, $2.5 billion on Polaris, and $2.1 billion on Minuteman I. Even the development of a new torpedo can cost as much as $75 million. Therefore, we need to know in advance the likely cost of completing any major weapon development.

Obviously, before we go ahead with the next phase, production and deployment, we need to know the investment cost of providing initial equipment for the proposed forces. And, finally, we need to know the cost of operating the proposed forces each year. In many cases, for example a B-52 wing, the 5-year operating costs are about equal to the initial equipment costs, and in some few cases, for example an infantry division, the operating costs for just 1 year are actually greater than the initial investment costs.

To facilitate the conversion of program costs to the budget, and vice versa, we also had to break down the costs of each program element by the various budget appropriation accounts in which it is financed. Operating costs typically are financed in the "military personnel" and "operation and maintenance" appropriations and, where operating spares are involved, in the "procurement" accounts as well. Initial investment costs typically are financed in the "procurement" and "military construction" appropriations.

We have nearly 1,000 program elements. Where military forces are involved, they are projected 8 years ahead in order to provide the necessary lead time for the determination of the procurement programs. All other program data, both physical and financial, are projected 5 years ahead. For purposes of continuity, all program data are shown for each year beginning in fiscal year 1962; thus, our present "Five-Year Force Structure and Financial Program" extends from fiscal year 1965 through fiscal year 1970, with forces projected through fiscal year 1973. The entire program is subject to continual change and is, therefore, updated every other month. Whenever a change is made in the cost of a program element in the current fiscal year, it must also be reflected in the budget for the same year, and vice versa. Considering the vast quantities of data involved in the planning-programming-budgeting system, the only practical solution was to computerize the entire operation. This we have now accomplished.

The next task was to relate the program elements to the major missions of the Defense Department. The objective here was to assemble related groups of program elements that, for decision purposes, should be considered together either because they supported one another or because they were close substitutes. The unifying principle underlying each major program is a common mission or set of purposes for the elements involved. We now have nine major programs: (1) strategic retaliatory forces, (2) continental defense forces, (3) general-purpose forces, (4) airlift and sealift, (5) reserve and guard, (6) research and development, (7) general support, (8) retired pay, (9) military assistance. . . .

All the program data, together with the description of the forces, their tasks and missions, procurement lists, facility lists, and so forth, constitute, collectively, what we term "The Five-Year Force Structure and Financial Program." Since the data are machine processed, they can be summarized in different ways. For the use of top management in the Defense Department, we prepare and update at regular intervals a special summary volume which displays in tabular form the forces, financial, manpower, and procurement programs. The "Five-Year Force Structure and Financial Program" is formally approved by the secretary of defense and is binding for programming purposes on all components of the departments.

We recognized from the beginning that the defense program is extremely dynamic and that changes would be required at various times during the year. Accordingly, we established a formal program-change control system. The basic elements of this system involve the submission of program-change proposals by any major component of the Defense Department, their review by all interested components, the secretary's decision on each proposal, and, finally, the assignment of responsibility for carrying out this particular decision to the appropriate military department or agency. Hundreds of program-change proposals are submitted each year requesting changes involving billions of dollars.

BUDGETING PHASE

This brings us to the third phase of the planning-programming-budgeting process. It may be worth emphasizing at this point that the programming review is not intended as a substitute for the annual budget review. Rather, it is designed to provide a Defense Department-approved program to serve as a basis for the preparation of the annual budget as well as guidance for future planning. In the budget review we go into greater detail, for the next year of the 5-year program, on procurement lists, production schedules, lead times, prices, status of funds, and all the other facets involved in the preparation of an annual budget. And, as I pointed out earlier, we still manage our funds in terms of the appropriation accounts as well as in terms of the program structure. Essentially, the annual budget now represents a detailed analysis of the financial requirements of the first annual increment of the approved 5-year program.

Thus we have provided for the secretary of defense and his principal military and civilian advisors a system which brings together at one place and at one time all the relevant information which they need to make sound decisions on the forward program and to control the execution of that program. And we have provided the necessary flexibility in the form of a program-change control system. Now, for the first time the largest business in the world has a comprehensive Defense Department-wide plan that extends more than 1 year into the future. And it is a realistic and responsible one – programming not only the forces, but also the men, equipment, supplies, installations, and budget dollars required to support them. Budgets are in balance with programs, programs with force requirements, force requirements with military missions, and military missions with national security objectives. And the total budget dollars required by the plan for future years do not exceed the secretary's responsible opinion of what is necessary and feasible.

With this management tool at his command, the secretary of defense is now in a position to carry out the responsibilities assigned to him by the National Security Act, namely, to exercise "direction, authority, and control over the Department of Defense" – and without another major reorganization of the defense establishment.

REFERENCES

[1] "Special Message to the Congress on Reorganization of the Defense Establishment, April 3, 1958," *Public Papers of the Presidents, Dwight D. Eisenhower, 1958* (Washington, D.C.: Government Printing Office, 1959), p. 278.

[2] For an interesting example of this technique, see "Memorandum for the Secretary of the Army," June 8, 1960, signed by the Special Assistant to the

Secretary of Defense, Mr. O.M. Gale, which outlines the budget guidelines set forth by President Eisenhower at a cabinet meeting on June 3, 1960. Reprinted in the *Congressional Record,* June 30, 1960, vol. 106, Part II. pp. 15100-15101.

[3] *Hearings on S. 1269 and S. 1843,* U.S. Senate, Committee on Armed Services, 81st Cong., 1st Sess., April 6, 1949 (Washington, D.C.: Government Printing Office, 1949), p. 79.

[4] Letter from Representative George H. Mahon, Chairman of the House Subcommittee on Defense Appropriations, to Secretary of Defense Neil McElroy, Aug. 18, 1959 (unpublished).

[5] Letter from Representative Mahon to Secretary of Defense Thomas S. Gates, Sept. 6, 1960 (unpublished).

[6] Charles J. Hitch and Roland N. McKean, *The Economics of Defense in the Nuclear Age* (Cambridge, Mass.: Harvard University Press, 1960).

[7] Extract from the transcript of an interview with Secretary of Defense Robert S. McNamara on the National Broadcasting Company program, "Today," Feb. 17, 1961.

[8] Statement of Gen. Maxwell D. Taylor, *Hearings on Organizing for National Security,* Subcommittee on National Policy Machinery, Committee on Government Operations, U.S. Senate, 86th Cong., 2d Sess., June 14, 1960 (Washington, D.C.: Government Printing Office, 1960), p. 769.

[9] For example, see House of Representatives Report No. 1607, 87th Cong., 2d Sess., (House Committee on Appropriations Report on the Department of Defense Budget for Fiscal Year 1963), Apr. 13, 1962, pp. 4-7.

[10] For further discussion of program element costs as well as a description of the defense programming system in 1962, see *Programming System for the Office of the Secretary of Defense, 25 June 1962* (Washington, D.C.: Government Printing Office, 1963).

After we have solved the problems in the planning process . . . we must still actually implement the planned project or program. There remains the problem of actually scheduling the components of the program over time, of allocating program resources to the parts, and of keeping track of progress and rescheduling activities as conditions change and feedback accumulates. In community development and urban planning . . . [we] are most often in a situation where what we are going to do has not been done before, at least not in quite the way that we want to do it or under the same conditions and constraints. Therefore, we need a technique to estimate the best way of proceeding. . . .

PERT [Program Evaluation and Review Technique] involves the definition of the various tasks that must be executed as part of a project or plan, linking these together in sequences, and then using time estimates for the completion of each task to study the possible ways of reallocating time resources among the tasks to expedite completion of the whole project. CPM [Critical Path Method] works with both time and cost estimates to adjust schedules efficiently under crash conditions. . . .

Donald A. Krueckeberg and Arthur L. Silvers, Urban Planning Analysis: Methods and Models, *New York (John Wiley) 1974, pp. 231, 232.*

20

P. Wood

THE USE OF A PROGRAM PLANNING AND BUDGETING SYSTEM TO IMPROVE THE COORDINATION AND IMPLEMENTATION OF PHYSICAL, ECONOMIC, AND SOCIAL PLANNING

Program planning and budgeting (PPB) is concerned with decision making. It is not a technique but a systematic approach – a sequence of steps – that can be adopted by management to improve the decision-making process. Local government is big business, for a city the size of Liverpool is likely to spend £2,000 million in providing services over a 10-year period. This is not to say that money is plentiful. Indeed the shortage of resources has become especially acute in recent years, partly as a result of the general economic situation and partly due to the increasing demands for local authority services. The range of problems which confront an authority are often new, complex and closely interrelated. Their solutions may be unclear, but are likely to be costly, long-term, and have a variety of side effects. This is a situation which is ripe for the introduction of new management methods that meet the needs of the area by providing better quality services in the most cost-effective way.

Selected from Papers and Proceedings, Volume II, *International Congress, International Federation of Housing and Town Planning, Copenhagen, Denmark, 527-535 (Sept. 1973).*

A NEW MANAGEMENT STRUCTURE

The introduction of PPB in Liverpool dates back to 1968 when McKinsey & Company Incorporated were commissioned by Liverpool Corporation to develop a new management system for the authority. Attention was concentrated initially on the structure of committees and departments, where the aim was to provide a more positive relationship between elected member and officer, and to develop a structure which would be better able to meet the requirements of the new management system. Whereas the council had operated through some 27 committees and 32 subcommittees, it adopted a new structure comprising only 10 main committees. In a similar way the number of departments was dramatically reduced, bringing together those functions with common overall objectives. A distinction was drawn between program (or spending) departments and functional departments, which provided professional services.

Six program departments were created: housing, personal health and social services, education, transportation and basic services, environmental health and protection, and recreation and open space.

Six functional departments provide a range of services to the organization: city treasury, city planning, land and property services, administrative services, medical officer of health, and program planning – a new department specifically created to service the management system.

The heads of the 12 departments form a management team headed by a chief executive. All chief officers meet together on a monthly basis, although a smaller group comprising the chief executive, the treasurer, the planning officer and the director of program planning meet more frequently.

THE MANAGEMENT SYSTEM

The coordination of services at chief officer level is insufficient unless it relates to a policy-making framework. PPB provides that framework by assisting elected members and chief officers in drawing up a corporate plan, in taking decisions on the use of resources, and in monitoring performance and revising the plan. The basis of PPB is a logical sequence of steps (Figure 1), although decisions at each stage can lead to modifications of earlier stages.

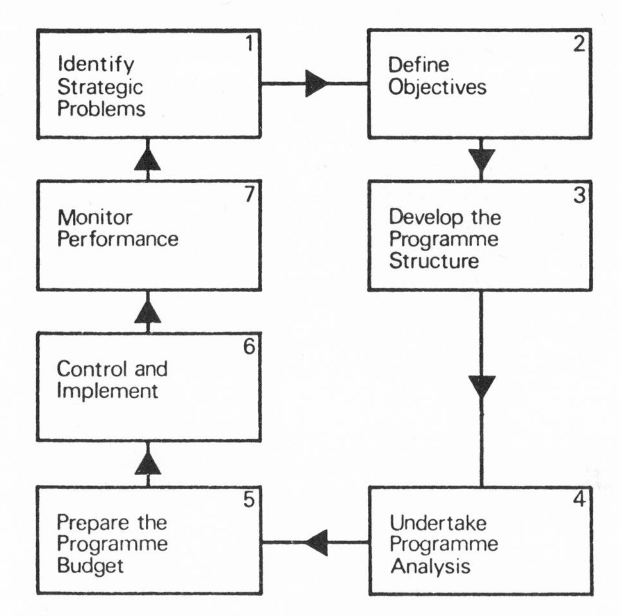

Figure 1
The sequence of the Program Planning and Budgeting System (PPBS).

The elements of the process are the following:

1. *Identify strategic problems:* information is needed on the nature and scale of problems facing the area, reflecting both needs and aspirations. It is often important to distinguish between symptoms and the underlying causes of problems, although both may require treatment.

2. *Define objectives:* objectives need to be defined explicitly and in a quantifiable way, as a yardstick against which the effectiveness of alternative policies can be measured.

3. *Develop the program structure:* broad objectives can be broken down into a series of subobjectives, each of which may have a series of activities which contribute to their attainment. The grouping of objectives, subobjectives, and activities into a hierarchy forms the program structure. The program structure should also include the activities of other organizations which contribute to the objectives of the local authority.

4. *Program analysis:* it is necessary to examine alternative courses of action before a policy decision

can be made. Having defined the objectives in a quantified form, it is possible to access the effectiveness of each alternative in meeting them. Program analysis may take the form of an issue paper which demonstrates, say, the best level of service that can be provided for a given sum of money. This provides the decision maker with relevant information on levels of performance, impact, and the consequences of each alternative course of action.

5. *Prepare the program budget:* this is the link to the budgetary process and is the stage when major decisions are taken, when the demands and supply of resources must be reconciled. Expenditure is classified according to the program structure, with costs and outputs identified for the next year and the subsequent 5-year period. The annual budgetary process then involves rolling the plan forward each year.

6. *Control and implementation:* it is necessary to ensure that the proposals in the plan are translated into action. This may be achieved through a variety of management techniques, such as critical path analysis.

7. *Monitor performance:* the assessment of need (for local authority goods or services), input (costs and other resources), and output (the achieved performance) requires careful and continuous measurement. The PPB approach, relying heavily on quantified methods, needs to be supported by a management information system.

THE ANNUAL CYCLE

When the stages of the program budgeting system are related to the budgetary process of local government, they result in an annual cycle of events (Figure 2). In Liverpool the cycle commences during the spring when the chief executive and other functional officers advise the policy committee on overall objectives and spending priorities. During the same period, program departments review past performance and revise their program structures. These are

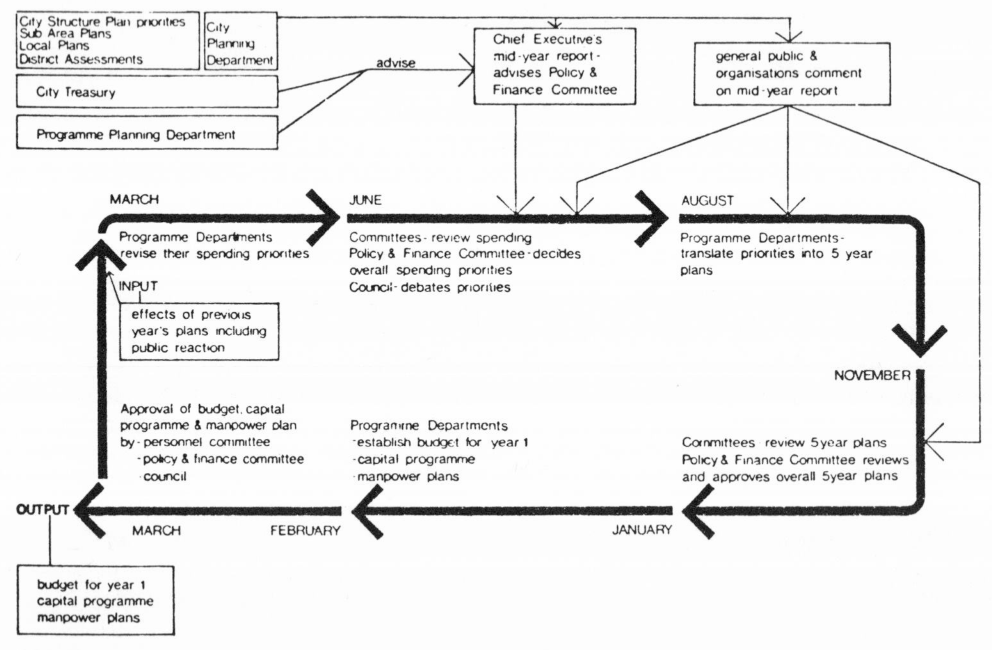

Figure 2
The program planning and budgeting cycle.

subsequently modified to take into account the decisions of the policy committee, and are then further refined into the detailed capital programs and manpower plans for each department.

CORPORATE PLANNING AND THE ROLE OF FUNCTIONAL OFFICERS

So far the description of program budgeting has concentrated on how the process is applied to program departments. There its attention is focused on improving the quality and efficiency of services such as housing, education, and transport. Their strands of policy form a series of parallel threads which must be woven together to form the fabric of the corporate plan. It is the responsibility of the functional officers to perform that weaving operation, ensuring that individual program department proposals are properly integrated into the total policy plan, and taking into account factors external to both the program department and the local authority.

The basic factors used in synthesizing these programs are the resources of land, labor, and capital. The functional officers most intimately concerned are the treasurer on the availability and use of finance, the head of administration on staff resources, and the planning officer in relation to the availability and use of land. His concern extends to the effects of economic and social activities which take place on land, both at the citywide scale and at the smallest local level of action.

THE PARTICULAR ROLES OF THE PLANNING OFFICER

The planner wishes to influence development from a number of standpoints. He needs to take a synoptic view of the elements in the process and relate them to broader or higher-level objectives. His concern may extend beyond physical planning; for example, in Liverpool the city is already heavily built up, and the main issue is not the design of new developments but the improvement in the quality of life for people already there. The planner is concerned therefore with a range of criteria extending across program areas, which I shall loosely describe as "locality" objectives. These would include the following:

1. The efficient use of land and buildings, for example, the multiple use of school facilities for community purposes.
2. Choice and opportunity within localities, advising on the range and timing of facilities to be provided.
3. Balance between districts, by identifying the special priorities of multiply deprived areas.
4. Environmental quality, by minimizing the harmful effects of development and seizing opportunities to improve and enhance the present environment.
5. Encouraging contributions from the private sector, through promotional activities, participation, and the control of development through planning applications.

The planner must assess the proposals of each program department in the light of these "locality" objectives. The *cross effects* must be identified for they may have fundamental repercussions on the program priorities. For example, the phasing of a school rebuilding project may be amended to relate more closely to the needs generated by the housing redevelopment program, and its design may be modified so that it is able to accommodate wider public use outside school hours. Cross effects may be of such significance that the original proposal may be deleted from the program. The obverse effect may occur through the creation of opportunities which induce new proposals. Slum clearance may provide an opportunity to improve open-space standards, while a road improvement scheme may open up land for development. Opportunities such as these must be reflected in the program budgets.

In Liverpool, the city planning officer has sought to influence program departments both during and after the preparation of the program budgets. In addition to providing an information service, he has prepared planning guidelines for each district; these are brought up to date annually, and are available to program directors when compiling their programs. After the program structures have been worked out in detail, the proposals are vetted by the planning officers as a district assessment exercise. Each district in the city is examined in turn, to identify deficiencies and to indicate where individual proposals may be getting out of step.

AREA MANAGEMENT

The process of influencing program departments through "locality" objectives has been given further impetus through initiatives from central government. A number of districts containing problems of multiple deprivation have been selected for special assistance from the government in the form of community development and urban aid projects. Of special interest is the government-sponsored "Inner Areas Study" in which an area has been selected for intensive treatment – the so-called "total approach" to the problem of the urban crisis.

The logical extension of this approach would result in area budgeting, giving the city council the problem of deciding how much money should be allocated to each district. Such an approach would give greater opportunities for community involvement to help determine local spending priorities.

SETTING OVERALL OBJECTIVES

Returning to the city scale, the setting of overall objectives and the definition of priorities among services is fundamentally a political decision. No techniques are available to show whether money is best spent on building more motorways or employing more social workers. Benefits are not quantifiable, and in any case affect different sections of the community. In an attempt to assist the policy committee in this most crucial decision, the city planning officer has undertaken a number of consultations with the general public to help identify key issues to be tackled. A variety of methods of consultation has been attempted, including the distribution of a booklet "Liverpool Prospects" to organizations and other interest groups for their reaction. More recently, in connection with the Merseyside Structure Plan, aims for the area are being identified through home interview techniques and by questionnaire surveys of local elected representatives.

CONCLUSION

Many lessons can be learned from the Liverpool experience. The consultant's recommendation was for quick and widespread adoption of the system – the "big-bang" approach – rather than partial introduction. This has some logic because it is a system of "corporate" management, but it was inevitable that teething troubles would occur. Departmental reorganization took place with almost indecent haste but was broadly successful. Familiarization with the new system by both elected members and officers was less successful, and this points to the need for extensive educational/training programs when introducing new management systems. In particular, more care is needed in explaining to elected members their role in the process, which differs significantly from their past experience. Rather than being concerned with details, they now have a crucial part to play in setting the objectives for the authority. This requires a new type of dialogue between member and officer to ensure that issues are fully understood, and that policy alternatives are explored in depth. In the absence of such involvement many elected members in Liverpool have felt remote from the system, and feel that key decisions are being taken by officials. This has led to some disenchantment with the system, and some reversion to "input" budgeting.

In concluding, it is necessary to repeat that a program budgeting system will not make decisions. It is a process to help the decision maker, the elected member. The planning officer can have a major influence in that process, both by advising the elected member, and by guiding and synthesizing the proposals of individual program departments. His task will become even more important after local government reorganization, which will replace the "all-purpose" authority (like Liverpool) with a two-tier system of local government. In addition, separate bodies will deal with water supply, health, gas, and electricity services. The presence of such a variety of organizations makes the introduction of PPB systems all the more difficult. However, without such coordination and planning of activities, the achievement of corporate objectives for such an area will never be accomplished.

A SHORT BIBLIOGRAPHY ON PPBS

Planned Programme Budgeting – A System of Management for Local Authorities. LAMSAC, 35 Belgrave Square, London.

The Planning-Programming-Budgeting Approach to Government Decision Making. Harold A. Hovey (Praeger Special Studies in U.S. Economic and Social Development). Praeger, New York, 1968.

Planning Programming Budgeting for City State County Objectives. Notes 1-8 (bound as one volume), 9, 10, and 11, Local Finance Project, George Washington University.

The Analysis and Evaluation of Public Expenditures: The PPB System. Government Printing Office, Washington, D.C., 1969.

Liverpool Prospects. F.J.C. Amos, City Planning Officer, Wilberforce House, 25 The Strand, Liverpool.

Master Plan. DEAD AT THE AGE OF 3, UNLOVED.

After years of hullabaloo and hearings, the controversial Master Plan for New York City was quietly buried by the agency that proposed, pushed, and praised it, the City Planning Commission. . . . The ambitious plan, developed over a period of three years at a cost to the city of nearly a million dollars was designed to be a guideline to rejuvenate the city's dying areas and to spur commercial development. . . . In addition to community criticism . . . two factors seem to be behind the reversal: New York has changed since the plan was conceived . . . necessary federal financing, $5.5 billion a year for 10 years, is not forthcoming.

Pranay Gupte, The New York Times, *10 June 1973.*

Superior Judge David A. Thomas summed up what's wrong with land-use planning in Los Angeles County when he voided the county's first general plan. Thomas ruled that the environmental impact report (EIR) on which the plan was based consisted on "no more than a sterile declamation of unsupported generalities almost entirely failing to convey any factual information . . . almost the entire EIR consists largely of pretentious statements of the obvious" . . . in preparation for nearly a decade . . . the plan was supposed to serve as a 20-year guide for the granting of zoning ordinances and building permits, and the setting of important density requirements. And it would have continued being used as a badly flawed planning yardstick. . . .

Editorial, The Los Angeles Times, *18 March 1975.*

PART VI ANALYSIS

an'a·lyze, an'a·lyse *(ăn'a·līz), v.t.; AN'A·LYZED or -LYSED (-līzd); AN'A·LYZ'ING or -LYS'ING (-līz'ĭng). [F.* analyser. *See ANALYSIS.]*
1. *To decompose or resolve into elements or constituent parts; as to* analyze *light by means of a prism; to* analyze *a fossil substance. "No one, I presume, can* analyze *the sensations of pleasure or pain."* Darwin.

2. *To separate mentally the parts of (a whole) so as to reveal their relation to it and to one another; to distinguish the elements of (a thing) in showing its structure or elucidating its meaning or essential nature; to examine critically for the sake of understanding the nature or organization; as, to* analyze *a personal impression, an economic theory.*

3. *To study the factors of (a situation, problem, or the like) in detail, in order to determine the solution, outcome, etc.; as, to* analyze *a position in chess. . . .*

Webster's New International Dictionary of the English Language, *Second Edition, Volume 1, Springfield, Massachusetts (Merriam) 1960, p. 95.*

21

Melville C. Branch

GOALS AND OBJECTIVES IN CIVIL COMPREHENSIVE PLANNING

Since the purpose of planning is to produce or favor predetermined consequences, establishing objectives is part of all planning and indeed its core consideration. Surprisingly little attention, however, has been devoted to the *analytical process* of identifying, formulating, and selecting objectives in planning. Many diverse objectives are suggested for corporate or city planning, but all too often the requirements for their progressive attainment have not been examined. Ends and means are discussed separately, not as one analytical continuum. Because of delayed recognition of the central importance of planning analysis as a basis for setting objectives, and because a method of conducting this difficult analysis has not yet crystallized, a theory and principles of establishing objectives in different forms of planning are lacking.[1]

This discussion seeks to clarify consideration of goals and objectives in planning by their further definition. By careful reference to the nature of the planning process, its function and practice in business

---

 Selected from OR 69, *John Lawrence, ed., Proceedings of the Fifth International Conference on Operational Research, Tavistock, London, 55-63 (1970).*

and other institutions, an underlying approach and general methodology are identified. These constitute a possible theoretical basis and suggest certain rudimentary principles for the formulation of objectives in comprehensive planning.

Just as most scientific inquiry starts with knowledge or speculation concerning the physical universe, planning theory and principles are suggested and confirmed by relating them to the real world. Abstract theory is intellectually helpful, but planning is an applied art as well as an applied science in the making. Its purpose and practice are reflections of the organism whose development it is intended to direct.

The literature and practice of planning are still characterized by wide differences of meaning in the use of even the most fundamental terms commonly employed to describe and discuss the process. This is the case today with the word *objective* and similar words such as *goal,* target, or aim. For the substantive content of planning to continue to advance, some order and uniform usage must evolve from the present semantic diversity. Since this paper intends to clarify the meaning and implied content of the terms goal and objective in planning, by relating them to the planning process, the words used to discuss this process require definition.

USE OF TERMS

1. *Organism* means the entity to which planning is applied: business enterprise, city, region, nation, military activity, or other unity seeking to shape its own future within the large context of which it is part.

2. *Planning* is the deliberate, organized, continuous process of identifying different elements and aspects of an organism, determining their present state and interaction, projecting them in concert throughout a period of future time, and formulating and programming a set of actions and plans to attain desired results. "The process itself is the actuality." Planning involves *internal* actions and events subject to partial or complete control within the organism, and those *external* to the organism over which it has limited or no control. Although this distinction is not absolute for many considerations and occurrences, it serves to identify what the organism can bring about on its own initiative. A wide range of probabilities and judgments must be developed for the diverse actions and events involved in planning: some physically tangible, others comparatively intangible, some subject to meaningful quantification, others difficult or impossible to measure at present in mathematical terms.

3. *Planning analysis* is the approach, method, procedure, and body of techniques used to investigate the past and present condition of the organism, determine the basic way in which it functions, project it into the future, formulate what is to be accomplished and attempted, and finally specify the actions to be taken. Numerous techniques are employed, many of them developed originally by other fields of concentration, but the total process of analysis, conclusion, decision, and implementation constitutes the substantive core of planning. Although different kinds of organisms require special knowledge and analysis fitted to their particular characteristics and requirements, there are basic similarities of approach, a sequence of study, techniques of analysis, and methods of programming effectuation which are common to all forms of planning.

4. *Comprehensive planning* embodies conceptually and analytically as many as possible of those most essential elements within the primary control of the organism itself which determine its development. It is planning for the totality rather than one or several of its parts, system rather than subsystem planning. It incorporates the best estimates that can be derived concerning pertinent events external to the organism.

5. A comprehensive *corporate plan* in business, and a *comprehensive, master,* or *general plan* in civil governmental planning, is the formulation into a set of interrelated policies and sequential actions of the results of continuing analysis and decision with respect to the future development of the organism. It is the current, adopted statement of analytical resolution, intent, strategy, programmed accomplishment, and expected results periodically reexamined to determine what modification is necessary or desirable – but subject to immediate revision whenever called for by emergency conditions or unexpected events of major import. It is a principal measure of institutional and managerial performance.

6. *Operational plans* are the specific and more detailed instruments required to realize part of a corporate or master plan. They are shorter-range and finite, with a predetermined beginning and outcome. Examples are a program and procedure of management reorganization within a certain time, or plans and specifications for manufacturing or service production of a given quantity, quality, and cost. Operational plans provide the more precise instruction, detail, and specification necessary to carry out portions of the comprehensive plan for the organism as a whole.

7. Planning *programs* show the precise sequence and interrelationship of actions and consequences which lead to a desired planning objective. Critical-path programs or schedules are familiar examples. Operational plans may be components of a planning program, or a planning program may specify the sequence of actions which will effectuate an operational plan. The word *target* is often used to designate specific accomplishments at particular points in time within a planning program.

PLANNING ANALYSIS AND THE DETERMINATION OF OBJECTIVES

In comprehensive planning, organism, ends, and means are one totality. Objectives for such planning emerge from what thorough analysis reveals concerning the nature, needs, and potentialities of the entity for which they are developed, and the feasibility of their progressive attainment. The general content of this examination is represented by the following sequence of analysis and resolution:

1. Nature, structure, and dynamics of the organism.
2. Historical facts and trends descriptive of its past and indicative of its probable future.
3. Socioeconomic, institutional, cultural, technological, and other internal and external limitations and potentialities.
4. Human and material resources available for directed change.
5. Existing realities which are desirable for retention or impractical to change within the planning period.
6. Commitments for the future, established by previous actions and events, available resources, and existing realities.
7. Tentative selection of objectives – immediate, short term, near term, intermediate, and long range.
8. Projection of the probable consequences of the application of available resources to different sets of objectives.
9. Selection among or amalgamation of several most desirable and feasible courses of future action and intent.
10. Derivation of a current statement of strategy, policy, specific objectives, and accomplishment for the planning period – constituting, together with the programs immediately below, the comprehensive, corporate, master, or general plan.
11. Programs of priority action and progressive implementation.
12. Periodic and emergency review and reanalysis – modification or revision of intent, objectives, and comprehensive plan.

The period of time covered by the comprehensive plan varies with the kind of organism, its situation, and the extent of directed development which experience and planning analysis reveal to be desirable and possible of attainment. Normally, different elements in a comprehensive plan are projected over different periods of future time or with different degrees of certainty. For example, in corporate planning, loss of personnel through mandatory retirement and depreciation ordinarily are projected further into the future than profits. Debt is projected more accurately than sales or management efficiency.

TYPES OF PLANNING OBJECTIVES

From an analytical viewpoint, comprehensive planning objectives are of several different types:

1. *Established:* for example, the necessity for a corporation to retire or refinance debt, or for a city to maintain essential municipal services.
2. *Alternative choice:* continued investment by a business in prolonging sales of an existing product versus its development of a new product, or the improvement of transportation facilities in a city versus expanded social services.

3. *Possible:* assuming a combination of favorable events, not now existing and unlikely for some time to come: dominance of a product market by a private enterprise currently accounting for a small fraction of the total market, or eradication of substandard housing in a city.

Planning objectives also differ in the probable time required for their attainment (lead times):

1. *Short term,* 1 to 5 years: for example, a corporation or city managing its liquid cash reserves so as to optimize percentage return.
2. *Near term,* 5 to 10 years: completing a large physical project or basic management reorganization.
3. *Intermediate,* 10 to 15 years: developing, manufacturing, and marketing a complex new product such as a new aircraft or next-generation electronic computer; acquiring right-of-way and constructing a new urban freeway.
4. *Long range,* 15 to 25 years or longer: extensive product diversification by a large corporation without merger or acquisition, first yield from commercial tree farming, or extensive change in the industrial base of a metropolitan city.

DETERMINATION OF OBJECTIVES AS PART OF COMPREHENSIVE PLANNING ANALYSIS

The development of an official master plan for an existing organism involves goals and objectives at each simulative step:

1. *Representation of the current state:* an expression, as concise as consistent with necessary coverage, which depicts the recent development of the organism and its current state within the environment in all essential respects. It embodies explicit and implicit objectives established in the past.
2. *Projection of current state:* showing probable development of the organism at successive stages in the future, assuming its internal situation, external environment, present objectives, policies, and planning remain the same. This provides an analytical base for identifying the effects of the modification described immediately below.
3. *Modification of projected current state:* incorporates known circumstances both within and external to the organism which will affect its future significantly. This simulative step determines directions of necessary and desirable action, and presents a partial picture of probable and possible change by comparison with step 2.
4. *Range of possible development:* optimistic representation of the private enterprise or other organism under exceptionally favorable and unlikely conditions 15 to 20 years hence. This encourages a greater range of imaginative consideration, and identifies unusual paths of opportunity to be followed if the opportunity occurs. A pessimistic projection portraying the situation under a combination of adverse circumstances can be made to complete the range of possible occurrence.
5. *Comprehensive plan:* an adjustment between steps 3 and 4. Incorporates a combination of established, necessary, desirable and possible objectives at successive future times. It is the formal statement of prescribed and accepted activity, direction, and intention for the organizational entity – periodically reviewed and reextended into the future, but subject to revision whenever indicated. It incorporates short- and near-term, intermediate, and long-range objectives which are often expressed in the form of a 1-year budget, 2-year operations plan, and progressively longer range extensions of internal activity and forecasts of external developments.

Planning objectives are those intentions which have been identified, examined, and adopted by such an analytical process. For small organisms with limited resources, sophisticated and expensive analysis is not necessary. What is essential is that the approach, analysis, and final formulation reflect this progressive sequence of consideration.

When the organism and its planning are analyzed in this way, objectives are usually limited; the range of possible change and attainment is greatly narrowed. The extent of previous commitment and investment reveal that a relatively small proportion of the total resources of the organism is available for new endeavors. Most are absorbed in maintaining existing activities, completing developments already underway, or undertaking new projects previously determined to be desirable whenever they can be initiated. The further the organism is projected analytically into the future, however, the greater the possibility of effecting extensive change by the cumulative application of surplus resources to this end, as they become available and provided interim developments permit continuation of the program of change.

There are, of course, exceptions to this general rule. Planning objectives representing rapid or radical changes in the organism and its achievements may be adopted if the higher risk is necessary or acceptable, such as a large corporation deciding to completely change the nature of its business within several years. Or internal resources may be withheld or drawn from some objectives and concentrated on others, if this is feasible and does not endanger the effective functioning or survival of the organism. There are circumstances when such departure from the usual analytical conclusion constitutes the soundest planning, such as a technological advance which quickly obsoletes the principal product of a business, or a serious socioeconomic situation or natural catastrophe in a city which demands immediate concentrated attention and a disproportionate share of available resources for a considerable time.

PLANNING GOALS

Purposes which cannot be specified precisely enough to be treated as planning objectives are differentiated by use of the word goals. *Goals are desires or intentions, whose nature is so general and hopeful, and whose attainment is so indefinite and distant, that they cannot be expressed and programmed as part of the comprehensive plan.* They are sufficiently fundamental to preclude their sudden change or abandonment, but opinion concerning their relative importance can shift more rapidly. They set the direction toward which planning objectives are oriented.

Goals are exemplified in corporate business planning by such general or distant intentions as the most successful enterprise in the industry, an excellent public image, satisfied employees, outstanding management, or the development of new technologies and very different products. Illustrative goals for comprehensive city planning might be a municipal economy providing full employment within the city, a healthy urban environment, widespread public participation in local government, rapid and inexpensive public transportation in all parts of the city, a beautiful community, or the elimination of excess outdoor advertising.

GOALS AND OBJECTIVES, UNCERTAINTY, AND COMPREHENSIVE PLANNING

Since planning analysis is so much easier and more conclusive when longer-range objectives are clear-cut and firmly set, it is tempting to search for absolutes. Unfortunately, absolute goals are few and far between. Because of the great diversities among people and continually changing conditions in the real world, there are few specific goals shared by a majority which can be stated with sufficient precision for inclusion in a corporate or urban master plan and program. As a consequence, it is important to remember that elegant quantitative analysis for comprehensive planning purposes is meaningful and usable in practice only if the underlying assumptions concerning human behavior, needs, value judgments, and decisions are valid. Awareness of goals helps to avoid analytical overelaboration.

Another form of searching for certainty which continually confronts the professional planner springs from the logically "open-ended" context of planning. To illustrate: a sizable business corporation is inseparably linked with one or more market levels; its planning is clearly related to population growth, economic and financial conditions generally, prices, interest rates, consumer-customer requirements, tastes and attitudes, employee skills, technology, government policies and politics, war and peace. For most companies, actual or potential competition anywhere in the world must be taken into account. Similarly, in city planning, a municipality is part of a metropolitan area, which is part of a larger local governmental unit and larger geographical or socioeconomic region, which in turn is within one or more even larger political-governmental jurisdictions. The latter are certainly affected by the nation, which in its turn is subject to international influences.

Therefore, for an absolute answer to many of the basic assumptions underlying comprehensive planning for a corporation or city, analysis should logically include all activities and events throughout the background spectrum of cause and effect. For a long time to come, our knowledge and techniques will be far too limited to encompass this scope. Furthermore, by the time such a complete sequence of analysis was accomplished, changing circumstances in the real world would have invalidated the results for current planning purposes, as well as any justification for such exhaustive and prolonged study from a cost-benefit point of view. Complete real-time analysis of the entire context logically related to corporate or comprehensive city planning is difficult to imagine because of its enormous intellectual difficulty. Not only is it difficult to handle a multitude of diverse interdependent variables and probabilities simultaneously, but we cannot treat mathematically the wide range of dissimilar elements and considerations needed for comprehensive planning. Some cannot yet be meaningfully quantified. When the relevance, reliability, and processing of necessary information are also taken into account, together with the complexities and subtleties of the decision-making process itself, sound analysis for comprehensive planning is indeed a formidable intellectual challenge. We do not now have even a rudimentary theory and principles for planning complex organisms at this level of consideration. Even if such analysis were possible, it is most doubtful whether we have at this time the human-institutional capabilities to make use of it.

By its nature, therefore, comprehensive planning is conceived and conducted in terms of uncertainty, even in highly controlled societies and institutions. Besides fitting both the characteristics of the organisms which would benefit most from sound comprehensive planning and the process itself, the use of probabilities permits delineation or "parameterization" of the open-ended context by reducing the logical necessity of a progressively endless search for the ultimate and absolute input. On the one hand, establishing a set of long-range planning objectives which cannot be changed or whose change involves great waste, is invalid in the real world. On the other hand, objectives cannot be changed continuously, since a minimum stability or "staying on the charted course" is essential to realize a planning purpose.

There is an underlying principle of indeterminacy intrinsic in comprehensive planning. Planning can provide the combination of sufficient accuracy and enough flexibility which justifies formal application of the process. But by its nature, planning cannot provide at the same time the highest order of accuracy *and* the greatest flexibility which could be attained for each if they could be considered separately. Therefore, a comprehensive plan which purports to define the organism as it should be 20 to 25 years hence, and to program the actions during the intervening years which will produce this end result, presupposes complete a priori information and knowledge, capabilities of quantitative statement and of combined correlation and projection which do not now exist, no unanticipated developments or events, and an organism which can be tightly controlled for many years regardless of what happens.

Unfortunately, just such a basically mistaken approach or failure to cope with society and cities as they really are has produced hundreds of master plan brochures in municipalities throughout the United States during the past half-century, which have been largely ignored because they were conceptually invalid or unrealistic to begin with. Corporate planning is much younger than city planning, and the necessity to produce products or services, general economic pressures, and business competition should prevent the same mistake. If such fundamental errors of comprehensive planning do occur in business, it is likely to be by large enterprises which have become inefficient because of size, some artificial advantage, or monopolistic position.

As can be seen, setting objectives in comprehensive planning is an inherently difficult and complex task because it involves available resources, existing commitments for the future, internal operating requirements, external trends and events, numerous wants and wishes – and their many interrelationships. It is especially unsatisfactory today because as yet there are few if any generally accepted theories and principles of comprehensive planning, common definition or description of goals and objectives, uniform method for their expression, or analytical techniques which permit consistent quantitative correlation and comparison among the many different considerations which must be taken into account in comprehensive corporate and city planning. Until these are developed, understanding the dynamics of the organism must be achieved by partial quantitative-statistical and cost-benefit subsystems analysis, logical reasoning, confirmed experience, structured judgment, and continuous observation. Human decision makers and the people who comprise the organism are incorporated in the analytical system because their appraisal, acceptance, value judgments, or imagination are critical – as individuals, groups, or collective consensus among many. They are the *raison d'être* for planning, its protagonists and beneficiaries.

---

[1] *Since goals and objectives are part of all planning, the professional and technical literature is filled with material* relating *to them: general and specific ends suggested for a vast array of organisms and activities, means and methods of attaining particular purposes, and other matters having to do in some way with goals or objectives. However, in the United States at least, there are comparatively few materials which treat goals or objectives* sui generis, *from a universal point of view, or with a general theory or methodology for their formulation in mind. Examples of the limited literature of this type include, from business, references [1-4], from city planning, references [5-7], and from public administration, reference [8].*

REFERENCES

1. Igor H. Ansoff (1957) Strategies for diversification. *Harvard Business Review* **35** (5) Sept.-Oct., 113-24.
2. Charles H. Granger (1964) The hierarchy of objectives. *Ibid.*, **42** (3) May-June, 63-74.
3. Bertram M. Gross (1965) What are your organization's objectives? a general systems approach to planning. *Human Relations* **18** (3) Aug., 195-216.
4. Cyril O'Donnell (1963) Planning objectives. *California Management Review* **46** (2) Winter, 3-10.
5. Alan Altshuler (1965) The goals of comprehensive planning. *Journal of the American Institute of Planners* **31** (3) Aug., 186-95.
6. Brian J. L. Berry and Jack Meltzer, eds. (1957) *Goals for America.* Prentice-Hall, Englewood Cliffs, N.J., 152.
7. Robert C. Young (1966) Goals and goal-setting. *Journal of the American Institute of Planners* **32** (2) Mar., 76-85.
8. Charles E. Lindblom (1958) The science of "muddling through." *Public Administration Review* **19** (2) Spring, 78-88. (Reading No. 45)
9. Melville C. Branch and Ira M. Robinson (1968) Goals and objectives in civil comprehensive planning. *The Town Planning Review* **38** (4) Jan., 261-74.

22

James Hughes and Lawrence Mann

SYSTEMS AND PLANNING THEORY

The recent appearance of two major books in systems research and systems theory[6b, 2j] provides a landmark in the growing influence of "system" on fields closely related to planning. It may be an appropriate moment then to review some of the literature that has contributed to the convergence of systems theory with planning theory.

"The Social System" of Talcott Parsons has been part of the intellectual climate for nearly 20 years, and there have been a number of attempts to relate this approach to cities and urban communities[3,11,21]. A somewhat different systems perspective has grown out of operations research (systems analysis or engineering), and this too has had considerable impact on planning research. Britton Harris has attempted some highly ambitious extensions of these approaches to broader planning methodology. Systems analysis probably finds its broadest possible framework in what has become known as "general systems theory," and a number of planners closely

---

follow the various publications that appear in that framework[17]

The Concept of System

The idea of "system" has been centrally placed in human science for nearly 40 years. Morris Cohen wrote in 1931 that

> *Ordinary, pre-scientific, or common-sense knowledge is disconnected, fragmentary, and chaotic or illogical. Science is devoted to the ideal of system, in which these defects are to be overcome. Indeed, instead of saying that system is a characteristic of science like certainty, evidence, and proof, definiteness and accuracy, or abstract universality and necessity, we may well maintain that the one essential trait of developed science is system, and that all these other traits are incidental to it.*[12]

Cohen points to three main traits of scientific systems: (1) interconnectedness of parts, (2) completeness, and (3) logical order. As an illustration of this last characteristic, he notes that

> *The principle of Occam's Razor ultimately implies that systems themselves should not be unnecessarily multiplied (where they are not mutually alternatives). Indeed common-sense may be said to be less systematic than science because it consists of a greater number of (uncoordinated) systems.*
>
> LUDWIG VON BERTALANFFY:
> Foundations of General Systems Theory

Even before Morris Cohen's formulations of systems in the philosophy of science, the germ of modern systems theory was planted in biology where a resistance to "mechanistic reductionism" (whereby life could be understood only as it was reduced to chemistry and physics) developed during the 1920s. In opposition, wholistic notions that considered the *organism* the basic element of the life sciences were advanced, and one of the early proponents of the organismic conception was Ludwig von Bertalanffy.

The probelms of the basic disrepute of analogies in science and the widespread misuse of the organismic analogy outside biology drove Bertalanffy into the philosophy of science. His first major contribution to the theory of general systems in a German philosophy journal[2a, 1945] was followed a few years later by his major theoretical book[2b], which was quickly translated and started a major ferment in philosophy. The outline of general systems theory was published in a British philosophical journal in 1950[2c], and six years later the *General Systems Yearbook* made its first appearance[2h].

Bertalanffy describes general systems theory as follows:

> *It is a logico-mathematical field, the subject matter of which is the formulation and derivation of those principles which hold for systems in general. A "system" can be defined as a complex of elements standing in interaction. There are general principles holding for systems, irrespective of the nature of the component elements and of the relations of forces between them.*

He wrote of structural conformity or "logical homology" of scientific laws in different realms. Bertalanffy also made it clear that he sought no vague qualitative analogies:

> *The principles that hold for systems in general can be defined in mathematical language. . . . Notions such as wholeness and sum, progressive mechanization, centralization, leading parts, hierarchical order, individuality, finality, equifinality, etc., can be derived from a general definition of systems; notions that hitherto have often been conceived in a vague, anthropomorphic, or metaphysical way, but actually are consequences of formal characteristics of systems, or of certain system conditions*[2b]

Bertalanffy was, of course, careful to argue that systems theory transcends mere analogy ("superficial similarities in phenomena that correspond neither to the factors operating in them nor in the laws applying to them") and concentrates on logical *homologies* ("phenomena differ in the causal factors involved, but are governed by structurally identical laws"). One of the main claims is that general systems theory can help distinguish analogies from homologies and "lead to legitimate conceptual models and transfer of laws from one realm to another." It is argued further that

> *In sciences that are not within the framework of physico-chemical laws, such as demography and sociology . . . exact laws can be stated if suitable model conceptions are chosen. Logical homologies result from general system characters, and the reason for this is the reason why structurally similar principles appear in different fields. . . .*

Of the logical homologies Bertalanffy is talking about, the principle of least action is probably the best known; but there are a number of other examples. The many applications of the exponential law in both positive and negative directions are detailed. Another is Verhulst's logistic law where an originally exponential increase is limited by restricting conditions. This has a variety of applications, including human population growth in a limited living space. Still another logical homology is found in the parabolic law of competition within a system whereby allocation to component elements is according to capacity expressed as a specific constant. (Pareto's law of income distribution is the best known example in the social sciences.) Finally, there is the relaxation oscillation principle of neon light, nerve physiology, and biocenosis of organic communities.

Bertalanffy specifically rejects the use of concepts such as "equilibrium" outside physics and chemistry, calling them "loose, if ingenious, metaphors." He also rejects the casual uses of concepts such as "system," "gestalt," "organism," "interaction," and "the whole is more than the sum of its parts," unless they can be formulated in an exact language such as general systems theory. To provide some understanding of this language, it is worthwhile to sketch verbally several key concepts.

A *system* is a complex of interacting elements. *Wholeness,* as a system attribute, means that the system behaves as a whole, changes in every element depending on all others, as opposed to *independence.*

A characteristic of many biological, psychological, and sociological systems is *progressive segregation and mechanization* in which interactions between elements decrease with time and the system passes from a state of wholeness, to a state of independence of elements, to a series of independent casual chains. This implies a loss of regulatory ability, since no longer will a disturbance be followed by the attainment of a new steady state as a result of interaction within the system.

A major attribute of systems is *centralization,* in which an element may be a *leading part* of a system centered around it. When a small change in a leading part causes great change (*amplification*) in the total system, the part may be called a *trigger.* In this case, the system is characterized by *instigation causality.* Centralization and instigation causality may be especially prevalent in systems characterized by *hierarchical order,* in which the components of a system are themselves systems of a lower order.

Competition is universal; there is always a struggle between system parts according to rules. In the simple case these rules are simply exponential. In more complex situations, such as bioecology and probably socioecological systems, a distinction is necessary.

Open systems behave differently in time than do *closed systems.* Closed systems must eventually reach a state of *equilibrium.* Open systems may, in certain conditions, reach a *steady state* that appears constant but involves a continuous inflow and outflow of materials.

Finality, or the way a system reaches its ultimate state, is another way in which different kinds of systems behave differently. *Equifinality* is characteristic of living systems, in that the final state may be reached from various initial conditions and in different ways. Closed systems cannot behave equifinally. Even open systems have limitations on equifinal behavior, as in the case of irreversible disturbances or hierarchical organization and progressive mechanization. One other kind of finality discussed by Bertalanffy is what he calls *planning intelligence:*

> *. . . the true finality of purposiveness, meaning that the actual behavior is determined by the foresight of the goal. . . . It presupposes that the future goal is already present in thought and directs the present action.*

Systems Theory and Analysis

Recent developments involve application and diffusion of general systems theory into public policy and social science. This spread has relevance to planning in that, first, it can eventually provide greater understanding of specific material peculiar to urban and regional planning. In other words, the complex, dynamic, and highly interrelated cultural, social, political, and spatial phenomena that define the environment of planning may be modeled or formulated into quantified, understandable systems. Second, it can aid understanding of the process of public policy formulation and possibly provide the basis or model of an optimal policy formulation procedure.

Recent applications see general systems theory as a "level of theoretical model building which lies somewhere between the highly generalized constructions of pure mathematics and the specific theories of specialized disciplines . . . a body of systematic theoretical constructs which will discuss the general relationships of the real world"[8]. More specifically, "the systems approach is one in which we fit an individual action or relationship into the bigger system of which it is a part, and one in which there is a tendency to represent the system in a formal model"[25]. A system, according to Buckley, is "a complex of elements or components directly or indirectly related in a causal network, such that each component is related to at least some others in a more or less stable way within any particular period of time"[6a]. According to Churchman, the following considerations must be kept in mind when defining a system:

1. The total system objectives and, more specifically, the performance measures of the whole system.
2. The system's environment: the fixed constraints.
3. The resources of the system.
4. The components of the system, their activities, goals, and measures of performance.
5. The management of the system[6a].

"The core of the systems approach lies in the attempt to describe the problem in relation to a total structure of objectives, costs, and benefits. The systems analyst looks at his system not as an end in itself but as a means toward the achievement of certain objectives of some broader entity in which he is operating"[23]. Systems analysis may be considered the "touchstone of the systems approach . . . undertaken with a view to identifying rational decisions as to the design, selection, or operation of a system"[4]. The first phase of a systems analysis is creating, from basic objectives, quantifiable analytic terms "that are sought by some as-yet-undefined complex of equipment and/or activity, taking into account the environment in which it is to operate"[4]. After this is accomplished, the analysis consists of the following:

1. The creation of an analytically manageable model of the interrelationships between major elements of the system and the external world.
2. Quantification of functional relationships between rate of system operation and system "outputs."
3. Quantification of functional relationships between rate of system operation and system "inputs."
4. The combination of (2) and (3) into an overall input-output relationship that, submerging intermediate relationships within the structure of the model, expresses system outputs as a function of system inputs.
5. The determination from the input-output relationship of optimum system design and rates of inputs and outputs that correspond to an optimum operation of that system[4].

Thus "the essence of the systems approach in organizations is to construct a model of the means-end configuration by comparing optional courses of action systematically (quantitative when possible), using a generalized sequence that can be retraced by others"[18]. The closeness of the above statements defining "systems" hints at a close relationship with concepts of economic rationality.

SYSTEMS AND POLICY

With this background in mind, it is desirable to examine systems and public policy. "The general theory of systems implies methodology in its strict philosophical sense, and the methodology of systems is represented symbolically by the design of models. . . . Models of organizations have been developed and administrative theories formulated in the area of management science"[18]. "Concepts such as wholeness, hierarchical differentiation, orderliness, interdependence, open systems, and steady state behavior seem especially applicable to those formal social organizations that are governed by public policy"[7]. One important model in the above sense is the PPB system, a model for planning and executing public policy. PPBS "transforms the conceptual clarity of general systems theory into a methodological, policy making strategy for use in numerous and diverse contexts ranging from defense planning to the management of elementary schools. . . . It can best be described as a general theory of planology because it is appropriate for planning in nearly any type of organization"[18].

PPBS, planning, programming, and budgeting systems, include programs designed around long-range goals broken down into measurable objectives. All relevant agency activities are included in these programs according to the mission or objective to which they contribute: "The need for a wholistic approach arises from the indissoluble connection between the allocation of human and material resources, or budgeting, and the formation of public policy"[22]. Alternative means to achieve the desired ends are examined, and in PPBS's fully refined state, a model is available to combine policy making and budgeting. PPBS thus constitutes "the process by which objectives and resources (ends-means), and the interrelations among them, are combined to achieve a coherent and comprehensive program of action for any organization conceived as a whole"[18]. A basic assumption "is that a unit of government can determine its policies most effectively if it chooses *rationally* among alternative courses of action that are placed in order of priority on the basis of anticipated benefits and costs for each alternative"[18]. PPBS bears a remarkable similarity to the means-end approach of Meyerson and Banfield as outlined in the appendix of *Politics, Planning, and the Public Interest.*

PPBS, however, does have its limitations:

1. Political elements may act as a barrier to systems procedures.
2. Systems analysis may not be able to provide much assistance in solving problems of social change.
3. The systems approach is not a "mathematical messiah" that can quantify and analyze the entire output of a social organization.
4. Generic models must be altered to fit specific situations.
5. Elaborate analysis may be based upon poor data or questionable premises.
6. Systems procedures may foster a centralizing bias that induces individual freedom and privacy.
7. Most government units presently have inadequate staffs and insufficient resources for comprehensive systems planning.
8. Present efforts suffer from a strong tendency to measure only what is easily measured (goal distortion)[18].

Besides these limitations, most of the criticism of the pure rationality models seems valid. However, the development of systems approaches to the entire field of urban phenomena clearly brings alternative proposals — determinations of subsequent consequences — closer to realization than ever before. Equally as important is the ultimate *testability* of this theory of planning. It is a workable process that can be tested empirically as it is implemented at various levels of goverment. Several criticisms that were an inherent part of the rational means-end scheme are mollified by this approach.

PPBS recalls Friedmann's definition of planning: "Planning may simply be regarded as reason acting on a network of ongoing activities through the intervention of certain decision structures and processes. . . . 'Introducing' planning, then, means specifically the introduction of ways and means for using *technical intelligence* to bring about changes that otherwise would not occur"[16]. Assuming a high degree of correlation between intelligence and knowledge, then the most effective planning will be based upon the most effective supporting knowledge. This is, of course, the central hypothesis of Mann's theory: "The probability that a planning agency will attain a given level of successful accomplishment depends to an important extent on the degree of effort made to base proposals on supportive knowledge of specific kinds — this to hold for agencies with like size, workloads, and resources over a defined period of time"[19].

Another hybrid approach combining systems analysis and city planning is that of Catanese and Steiss: "Systemic planning, a framework for structuring the application of the systems approach to urban systems, seeks the formulation of maximal, minimal, optimal, and utopian level alternatives to urban growth and development"[8]. Combining systems analysis, operations research, and comprehensive planning, each of which is, or should be, concerned with methodological process rather than the substantive content of the application area, the systemic planning process fits systems and operations procedures into the political process so as not to produce unimplementable proposals. According to the authors, "The Systemic Planning process does not apply *strict* optimization techniques nor does it restrict freedom of choices. Rather it attempts to make planning a more meaningful procedure by evaluating the likely functional or performance characteristics to be derived from alternative policies, plans, and programs"[13].

Dror also uses "a systems approach that applied qualitative ideas and methods from modern management sciences to a complex social phenomenon, namely, public policy-making"[14]. He considers the policy-making system a "subsystem of the social system whose subsystems, such as the power subsystem, partly overlap the policymaking subsystem"[14]. Thus urban planning is a subsystem of policy making or rather a specific form of policy-making types. He creates an optimal model of public policy making with the following characteristics[14]:

1. It is qualitative, not quantitative.
2. It has both rational and extrarational components.
3. Its basic rationale is to be economically rational.
4. It deals with meta policy making (policy making on how to make policy).
5. It has built-in feedback.

The optimal model consists of three major stages that are closely held together by communication and feedback channels: (1) meta policy making (policy making on how to make policy), (2) policy making, and (3) post policy making.

The first major stage, meta policy making, sets down guidelines for specific situations for which policy makers will employ a certain form of decision making; that is, in what areas is it profitable to employ pure rationality if it is feasible, or will incrementalism suffice? Also, methodological instructions for policy making are set up – a planning-programming-budgeting system may be appropriate here. Realistic long-range goals, necessary to the second stage, are established. The main function of this first stage is to design and manage the system of policy making as a whole in order to establish basic guidelines and rules for decision making.

The second state is policy making, the actual process for formulating policy. Various phases of this stage essentially correspond to the economically rational model as conceived by Dror.

Third is the post-policy-making stage which is concerned with executing and evaluating policy. This final stage includes a comprehensive system of feedback to all levels of the model.

One must look at Dror's efforts in the light of his earlier statements concerning the entry of the economic approach into decision making through the introduction of PPBS at the federal level. In his article concerning the creation of policy analysts, he seems to express a fear of systems analysts moving into public administration.

As a result of these weaknesses, systems analysis as such is of doubtful utility for dealing with political decisions, overall strategic planning, and public policy-making. . . . What is needed is a more advanced type of professional knowledge, which can be used with significant benefits for the improvement of public decision making[15]. This knowledge would be based in a new professional discipline called policy analysis, which seeks to shift emphasis within public systems from quantitative methods to political and qualitative analysis. Dror states:

> *The one-sided invasion of public decision-making by economics was caused largely by the inability of modern political science and public administration to make contributions to government decision making. Economics developed a highly advanced action-oriented theory and put it to the test of innovating economic policy making. At the same time, the modern study of political science and public administration became sterilized by an escape from*

> *political issues into behavioral value-free research and theory, or exhausted itself in suggestions for insignificant incremental improvements on the technical level*[15].

Thus Dror's book may be considered as an effort to keep public administration and political science in the mainstream of public decision making or as a response to an invasion by a somewhat alien (to political science) and threatening animal, PPBS. However, this should not affect his contribution including consideration of political aspects of public decision making, extrarationality, creativity, and qualitative methods. Additionally, his concept of policy analysis is considered "a variant of planning" by Wildavsky, who defines policy analysis as "the sustained application of intelligence and knowledge to social problems"[24].

It seems that the concept of "system" has, as predicted by Bertalanffy, "become a fulcrum in modern scientific thought"[2].

> *The tendency to study systems as an entity rather than as a conglomeration of parts is consistent with the tendency in contemporary science no longer to isolate phenomena in narrowly defined contexts, but rather to open interactions for examination and to examine larger and larger slices of nature*[1].

This concept can help illuminate the dynamic, complex, and highly interrelated cultural, social, and political phenomena that comprise the substantive concerns of urban planning. If the phenomena are better understood in terms of quantifiable "systems" that facilitate prediction, then the consequences of alternative actions may be brought into clearer focus. This, then, makes decision making, and therefore planning, that much more effective. As a result, scientific approaches to planning and policy making become much more realistic.

REFERENCES

[1] Russell L. Ackoff. "Games, Decisions, and Organizations," *General Systems*, 1954.

[2] a. Ludwig von Bertalanffy. "Au einer allgemeinen Systemlehre," *Blatter fur Deutsche Philosophie,"* XVII Nos. 3 and 4 (1945).

b. ——. *Das Biologische Weltbild: Die Stellung des Lebens in Natur und Wissenschaft.* Bern: A. Francke A.G., 1949. English translation, *Problems of Life.* New York: Wiley, 1952.

c. ——. "An Outline of General Systems Theory," *British Journal of the Philosophy of Science*, I (1950).

d. ——. *Human Biology*, XXIII (1951), a special issue on general systems theory.

e. ——. "Philosophy of Science in Scientific Education," *Scientific Monthly*, LXXVII (1953).

f. ——. "Essay on the Relativity of Categories," *Philosophy of Science*, XXII (1955).

g. ——. "A Biologist Looks at Human Nature," *Scientific Monthly*, LXXXII (1956).

h. ——. "General Systems Theory," *General Systems Yearbook*, 1956.

i. ——. "General Systems Theory: A Critical Review," *General Systems*, 1962; and in Walter Buckley, *Modern Systems Research for the Behavioral Scientist.*

j. ——. *General System Theory: Foundations, Development, Applications.* New York: George Braziller, 1968.

[3] John E. Bebout and Harry C. Bredemeier. "American Cities as Social Systems," *Journal of the American Institute of Planners*, XXIX (May 1963).

[4] Guy Black. *The Application of Systems Analysis to Government Operations.* New York: Praeger, 1968.

[5] Kenneth Boulding. "General Systems Theory: The Skeleton of Science," *Management Science* (1956).

[6] a. Walter Buckley. *Sociology and Modern System Theory.* Englewood Cliffs, N.J.: Prentice-Hall, 1967.

b. ——. *Modern Systems Research for the Behavioral Scientist.* Chicago: Aldine, 1968.

[7] Rocco Carzo, Jr., and John N. Yanouzas. *Formal Organizations: A Systems Approach.* Homewood, Ill.: Irwin, 1967.

[8] Anthony J. Catanese. "Urban System Planning: Retrospect and Prospect," *High Speed Ground Transportation Journal*, III (Jan. 1969).

[9] Anthony J. Catanese and Alan W. Steiss. "Systemic Planning," *Journal of the Town Planning Institute*, LIV, No. 4.

[10] West Churchman. *The Systems Approach.* New York: Delacorte Press, 1968.

[11]Terry N. Clark. *Community and Decision-Making: Comparative Analyses.* San Francisco: Chandler, 1968.
[12]Morris Cohen. *Reason and Nature.* New York: Harcourt Brace, 1931.
[13]John W. Dickey and Alan W. Steiss. "Programming Models in the Systemic Planning Process," *High Speed Ground Transportation Journal,* III, No. 1 (Jan. 1969).
[14]Yekezkel Dror. *Public Policymaking Reexamined.* San Francisco: Chandler, 1968.
[15]Yehezekel Dror. "Policy Analysts: A Professional Role in Government Service," *Public Administration Review,* XXVII, No. 3 (Sept. 1967).
[16]John Friedmann. "A Conceptual Model for the Analysis of Planning Behavior," *Administrative Science Quarterly,* XII, No. 2 (Sept. 1967).
[17]*General Systems.* Yearbook of the Society for General Systems Research, Chicago: Annually, 1956 to present.
[18]Harry Hartley. "PPBS: The Emergence of a Systemic Concept for Public Governance," *General Systems,* XIII (1968).
[19]Lawrence D. Mann. *Practicable Planning Theory.* (Processed).
[20]Talcott Parsons. *The Social System.* New York: Free Press, 1950. Much of this work dates from the 1930s and it has continued through a number of publications to date.
[21]Irving T. Sanders. *The Community: Introduction to a Social System.* New York: Ronald, 1958.
[22]Arthur Smithies. *The Budgeting Process in the United States.* New York: McGraw-Hill, 1955.
[23]Michael Tietz. "Cost Effectiveness: A Systems Approach to Analysis of Urban Services," *Journal of the American Institute of Planners,* XXIV, No. 5 (Sept. 1968).
[24]Aaron Wildavsky. "Rescuing Policy Analysis from PPBS," *Public Administration Review,* XXIX, No. 2 (Mar.-Apr. 1969).
[25]Charles J. Zwick. *Systems Analysis and Urban Planning.* Chicago: The Rand Corporation, 1963.

. . . the project leadership had some persistent need to be shown how and why sophisticated analysis could do the job better, and actual management procedures in the context of planning a large, complex project were fluid, rough and ready, and therefore not amenable to rationalized, systematic analysis. The subtle dictum "Model simple, but think complex" is worth mentioning . . .

One must begin to take seriously the profound mismatches that exist between available analytic tools, the current quality and quantity of theory and supportive data, and what policy makers desperately require and are capable of understanding and utilizing.

Garry D. Brewer, quoted in Book Review of New Tools for Urban Management, Science, *Vol. 176, No. 4035, 12 May 1972, p. 648.*

23

Richard A. Johnson, Fremont E. Kast and James E. Rosenzweig

SYSTEMS THEORY AND MANAGEMENT

The systems concept can be a useful way of thinking about the job of managing. It provides a framework for visualizing internal and external environmental factors as an integrated whole. It allows recognition of the proper place and function of subsystems. The systems within which businessmen must operate are necessarily complex. However, management via systems concepts fosters a way of thinking which, on the one hand, helps to dissolve some of the complexity and, on the other hand, helps the manager recognize the nature of the complex problems and thereby operate within the perceived environment. It is important to recognize the integrated nature of specific systems, including the fact that each system has both inputs and outputs and can be viewed as a self-contained unit. But it is also important to recognize that business systems are a part of larger system – possibly industrywide, or including several, maybe many, companies and/or industries, or even society as a whole. Further, business systems are in a constant state of change –

they are created, operated, revised, and often eliminated.

What does the concept of systems offer to students of management and/or to practicing executives? Is it a panacea for business problems which will replace scientific management, human relations, management by objective, operations research, and many other approaches to, or techniques of, management? Perhaps a word of caution is applicable initially. Anyone looking for "cookbook" techniques will be disappointed. In this article we do not evolve "ten easy steps" to success in management. Such approaches, while seemingly applicable and easy to grasp, usually are shortsighted and superficial. Fundamental ideas, such as the systems concept, are more difficult to comprehend, and yet they present a greater opportunity for a large-scale payoff.

SYSTEMS DEFINED

A system[1] is "an organized or complex whole; an assemblage or combination of things or parts forming a complex or unitary whole." The term system covers an extremely broad spectrum of concepts. For example, we have mountain systems, river systems, and the solar system as part of our physical surroundings. The body itself is a complex organism including the skeletal system, the circulatory system, and the nervous system. We come into daily contact with such phenomena as transportation systems, communication systems (telephone, telegraph, etc.), and economic systems.

A science often is described as a systematic body of knowledge; a complete array of essential principles or facts, arranged in a rational dependence or connection; a complex of ideas, principles, laws, forming a coherent whole. Scientists endeavor to develop, organize, and classify material into interconnected disciplines. Sir Isaac Newton set forth what he called the "system of the world." Two relatively well known works which represent attempts to integrate a large amount of material are Darwin's *Origin of the Species* and Keynes's *General Theory of Employment, Interest, and Money.* Darwin, in his theory of evolution, integrated all life into a "system of nature" and indicated how the myriad of living subsystems were interrelated. Keynes, in his general theory of employment, interest, and money, connected many complicated natural and man-made forces which make up an entire economy. Both men had a major impact on man's thinking because they were able to conceptualize interrelationships among complex phenomena and integrate them into a systematic whole. The word system connotes plan, method, order, and arrangement. Hence it is no wonder that scientists and researchers have made the term so pervasive.

The antonym of systematic is chaotic. A chaotic situation might be described as one where "everything depends on everything else." Since two major goals of science and research in any subject area are explanation and prediction, such a condition cannot be tolerated. Therefore, there is considerable incentive to develop bodies of knowledge that can be organized into a complex whole, within which subparts or subsystems can be interrelated.

While much research has been focused on the analysis of minute segments of knowledge, there has been increasing interest in developing larger frames of reference for synthesizing the results of such research. Thus attention has been focused more and more on overall systems as frames of reference for analytical work in various areas. It is our contention that a similar process can be useful for managers. Whereas managers often have been focusing attention on particular functions in specialized areas, they may lose sight of the overall objectives of the business and the role of their particular business in even larger systems. These individuals can do a better job of carrying out their own responsibilities if they are aware of the "big picture." It is the familiar problem of not being able to see the forest for the trees. The focus of systems management is on providing a better picture of the network of subsystems and interrelated parts which go together to form a complex whole.

Before proceeding to a discussion of systems theory for business, it will be beneficial to explore recent attempts to establish a general systems theory covering all disciplines or scientific areas.

GENERAL SYSTEMS THEORY

General systems theory is concerned with developing a systematic, theoretical framework for describing general relationships of the empirical world. A broad spectrum of potential achievements for such a framework is evident. Existing similiarities in the theoretical construction of various disciplines can be pointed out. Models can be developed which have applicability to many fields of study. An ultimate but distant goal will be a framework (or system of systems) which could tie all disciplines together in a meaningful relationship.

There has been some development of interdisciplinary studies. Areas such as social psychology, biochemistry, astrophysics, social anthropology, economic psychology, and economic sociology have been developed in order to emphasize the interrelationships of previously isolated disciplines. More recently, areas of study and research have been developed which call on numerous subfields. For example, cybernetics, the science of communication and control, calls on electrical engineering, neurophysiology, physics, biology, and other fields. Operations research is often pointed to as a multidisciplinary approach to problem solving. Information theory is another discipline which calls on numerous subfields. Organization theory embraces economics, sociology, engineering, psychology, physiology, and anthropology. Problem solving and decision making are becoming focal points for study and research, drawing on numerous disciplines.

With these examples of interdisciplinary approaches, it is easy to recognize a surge of interest in larger-scale, systematic bodies of knowledge. However, this trend calls for the development of an overall framework within which the various subparts can be integrated. In order that the *interdisciplinary* movement does not degenerate into *undisciplined* approaches, it is important that some structure be developed to integrate the various separate disciplines while retaining the type of discipline which distinguishes them. One approach to providing an overall framework (general systems theory) would be to pick out phenomena common to many different disciplines and to develop general models which would include such phenomena. A second approach would include the structuring of a hierarchy of levels of complexity for the basic units of behavior in the various empirical fields. It would also involve development of a level of abstraction to represent each stage.

We shall explore the second approach, a hierarchy of levels, in more detail since it can lead toward a system of systems which has application in most businesses and other organizations. The reader can undoubtedly call to mind examples of familiar systems at each level of Boulding's classification model.

1. The first level is that of static structure. It might be called the level of *frameworks;* for example, the anatomy of the universe.
2. The next level is that of the simple dynamic system with predetermined, necessary motions. This might be called the level of *clockworks.*
3. The control mechanism or cybernetic system, which might be nicknamed the level of the *thermostat.* The system is self-regulating in maintaining equilibrium.
4. The fourth level is that of the "open system" or self-maintaining structure. This is the level at which life begins to differentiate from not-life: it might be called the level of the *cell.*
5. The next level might be called the genetic-societal level; it is typified by the *plant,* and it dominates the empirical world of the botanist.
6. The animal system level is characterized by increased mobility, teleological behavior, and self-awareness.
7. The next level is the "human" level, that is, of the individual human being considered as a system with self-awareness and the ability to utilize language and symbolism.
8. The social system or systems of human organization constitute the next level, with the consideration of the content and meaning of messages, the nature and dimensions of value systems, the transcription of images into historical record, the subtle symbolizations of art, music, and poetry, and the complex gamut of human emotion.
9. Transcendental systems complete the classification of levels. These are the ultimates and absolutes and the inescapables and unknowables, and they also exhibit systematic structure and relationship.[2]

Obviously, the first level is most pervasive. Descriptions of static structures are widespread. However, this descriptive cataloguing is helpful in providing a framework for additional analysis and synthesis. Dynamic "clockwork" systems, where prediction is a strong element, are evident in the classical natural sciences such as physics and astronomy; yet even here there are important gaps. Adequate theoretical models are not apparent at higher levels. However, in recent years closed-loop cybernetic, or "thermostat," systems have received increasing attention. At the same time, work is progressing on open-loop systems with self-maintaining structures and reproduction facilities. Beyond the fourth level we hardly have a beginning of theory, and yet even here system description via computer models may foster progress at these levels in the complex of general systems theory.

Regardless of the degree of progress at any particular level in the above scheme, the important point is the concept of a general systems theory. Clearly, the spectrum, or hierarchy, of systems varies over a considerable range. However, since the systems concept is primarily a point of view and a desirable goal, rather than a particular method or content area, progress can be made as research proceeds in various specialized areas but within a total system context.

With the general theory and its objectives as background, we direct our attention to a more specific theory for business, a systems theory which can serve as a guide for management scientists and ultimately provide the framework for integrated decision making on the part of practicing managers.

SYSTEMS THEORY FOR BUSINESS

The biologist Ludwig von Bertalanffy has emphasized the part of general systems theory which he calls open system [1]. The basis of his concept is that a living organism is not a conglomeration of separate elements but a definite system, possessing organization and wholeness. An organism is an open system which maintains a constant state while matter and energy which enter it keep changing (dynamic equilibrium). The organism is influenced by, and influences, its environment and reaches a state of dynamic equilibrium in this environment. Such a description of a system adequately fits the typical business organization. The business organization is a man-made system which has a dynamic interplay with its environment — customers, competitors, labor organizations, suppliers, government, and many other agencies. Furthermore, the business organization is a system of interrelated parts working in conjunction with each other in order to accomplish a number of goals, both those of the organization and those of individual participants.

A common analogy is the comparison of the organization to the human body, with the skeletal and muscle systems representing the operating line elements and the circulatory system as a necessary staff function. The nervous system is the communication system. The brain symbolizes top-level management, or the executive committee. In this sense an organization is represented as a self-maintaining structure, one which can reproduce. Such an analysis hints at the type of framework which would be useful as a systems theory for business — one which is developed as a system of systems and that can focus attention at the proper points in the organization for rational decision making, both from the standpoint of the individual and the organization.

The scientific-management movement utilized the concept of a man-machine system but concentrated primarily at the shop level. The "efficiency experts" attempted to establish procedures covering the work situation and providing an opportunity for all those involved to benefit — employees, managers, and owners. The human relationists, the movement stemming from the Hawthorne-Western Electric studies, shifted some of the focus away from the man-machine system per se to interrelationships among individuals in the organization. Recognition of the effect of interpersonal relationships, human behavior, and small groups resulted in a relatively widespread reevaluation of managerial approaches and techniques.

The concept of the business enterprise as a social system also has received considerable attention in recent years. The social-system school looks upon management as a system of cultural interrelationships. The concept of a social system draws heavily on sociology and involves recognition of such elements as formal and informal organization within a total integrated system. Moreover, the organization or enterprise is recognized as subject to external pressure from the cultural environment. In effect, the enterprise system is recognized as a part of a larger environmental system.

Since World War II, operations research techniques have been applied to large, complex systems of variables. They have been helpful in shop scheduling, in freightyard operations, cargo handling, airline scheduling, and other similar problems. Queuing models have been developed for a wide variety of traffic- and service-type situations where it is necessary to program the optimum number of "servers" for the expected "customer" flow. Management science techniques have undertaken the solution of many complex problems involving a large number of variables. However, by their very nature, these techniques must structure the system for analysis by quantifying system elements. This process of abstraction often simplifies the problem and takes it out of the real world. Hence the solution of the problem may not be applicable in the actual situation.

Simple models of maximizing behavior no longer suffice in analyzing business organizations. The relatively mechanical models apparent in the "scientific management" era gave way to theories represented by the "human relations" movement. Current emphasis is developing around "decision making" as a primary focus of attention, relating communication systems, organization structure, questions of growth (entropy and/or homeostasis), and questions of uncertainty. This approach recognizes the more complex models of administrative behavior and should lead to more encompassing systems that provide the framework within which to fit the results of specialized investigations of management scientists.

The aim of systems theory for business is to develop an objective, understandable environment for decision making; that is, if the system within which managers make the decisions can be provided as an explicit framework, then such decision making should be easier to handle. But what are the elements of this systems theory which can be used as a framework for integrated decision making? Will it require wholesale change on the part of organization structure and administrative behavior? Or can it be woven into existing situations? In general, the new concepts can be applied to existing situations. Organizations will remain recognizable. Simon makes this point when he says that

> 1. *Organizations will still be constructed in three layers; an underlying* system *of physical production and distribution processes, a layer of programmed (and probably largely automated) decision processes for governing the routine day-to-day operation of the physical* system, *and a layer of nonprogrammed decision processes (carried on in a man-machine system) for monitoring the first-level processes, redesigning them, and changing parameter values.*
>
> 2. *Organizations will still be hierarchical in form. The organization will be divided into major subparts, each of these into parts, and so on, in familiar forms of departmentalization. The exact basis for drawing departmental lines may change somewhat. Product divisions may become even more important than they are today, while the sharp lines of demarcation among purchasing, manufacturing, engineering, and sales are likely to fade.*[3]

We agree essentially with this picture of the future. However, we want to emphasize the notion of systems as set forth in several layers. Thus the systems that are likely to be emphasized in the future will develop from projects or programs, and authority will be vested in managers whose influence will cut across traditional departmental lines. This concept will be developed in more detail throughout this article.

There are certain key subsystems and/or functions essential in every business organization which make up the total information-decision system, and which operate in a dynamic environmental system subject to rapid change. The subsystems include the following:

1. A *sensor subsystem* designed to measure changes within the system and with the environment.
2. An *information processing subsystem* such as an accounting or data processing system.
3. A *decision-making subsystem* which receives information inputs and outputs planning messages.
4. A *processing subsystem* which utilizes information, energy, and materials to accomplish certain tasks.
5. A *control component* which ensures that processing is in accordance with planning. Typically this provides feedback control.

6. A *memory* or *information storage subsystem* which may take the form of records, manuals, procedures, computer programs, etc.

A goal setting unit will establish the long-range objectives of the organization, and the performance will be measured in terms of sales, profits, employment, etc., relative to the total environmental system.

This is a general model of the systems concept in a business firm. In the following section a more specific model illustrating the application of the systems concept is established.

AN ILLUSTRATIVE MODEL OF THE SYSTEMS CONCEPT

Traditionally, business firms have not been structured to utilize the systems concept. In adjusting the typical business structure to fit within the framework of management by system, certain organizational changes may be required. It is quite obvious that no one organizational structure can meet operational requirements for every company. Each organization must be designed as a unique system. However, the illustrative model set forth would be generally operable for medium- to large-sized companies which have a number of major products and a variety of management functions. The primary purpose of this model is to illustrate the application of systems concepts to business organizations and the possible impact upon the various management functions of planning, organizing, communication, and control. The relationships which would exist among the top management positions are shown in Figure 1.

The master planning council would relate the business to its environmental system, and it would make decisions relative to the products or services the company produced. Further, this council would establish the limits of an operating program, decide on general policy matters relative to the design of operating systems, and select the director for each new project. New project decisions would be made with the assistance and advice of the project research and development, market research, and financial groups. Once the decision was made, the resource allocation committee would provide the facilities and manpower for the new system and supply technical assistance for systems design. After the system had been designed, its management would report to the operations committee as a major project system, or as a facilitating system.

Facilitating systems would include those organized to produce a service rather than a finished product. Each project system would be designed toward being self-sufficient. However, in many cases this objective may not be feasible or economical. For example, it may not be feasible to include a large automated mill as a component of a major project system, but the organization as a whole, including all the projects, might support this kind of a facility. A facilitating system, would be designed, therefore, to produce this kind of operating service for the major project systems. The output of the facilitating system would be material input for the project system and a fee should be charged for this input, just as if the input had been purchased from an outside source.

A soap manufacturer could have, for example, major project systems in hand soap, laundry soap, kitchen soap, and tooth paste. A facilitating system might be designed to produce and *sell* containers to the four project systems.

Operating Systems

All operating systems would have one thing in common – they would use a common language for communicating among themselves and with higher levels of management. In addition, of course, each system designed would be structured in consideration of companywide policies. Other than these limits, each operating system would be created to meet the specific requirements of its own product or service. A model of an operating system is shown in Figure 2.

Figure 2 illustrates the relationship of the functions to be performed and the flow of operating information. The operating system is structured to (1) direct its own inputs, (2) control its own operation, and (3) review and revise the design of the system as required. Input is furnished by three different groups: technical information is generated as input into the processing system; in addition, technical information is the basis for originating processing information. Both technical and processing information are used by the material input system to determine and supply materials for processing. How-

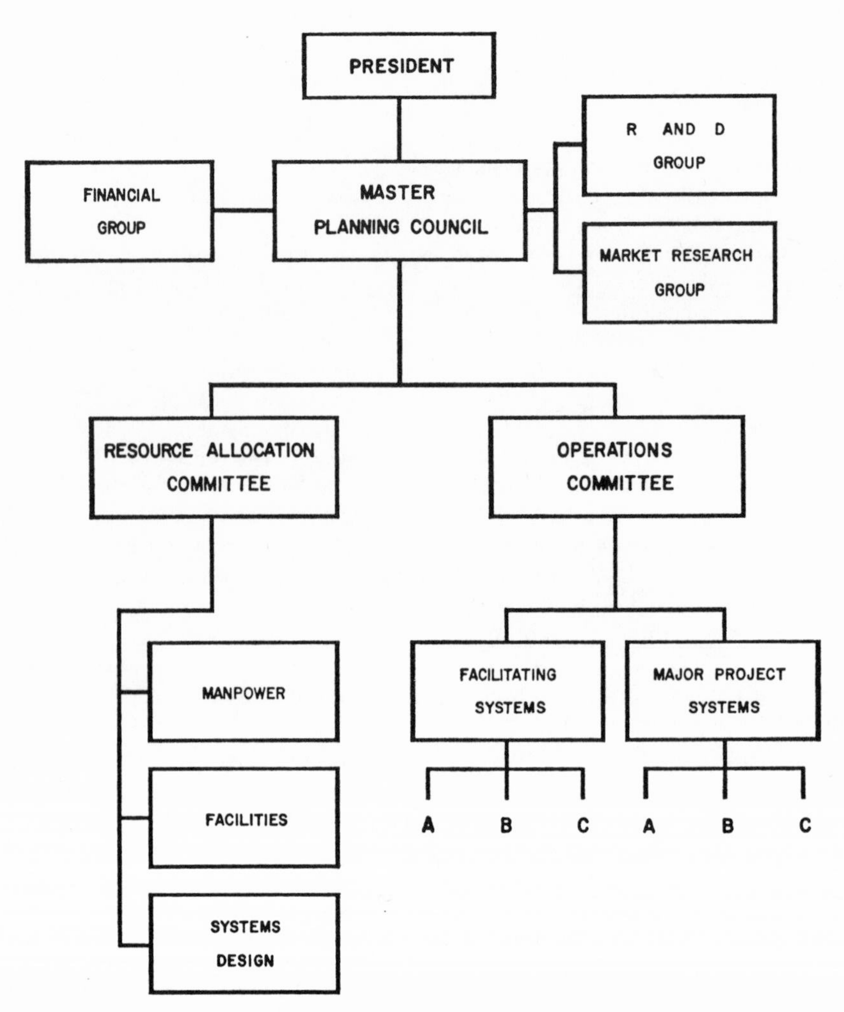

Figure 1
The Systems Model: Top Management.

ever, corrective action, when necessary, would be activated by input allocation.

This model can be related to most business situations. For example, if this represented a system to produce television sets, the technical information would refer to the design of the product, processing information would include the plan of manufacture and schedule, and the material input would pertain to the raw materials and purchased parts used in the processing. These inputs of information and material would be processes and become output. Process control would measure the output in comparison to the standard (Information Storage) obtained from input allocation, and issue corrective information whenever the system failed to function according to plan. The design of the system would be reviewed continually and the components rearranged or replaced when these changes would improve operating efficiency.

Basically, the operating systems would be self-sustaining with a high degree of autonomy. Therefore, they could be integrated into the overall

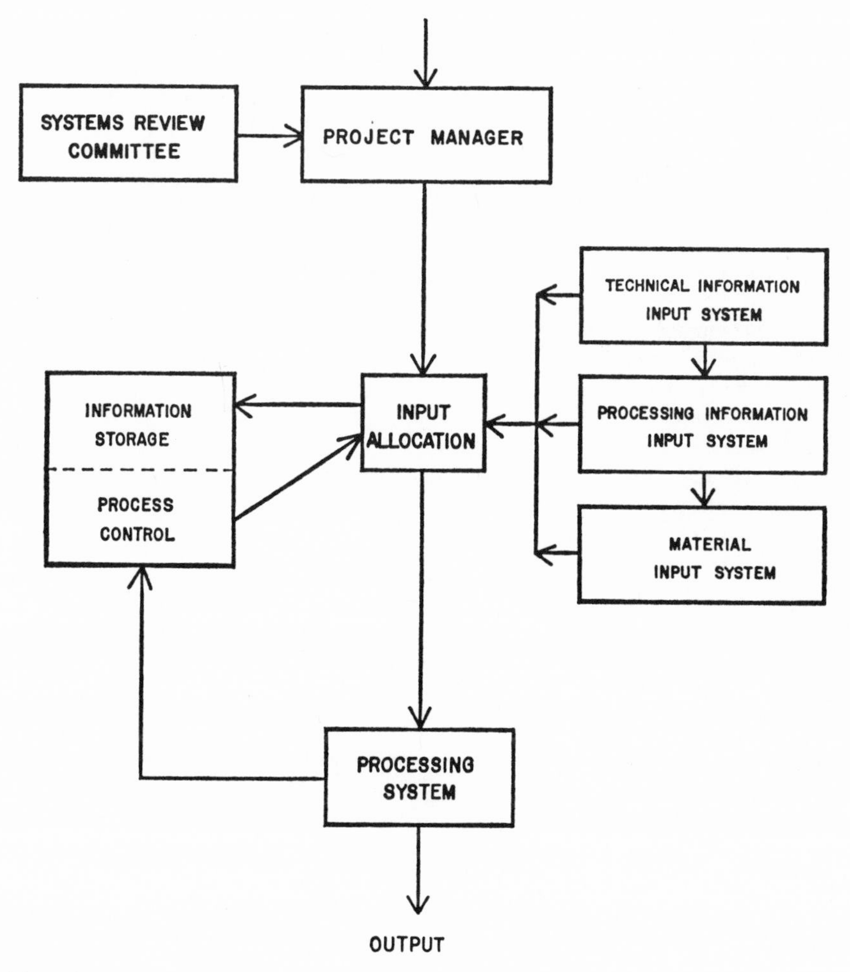

Figure 2
An Operating System Model.

organizational structure (Figure 1) with a minimum of difficulty.

SYSTEMS CONCEPTS AND MANAGEMENT

Managers are needed to convert the disorganized resources of men, machines, and money into a useful and effective enterprise. Essentially, management is the process whereby these unrelated resources are integrated into a total *system for objective accomplishment.* A manager gets things done by working with people and physical resources in order to accomplish the objectives of the system. He coordinates and integrates the activities and work of others rather than performing operations himself.

Structuring a business according to the systems concept does not eliminate the need for the basic functions of planning, organization, control, and communication. However, there is a definite change of emphasis, for the functions are performed in conjunction with operation of the system and not as separate entities. In other words, everything revolves around the system and its objective, and the function is carried out only as a service to this end. This point can be clarified by reviewing each of the functions in terms of their relation to the model of the systems concept illustrated previously.

Planning

Planning occurs at three different levels in the illustrative model. These levels are shown in Figure 1. First, there is top-level planning by the master planning council. Second, the project and facilitating systems must be planned and resources allocated to them. Finally, the operation of each project and facilitating system must be planned.

The master planning council establishes broad policies and goals and makes decisions relative to the products or services the company produces. It decides upon general policy matters concerning the design of the operating systems and selects the director for each new program. It is the planning council which receives informational inputs from the environmental and competitive systems. It combines these inputs with feedback information from the internal organizational system and serves as the key decision-making center within the company. Much of the decision making at this level is nonprogrammed, unstructured, novel, and consequential. While some of the new techniques of management science may be helpful, major reliance must be placed upon mature assessment of the entire situation by experienced, innovative top executives.

Once these broad decisions have been made, the planning function is transferred to the resource allocation and operating committees. They plan and allocate facilities and manpower for each new system and supply technical assistance for individual systems design. At this planning level it is possible to utilize programmed decision making – operations research and computer techniques.

The third level, planning the operations of each project or facilitation system, is concerned primarily with the optimum allocation of resources to meet the requirements established by the planning council. This planning can most easily be programmed to automatic decision systems. However, the project director would still have to feed important nonquantifiable inputs into the system.

Under the systems concepts of planning there is a direct relationship between the planning performed at each of the three levels. The first planning level receives informational inputs from the environment and competitive system and feedback information from within the organization. It translates this into inputs for the next planning level, which in turn moves to a more detailed level of planning and provides inputs for the third or project level. One of the major advantages of this sytems model is to provide a clear-cut delineation of the responsibility for various types of planning.

This concept facilitates integrated planning on a systems basis at the project level within the organization. Given the inputs (premises, goals, and limitations) from the higher levels, the project managers are delegated the function of integrated planning for their project.

Organization

Traditional organization theory emphasized parts and segments of the structure and is concerned with the separation of activities into tasks or operational units. It does not give sufficient emphasis to the interrelationships and integration of activities. Adapting the business organization to the systems concept places emphasis upon the integration of all activities toward the accomplishment of overall objectives but also recognizes the importance of efficient subsystem performance.

The systems basis of organization differs significantly from traditional organization structures such as line and staff or line, staff, and functional relationships. As shown in Figure 1, there are three major organizational levels, each with clearly delineated functions. The master planning council has broad planning, control, and integrative functions; the resource allocation committee has the primary function of allocating manpower and facilities, and aids in systems design for the facilitating or project systems. One major purpose of this type organization is to provide an integration of activities at the most important level – that is the individual project or program.

Staff specialization of skills is provided for the master planning council through such groups as financial, research and development, and market research. Their activities, however, are integrated and coordinated by the planning council. There are specialists at the operating level who are completely integrated into each project system. Thus the activities of these specialists are geared to the effective and efficient performance of the individual project system. This type organization minimizes a major problem associated with staff and functional personnel — their tendency to associate their activities with specialized areas rather than with the optimum performance of the overall operation. Yet, under the model the importance of initiative and innovation are recognized. In fact, the major function of the master planning council is planning and innovation. Specific provision for receiving information inputs from product and market research are provided in the model.

There are other advantages of the systems concept. Business activity is dynamic, yet the typical organization is structured to perpetuate itself rather than change as required. There is generally resistance by the various specialized functions to change in order to optimize organization performance. For example, Parkinson's law states that there is an ever-increasing trend toward hierarchies of staff and functional personnel who are self-perpetuating and often do not contribute significantly to organizational effectiveness, or in extreme cases may be dysfunctional. In contrast, a system is designed to do a particular task. When the task is completed, the system is disbanded.

Systems are created from a central pool of resources. Facilities, machines, and manpower are assigned to specific projects or programs. The approach is to create and equip the project system with a complete arrangement of components to accomplish the job at hand. This may result in the duplication of certain activities in more than one operating system; however, this disadvantage is not as serious as it may seem. For example, it may be more efficient to have several typewriters assigned to each system rather than a center pool of typewriters. In the first instance, the typewriters may be utilized less than 100 percent of the time, but the problems of scheduling the work at the central pool, delays, accountability, measurement of contribution, etc., would soon offset the advantages of centralizing equipment. Too much effort may be spent in creating processing information which accomplishes no objective other than keeping the machines utilized. A reasonable amount of redundancy or extra capacity will provide more flexibility, protect against breakdowns, reduce flow time, require less planning, eliminate many problems associated with interdepartmental communication, and reduce the amount of material handling.

Obviously, there are situations when it is impractical to decentralize a particular facility, because individual systems cannot utilize it sufficiently to warrant its incorporation into each separate operation. In these instances, a facilitating system would be created which would sell its services to any or all of the major project systems. These service systems would have to justify their existence and compete with outside vendors as suppliers to the major project system.

One great advantage of the systems concept for organizing pertains to the decentralization of decision making and the more effective utilization of the allocated resources to the individual project system. This has the merit of achieving accountability for performance through the measurability of individual systems of operation.

Control

The systems concept features control as a means of gaining greater flexibility in operation, and, in addition, as a way of avoiding planning operations when variables are unknown. It is designed to serve the operating system as a subsystem of the larger operation. Its efficiency will be measured by how accurately it can identify variations in systems operation from standard or plan, and how quickly it can report the need for correction to the activating group.

We must conclude that error is inevitable in a system which is subject to variations in input. When the lag in time between input and output is great, more instability is introduced. Feedback can reduce the time lag; however, corrective action which is out of phase will magnify rather than overcome the error. Every system should be designed to make its own corrections when necessary. That is, a means should be provided to reallocate resources as conditions change. In our model the Systems Review Committee (see Figure 2) should be aware of any change in operating conditions which might throw the system "out of control." Replanning or redesign may be required.

In controlling a system it is important to measure inputs of information, energy, and materials, and outputs of products and/or services. This will determine operating efficiency. In addition it may be important to establish points of measurement during critical or significant stages of processing. Such measurements would be used principally to help management analyze and evaluate the operation and design of individual components. The best approach is to spotlight exceptions and significant changes. Management can focus their attention on these areas. One important thing to remember is that the control group is not a part of the processing system – it is a subsystem, serving the operating system. Cost control can be used as an example to illustrate this point. The cost accountant must understand that his primary objective is to furnish managers with information to control costs. His task is to inform, appraise, and support; never to limit, censure, or veto. The same principle applies to every control group serving the operating system.

Communication

Communication plays a vital role in the implementation of the systems concept. It is the connecting and integrating link among the systems network. The flow of information, energy, and material – the elements of any processing system – are coordinated via communication systems. As shown in the model (Figure 2), the operating system requires information transmission to ensure control. Communication systems should be established to feed back information on the various flows – information, energy, and material. Information on the effectiveness of the planning and scheduling activities (as an example of information flow) would be helpful in adjusting the nature of this activity for the future. Similarly, reports on absenteeism are examples of communication concerning the energy flow (the people in the system) to the processing activity. Information on acceptance inspection is an example of information stemming from the material flow aspect of an operating system. All these feedback communication systems provide for information flow to a sensor and a control group. Comparison between the information received and the information stored (the master plan for this particular operating system) would result in decisions concerning the transmission of corrective information to the appropriate points.

Relationships within and among various project systems and between the levels of the system as a whole are maintained by means of information flow, which also can be visualized as a control device. Moreover, any operating system maintains contact with its environment through some sensory element. Referring to Figure 1, the sensory elements in this case are the groups reporting to the master planning council. The master planning council makes decisions concerning the product or service the organization will produce based on information gained from market research, research and development, and financial activities. In a sense, these activities function as the antenna of the organization, maintaining communication with the external environment. The master planning council melds the information received through these activities with other premises covering the internal aspects in order to make decisions about future courses of action. Here, again, communication or information flow can be visualized as a necessary element in controlling the course of action for the enterprise as a whole. Based on the feedback of information concerning the environment in general, the nature of competition, and the performance of the enterprise itself, the master planning council can continue its current courses of activity or adjust in light of developing circumstances. Thus communication or information flow facilitates the accomplishment of the primary managerial functions of planning, organizing, and controlling.

Communication by definition is a system involving a sender and a receiver, with implications of feedback control. This concept is embodied in the lowest-level projects or subsystems, in all larger systems, and in the system as a whole. Information-decision systems, regardless of formal charts or manuals, often flow across departmental boundaries and are often geared to specific projects or programs. The systems concept focuses on this approach and makes explicit the information-decision system which might be implicit in many of today's organizations.

The systems concept does not eliminate the functions of management, i.e., planning, organizing, control, and communication. Instead, it integrates these functions within a framework designed to emphasize their importance in creating more effective systems. Because of the great diversity of operations and environments, particular missions of organizations differ and each system must be unique or at least have some unique elements. Nevertheless, the illustrative model and its application to the management functions of planning, organizing, controlling, and communication can serve as a point of departure in systems design.

PERVASIVENESS OF SYSTEM CONCEPTS

Many of the most recent developments in the environment of businessmen and managers have involved systems concepts. For example, the trend toward automation involves implementation of these ideas. Automation suggests a self-contained system with inputs, outputs, and a mechanism of control. Yet the concept also recognizes the need to consider the environment within which the automatic system must perform. Thus the automated system is recognized as a subpart of a larger system.

The kinds of automation prevalent today range in a spectrum from sophisticated mechanization to completely automatic, large-scale production processes. Individual machines can be programmed to operate automatically. Large groups of machines also can be programmed to perform a series of operations, with automatic materials-handling devices providing connecting links among components of the system. In such a system, each individual operation could be described as a system and could be related to a larger system covering an entire processing operation. That particular processing operation could also be part of the total enterprise system, which in turn can be visualized as a part of an environmental system.

Completely automated processing systems such as oil refineries are also commonplace today. In such cases the entire process from input of raw material to output of finished products is automated with preprogrammed controls used to adjust the process as necessary, according to information feedback from the operation itself.

The systems concept is also apparent in other aspects of automation. The above examples deal with physical processing; another phase which has been automated is information flow. With the introduction of large-scale, electronic data-processing equipment, data-processing systems have been developed for many applications. Systems concepts are prevalent, with most applications built around the model of input-processor-output and with feedback control established within the instructions developed to guide the processing of data. Here, again, there is an entire spectrum of sophistication leading from simple, straightforward data-reduction problems to elaborate, real-time data-processing systems.

Physical distribution systems have received increasing attention on the part of manufacturers and shippers. The concepts of logistics, or materials management, have been used to emphasize the flow of materials through distribution channels. The term *rhochrematics* has been coined to connote the flow process from raw-material sources to final consumer.[4] In essence, these ideas embrace systems concepts because emphasis is placed on the total system of material flow rather than on functions, departments, or institutions which may be involved in the processing.

In recent years increasing attention has been focused upon massive engineering projects. In particular, military and space programs are becoming increasingly complex, thus indicating the need for integrating various elements of the total system. Manufacturing the product itself (a vehicle or other hardware) is quite complex, often involving problems of producibility with requirements of extremely high reliability. This is difficult to ensure for individual components or subsystems. In addition, each subsystem also must be reliable in its interrelationship with all other subsystems. Successful integration of subcomponents, and hence successful performance of a particular product, must also be integrated with other elements of the total system. For example, the functioning of the Nike-Zeus antimissile missile must be coordinated with the early warning system, ground facilities, and operating personnel. All elements must function as an operating, integrated whole.

The previous discussion has emphasized the mechanistic and structural aspects of the systems concept. Yet, we cannot forget that business organizations are social systems; we are dealing with man-made systems. Obviously, a great deal could be said about the possible consequences of applying systems concepts to human relationships, but such a task is beyond the scope of this article. However, in discussing the impact of the systems concept it should not be assumed that people basically resist systems. Much of man's conscious activities since the dawn of history has been geared to creating system out of chaos. Man does not resist systematization of his behavioral patterns per se. Rather, the normal human being seeks satisfactory systems of interpersonal relationships which guide his activities. Without systematization, behavior would be random, non-goal-oriented, and unpredictable. Certainly, our complex, modern, industrial society demands more systemized human behavior than older, less-structured societies. A common characteristic in a rapidly advancing society is to make systems of interpersonal relationship more formal. While many of these systems have been implicit in the past, they are becoming more explicit. This remains one of the basic precepts of our systems model; systematic interpersonal relationships are necessary for accomplishing group objectives, and an effective organizational system should be designed to meet this need.

SUMMARY

General systems theory is concerned with developing a systematic, theoretical framework for describing general relationships of the empirical world. While a spectrum or hierarchy of systems can be established over a considerable range, the systems concept is also a point of view and a desirable goal, rather than a particular method or content area. Progress can be made as research proceeds in various specialized areas but within a total system context.

The business organization is a man-made system which has a dynamic interplay with its environment – customers, competitors, labor organizations, suppliers, government, and many other agencies. In addition, the business organization is a system of interrelated parts working in conjunction with each other to accomplish a number of goals, both those of the organization and those of individual participants. This description parallels that of open systems in general which maintain a constant state while matter and energy which enter them keep changing; that is, the organisms are influenced by, and influence, their environment and reach a state of dynamic equilibrium within it. This concept of the organization can be used by practicing managers to integrate the various ongoing activities into a meaningful total system. Regardless of specific adjustments or organizational arrangements, there are certain subsystems or essential functions which make up a total information-decision system. However, the exact form utilized by a particular organization may depend upon the task orientation. We have presented a generalized illustrative model which indicates an approach that may be appropriate for a large segment of modern business organizations.

Managers are needed to convert disorganized resources of men, machines, and money into a useful, effective enterprise. Essentially, management is the process whereby these unrelated resources are integrated into a total *system for objective accomplishment.* The systems concept provides no cookbook technique, guaranteed to provide managerial success. The basic functions are still planning, organization, control, and communication. Each of these activities can be carried out with or without emphasis on systems concepts. Our contention is that the activities themselves can be better accomplished in light of systems concepts. Furthermore, there can be a definite change in emphasis for the entire managerial process if the functions are performed in light of the system as a whole and not as separate entities.

The business organization as a system can be considered as a subsystem of a larger environmental system. Even industry or interindustry systems can be recognized as subelements of the economic system, and the economic system can be regarded as a part of society in general. One major change within business organizations of the future may be the breakdown of traditional functional specilization geared to optimizing performance of particular departments. There may be growing use of organizational structures designed around projects and information-decision systems. The systems concept calls for integration into a separate organizational system of activities related to particular projects or programs. This approach currently is being implemented in some of the more advanced technology industries.

The breakdown of business organizations into separate functional areas has been an artificial organizational device, necessary in light of existing conditions. Management science techniques, computer simulation approaches, and information-decision systems are just a few of the tools which will make it possible for management to visualize the firm as a total system. This would not have been possible two decades ago; it is currently becoming feasible for some companies; and it will become a primary basis for organizing in the future.

---

[1] *For a more complete discussion see Johnson, Kast, and Rosenzweig [3], pp. 4-6, 91, 92.*
[2] *Boulding [2], pp. 202-205.*
[3] *Simon [4], pp. 49-50. (Italics by authors.)*
[4] *Rhochrematics comes from two Greek roots; rhoe, which means a flow (as a river or stream), and chrema, which stands for products, materials, or things (including information). The abstract ending -ics has been added, as for any of the sciences.*

REFERENCES

1. Bertalanffy, L. von, "General System Theory: A New Approach to Unity of Science," *Human Biology*, Dec. 1951, pp. 303-361.
2. Boulding, K., "General Systems Theory: The Skeleton of Science," *Management Science*, Apr. 1956, pp. 197-208.
3. Johnson, R. A., Kast, F. E., and Rosenzweig, J. E., *The Theory and Management of Systems*, McGraw-Hill Book Company, Inc., New York, 1963.
4. Simon, H. A., *The New Science of Management Decision*, Harper & Row, Inc., New York, 1960.

We are in the first stages of the "post-industrial society," which will be characterized by the centrality of theoretical knowledge . . . as the source of innovation and policy formulation; control of technology and technological assessment; and the creation of a new "intellectual technology" — methods such as systems analysis, operations research, decision theory — for dealing with problems of organized complexity.
Daniel Bell, as reported by Nicholas Wads, "Daniel Bell: Science as the Imago of the Future Society,"

Science, *Vol. 188, No. 4183, 4 April 1975, p. 35.*

24

Jay W. Forrester

THE IMPACT OF FEEDBACK CONTROL CONCEPTS ON THE MANAGEMENT SCIENCES

This evening I wish to discuss with you a new frontier in our society. You may choose to look upon this as the new frontier in control technology. Equating the two implies that information feedback system concepts are about to emerge from the strictly technological world into the main stream of international social and professional change.

Each major step in human advancement has resulted from an assault on a new frontier. In our history, we have passed through the periods of clarifying religious and ethical beliefs, great literary development, governmental organization first into monarchies and then followed by evolution into democracy, geographical exploration radiating outward from Europe after the fifteenth century and ending in this country only 60 years ago, the industrial revolution, the development of the professions of law and medicine, and the scientific and technological revolution which we have experienced in the last three decades. These frontiers come, they

---

Reprinted from The Impact of Feedback Control Concepts on the Management Sciences, *by Jay W. Forrester, Foundation for Instrumentation, Education and Research, New York, 24 pp. (1959).*

are exploited, and they recede into the generally accepted fabric of society.

What is the next frontier of our civilization? To look upon space exploration as the next great human endeavor is, I think, to be blinded by the crest of the recent scientific wave of progress. Space exploration is an extension and merging of past geographical and technological explorations. To me, the great opportunities and the great advances of the next three decades appear to lie in the fields of management and economics.

This shift from the technical to the social frontier is clearly visible in the recent change of emphasis on the international scene. For the last 20 years, international competition has been in the arena of force supported by scientific advances. The rules of the contest are now changing. The test is now of leadership in showing the way to economic development and political stability. Do we understand the dynamics of change in industrial and economic systems well enough to pioneer this new frontier?

A new frontier is opened when there is a coincidence of a great need existing simultaneously with means for filling the need. Let us see if circumstances are timely for making new advances in management and economic systems the next frontier. Is there a great need? If so, has anything happened to provide new tools for attacking the great problems?

MANAGEMENT AS AN ART

Management of industrial and economic activity has evolved over the centuries as an art – a very skilled art. No matter how skilled is an art, it is greatly handicapped by the absence of a unifying, underlying structure of fundamental principle. Such is the condition of management today. The effective practice is executed as an art. The associated science has been notably ineffective.

Management as an art has progressed by the acquisition and recording of human experience. But, as long as there is no orderly underlying scientific base, the separate experiences remain as special cases. The lessons are seldom transferrable either in time or in space.

The corporate manager today is but little helped by experiences recorded in the literature and carried forward from a generation ago. The descriptions are incomplete, lack precision, and arose from circumstances that cannot be properly related to today's events.

Thus far, in the teaching of management, this verbal and descriptive recording of the past as "cases" representing selected experiences has had to serve as the vehicle for transmitting knowledge. It was the best available, although it was far from adequate. Were engineering still to rest on the same descriptive transmittal of experience, we would not have today's advanced technology. The liberal arts training through multiple exposure to recorded incidents of the past presumes that the student will distill an intuitive structure of human and social behavior around which to assemble and interpret his own experiences. The rapid strides of professional progress come when this structure can be identified and taught explicitly rather than by indirection and diffusion. The student can then inherit an intellectual legacy from the past and build his own experiences upward from that level, rather than having to start over again at the point where his ancestors began.

Not only is descriptive transmittal of an art ineffective in going from the past to the present, but it serves almost as poorly in moving experience and judgment from one point to another in the present. We still find each company and each industry believing its problems are unique. A discussion of a set of experiences in the context of another situation often elicits the rejoinder, "Yes, but my company is different." Because of the lack of a suitable fundamental viewpoint, we fail to see how inudstrial experiences all deal with the same material, financial, and human factors – all representing variations on the same underlying system.

Without an underlying science, advancement of an art eventually reaches a plateau. The human lifetime limits the knowledge that can be gained from personal experience. In an art, experiences are poorly transferred from one location to another or from the past to the present. Historical experiences lack a framework for relating them to current problems. One company's experiences cannot be directly and meaningfully used by another.

Management has now reached such a plateau. If progress is to continue, there must arise an applied science as a foundation to support further development of the art. Such a base of applied science will permit experiences to be translated into a common frame of reference from which they can be transferred either in time or space to be effectively applied in new situations by other managers.

CONVENTIONAL MANAGEMENT SCIENCE INADEQUATE

But, you may ask, are not the fields of operations research and mathematical economics providing this underlying foundation of principle? I believe they are conspicuous in their failure to do so. Past analytical methods fall so far short of the practical requirement that they have become a field of activity unto themselves.

Mathematical economics and management science have often been more closely allied to formal mathematics than to economics and management. The difference in viewpoint is evident if one compares the business literature with publications on management science, or compares descriptive economics books with texts on mathematical economics. In many professional journal articles the attitude is that of an exercise in formal logic rather than that of seeking useful help in real problems. In such an article, assumptions of unlikely validity are stated in an introductory paragraph and adopted without justification. On this formal but unrealistic foundation is then constructed a mathematical solution for the behavior of the assumed system.

Another evidence of the bias toward the mathematical rather than the managerial motivation is seen in a preoccupation with "optimum" solutions. For most of the great problems, mathematical methods fall far short of being able to find the "best" solution. The misleading objective of trying only for an optimum solution often results in simplifying the problem until it is devoid of practical interest. The lack of utility does not, however, detract from the elegance of the analysis as an exercise in mathematical logic.

A conspicuous dichotomy exists between the practicing managers and economists who reside in what I will call Region A and the mathematical analysts of economic and management phenomena who reside in what I will call Region B. In business, Region A is inhabited by men responsible for decisions and policy and action, and Region B by staff specialists who advise but take no responsibility. In academic circles, Region A is the home of the descriptive social scientist whose skill is measured by his acuteness in perceiving the motivations and interrelationships in economic and managerial affairs; in contrast, Region B is more apt to include those searching for problems to fit available mathematical tools. In Region A the goal is improvement of real situations; in Region B the goal is often the explicit or optimum solutions to unrealistically simplified hypotheses. In Region A the manager acts on such information as he can obtain; in Region B the analyst often ignores phenomena that he admits are crucial but which cannot be precisely measured. In Region A, success is measured by financial results and by economic development; in Region B, reputations are based on published papers and mathematical elegance. In Region A the literature is the business press and the descriptive books on economics and management; in Region B the literature is the journals and books on operations research and mathematical economics; and not often do the inhabitants of one region read the literature of the other. In Region A, system nonlinearities are recognized as primary causes of important occurrences; in Region B, nonlinear behavior is usually ignored. In Region A, data comes from personal observation and participation in economic and business affairs; in Region B, inputs are often limited to those for which statistical measurements are available. In Region A, opinions are more apt to be built up from individual incidents, such as how the individual person reacts, how the actual production process is designed, and how long it takes to build a factory; in Region B, relationships are extracted from averages and aggregates in which the nature of individual action is often lost. In Region A the manager and the government policy maker deal with the dynamic interactions between men, materials, decisions, equipment, and money; in Region B, simplified money-flow relationships have often claimed the exclusive attention of the mathematical analyst.

These distinctions between Regions A and B have been sharply etched and somewhat exaggerated. However, the proper impression emerges — the "art" of Region A is still better able to deal with decisions of great consequence than the "science" of Region B. The overlap between the two is slight. The manager has often found that management science did not deal with his most urgent problems; it has not learned to take into account the variables that he knows to be important; it is not cast in a language with which he is familiar.

REASONS FOR GULF BETWEEN APPROACHES

The reasons for this gulf between the art and the present embryonic attempts at a science have arisen naturally enough and are easy to trace. Almost all the present management science work traces its ancestry back to the fields of physics, applied mathematics, statistics, and the agricultural and genetic sciences. This evolution of methodology began several decades ago when there was no possibility of dealing with complex nonlinear systems. The emphasis has been on elementary formulations in which explicit analytic solutions have been possible. This high degree of oversimplification has kept the work sufficiently separated from the practical world that the field has remained uncomtaminated by demands for practical utility. Students of the area have become the future teachers.

The failures of management science have been explained away on the basis that suitable data were not yet available, rather than letting these failures raise questions about the foundations of the approaches being used. Much of the statistical methodology has been carried over from agriculture and genetics where it was developed as a tool in the study of unidirectional, open-loop systems wherein causes produced effects that did not react on causes. The hereditary characteristics of the offspring arise from the parents, but no characteristics of the offspring influence the heredity of the parents. A methodology which is suitable for open-loop systems may be quite unsuitable for use in complex, closed-loop, information feedback systems. If these systems are sufficiently complex and noisy and contain frequency sensitive delays and amplification, it becomes possible to attribute failures to an insufficient application of a method rather than to a fundamental unsoundness in its approach.

A number of "obvious truths" seem to have been accepted in varying degrees as the philosophical guidelines for much of the search for a scientific foundation underlying management and economics. All the following appear to be given at least some credence, and all seem to me to be misleading:

1. *Linear analysis:* That a linear analysis is an adequate representation of industrial and economic systems. Almost every factor in these systems is nonlinear. Much of the important behavior is a direct manifestation of the nonlinear characteristics. The amplitude of excursion of system variables is so large that "small-signal" linear analysis is not suitable.
2. *Stable systems:* That our social systems are inherently stable and can be attacked with methods that are valid only for stable systems that tend toward equilibrium. There seems ample evidence that much of our industrial and economic behavior shows the characteristics of an unstable system. Many industries are characterized by an unstable, nonlinear, self-limiting systems behavior.
3. *Prediction function:* That the obvious purpose and test of a model of an industrial system is its ability to predict specific future action. We should use a model to predict the character and the nature of a system and for the design of the kind of system that we desire. This is far different, less stringent, and much more useful than the prediction of the specific future times of peaks and valleys in a sales curve.
4. *Data:* That the construction of a model must be limited to those variables for which numerical time-series data exists. A model of system behavior must deal with those variables which are thought to control system action. If data have not been collected in the past, best guesses must be substituted until measurements are taken.
5. *Accepted definitions:* That a model must be limited to considering those variables which have generally accepted definitions. Many underfined concepts are known to be of major importance. Integrity, hope, research output, quality, customer satisfaction, and confidence must all be given definitions and be incorporated in those system models where they are presumed to be important.
6. *Descriptive knowledge:* That our vast body of descriptive knowledge is unsuitable for use in model

formulations. Just as formal numerical data have been the preferred ingredient for model making, so has the wealth of information in the business press been rejected. *Business Week* and the *Wall Street Journal* lack academic stature even though they may contain the clearest and most perceptive published insights into the reasons for industrial managers' decisions.

7. *Exact versus social:* That there is a sharp distinction between the "exact" and the social sciences. At the most this is but a quantitative distinction and at the least there is no distinction at all. Exactness and accuracy must be measured not in terms of the number of decimal places but in terms of the requirements. The design and construction of a machine tool may require measurements to five significant decimal digits. This accuracy of measurement has been developed because it was needed. The next advancement is based on the preceding level of accuracy. Dramatic progress is possible in the dynamic behavior of industrial systems using parameters which may be in error by a factor of 3 – which are not even correct in the first decimal place.

8. *Source of Analogy:* That the physical and genetic sciences provide the proper analogy for model building in the social sciences. The "laws" of physics usually relate to open systems rather than information feedback systems. Furthermore, they relate to fragments of systems rather than to entire systems. A much better analogy exists in the engineering and military models – models of telephone systems, of aircraft, of military systems, and of missile controls.

9. *Accuracy of structure:* That accuracy of parameters is more important than system structure. A great deal of time and effort in social science is devoted to the measurement of parameters. Yet these parameters are put into models which I believe do not belong to the general class to which the actual systems themselves belong. Correct parameters can hardly succeed in a grossly incorrect model structure. Here I refer to the failure to deal adequately with those factors that give information feedback systems their characteristic behavior.

10. *Accuracy versus precision:* That accuracy must be achieved before precision is useful. The ability to precisely state a hypothesis and to examine its consequences can be tremendously revealing even though the accuracy of the statement is low. A precise and explicit statement with assumed numerical values will tell us the kinds of things which can happen. Should these things be important we can later devote attention to improving the accuracy of their statement.

11. *Optimum solutions:* That it is necessary to find optimum solutions to managerial questions. Tremendous gains lie ahead in systems management. Mere improvement will often be dramatic even when it falls far short of some optimum. Optimum solutions are generally possible only for naively simply questions. More is to be gained by improving areas of major opportunity than by optimizing areas of minor importance.

12. *Controlled experiments:* That the social sciences differ fundamentally from the physical sciences by inability to conduct "controlled experiments." Again, this is nothing more than a matter of degree. Controlled experiments in engineering are done with models. The models are as complete and realistic as our knowledge permits. The same concepts in using a model are possible in management and economics. Effective models are entirely feasible with present technical resources. We have only to construct the models and then the laboratory experimental stage can begin.

13. *Human decisions:* That human decision making is obscurely subtle and impenetrable. The major factors to which a decision is responsive are relatively few in number. They are usually subject to clarification if properly approached. Once we have dealt with a relatively few properly selected factors, the remaining can be relegated to a noise and uncertaintly category.

14. *Decision versus policy:* That emphasis in models should be on decision making. The sharp distinction between policy and decision has been obscured. Too much attention has been concentrated on the individual decisions and not enough on the policy which governs how the decisions are made. Models of industrial systems should be directed toward policy. In other words, what are the rules by which information sources are converted into a continuous flow of decisions? The present-day emphasis on the management game as a training device arises from a misplaced concentration on decisions rather than on policies.

FOUNDATIONS FOR PROGRESS

If the past conventional approaches to a management science have been inadequate, what is to be the foundation for a better future understanding of industrial and economic dynamics? The need for a better understanding of industrial systems has long existed. Why is a major breakthrough now timely? Why is the opportunity now at hand when it was not previously? We can move ahead now because four essential foundations for progress have come into existence in the last two decades.

The first and most important foundation is the concept of servomechanisms (or information feedback systems) as evolved during and after World War II. Until recently, we have been insufficiently aware of the effect of time delays, amplification, and structure on the dynamic behavior of a system. We are coming to realize that the interactions between system components can be more important than the components themselves. I will return later to a discussion of information feedback aspects of industrial systems.

The second foundation is the realization of the importance of the experimental approach to the understanding of system dynamics. No longer do we limit our attention to uselessly oversimplified analysis simply to achieve analytical solutions. Mathematical models have been extensively used to simulate time relationships in systems and to yeild empirical answers to special cases. Instead of going from the general analytical solution to the particular special case, we have come to appreciate the great utility, if not the elegance, of the empirical approach. In this we study a number of particular situations and from these we generalize as far as we dare.

The third foundation for progress is the electronic digital computer that has become generally available during recent years. Without it the vast amount of work to get specific solutions to the characteristics of complex systems would be prohibitively expensive. In the last 15 years the cost of arithmetic computation has fallen by a factor of 10,000 or more in those areas where digital computers can be used in their most efficient modes of operation. The simulation of information feedback models of industrial behavior is such an area of high efficiency. A cost reduction factor of 10,000 or even 100,000 places one in a totally different research environment than existed even a decade ago.

The fourth foundation is the better understanding of policy and decision making which has been achieved while developing modern military control systems. During World War II fire control prediction decisions were made automatically by machine, but before 1950 there was almost no acceptance of automatic threat evaluation, weapon selection, friend and foe identification, alerting of forces, or target assignment. In a mere 10 years, these automatic decisions have been pioneered, accepted, and put into practice. In so doing, it has been necessary to interpret the "tactical judgment and experience" of military decision making into formal rules and procedures. This change has been forced because the pace of modern military operations has exceeded the ability of a human organization to respond. Ten years and thousands of people have been involved in this interpretation of the decision process and in automatizing the operational policies that are the basis for tactical military decision making. We now see a migration of men from military systems research into the study of industrial and economic systems. These men started in 1950 in the environment of "You can't make a machine substitute for my military training and command experience," and in 10 years have seen the same critics accept as better and as commonplace the automatic execution of front-line military judgment. Likewise, we are seeing that there is an orderly basis that prescribes most of our present managerial decisions, that these decisions are not entirely "free will," but are strongly conditioned by the environment. This being true, we can set down the policies govering such decisions and determine how those policies are affecting industrial and economic behavior.

AN ANALOGY AND PRECEDENT

I would like to illustrate the changes which I foresee in the fields of management and economics in the next three decades by citing an analogy. This precedent is the great change which has occurred in engineering and technology during the last three decades. The changes in the status and the world position of science and technology since 1935 are of the same nature and have occurred for the same fundamental reasons as the changes we can expect in management and economics between now and 1990.

Over the last 30 years, technology has held the spotlight in the center of the world stage, just as management and economics will during the next three decades. Furthermore, the fundamental reasons for great advances in management will be essentially the same as those which have thrust science and engineering to a dominant international position.

The first of these reasons is the development of a scientific base underlying the empirical practice. Prior to 1935, engineering tended to be an empirical art following handbook procedures, precedent, and experience. In the same way, management today is an empirical art, following the precedent of past experience. Before World War II, basic scientific developments in the world's universities lacked close ties to the practice of engineering. There was no strong applied science intermediate between basic information and its practical application; industrial research laboratories were the exception rather than the rule. In the same way, we now see developing a body of underlying principles of industrial systems behavior which has not yet found its way into the practice of management. There is thus far only an awakening of industrial awareness to the need for in-house applied research in the problems of the managerial process. Since 1940, the practice of engineering began to converge with the underlying science. Engineering advancement was founded more closely on mathematics, physics, and chemistry. The methods, instruments, and attitudes of research were no longer foreign to the practical fields of application. Research came to be recognized as an essential part of the technological process. Likewise, we are now beginning to see practicing managers recognizing the importance of research into management problems. Just as we now expect a percentage of the sales dollar to go to product research, so we may come to expect in the future that a certain fraction of the management and white collar payroll should be devoted to research toward managerial innovation.

A second reason for the upsurge in technology was a change in the concept of how to organize for scientific research. Prior to 20 years ago, research was most often a one-man activity. Now team research is recognized as essential if resources are adequately to match the task. Until recently, research in the social sciences has been largely at the individual level in the form of doctorate theses and university faculty research. The attitude toward management research is changing, and we already begin to see problems yielding to organized group effort.

The third reason for the upsurge in engineering revolved around an articulate recognition of the importance of "systems engineering." Systems engineering is a formal awareness of the importance of interaction between the parts of a system. Systems engineering is the central theme in the information feedback concepts. Systems engineering implies integration. It says that the whole is more than the sum of the parts. The systems concept is only now taking on some real meaning in the management picture. Until now, much of management education and practice has dealt only with components. Management schools have taught accounting, production, marketing, finance, human relations, and economics as if they were separate unrelated subjects. Only in the topmost managerial positions has it been necessary to integrate the separate functions. Our industrial systems have become so large and sophisticated that a knowledge of the parts taken separately is not sufficient. In management as in engineering, we can expect that the interconnections and interactions between components of the system will be more important than the separate components themselves.

INFORMATION FEEDBACK SYSTEMS

I have already stated that the concepts of information feedback systems will become a principal basis for an underlying structure to integrate the separate facets of the management process. What is an information feedback system? I would like to give an unusually broad definition:

> *An information feedback system exists whenever the environment leads to a decision that results in action which affects the environment.*

This is a definition which encompasses every conscious and subconscious decision made by human beings, as well as those mechanical decisions made by devices called servomechanisms. The condition of sales and inventories leads to decisions on the placing of orders that adjust inventories and the ability to make sales. The level of unemployment leads at one extreme to public works to boost the activity of the economy, and at the other to fiscal and monetary controls to cope with inflation. We are able to stand without toppling over because our environment, as manifested by our state of balance, leads to a subconscious decision to shift the body weight and correct any tendency toward falling. Everything we do as an individual, as an industry, or as a society is done in the context of an information feedback system. The definition is so common and all inclusive as to seem meaningless. Yet, we are only now becoming sufficiently aware of the tremendous significance of information feedback system parameters in creating the behavior of these systems. We are only beginning to realize the way in which structure, time lags, and amplification combine to determine behavior in our social systems.

Why has the fundamental nature and the importance of information feedback systems escaped notice until the last three decades? I think it is because of the peculiar classes into which these systems fell before 1940. On the one hand, we had the biological information feedback systems that regulate body temperature, muscular coordination, and other processes. These systems had been so ideally perfected for their purposes, and at the same time we had come to completely accept their shortcomings, that the systems and their information feedback character went unnoticed. On the other hand, social, economic, and industrial systems developed in recent centuries on so large a scale compared to the individual that the fundamental information feedback characteristics of the systems were most difficult to discern. Furthermore, hosts of other causes have been advanced to explain the behavior that arises directly from the information feedback character of these systems. Explanations have been in terms of the superficial specific manifestations of the particular problem rather than in the more fundamental terms of generalized closed-loop systems.

Our social systems are a great deal more complex than the information feedback systems that have already been mastered in engineering. Are we ready to tackle them?

Our knowledge of information feedback systems has been growing in that exponential manner so characteristic of human knowledge. The ability to deal with information feedback systems seems to have been progressing by about a factor of 10 per decade. Let me illustrate. During my first contact with the field of servomechanisms in 1939, one of the recent papers then read by the student was by Harold Hazen in the *Journal of the Franklin Institute* dealing with the dynamic characteristics of a control system described by a second-order, linear differential equation. By the early 1940s the field had developed into the concepts of Laplace transforms, frequency response, and vector diagrams, as later presented in books such as *The Principles of Servomechanisms* by Gordon Brown and Donald Campbell. But as usual, the forefront of mathematics was unable to cope with the problems of major engineering interest. Military necessity exerted pressure. Engineers did not long linger waiting for analytical solutions to information feedback system behavior. Linear and nonlinear mathematical models were constructed for solution on analogue computing machines.

By 1945, systems of 20 variables were easy to handle — a tenfold increase over 1935. By the end of the second decade in 1955, new methods and new areas were again being pioneered. The digital computer had appeared, opening the way to the simulation of systems far beyond the capability of analogue machines. With the new tools, attention began to center on the information feedback characteristics of military combat systems, incorporating both equipment and people. Systems of 200 variables could be feasibly studied. The pace has not slackened. We are now entering the 1965 era with another factor of 10 within reach. Models of 2,000 variables with no restrictions on representing nonlinear phenomena put us within a vast area of important managerial and economic questions.

PRESENT AND FUTURE

I have sketched for you my impression of the histories of management development and of control systems. I have suggested that as these converge there will be a basis for a great upsurge in our understanding of industrial and economic systems. Where do we stand today? Has this change begun to occur? It has. Persons in many places are beginning to study the dynamic information feedback character of our social systems. I would like to give you some tangible feeling for this, not by a general survey, but by reporting to you on our own specific activities at the M.I.T. School of Industrial Management.

Four years ago I moved from the development of military information handling systems to the Management School. For the last 3 years our work in this area has been financed by a grant from the Ford Foundation for the purpose of launching the field of Industrial Dynamics. In addition, there has been support from the Sloan Foundation and from several industrial companies. We have 12 full-time staff members, many of whom are also carrying on graduate study toward an advanced degree.

We are nearing the end of the first phase of our program, which has been to establish a general basis of methodology on which to build. This basis seems to consist of three parts:

1. Objectives and a guiding philosophy.
2. A generalized model structure that serves efficiently for representing any dynamic systems interrelationships.
3. A digital computer compiler for automatically programming computers to carry out the necessary computation.

As the first of the above results, we have thought through a guiding philosophy for objectives and approach. The goal is to design better systems, not merely explain what is now done. Models are not primarily for predicting specific future events, but are for displaying the general performance characteristics of a system. Models must be based on our vast body of descriptive knowledge and not merely on that for which numerical data exist. The approach must be completely amenable to any kind of nonlinear factors and to the inclusion of noise and uncertainty. Perhaps surprisingly, it has become apparent that the nonlinear model is much easier to develop successfully than a linear one because a great deal of our descriptive knowledge relates to the nonlinear conditions. We know what would happen under many limiting conditions even though data have never been collected. Functional relationships which meet all the known boundary conditions are constrained so that they cannot be far wrong in the normal operating regions. The purpose of a model is to give the better understanding of a system which can lead to improvement in system performance. Dramatic improvements are possible, and we need not at this time undertake a search for optimum solutions. In short, systems will be studied with the same philosophy that has succeeded in complex engineering and military systems. Models of industrial and economic systems will be evolved in the same ways and for the same purposes as they have in engineering systems. The greater complexity by a factor of 10 over what has been handled before is merely the next exponential step in knowledge beyond the background which now exists.

The second result from our work so far has been the establishment of a fundamental concept of systems structure, which realistically characterizes information feedback systems, be they technical, industrial, or economic systems. This is a simple, straightforward model structure into which we find we can cast any combination of flows, time delays, amplification, structure, and nonlinear decisions. This model structure lends itself to an orderly, straightforward mechanization on today's digital computers. I will say only a word about the actual structure. We find that two fundamental types of equations are required. One type we call a level equation; the other we call a rate or a decision equation. A level equation is a simple integration which gives the content of a level, such as an inventory, by accumulating the net difference between inflow and outflow rates. The rate equations are the decision functions of the system and control the rates of flow between the levels. These rates depend only on the levels in the system. There are no simultaneous algebraic equations. At every step in time for which the system is to be evaluated, each equation can be evaluated individually. When its time for computation comes, all values necessary for the determination of the new value of a variable are available. The computation is efficient, and the amount of digital computer time required is not an important consideration in conducting a systems study project.

The third accomplishment has been the design of

an automatic computer compiler especially for the efficient handling of this class of complex information feedback system. The compiler which we now have and call DYNAMO operates with the IBM 704 and 709 computers and will cope with systems up into the range of a few thousand variables. It takes the elementary system equations in the form prepared by the manager making a study. It applies extensive logical checking to find those errors which are oversights and logical mistakes. It arranges the equations in the proper order. It automatically converts them into a digital computer running program, and runs this program through the specified number of time steps. It then prints out the values of all specified variables at the desired time intervals. It also plots with respect to time such variables as are desired. It can do all of this for a system of 100 variables solved at 1,000 successive time steps in a total of 3 minutes of computer time for automatic programming, running, and generation of printed and plotted information.

With this philosophy and methodology available, effort can be concentrated where it belongs, on the system under study rather than on the routine drudgery so often associated with empirical experiments with models. Even within the time span of a master's degree thesis, we regularly have students formulate and study the characteristics of systems involving from 100 to 300 variables.

SOME RESEARCH PROJECTS

You might be interested in some of the kinds of systems which have been explored thus far. Some have been staff research projects carried on under the Ford Foundation support. Some are sponsored by industrial organizations with whom we are working cooperatively. Some have been student theses.

Effect of Management Policies - We have been working with a manufacturer of electronic components on one product line which has been set up as a dynamic information feedback system of some 150 variables. These variables describe the processes involved in order filling, inventory reordering, manufacturing, the labor sector with its hiring and firing, product delivery delay, and customer ordering. The phenomenon under study is fluctuation of employment and manufacturing rate that arises from a fundamental instability of the interacting policies within the system. The interaction includes the tendency for customers to order ahead when deliveries become slower. This builds up backlogs of unfilled orders and consequently increases still further the delivery delay. This ordering increase in response to rising delay, and rising delay caused by greater ordering, can regenerate until after the delayed build-up of factory labor and production rate has occurred. By this time, sales seem to be at a new high rate and order backlogs have become excessive. Manufacturing rate then overshoots in an effort to meet the apparent increased demand and to reduce the excessive backlog of unfilled orders. As soon as production equals the sales rate and begins to reduce the backlog, delivery delay begins to improve. The customer can reduce his ordering rate and live on the accelerating flow of previously ordered goods. This typical industrial system structure in combination with common and ordinary management policies here creates a production fluctuation of about 2-years interval from peak to peak in the production and sales curves. This occurs even when the primary demand for the product is constant.

Dynamics of Growth - Another cooperative study is with a small corporation which is only 3 years old. Here we are formulating the interactions of those factors which, seem to determine the transient dynamics of corporate growth. These include the characteristics of the products being developed, the nature of the market, the technology of the development and manufacturing processes, and financing and tax policies. The study is aimed at showing those combinations which on the one hand can give us corporations with a 30 or 50 percent annual growth rate and, on the other hand, those which lead to stagnation and failure.

Stability Versus Weather - Another cooperative study just now being started deals with an agricultural product. It will represent in one dynamic system the motivations for planting, the uncertainties of weather, the inventory carryover from preceding years, price generation, manufacturing processing rates, advertising expenditures, promotional policies, inventories in the distribution system, profitability, and the effect which profitability and other factors have back on the decisions in planting. This is a system of many major and minor closed loops. It will be aimed at finding those inventory, price, production, and marketing policies which will best stabilize a system which is constantly being disturbed by deviations in the weather.

Instability in an Industry - One of our staff members is working on the dynamic instability of the electric generator and turbine industry. Here we have a situation where the demand for electricity grows very smoothly and yet the manufacturing of generating equipment is one of our most unstable industries. Orders fluctuate over a 10 to 1 ratio with a typical period of some 5 or 6 years. Manufacturing rate fluctuates by as much as 4 to 1. Informal estimates place the excess product cost arising from this instability as high as 20 percent of the product price. It is easy to show how this kind of behavior will result from a combination of the load forecasting methods of the utilities, the long lead time enforced by the technological nature of the product, the existence of excess manufacturing capacity at the equipment manufacturers, and the variability of delivery lead time which occurs as the manufacturers become loaded with orders. The trouble is accentuated rather than alleviated by a tendency which has developed toward price softening and discounts occurring near the time when order backlogs are reaching a minimum.

Research by Sloan Fellows

Some very interesting systems studies have been made by our Sloan Fellows. These are men between 30 and 40 years old, usually with some engineering background, who are reaching management levels in their companies. They are sent to the M.I.T. School of Industrial Management for a full calendar year where they work toward a master's degree in management. Several of these men have done very interesting theses in Industrial Dynamics.

One is Edward Kinsley of the Texas Instruments Company. Using 160 variables he set down the interrelationships that he felt described the life cycle of a new product as it was affected by six sectors internal to the company (manufacturing, personnel, capital equipment, money, marketing, and research and development) and two external sectors (the market, and pricing and competition). He dealt with the generalized characteristics of a typical product rather than a specific product. Treating the general case forces attention onto the fundamental factors and prevents diversion by the more superficial details of a specific product. An example is the ratio of development time to market education time. This is a fundamental dynamic parameter. You can see immediately that a long development time combined with a long marketing education time and a short length of time necessary to start production defines a product in which one cannot expect a period of high profitability created by technological monoply. Full information and samples must be released early to the market. Before the market demand develops, competitors can enter to produce the product. This is characteristic of the transistor industry in which the organizations doing early development are not the major producers today. By contrast, a long production preparation time and a short market education interval makes it possible to enjoy a competition-free period with a new product. Simultaneously, price, quality, capital equipment, manpower, and other considerations would thread their interacting ways through the system dynamics.

Ray Ballmer from the Kennecott Copper Company developed a model of the international copper industry involving some 150 variables. This dealt with mining, concentrating, smelting, refining, fabricating, and manufacturing. Historically, copper price and supply have been notably unstable. The dynamic model built by Ballmer shows that the accepted and recognized policies of the industry can cause this instability even with a constant average final usage rate of copper. Even more exciting are the indications that one can hope to find corporate policies that will alleviate the problems.

Others have dealt with the relationship between design policies and market penetration in the automobile industry, and still other studies have begun to set up the policies and the flows of information which seem to represent the dynamic variables in military research and development management.

BACKGROUND FOR SUCCESS

In this work we have found two kinds of men to be the most successful. One kind has had formal training in the analysis of information feedback systems and preferably has also had laboratory work and firsthand experience with such systems. The second kind is the competent industrial manager. He also has dealt with complex information feedback systems in the true-life environment. Even better, there are many men with both kinds of training. To be successful, a man needs a good feeling for system dynamics and also he must understand the factors which go into managerial decision making. He needs a strong motivation toward finding helpful insights into better management. So far, we have had much less success with those trained in mathematics, in conventional operations research, and in mathematical economics. Men who are interested primarily in these latter areas seem often to lack one or all of the essential qualifications of dynamic systems intuition, motivation, and understanding of the business processes.

Formal training in servomechanisms theory is very valuable as a background but not so that the man shall use it for formal analysis. The problems are too complex for analytical treatment. However, a good foundation in linear systems analysis develops an alert, intuitive understanding and keen perception of the kinds of factors which can give information feedback systems their typical behavior characteristics. It is my feeling that formal theory will continue to be an educational rather than an operating tool.

While the fundamental principles of ordinary information feedback systems are applicable, the quantitative nature of industrial and economic systems can be quite different from what the engineer ordinarily encounters. We do not have the simple, single-loop control systems in which there is one point of error detection and one place where correcting control is exercised. Error detection and control are both distributed throughout the system. Every decision point constitutes both. Many of the practical solutions to stability problems which have been worked out for engineering systems are not applicable in social systems. Many of the methods which have been found of practical value in engineering systems work because of the special characteristics of those systems. Engineering systems tend to be so designed and so used that the normal operating frequencies, the natural frequencies of the system, and the impressed noise frequencies tend to fall into different frequency bands. This seems not to be true in our industrial systems. The disturbing noise, the natural frequencies of the systems, and the frequencies to which they are expected to respond all fall within the same frequency band.

Out of this work there is developing a central framework for management education. Management education in the past has been split into various unrelated functional studies. Courses have tended toward mere description of industrial practice. From the new viewpoint, industrial activities emerge not as the seperate functional specialties, but as a system of flows dealing with materials, money, manpower, orders, and capital equipment, all of which are integrated by an information flow and decision-making network. This is a dynamic information feedback system viewpoint. It provides a vehicle for management policy education and for the transfer of knowledge and experience from one point and from one time to another.

25

Jay W. Forrester

A DEEPER KNOWLEDGE OF SOCIAL SYSTEMS

When considering the computer in society, one could ask, "How can man improve his understanding of the computer so as to make the computer better serve society?" But here the perspective will be reversed to ask, "How can the computer improve our understanding of society, so that society can better serve man?"

Man neither understands nor has mastery of his social systems, as a glance at the newspapers will testify. More technically, we fail to understand social dynamics – how our social systems change with time, and how their behavior depends on their structures and on the policies at work within them. Toward a better understanding of human systems, we have at the Sloan School for the last 12 years been developing "industrial dynamics." But that title has become a misnomer. The theory and concepts we employ describe the dynamics of systems in general – not

---

 Selected from Technology Review, *71(6), 22-31 (April 1969). To appear in* Collected Papers of J.W. Forrester, *J.W. Forrester, ed., Wright-Allen Press, Inc., Cambridge, Massachusetts (1974).*

merely managerial systems but all systems. Our work focuses on the principles of systems, principles that apply to any system whose condition changes with time. The principles cover systems in physics, engineering, management, economics, medicine, and politics — wherever interactions between components change the condition of the system as it progresses from the present into the future. The universality of the principles underlying the structure and dynamics of systems gives our students a very remarkable mobility. They can readily move between fields as different as engineering, public health, ecology, and management. This mobility is possible because important and useful principles apply equally well to controlling chemical processes, guiding spaceships, and working out policies within our social systems.

To illustrate this generality, let me now describe some principles governing structure in systems and then show how these principles can be used to examine the dynamics of growth, decline, and revival of an urban area.

Orderly and universal concepts of structure apply throughout all systems. Most fundamental is the concept of the feedback loop. Indeed, all decisions are embedded in feedback loops. All processes of growth and goal seeking are governed by feedback loops. All interactions that cause change through time occur within feedback structures. The simplest possible feedback loop (Figure 1) illustrates the basic structure. Here the "rate" is a statement of policy that tells how the available information is processed to generate a flow that changes the "level" or condition of the system. As the condition (level) changes, it presents new information to the policy (rate) to change the flow to further change the level. The process is continuous and circular.

To illustrate, Figure 1 can describe the process of filling a glass under a faucet. The rate variable describes the water flow. The level variable is the water level. The level of water is being controlled by the rate of flow, and the rate of flow is being controlled by the level. As one watches the level, he adjusts the faucet to control the rate of flow based on the level and how full he wants the glass. The information part of the loop from glass to eye to hand to faucet is usually overlooked. We are often not aware of the circular structure within which our actions take place. All decisions occur within a feedback loop structure, whether the decisions are public or private, conscious or subconscious.

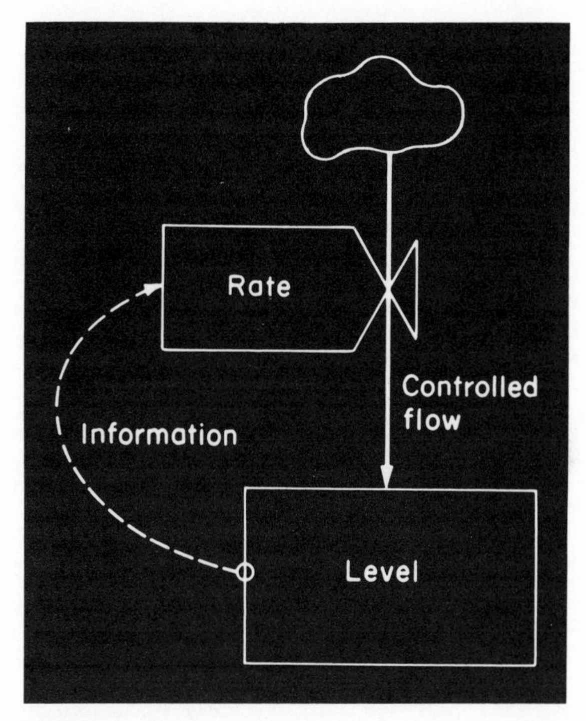

Figure 1
The simplest possible feedback loop, having one rate and one level.

Figure 1 illustrates a general truth about systems. They are composed of two and only two kinds of variables — the rates and the levels. The rates are instantaneous, algebraic relationships that describe how the levels influence the flows. The levels are the accumulations of the rates of flow, and are generated by integration over time.

DIFFERENTIATION IS UNREAL

Integration is the fundamental process by which change occurs in all systems. Integration, not differentiation, is the time-dependent process in dynamic systems. This may be surprising to engineering readers; in our educational system differential equations and the mathematical notation for differentiation are deeply entrenched. But there is in fact no place in nature or society where the process of differentiation occurs. Differentiation is a figment of the mathematician's imagination; it has no reality. All the processes of physics and of social systems integrate or accumulate. In a swinging pendulum, acceleration is integrated into velocity, and velocity is integrated into position, even though the equations are usually written as derivatives. Indeed, the concept of differentiation, when its unreality is not fully understood, misleads many students. It tends to create a reversed sense of causality. Returning to the filling of a glass of water, if one expresses the rate of flow as the derivative of the water level, as is commonly done, one is perilously close to saying that the water is flowing because the water level is changing. As more complex systems are encountered, the danger of failing to see the direction of causality increases, unless the flow-diagram structure and mathematical notation are made to agree with the processes in the actual system.

That two kinds of variables (levels and rates) are necessary and sufficient in a system has been recognized in several fields. For example, the difference between levels and rates is clearly recognized in financial accounting; the balance sheet variables are kept strictly separate from the profit-and-loss sheet variables. The balance sheet variables are levels (or state variables); the profit-and-loss variables are rates. Also, the distinction between rates and levels was formulated in psychology by Kurt Lewin when he wrote of the "psychological field variables," the set of mental conditions representing the state of a person's psychological system. These psychological field variables (levels) and the rates of change (rates) in the field variables were described as a complete statement of the psychological system.

Figure 1 illustrates also a second general truth about systems. Rate variables depend only on levels. Level variables are changed only by rates No rate depends on another rate. No level depends directly on another level — there has to be an intervening rate. . . .

These principles (that two kinds of variables are necessary and sufficient, and that neither type of variable depends on its own kind) are but examples of many that can be identified in systems. . . . Other principles relate dynamic behavior to system structure. Such principles, when one becomes convinced of their universal applicability, are powerful aids in dealing with an unfamiliar system. It is far easier to identify a quantity as either a level or a rate than to ponder which of an unknown number of categories it might belong to. Knowing the relationship between levels and rates, and anticipating the feedback loop structure (even though the structures are far more complex than Figure 1) helps to organize our observations about systems. Principles that are universal from physics to psychology and from economics to ecology give a man mobility by providing a common foundation from which to perceive all dynamic behavior.

Engineers have studied and worked with feedback (or control, or servomechanism) systems for several decades. But engineers have usually not seen the full generality of the feedback process. Feedback control has often been defined as limited to physical equipment where small signals control large flows of energy, without recognizing that the same structures, the same modes of behavior, and the same dynamic principles apply wherever changes occur through time.

THINKING TOO SIMPLY

Because of inability to obtain mathematical solutions to problems relating to complex systems, engineers developed the differential analyzer and later the digital computer. These computers manipulate models which represent the real system in which we are interested. But models were not first created by or for computers. Models have been used as long as man has been able to think. Mental images in thinking are models. Such images are not the real system, they are abstractions that stand for, represent, or simulate the real system. Computer models differ by being more explicit than mental models. They are more useful for communication because they are less ambiguous than descriptions in words.

In addition, computer models of feedback systems can show the dynamic consequences of interactions between components of the system — an ability almost totally lacking in our mental processes. We can be sure that a computer is showing correctly the consequences that follow from the assumptions it is fed. But the mind refuses the task, or, worse, is usually wrong, when it attempts to estimate the dynamic behavior of a system having a given structure.

It technology, where systems are comparatively simple, engineers have long ago recognized that understanding must be based on models, now usually computer models. But in social systems, with their far greater complexity, we still attempt intuitive solutions to give dynamic behavior.

However, intuition is unreliable. It is worse than random because it is wrong more often than not when faced with the dynamics of complex systems. There are reasons why intuition leads to error. Our intuition has been built up over a lifetime of immersion in the simplest possible feedback loops. A baby in turning his head toward a light source gains his first experience with the simple type of feedback loop which operates on a discrepancy between a goal – where the light is – and a condition of the system – where the head is. From then on, every learning process takes place within the context of simple feedback loop structures. But this learning process is misleading in two ways. First, even the simple feedback structures are perceived incorrectly. Second, the behavior of the simple systems that we learn to understand is in many ways quite the opposite of that of complex systems. We tend to perceive incorrectly even the simplest loop structures. We think of A as causing B without realizing that B is also causing A. That is, we think of unidirectional cause and effect linkages rather than the feedback loop structure that actually exists. For example, we think of the faucet controlling the water in the glass but do not think of the faucet at the same time being controlled (through our sight and action) by the water in the glass.

The conditioning acquired from simple systems misleads us in the realm of complex social systems. We do not realize that the behavior of a system is determined primarily by its structure and (rates) policies, and not by incidents of the A-causes-B variety. Complex systems are diabolical. We approach them expecting to find simple cause-effect relationships to explain the troubles in a system. Not only is there no simple cause – the "cause" is in the structure, and in the interactions of components interrelated in complex ways – but even worse, the system deludes us into a sense of false security by presenting apparent cause-effect pairs that match our expectations. Such apparent causes are usually only coincident symptoms; they are not leverage points through which the behavior of the system can be corrected. In fact, the "obvious" correction exerted on the apparent cause will often make matters worse.

This counterintuitive behavior in complex systems appears repeatedly in management systems. In our research we often examine the interactions within a corporation with a clearly recognized set of troubles – like fluctuating employment, falling market share, or low profitability. In such situations, one usually finds that people are quite aware of their actions, their response to pressures, and their efforts to correct the troubles. By careful examination, we ordinarily find that people are in fact doing what they believe they are doing. One can then build a system model from these known practices that are governing behavior and show that the known practices, in fact the very things being done to solve the problems, are sufficient to cause the troubles. The consequences of the practices as they interact with one another are very different from those predicted by intuition and judgment.

HOW TO LOOK AT CITIES

The contrary nature of complex systems is illustrated by my recent examination of the growth, stagnation, and revival of a city (*Urban Dynamics*, MIT Press, Cambridge, Mass., 1969). The study shows that the depressed areas of our cities are generated by the policies that govern urban areas. Furthermore, the actions intended to relieve the symptoms of urban distress can often make matters worse. In examining a social system such as a city, one should not begin by attempting to solve its problems. First, one must identity the system structure that creates the undesirable symptoms. Unless the underlying casual relationships are understood and altered, any efforts to relieve symptoms struggle against the still-present forces that continue to work toward undesirable ends. In fact, efforts merely to treat symptoms usually increase the counterpressures within the system and neutralize the corrective efforts.

So if one wishes to redesign urban policies with

the aid of a computer simulation model, the model should contain the processes that generate urban blight. Only then, with the causal interactions present, can corrective policies be reliably evaluated. A proper model will be able to start with empty land, show the growth of a city, and reproduce the aging sequences as slums develop.

In addition to incorporating the processes that create the troubles, the model's construction should follow another guiding principle. It should be a model of the *general* urban life cycle, not a model of a specific city. The general model will be simpler and clearer than a specific model. It focuses on the processes that are common to all slum areas whether now or in the past, in the United States or elsewhere, New York or a gold rush camp. It avoids the peculiar facets of individual situations which, because they can sometimes be missing, must not be an essential part of the urban decay process.

Figure 2 shows a model structure for studying the life cycle of an urban area. The model contains two kinds of variables – nine level variables represented by rectangles and 22 rate variables represented by valve symbols. The rates cause the levels to change as shown. The levels control the rates of flow through "information" connections (in the sense of Figure 1) too complex to show on this diagram.

This representation of a city contains three main

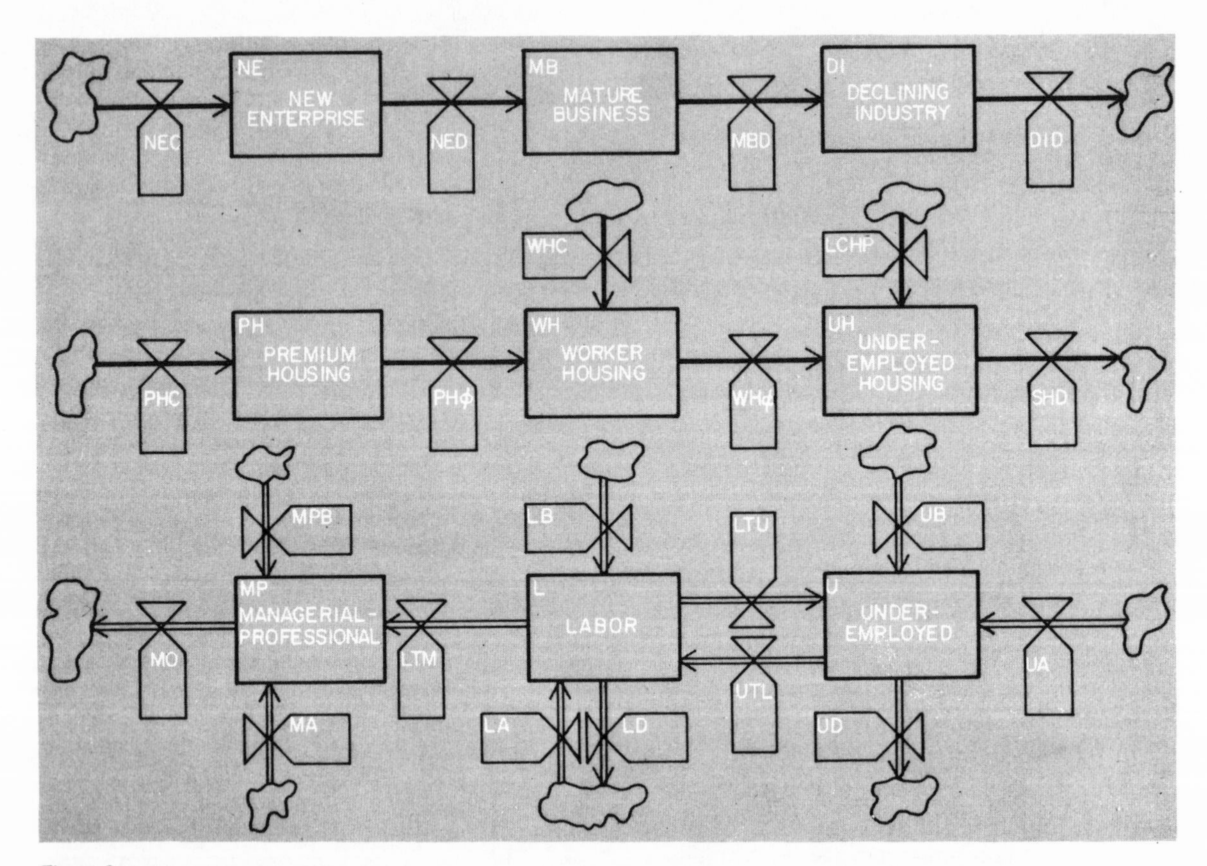

Figure 2
Major levels (rectangles) and rates (valve symbols) for the model of an urban area. Reading from the top, the levels modeled are the amounts of different classes of business, housing, and people. Flow rates are shown, but not the influences controlling them.

subsystems, shown one below another. Across the top is the industrial sector that generates new enterprise and causes it to age progressively into mature business, then declining industry, and finally demolition. In the center is the housing sector, which builds three categories of housing and, by aging, causes structures to move from the left toward the right categories and finally to be demolished. The lower subsystem generates the movement of people to and from the urban area and between economic categories of population.

Figure 3 shows the kind of information linkages that were omitted from Figure 2. In Figure 3 the dotted lines connect the system levels to just one of the 22 flow rates, here the rate of arrival of underemployed into the urban area. A similar net-

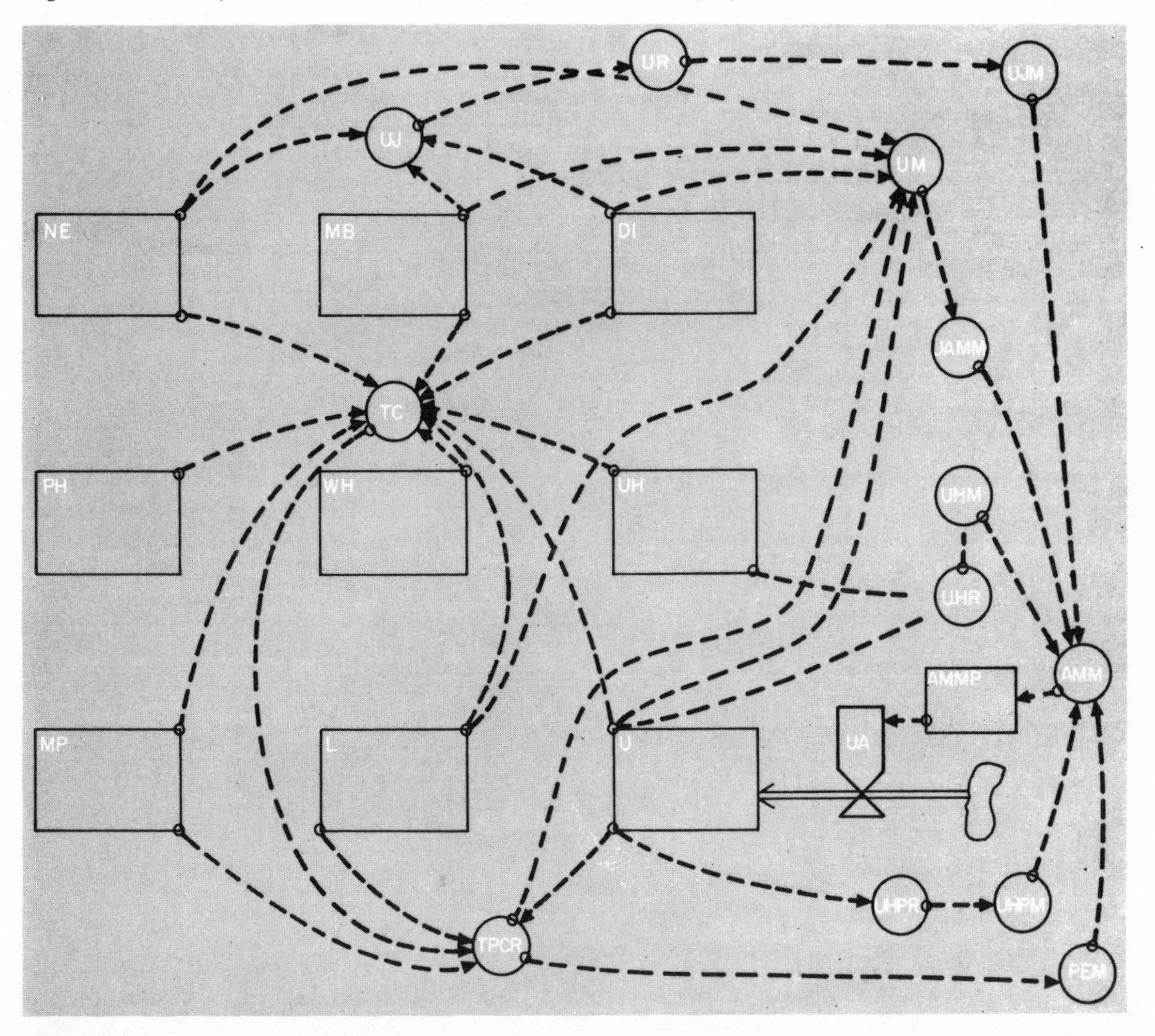

Figure 3
Typical set of influences controlling one of the flow rates in Figure 2. This diagram shows how the levels in all parts of the system govern the rate of arrival of underemployed people (lower right).

work controls each of the other 21 rates of flow. In the figure the arrival rate depends on the underemployed/job ratio (U.R.), the upward economic mobility from the underemployed to the labor class (U.M.), the underemployed housing ratio (U.H.R.), the low-cost-housing construction rate (U.H.P.R.), and the public expenditure ratio (P.E.M.). The underemployed-arrival rate depends on the condition of the industrial sector describing the available jobs, on the housing sector, and on the conditions within the population sector.

THE CITY'S HISTORY

Figure 4 shows a 250-year urban life cycle generated by the computer model. It starts with nearly empty land, generates growth and the filling of the particular land area, and then emerges into a final equilibrium stagnation. The 100th year is about the end of the growth period with the peak in new enterprise about equal to the amounts of mature business and of declining industry. At that time the underemployed population is half the labor population. But at the 250th year, underemployed population has risen and labor has fallen until the two are equal. Declining industry has risen to dominate the industrial scene compared to new enterprise or mature business.

Figures 2 and 4 help show the underlying reason for decline of an urban area. During the first 100 years construction is predominantly of new enterprises, premium housing, and worker housing. As the

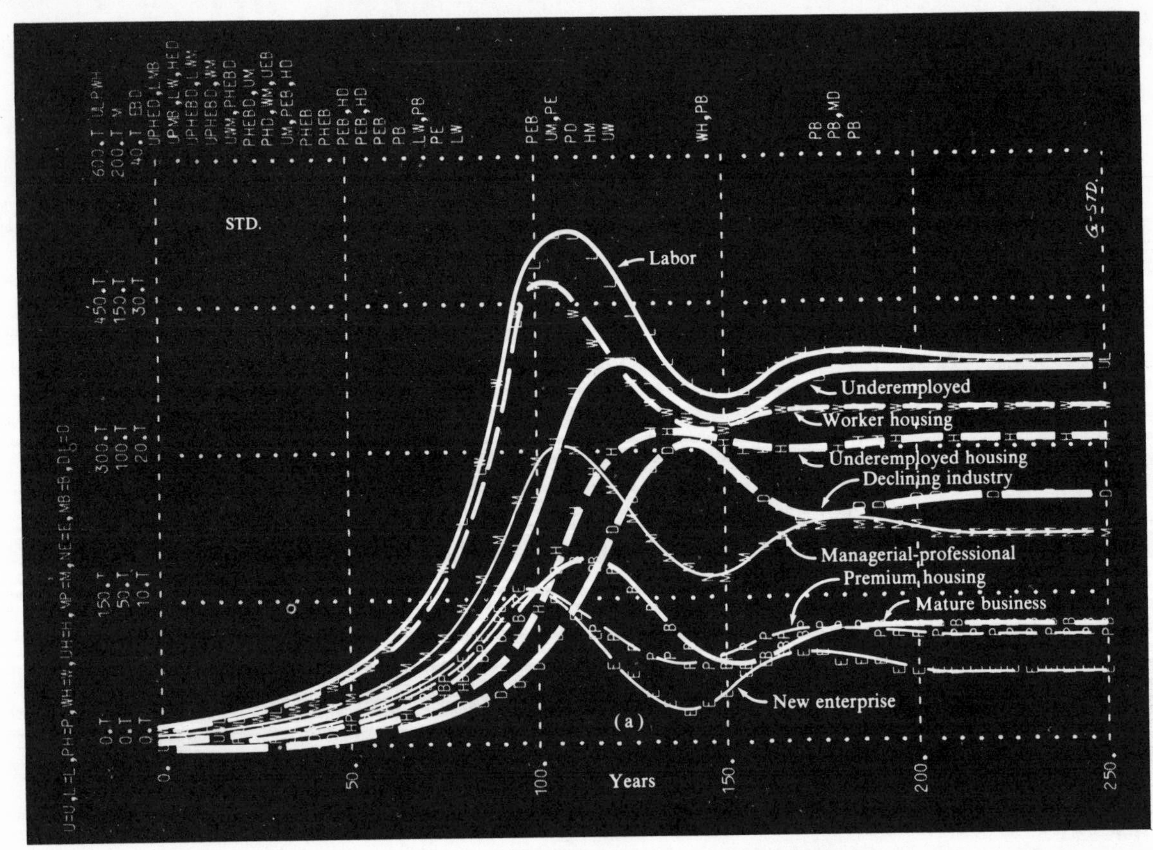

Figure 4
Urban development, maturity, and stagnation, showing the major levels during the urban life cycle.

particular land area (a fixed area representing part of a large city) becomes full, new construction is reduced to the replacement of demolished structures. At the end of the growth phase, buildings are mostly new and are occupied by successful industry and economically self-sufficient population. But as they age, employment in the industrial buildings decreases while occupancy in the housing increases. The increased population density in housing occurs because the aging housing is acceptable only to lower-income families who must use the space more intensively than those who occupied it when the buildings were new. The declining area is then characterized by falling job opportunities and rising population. The living standards of the area fall as the per capita income declines. When the standard of living falls far enough, the depressed conditions limit further migration into the area even though there may be empty housing. A balance is established where available housing beckons the underemployed while lack of available work repels them.

Figure 5, showing additional variables over the same time interval as Figure 4, tells how internal conditions change as the area shifts from growth to stagnation. The underemployed/job ratio, a measure of unemployment, was low during the first 80 years of growth. But between 80 and 130 years this ratio rises suddenly as the underemployed population continues to rise while jobs decline. At the same time the underemployed/housing ratio, a measure of housing scarcity, falls. During growth, population balance was maintained by high job opportunity counteracted

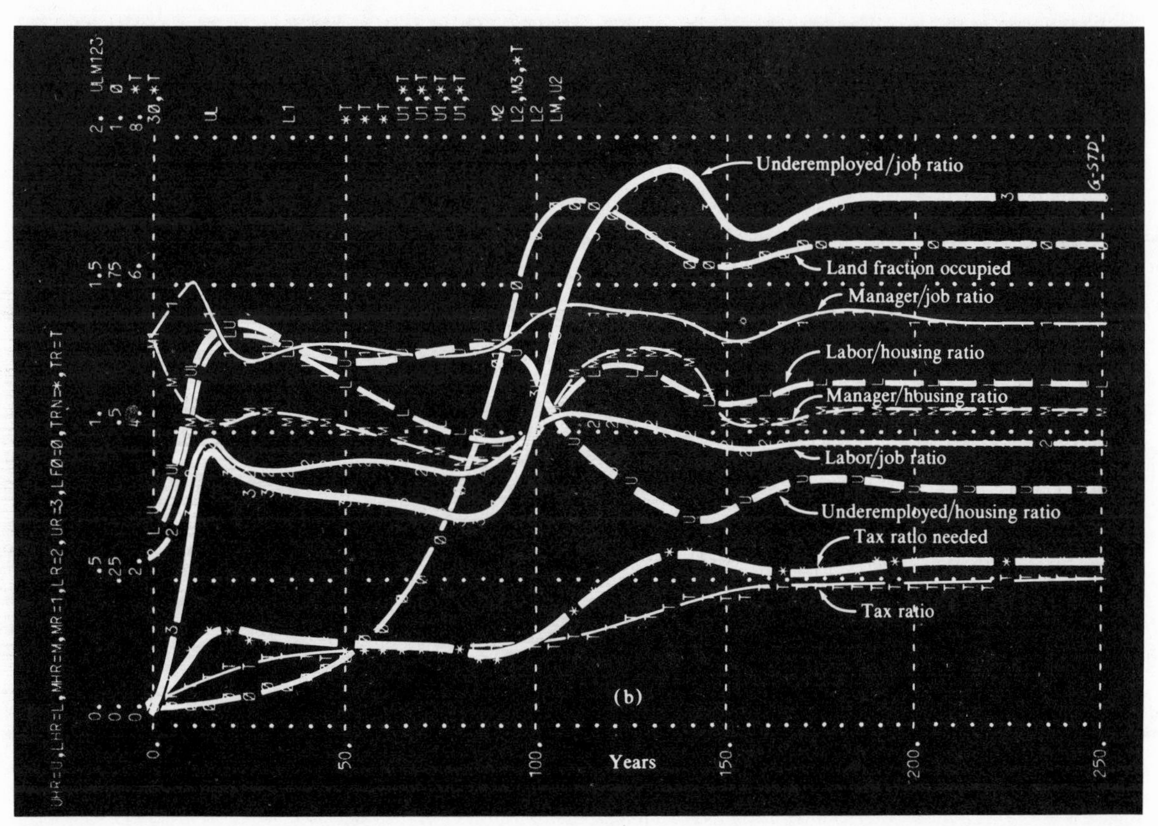

Figure 5
Important ratios as they change during the life cycle in Figure 4.

by a housing shortage. During stagnation, the roles reverse, with a job shortage and a housing excess. Few people think of the slum areas as having excess housing, but housing is excess compared to the available work and even excess compared to the population (given the housing density that their limited means forces them to tolerate). Empty buildings exist adjacent to overcrowded buildings. This is caused by a depressed economic condition, not a housing shortage. The income of the area is inadequate to maintain and use all the housing.

LOW-COST HOUSING?

Figure 6 shows how the computer model can be used to examine policy changes. The 50-year span starts with the conditions as at the end of Figure 4. A low-cost-housing program is started and sustained throughout the 50 years. It builds housing each year for about 2.5 percent of the underemployed population. The amount of housing available to the underemployed rises, but most other conditions in the area are further depressed. The housing attracts more underemployed population until economic conditions decline far enough to counteract this inflow. During the 50 years, new enterprise falls by 49 percent and mature business 45 percent. Premium housing declines by 34 percent and worker housing 31 percent, while housing for the underemployed rises 45 percent. The managerial-professional and labor populations fall by over 30 percent. The underemployed/job ratio rises 30 percent – there is more unemployment than before.

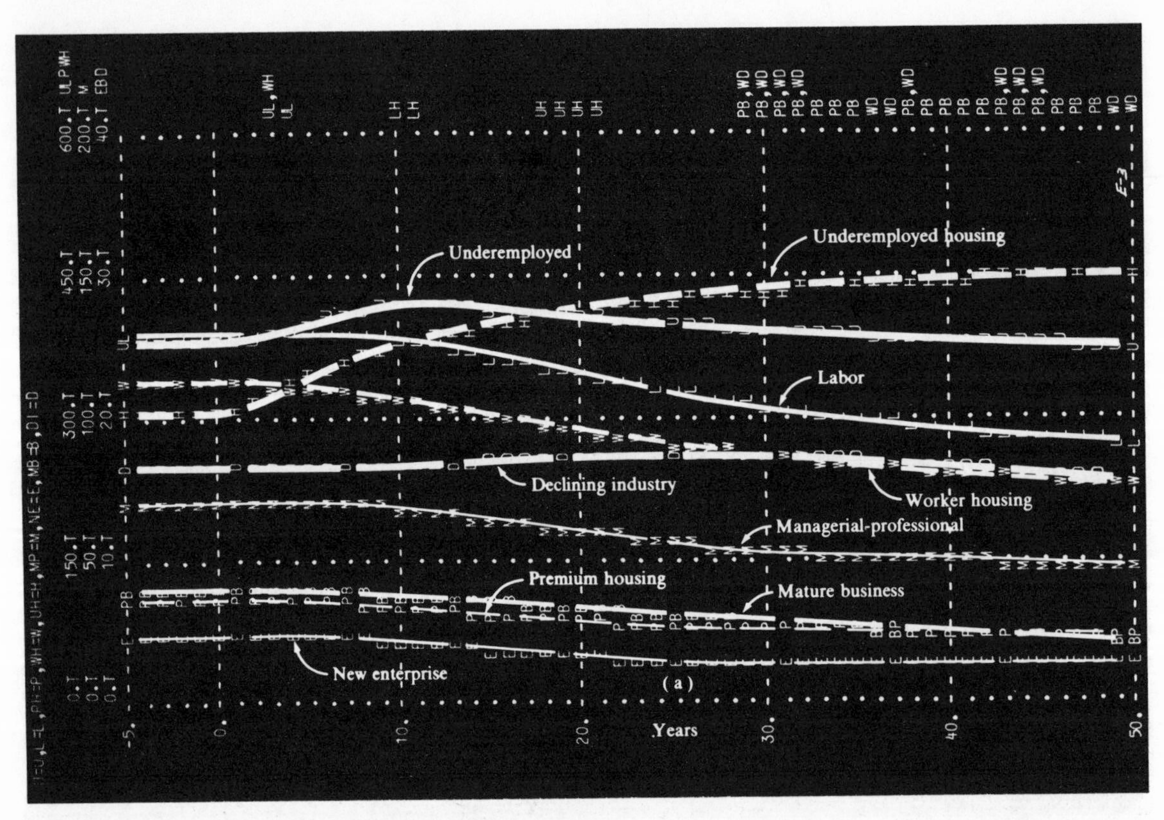

Figure 6
Construction of low-cost housing each year for about 2.5 percent of the underemployed (starting at the end of the life cycle already shown) causes deterioration in most of the system's conditions.

Of the many policies examined that are commonly assumed to help an urban area, construction of low-cost housing appears the most detrimental. It draws people into a social trap from which there is little economic escape. As explained in *Urban Dynamics,* other common proposals for alleviating the condition of the cities – job training, government-created jobs, and national financial subsidy to the cities – all appear to lie between neutral and detrimental.

REVIVAL FROM WITHIN

Figure 7 suggests an approach for urban revival from within. Again, a new set of urban management policies operate for 50 years, starting from the final stagnation conditions of Figure 4. Here a slum-demolition program removes 5 percent per year of the most deteriorated category of housing. At the same time, incentives have been added and hindrances removed to favor a more rapid construction of new enterprise. In the 50 years, we see that new enterprise and mature business rise by over 60 percent; professional and labor populations are up by over 50 percent while the underemployed population has fallen 11 percent.

Figure 8 accompanies Figure 7 and shows the compensating changes in housing and jobs. The underemployed/housing ratio has risen by 58 percent,

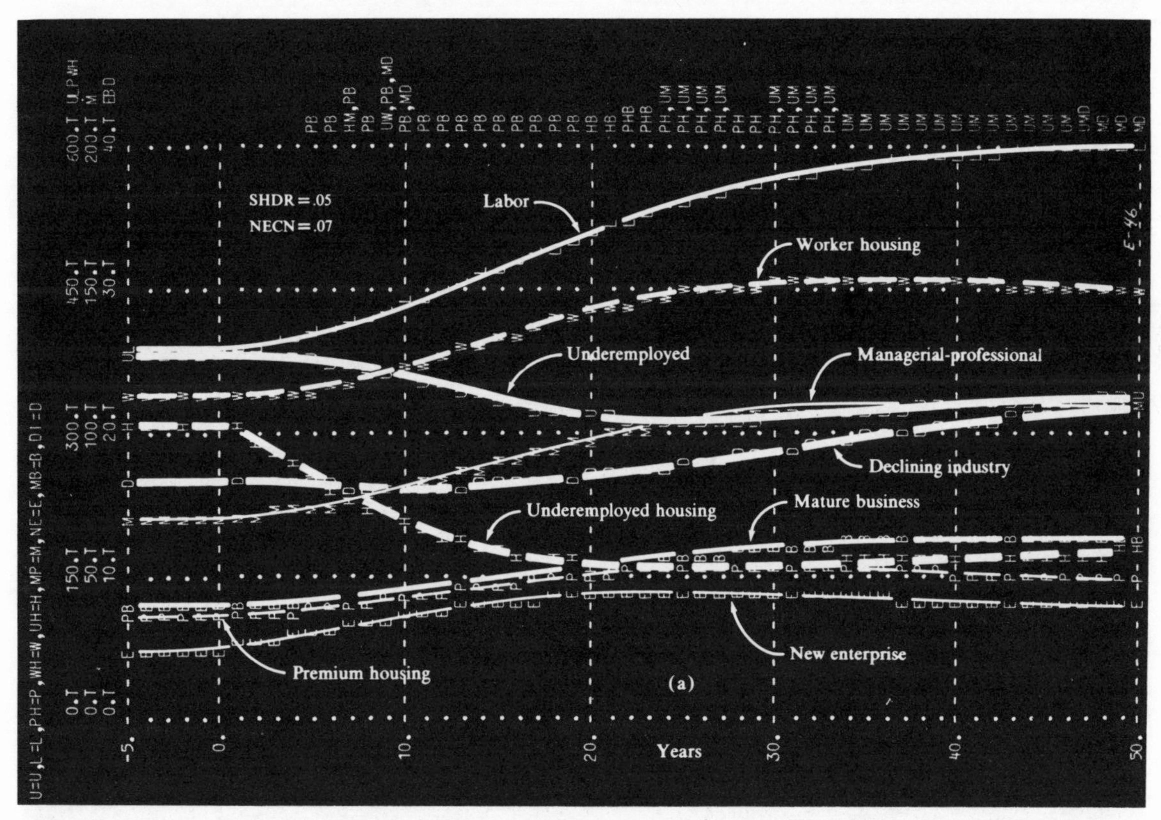

Figure 7
Again starting at the end of the life cycle, the demolition of 5 percent of the underemployed housing each year, together with business encouragement that increases the construction of new enterprises by 40 percent above that which would otherwise have occurred for the same conditions, causes an economic revival, with improved job opportunities for the underemployed.

meaning that less housing is available. But the underemployed/job ratio has fallen by 41 percent, to a point where people and jobs are about in balance. If the area is to revive economically, it is essential that the excess housing at the lowest level by removed. Otherwise anything that tends to improve living conditions will attract enough lowest-income people into the available housing to pull the standard of living back down far enough so that economic distress limits further inward migration.

Such a proposal for reducing the slum housing might seem antisocial and detrimental to the interests of the underemployed population, if one did not look deeper into what the system is doing. Before the new policies for revival, the run-down housing attracted people into a situation that had little upward economic mobility, so they became stranded in economic despair. By contrast, in Figures 7 and 8, additional information from the simulation model shows that after 50 years of the new policies more underemployed are entering the area each year than before, fewer are leaving, and the upward economic mobility of the residents has risen by 67 percent. The area has become a much more effective socio-economic converter for the poor, as well as having a higher internal standard of living.

The convergence of two areas of M.I.T. pioneering – feedback theory and computers – may product far greater impact on society than more popularized developments like atomic energy and space flight.

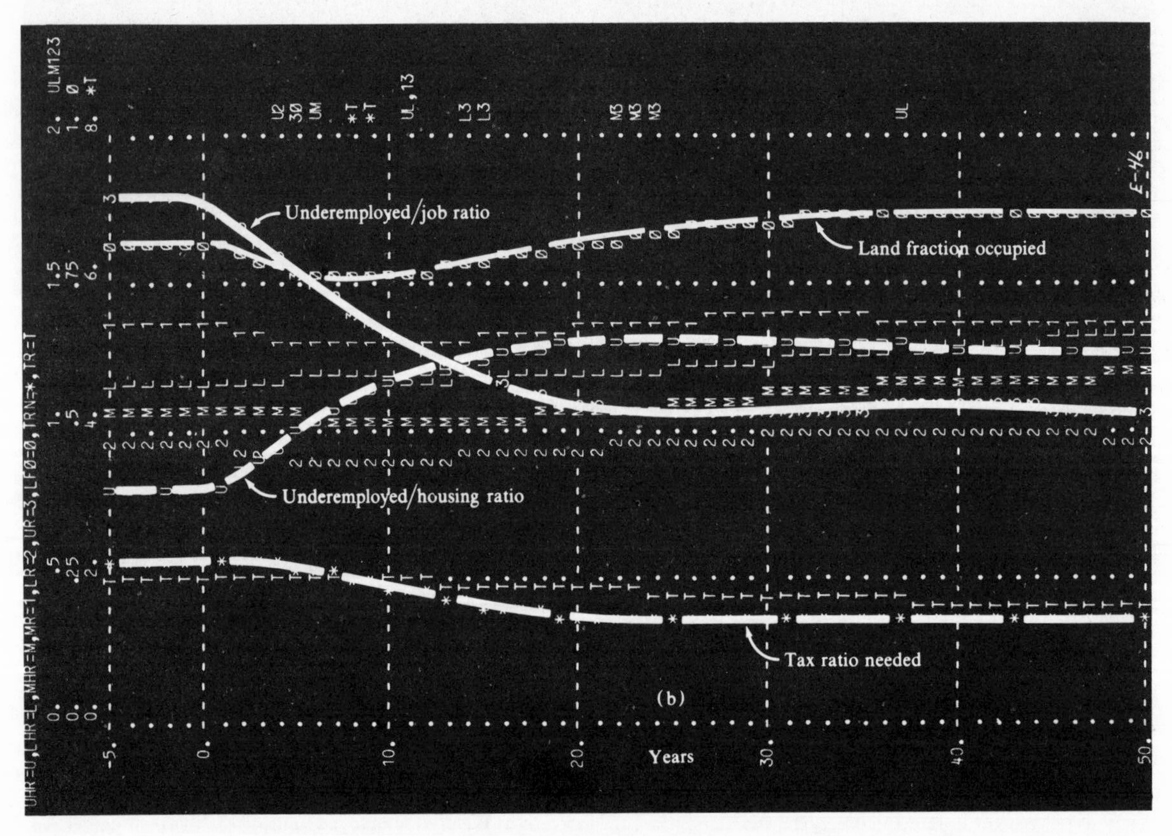

Figure 8
Changes in underemployed/job ratio, the underemployed/housing ratio, and the tax ratio that result from the changes shown in Figure 7.

Economic crises, inflation, international conflict, failures in developing nations, price instability in world commodities, urban deterioration, student unrest, population pressures, and the stresses between the individual and the organization all reveal our failure to master the social dynamics of our environment.

It is now clear that such complex systems can be understood and that much improved legal, financial and organizational structures can be devised. But the improved understanding will come slowly. The professional training needed to deal with such systems is at least as extensive as training for the professions of engineering or medicine. The student must acquire theory, case studies, laboratory work, and an apprenticeship. Only parts of an appropriate educational program now exist; the remainder must still be created. Here lies the most challenging frontier in research and education.

Much of the decision-making in the real world takes place in an environment in which the goals, the constraints and the consequences of possible actions are not known precisely. To deal quantitatively with imprecision, we usually employ the concepts and techniques of probability theory and, more particularly, the tools provided by decision theory, control theory and information theory. In so doing, we are tacitly accepting the premise that imprecision – whatever its nature – can be equated with randomness. This, in our view, is a questionable assumption.

Specifically, our contention is that there is a need for differentiation between randomness *and* fuzziness, *with the latter being a major source of imprecision in many decision processes. By fuzziness, we mean a type of imprecision which is associated with* fuzzy sets . . . *that is, classes in which there is no sharp transition from membership to nonmembership. . . . Actually, in sharp contrast to the notion of a class or set in mathematics, most of the classes in the real world do not have crisp boundaries which separate those objects which belong to a class from those which do not. . . .*

R.E. Bellman and L.A. Zadeh, "Decision-Making in a Fuzzy Environment," Management Science, *Vol. 17, No. 4, December 1970, p. B-141.*

26

Harper Q. North and Donald L. Pyke

TECHNOLOGICAL FORECASTING TO AID R & D PLANNING

There is probably no segment of society which has been left untouched by the rapid growth of the world's technologies. Their growth has certainly revolutionalized our industries. The consumer sector of our economy has felt the tremors from the explosion in technology, and their indirect effects have become evident in the social sector, where we see unrest among groups protesting events which are engulfing us, or movements toward dropping out of the complex society which we have created. The quest is for a more peaceful and manageable existence. The technological explosion has probably had its greatest impact in national defense with the advent of nuclear warfare, and in electronics with the invention and exploitation of the transistor. In electronics, a normalized curve from 1940 on reveals that sales have literally doubled every 5 years, a fantastic rate of growth!

Most of us have heard the observation that 80 to 90 percent of all the scientists who have ever lived are alive today.[1] The same observer prophesies that

---

today's technical college graduate, during his professional career, will have witnessed 80 percent of the major scientific discoveries and inventions which are common knowledge at his retirement. This is the pace with which industry must cope during the years ahead. A technically based company is as likely to miss the proper targets with short-range opportunistic product planning as a ballistic missile would be with only a magnetic compass for guidance.

Today's diverse company needs far-sighted technical vision and guidance which can be based either upon extrapolations of the past or, where perturbations are anticipated, upon expert opinion. TRW has chosen to supplement trend extrapolation by seeking the opinions of in-house technical experts to forecast events which should be considered in product planning and in the preparation of medium- and long-term business plans.

Technological forecasting is only one of a number of tools available to aid in the product selection process, but as the rate of expansion in technology increases, so does the importance of some form of technological forecasting. Such forecasting is far from an exact science. It is not a panacea to product planning, and the route is fraught with uncertainties, but a great deal can be learned in the preparation of a technological forecast. In short, we must use the tools at hand and make the best of them.

In mid-1965, we noted a report[2] on a long-range forecasting study by Dr. Olaf Helmer and Mr. T. J. Gordon, now associated with the Institute for the Future, which summarizes work done earlier in the development[3] and use[4] of a technique called the Delphi method. This technique seeks to take full advantage of the committee approach, while at the same time eliminating some of the disadvantages.

Those who have dealt with committees of experts may have run into problems involving one or more of the following people: (1) the expert with a reputation who feels called upon to defend his publicly stated opinion, (2) the senior executive with whom subordinates are reluctant to differ, or (3) the "silver-tongued salesman" who can "sell refrigerators to Eskimos." The Delphi technique attempts to avoid these problems and others by dealing individually with each member of the "committee" and protection his anonymity.

PROBE I USED DELPHI TECHNIQUE

TRW's experimental study, which we now call Probe I, was completed in June of 1966. We used our own modified Delphi technique in this exercise, which resulted in the publication, for internal use only, of a 50-page document[5] containing a forecast of over 400 technical events which our panel of 27 experts felt would occur during the next 20 years and which would have a significant impact on TRW – its products, services, or processes.

We recognized some time ago that corporate executives, even those who are highly trained and broadly experienced in technical fields, find it impossible to remain current in all technologies of interest to their companies. Thus Probe has been developed to aid corporate executives in their perception of the need for and feasibility of new products or services and the appropriate modification of long-range corporate plans. In addition, along with the revised corporate plan, Probe has been designed to aid R&D managers in their modification of long-range R&D plans, which in turn are subject to the usual executive evaluation of the match with corporate plans and objectives, the results of which lead to a decision to proceed with specific programs.

However, "You can lead a horse to water but cannot make him drink," as the saying goes; so it was with TRW's R&D people. A review of R&D activities at TRW during the latter half of 1966 revealed that, whereas Probe I had been a useful checklist and a vision-extending exercise for the panelists and others, the principal benefit which the company received was a great deal of publicity concerning its foresight. There was no real evidence that it had been used directly in R&D planning.

We concluded that many of the final events predicted by our panelists, while interesting, were not events in which the company could participate directly. Some were too formidable to match available research and development resources while others were too distant in time for immediate attention.

MORE DETAILED MAP NECESSARY

We have often compared our exercise to the printing of a road map which showed only the names and locations of large cities. Roads and smaller cities were omitted and the traveller was at a loss to plot his route. So it was with many of the technological events anticipated by our Probe I panel of experts. It was interesting, for example, to contemplate that "3-D color movies utilizing holographic techniques will be technically feasible by 1972"; but unless one considers the technological developments which must precede that event, he has no "road map" to use as a basis for planning an R&D program in this or any related technical area. Moreover, the "terminal city," the 3-D holographic color movie, is too big for any one company to contemplate. This led us to experiment with the preparation of technological road maps associated with this event.

The first step was a logic network which contained a "map" of developments to precede the technical feasibility of 3-D holographic color movies. Creating the network (which resembled a PERT chart) was a time-consuming task requiring the counsel of experts in many fields. These experts, however, became intrigued with the "fallout" which might be possible once certain milestones were passed. Thus it was that our logic network was expanded to include such fallout items of interest, and a technique was developed for the creation of more detailed charts.

These charts are so complex that one hesitates to produce them unless his basic data provide a sound foundation. In view of the experimental nature of Probe I and of the possible improvements which were evident, we decided that it did not provide the desired foundation and that predictions were lacking in some of the areas where they were most needed. Accordingly, Probe II was launched and is under way at present (its events have already been studied in connection with future plans of two TRW groups).

TRW HIGHLY DIVERSIFIED

The current study will provide a useful vehicle for describing our method in detail. First, however, some information about TRW will make what follows more meaningful. As some will know, it is a diversified corporation of some 80,000 employees spread throughout nearly 250 worldwide locations. In 1967, it passed its former 1970 goal of $1 billion in sales and is now operating at an annual level of $1.5 billion. TRW is decentralized into five highly autonomous groups to which two subsidiaries have been added recently:

1. Our Automotive Group manufacturers valves, pumps, steering linkage, and other chassis parts for original equipment manufacturers and for the after market.

2. The Electronics Group specializes in electronic components from sophisticated semiconductor devices and integrated circuits through capacitors and resistors to coils for radio and TV receivers.

3. The Equipment Group produces aircraft parts, including jet engine components, pumps, and actuating units; torpedo propulsion systems; nuclear reactor components; and high-temperature alloys for outside customers as well as for internal use.

4. The Systems Group has been involved in over 90 percent of all U.S. space missions, having previously provided systems engineering and technical direction for the four initial Air Force intercontinental ballistic missile programs. These people are currently directing their attention toward civil systems, including such things as high-speed ground transportation and hospital design, among others. The group also includes Hazelton Laboratories, which deals in the life sciences.

5. The newly formed Industrial Operations consist of TRW Instruments (supplier of electro-optical instruments, ultra-high-speed cameras, and spectrographic equipment), Mission Manufacturing (a company which furnishes oil field equipment), and Reda Pump (which manufactures submersible pumps for the oil industry). Each of these is related to TRW's other activities through common technologies.

6. United-Carr, a TRW subsidiary, originally focused upon punch press operations, making fasteners for the auto industry, but it has added a Cinch Group, a leader in electrical connectors and multilayer printed circuit boards.

7. United Greenfield, our newest subsidiary, is the world's largest producer of cutting tools and hand service tools. It also produces gages, springs, forgings, extrusions, and a variety of special equipment for the automobile servicing industry.

Our purpose in including this information about TRW is to provide an indication of the scope of the forecast with which we have been struggling.

In reviewing our experience with Probe I, we were able to list a number of areas in which improvements could be made in our use of a modified Delphi technique:

1. Probe II assumes a socioeconomic environment left to the panelists in Probe I. That environment is essentially the one assumed for TRW's 1975 long-range corporate business plan. Supplementary publications, which elaborated on various aspects of the assumptions, were also furnished to the panelists.

2. In Probe I, no restrictions had been placed upon our panelists with respect to subject matter except for a request that predicted events should have some bearing upon TRW's present and envisioned future. This time, we have asked for predictions in 15 specific areas, identified as those in which events are likley to lead to opportunities for our company or to threaten our existing product lines.

3. Participation in Probe I had extended the vision of 27 staff members who contributed. To expand this benefit, we decided to increase the number of participants to be drawn from key members of TRW's technical staff of over 5,000 graduate engineers and scientists. A group captain, appointed by the executive vice-president of each of the then existing TRW groups, was asked to indicate which of his divisions had interest in the different categories. Each group captain then arranged for participation by one or more experts from each organization likely to be affected by the occurrence of events in the specific categories. This led to the selection of 140 experts.

4. Whereas Probe I was limited to a quest for the most probable date of occurrence, in Probe II we are asking more penetrating questions concerning each event. Probe II is being conducted in two rounds plus a follow-up as required.

In Round 1, panel members were asked to list their forecast events considering *desirability*, *feasibility*, and *timing*. Each event was to be weighed in accordance with these factors. Desirability was to be considered from the viewpoint of the customer. Accordingly, this rating was to reflect an estimate of the potential demand which would indicate the importance of the event from a marketing standpoint. Feasibility, on the other hand, was to be considered more from the viewpoint of the producer. It was to reflect an estimate of both the technical feasibility and the difficulty likely to be encountered in prerequisite developments. Timing was to reflect both an estimate of the date by which the probability is 0.5 that the event will have occurred and the degree of uncertainty associated with that estimate, i.e., the date by which there is a "reasonable chance" that the event may have occurred ($P = 0.1$), and that date by which the event is almost certain to have occurred ($P = 0.9$).

In Round 2, each panel member was provided with a list representing a composite of all predictions of his own panel plus those of other panels which are also related to his category. In this round, he was asked to evaluate all events with respect to the same three factors as in Round 1. Estimates made by the originator were not included; each panelist was on his own. Self-consistency in the list, particularly with respect to order and anticipated date of occurrence, was an important objective.

PILOT STUDIES BROUGHT MODIFICATIONS

Pilot studies, conducted in the categories of "Oceans" and "Personal and Medical," led us to modify several features of the study. In recognition of the fact that the expertise of a specialist varies according to the issue, we asked our panelists to indicate their familiarity with the technologies relevant to each event in a scale ranging from a layman's knowledge at one end to a specialist's knowledge at the other. Experience without pilot panels led us to suggest further that panelists ignore those events about which they know nothing. Another refinement emerging from the pilot studies permitted panelists to record their estimate of the probability of the event ever occurring.

In an informal third round, resolution of wide differences of opinion will be sought on an individual basis. Panelists considering themselves specialists in a relevant technology and whose responses deviate too far from the mean will be contacted by their chairmen to determine whether the deviation was due to a misunderstanding or the fact that the panelist was aware of information not considered by other panelists.

The first round of Probe II has been completed, yielding over 2,100 events. Of these, about 1,750 survived initial editing in which obvious duplications, vague statements, and descriptions of trends rather than events were eliminated. Editing by panel chairmen and one final "editorial pass" reduced the list to about 1,200 events. Printing of Round 2 questionnaires and the processing of subsequent data are being done by computer. At this writing, 60 percent of the responses to Round 2 have been received. Ultimately, we expect that, on a statistical average, each event will have been evaluated by approximately 40 experts representing three or more different viewpoints.

We are concerned about the practicality of formally charting a large number of events to the degree of detail typical of earlier experiments. The task appears to be too formidable for the time available to planners. Accordingly, we are considering introduction of a computer-based approach to achieve most of the results which might be obtained from "mapping" each event separately – an approach by which scientists and engineers familiar with fields associated with a particular end event would relate it to the major technological developments required to reach essential milestones with dates of achievement consistent with the predicted date of the end event. These would be interrelated in a composite computerized network which would provide us with an ability (1) to make an initial cross-check on the accuracy of forecast dates for different end events which are dependent upon the same technological development and (2) to maintain a continuing surveillance over the accuracy of the network. Thus we should be able (1) to monitor, through time, development of those technologies of importance to TRW and, for that matter, to its competitors, and (2) to avoid the necessity for a Probe III by updating the network as required to conform with experience and changes in outlook. The extension of this technique to other variables is, of course, possible.

It is probably necessary, in closing, to reiterate that technological forecasting is not a panacea which will assure corporate success for its user. Nevertheless, we feel that any advance notice of perturbations in long-range technical trends is worth seeking and that the Delphi forecasting is a more useful approach to identifying those perturbations than other methods we have encountered. Imperfect though it is, technological forecasting seems essential in the face of the "technology explosion" we've experienced which, in retrospect, may prove to have been merely a "burning fuse" with the real explosion yet to come.

NOTES

[1] Derek J. DeSolla Price, *Little Science, Big Science,* Columbia University Press, New York, 1963.

[2] Olaf Helmer and T. J. Gordon, "Probing the Future," *News Front,* Vol. 9, No. 3, April 1965.

[3] N. Dalkey and Olaf Helmer, "An Experimental Application of Delphi Method to the Use of Experts," *Management Science,* 1963, 9, 458, 467.

[Donald L. Pyke, "A Practical Approach to Delphi," *Futures,* June 1970, 143-152.]

[4] T. J. Gordon and Olaf Helmer, "Report on a Long-Range Forecasting Study," RAND Corporation, P-2982, September 1964.

[5] *A Probe of TRW's Future/The Next 20 Years,* a proprietary document, July 1966.

[Harper Q. North and Donald L. Pyke: *Technological Forecasting in Planning for Company Growth,* IEEE Spectrum, Vol. 6, No. 1, January 1969, 30-36; " 'Probes' of the Technological Future," *Harvard Business Review,* May-June 1969, 68-78.]

PART VII SIMULATION

. . . A big advantage of a model is that it provides a frame of reference for consideration of a problem. This is often an advantage even if the preliminary model does not lead to successful prediction. The model may suggest informational gaps which are not immediately apparent and may suggest fruitful lines for action. When the model is tested the character of the failure may sometimes provide a clue to the deficiencies of the model. . . .

Another advantage of model-making is that it brings into the open the problem of abstraction. The real world is a very complex environment indeed. . . . Some degree of abstraction is necessary for decision. . . .

By making this process of abstraction deliberate, the use of a model may bring [important] questions to light. Moreover, it may suggest which characteristics are relevant to the particular decision problem under consideration.

Once the problem is expressed in symbolic language there is the advantage of the manipulative facility of that language. The symbolic language also offers advantages in communication. It allows a concise statement of the problem which can be published. Moreover, it is more easily integrated with the other scientific work which is also in symbolic language.

Another advantage of mathematical models is that they often provide the cheapest *way to accomplish prediction. . . .*

Irwin D.J. Bross, Design for Decision, *New York (Macmillan) 1953, pp. 169-170.*

27

Alan J. Rowe

SIMULATION: A DECISION-AIDING TOOL

One can hardly question the increase in complexity of modern organization and the dynamic requirements for management decision making. In view of shorter lead time, greater number and complexity of products, wider geographic distribution of customers, tax considerations, government requirements, etc., there are increased requirements for improved management decision making. Not only do organizations have to respond rapidly to changing conditions, but the decisions must take into account the varying constraints imposed upon the organization. It is no accident, therefore, that computers are assuming an ever-increasing role in what are called real-time management control systems. These computer-based systems can readily incorporate simulation models to provide managers with a means for pretesting ideas and aid in evaluating the effects of their decisions.

THE MANAGER AS A DECISION MAKER

Before one can answer the question of how simulation is used as an aid in the decision-making function, it is necessary to review the management function in detail. Since the manager primarily deals with people, the mechanization of decision making appears highly unlikely. Rather, the small percentage of time that the manager spends in making decisions will be improved by the use of disciplines such as operations research, management science, behavioral science, and econometrics, and supported by improved information supplied by computers.

A review of the decision-making function in an organization reveals that there is a clear-cut dichotomy between policy making at the top management level and decisions at the operating level which are concerned with the physical processes of producing goods to meet customer demands. Figure 1 illustrates the time span of decisions as a function of level of management in the organization. In a sense, middle management's function can be defined as assuring the implementation of the policy decisions as defined by top management and which have to be carried out at the operating level. To this extent, they exercise a management control function which involves the measurement, evaluation, and feedback of information.

Policy decisions which are often considered at the top management level are the following:

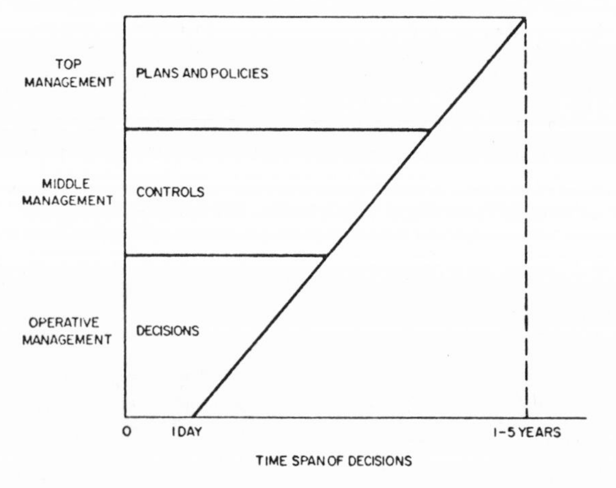

Figure 1

1. Assure a consistent set of objectives for all divisions of an organization.
2. Determine the financial support required and risks involved in achieving long-range goals.
3. Assert effective control via accurate and timely performance measures based on a continuous reappraisal of the relationship of operating decisions with corporate objectives.
4. Evaluate new products, new businesses, acquisitions, etc., in relation to growth objectives and economic analysis of business cycles.
5. Provide research support for product development, management techniques, computer applications, operations research, etc.
6. Determine basic strategies, such as pricing, in respect to industry trends, competition, legal and tax considerations, etc.
7. Establish an appropriate corporate image through public relations and employee benefits.

Since top management is primarily concerned with long-range decisions, there is a loose coupling with the actual business processes. Thus the amount and frequency of information used for decision making depends on the measurement system. Rather than continuous reporting of operating status to top management, periodic reports are generally sufficient. However, computers provide the capability of random access to updated information on detail system status as the need arises. In this sense, then, top management can have a real-time information system as the base for decisions.

It is fairly obvious that the use of simulation as a decision-aiding tool will have its major impact at the operating level. However, there is considerable effort being expended to develop models of the total business system. Several efforts along these lines have been undertaken at Systems Development Corporation[1], Stanford Research Institute[2], IBM Corporation[3], and M.I.T. A notable example of work being undertaken in this area is by the industrial dynamics group at M.I.T. which is concerned with studying total system behavior. The basic premise of this latter simulation is that the dynamic interaction of system variables and information feedback leads to amplification, oscillation, and delays in system performance. The behavior of a system, then, is the result of desired objectives and the decision rules employed to carry out these objectives. Thus, where there is an attempt to make corrections and adjustments in flow rates, there is the possibility of delays or amplification or there may be conflicts between short- and long-term objectives. Forrester in his recent book on industrial dynamics[4] describes a number of studies that have been undertaken and describes future studies of total management systems, as well as broad economic systems.

CRITICAL MANAGEMENT DECISIONS

One important area in management decision making is the allocation of the decision maker's time so that the important problems are being tackled. An analysis of the decisions made by managers[5, 6] indicates a Pareto distribution; that is, if we look at the whole set of decisions for which a manager is responsible, some small percentage of these constitute the really important aspects of his job. This is shown in Figure 2. The remaining decisions, although necessary, nonetheless have a much smaller operational impact. In a sense, therefore, we can think of applying two concepts on this area, that of continuous monitoring in real time, and exception reporting. The critical decision area, then, would be under continuous surveillance by a computer to provide current status and operational information so that the decision maker has the maximum visibility possible. The remaining decisions would be updated periodically to indicate trends or deviations, but only when these appeared out of control would they be reported. We can consider the problem of decision making, then, from the following point of view:

1. Define that set of critical decisions over which management must maintain continuous surveillance. It is in this area that simulation might be most appropriate.
2. Define a set of operating decisions which can be controlled on an exception basis. No need for simulation.
3. Determine the effects of interaction of one decision on another. This area appears suitable for simulation.
4. Examine the effect of timing upon decisions; i.e., how rapidly must information be updated in order to maximize the effectiveness of decisions being made; or what is the effect of delays on decisions on response characteristics of a system?

ORGANIZATIONAL DECISIONS

In addition to the normal decisions confronting managers in the operational area, there are major

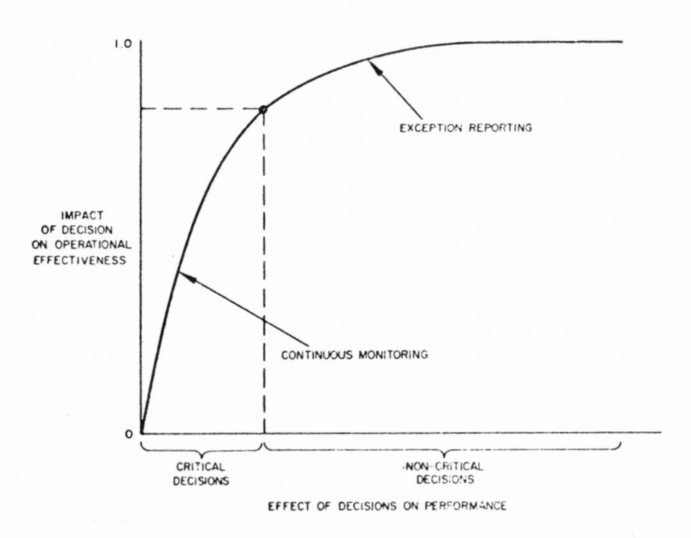

Figure 2

organizational questions which should be considered. Although at the present time these are difficult to examine by the use of simulation, nonetheless there is increasing effort being applied which should yield insights into these important problem areas. Some of these problem areas are the following:

1. Management amplification: determination of the correct manner in which the manager can extend his own capability by delegating decisions within the organization.
2. Balanced control: determination of the basis for centralized control to provide direction and unity while maintaining decentralization decision making for the purposes of growth, motivation, and operational effectiveness.
3. Matching requirements: establishment of the needs and drives for individuals and relating these to the organization as well as to the supervisor.
4. Planning requirements: determination of work definition based on time-phased definition of elements of work and their interaction, clearly indicating organizational responsibility and authority.
5. Management control: establishment of closely coupled feedback to provide timely information which permits an organization to adapt to dynamic changes.
6. Operational strategies: determination of the application of advanced techniques to evaluate decision alternatives and help to establish policies.
7. Performance measures: need to establish a basis to identify technical performance in terms of results rather than budgets or dollar expenditures.
8. Resource allocation: establish a basis for value of work to be performed and estimate of time and resources, using priorities based on utility rather than arbitrary percentage allocation.
9. Forecasting: determine the basis for coping with dynamic changes in an environment which permits growth and a viable organization.

In addition, there should be a clear distinction among ownership requirements, staff responsibilities, operating management, and business functions. With a clear understanding of the division of responsibility among these activities, an analysis of the application of simulation to the decision-making problem can be properly undertaken. Appendix A lists some of the basic management decision areas which should be considered.

MANAGEMENT DECISION MAKING AND THE COMPUTER

One further consideration which should be borne in mind is the relationship of managers to computers. Management is part of a man-machine system. As such, they use computers and have programmed decisions. Unquestionably, computers have played an important role in business and appear to have increasing importance. Computers have large rapid memories which retain information indefinitely; they operate at microsecond speed on a variety of abstract problems. On the other hand, the human has some unique capabilities. He is adaptive and can accommodate unknown and new requirements, is creative in terms of providing useful and purposeful products, and is inspired by insight, hunches, experience, and emotion. The implication here is that there is a serious gap between what might be called rational decision making and emotional decision making. Until the time is reached that all decision making can be reduced to a purely rational basis, it appears that the computer will at best provide input to decision making in the form of data and analysis, utilizing techniques like simulation. However, it does not appear that the computer will be replacing the decision maker.

THE SIMULATION TECHNIQUE

A considerable body of literature exists covering the use of simulation in various applications[7,8,9]. As shown in Figure 3, simulation should be thought of as a continuum, starting with exact models or replication of reality at one extreme, with completely abstract mathematical models at the other. When viewed in this manner, the breadth of simulation can be appreciated.

For many years engineers have used reduced or scaled models for testing. The Armed Services have used exact models for training. In addition, there have been many laboratory studies which, in fact, constitute what might be thought of as a similitude or an attempt to duplicate reality in a laboratory environment. This has been extended into the area of management games, where the human interacts with

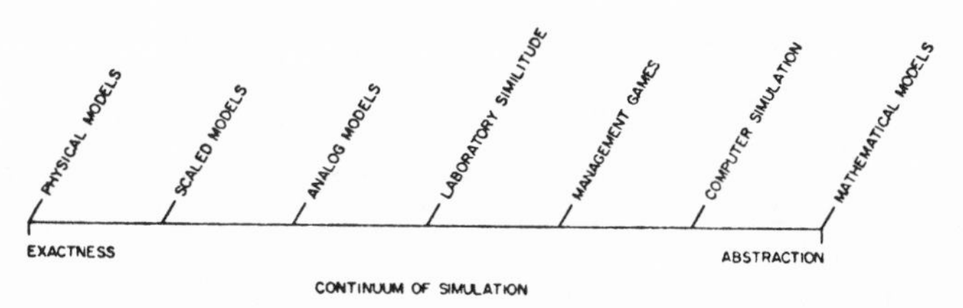

Figure 3

the output of the computer and makes decisions on information received. Extending this further, there is the area of computerized simulation. Here the attempt is to model the behavior of a system so that the results obtained correspond fairly closely to the problem being studied. It is in this area that there appears to be a maximum advantage for the use of simulation as an aid in decision making. Continuing on, abstract mathematical models, which often use Monte Carlo or other techniques, are used to study problems which may not correspond exactly to reality. However, they produce useful results for solution of problems.

Not only has simulation increased in use as a means for studying and understanding new problem areas and providing useful insights which might otherwise not be available, but it has a number of distinct advantages. Once a simulation model is completed, the time for experimentation is considerably reduced, since it is possible to examine 1 year of simulated activity in as short a time as 15 minutes. The cost of simulation models is being reduced to the point where for larger problems it is now an extremely economical tool. The fact that all the work is done in a computer rather than a laboratory or actual operation provides better experimental design and control in addition to avoiding disturbance of an actual operation. The ability to explain a simulation model in terms of the real problem is a far more powerful tool than some of the mathematical or analytic techniques which, although they may provide improved solutions, nonetheless cannot be described to management or the potential user.

USES OF SIMULATION

Simulation has been used in a wide variety of ways as well as in a broad range of application. The uses of simulation may be categorized in the following manner:

1. Research: it is an important tool for experimentation and is being used to an ever-increasing degree as a policy laboratory for management.
2. Gaining insights: testing of new areas, stressing the system or extrapolating its behavior, all of which cannot be done in a real-world situation, can be studied with simulation.
3. Sensitivity: simulation provides the capability of testing for the importance of system variables and the effect of their interactions.
4. Decision rules: based on a systematic formulation, simulation can be used to test a wide variety of decision rules within a well-defined situation.
5. System design: based on formulation of the operating characteristics of new systems, simulation provides the basis for applying decision rules or management principles in the design of operational systems.
6. Pretest: simulation provides a capability similar to prototype or breadboard testing, wherein new designs can be examined prior to their installation.
7. Evaluation of alternatives: because of rapid speed, a large number of alternative designs can be examined as well as comparing the new system with current system operation.
8. Demonstration: simulation provides an effective vehicle for demonstrating to management and operating people the effects of new decision rules, new designs, or impact of system variables.
9. Real-time control: without simulation, systems such as the SAGE Air Defense or the SABRE system would not be possible owing to the necessity of extrapolating information received.
10. Training: simulation requires participation, which is an effective means for training. In addition,

it forces the consideration of formalized decision rules on the part of the participants.

11. Develop new theories: techniques which are difficult to examine mathematically can be studied by the use of simulation, such as queuing.

12. Solution of complex problems: where analytic techniques are not available, Monte Carlo simulation can be used to find solutions to a variety of complex problems.

PROBLEMS IN SIMULATION

Although simulation has many advantages, one should not overlook the difficulty involved in developing a model, programming it on a computer, and utilizing the results. Although computer simulation has been used for a number of years, there are still many pitfalls that must be avoided. One of the greatest difficulties is that of developing a suitable model. Another is the use of computers in the simulation process, which poses a number of problems, including computer programming, search techniques, data storage and retrieval, function generators, etc. The computer programming problem has in many instances proved to be a major stumbling block.

In recent years there have been a number of approaches taken to minimize the programming problem. One is the development of models to study a specific area, such as Job Shop Simulation[10]. Using this type model, the user is required to provide appropriate data and description of the facility to be studied, and the computer program needs little modification. A similar approach has been taken in the Gordon General Purpose Simulator[11]. A somewhat more general approach to this problem has been tackled by the use of the DYNAMO Compiler[12], in which a set of equations is submitted to the computer, which in turn compiles these and generates a computer program. Therefore, once the model is completed, no further programming is required. As an alternative to writing directly in machine computer language, a new simulation language has been developed called Simscript[13]. This language is similar to the DYNAMO Compiler in that, once the model is written, no further programming is required. The Simscript language, since it is based on the use of FORTRAN, has all the flexibility of straight computing language but much of the simplicity of a special-purpose approach. Thus, depending on the type of problem being undertaken, it is possible to use a variety of approaches to obtain a computer program. Several of the computer manufacturers have developed standard programs which are readily available and require no further computer programming effort[14].

A second problem is in the area of experimental design. Considerable effort is often expended in an attempt to obtain information in an inefficient manner based upon poor input data. It is therefore necessary to consider a computer simulation for research as an equivalent to a laboratory experiment. Before any simulation model is undertaken, areas of payoff or urgency should be established, and the feasibility of completion with estimates and budgets should be provided. In the area of defining the problem, there should be careful observations and correct statements concerning what is being studied, and discussions held with experienced personnel. Preliminary approaches or brainstorming should be undertaken in order to attempt to define solutions to the problems being studied. Organization of the data, the use of sample versus exhaustive representation, and the use of statistically designed experiments should all be incorporated. This becomes particularly important when trying to state on a rigorous basis the comparison of one system design to another. Simply because a problem is run on a computer does not mean it is either valid or statistically significant.

A number of fairly significant techniques have been developed for analysis and evaluation of data[15]. Some of these are referred to as Monte Carlo sampling or important sampling. In these techniques the data are handled in such a way as to minimize the amount of data required and to maximize the information that can be derived from the manipulation of the data. In some complex areas involving time series, appropriate analytic techniques are still not available. However, in many other applications the use of analysis of variance or regression analysis is very important. It is necessary in evaluating the results of a simulation to have the appropriate criteria and measures of system performance. These, of course, do not depend on the simulation but rather on the user.

The problem of modeling is another important area since the results of simulation are no better than the model provided. A model provides a formal statement of system behavior. The model may be symbolic, mathematical, or merely descriptive. The model should be constructed so that the parameters, variables, and forcing functions correspond to the actual system. The parameters should include properties which are sufficient to define the behavior of the system; whereas the variables are the quantities which describe the behavior for a given set of parameters. The forcing function provides the stimulus, external to the system, which causes the system to react. For example, job orders which enter a production system cause men to work, machines to run, queues to form, etc. In this way, job orders become the forcing function for the system. Whatever particular form is used, a model provides the frame of reference within which the problem is considered.

A model often indicates relationships which are not otherwise obvious. However, a model need not duplicate actual conditions to be useful. The model should be designed to predict actual behavior resulting from changes in the system design or application of new decision rules. Prediction implies and understanding of the manner in which the system reacts, that is, being able to specify the outputs for a given set of inputs. This approach differs from the conventional concept of the "black box."

Models are merely the basis for testing new ideas and should not become ends in themselves. The simpler the model, the more effective for simulation purposes. Tests should be made prior to model building to determine the sensitivity of the characteristics which are incorporated. Typically, certain key characteristics contribute the majority of the information to be derived from simulation. Many of the other characteristics, although more numerous, do not contribute much to the final systems design. In a sense, simulation can be considered as sampling the reaction of a system to a new design. It is imperative, therefore, that a representative sample be taken, rather than an exhaustive sample. To this end, the number and type of characteristics to be included should be carefully selected.

One approach, then, is to start with a simple model which can be easily modified to incorporate new factors or eliminate undesirable ones. Starting with paper-and-pencil models often helps to clearly define the problem and specify the system design. Conventional flow charting can be used to obtain information with which to start the study. These data can be refined and expanded as the model develops. At the completion a "logical" model is available which should appropriately describe the system characteristics.

The major task of simulation is reached at this point. A logical model, which is merely descriptive, is not suitable for computer simulation. The model must be modified to suit the particular computer on which it will be programmed. Factors such as kind of memory, manner of indexing, speed of computation, and errors due to rounding must all be taken into account. Simplification is often necessary owing to speed of computation or limitation of the computer memory. The method of filing information and representing time are also significant problems. Which data to accumulate and at what point in time often are difficult to decide beforehand. Thus the program must be flexible and easy to change.

As computer programming proceeds, there is generally feedback which provides the basis for further modification of the model. At the outset, the decision must be made whether to make the program general or special purpose. The type of programming changes radically, depending upon the end use of simulation. Modular programming which treats each section independently provides flexibility at a small cost in computation time and storage. In view of the many logical relations which exist in systems, computer programming represents an important and often difficult aspect of the problem.

The design aspect of the problem can be accomplished either by modifying the parameters of the model and examining the behavior of the system or by specifying a given design and examining behavior under varying conditions. The first type of simulation is exploratory in that the new design will result from information obtained by simulation. The second type is a form of statistical sampling. That is, a given design is subjected to many conditions in order to determine its suitability. In most instances, this latter form of simulation is the one used in industry.

SUCCESSFUL APPLICATIONS OF SIMULATION

The wide variety of simulation applications is somewhat astounding. Not only has simulation been extremely successful for purposes of studying physical systems, but it has been used for such diverse applications as the study of personality[16], election results, gross economic behavior, etc. To evaluate where simulation is most effective, it is probably best to categorize problems as involving physical and nonphysical systems with high and low risk decision alternatives. As shown in Figure 4, the area for greatest success is physical problems having very low risks. The poorest applications are nonphysical problems having high risk or little data. This situation, of course, may change as the simulation technique is applied to a broader class of problems.

A review of the literature indicates many successful applications of simulation in the business area[17,18,19,20]. A publication[21], "Simulation – Management's Laboratory," based on a fairly extensive survey of simulation applications, shows the applicability to a wide variety of management problems. Some of the applications include the following:

Company	Management decision area simulated
1. Large paper company	Complete order analysis
2. U.S. Army Signal Supply	Inventory decisions
3. Sugar company	Production, inventory, distribution
4. British Iron & Steel	Steelworks melting operation
5. General Electric	Job shop scheduling
6. Standard Oil of California	Complete refinery operation
7. Thompson Products	Inventory decisions
8. Eli Lilly & Company	Production and inventory decisions
9. E. I. duPont	Distribution and warehouse
10. Bank of America	Delinquent loans

In another survey by Malcolm[22], the following applications are described:

Company	Problem simulated
1. Eastman Kodak	Equipment redesign, operating crews
2. General Electric	Production scheduling, inventory control
3. Imperial Oil	Distribution, inventory
4. United Airlines	Customer service, maintenance
5. Port of New York	Bus terminal design
6. Humble Oil	Tanker scheduling
7. U.S. Steel	Steel flow problems
8. IBM	Marketing, inventory, scheduling
9. SDC	SAGE Air Defense
10. Matson	Cargo transportation

These lists are not mean to be all-inclusive, but rather indicate the variety and type of problems that have been solved by simulation.

A CASE STUDY

To illustrate the use of simulation in an actual application, the history and development of a specific case will be described. Starting approximately in 1952 the problem of scheduling was considered a critical area. Extensive work was begun, using analytic techniques to find a suitable solution. However, in view of the large-scale combinatorial nature of this problem, no solution was found except for extremely small cases. It soon became apparent that some alternative approach should be tried. The first attempt in the direction of simulation was the use of computers to examine the effect of alternative schedules by complete enumeration. From this simple beginning, more extensive models were developed and over a period of years these evolved into what today is known as the Job Shop Simulator. Although the computer program cost a large sum of money and took almost 2 years to develop, the Job Shop Simulator has been used successfully in a large number of companies. Notably, it has become an integral part of the manufacturing function at General Electric and has been used extensively in many other companies, including the Hughes Aircraft Company.

The kind of decisions that can be aided by the use of this type of simulation are the following:

1. Establishing required capacity in terms of equipment, facilities, and manpower in order to meet unpredictable customer demand.
2. Examination of alternative types of demands and the capability of the system to respond.
3. Examination of the inventory problem relat-

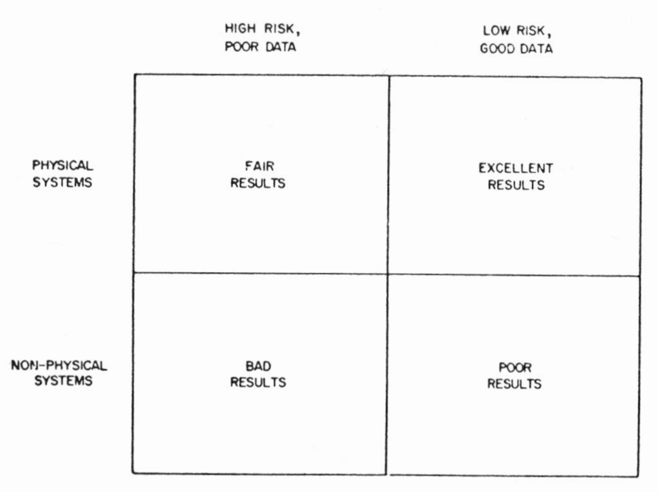

Figure 4

ing equipment utilization to cash requirements and customer demand. (It is possible to meet customer demand by maintaining large inventories.)

4. Development of appropriate scheduling decision rules to maintain a minimum inventory and meet specified delivery requirements.

5. Study the operation of a physical facility through the appropriate use of forecasting techniques, load level techniques, scheduling decision rules, and priority decision rules.

In addition to specific decision areas, there is the information generated from the simulation which provides the basis for feedback on performance so that management can make decisions on a number of shifts to run, need for additional equipment or capacity, or amount of cash to maintain for adequate inventory. The use of this particular program has been extended to an operational system at the Hughes Aircraft Company for real-time manufacturing control. The Job Shop Simulator was first used to examine alternative scheduling decision rules. These rules, in turn, provided the basis for developing a supplemental computer program which is used to generate the factory job order status on a daily basis. This computer program, by application of priority decision rules, is used to generate new priority lists each day, taking into account all occurrences for the given day. Thus the system operates on essentially a daily cycle with all information current and correct as of that point in time. This type of real-time application appears to offer considerable opportunity for the use of simulation in industry[23,24]. In addition to its use as the basis for decision aiding in system design, simulation is thus also useful for operational decisions.

FUTURE OF SIMULATION

It is apparent from the many successful applications that simulation will continue to grow in importance and become a truly operational tool for management decisions. There is still a vast area that can be tackled, ranging from the study of broad national economic problems to total company system problems to specific problem areas. Because of its many advantages and because of the need for improved techniques in management, simulation appears as one of the most exciting tools that has come on the horizon. There is still much required in the way of improved modeling, reduced cost of programming, improving the outputs, etc. However, none of these problems is insurmountable, and the evidence is quite clear that there is continued improvements on all fronts. Thus we can expect to see the use of simulation as a normal part of business operations in the not too distant future. The challenge is ours to learn how to best utilize this important management tool.

APPENDIX A

To provide the proper context for the use of simulation in the decision-making function, an understanding of basic management activities is required.

I. *Management Functions*
   1. Ownership: provides the purpose, direction, and design of the enterprise, in addition to capital and risk taking.
   2. Staff is responsible for the following:
      (a) Planning and policy making.
      (b) Control through evaluation of timely and accurate information.
      (c) Uniform measurement based on total business considerations.
      (d) Coordination of the business through programs, communication and organization structure.
      (e) Research and development of advanced techniques.
      (f) Initiating new ventures and business activities.
      (g) Providing leverage in negotiations by representing the combined strength of the business.
      (h) Providing financial stability through borrowing power and allocation among divisions.
      (i) Maintaining corporate image, legal services, etc.
      (j) Establishing staffing and resource requirements.
      (k) Assisting in carrying out new programs.
   3. Operating management is responsible for the following:
      (a) Developing plans to carry out specified objectives.
      (b) Developing personnel and staffing.
      (c) Determining effective operating procedures and decision rules.
      (d) Developing the capability to respond to varying demands.
      (e) Improving processes, product, and delivery capability.
      (f) Providing means to evaluate cost and quality effectiveness.
      (h) Maintaining a stable work force under variable demand.
      (i) Using real-time feedback control mechanisms.

II. *Management Decision Areas*
   1. Analysis of demand (profit and cost):
      (a) Pattern of demand and forecasting: backlog, sales effort, proposal effort.
      (b) Number and type of customers and position or status.
      (c) Forecast of customer needs and sales effort – where used.
      (d) Basis for customer satisfaction.
      (e) Inventory and spare parts requirements, distribution.
      (f) Pricing and bidding policies.
      (g) Quality and reliability requirements.
      (h) Number and position of competitors.
   2. Meeting demand:
      (a) Business strategies: investment, allocation, system design.
      (b) Product planning and scheduling: capacity by equipment – make or buy.
      (c) Allocation and capital requirements.
      (d) Skills required: number, type, flexibility, shifts.
      (e) Measurement and control of performance.
      (f) Organizational requirements – key personnel.
      (g) Budgeting, estimating, manufacturing cycle time.
      (h) Raw material procurement policies.
      (i) Production leveling possibilities, load analysis, stability.
      (j) Measuring impact of new products.
      (k) Utilization of resources – idle time.
      (l) Delivery distribution, rejects, reworks, breakdowns.
      (m) Maintenance of quality and reliability.
      (n) Information requirements.
   3. Provision for new product:
      (a) Analytic techniques for evaluating alternative strategies.
      (b) Attracting and keeping high-calibre personnel.

(c) Research and development requirements.
(d) Long-range planning, business or defense cycles.
(e) Diversification, acquisition, mergers.
(f) Economic analysis of trends.
(g) Productivity improvement through automation.

4. Company growth:
   (a) Establish company goals.
   (b) Establish plan for growth.
   (c) Recognize skills and limitations.
   (d) Analyze product lines.
   (e) Investigate opportunities:
      (1) Growth industry is twice national average (8 percent).
      (2) Means to reduce costs.
   (f) Product objectives:
      (1) How to enter new industries: acquisition, research.
      (2) Integration versus diversification.
      (3) Product redesign.
      (4) Corporate requirements: capital, executive talent, and sales and manufacturing organization. . . .

REFERENCES

[1] A.J. Rowe, *Modeling Considerations in Computer Simulation of Management Control Systems,* System Development Corporation SP-156, March 1960.

[2] C.P. Bonini, *Simulation of Information and Decisions in the Firm,* Stanford University, April 1960.

[3] D.F. Boyd and H.S. Krasnow, "Economic Evaluation of Management Information Systems," *IBM System Journal,* Vol. 2, March 1963.

[4] J.W. Forrester, *Industrial Dynamics,* MIT Press, 1961.

[5] A.J. Rowe, "Research Problems in Management Controls," *Management Technology,* No. 3, December 1961.

[6] P.F. Drucker, "Managing for Business Effectiveness," *Harvard Business Review,* May-June 1963.

[7] D.G. Malcolm (ed.), *Report of System Simulation Symposium,* American Institute of Industrial Engineers, May 1957.

[8] W.E. Alberts and D.G. Malcolm, *Report of the Second System Simulation Symposium,* American Institute of Industrial Engineers, February 1959.

[9] Report #55, *Simulation and Gaming: A Symposium,* American Management Association, 1961.

[10] A.J. Rowe, "Toward a Theory of Scheduling," *Journal of Industrial Engineering,* Vol. XI, No. 2, March-April 1960.

[11] G. Gordon, "A General Purpose Systems Simulator," *IBM Systems Journal,* Vol. I, September 1962.

[12] E.B. Roberts, *Simulation Techniques for Understanding R&D Management,* MIT, March 1959.

[13] H. Markowitz, B. Hausner, and H. Karr, *Simscript: A Simulation Programming Language,* Prentice Hall, Inc., 1963.

[14] "Inventory Management Simulation," *IBM Data Processing Information,* April 1961.

[15] S. Ehrenfeld, and S. Ben Tuvia, "The Efficiency of Statistical Simulation Procedures," *Technometrics,* May 1962.

[16] S.S. Tomkins and S. Messick, *Computer Simulation of Personality,* John Wiley & Sons, Inc., June 1962.

[17] J. Moshman, "Random Sampling Simulation as an Equipment Design Tool," *CEIR,* May 1960.

[18] A. Rich, and R.T. Henry, *A Method of Cost Analysis and Control Through Simulation,* Linde Company.

[19] H.N. Shycon and R.B. Maffei, "Simulation – Tool for Better Distribution," *Harvard Business Review,* December 1960.

[20] D.G. Malcolm, "System Simulation – A Fundamental Tool for Industrial Engineering," *Journal of Industrial Engineering,* June 1958.

[21] *Simulation – Management's Laboratory,* Simulation Associates, Groton, Connecticut, April 1959.

[22] D.G. Malcolm, *The Use of Simulation in Management Analysis – A Survey and Bibliography,* System Development Corporation SP-126, November 1959.

[23] A.J. Rowe, "Management Decision Making and the Computer," *Management International,* Vol. 2, No. 2, 1962.

[24] A.J. Rowe, *Real Time Control in Manufacturing,* American Management Association, Bulletin #24, 1963.

28

Richard L. Van Horn

VALIDATION OF SIMULATION RESULTS

Simulation models change the state of our knowledge or, at least, beliefs about some process. Or, in other words, simulators are designed and used with a goal of learning something. By all generally accepted definitions, a simulation is a symbolic or numerical abstraction of the process under study and is not the process itself. Thus "learning" from a simulation requires two stages. First, understand the behavior of the simulator itself in terms of the relations that exist between inputs and results. The second, and often more difficult task, is to translate "learning" from the simulation to "learning" about the actual process. The second task, the translation of learning from simulator to actual process is generally viewed as the focus of the validation process.[a]

This paper starts with the assumption that a set of "statistically significant" inferences is available from a simulation. The random-number generator is truly random and the computer program correctly executes the logic desired by the modeler. The modeler has

---

examined run length, replication, variance-reduction techniques, and related problems of operation. Finally, the results have passed adequate tests of statistical significance. These areas are clearly difficult and important, but they are outside the scope of the current discussion. Good discussions and surveys of these problems are found in Conway[5], Fishman and Kiviat[11,12], and Naylor[15,16].

Validation, in this paper, is the process of building an acceptable level of confidence that an inference about a simulated process is a correct or valid inference for the actual process. Seldom, if ever, will validation result in a "proof" that the simulator is a correct or "true" model of the real process. A simulator on a digital computer is a particular instance of a finite-state machine that will transform inputs into outputs. Thus Turning's proof[16] on the equivalence of finite-state machines implies that one can never prove that two machines are identical just by comparing input-output transformations, no matter how large a (finite) sample is used.

Fortunately, the users of simulators are seldom concerned with proving the "truth" of a model.[b] Instead, the simulator produces some specific insight which needs validation. Often, a large number of actions – for example, the undertaking of various statistical tests, special data-collection efforts, complementary studies and field tests – will exist each of which may increase (or decrease) our confidence in the specific insights. One approach to validation is to list a number of possible actions; however, no one wants the experimenter to take all possible actions. Some are inapplicable; others are duplicative, and all involve a cost. The experimenter is supposed to select a set of actions. Thus, in concept at least, validation reduces to a standard decision problem – to balance the cost of each action against the value of increased information about the validity of an insight. In this view, two important characteristics of the validation problem are the following:

1. The objective is to validate a specific set of insights, not necessarily the mechanism that generated the insights.
2. There is no such thing as "the" appropriate validation procedure. Validation is problem-dependent.

Validation clearly applies to a far more general environment than simulation; validation is a problem associated with all modeling. It seems appropriate to ask, "Why even discuss validation (or statistical analysis, etc.) in the context of simulation?" Several reasons come to mind. Simulations tend to become far more complex than other management science models. Most analytic models either deal with small problems – for example, queuing models – or deal with one aspect of large problems – input-output models. Simulators allow the modeler to include many different parts and processes in one model and allow the parts to interact in nonlinear, nonstationary modes.

In addition, simulators conceal their assumptions and processes, certainly from the casual observer, and often from their designer. The simple statement that model x is a linear programming model conveys a great deal of information about its structure, assumptions, and limitations. The statement that model y is a simulation conveys virtually no information. Finally, simulators, either explicitly or implicitly, often claim to represent "reality." Economists would not claim nor would managers believe that the set of differential equations from theory of the firm represent the firm's actual decision process. But simulators look real and both modelers and managers find them easy to believe. For all those reasons, validation holds a special and important role in simulation.

ASPECTS OF VALIDATION

Two broad questions appear relevant to building a framework for validation. First, what are the characteristics of the processes or systems that are common to the management and social sciences? Second, what methodologies or approaches should enter into validation?

Earlier in this paper validation was characterized as problem-dependent. Thus the major attributes of the processes that are simulated should guide any general discussion of validation. Clearly, the entire spectrum of problems of interest to simulation modelers in the management and social sciences is very broad. For example, models frequently are built to simulate the behavior of an abstract problem such as queue behavior under a set of explicit mathematical assumptions. When the actual process is specified by the modeler, the validation problem as defined here does not exist. Simulation of intelligent behavior represents the other extreme. Here the task of finding a satisfactory description for the actual process is immensely more difficult and controversial than comparing the actual to a simulator.

Most management science simulations involve a production or service facility – for example, hospitals, job shops, transit systems, and air traffic – or perhaps involve some aspect of the economy. In general these systems are characterized by the following:

1. The structure and parameters of the process are determined by the environment, not by the modeler.
2. Part of the process depends upon physical phenomona – the behavior of aircraft, trains, or drill presses.
3. People are part of the process either directly as information processors and decision makers or indirectly as consumers.
4. The process tends to consist of many parts, and the behavior of the process depends on interaction among the parts.

Much of this discussion is relevant to other forms of simulation – for example, engineering or physics simulation of atomic particles, satellite orbits, etc. – but the primary focus is on problems suggested by the above environment.

After examining philosophical views on validation, Naylor and Finger[15] suggest a three-stage approach. In a slightly generalized form, the three phases are the following:

1. Construct a set of hypotheses and postulates for the process using all available information – observations, general knowledge, relevant theory, and intuition.
2. Attempt to verify the assumptions of the model by subjecting them to empirical testing.
3. Compare the input-output transformations generated by the model to those generated by the real world.

These three phases appear to capture the major ways to build confidence in a model, and the subsequent discussion will follow this structure.

MODEL CONSTRUCTION

Management science processes, as described, consist of three main components: people, physical processes, and an organization structure. The simulator must find some way to represent these components in the model, and these models will possess varying degrees of a priori confidence. When a process is easy to observe and measure, the confidence in its representation is high. Some representations will be "well known" in the sense that previous validation has occurred. For example, it is often possible to represent machines and physical processes by production functions with a substantial degree of confidence.

For many processes, model confidence is increased by the existence of an extensive body of research. Some examples are highway traffic, air traffic, job shop, elevator operation, machine failures, and telephone switching. For many of these activities, the mathematics of stochastic processes provide a strong theoretical base for modeling, even though direct analytic solutions are not known.

Clearly, people are harder to model. In a stationary world, probability distributions provide reasonable models for arrivals at a supermarket or subway stop. Models to describe behavior in a nonstationary situation – the opening of a new supermarket or a subway fare raise – are harder to find. Now the modeler faces questions of utility or preference. In many systems, overall system behavior is strongly influenced by people acting as decision makers or information processors and, at this point, models become extremely scarce or at least controversial. The satisficing, limited-capability man of March and Simon[18] appears, on the surface, to differ greatly from the rational, optimizing Economic man. If the modeler accepts the March-Simon view, he can with great effort construct a model of a specific type of man; but an operable general model has yet to appear.[c]

Unfortunately, man's intellect beclouds even the

production functions for his physical activities. A number of researchers have observed that jobs take longer when there is not much work to be done – the job-shop effect. Cyert and March[7] generalize this effect to a concept that they call organizational slack. However, the problem remains the same; good models for human behavior are hard to find.

The above implies that the initial confidence attached to most representations of human behavior will tend to be low. Subsequent validation by empirical testing of assumptions or input-output transformations tends to involve unwieldy statistical properties and, in general, is difficult and costly. The scarcity of empirical work testing the March-Simon postulates offers some evidence of the difficulty of verifying models of man as a complex information processor.

These problems of finding adequate symbolic models of human behavior have led a number of investigators to man-machine or game simulations. Man-machine simulations attempt to solve the representation of human behavior by inserting a person directly in the simulation. The SAGE research by RAND (and later by the Systems Development Corporation) is now a classic example of this solution. The SAGE man-machine simulation subsequently was used for training, but its original purpose was research with a high confidence representation of human behavior. In the man-machine simulations of the RAND Logistics Simulation Laboratory, men again were introduced explicitly because an adequate model of their behavior was unavailable[9].

Although one can question whether a man in a simulation is a valid representation of a man in a different process, most people will agree to placing higher a priori confidence on a man than on most models of him. However, this solution raises a host of new problems. Substantial time compression is ruled out; most people find microsecond operating cycles beyond their abilities. Since data produced by people are costly and noisy, severe problems arise when one attempts to draw "statistically significant" inferences from the simulation itself. Thus man-machine simulation trades increased validity for a loss in analysis capability.

In most situations people and physical processes are linked by information and decision flows – an organization structure. A common simulation model for an organization is a simple noise-free network for instantaneous transmission of discrete messages. There is general agreement that real organizations are more complex, but again there are no widely accepted complex models. One obviously can improve face validity by adding error and delay mechanisms to the simple network model. Bonini[1] constructed an elaborate organizational model with motivational factors and interactions between human behavior and physical processes. However, Bonini used his model to explore the consequences of a set of postulates; he did not argue that it was a valid general representation of an organization.

EMPIRICAL TESTING OF ASSUMPTIONS

The notion of subjecting assumptions, parameters, and distributions to empirical testing appears eminently reasonable. The statistical theory of estimation and hypothesis testing provides a rigorous approach to this task. A model with untested, untestable, or refuted assumptions is at least disturbing. Most articles on simulation applications report some form of assumption testing, even if it is only an eyeball comparison of means and ranges. And articles on validation (the few that exist) offer a list of appropriate statistical tests. Some degree of assumption testing appears essential to validation.

Two qualifications to this testing deserve mention. First, finding a "genuine" underlying distribution or assumption to observe is a nontrivial task. For example, consider a manual order-processing system that is scheduled for conversion to a computer system with remote terminals. It is reasonable to assume that orders arrive randomly and are simulated by a model that uses the average arrivals per time unit as the mean for a Poisson distribution. If the underlying arrival distribution is essentially random but the manual system collects and forwards orders in batches, the data collected at the order receipt point may show a variance to mean ratio that is significantly larger than 1. The data thus refute the Poisson assumption. However, if the computer system eliminates batching, the Poisson assumption is valid; and the data (as interpreted) are wrong. This type of problem is common in inventory and service systems.

Second, empirical testing of assumptions often has a lower cost substitute – sensitivity testing. A number of results exist in statistics and probability that are true for classes of distributions or even for a general distribution. It seems reasonable to expect that often the insight gained from a simulation will not depend on a specific distribution. In similar fashion, an insight normally is relevant for a range of parameter values. Sensitivity testing can establish the set of distribution and parameter values for which a set of insights is relevant. In this way, the requirement for and cost of testing assumptions empirically is reduced. In addition, the model insights now apply to a much broader set of processes and presumably are of greater value to the research community. Conway[6] raises the question of whether researchers (not implementers) should ever become involved with empirically based simulators.

Since many simulators do focus on an actual process, the need to test assumptions does arise. The papers of Naylor and Finger[15] and Fishman and Kiviat[11] provide a good review of statistical tests of means and variances, analysis of variance, regression, factor analysis, spectral analysis and autocorrelation, chisquare, and nonparametric tests. Both articles point out that all statistical tests make some assumptions about the nature of a process. Thus tests themselves are subject to questions of validity.

Some tests require fewer assumptions than others, but in general the power of tests will decrease as one relaxes assumptions. Mood and Graybill[14] point out that for comparing sample means from a normal population, the (nonparametric) Mann-Whitney test has an asymptotic relative efficiency equal to 95 percent that of a *t* test, "a small price to pay when the assumption of normality is suspect." Thus, for large samples at least, some validity problems again can be reduced by reducing the dependence on assumptions. In the same spirit, Fishman and Kiviat suggest a variance test described by Cochran[3] to test goodness of fit without the need to assume class intervals. Perhaps this area is summed up best by a Fishman and Kiviat statement that "(statistical) validation, while desirable, is not always possible."

COMPARISON OF INPUT-OUTPUT TRANSFORMATIONS

A digital simulator, as referenced earlier, is a finite-state machine for transforming an input set into an output set. Since insight comes from observing and analyzing the transformation, overall confidence in the insight clearly depends greatly on confidence in the transformation process. One obvious way to gain confidence is to compare the output of the simulator and an actual process using, if possible, identical input. Many of the statistical tests discussed in the previous section are again relevant for this comparison, as are the limitations. Often simple comparisons of means, ranges, and variances and graphical comparison of distribution or time behavior will capture most of the available information.

Since simulation produces a set of time series, methods that look at time series appear particularly appropriate. One of the more interesting suggestions is the use of spectral analysis[10,13,15]. Many simulation outputs are autocorrelated, and spectral techniques provide the autocorrelation characteristics in a convenient form for analysis and comparison. Jenkins[13] and Fishman and Kiviat[10] describe procedures to test the equivalence of two spectra. If the spectra are equivalent, the modeler certainly increases his confidence in the model. If the spectra are not equivalent, the interpretation is more difficult. Often the relation between any deviation in spectra and confidence in a particular insight is unclear.

Spectral analysis faces several other problems. First, it requires a large number of observations. The cost of data collection on an actual process or a man-machine simulator may preclude obtaining a sufficient sample for the use of spectral techniques. Another requirement is even more restrictive. The procedures described apply to "covariance stationary" processes. But many simulators are designed precisely because the process under study is nonstationary. For example, consider a simulation of a computer center. At 8:00 A.M. the system starts off with a zero backlog of priority jobs (jobs submitted by the research staff). The expected length of the backlog may increase during the day until 5:00 P.M., when the staff goes home. The center reduces the backlog to zero and then runs background or sells time.

Thus the process never reaches steady state. The backlog statistics are direct functions of clock time and are not stationary. Problems with some sort of start-stop phenomena or time-varying parameters abound in the simulation environment.

Ideally, a comparison test should handle nonstationarity, compensate for noisy data, simultaneously evaluate a number of output measures, and work for small samples. Does such a test exist? The answer is yes if one is willing to define test very broadly. The test is simple. Find people who are directly involved with the actual process. Ask them to compare actual with simulation output. To make the test a little more rigorous, one might offer several sets of simulated data and several sets of actual data and see if the "experienced" people can tell which is which. One might even test the classification for statistical significance. If people can discriminate, ask them how they do it. The experimenter can then decide if the detectable difference affects the inferences that he wishes to make.

This test is sometimes attributed to Turing, although Turing[20] actually was trying to find an operational definition of human intelligence when he suggested a similar procedure. The idea is certainly appealing and deserves further exploration. It is probably a great improvement over having the modeler use his intuition to validate his model. However, whether one can make meaningful statements on the power of such a test is an open question.

Assume that a model has passed a reasonable set of tests for input-output equivalence with an actual process. Often the modeler has little interest in the tested situations – those represented by the empirical data. One reason for building simulators is to explore situations for which no empirical data exist. In this event, the inferences represent extrapolation from the experience base. The experimenter must now ask whether his insight applies to a property of the actual process or merely to a peculiarity of the simulation. There is no answer to this question in the simulation situation. If the modeler wishes to further increase confidence, he must look outside.

Complementary research offers one path to further confidence. The basic idea is to ask the questions that led to the simulation result in a different context. For example, if the simulation leads to insight that appears strongly dependent upon human behavior, the next step might be to conduct psychological experiments. Physical scientists and engineers are long accustomed to this sequence. Theoretical or abstract model results are tested in a series of small experiments. If all goes well, large-complex experiments are attempted. Despite extensive, rigorous, validity testing, the design for a new aircraft does not go from paper to production. Instead, it is subjected to a great deal of complementary research.

Complementary research appears less common in the management and social sciences. For example, the notion of building prototype information or management systems specifically for research deserves serious attention in view of the large expenditures that go into such systems. Dunlop[8] reports that IBM is engaged in some activities of this nature but certainly no widespread pattern is visible. For this discussion, a prototype is defined as an iconic model – often a simplified version of the actual process – operated explicitly for research and testing purposes.

A more common activity is the field test. The field test places an "actual process" in an operational situation and tries to measure performance. In many field-test situations, the operational decisions are dominant. At the first sign of real or imagined difficulty, experimental controls and data collection are abandoned or seriously compromised. As a prelude to implementation, field tests undoubtedly are valuable. But given their many problems, their usefulness as a representation of reality is open to question.

A VALIDATION EXAMPLE

Validation clearly does pose a large number of problems. The experiences encountered during Laboratory Problem Four (LP-IV) in the RAND Logistics Simulation Laboratory illustrate a number of the points discussed in the previous sections.[d] LP-IV centered around a man-machine simulation of aircraft recovery operations at an Air Force base. In many aspects, the problem resembles a complex job shop. Aircraft, as a result of alert and training activities, generate both planned and unplanned maintenance demands. Planned activities include preflight and postflight inspections and fueling. Unplanned jobs result from problems encountered inflight or during ground inspections. Performance of the jobs requires men from over 20 different skill areas plus extensive equipment and facilities.

The objective of the project was to achieve more effective use of aircraft by reducing turnaround time – the time from landing of an aircraft until it is ready for its next flight. The mechanism for this reduction was limited to changing the management system – scheduling and control procedures. Resource levels and the production functions for resources were viewed as fixed.

A number of previous RAND studies had examined the maintenance and operating characteristics of aircraft so that a reasonable research base was available for construction of environment models. Preliminary studies of existing Air Force scheduling and control indicate a high degree of complexity and a lack of any recognizable structure. The process has a very strong human information-processing and ad hoc decision-making element. None of the common "dispatch rules" from job-shop studies appeared to capture more than a small part of the process. Initial runs of a computer simulation of the process strengthened this view. In addition, Air Force controllers, who were contacted during a number of field visits, expressed great doubt that control people would or even could respond to some of the contemplated changes. At this point, the LP-IV staff decided that introducing actual Air Force controllers into a man-machine simulation was required to achieve reasonable confidence in any insights.

The staff further agreed to produce a special validation run – the Benchmark. The Benchmark combined a simulated (iconic) version of the existing information system and control procedures with the computerized environment models of aircraft and maintenance men, equipment, and facilities. This step provided the mechanism for direct comparison of output between the simulation and the actual process. In contrast to a similar all-computer run of this nature, the cost was high. About 4 man-years of our development effort, 2 man-months of Air Force time, and $20,000 of computer time went into this effort. (Exact cost apportionment is difficult because a substantial part of the effort carried over to later phases of the experiment.) Previous RAND Laboratory Project studies (LP-I, -II, and -III) did not include a Benchmark. One modeled a nonexistent future environment and the other two modeled inventory processes for which initial confidence in the model and estimation processes was believed high enough that a Benchmark was not needed.

The first organization structure for the Benchmark was a simple noise-free network with constant delays. For example, the controller received information on maintenance needs 15 minutes after an aircraft landed. This model subsequently failed several validity tests – it was simply too good. After another round of field observation and data collection, it was replaced with a model that introduced random delays and random errors into the information and decision flows. Under the new mode, some jobs were reported long after the aircraft landed; occasionally a team that was sent out to work on a job did not go; and so on.

Despite the extensive experience, even the aircraft maintenance environment models presented a challenge due to inadequate data. The Air Force at that time collected "total man hours" for each job. The models used team size and elapsed time. Part of the problem was resolved by special data collection, but some areas were not covered. For these areas, senior Air Force maintenance people were asked to convert a stratified set of total-time observations into their elapsed time and team size equivalents. These estimates were then used by the LP-IV staff to construct the sampling distributions.

The LP-IV environment did not provide much basis for testing assumptions. For some parameters, such as ones described, empirical data were unavailable. When data were available, the empirical distributions (with minor smoothing and truncation) were used directly. However, a critical environment assumption – the relation between various aircraft activities and resulting maintenance workload – did look testable. Existing Air Force planning policies relate workload to flying hours. Some convincing a priori arguments and our discussions with our Air Force advisors suggested that the act of flying – the sortie – regardless of flight length, was the prime generator of workload.

In the LP-IV model, workload does generate directly from the sortie. Aside from some limited special tests, only aggregate data – total man-hours, sorties, and flying hours – were available. Regressions on this data produced only discouragement. The constant term was large and r^2 was small. Furthermore, sorties and flying hours are highly correlated. Part of the problem is explained by the fact that the correlation between available man-hours (a man-hour measure in the accounting system) and expended man-hours was high. One inference of these results is systematic bias in the data. Reported man-hour expenditures apparently are tailored to fit available man-hours – a not surprising phenomenon. One might at this point feel forced to reject the model assumption. However, it is important to look back at the insight that LP-IV is after – the prediction and improvement of aircraft turnaround. For this purpose, a model that relates workload to time available is useless and inappropriate. This relation probably exists only in the accounting system.

Fortunately, LP-IV obtained (at high cost to several staff members) a special set of data with two sortie types, one approximately three times the length of the other. These data show a much stronger relation of workload with sorties than with flying hours. The sorties with a three times increase in flying show only a 25 percent increase in workload. Although the true situation is far from clear, we concluded that the assumption of workload related only to sorties was reasonable for our particular purpose. We also concluded that one should use empirical data with extreme care and a large amount of skepticism.

The Benchmark model ran for 5 simulated weeks. For input-output comparison, the staff obtained 4 months of data from an Air Force base with a flying program that was almost identical to the laboratory. These data encompass 130 sorties for the laboratory and over 500 for the actual process. Unfortunately, the actual workload data are four observations of average man-hours per sortie for each of 20 skill groups. The strong tendency of real-world data systems to aggregate data has repeatedly frustrated our attempts to conduct a reasonable validation.

Within the limits of the data, the comparisons of the workload for the laboratory run with the four actual observations look reasonable. The largest deviation is 2σ for one of the low workload skill groups, and over half the skill group means agree within one standard deviation. Air Force bases place considerable emphasis on meeting the takeoff schedule and record both missed and late takeoffs. Takeoff deviations in the environments we were examining are rare but do exist; the lab model showed the same behavior, but the sample is too small for any meaningful statistical analysis.

The primary focus of LP-IV was turnaround. Unfortunately, aircraft turnaround data were not collected. Special data collection for turnaround, which would have required three people for at least a month at a base, was rejected as too expensive.

Since LP-IV was a man-machine simulation, with experienced Air Force participants, some form of "Turing" test had great appeal. Thus, at the end of the Benchmark, the 10 Air Force participants were shown the data and extensively questioned on their reactions to the data and to the experiment itself. They strongly reported that turnaround times were consistent with their actual experience. They also pointed out many problems including (1) the lack of noise in the organization structure, (2) restriction with concurrent work on aircraft, and (3) inadequate representation of service activities – fueling, towing, and inspecting aircraft. These were corrected to their satisfaction.

Subsequent runs examined policy innovations centered around the introduction of new scheduling procedures and a modified information system. Results were examined by analysis of variance using an F test. The mean time for turnaround dropped by 39 percent below the Benchmark (significant at the 0.001 level). The same was true for several other important measures. Thus the simulation itself did generate statistically significant results. But in view of the validation difficulties, LP-IV concluded that the confidence that these results held for the actual process was still too low.

Two efforts to increase confidence started. The first was a series of psychological experiments to determine whether the dramatic improvement in scheduling was related to some accident of the simulation or whether it was reproducible in a different environment. The psychological experiments used 40 college students with no knowledge of the Air Force version of the scheduling task. The same results appeared and the difference between scheduling modes is again statistically significant. This experience certainly did increase the confidence that the difference in scheduling systems is valid for the actual process.

For a second complementary research effort, an Air Force group and the LP-IV staff decided to conduct a field test. In terms of the earlier discussion, LP-IV preferred a prototype test, but the Air Force could not obtain agreement for any control over or interference with operational requirements. During the first month, the field test did show a large (and significant) reduction in turnaround. At this point, the controllers complained about their workload (the requirement to preserve operational integrity resulted in much duplicate work) and the procedures were modified. The test went on and a great deal of useful data was generated. However, the procedural changes plus several major changes in operations make any direct comparisons of questionable value.

One suspects that LP-IV does not represent an isolated incident in the simulation world. The difference between looking for a problem which fits a technique and fitting techniques to a given problem is well known. LP-IV started with a problem and tried to solve it. Many validation techniques, particularly statistical ones, appeared highly desirable, but practical limitations precluded their use. One, of course, can greatly improve the picture presented here by selective reporting; but perhaps this view is more indicative of the world. Should one give up at this point? LP-IV chose to go ahead and tried a variety of ways to achieve confidence in the results. In retrospect, perhaps too little of the total effort went into validation. For example, the decision not to incur the cost of collecting turnaround data was reversed for subsequent work in this area. So, hopefully, some of the experiences in LP-IV will have long-run benefit toward improving validation abilities.

CONCLUDING REMARKS

Simulation offers the most flexible and realistic representation for complex problems of any quantitative technique. Its look of realism makes it a frequently preferred technique for large, significant problems. Thus many of the aspects that make validation difficult for simulation also give validation a great deal of importance. Decisions, often major decisions, are made on the basis of simulation results.

Knowledge about appropriate statistical tests for validating simulations is increasing. However, testing suffers from the standard problems of empirical research: (1) small samples due to the high cost of data, (2) too aggregate data, and (3) data whose own validity is questionable.

When adequate data are available, statistical tests are an essential part of validation, but the overall validation process should encompass much more. In rough order of decreasing value-cost ratios, some of the possible validation actions are the following:

1. Find models with high face validity.
2. Make use of existing research, experience, observation and any other available knowledge to supplement models.
3. Conduct simple empirical tests of means, variances, and distributions using available data.
4. Run "Turing"-type tests.
5. Apply complex statistical tests on available data.
6. Engage in special data collection.
7. Run prototype and field tests.
8. Implement the results with little or no validation.

The real task of validation is finding an appropriate set of actions.

---

[a] *Fishman and Kiviat*[11] *divide simulation testing into three categories. "(1)* Verification *insures that a simulation model behaves as an experimenter intends. (2)* Validation *tests the agreement between the behavior of the simulation model and a real system. (3)* Problem analysis *embraces statistical problems relating to (the analysis) of data generated by computer simulation."*

[b] *Some experimenters, of course, are concerned with proving the "truth" of their model. One example is the work of Newell and Simon*[17] *on simulation of human thought. The difficulties suggested by Turing's*

work become very meaningful if one wishes to use a model as a general theory.

[c] *For example, Clarkson*[2] *devised a model of a trust investment officer.*

[d] *Cohen and Van Horn*[4] *provide a fuller description of the LP-IV experiment.*

REFERENCES

[1] Bonini, Charles P., *Simulation of Information and Decision Systems in the Firm,* Prentice-Hall, Englewood Cliffs, N.J., 1963.

[2] Clarkson, G. P. E., *Portfolio Selection: A simulation of Trust Investment,* Prentice-Hall, Englewood Cliffs, N.J., 1962.

[3] Cochran, W. G., "Some Methods for Strengthening the Common χ^2 Test," *Biometrics,* Vol. 10, No. 4 (Dec. 1954).

[4] Cohen, I. K., and Van Horn, R. L., "A Laboratory Experiment for Information System Evaluation" in J. Spiegel and D. Walker, *Information System Sciences,* Spartan, 1965.

[5] Conway, R. W., "Some Tactical Problems in Digital Simulation," *Management Science,* Vol. 10, No. 1 (Oct. 1963), pp. 47-61.

[6] ———, *An Experimental Investigation of Priority Assignment in a Job Shop,* RAND Corporation, RM-3789-PR, 1964.

[7] Cyert, R. M., and March, J. G., *A Behavioral Theory of the Firm,* Prentice-Hall, Englewood Cliffs, N.J., 1963.

[8] Dunlop, R. A., "Some Empirical Observations on the Man-Machine Interface Question," Proceedings of the 1968 Carnegie-Mellon Symposium on Management Information Systems (to be published).

[9] Geisler, M. A., Haythorn, W. W., and Steger, W. A., *Simulation and the Logistics Systems Laboratory,* The RAND Corporation, RM-3281-PR, 1962.

[10] Fishman, George S., and Kiviat, P. J., "The Analysis of Simulation-Generated Time Series," *Management Science,* Vol. 13, No. 7 (Mar. 1967), pp. 525-557.

[11] ——— and ———, *Digital Computer Simulation: Statistical Considerations,* RAND Corporation, RM-5387-PR, 1967.

[12] Fishman, George S., *Digital Computer Simulation: Input-Output Analysis,* RAND Corporation, RM-5540-PR, 1968.

[13] Jenkins, G. M., "General Considerations in the Analysis of Spectra," *Technometrics,* Vol. 3, No. 2 (May 1961).

[14] Mood, A. M., and Graybill, F. A., *Introduction to the Theory of Statistics,* 2nd ed., McGraw-Hill, New York, 1963.

[15] Naylor, T. H., and Finger, J. M., "Verification of Computer Simulation Models," *Management Science,* Vol. 14, No. 2 (Oct. 1967), pp. B-92 to B-101.

[16] ———, Wertz, K., and Wonnacott, T. H., "Methods of Analyzing Data from Computer Simulation Experiments," *Communication of the ACM,* Vol. 10, No. 1 (Nov. 1967).

[17] Newell, Allen, and Simon, H. A., "GPS, A Program That Simulates Human Thought," in E. A. Feigenbaum and J. Feldman (eds.), *Computers and Thought,* McGraw-Hill, New York, 1963.

[18] Simon, H. A., and March, J. G., *Organizations,* Wiley, New York, 1958.

[19] Turing, A.M., "On Computable Numbers, with an Application to the Entscheidungsproblem" *Proceedings of the London Mathematics Society,* Part 1, No. 42 (1936), pp. 230-265. Part 2, No. 43 (1937), p. 544.

[20] ———, "Computing Machinery and Intelligence," *Mind,* Vol. 59 (Oct. 1950), pp. 433-460; reprinted in E. A. Feigenbaum and J. Feldman (eds.), *Computers and Thought,* McGraw-Hill, New York, 1963.

29

Ira S. Lowry

A SHORT COURSE IN MODEL DESIGN

The growing enthusiasm for the use of computer models as aids to urban planning and administration derives less from the proven adequacy of such models than from the increasing sophistication of professional planners and a consequent awareness of the inadequacy of traditional techniques. As Lowdon Wingo has put it, planners are now prisoners of the discovery that in the city everything affects everything else:

> *In the good old days we tackled the slum in a straightforward way by tearing it down. Now we know the slum to be a complex social mechanism of supportive institutions, of housing submarkets, of human resources intertwined with the processes of the metropolitan community as a whole. . . . To distinguish favorable policy outcomes from unfavorable ones is no longer a simple matter. Decisions by governments, firms, and individuals in metropolitan areas turn on the state of such interdependent*

---

spatial systems as use of recreation facilities, transportation and communication nets, and the markets for land, housing, and even labor, rather than on the highly localized consequences directly elicited by policy actions. The rapid evolution of a genus of mathematical techniques, or models, to conditionally predict certain locational aspects of the behavior of urban populations has been both cause and consequence of these developments.[1]

During the coming decade, it is safe to predict, many of our readers will be called upon to evaluate proposals for such models or to participate in their construction. In this essay, I hope to provide some orientation to the model builder's way of thinking, interpret the jargon of his trade, and suggest a few standards for the evaluation of his product.[2]

Granted the complexity of the urban environment and the potentially extensive ramifications of planning decisions, we may ask, first of all, how computer models improve the planner's ability to generate sound policy and effective programs. The answer is certainly not that computers are wiser than their masters, but rather that they perform the most monotonous and repetitive tasks at high speed and with absolute mechanical accuracy. The model builder can make use of this capacity only insofar as he is able to perceive repetitive temporal patterns in the processes of urban life, fixed spatial relationships in the kaleidoscope of urban form.

If he can identify such stable relationships, he may then find it possible to use them as building blocks or elements of a computer model. These elements, replicated many times, can be combined and manipulated by the computer (according to rules specified by the model builder) to generate larger, quasi-unique patterns of urban form and process which resemble those of the real world. The model literally consists of "named" variables embedded in mathematical formulas (structural relations), numerical constants (parameters), and a computational method programmed for the computer (algorithm). The pattern generated is typically a set of values for variables of interest to the planner or decision maker, each value tagged by geographic location and/or calendar date of occurrence.

THE USES OF MODELS

The model thus constructed may fall into any of three classes, depending on the interest of the client and the ambition of the model builder. In ascending order of difficulty, these are descriptive models, predictive models, and planning models.

Descriptions

The builder of a descriptive model has the limited objective of persuading the computer to replicate[3] the relevant features of an existing urban environment or of an already observed process of urban change. Roughly speaking, the measures of his accomplishment are (1) the ratio of input data required by the model to output data generated by the model, (2) the accuracy and cost of the latter as compared to direct observation of the variables in question, and (3) the applicability of his model to other times and places than that for which it was originally constructed.

Good descriptive models are of scientific value because they reveal much about the structure of the urban environment, reducing the apparent complexity of the observed world to the coherent and rigorous language of mathematical relationships. They provide concrete evidence of the *ways* in which "everything in the city affects everything else," and few planners would fail to benefit from exposure to the inner workings of such models. They may also offer a shortcut to fieldwork, by generating reliable values for hard-to-measure variables from input data consisting of easy-to-measure variables.[4] But they do not directly satisfy the planner's demand for information about the future, or help him to choose among alternative programs. For these purposes, he must look to the more ambitious predictive and planning models.

Predictions

For prediction of the future, an understanding of the relationship between form and process becomes crucial. In a descriptive model it may suffice to note that X and Y are covariant (e.g., that the variable Y consistently has the value of $5X$, or equivalently, that $X = 0.2Y$); but when the aim is to predict the value of

Y at some future time, the model must specify a causal sequence (e.g., that a 1-unit change in the value of X will *cause* the value of Y to change by 5 units). If one is able to postulate the direction of causation, knowledge of the future value of the "cause" enables one to predict the future value of the "effect."[5]

Thus the first task of the builder of a predictive model is to establish a logical framework within which the variables of interest to his client stand at the end rather than at the beginning of a causal sequence. (Variables in this terminal position are often described as "endogenous.") His second task is to make sure that those variables which stand at the beginning (prime causes, often called "exogenous") can be plausibly evaluated as far into the future as may be necessary. These requirements may enlarge his frame of reference far beyond that which would serve for a merely descriptive model.[6]

The second requirement is partly relaxed in the case of *conditional* predictions, which are in any case of greater interest to planners than the unconditional variety. The planner is ordinarily interested in the state of the world following some contemplated act on his part, or following some possible but uncertain event outside his control. The model may then be allowed to respond in the form, "if X occurs, then Y will follow," without explicitly asserting the likelihood of X's occurrence. But explicit predictions must still be made for other exogenous events, since these may reinforce or counteract the effects of the hypothetical change in X.

A special case of conditional prediction is called "impact analysis." Here the interest is focused on the consequence that should be expected to follow a specified exogenous impact (change in X), if the environment were otherwise undisturbed.

Planning

Finally, there are planning models, a class whose technology is not far developed. A planning model necessarily incorporates the method of conditional prediction, but it goes further in that outcomes are evaluated in terms of the planner's goals. The essential steps are as follows: (1) specification of alternative programs or actions that might be chosen by the planner, (2) prediction of the consequences of choosing each alternative, (3) scoring these consequences according to a metric of goal achievement, and (4) choosing the alternative which yields the highest score.

The best-known species of planning model executes these steps by means of a "linear program," a computational routine allowing the efficient exploration of a very wide spectrum of alternatives – albeit under rather special restrictions as to permissible cause-effect relationships, and assuming complete information about alternatives and their consequences at the time of choice. Perhaps more relevant to urban planners is the problem of making a sequence of choices, the effects of each choice conditioning the alternatives available for subsequent choices. Since, at each decision point, there are as many "branches" as alternatives available, the spectrum of possible final outcomes can easily become astronomical. If steps 3 and 4 are programmed for the computer, it is feasible to trace a fairly large number of alternative decision sequences through to their final outcomes; and mathematicians have reported some success with "dynamic programs" for identifying optimal sequences more efficiently than by trial and error.[7]

THEORIES AND MODELS

I have indicated that the model builder's work begins with the identification of persistent relationships among relevant variables, of causal sequences, of a logical framework for the model. In so doing, he must develop or borrow from theories of urban form and process. Although "theory" and "model" are often used interchangeably to denote a logicomathematical construct of interrelated variables, a distinction can be drawn. In formulating his constructs, the theorist's overriding aims are logical coherence and generality; he is ordinarily content to specify only the conceptual significance of his variables and the general form of their functional interrelationships. The virtuosity of the theorist lies in rigorous logical derivation of interesting and empirically relevant propositions from the most parsimonious set of postulates.

The model builder, on the other hand, is concerned with the application of theories to a concrete case, with the aim of generating empirically relevant output from empirically based input. He is constrained, as the theorist is not, by considerations of cost, of data availability and accuracy, of timeliness, and of the client's convenience. Above all, he is required to be explicit where the theorist is vague. The exigencies of his trade are such that, even given his high appreciation of "theory," his model is likely to reflect its theoretical origins only in oblique and approximate ways. Mechanisms that "work," however mysteriously, get substituted for those whose virtue lies in theoretical elegance.

The theoretical perspective of the model builder is most clearly visible in the set of structural relations he chooses as the framework of his model. A neatly articulated model will consist of a series of propositions of the general form $Y = f(U, V, X, Z, \ldots)$.[8] These propositions embrace the variables in which he is interested and specify the ways in which these variables act on one another. For most models relating to policy issues, it is useful to classify the propositions in terms of their content as technological, institutional, behavioral, or accounting.[9] While there may well be alternative sets of such propositions that convey the same meaning, the model builder is at least bound by rules of consistency (no contradictory propositions) and coherence (as many independent propositions as there are variables). Within these rules, his choice of structures is guided mostly by his sense of strategic advantage.

The pure theorist is often satisfied with the general forms indicated above, or with these forms plus a few constraints or restrictions. The model builder must be much more explicit, detailing the exact functional forms of his structural relations (e.g., $Y = \log U + a(V/X) - Z^b$); he must also fit his variables (Y, U, V, X, Z) and parameters (a, b) from empirical sources.

THE STRATEGY OF MODEL DESIGN

The "dirty work" of transforming a theory into a model is further discussed below (Fitting a Model). At this point I want to review some strategic alternatives of design open to the model builder, choices which demand all his skill and ingenuity since they bear so heavily on the serviceability of his model to its predetermined purposes. Typically, these decisions must be made in an atmosphere of considerable uncertainty with respect to problems of implementation and eventual uses, and there are no clear canons of better and worse. Though the model builder can profit from the experience of others who have dealt with similar problems, he is to a large extent thrown back on his intuitive perceptions and his sense of style.

The Level of Aggregation

Perhaps his most important choice concerns the level of aggregation at which he finds it profitable to search for regularities of form and process. While there is an accepted distinction between macroanalysis and microanalysis, the differences between these modes of perception can be elusive. Neither is the exclusive property of a particular academic discipline, but in urban studies, macroanalysis is closely associated with urban geography, demography, social physics, and human ecology, while microanalysis is typically the metier of economics and social psychology.

The geographers, demographers, ecologists, and social physicists prefer to deal with statistics of mass behavior and the properties of collectivities. The elements of a model based on this tradition are likely to be stock-flow parameters, gravity or potential functions, matrices of transition probabilities.[10] Faced with the same *explanandum,* the economist is much more likely to think in terms of a market model, in which resources are allocated or events determined through competitive interaction of optimizing individuals whose behavior is predicated on a theory of rational choice. The social psychologist also works from a theory of individual choice, and has his own version of the market model – though it is less articulate because it embraces a much wider variety of transactions.[11]

The principal criticism of the macroanalytic approach is that its "theory" consists in large part of descriptive generalizations which lack explicit causal structure. Thus a macro model of residential mobility may consist essentially of a set of mobility rates for population subgroups classified by age, sex, or family status, rates based on historical evidence of the statistical frequency of movement by the members of such groups. For purposes of prediction, one may assume that these rates will apply to future as well as past populations; but since the reasons people move are not explicit in such a model, the assumption of continuity is behavior cannot be easily modified to fit probable or postulated future changes in the environment of this behavior.

A second objection to macroanalytic approaches is that they do not lend themselves easily to financial accounting schemes. These are of particular relevance to planning models whose purpose is to distinguish among better and worse alternatives of policy or program. Strictly speaking, such distinctions can only be made if goal achievement is reducible to a single metric, and the most comprehensive metric available in our society, whether we like it or not, is money.[12] Thus, in choosing among alternative transportation plans, the objective may be to maximize net social return to transportation investments – for example, to maximize the difference between benefits to be derived from the investment and costs allocable to it. Even though a gravity-model representation of the journey to work/residential location relationship may "work" in the sense of generating accurate predictions of population distribution and travel patterns, it will not yield financial data so easily as a market model of travel behavior and residential site selection, since the latter operates throughout in terms of price-defined alternatives faced by households.

The microanalytic approach also has its problems. Chief among these is that a model based on the theory of rational choice can be implemented only if the chooser's system of relative values – technically, his "preference system" – can be specified in considerable detail. The search for an empirical technique to achieve this detailed specification has frustrated generations of economists, and approximations to date are both crude in detail and based on highly questionable operating assumptions. Lacking the ability to observe these preference systems directly, the modeler is restricted to a very meager menu of empirically relevant propositions concerning the complementarity and substitutability of economic goods, propositions deducible from general theoretical principles.

The second problem of the microanalytic approach is the implementation of a comprehensive market model – one embracing the entire range of transactions which substantially affect the patterns of urban development and land use. Given complete information about the demand schedules of buyers and the supply schedules of sellers, the classical theory of a perfectly competitive market for a homogeneous commodity is simple enough, having a determinate solution for both the volume of transactions and the emergent price of the commodity. But the model builder is faced, empirically, with a congeries of interrelated markets, subtly differentiated commodities, imperfections in communication, and inequities of bargaining position, all of which rule out the easy mathematical resolutions of the classical case. The fact is that we are presently able to implement only quite crude and tenuous approximations of market models.[13]

The Treatment of Time

Except for the simple descriptive case, a model usually purports to represent the outcome of a process with temporal dimensions. Beginning with the state of the (relevant) world at time t, it carries us forward to the state of that world at $t + n$; thus a land use model may start with a 1960 land use inventory in order to predict the 1970 inventory. The way in which this time dimension is conceived is a matter of considerable strategic significance; the choice lies among varying degrees of temporal continuity, ranging from comparative statics at one extreme, through various types of recursive progression, to analytical dynamics at the other extreme.

At first glance, the choice seems to hinge merely on the question, how often need results be read out? But the issues go deeper, involving the model-builder's perception of the self-equilibrating features of the world represented by his model, the empirical evaluation of response lags among his variables, and his interest in impact analysis as distinguished from other types of conditional or unconditional prediction.

The method of comparative statics implies a conviction that the system is strongly self-equilibrating, that the endogenous variables respond quickly and fully to exogenous changes. The model's parameters, fit from cross-section data, represent "equilibrium" relationships between exogenous and endogenous variables; a prediction requires specification of the values of the exogenous variables as of the target date. The process by which the system moves from its initial to its terminal state is unspecified.[14]

Alternatively, comparative statics may be used for impact analysis, where no target date is specified. Assuming only one or a few exogenous changes, the model is solved to indicate the characteristics of the equilibrium state toward which the system would tend in the absence of further exogenous impacts.

Self-equilibration is not a necessary assumption for analytical dynamics, an approach which focuses attention on the processes of change rather than on the emergent state of the system at a specified future date. Technically, this type of model must be formulated as a set of differential equations, at least some of which include variables whose rates of change are specified with respect to time.

Implementation of such a model requires only specification of its structural parameters and the "initial conditions" of its variables. Thereafter, all processes are endogenous except time, and the time path of any variable can be continuously traced. The state of the system can be evaluated at any point in time. If the system *is* self-equilibrating, the values of its variables should converge on those indicated by analogous comparative statics; but without self-equilibrating properties, the system may fluctuate cyclically, explode, or degenerate.

Because comparative statics requires such strong equilibrium assumptions (seldom warranted for models of urban phenomena), and because analytical dynamics requires virtually complete closure (all variables except time are endogenous), most model builders compromise on recursive progressions. This method portrays the system's changes over time in lock-step fashion by means of lagged variables, for example:

$$Y_{t+1} = a + bX_t \qquad (1)$$

$$X_t = c + dY_{t-1} \qquad (2)$$

Starting with initial values for either X or Y, one carries the system forward by alternately solving equations (1) and (2). Of course in this example, a bit of algebraic manipulation suffices to evaluate Y_{t+n} directly from a given Y_t; but the case is seldom so simple – and the model builder is likely to want to inject periodic exogenous changes into this recursive sequence.

The Concept of Change

Any model dealing with changes over time in an urban system must distinguish (at least implicitly) between variables conceived as "stocks" and variables conceived as "flows." A stock is an inventory of items sufficiently alike to be treated as having only the dimension of size or number – for example, dwelling units, female labor force participants, acres of space used for retail trade. This inventory may change as items are added or deleted; such changes, expressed per unit of time, are called flows. A model builder may choose to focus either on the factors which determine the magnitude of each stock, or on the factors which determine the magnitude of each flow.[15]

Since a stock is by definition the integral over time of the corresponding flow, it must also have the same determinants as the flow. But if the model builder limits his attention to flows which occur over any short span of time, he can afford to take a number of shortcuts. Exogenous variables whose effects on stocks are visible only in the long run can be ignored or treated as fixed parameters. Whereas nonlinear expressions may be necessary to represent the long-run growth of a stock, marginal increments in the short run can often be represented by linear expressions. By accepting the initial magnitude of a stock as historically "given," one avoids the necessity of replicating the past and can devote himself to modeling the events of the present and near future.

Consider a model of retail location whose eventual application will be a 5-year projection of the distribution of retail establishments within an urban area. The existing pattern (initial stock) of retail establishments in a large city reflects locational decisions made over the course of a century or more, during which time the transportation system, merchandising techniques, and patterns of consumption all have changed slowly but cumulatively. Most of the present stock of retail establishments will still be in operation at their present sites 5 years hence.

If the model builder is willing to organize his design around the *present* characteristics of the transportation network, of merchandising methods, of consumption patterns, his task may be greatly simplified. And the resulting model may be quite adequate for the prediction of short-run *changes* in retail location (say, as a consequence of population growth), even though it would not be able to recapitulate the city's history of retail development.

Clearly, the model builder must weigh the advantages of such simplifications against the fact that his model will have a shorter useful life. Since its structure postulates stability in a changing environment, the model will soon lose its empirical relevance.[16] By way of compromise, many model builders make use of "drift parameters": structural "constants" which are programmed for periodic revision to reflect changing environmental conditions, conditions which cannot conveniently be made explicit in the model.

Solution Methods

An integral part of the strategy of model design is a plan for operating the model – an algorithm or method of solution. This plan describes the concrete steps to be taken from the time that input data are fed to the computer until final results are read out. Four general methods are prominent; the choice among them is largely governed by the degree of logicomathematical coherence of the model itself.

The neatest and most elegant method is the analytic solution. Ordinarily, this method is applicable only to models which exhibit very tight logical structures and whose internal functional relationships are uncomplicated by nonlinearities and discontinuities. In substance, the set of equations constituting the model is resolved by analysis into a direct relationship between the relevant output variables and the set of input variables; intervening variables drop out of the "reduced form" equations. The paradigm system used above to illustrate recursion [equations (1) and (2)] can be solved analytically; for example,[17]

$$Y_{t+4} = (a + bc)(1 + bd) + (bd)^2 Y_t \qquad (3)$$

For models lacking complete logical closure, or whose structures are overburdened with inconvenient mathematical relationships, an alternative to the analytic solution is the iterative method. This method comprises a search for a set of output values which satisfy all the equations of the model; it proceeds initially by assuming approximate values for some of the variables and solving analytically for the remainder. These first-round solutions are then used as the basis for computing second approximations to replace the initially estimated values, and so on. Except for various degenerate cases, the solution values eventually "converge" – that is, further iterations fail to result in significant changes in the solution. Mechanically, the process is quite similar to recursive progression of a self-equilibrating system, but the iterative process need not imply either a sequence over time or a causal sequence. A drawback of this method is that it fails to signal the existence of alternative solution sets, a possibility that may have considerable importance for the interpretation of results.[18]

Ambitious models of urban processes may not meet the requirements for either the analytic or iterative methods of solution because of their scope: in the attempt to embrace a wide range of obviously relevant phenomena, one easily loses mathematical rigor and logical closure. For models of this class – loosely articulated "system analyses" – machine simulation may be the best resort. The model specifies an inventory of possible "events" and indicates the immediate consequences of each event for one or more variables representing a "stock" or population. A change in the magnitude of a stock has specified (endogenous) consequences in the form of inducing new events; but, characteristically, the major source of new events is exogenous. Indeed, the more sophisticated simulations (Monte Carlo or stochastic models) generate exogenous events by random choice from a given frequency distribution of possibilities. The computer's principal task is to keep a running account of all stocks and to alter them in response to events. This method is less appropriate for explicit projections than for tests of the sensitivity of the model (and, by implication, of the real-world system represented by the model) to various possible constellations of exogenous events.

Finally, there is the method of "man-machine simulation," in which computer processing of input data is periodically interrupted, and the intermediate state of the system is read out for examination by a human participant. He may adjust intermediate results to correspond with his judgment as to their inherent plausibility, or he may use these intermediate results as a basis for a "policy" decision which is then fed back to the computer model as an exogenous change in values for specified variables or parameters. The human participant is ordinarily included for educational reasons – to give him practice in responding to planning problems – but on occasion he is there simply because the model builder does not fully trust his model to behave "sensibly" under unusual circumstances.[19] (See Parameters.)

FITTING A MODEL

Once the model builder has selected a theoretical perspective, designed a logical framework large enough to encompass his objectives, and postulated the existence of enough empirical regularities to permit the resolution of his problem, his next task is to "fit" or "calibrate" the model. This task involves two types of transformation: the variables mentioned in the model must be given precise empirical definition, and numerical values must be provided for the model's parameters.

Variables

The first transformation always involves compromise. A variable conceived in general terms (household income) must be related to an available statistic (median income of families and unrelated individuals as reported by the U.S. Census of 1960 on the basis of a 25 percent sample), and the restrictions and qualifications surrounding the data must be carefully explored to be sure they do not seriously undermine the proposed role of the variable in the model (aggregation of medians is difficult; response errors may create serious biases in the data; sampling variability of figures reported for small areas may be uncomfortably large).

A variable included in the model because of its theoretical significance may not be directly observable in the real world, so that some more accessible proxy must be chosen. Thus many land use models deal in "location rents" (defined as that portion of the annual payment to an owner of a parcel of land which is attributable to the geographical position of the parcel as distinct from its soil or slope characteristics, existing structural improvements, or services provided by the landlord), but empirical sources tell us only about "contract rents" (the total contractual payment of tenant to landlord). Can contract rents be statistically standardized to serve as a reliable proxy for location rents?

I know of no formal canon of method for fitting variables, although I can think of some scattered principles to be observed.[20] More frequently than not, the problems encountered at this step force the model builder to backtrack and revise parts of his logical structure to lessen its sensitivity to bad data or to make better use of what data are actually available. Since few published statistics are exactly what they seem to be from the table headings and column stubs, it is very easy for one inexperienced in the generation of a particular class of data to misinterpret either its meaning or its reliability.

Parameters

The fitting of parameters – numerical constants of relationship – is necessary for two reasons: (1) theoretical principles and deductive reasoning therefrom are seldom sufficient to indicate more than the appropriate sign (positive or negative) and probable order of magnitude for such constants; and (2) since these constants are measures of relationship between numerical variables, the precise empirical definition of the variable affects the value of the parameter. For instance, the appropriate value of a labor force participation rate depends among other things on whether the pool from which participants are drawn is defined to include persons 15 to 60 or persons 14 to 65.

Parameter fitting is a highly developed branch of statistical method.[21] The most common tool is regression analysis, the simplest case being the estimation of parameters for a linear function of two variables, $Y = a + bX$. From a set of coordinate observations of the values of X and Y, one can estimate values for a and b in such a way as to minimize the expected error of estimate of Y from known values of X.

If the model can be formulated as a set of simultaneous linear equations, an elaboration of this method can be used to locate "best fit" values for all parameters in the system.[22] Models fitted in this way are often described as "econometric," although the method is equally applicable to noneconomic variables. A significant drawback of econometric fitting is that the criterion of selection for the values assigned to each parameter is the best *overall* fit of the model to a given array of data. The values generated for individual parameters are often surprising, yet it is difficult to look "inside" the fitting process for clues of explanation.

Alternatives to a comprehensive econometric fit can be described generally as "heuristic" methods. The model is partitioned into smaller systems of equations – some perhaps containing a single parameter – so that the parameters of each subsystem can be fit independently. This is in fact the typical approach, since few large models or urban form and process can be formulated as a single system of linear equations and still meet the objectives of the client.

Methods for obtaining estimates of the various parameters in these subsystems may vary considerably. A model ordinarily contains parameters whose function is nominal, and a model builder anxious to get on with his job may simply assign an arbitrary but plausible value to such a parameter. Where the context rules out direct methods for deriving simultaneous "best fits" even for the parameters of a limited subsystem, trial-and-error methods can be used to find a set of parametric values which seem to work. Or parametric values may be taken directly from empirical analogues, without regard for "best fit" in the context of the model.[23]

Finally, I should mention that model builders sometimes despair of finding a mathematically exact expression of relationship among certain of their model variables, so resort instead to "human" parameters. At the appropriate point in the operation of the model, intermediate or preliminary results are scanned by persons of respected judgment, who are asked to alter these outputs to conform to an intuitive standard of plausibility based on their experience in the field. The altered data are then fed back to the computer for further processing.

TESTING A MODEL

Fitting a model is analogous to the manufacture and assembly of a new piece of electrical machinery. A work team, guided by engineering drawings, fabricates each component and installs it in proper relation to other components, connecting input-output terminals. Along the way, considerable redesign, tinkering, and mutual adjustment of parts in inevitable; but eventually the prototype is completed. However carefully the individual components have been tested and their interconnections inspected, a question remains about the final product: will it really work?

Industrial experience indicates that the best way to answer this question is to turn the machine on and apply it to the task at hand. This precept applies also to computer models of urban form and process, with the important reservation that it is extremely difficult to select a "fair" but revelatory task, or to establish clear and objective standards of performance.

The appropriate test for a model depends, of course, on its predetermined function. It is unfair to ask a descriptive model to make a prediction, or a predictive model to find the optimal solution to a planning problem. But it is unfortunately the case that even an appropriate test may be infeasible.

The easiest model to test is the descriptive variety. Thus, for a model of urban form, the appropriate test would be its ability to replicate the details of an existing urban pattern on the basis of limited information concerning the area in question. Since most such models are built with a particular urban area in mind and fitted with reference to this area's characteristics, one ordinarily has detailed observations (for example, concurrent and otherwise compatible inventories of land use, structures, human populations, business enterprises, transportation facilities, and so forth) against which the model's output may be checked. The limitations of the test should also be apparent: the model's structure and parameters may be so closely locked into the patterns evident in this particular area and time that its descriptive abilities may have no generality; applied to another city the model may fail miserably.

The appropriate test for a predictive model is to run a prediction and verify the details of its outcome. The more distant the horizon of forecast, the more stringent the test; it would be easy to predict the distribution of workplaces in Boston tomorrow if one were given today's inventory. But few clients have the patience to finance several years of model building, then wait several more years to verify the model's first predictions. And even if one were willing to wait, there is the further problem that the model will almost certainly be designed for *conditional* predictions, and it would be remarkable indeed to discover in retrospect that all postulated conditions had been fulfulled.

The more accessible alternative is *ex post facto* prediction: take the state of the world in 1950 as a starting point and apply the model by forecasting for 1960; then compare the forecast values to the observed values for 1960.[24] This procedure is likely to suffer from the same limitation of semicircularity that plagues the testing of descriptive models. More likely than not, the predictive model was *fitted* to the recorded processes of change, 1950-1960. And if not, the reason is likely to be that comparable data are not available for the two dates. A predictive model is oriented to the problems of the future, and the model builder is anxious to feed his model the most recent additions to the menu of urban data – indeed, he may well initiate field work on a new series to provide it with a balance diet. Why limit his freedom by insisting that his model be able to subsist on the more limited menu available a decade ago?

The test of a planning model has two distinct phases. The first is a check on its ability to trace through the consequences of a given planning decision or set of decisions; this phase is a form of conditional prediction, and subject to all the hazards described above. The second phase is a check on the ability of the model to select an optimal result from a spectrum of alternative outcomes. It may fail to do so because (1) shortcut methods may eliminate as suboptimal some outcomes which have more promise than they immediately show; (2) the evaluation of outcomes may be very sensitive to engineering estimates of cost or imputation of benefits, and these are intrinsically nebulous; or (3) the criteria of selection may be poorly stated, so that an outcome which would in fact be acceptable to the client is classified as unacceptable by the model.

"Sensitivity testing" is sometimes urged as a more accessible substitute for the performance tests discussed above; although it is easy to perform and applicable to a wide variety of models, sensitivity testing elicits indications of the "strength" of a model's design rather than of its descriptive or predictive or evaluative accuracy. The procedure is as follows: by varying the value of a single parameter (or even of an input variable) in successive runs of the model, one can measure the difference in outcome associated with a given parametric change. If the model's response to wide differences in parametric values is insignificant, this may be an indication that the parameter and the associated network of functional relations – is superfluous. On the other hand, extreme sensitivity of outcomes to parametric changes indicates either that the parameter in question had better be fit with great care, or that some further elaboration of this component of the model is in order – on the grounds that the analogous real-world system must in fact have built-in compensations to forestall wild fluctuations in outcome.

EVALUATION

The picture I have painted above is rather grim, but I think it is accurate. The truth is that the client ordinarily accepts from the model builder a tool of unknown efficacy. The tests that the client can reasonably insist upon are at best partial and indecisive. Perhaps worst of all, those who must make the major decisions about sponsoring a model building project are unlikely to have the time or training to evaluate a proposal, and later, having footed a large bill, have a vested interest in the model hardly second to the professional stake of its builder. In the absence of incontrovertible evidence to the contrary, the builder and sponsor will agree that the model "works."

In the face of such ambiguities, it is not hard to imagine a reasonable man's refusal to participate in such a probable boondoggle. But for the reasons indicated at the beginning of this article, I do not anticipate any shortage of sponsorship for model building projects: it is better to try something – anything – than to merely wring one's hands over the futility of it all. Sponsors and model builders too can take comfort in the thought that they are building for the distant if not the near future.

Above all, the process of model building is educational. The participants invariably find their perceptions sharpened, their horizons expanded, their professional skills augmented. The mere necessity of framing questions carefully does much to dispel the fog of sloppy thinking that surrounds our efforts at civic betterment. My parting advice to the planning profession is: If you do sponsor a model, be sure your staff is deeply involved in its design and calibration. The most valuable function of the model will be lost if it is treated by the planners as a magic box which yields answers at the touch of a button.

NOTES

[1] Wingo (Ref. 15), p. 144. Model building has also been greatly encouraged by the electronic revolution in data processing and computation; mathematical models have an insatiable appetite for numbers.

[2] An immensely important topic in the field of model design which is *not* covered by this article is the joint effort of model builder and client to define the "problem" to which a model offers a possible "solution." On this point, I know of no better reference than a RAND book on systems analysis (Ref. 21), particularly the essays of R. D. Specht ("The Why and How of Model-Building"), Roland McKean ("Criteria"), and E. S. Quade ("Pitfalls in Systems Analysis").

[3] Some model builders would freely substitute "simulate" for my "replicate." All models are intended in some sense to simulate reality, but this usage is the source of some confusion in the literature since "simulation" has acquired another more technical meaning, descriptive of a class of algorithms. In this essay, I use the term *only* in the latter sense. See below, "Solution Methods."

[4] For example, traffic analysts use zonal interchange models to generate estimates of zone-to-zone traffic flows from inventories of the land uses in each zone. One prominent member of the profession is so convinced of the descriptive reliability of these models that he sees no further need for direct surveys of traffic movements (O&D) studies).

[5] Philosophers of science view the concepts of "cause" and "effect" with jaundiced eyes. For lesser mortals these concepts are most helpful and not at all dangerous so long as they are applied within the framework of a system of interdependence. Cf. Simon (Ref. 23), Ch. 1-3.

[6] No variable is intrinsically endogenous or exogenous. These terms, like the statistician's "dependent" and "independent," merely define the position of a variable within a particular model. A further useful distinction can be made between exogenous variables subject to policy control and those which are not; and between endogenous variables of direct interest to the planner and those which are included only because they are necessary to complete the logical structure of the model. Cf. Sonenblum and Stern (Ref. 24), pp. 112-114.

[7] The fundamentals and applications of linear programming are summarized in very readable form by Baumol (Ref. 2), pp. 837-853. I cannot find any simple exposition of dynamic programming, but see Bellman (Ref. 4), pp. vii-xi, for a brief account of the class of problems to which the technique is applicable.

[8] "The value of Y is a function of (depends on) the values of U, V, X, and Z, and so forth." For a gentle introduction to the notation and methods of mathematical modeling, Beach (Ref. 3) is an excellent source.

[9] Some examples, in prose rather than symbols:

Technological: the maximum vehicular capacity of a roadway is a function of the number of lanes, the average distance between signals, and the weather.

Institutional: disposable family income is a function of gross family earnings and the tax rate.

Behavioral: the level of housing density chosen by a family depends on disposable family income, the average age of family members, and the location of the workplace of the principal wage earner.

Accounting: total land in use is the sum of land in residential use, in retail use, in manufacturing use, and so forth.

[10]The essays in Part II of Zipf (Ref. 26) should give the reader a "feel" for the macroanalytic perspective in urban models. See also Carrothers (Ref. 8), and Berry (Ref. 5).

[11]Dyckman (Ref. 9) provides an excellent review of the theory of rational choice in a planning context. Any introductory text in economics will describe the microanalytic underpinnings of demand and supply schedules and will also review a family of market models. The most ambitious microanalytic model ever undertaken in the social sciences is described in Orcutt (Ref. 20). For models that embrace more than "economic" man, see Simon (Ref. 23) or Lazarsfeld (Ref. 16).

[12]Cf. Lichfield (Ref. 17).

[13]It is my personal conviction – not shared by all members of the fraternity of model builders – that the macroanalytic approach to modeling urban form and processes shows the greater promise of providing reliable answers to concrete problems of prediction and planning. For a contrary view, see the forceful statement by Harris (Ref. 13), p. 16.

[14]Descriptive models of urban form are nearly always static or "equilibrium" models, and are sometimes used for quasi-predictions (comparative statics). For convenient examples, see Harris (Ref. 12) or Lowry (Ref. 18).

[15]Contrast the emphasis on stocks in the San Francisco CRP model designed by A. D. Little, Inc. (Ref. 1) with the emphasis on flows in Bolan et. al. (Ref. 7) or with the several "growth allocation" models described in the *Journal of the American Institute of Planners,* XXXI(2) (May 1965).

[16]Cf. Black (Ref. 6).

[17]The reader is warned that equation (3) is not a general solution for any Y_{t+n}, but merely the simplest expression for Y_{t+4}.

[18]An example of the iterative technique is given in some detail in Lowry (Ref. 18), pp. 12-19.

[19]A good bibliography in simulation methods is Shubik (Ref. 22). Geisler et al., (Ref. 10) offer a quick and readable review of the field, with emphasis on man-machine simulation or "gaming." Grundstein (Ref. 11) describes a "community game" for the training of planners and municipal administrators.

[20]Special data problems encountered in modeling urban form and process are discussed by Britton Harris, "An Accounts Framework for Metropolitan Models," in Ref. 15, pp. 107-127. Also see Steger (Ref. 25), pp. 1-6.

[21]Beach (Ref. 3), Part II, provides an especially good introduction to statistical and econometric methods.

[22]The convenience of this method is so great that it is often applied to systems containing known nonlinearities, on the grounds that a linear approximation is better than nothing. Simultaneous estimation of the parameters of nonlinear systems is possible, but more difficult; the outstanding example among land use models is Kark Dieter's Program Polimetric for fitting an exponential model with a great many parameters. (The model, but not the fitting method, is described in Ref. 7.)

[23]Cf. Niedercorn (Ref. 19). His model is partitioned into three subsystems, each of which was fit independently. The discussion on pp. 14-15 illustrates the variety of estimating methods ordinarily required to fit a model. See also Harris (Ref. 12) for a discussion of the "gradient search" method of estimating parameters.

[24]Hill (Ref. 14) reports with unusual thoroughness on a test of this type for the *EMPIRIC* Model developed for Boston by Traffic Research Corporation.

REFERENCES

1. Arthur D. Little, Inc. "A Simulation Model of the Residential Space Market in San Francisco." Paper prepared for the Seminar on Models of

Land Use Development, Institute of Urban Studies, University of Pennsylvania, Philadelphia, Oct. 1964.
2. Baumol, William J. "Activity Analysis in One Lesson," *American Economic Review*, XXXXVIII (Dec. 1958), 837-873.
3. Beach, E. F. *Economic Models: An Exposition.* John Wiley & Sons, New York, 1957.
4. Bellman, Richard. *Dynamic Programming.* Princeton University Press, Princeton, N.J., 1957.
5. Berry, Brian J. L. "Cities as Systems Within Systems of Cities." Paper presented at the Annual Meeting of the Regional Science Association, Chicago, Nov. 1963.
6. Black, Russell VanNest. "Scientific Versus Empirical Projections," *Journal of the American Institute of Planners,* XXVI (May 1960), 144-145.
7. Bolan, Richard S., Willard B. Hansen, Neal A. Irwin, and Karl H. Dieter. "Planning Applications of a Simulation Model." Paper prepared for the New England Section, Regional Science Association, Fall Meeting, Boston College, Oct. 1963.
8. Carrothers, Gerald A. P. "An Historical Review of the Gravity and Potential Concepts of Human Interaction," *Journal of the American Institute of Planners,* XXII (May 1956), 94-102.
9. Dyckman, John W. "Planning and Decision Theory," *Journal of the American Institute of Planners,* XXVII (Nov. 1961), 335-345.
10. Geisler, M. A., W. W. Haythorn, and W. A. Steger. *Simulation and the Logistics Systems Laboratory.* RM-3281-PR, The RAND Corporation, Santa Monica, Calif., 1962.
11. Grundstein, Nathan D. "Computer Simulation of a Community for Gaming." Paper prepared for the Annual Meeting of the American Association for the Advancement of Science, Denver, Dec. 1961.
12. Harris, Britton. "A Model of Locational Equilibrium for Retail Trade." Paper prepared for the Seminar on Models of Land Use Development, Institute of Urban Studies, University of Pennsylvania, Philadelphia, Oct. 1964.
13. Harris, Britton. "Some Problems in the Theory of Intra-Urban Location." Paper prepared for a seminar sponsored by the Committee on Urban Economics of Resources for the Future, Washington, D.C., Apr. 1961.
14. Hill, Donald M. "A Growth Allocation Model for the Boston Region – Its Development, Calibration and Validation." Paper prepared for the Seminar on Models of Land Use Development, Institute of Urban Studies, University of Pennsylvania, Philadelphia, Oct. 1964.
15. Hirsch, Werner Z., ed. *Elements of Regional Accounts.* Johns Hopkins Press, Baltimore, 1964.
16. Lazarsfeld, Paul F., ed. *Mathematical Thinking in the Social Sciences.* The Free Press, New York, 1954.
17. Lichfield, Nathaniel. "Cost-Benefit Analysis in City Planning," *Journal of the American Institute of Planners,* XXVI (Nov. 1960), 273-279.
18. Lowry, Ira S. *A Model of Metropolis.* RM-4035-RC, The RAND Corporation, Santa Monica, Calif., 1964.
19. Niedercorn, John H. *An Econometric Model of Metropolitan Employment and Population Growth.* RM-3758-RC, The RAND Corporation, Santa Monica, Calif., 1963.
20. Orcutt, Guy H., M. Greenberger, J. Korbel, and A. M. Rivlin. *Microanalysis of Socioeconomic Systems: A Simulation Study,* Harper & Row, New York, 1961.
21. Quade, E. S., ed. *Analysis for Military Decisions.* R-387-PR, The RAND Corporation, Santa Monica, Calif., 1964.
22. Shubik, Martin. "Bibliography on Simulation, Gaming, Artificial Intelligence and Allied Topics," *Journal of the American Statistical Association,* LV (Dec. 1960), 736-751.
23. Simon, Herbert A. *Models of Man.* John Wiley & Sons, New York, 1957.
24. Sonenblum, Sidney, and Louis H. Stern. "The Use of Economic Projections in Planning," *Journal of the American Institute of Planners,* XXX (May 1964), 110-123.
25. Steger, Wilbur A. "Data and Information Management in a Large Scale Modelling Effort: The Pittsburgh Urban Renewal Simulation Model." Paper prepared for the Seminar on Models of Land Use Development, Institute of Urban Studies, University of Pennsylvania, Philadelphia, Oct. 1964.
26. Zipf, George Kingsley. *Human Behavior and the Principle of Least Effort.* Addison-Wesley Publishing Co., Reading, Mass., 1949.

30

Martin Shubik

ON THE SCOPE OF GAMING

There are many forms of gaming, stretching from complex mathematical models to free-form verbal interchanges. Individuals whose world view and professional backgrounds are utterly different may all regard themselves as being involved in "gaming."

The subjects are different, their purposes are different, and the criteria of validation differ, but the name is the same. In this paper, an attempt is made to sort out these major differences.

In a companion paper, definitions of the words gaming, game theory, and simulation are given to provide a context both for the discussion here and there.

The prime purposes of this classification are the following:

1. To call attention to the important prevalidation problems of *specification,* i.e., stating purpose and devising criteria by which to judge the attainment of one's goals.

---

2. To indicate the possibility that *in spite* of the diversity there may be a common core of knowledge and professional skills of importance to all gamers.

3. To suggest that all specialists stand to benefit from an understanding of the diversity of gaming, because frequently different types of gaming overlap and errors or important phenomena that may be completely ignored by one specialist may be obvious to another who sees the same game from a somewhat different viewpoint.

THE MANY GOALS OF GAMING

Teaching

Figure 1 shows the six main divisions of the goals of gaming, together with a finer breakdown of the categories of teaching and training. The breakdowns of the other categories are given subsequently.

In teaching and training, the audience for different games is extremely varied with respect to age, occupation, and reasons for using a game. A useful breakdown which correlates well (but not perfectly) with age is the type of educational operation: preschool, elementary school, high school, undergraduate college, graduate, and adult educational programs.

An individual's occupation and his reasons for using a game are highly correlated. Without going into great detail, four reasons are suggested which broadly describe why most players are involved with teaching or training games: they volunteer to play, they are advised to play and follow the advice, they are ordered to play by a superior, or bureaucratic or organizational rules require that they play.

Most games in most educational institutions are parts of courses or programs. There may be electives prior to registering for the program; however, once a student is in a program the organizational rules will require that he participate. In many colleges and universities in the United States there is a considerable amount of voluntary gaming.

Where the participants are members of large bureaucratic organizations, such as the military, other parts of government service, or private corporations, they have, for the most part, been advised or ordered to participate. On occasion they may be volunteers. When this is the case, the type of volunteering is usually of the type where a department head is told to supply three out of his twenty men for a game. He may call for volunteers. It is not uncommon that the volunteers may be the three least busy or most junior men in the department.

Concerned citizens groups, curious students, and

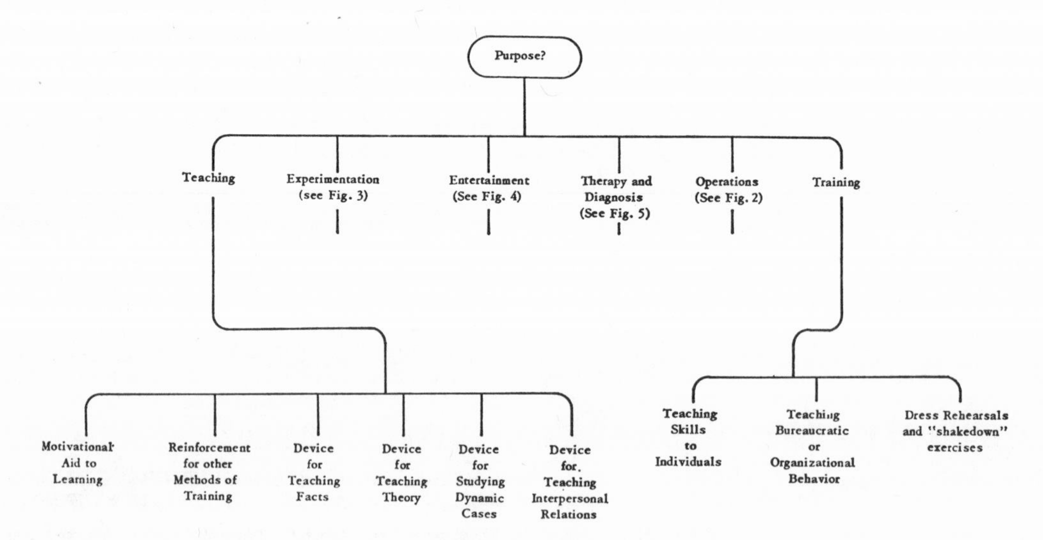

Figure 1

"buffs" form the hard core of volunteer gamers. A crude estimate indicates that in 1970 there were between 15,000 to 25,000 war-gaming amateurs in the United States.[a] Currently there is a trend towards games stressing social interaction and the problems of society. This is manifested in the growth of a number of board games in the penumbra between education and entertainment; thus we have had a progression from Monopoly to Smog. Even with war games there appears to have been an upswing in the last 10 years of games calling for diplomacy, negotiations, and grand strategy, such as Diplomacy and Summit, as contrasted with straight war games. From a technical game theoretic point of view there has been a shift from two-person zero-sum games or noncooperative individualistic enterprise games to nonconstant sum games where coalitions are of importance.

The overall trend in voluntary gaming in the last 30 years has been from an almost exclusive emphasis on military games to military-diplomatic games and to business games, and now more recently to games concerning society.

Different Roles in Gaming for Teaching - Before questions concerning validation can be asked with respect to a single game, it is desirable to consider goals and criteria of success from several different points of view.

In particular, any game should be considered in the context of its impact on individuals engaging in four activities related to it. They are the players, the builders, the controllers or directors, and the sponsors. Frequently, an individual may play more than one role. Furthermore, the roles are often more finely differentiated than the breakdown noted above. For instance, the game direction may consist of a team which contains not only umpires or teachers who direct the game, but also experts who are called upon to judge the feasibility of certain acts while otherwise having no control role.

At the university level, especially with graduate students, more may be learned by the students in constructing games than in playing them. The locus of the learning experience is by no means centered with the players.

In gaming for teaching purposes, especially at the high school level or younger, the worth of a game is frequently no more than that of the teacher. An inspired teacher can direct a mediocre game with good results, and the best of teaching games can be of little use if it calls for considerable direction from an inadequate teacher.

The breakdown of roles noted above applies to gaming used for purposes other than teaching. It is referred to again later.

Motivational Aid to Learning - One major attraction of gaming has been as a motivational device. It appears to attract the attention of and involve the players deeply where other methods have far less impact. There is reasonable consensus on this point among those who have used games and a small amount of experimental evidence as shown by the work of Wing[29] and others[5]. Creators of educational games, such as Layman Allen[21], stress the positive motivational features of educational games. However, it is easy to slip from conjecture to unsubstantiated advocacy, as is exemplified by the writings of Clark Abt[1]. Coleman has stressed the value of games in teaching disadvantaged children[9].

Reinforcement for Other Methods of Teaching - In the universities and schools, games are frequently used as part of a program along with more traditional methods of teaching. This is also true of the business schools and military academies. Gaming proponents claim that the mix of methods is most effective.

A Device for Teaching Facts - In virtually every type of gaming, including the diplomatic-military games of the Studies, Analysis and Gaming Agency,[b] and business games such as the Carnegie Tech Game or INTOP[28], gaming practitioners and players have claimed that gaming is an extremely useful way to learn and organize facts. A game usually provides a handy scheme for supplying associative links between facts, and as such it may aid both learning and remembering; although to date there is little hard evidence substantiating these claims.

A Device for Teaching Theory - At the advanced undergraduate and graduate level the building of games appears to be extremely useful in encouraging students to think in terms of models and abstractions. This improves their ability to theorize. In the social sciences especially, the importance of improving the ability of an individual to enable him to construct abstract representations of complex systems cannot be overemphasized. The discipline in constructing a playable game provides a deep appreciation of logical consistency and completeness, as well as stressing the connection between the model and its subject matter.

On the other hand, it is important to stress that before a game can be used with any success to teach theory it is rather desirable that the theory exists to be taught. In the exploitation of business games over the last decade this has not always been the case. A flagrant example of potential misuse has been in the modeling of advertising in business games. Even a brief glance at the literature on how advertising affects sales is sufficient to indicate that there is little substantiated theory in advertising; yet in many of the business games played both at universities and in business training programs, advertising has been thown in as an ad hoc modification on demand with teaching results which could be damaging were it not for the basic skepticism of most of the players. It is critically important that players be warned against learning false or unsubstantiated principles.

A Device for Studying Dynamic Cases - Several business schools, especially the Harvard Business School, favor the use of the case method. A specific historical case may be taken up, a "scenario" written describing it, and the class is required to consider the problems it poses and the ways in which they were handled or might have been handled.

A game lends itself with great ease to providing a dynamic context to a case. Furthermore, like the Czech experimental theater at Expo '67, it provides a natural means whereby alternative histories can develop.

A formal game, especially a large and complex one, has both the advantages and disadvantages of an institution. It may take on the inertia of an institution itself, as is exemplified by the Carnegie Tech game[7]. However, this may be an advantage, as it is extremely difficult to explain or reproduce in the classroom the ambience of decision making within a bureaucracy.

A Device for Teaching Interpersonal Relations - Many of the basic games for younger children and disadvantaged groups, as well as community action games to study urban redevelopment or other social problems, stress interpersonal relations both from the viewpoint of the individuals and their roles. In many of the uses of gaming, "seeing the other individual's point of view" by role playing his position appears to be of value. Thus, for example, a slum child may begin to appreciate that a policeman's lot is not a happy one. Furthermore, it might even be possible for a U.S. official to appreciate that to a North Vietnamese he does not necessarily appear as the epitome of sweetness, light, reason, and democracy.

At the more direct level an appreciation of the need for bargaining, communication, and compromise can be obtained from many of these games. A good example of such a game is Democracy[8]. Some of the insights gained here do not pertain only to personality factors but to a basic game theoretic phenomenon that in an n-person, non-constant-sum game there is no neat unique way of defining socially rational behavior. There are many different criteria for social rationality, and (as evinced by the lack of a core)[22] it is frequently not possible to satisfy the demands of all groups, even if each group can show that its demand is within the scope of its own power if it fails to cooperate with the remainder of society.

Training

Teaching blends into training, training into operational uses and so forth. Nevertheless, it is useful to make the distinctions among different goals for gaming although they may blend together at the boundaries between them. In particular the major distinction between teaching and training concerns the emphasis placed on the *why* of the process. There are several quite effective small games which can be of use in improving an individual's performance in production and inventory scheduling without ever going into the depths of why certain methods work. An operator does not have to get a course in dynamic or in integer programming to become a better manager of production and inventory scheduling.

Many individuals can be taught to drive safely by means of analogue device trainers without having to learn much about Newtonian mechanics or how an automobile works. Training games for simple manual skills, especially those requiring a fair amount of coordination, are not particularly exciting, but they can be of tremendous use and can provide valuable simulated experience that would be costly in the extreme to obtain from the field.

In general, when games are used for training, the only role occupied by the individual being trained in that of player. This contrasts with gaming for teaching where because the *why* is so important it is highly desirable in some instances to have students build or supervise as well as play games.

Bureaucratic and Organizational Behavior - In a complex society, licenses must be obtained, permits granted, rules checked, expectations examined, accounts audited, telephone calls made and routines for processing torrents of communication must be established. Training games offer the possibility not only of training individuals to acquire individual skills but also to learn bureaucratic routines.

Dress Rehersals and Shakedown Exercises - Rehearsals in the theater, field maneuvers, and battle exercises are all examples of operations devoted to seeing that individuals know "their lines" and are able to cooperate in team action. They differ from the previous category only inasmuch as they are usually aimed at preparing for coordination in a temporary context such as a specific play or a projected offensive. The phrase "shakedown" appears to come from the naval usage "shakedown cruise," which is the original cruise of a ship devoted to getting the crew to coordinate and to check to see if the equipment works.

Operational Gaming

The different goals of operational gaming are indicated on Figure 2. In contrast with gaming for teaching, operational gaming is used almost exclusively by adults in military, governmental, or corporate organizations.

There is an overlap between operational and training games in the domain of field exercises. It is difficult to say where the dress rehersal and coordination aspects of an exercise cease and where planning, strategy testing, and exploration begin. In Figure 2 the category "shakedown" has been included under operational gaming as well as in Figure 1 under training.

By far the largest use of operational games to this day is military or diplomatic-military. Relative to these uses corporate operational gaming is insignificant and the use of operational gamings for social planning is in its infancy.

Because of the nature of the bureaucratic structure of decision making, a clear understanding of the roles and goals of the players, builders, controllers, and sponsors of operational gaming exercises is far more important to the professional who wants to know "what is going on" than is such detailed understanding of the use of gaming for teaching.

Operational gaming is "where the money is" currently and the goals of a consulting firm wanting to build a large game, a general wishing to advocate a weapons system, and a colonel assigned to play in or operate the game can be sufficiently diverse that the mismatch makes an objective evaluation of such a game harder than reading the Rosetta stone.

Cross-Checking and Extra Validation for Other Methods - A game may be used as a backup procedure to provide an extra insight into a process that has

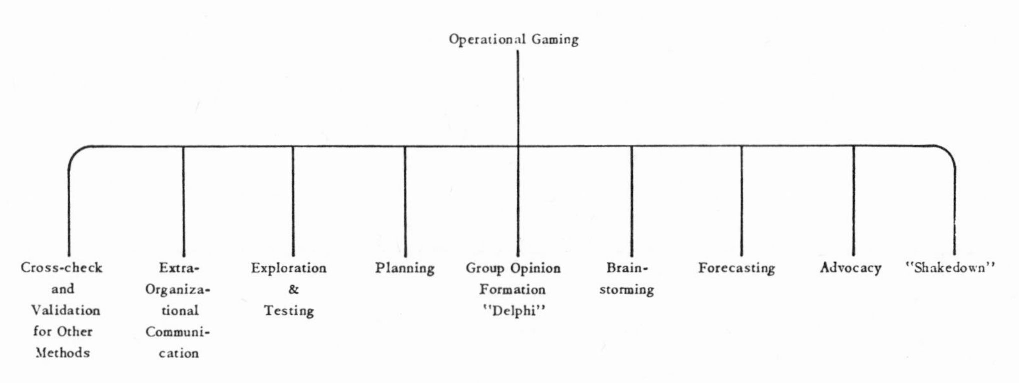

Figure 2

been investigated by other means. For example, a recommendation may be presented in report form. The basis for the recommendation may be expert opinion and/or empirical evidence. A gaming study of the same problem may turn up insights or raise questions overlooked by the approach. As operational games in general tend to be somewhat expensive in both time and money, the problem has to be of sufficient importance to merit the extra effort.

There is also the danger that a game may be employed to give a pseudoscientific window dressing to a recommendation.

Extraorganizational Communication - There may be a game outside of the game being played. With operational games it is critical to understand both the stated and the unstated purposes of gaming by the individuals involved in the exercise. In particular, gaming, along with short courses and seminars, is used to establish informal means of communication. In some instances the main objective may be to arrange to get two or three ranking individuals trapped together for 2 or 3 days on neutral ground.

Participants in diplomatic-military war games frequently comment on the value of being able to watch the decision-making styles of different high-ranked individuals.

The use of a game as a means for establishing informal communication will vary heavily with the style of play. If the game is held in an isolated locale over an intense period of play for 3 or 4 days or more, the effect may be quite striking. If, on the other hand, it is played in an intermittent manner over several weeks or months, then it is easy for most of the participants to minimize the disturbance to their set patterns.

Exploration, Testing, and Planning - The strict meaning of a strategy in the sense of game theory, while precise and worthy of note to a gamer, is not particularly useful to a planner. Planning involves the selection and aggregation of information. Even with the aid of high-speed digital computers the number of alternatives which can be explored is miniscule. Games such as those played by the SAGA[c] operation or the SIERRA series[d] of The Rand Corporation and many other have been used for planning, exploration, and the testing out of a limited number of alternatives.

An intense amount of preparation goes into a game of this type. The preparation is in general far more extensive than the play. Two or three moves on each side may be taken, and in a debriefing session after the game there will be an attempt to summarize and note the consequences, alternatives, or facts that had been overlooked prior to the commencement of play.

A planning game to be of use must utilize individuals sufficiently involved in the process that they can be privy to the actual problem and the major considerations. In military and governmental games these may range from colonels to five-star generals and cabinet officers.

There is some evidence that some high-ranking officials enjoy participating in gaming exercises; but there appears to be little evidence beyond the occasional testimonial as to what was accomplished. This last comment applies to gaming regarded as a "brainstorming" exercise as well.

Group Opinion Formation and "Delphi" - In the behavioral sciences and in the study of organizations, in evaluating many aspects of the present, and in forecasting the future, we have very little "hard" knowledge in the sense of the sciences in which experiments are performed and *replicated* frequently. In most professions much use is made of expert opinion. Up until recently little systematic thought was given to the study of how expert opinion is used and what the techniques are for optimizing the use of this scarce resource. Furthermore, little was known of the relative worth of using the opinion of more than one expert. When do diminishing returns set in? What sort of controls should there be over the interaction, and so forth?

An operational game may be regarded as a formal structure to elicit group planning – a process which involves both evaluation and prediction of the likelihood of contingencies.

Helmer[17] and Dalkey[10] have advocated the use of Delphi techniques, which consist of having a group of experts who are anonymous to each other respond to questionnaires, after which the results of their responses are processed and returned to them so that they can adjust their estimates in the light of the new information. Dalkey currently is engaged in large-scale experimentation[11] on the properties of the Delphi method.

One important feature that differentiates a formal operational game from Delphi is that there has been less emphasis on the aspects of motivation in relation to performance with the former than with the latter. To date there has been little effort to blend these two approaches. However, the potential appears to be worthwhile.

Forecasting - In general a game is *not* a forecasting device. A good operational game may make use of good forecasting procedures but it is not in itself aimed at providing forecasts. This should not be confused with its use in discovering new alternatives and in helping to evaluate future possibilities. *Forecasting* and *contingency planning* are related but extremely different activities. In particular, a good forecaster may not be in the slightest interest in the importance or worth of his forecast. Accuracy may be a goal for the forecaster in and of itself, not because of its relevancy to the planning process.

A game may be a useful device for stressing the need for coordination of forecasting activities with planning and decision-making processes. In this sense the involvement of forecasters in the design and play of operational games may be of considerable use.

Advocacy - Last, but not least, we must note the use of operational games for advocacy. A competent game designer can build biases of almost any size into a game. Advocates for specific policies or weapons systems can load the dice so that the game has a great probability of producing the results they want to see. Games are fun. They are great propaganda devices. The exploitation of the AMA business game provides one such example[3]. Action groups of nonprofessionals can easily be hornswoggled by a latterday snakeoil salesman peddling a game to cure all ills.

Smog, fog, the crime rate, central city decay, impotency, war, lack of understanding among nations, the evils of unemployment and the drug culture, the curse of the automobile, and the lack of a good 5-cent cigar will all be cured if we only have a big enough data bank tied into a game room with large fancy maps.

Recently there has been a move for the building of a "World Game" by several extremely well-meaning individuals[13]. As a mild advocate of gaming this author believes that there are many good reasons to proceed with the use and building of large games for operational purposes, especially in areas dealing with social policy. However, one must not confuse conversational feasibility with operational feasibility.

In some instances a game can be used as a euphemistic way for informing others of a change in policy by asking them to participate in an exercise whose outcome is a foregone conclusion. The Japanese war gaming prior to Pearl Harbor could be interpreted in the manner.[e]

Experimental Gaming

Human beings fortunately are more difficult to experiment with than rats or guinea pigs. Even so, there is now a fast-growing literature on experimental gaming in which human decision-making behavior is studied by observing the performance of individuals in formally structured games. To pursue this type of work fruitfully, it is important that the experimenters have at least a basic elementary understanding of game theory and social psychology. In a companion article, a background of game theory relevant to gaming has been presented[25].

Much experimentation has been done with simple 2 by 2 matrix games under relatively restricted conditions. The experimental subjects have been, for the most part, undergraduates at various universities; some army personnel have been used, as have been some inmates of local jails and some middle- and upper-level corporate personnel.

These experiments are psychological-light-years distant from preschool educational games or from military-diplomatic free-form war games. The criteria for validation belong to more or less accepted statistical methods familiar to physical scientists, econometricians, and experimental psychologists.

Some experimentation has been performed with business games of middling or of considerable complexity[19] and with political, diplomatic, and war games[18]. In general, owing to the greater complexity and smaller degree of control on these games, they have been harder to control, and hypotheses have been difficult to test. In some instances (Hoggatt[19], Shubik, Wolf, and Lockhart[27]) players have been faced with artificial players as competitors.

Validation of Hypotheses - In general, although the goals of the game designers are usually clear in experimental gaming, the goals of the players are by no means clear. There exists an enormous, and frequently poorly handled, problem in specifying, controlling, and measuring the goals and motivations of players in simple as well as in complex experimental games.

A separate article is needed to do justice to the literature on experiments with 2 by 2 matrix games, and another article is needed to discuss experimental work on the analysis of human factors in complex competitive systems. Nevertheless, without going into detail, several disturbing features of work in gaming can be seen. Specifically, much of the work with operational games presupposes that a considerable number of problems that belong to the domain of experimental gaming, i.e., basic research, have been solved, whereas, in fact, the expenditures and activities in experimental gaming are miniscule as compared with operational gaming.

Furthermore, although the word "validation" is popular and takes on a particular scientific flavor when applied to experimental games, if you do not know what you are trying to validate then all the statistical apparatus you have may not help you (unless you are consciously searching your data to generate hypotheses). *Control* and *specification* are prevalidation procedures which even at this time are not yet carried out adequately on many of the experiments. The major contributing factors to the failure of control and specification are lack of cooperation among specialists (i.e., social psychologists who know no game theory, misunderstanding the competitive structure, or game theorists knowing no social psychology, failing to allow for simple explanations of behavior) and lack of sufficiently automated laboratory facilities to enable the careful experimenters to obtain detailed observations and to run standard analyses at a reasonable cost.

Artificial Intelligence - Figure 3 has three branches. The first covers the type of experimentation that is more or less familiar in other disciplines. In the past decade there has been a considerable upsurge in the study of *artificial intelligence*, or in the study and the construction of computer programs which perform tasks that are usually regarded as requiring intelligence. No distinction has been made, in general, between the sort of intelligence required to solve difficult problems, such as playing chess, and

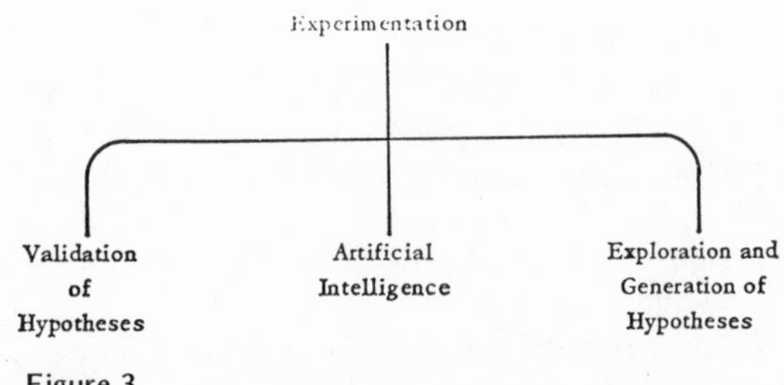

Figure 3

to resolve interpersonal problems, such as those which arise in nonconstant sum games – bargaining, for instance.

Frequently, the gamer is more interested in *social intelligence* than in individual intelligence. The problems in the construction of a good problem solver or a socially intelligent player differ inasmuch as the criteria for the performance of the former are relatively easy to construct, whereas there are no such easy criteria that can be constructed to judge group or social rationality.

In particular, it appears that a good problem solver, a program which can play chess well, for instance, requires efficient searching and calculating abilities and other features usually associated with intelligence and intellect. By the very nature of the game it need not have any "personality." A good chess playing program has to be a "smart" or intelligent program, not a pleasant or nice one. This is not the case when we turn to non-constant-sum games. It is possible to build an artificial player for a business game[19] which plays in a manner comparable to human players. The rules or "heuristics" needed to construct such a player call more for an emphasis on his interpersonal relationships than on his ability to compute. A "nice," moderately cooperative and not particularly aggressive artificial player in a business game may elicit cooperation from his competitor and will do quite well.

The literature on artificial intelligence has very little on the subject of social intelligence. There has been and there is currently an extreme division of opinion on the nature of problem solving, leaving aside the extension to social interaction. Simon, Minsky, and Pappert[23], and many others, are the proponents, whereas considerable criticism of the basis of artificial intelligence work has been offered by Bar Hillel and H. Dreyfus[12].

To enter into the debate on the pros and cons of artificial intelligence would take us too far astray from the work relevant to gaming; hence we confine our remarks only to those aspects of the subject relevant to those interested in gaming.

Along with the growth of interest in artificial intelligence has come a considerable growth in the design of protocols and ways to describe decision-making processes. Much work has been addressed to analogies between how one teaches a machine and how one teaches a child[23]. In particular those interested in experimentation with computer-aided instruction[f] need to be aware of the developments in artificial intelligence.

The experimental gamer is usually more interested in games which are more than problem-solving exercises. Many war games and games such as chess can be modeled as two-person zero-sum games; hence the main analytical problems they pose are in the domain of information processing and problem solving. Diplomatic-military, business, social development, and most other games do not fall under the zero-sum rubric. Social, political, or economic behavior all call for attention to interpersonal interaction. The construction of robots or artificial players in these games both provides opportunities to attempt to model sociopsychological processes in the building of the players and gives the experimenter greater control over his experiments, especially when he is able to replace a set of two-person experiments with a set of experiments consisting of a group of individual human players playing with the same artificial competitor.

Exploration and Generation of Hypotheses - Frequently, experimental games are used to explore decision-making processes and to generate hypotheses rather than to test specific hypotheses. Sometimes this is not the way things were planned, but this is how it works out. Prior to the experiment, several hypotheses may be suggested. After the experiment it appears that hypotheses can neither be accepted nor rejected, owing to insufficient definition or complications in the control of the experiment. Nevertheless, the running of the experiment clarifies the definition of the hypotheses, locates others, and locates the control difficulties.

The above reasoning is often used as an excuse or self-justification after an ill-conceived experiment has been run. However, this is not always the case and pilot experiments play an extremely useful role when the topic being studied is both complex and ill defined.

Games for Entertainment

The Theater - It is important to remember the deep interconnection between gaming and theater. For example, many war exercises, fleet maneuvers, and "dry-runs" are identical in purpose with dress rehearsals. Huizinga[20], Callois[6], and many others have discussed the relationship between plays and games. It is not the purpose of this paper to explore the historical, anthropological, and religious aspects of this interconnection. They form a fascinating subject in themselves. However, those who wish to use games for more mundane purposes should at least be aware of the interrelationship among the games, plays, theater in general, mass spectacles, and ceremonial parades. The military parade itself is an extremely complex phenomenon, being part entertainment, part training, part a signaling process in a diplomatic dialogue, and a device for influencing morale.

An important but open question is what are the basic features that differentiate good theater from good operational gaming? For example, how does the "realism" of the scenery affect both of these activities? The audiences are different, the role playing is different, and the stated purposes are different. Nevertheless, an analytical categorization of these differences is not an easy task.

Gambling - Three categories of individuals involved in gambling must be distinguished. There are those businessmen who run gambling ventures, professional gamblers who make a living from playing, and those who play for other reasons.

The individuals who run gambling establishments are in many senses not particularly distinguished from other businessmen except that possibly gambling as a business tends to be an enterprise with not very large components of risk as compared with a high innovation technology enterprise. The professional gambler such as the poker player (see for example H. O. Yardley[31], *The Education of a Poker Player*) does not seem to be far different from the professional arbitrageur. They both take risks but they are in the

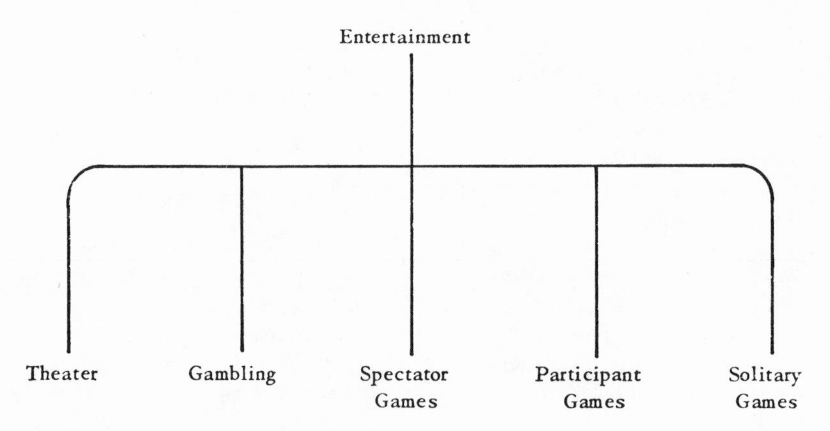

Figure 4

true sense of the word calculated risks, and the individuals who devote professional attention to these occupations are usually skillful enough that they are able to earn a good (but in general not spectacular) living from their professional skills.

Although there is a large element of chance in a game such as poker in contrast with a game such as roulette, it is primarily a game of skill and not of chance. The calculation of probabilities is one of the key aspects of good poker playing. There are obviously enormously important psychological factors in one's ability to judge the competence and style of the other players. Betting on horse races stands somewhere between a roulette game and a poker game in its skill component. The element of chance is extremely large. However, there are some useful calculations to be made concerning the odds being offered and the probable performance of the horses. The main factor, however, does come in the judgment of the horses and their performance on a specific track under the appropriate weather conditions.

In general, especially in large organizations, when someone states "we have taken a calculated risk," that frequently means that individuals have made a decision without doing the calculations necessary. In the case of the professional gambler, the very reverse is true. In general they have no bureaucratic structure around them and they are in a position where fast and explicit calculation of the odds is a central aspect of their very living.

In contrast with the fighter pilot, the poker player thinks in terms of odds explicitly. It is unlikely that the pilot calculates a probability of 0.15 of success by one avenue or 0.3 of success by another approach. There is undoubtedly an important difference in calculating explicit probabilities involving death and explicit probabilities involving money. Furthermore, the very nature of many gambling games of skill makes the calculation of probabilities a natural and explicit way of evaluating one's position. It is unlikely, however, that these features are in themselves sufficient to explain the fundamental difference in approch to thinking in terms of explicit probabilities evinced by professional gamblers as compared with, say, middle management or army colonels or, even more so, the citizen on the street. The literature on the social and personality aspects of the professional gambler appears to me to be surprisingly slight. This also holds true for the handful of special professions in which the risk-of-life component is sufficiently important to make the gambling aspect explicit, for example, test pilots and steeplejacks.

The interests of the individuals not professionally involved in gambling run the gamut from mild entertainment to deep addiction. Many individuals who lose \$20, \$30, or \$100 at the tables in Las Vegas or Monte Carlo are paying an entertainment fee. For the most part they know that they are paying this fee and have decided that the entertainment is worth it. It is worth noting that the mere location and decor of main major gambling towns and main casinos lay stress on the theatrical aspects and the role-playing features of gambling. Las Vegas is designed so that the perfectly ordinary middle-class dweller of suburbia can lose his $100 or so in a socially acceptable manner in surroundings ranging from pseudo-luxury to pseudo-wickedness.

What are the risk-taking features of the ordinary individual who is not addicted to gambling, who plays small-stake roulette at a casino, or who buys the occasional ticket for the races? There exists a certain amount of literature in economics and psychology concerning gambling and the buying of insurance where the odds are in general extremely small for an event to occur. However, there is virtually no analytical literature on ordinary gambling behavior. Erving Goffman has several highly stimulating articles on con games, where the otherwise prudent and nonaddicted individual is taken for a sucker[14].

There are many individuals for whom gambling is an addiction. Dostoevski was a good example on one of these. There is a small psychopathological literature on gambling as is evinced by the somewhat unsatisfactory book of Dr. E. Bergler[4]. One difficulty in studying a subject such as pathological gambling is that it requires a multidisciplinary approach. Psychiatrists will tend to see only the psychiatric aspects whereas, for example, those trained in a theory of games will undoubtedly lay heavy emphasis on the structural differences among various games.

From the viewpoint of those interested in operational games, especially games of a military or social variety, the study of addiction and extreme risk-taking would appear to be critical. The distance between the gambling addict and the drug addict may not be great. There also appears to be an important psychopathological risk-taking component in assassinations, in some forms of exploratory behavior, and in the actions of some extremist groups.

In summary, it appears to me that gambling behavior of virtually all types is a critical phenomenon in the understanding of many important features of risk taking. Those who argue for operational games as a means for studying extremely original or surprising alternatives should also consider the need to explore the genesis of both "reasonable and pathological risk behavior."

Spectator Games - Many sports, such as football, baseball, hockey, basketball, cricket, etc., are primarily spectator sports. The vast majority of the participants are in the audience and derive vicarious pleasure from the play. There the analogy between the game and theater is possibly at its closest. There are the actors, and the great majority are spectators. The sports event is far more of a free-form play than is a theatrical performance. In the former, although the rules are given, the actual path of the play is not completely known in advance. In the latter, the complete path of the play has been specified except for the acting that has not been controlled by the direction. Spectator games may have a small advocacy and teaching component to them, inasmuch as they may inculcate an appreciation of teamwork and an ability to judge and understand the qualities of effective performance. However, for the most part they are pure entertainment. For a discussion of the vicarious pleasure and rule identification aspects of spectator entertainment see Callois[6].

Participant Games - Bridge, poker, tennis, chess, football, charades, monopoly, and many board games, many of which can be played as spectator games, are most frequently played only by the active participants for their own amusement.

The distinction between participation in a poker game for amusement and for gambling purposes may easily vary as the size of the stakes. The importance of the payoffs to the players as an influence on the nature of the game cannot be overstressed. When an individual participates in a game whose stated purpose is operational or educational, but which nevertheless is formulated in such a way that the payoffs to him are not particularly clear, it becomes absolutely crucial to investigate the possibility that he has turned the exercise into a game for his entertainment.

It is a safe rule to apply when using games for teaching, experimentation, operations, or therapy to have as a null hypothesis that in fact the game was primarily theater or participant entertainment.

Solitary Games - Possibly one of the greatest sinks for the use of man-hours in gaming is the solitary game. Crossword puzzles, jigsaw puzzles, and solitaire are major examples of games "played to while away the time," although it can be argued that they may have an educational component. The

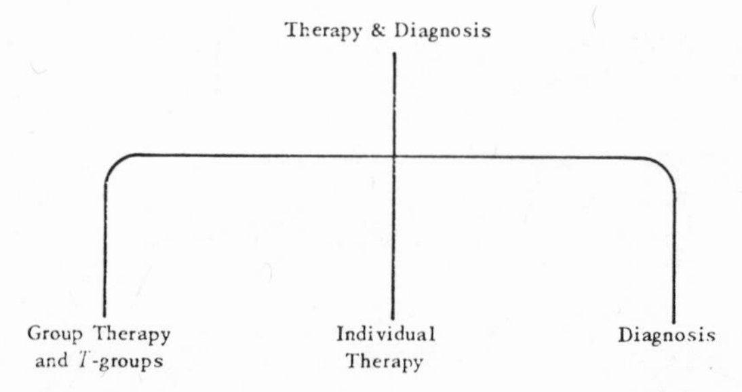

Figure 5

origins of both the crossword puzzle and the jigsaw puzzle are relatively recent (within the last 100 years). Precisely what makes them so popular? Will they be supplanted by other solitary games? Could solitary games be designed that would be fun and more explicitly educational or experimental?

Games may well have both diagnostic and therapeutic value. Although these areas lie well beyond my own training and competence, it would be a glaring omission not to call them to the attention of the reader.

Group Therapy and T-Groups - In some ways group therapy sessions and *T*-groups might be regarded as "antigames"; as such, the comparison between them and a formal operational game such as, say, a diplomatic-military game becomes of considerable interest. In the case of the latter, the individuals are encouraged to concentrate on certain aspects of role playing. Very frequently an individual is required to simulate the decision-making process of someone else. In contrast with this activity, in group therapy individuals are encouraged to find out who they are. The stress appears to be in the other direction. Individuals will hopefully be able to examine where they have been role playing in a manner not consistent with their comprehension of self.

It appears to me that the paradigm of the game offers an extremely fruitful basis for joint work by psychiatrists, social psychologists, and those interested in organizational decision making.

Diagnosis - It is not difficult to design games that focus on relatively narrow bandwidths of decision making and of interactive behavior. Informal experimentation with several games, such as "So-Long Sucker"[16] and "The Dollar Auction"[26], indicates that it is possible to obtain extremely strong participant reactions to relatively simply games. Experimentation with 2 by 2 matrix games of certain design has also indicated this. The use of small games for diagnosis might be relatively cheap and effective.

Individual Therapy - The use of games for therapeutic and corrective purposes is clearly closely related to, but somewhat different from, the use of games in teaching. Little appears to be known about the potentialities of this use.

CONCLUDING REMARKS

The scope of gaming is considerable. Many uses are utterly different from each other both in concept and purpose. Yet at the same time amid all the diversity a certain common thread is present. The game is a paradigm for competitive and/or cooperative behavior within a structure of rules. The rules vary in formality in free-form gaming or in rigid-rule gaming. They vary in portrayal of war situations, economics, social contract formulation, and so forth. But *all* games call for an explicit consideration of the role of the rules. A serious user of games is well advised to be broadly aware of the alternative uses and meanings of games as well as deeply specialized in his own type of gaming.

A key word concerning gaming that he hears frequently and from different sources is "validation." How can games be validated? Prior to asking this question it is imperative that we ask, "For what purpose and for whom?" A common vocabulary employed by different groups to mean different things is guaranteed to breed confusion. Around 30 considerably different purposes for gaming have been suggested. The criteria of validation for the success of a gaming endeavor are extremely different as we consider the different uses.

The size of the box office receipts is a good criterion for evaluating the success of a spectator sport from the viewpoint of the promoter. The number of people doing crossword puzzles is a good criterion for the owner of a newspaper. The criteria applied by a general, a zero-sum game theorist, and a military hardware expert to force posture and allocation games are planets apart from the criteria that might be used to judge the success of a political-military exercise run by a mixed group of political scientists and top government officials. This is turn would be different from the judgments applied to evaluate the worth of a teaching game designed to give high school students an appreciation of international relations.

In light of the many different types of gaming, the different goals of the various interested parties, and the problems of control, rather than talk about validation we are probably better off concentrating on four stages in the evaluation of gaming. They are (1) intention, (2) specification, (3) control, and (4) validation.

The first refers to finding out generally what the goals of the concerned parties are. The second involves translating these goals into well-defined measures and in establishing that the measures can in fact be obtained from the game. The third refers to the actual control of the game necessary to guarantee that extraneous elements do not confound the obtaining of the measurements called for by the specification. The process of validation calls for interpreting the significance of the measurements in terms of the specification.

A methodology and a theory of gaming are only beginning to emerge. It is certainly premature to believe that there is such a thing as "the method for evaluating or validating all games." The "prevalidation" steps are still not always done adequately. The surprising feature of the growth in gaming is not that there is no single method of validation, but that so little attention has been paid to sorting out the different uses of gaming and to developing criteria and methods that apply to the special uses.

The promise from many of the different types of gaming appears to be considerable. The proof of the promise is by no means empty for some of the categories of gaming. There are some reasonable criteria available for judging the success of a social psychology experiment, the teaching value of some elementary games, and of some business games; the worth of some formal game theoretic and simulation models for weapons evaluation; the profitability of sports; the entertainment value of the theater and entertainment games, and several other uses of gaming.

Our hard knowledge is extremely limited concerning how successful (and what are the criteria for success?) operational games are. What is really learned from political-diplomatic and military exercises? Who learns what from teaching games? The words "ad hoc" are frequently used in the pejorative sense. I would like to use them in a nonpejorative manner. It is my belief that the potentialities of gaming are considerable in many different fields of application. The ad hoc construction of specification, control, and validation procedures with extreme attention paid to special purpose at hand could yield valuable insights and results from which the broader generalizations called for by a general theory of gaming might be constructed.

---

[a] *Based on the number of subscribers to the Avalon Hill publication,* The General.

[b] *This agency is the successor to the Joint War Gaming Agency.*

[c] *The type of game used here was originally suggested by Goldhamer and Speier [15].*

[d] *See, for example, Northrop [24].*

[e] *See Wohlstetter [30].*

[f] *See for example [2].*

REFERENCES

1. Abt, C. C., *Serious Games,* The Viking Press, New York, 1970.
2. Atkinson, R. C., "Role of the Computer in Teaching Initial Reading," *Childhood Education* (1968).
3. Bellman, R., Clark, C. E., Malcolm, D. G., Craft, C. J., and Ricciardi, F. M., "On the Construction of a Multistage, Multiperson Business Game," *Operations Research,* Vol. 5, No. 4 (Aug. 1957).
4. Bergler, E., *The Psychology of Gambling,* Hill and Wang, New York, 1957.
5. Boocock, S. S., and Schild, E. O., *Simulation Games in Learning,* Sage Publications, Beverly Hills, Calif., 1968.
6. Callois, R., *Man Play and Games,* Thomas & Hudson, London, 1962.
7. Cohen, K. J., Dill, W. R., Kuehn, A. A., and Winters, P. R., *The Carnegie Tech Management Game,* Richard D. Irwin, Homewood, Ill., 1964.
8. Coleman, J., *Democracy,* The Johns Hopkins Uni-

versity Department of Social Relations and Academic Games Associates, Baltimore.

[9] ——, "Social Processes and Social Simulation Games," in Boocock, S. S., and Schild, E. O., *Simulation Games in Learning.*

[10] Dalkey, N. C., *The Delphi Method: An Experimental Study of Group Opinion,* The RAND Corporation, RM-5888-PR (June 1969).

[11] ——, and Roarke, D. L., *Experimental Assessment of Delphi Procedures with Group Value Judgments,* R-612-ARPA, The RAND Corporation (to appear).

[12] Dreyfus, H., *Critique of Artificial Reason,* 1971 (to appear).

[13] Fuller, B., *Presentations to Congress: The World Game,* Southern Illinois University, Carbondale, Ill., 1970.

[14] Goffman, E., "On Face Work," *Journal for the Study of Interpersonal Processes,* Vol. 18, No. 3 (Aug. 1955), pp. 213-231.

[15] Goldhamer, H., and Speier, H., "Some Observations on Political Gaming," *World Politics,* Vol. 12 (1959), pp. 71-83.

[16] Hausner, M., Nash, J. F., Shapley, L. S., and Shubik, M., "So Long Sucker," a four-person game in *Game Theory and Related Approaches to Social Behavior,* M. Shubik (ed.), John Wiley & Sons, New York, 1964.

[17] Helmer, O., *A Use of Simulation for the Study of Future Values,* The RAND Corporation, P-3443 (1966).

[18] Hermann, C. F., "Validation Problems in Games and Simulations with Special Reference to Models of International Politics," *Behavioral Science,* Vol. 12 (May 1967), pp. 216-231.

[19] Hoggatt, A. C., "Measuring the Cooperativeness of Behavior in Quantity Variation Duopoly Games," *Behavioral Science,* Vol. 12, No. 2 (Mar. 1967).

[20] Huizinga, J., *Homo Ludens,* Beacon Press, Boston, 1955 (translation).

[21] Layman Allen, *WFF'N PROOF,* New Haven, Conn.

[22] Luce, R. D., and Raiffa. H., *Games and Decisions: Introduction and Critical Survey,* John Wiley & Sons, New York, 1957.

[23] Minsky, M., and Pappert, S., "Artificial Intelligence Memo No. 200, Progress Report 1968-1969," Massachusetts Institute of Technology, Cambridge, Mass., (1970).

[24] Northrop, G. M., *Use of Multiple On-Line, Time-Shared Computer Consoles in Simulation and Gaming,* The RAND Corporation P-3606 (1967).

[25] Shubik, M., "On Gaming and Game Theory," *Management Science* (Jan. 1972).

[26] ——, "The Dollar Auction Game: A Paradox in Non-cooperative Behavior and Escalation," Yale University, Department of Administrative Sciences, Report No. 30 (1970).

[27] ——, Wolf, J., and Lockhart, S., "An Artificial Player for a Business Market Game," *Simulation and Games* (1971) (to appear).

[28] Thorelli, H. B., and Graves, R. L., *International Operations Simulation,* The Free Press, New York, 1964.

[29] Wing, R. L., *The Production and Evaluation of Three Computer-Based Economics Games for the Sixth Grade,* Board of Cooperative Educational Services, Westchester County, N.Y., 1967.

[30] Wohlstetter, R., *Pearl Harbor: Warning and Decision,* Stanford University Press, Stanford, Calif., 1962.

[31] Yardley, H. O., *The Education of a Poker Player,* Simon and Schuster, New York, 1957.

PART VIII MANAGEMENT

The difficulties of governing our large metropolitan cities are manifested in the increasingly severe survival problems of large-city mayors, the struggle to maintain service standards in a number of traditional municipal activities such as sanitation, the extension of blight despite sizeable renewal and housing programs in many communities, and the inability to respond effectively to demands for changes in programs and services. The disgovernance syndrome (as distinct from non-, mal- and un-governance) has developed during a decade when municipal governments as a group have broadened their responsibilities, have attempted to do more and better, and have in fact been more responsive to problems of human welfare than ever before. . . .

Some of the factors [making municipal management more difficult] are those of scale and expanse. Some are primarily political, and some administrative and managerial. Others are more clearly sociological or cultural in character. All are, however, many-sided and so interrelated with one another that they become difficult to group into a typology. Individually, the factors must be dealt with; it is their simultaneous occurrence which creates the disgovernance syndrome.

Henry Cohen, "Urban Disgovernance," Public Administration Review, *September/October, 1970, pp. 488-489.*

31

Melville C. Branch

SIMULATION, MATHEMATICAL MODELS, AND COMPREHENSIVE CITY PLANNING

BASIC REQUIREMENTS FOR SIMULATION[a] IN COMPREHENSIVE CITY PLANNING

A brief description of the intellectual task of planning comprehensively for a large city helps to delineate the problem and identify some of the difficulties and potentialities of methodological advancement. What does a person who is in the decision-making position in city planning need to satisfy himself that his actions are well founded? Since nothing approaching an established method exists for this difficult task of absorption, analysis, and conclusion, we must begin with individual experience. My observations in this paper are derived from direct experience as a city planning commissioner responsible for making decisions within a legal time limit on the variety of matters connected with physical planning for a large, dynamic municipality. These observations also reflect earlier experience as a member of the planning staff of a large corporation.

These experiences confirm the differences between

 Selected from Urban Affairs Quarterly, *1(3), 15-38 (March 1966).*

the roles of "line" and "staff" as discussed in management literature. Staff members seek to understand the position and point of view of the decision maker and accommodate their analysis to his needs. In examining the thought processes involved in comprehensive city planning, we must explain the human requirements for mathematical models as "decision-aiding tools," in addition to describing some of the executive attitudes which are covered in a later section of this paper. Something of the same analytical comprehension must be shared by the staff maker of mathematical models and the decision maker if these models are to simulate the same reality for both.

All matters which an individual reviews and decides upon are referred to his mind and memory — what might be called his personal internalized simulation, the product of accumulated observations, experience, formal knowledge, and thought. This memory incorporates unconscious as well as conscious material. It is the mental picture and related store of readily accessible information and conclusion which is called up to awareness from the sum-total synthesis in the mind of a lifetime of sensory inputs. This total includes highly specific and clear images, partly formulated thoughts, and vague impressions. Pictorial memories, visual constructs, data, correlations, judgments, intuitions, and feelings are all present — well ordered in the mind or confused, depending on the situation and stage of appraisal. Mental awareness is very broad and can shift rapidly among physical, economic, human, political, strategic, legal, personal, and many other considerations. No inanimate device can approach an equal *range* of sensory input and cognitive interconnection; the human decision maker must rely on his own mental-nervous system if he is not to abrogate his role, responsibility, and his ego. Although intuition is an important part of reasoning, the emphasis here is on deliberate thought.

How is urban planning conducted in the mind? At present, no definitive answer can be given to this difficult question, but it is likely that similarities exist among the mental processes of different people. To the extent this is so, there are common objectives and requirements for successful simulation that is absorbed, accepted, and used constructively by those responsible for reaching conclusions and making final decisions. While the following comments concerning the mental processes involved in comprehensive city planning are the observations of only one person, they illustrate the type of descriptive formulation which must be developed if we are to extend our understanding of the planning process. They also suppot conclusions concerning effective simulation which are reached later in this paper.

The accomplished decision maker in city planning visualizes a plan picture of the metropolitan region as a whole, as it would appear from an aircraft high above. This internalized view is derived from direct observation on the ground or from the air, from maps and photographs. A second portion of the mental image for planning comprises segmental views of the city: entire neighborhoods, city blocks, stretches of streets, groups of buildings, or single structures. A third portion incorporates a large volume and variety of separate items and detailed recall of particular visual features or minutia of every sort. Still another part consists of more abstract materials: data, indications of trends and change, concepts of interrelationships, opinions, convictions, ideas of many kinds, or imaginative speculations. In this fourth part are included considerations relating to the acceptance and implementation of plans and planning policy: politics, public and institutional attitudes, motivations, reactions, and other behavioral realities. This most abstract material is manifest in diverse memories; some of these expressly recall facts, figures, graphics, pictures, and other "photographic" recollection; others are much more general conclusions, judgments, or feelings which emerge from an unspecified accumulation of experience.

The four parts of this conscious material are interrelated in the mind by a process as much involuntary as intentional. Certain interconnections are made by deliberate reasoning; others come about with far less recognition of progressive sequence; some emerge already formulated without awareness of their specific derivation. The mind tends to generalize; in order to assimilate the stream of sensory inputs to which it is subjected, if filters, suppresses, simplifies, correlates, and combines into related parts and impressions. It incorporates the concept of time, remembering and visualizing the past and passing present, and projecting into or imagining the future. The resultant totality is a dynamic construct composed of pictorial and nonpictorial or abstract parts, subject to a constantly changing internalized scan of these components, but retaining a consistent connection or persistent memory core. The individual accumulates a conscious store of constructive information for city planning as long as he can continuously relate diverse inputs to a conceptual continuity within his mind. When this system of voluntary and involuntary selection and integration is interrupted, mental material is more confused and conclusions are fragmentary and not well reasoned.

The process of comprehensive planning involves conceptualizing a city from the broad to the particular. Consequently, the decision maker must see the city in its main physical outlines and elements: configuration, general geography and topography, basic transportation network, centers and subcenters of activity, main types and locations of land use – as much as he can retain in one coherent, unified image. To this core picturization is tied as much dynamic economic, social, political, quantitative, abstract information of all types, and as much detail as the mind can handle and consciously recall. Although time and continuous application permit enlargement and intensification of the complete mental image, there is a limit to the total breadth and depth which is available on call. When this conscious capacity is reached, further development should take place in the *quality* of the mental approach, attitudes, analytical synthesis, and conclusions.

Various decisions as they arise are related to the individual's mental image of the city and its planning. This means that material submitted in connection with a particular decision is most helpful when it relates easily to the person's internalized conceptualization. If the language or form of the material is strange or unclear, it is difficult or impossible for the decision maker to incorporate it, and he may tend to request reformulation of the information or to reject it. If the reason for rejection is recognized, further staff work is usually requested; unwillingness to admit that the information is not understood often leads to rejection without explanation; and if the reason is not consciously recognized, an "illogical" justification may be adopted for the rejection. The decision maker must be able to comprehend material *in his own terms*, to incorporate it within his mind and thereby make it "his own." If he accepts it on faith, he has not tested it with relation to his own knowledge and judgment; he has not included it as part of his comprehension, and probably he will continue to be dissatisfied with this area of consideration until it is further clarified.

Analogy is an important part of thinking. No man can know all or even consciously recall and manipulate at any one time but a fraction of the information he has absorbed. For both reasons there are inevitable gaps in the mental picture and abstract comprehension, temporary or permanent as the case may be. There is a natural tendency to fill these gaps as need be by analogy or extrapolation based on what are believed to be or imagined as similar situations or comparable circumstances. When a decision is required in a section of the city for which there is no visual memory, a concept of the three-dimensional reality is derived by analogous recall and mental construction. What is this section probably like, based on the city in general and similarly constituted sections? Or "facts" and trends concerning abstractions such as growth, attitudes, motivations, and law – only partly explained or understood – are developed from the closest parallels and indicators readily recalled. Therefore, some representation of the urban organism as a whole is not only essential for formal analysis, but as a continuous reference to refresh and fill out the internalized picture and abstract comprehension of the decision maker, in this way reducing the number of his questionable analogies and the uncertainties of his mental image in general.

We can see from this abbreviated description that simulation for comprehensive city planning must meet certain minimum requirements. It must encompass and incorporate *all types of information of primary import* whether or not it is tangible, quantifiable, physically visible in space, stable, reasonably understood, or projected. Such simulation must therefore be a simplified or generalized *distillation* to the point where it can be comprehended and recalled to mind as a whole. This requires selection and clear expression of a *basic conceptual structure* or chassis of mental interconnection to which other material is continually related. Matters which cannot be connected with this central referent as a mental continuum must be treated separately without benefit of correlation with the "whole."

The simulation must be in a form and language *comprehensible to decision makers.* When significant differences of background, training, or personality exist among those who decide, a form is adopted which represents the best common denominator of comprehension, requiring some special effort on the part of each decision maker and sufficient staff explanation for each individual's need of amplification. The amount of *information and analysis* presented is *restricted* to that which does not produce conceptual confusion for those who use it; and the *rate at which it is changed is limited* by the stability necessary for mental absorption and analysis. The simulation must include both *pictorial components and abstractions* that are meaningfully interrelated. It must incorporate *past, present, and future* in its content. It should lend itself to the visualization and understanding of different solutions in *space* and alternative *policies* of many kinds, for such choices are a key requirement of city planning. By its nature, simulation cannot be all-inclusive, but it is as soundly *representative* as inherent characteristics and limitations permit.

The *validity* of the model must be regularly tested by comparing its representative and predictive accuracy with the actual behavior of the organism it depicts; otherwise, the decision maker will not or should not accept and use it as a basis for his conclusions. These general specifications for successful simulation present a formidable methodological task of analytical expression and interrelation in themselves. To this must be added the enormous difficulty of reliably simulating the dynamics of a large metropolitan city, and in addition the process of comprehensive guidance itself.

Comprehensive city planning is at present second in its difficulty and complexity only to similar coordinative planning at the national level; for smaller communities it is less difficult in degree of complication rather than in kind. The elements to be considered are multitudinous and very diverse; they vary over time at different rates, in the extent and type of their interrelation, in their relative importance. There are never enough resources to accomplish what is needed or desired; their allocation is directed by the public-political process of representation, expression, pressure, and manipulation. Basic objectives evolve and emerge only gradually; they are most often clear and definite in emergencies. Consideration of the longer-range urban future is an interest or concern which must in general be stimulated; separate self-interest and the immediate present are the normal preoccupations. Historical tradition, cultural values, motivations and attitudes, politics and government, legal rights and equities are vital continuities, rarely subject to sudden deliberate change. The entire effort of city planning is by and for people, and the human condition of a municipality is an awesome responsibility to undertake beyond the signposts of public participation and the constant monitoring of delegated authority provided by the political-representative process. The directive forces and constraints of planning must be continually balanced with freedoms which are the essence of our national past and future purpose.

Our knowledge of cities and their comprehensive planning is very limited. We are not yet able to evaluate the costs and benefits of a plan to include those municipal elements that cannot be measured meaningfully in dollars or other quantitative terms, but are of major importance. Our capacities to forecast or project into the future several elements involving spatial, socioeconomic, and political-governmental forces are very limited; we are a long way from the complex multiple projection required for sound comprehensive planning. All too often the facts as they develop in the real world necessitate the continual adjustment of relatively simple "descriptive" models rather than confirm their predictive accuracy. We cannot yet calculate the relationship between the different metropolitan transportation systems and land uses in a dynamic economy with a spectrum of individual and institutional developmental actions and choices. More practically, we do not know now how to bring order and sound purpose to seriously confused municipal tax policies. We cannot develop widespread and continuing civic interest and local political concern about crucial decisions that continually confront a community, when we are preoccupied with special project and political-governmental detail. We have not established and organized comprehensive planning which includes all primary municipal functions within its analytical scope and coordinates effectively with abutting jurisdictions sharing a common watershed, air space, or major transportation route. We have no yet even found a way of ensuring that the best brains and knowledge available are applied to critical urban problems.

This is not to suggest that the task is so formidable that there is little prospect of constructive proposal and accomplishment. The ever-present momentum of current actions and events contains within itself directive forces for the future; to the extent that these can be soundly analyzed, correlated, and channeled toward agreed-upon objectives, man and society benefit. The alternatives of development by random chance or many separate fragments become less feasible as population growth and scientific-technological advance intensify the interdependency of an increasing number of activities. But the scope and difficulty of comprehensive city planning and decision making indicate that *simulation must appropriately and reliably reflect the nature of the urban organism it represents* if it is to eventually fulfill its potentialities as a key mechanism of planning. Either it must meet the requirements previously noted for comprehensive analogy, or treat a segment with the final combination achieved by a further process of resolution.

POTENTIALITIES OF MATHEMATICAL MODELS

Mathematical models are one of the many means and methods of simulation. They are, of course, unique in several ways. If properly formulated, the rules by which they are derived and tested produce descriptive statements more exact in themselves and less subject to differences of interpretation than is true for many other forms of simulation. Their foundation in a long accumulation of mathematical knowledge makes consistency possible within an established context of analytical expression. Since mathematical formulation requires quantification in numbers or other symbolic representation, precise statement with minimum ambiguity is possible. This statement can be modified or extended as long as logical-mathematical validity is maintained. Numeration promotes thinking in absolute terms; and the basic mathematical purpose of treating interrelationships is a central requirement of coordinative planning. The language of mathematics is understood by all trained in the field; they can utilize the work of others without substantive misinterpretation.

When mathematical models are translated into computer language, the superspeeds at which computations can be made is a significant advantage. Many people are now familiar with examples of data processed by electronic computer which would have taken lifetimes of laborious computation without them; and when self-checking routines are incorporated in the program of instructions, computers are many times more accurate than humans performing the *same* calculations. Mathematical models stored in a computer memory can be linked with the source and processing system of input information so that they are kept up to date quickly and automatically. Hypothetical values can be substituted for actual figures; in this way the model may be used to determine the consequences of whatever it describes when certain assumptions are made. Without the electronic computer such calculation of the effects of one action on several related elements may be difficult, prolonged, and often impossible in the time available for decision. Were a mathematical model available, for example, which accurately described a group of interconnected municipal costs, it would be possible to ascertain quickly how an increase or decrease in one or several of these would affect the others. The entire representation could be manipulated until an acceptable array of the constituent elements was attained. Parameters or limitations could be incorporated so that some particular element would not exceed a given amount (statutory debt limit), one would be less than half as large as another (policy decision), or the entire combination could retain certain proportionate relationships (to fit an overall tax distribution).

Another value of mathematical simulation and associated data processing for city planning is its focus on facts, figures, and careful thinking. The use of numbers to describe as many phenomena as possible, of equations to express interrelationships, encourages the accuracy and disciplined reasoning which are highly desirable in endeavors as difficult and far-reaching as city planning. With a background of more accurate information made available in this way, the inevitable gaps of memory, knowledge, and comprehension on the part of decision makers would more likely be filled with fact than questionable deduction from questionable analogy. Parenthetically, not every one necessarily considers this a gain; there are many with special interests or emotional involvements of some sort who much prefer to maintain without evidence, to argue a personal point of view, to exhort without the difficulties and constraints of factual analysis. Perhaps most of us tend to distort facts unconsciously in matters about which we feel deeply, but modern society is inexorably forcing facts upon us. To the extent that mathematical models or any other demonstrably representative statement are available and generally accepted as a factual base, the process of related consideration, judgment, and conclusion is made easier and more accurate. In particular, an accepted referent permits clarification of differences in interpretation and diverging opinions, thereby reducing the stalemate situations which so often occur when no one is sure where the discussion and thinking stand, of the positions of participants, or how to progress toward resolution and decision.

Quantitative models can serve the purposes of backward or forward projection. Most city planning problems today are closely tied to past action or inaction, and the consequences of mistakes made today will be felt for a long time to come. One problem in evaluating past experience is the lack of consistent municipal data permitting comparison through the years. For example, investigation of trends in the unit cost of public utility installations in the past to determine what factors have been most influential soon bogs down because of the dearth of consistent, compatible data. Better municipal housekeeping and an effective data-processing storage system will not alone solve the problem; some simulation or quantitative reference is needed to which various data can be meaningfully related and compared over time. This record would not only increase the useful lessons of experience by revealing trends and identifying causes, but in so doing probably would improve the reliability of projection into the future. To be sure, its most important benefits would be felt some years hence when a sufficient accumulation of comparable data interrelated through consistent statement would then be available for city planning at that future time. In considering this possibility, the methodological difficulties of allowing for the many changes which occur over time – such as monetary inflation, changed technologies, and different levels of service-demand – should not be underestimated.

Finally, there is the role of quantification and simulation in extending the knowledge of comprehensive city planning itself. Starting with a vastly oversimplified construct of a few principal elements and their interaction, which, hopefully, nevertheless conform reasonably with the actual behavior of the city, it should be possible to gradually add more elements and aspects into the mathematical construct while retaining its representative validity. Another research approach is to examine a part of the urban organism, such as the interaction of transportation and land use now under investigation in several places, to the limit possible with present mathematical methods, disregarding those factors known to affect the matter under study but not now mathematically manageable. Various subsystems could be studied with results which should improve the quality of the separate analysis of these parts of the urban organism and prove suggestively helpful in the judgmental process of decision making. As more subsystems are reliably treated in mathematical terms, it should be possible in time to combine several of them into an encompassing mathematical statement which retains descriptive and predictive validity. By such accumulation of parts coincident with advances in mathematics, it may be possible eventually to arrive at an overall mathematical simulation which can be employed directly for comprehensive city planning. An obvious advantage of this latter approach is the interim usefulness of each successive attainment.

Notwithstanding present and potential uses of models in certain simulation, their employment in comprehensive planning must be approached with caution. Mathematical models can incorporate only that which can be expressed and interrelated with the requisite quantitative and predictive rigor. At present they cannot encompass the range of interdependent variables which must be taken into account in comprehensive city planning; nor can they even approximately simulate the complex dynamics of a city. For example, except for the simplest events, very few mathematical models for traffic planning have been validated by comparison with the actual behavior of the phenomena they seek to predict. Models which have been adjusted to conform to the *present* disposition of vehicular traffic in one city do not accurately describe the situation in another municipality of comparable size.[1]

The searching critique of simulative models and insistence on their experimental validation that characterizes scientific method in fields such as physics or chemistry has been largely absent in the broader applications to human activities now being investigated and attempted. Admittedly, validation is difficult when highly complicated urban phenomena are involved, but unverified models cannot and should not be employed as a basis for city planning decisions usually with inportant, long-lasting effects. Mistaken faith, deceptive numerology, or the tactical advantages of superficial but seemingly conclusive quantification are pitfalls which the responsible decision maker avoids.

To infer therefore that mathematical models or any system of precise quantification will soon usher in a new era of scientific city planning could only come from serious misunderstanding of the nature of cities, people, and planning. As suggested previously, mathematical treatment of an organism such as a city will be some time coming. To the functional complexities of the city itself must be added those of planning and projecting into the future, and important interconnections between the city and other municipalities, its surrounding region, the state and national environment, and increasingly with international developments.

> *Today we preach that science is not science unless it is quantitative. We substitute correlation for causal studies, and physical equations for organic reasoning. Measurements and equations are supposed to sharpen thinking, but . . . they more often tend to make the thinking noncausal and fuzzy. They tend to become the object of scientific manipulation instead of auxiliary tests of crucial inferences.*
>
> *Many — perhaps most — of the great issues of science are qualitative, not quantitative, even in physics and chemistry. Equations and measurements are useful when and only when they are related to proof; but proof or disproof comes first and is in fact strongest when it is absolutely convincing without any quantitative measurement.*
>
> *Or to say it another way, you can catch phenomena in a logical box or in a mathematical box. The logical box is coarse but strong. The mathematical box is fine-grained but flimsy. The mathematical box is a beautiful way of wrapping up a problem, but it will not hold the phenomena unless they have been caught in a logical box to begin with.*[2]

THE DECISION-MAKING INTERFACE

Until quantitative simulation is available for practical application in comprehensive planning, those with the executive responsibility will continue to be confronted by a stream of decisions to be made. What can be done in the interim to improve decision making in city planning?

A problem to be overcome is the conceptual gap between staff specialists and decision makers. This has always existed but is accentuated by the current surge of knowledge. As technical information and analytical techniques advance rapidly in many of the substantive areas of concern in comprehensive city planning, the decision maker finds it more difficult to follow and comprehend an expanding spectrum of special knowledge. Since keeping up with all of it is intellectually impossible, assimilation of specialized material becomes increasingly marginal for the decider unless a deliberate effort is made to bridge the conceptual gap. It is for this reason that the special rooms, military command centers, electronic display devices, and other mechanisms to portray coordinative information more clearly are being developed and used. Of course, the mathematical model builder hopes that eventually he can supply a basic analytic referent which will be seen in the mind's eye and comprehended abstractly in the same way by all those actively concerned in final decision making.

For some years to come, however, the type of analytical representation *intermediate* between traditional methods and mathematical models will be employed in city planning. A map or map-model of the entire city and its immediate surroundings will be the base analytical referent, with as much additional information as practicable related directly to it by legend, color code, overlay, adjacent presentation, or small upright standards on the map-model which display key data by district, manually or automatically posted. Other representations and forms of simulation can be utilized to depict particular projects or plan components; these are interrelated with the master model by verbal exposition, discussion, and the connections noted within each executive mind.

Much depends on the working relationship between staff and line, for this constitutes the interface of different responsibilities, personalities, experience, knowledge, and points of view. There are various requirements for a successful relationship, some quite subtle. For the decision maker, probably most important is the *selection of immediately supportive staff.* Besides the necessary knowledge, experience, and capacity to continue learning, staff members must understand the realities of city planning. They will, for example, not only accept but recommend any strategic modification of the best analytical plan necessary for its implementation; nor will their studies be carried beyond the level of detail needed or desired for decision. They will find out what the executive has in mind rather than presume what he wants; they will recognize the primary obligation of always stating the limitations as well as the strengths of the material being processed and presented.

> *If properly employed . . . conditional forecasts can constitute a valuable tool. The danger lies in the fact that the details of the planning process, once the assumptions have been made, are identical in this type of imaginative planning with those that lead to a legitimate program of action in, for example, a major move [of a large institution from one location to another]. . . . It is probably this appearance of reality in the final planning document that accounts for the frequency with which not only the program planners but also members of top management ascribe more validity to the detailed conclusions than the oftentimes highly questionable nature of the assumptions would justify. Awareness of this danger causes some executives to be less than completely enthusiastic about the formally organized [comprehensive] planning process. This is a situation which the practitioners and proponents of . . . planning must remedy. They must take the lead in protecting themselves and others against improper use of projective studies. Every planning department should start with a statement of the assumptions upon which the analysis is based and, when summarizing the results of analysis, should restate these assumptions clearly as conditions upon the validity of the conclusions.*
>
> *If the staff planner and line executive discipline themselves in this way, the habits of thought thus encouraged will produce other benefits. In particular, it is less likely that the planning effort will be carried to unprofitable lengths in terms either of detail or of extrapolation into the future: the pyramiding of assumptions necessary to permit such extended analysis will warn the [staff] planner when he is approaching the point where what has hitherto been a solid piece of fiction is about to cross the line into meaningless fantasy.*[3]

For his part the decision maker must understand the characteristics of staff personnel. For example, a person with a fine mind for staff work may not be fluent verbally; or the personality best suited for staff support may not have the same kind of drive characteristic of many line executives. The decision maker cannot expect objective analysis from his staff if he imposes a straitjacket of foregone conclusion, premature opinion, or prejudice. If he demands a near-perfect record of analysis and recommendation, he is not only unrealistic but restricts the scope of staff consideration to certainties and safe generalizations, stifles the courage to explore, and inhibits activities in general. He does not display disinterest in the ideas of subordinates, disdain to listen constructively, except excessive personal loyalty, nor communicate so sparingly that it is well nigh impossible for the staff to be helpful.

The working relationship at this important operational interface between decision makers and staff is mutually most productive when there is the maximum *intellectual and emotional compatibility consistent with respective positions and responsibilities.*

THE JUDGMENTAL PROCESS

Major city planning decisions are made in the United States by commissions and the directors of planning departments. Although their conclusions are usually only advisory to the local legislature, they involve substantially the same problems of analytic resolution as if they were final. The main difference is that certain categories of political consideration are left more appropriately to elected municipal representatives.

While research and development progresses in the field of data processing and mathematical simulation, there is much that can be accomplished by decision makers themselves to improve the quality of analytical consideration and the soundness of conclusions. Normally, city planning commissions are composed of seven people. An increase in membership changes the character and operating dynamics of the body. Very likely there are greater differences of precise point of view; more time is required for presentation and discussion; staff problems of explanation and service to the commission, individually and collectively, are increased; it takes longer to complete official business and discuss new planning proposals; each item, therefore, on the average, probably receives less attention. There are likely to be more combinations of disagreement and tactical position, and therefore a more complicated voting pattern; consensus is more difficult to develop; and attaining an effective working relationship *as a group* is made more difficult. These consequences are progressively amplified as commission membership exceeds five.

The success of well-known precepts of organization, procedure, and behavior for small groups depends first on *individual* attitudes and deportment.

> *"Ideally, several forms of conscious review or mental monitoring operate throughout the formation of subjective judgments. The source, intent, derivation, and reliability of informational inputs are noted as they are pigeonholed in the mind, Throughout the process of analysis, a critical attitude is applied to foster a mental image of reality and reduce progressive distortion and wishful thinking. Recognized gaps in knowledge are filled by others, and known areas of emotional inaccuracy are compensated by appropriate self-adjustment. Realization of the necessity of simplifying complex, open-ended comprehensive planning questions to mentally manageable proportions is reflected in a continuous choice, elimination, and combination of information and deduction. Awareness of when a conclusion must be reached to be useful establishes the time and conceptual restrictions within which the evaluative effort is made. Understanding the nature of intuition helps in recognizing the value and limitations of this insight, whether intentionally applied in conscious analysis or immediately responsive as demanded by circumstance.*
>
> *To the extent possible, [the decision maker] subordinates self to the analytical problem at hand. His mind is receptive to the conclusions of scientific method and informed opinion, but he has the courage to proceed on his own convictions in the face of wide disagreement. Admittedly, this approach is not always easy to achieve, but [comprehensive] planning decisions in the longer-range interests of a [city] merit the maximum effort toward objectivity. By this approach implicitly and explicitly applied, the executive increases his own managerial effectiveness and sets a desirable precedent for the performance of subordinates.*
>
> *The group operates as a unit, each member deferring his prejudices . . . to realities, his personal preferences to the [city planning] purpose, his predilections to group action. Since the [commission] functions within operating limits of time . . . discussion is controlled by formal procedure or informal agreement. It is constructive and additive, rather than negative, argumentative, and self-justifying.* Interpersonal compability is as important as aggregate brainpower. *Overreaction to the opinions or mannerisms of others slows progress. Undue agressions, rivalry, or unconscious clash of personalities can destroy group effectiveness. Dissenting*

opinion is neither sacrificed to a lowest common denominator of concurrence nor allowed to become so pervasive that useful agreement and solution are impossible. . . . Individuality is not subdued but controlled. Any tendency to defer irresolutely to majority opinion is discouraged.

Some type of [simulation], as discussed earlier in this chapter, is maintained. . . . It serves as a factual referent which reduces wasteful digression because of lack of information, misunderstanding, or avoidance of matters of fact. It assists the group in conceptualizing the analytical problem and thereby advances its work. It embodies the consequences and current stage of analysis, available at all times to both membership [and staff]. It emphasizes the objective end to which personal interests are subordinated and related. It permits a structuring and expression of individual contributions otherwise more difficult. It reduces the need [of detail in] minutes and recapitulation between meetings. It facilitates continuity of effort and more rapid indoctrination as membership changes.[4]

Since the decision maker can seldom encompass all the fields of knowledge which relate to his decision, he functions in considerable measure as a catalytic agent in city planning analysis. He must comprehend, evaluate, interrelate, and judge the work of others without being able to absorb it in detail. Frequently, the validity of analysis and conclusion must be sensed through the person responsible, rather than by any self-deceptive and self-defeating attempt at precise discussion and logical argument in an unfamiliar field. The recommendations of a staff member may be influenced by professional pride, intellectual characteristics, prior commitment, emotional involvement, excessive caution in the interests of self-preservation, egoism, unexpressed motivations as well as stated intentions, or ulterior purposes such as organizational empire building.

Ordinarily, the decision maker must in each case identify the weight of such factors *through the person involved* by using perceptive questioning, derived from understanding people in general and the individual in particular, which unearths additional information concerning the methods of analysis employed, underlying assumptions, the degree of confidence the individual really holds in his material when challenged, and the general impression he gives of competence, frankness, and high standards *on the matter under consideration.* Prior experience with the individual is recalled to mind as a further check, remembering, of course, that there are times and extenuating circumstances which lower usual performance. Outside consultation is also available, but there are practical limitations to the frequency of this managerial resort as well as additional cost and delay. A substantial part of comprehensive planning decision is the instigation, evaluation, modification, and development of other people's conclusions and opinions. The decision maker soon learns that he must take a realistic, continuously questioning view because he is subjected to conscious and unconscious overstatement, intensive efforts to persuade, stratagems that are a natural part of institutional and human relationships, and inevitably some deliberate misrepresentation. When sound mathematical simulation is achieved some time in the future, many aspects of staff work can be checked quickly against this base referent.

Besides making the most of staff support, there is another path of possible improvement in the decision-making process. Because of the number and diversity of determinations there can develop over time a pattern of judgmental response unrecognized by those deciding. Few people in executive positions have the time or the inclination to keep a record of their formal and informal decisions that have major import. Their full concentration is given to the exacting task of assimilating, evaluating, synthesizing, and concluding.

Faced with crucial and sometimes agonizing choices, the decision maker develops a certain necessary dispassion. He cannot allow emotional attachment or overreaction to affect the essential objectivity of his conclusions; nor can he function successfully if he is torn by anxieties, overconcern with the problems of others, recrimination, or any illusion that he can perform perfectly. Consequently, the decision maker attains his own best balance of objectivity and subjectivity, detachment and sympathy, patience and decisiveness.

Naturally, those on the "other side" of the

interface – staff, subordinate line managers, others preparing and presenting material for executive consideration – have a different view. Ordinarily, they are concerned with only one aspect of the whole; therefore, it is much easier for them to discern continuities and discrepancies in the succession of decisions relating to their special interest. Since it is their main concern and preoccupation which hangs in the balance of decision, they are more intensely involved occupationally and emotionally. This group naturally watches the record and actions of the decision makers because the ultimate success of their work depends on them.

Lawyers, for example, are quick to note any predilection or exceptional pattern in juridical decisions; an effort is made to secure or avoid judges whose records indicate consistently unusual personal reaction and treatment. Under these circumstances lawyers believe they can predict the responses and decisions of particular judges in certain trial situations. Research is underway to explore this matter more fully by examining many decisions by the same judges to see if there are patterns of predictable behavior and conclusion.[5] Should this possibility surprisingly prove to be a fact, it would be theoretically possible to simulate in a computer the rationale of judicial determination discovered by the analysis of past case histories and decisions; and by programming into the machine "facts," arguments, and precedents, a decision could be rendered automatically in much less time and probably at lower cost than is required for the usual trial. More likely is the possibility that simulation of legal precedents and factual material be employed where trials are not necessary and only an administrative ruling is sought.

While it will probably be a long time before such simulation is successful enough to substitute for human judgment and the human process, the approach might be utilized in a limited way. A statistical analysis of the resolution of matters brought for decision over a period of several years would enable an official to review his own actions. He might discern a cumulative trend in his decisions on a particular type of city planning consideration which he had not recognized; he might find that his conclusions in a class of cases varied from those of his associates more than he thought. He would decide for himself whether there was any significance in the reportive analysis; from this "mirror" or feedback of his actions, he might choose to bear in mind certain apparent reactions and adjust his reasoning accordingly. If reliable statistical analysis could be carried further to show likely causes for the pattern of decision detected, each individual could identify his own dynamics more specifically. If successful for the decision maker, the same type of analysis could be used for review of staff conclusion.

Certainly, this kind of highly realistic self-appraisal is not easy. Because of their roles and personalities, decision makers are far more accustomed to looking at others than at themselves. Willingness to view oneself in a potentially revealing way requires inner security, and constructive compensation for any overreactions suggested by the statistical analysis is unlikely except by the emotionally mature. Conviction of the necessity of the soundest decisions in comprehensive city planning, or another motivation equally strong, is needed to sustain the effort. As with staff, the personalities of decision makers are as vital for effective planning as technical methodology, reaffirming the extreme importance of selecting the best people available with a balance of intellectual capability and emotional maturity. Perhaps in time mathematical simulation will join psychology and psychiatry in advancing more reliable measurement in these fields.

If in the future electronic computers are linked directly with the decision maker, his analytical capabilities might be further improved. For under these conditions he could ask questions of the machine and – provided answers could be derived from the stored information and calculative programs available in the computer – receive such rapid response that the mental process is not significantly interrupted. The decider could have before him a television-like display tube which would present information analysis from the computer in the graphic form most helpful to him: trend lines, bar chart, matrix, flow diagram, tabulation, and others. The individual could instruct the computer to modify these displays to facilitate conceptualization of the material by making it, for example, more or less detailed, changing the grouping of the data, or superimposing an array of related information for comparison. In this way, analytical displays would fit and follow individual thinking. If the person were also able to "type" into the computer elements of his judgmental process as he proceeded, there could be recorded immediately for him to see the sequence of his considerations, the logical accuracy of a series of interdependent premises and deductions, or perhaps the configuration resulting from spatial postulates expressed in descriptive terms. The computer could thus become in effect an analytical extension of the human mind.

---

[a]*Different kinds of models or "representations of a thing" have long been employed for many purposes including planning. Primitive man on occasion probably traced out on bare ground a plan of attack on on some enemy position. A model of a steam device is described by Hero of Alexandria early in the second century A.D. Many of the technically inventive and anatomically explanatory drawings of Leonardo da Vinci are "models" portrayed in two-dimensional form on paper; subsequent development of a geometric method of perspective drawing was another step in the simulation of the three-dimensional world on a flat surface.*

Every preliminary plan for a building or engineering structure is a form of analytical simulation. Small-scale or full-size models are part of the process of designing automobiles, aircraft, ships, and a multitude of simpler industrial products. Miniature replicas of machinery and equipment are sold commercially for the scale models used by industrial engineers in analyzing alternative layouts for manufacturing plants. For years prototype sections of highways have been built to full size to test their resistance to the destructive forces of heavy traffic. Flight simulators used to check out commercial airline pilots on emergency procedures are so realistic that the pilots react physically and emotionally as they would under crisis conditions.

Similarly, abstract models have been employed for centuries in intellectual and scientific developments. Mental pictures and concepts constitute simulative abstractions within the human brain. The translation of such mental pictures into symbols on paper constitutes most of the literature of science. An accumulation of thousands of mathematical formulations describe forces, conditions, and interrelationships of matter in physics, chemistry, or biology. Accounting systems are accepted simplifications of the financial operations of a business enterprise. And scientifically selected statistical samples are commonplace in national censuses, opinion surveys, product quality control, and other research directed at a distinct group of people, quantity of objects, or commercial market.

If the concept of simulation is extended further, literature and fine arts may be included, because they seek to describe in their particular way subtle, complex, unquantifiable aspects of the human condition and aesthetic qualities of the human environment.

REFERENCES

[1] The author is indebted to Dr. Robert Brenner and Walter W. Mosher, Jr., of the Institute of Transportation and Traffic Engineering, University of California, Los Angeles, for these examples and for constructive comment on this chapter in general.

[2] John R. Platt, "Strong Inference," *Science,* Vol. 146, No. 3642, Oct. 16, 1964, pp. 351-352.

[3] Dean E. Wooldridge (Member of the Board of Directors and Past President of Thompson Ramo Wooldridge Inc.) in Melville C. Branch, *The Corporate Planning Process,* New York (American Management Association), 1962, p. 15.

[4] Melville C. Branch, *ibid.,* pp. 167, 171, 173.

[5] Reed C. Lawler, "What Computers Can Do: Analysis and Predictions of Judicial Decisions," *American Bar Association Journal,* Vol. 49, No. 4, April 1963, pp. 337-344; "Foundations of Logical Legal Decision-Making," *Modern Uses of Logic in Law,* Committee on Electronic Data Retrieval, American Bar Association, Chicago (1155 East 60th Street), June 1963, pp. 98-114.

32

David L. Johnson and Arthur L. Kobler

THE MAN-COMPUTER RELATIONSHIP

Recently, Norbert Wiener, 13 years after publication of his *Cybernetics,* took stock of the man-computer relationship[1]. He concluded, with genuine concern, that computers may be getting out of hand. In emphasizing the significance of the position of the computer in our world, Wiener comments on the crucial use of computers by the military: "it is more than likely that the machine may produce a policy which would win a nominal victory on points at the cost of every interest we have at heart, even that of national survival."

Computers are used by man; man must be considered a part of any system in which they are used. Increasingly in our business, scientific, and international life the results of data processing and computer application are, necessarily and properly, touching the individuals of our society significantly. Increasing application of computers is inevitable and requisite for the growth and progress of our society. The purpose of this article[2] is to point out certain cautions which must be observed and certain paths which must be emphasized if the man-computer relationship is to develop to its full positive potential and if Wiener's prediction is to be proved false.

In this article on the problem of decision making we set forth several concepts. We have chosen decision making as a suitable area of investigation because we see both man and machine, in all their behavior actions, constantly making decisions. We see the process of decision making as being always the same: within the limits of the field, possibilities exist from which choices are made. Moreover, there are many decisions of great significance being made in which machines are already playing an active part. For example, a military leader recently remarked, "At the heart of every defense system you will find a computer." In a recent speech the president of the National Machine Accountants Association stated that 80 to 90 percent of the executive decisions in U.S. industry would soon be made by machines. Such statements indicate a growing trend – a trend which need not be disadvantageous to human beings if they maintain proper perspective. In the interest of making the man-machine relationship optimally productive and satisfactory to the human being, it is necessary to examine the unique capabilities of both man and machine, giving careful attention to the resultant interaction within the mixed system.

BASIC PARAMETERS

In any analysis of the types of problems which may, or should, be solved by automatic methods, the decision capability of the machine is fundamental to the entire solution. Whether the problem is the addition of a series of numbers or the firing of a retaliatory nuclear weapon, the computer can act only through the processing of a series of yes-no decisions. Much work has been done in the definition of decision structure. The fundamental decision element is one of binary choice with one or more inputs and two outputs (or at least a single output capable of bistable condition). Such basic decision elements may be combined to provide decision systems as complex as the application requires. The decision "Is A greater than B?" may be considered a single basic decision[3]. As input, we have two variables, the magnitudes of A and B. The decision element in this case can be a simple comparator. The output may be either a "yes" or a "no." The inputs must accommodate variables of specified or unspecified limits. The output is limited to a simple binary choice, the forms of which are fixed. There is no room for a "maybe" answer within the single decision unit. Of paramount significance, however, are the decision parameters which are neither input nor output but which determine the structure of the actual decision apparatus. Thus, in the foregoing example, such parameters might include the following considerations: (1) *greater* should be defined: (2) both quantities are (or are not) represented in the same number systems; (3) infinite magnitude is (or is not) allowed; and (4) magnitude relates only to the comparison (or signs must be considered). Clearly, these are just samples. Many other elements must be fixed before the decision structure is complete.

In cases of equipment design, such "basic" parameters usually exist within the discipline of a determining operation and may be resolved without extensive ambiguity. In more complex decision simulation, the parameters may vary from one decision to another in ways so subtle as to elude identification.

In considering decision characteristics in their relationship to man and the computer, a broad examination of the generally relevant field is required before adequate definition within restricted specific subfields is possible. One must recognize that the general field encompassing the environment and context of any decision determines to a large extent the type of decision process used, as well as the parameters of the decision structure. Although the field forces will affect both input and output forms, the most insidious effect will be upon the parameters relative to the decision itself.

In decision situations one important factor is the amount or degree of input information available. There may be little information about the choices, all the necessary information, or a confusing redundancy or superfluity of information. The evaluation of the output, of the decision made, will be influenced by the criteria available for judging it. We can have absolute, defined criteria or literally none at all. In the latter case, a number of "reasonable" men (or rational machines) may arrive at a number of equally satisfactory decisions: moreover, each of the choices, if implemented, may result in equal success-failure probabilities.

"ROUTINE" AND "SPECIAL" PROBLEMS

Today, computers are used most in dealing with what may be called "routine" problems, as contrasted with "special" problems. "Routine processing" can be used when the problems are subject to solution by specific, well-defined methods, when the validity of the solutions can be appraised, and when all parameters are defined. "Routine-direct" decisions are most often made in the physical sciences, in a system so bounded that the human response or cause is not considered. The decision structure is defined, as are the inputs, the outputs, and the decision parameters. In most cases the variables are measurable, or, at the least, probabilities are available, together with adequate information as to their reliability. Within a given solution predetermination of particular decision paths may be impossible, but implicit in the system is the characteristic that all possible decisions are recognized and considered within the rigid decision structure.

Different from the routine-direct solutions, but still within the defined "routine" category, are the routine-learning solutions, which involve training with, and use of, computers as learning machines. These are discussed later in this article.

Problems susceptible of routine-direct solution arise within limited environmental fields. Such problems lend themselves readily to automation. Our "special processing" category of decision problems includes all problems outside the rather restricted "routine" category. Most routine problems are part of systems which are themselves special in nature. Thus, to use and evaluate the meaning of a routine solution in its application within the total environment is a special problem which requires evaluation in the field of human reaction: how is the routine information output to the used?

Routine-direct processing may be applicable to dull, time-consuming. massive clerical jobs, or to problems requiring tremendous amounts of prescribed, iterative calculations. Sometimes, however, such a job may appear routine to some yet special to others. These variations in categorization are not, for the most part, variations in the means of calculation or solution but variations in the input parameters upon which the decision is to be based. The variations appear as soon as man is considered a part of the system to be examined. For example, in Ohio, computers "study possible rights of way, tot up the estimated property values involved in purchasing them, and pick out those which best combine cheapness and directness and ease of construction. Then they work out most of the engineering problems for the new highways to be built over them[4]. Some Seattle citizens feel that their beautiful city is being destroyed by a cheap, direct, easily constructed, but ugly freeway, which has taken over some of the most beautiful public park land. For them, then, the Ohio computer problem is not routine; all necessary and appropriate information is not available, and there is no clearly defined criterion of output. The problem is a special one; beauty, they feel, should be one of the parameters. How does one measure beauty? The problem here is not one of computer function; highway engineers could reach a like conclusion more slowly and less efficiently. The problem lies in the human response to a computer output: the computer has delivered results, and they have the aura of finality and correctness.

Clearly, the special area of decision making has been the unique property of human judgment. With the contemporary state of knowledge, this fact is both reasonable and proper. Inputs, parameters, and outputs for special problems are poorly defined and impossible to measure. Even on a probabilistic basis. Little information is available and statistics are inadequate. In most cases, elemental decision probabilities may be determined with some validity, but information as to the relationship of the decision elements in the total system are unique in each given environment and extremely difficult to fix by either joint or conditional probabilities. Clearly, more definitive understanding of the human complex is required before all phases of decision systems which include human parameters may be resolved. It is not surprising that it is currently difficult to find decisions involving human reactions within a system which can be adequately and generally treated by mechanized simulation.

In this context, then, we see dangers which fall into two categories: first, in the present state of our knowledge we may too easily overlook crucial parameters in the decision situation, parameters which do not permit processing – for example, values; second, we must be aware of the frailty of man qua man, particularly of man in our complex world. Man exerts a dominant influence on the use of computers and on the man-machine relationship. In the remainder of this article we will consider these two general areas, which often overlap.

PARAMETERS OF VALUE

Values, broadly conceived, are required for the solution of any decision problem. If the problem is dominated by a rigid and well-measured scientific discipline, the values may simply be mathematical – for example, that 3 is greater than 2, and that 5 times 4 equals 20. The discipline itself has defined and fixed the necessary assumptions as to the meaning of the operations and the number system intended and has set a scale of values under which the solution is to be obtained. Thus, for even the most routine tasks, values are programmed in; every problem statement or program inherently contains what is wanted, or valued, by the programmer. In cases of routine-direct solutions, the parameters are fixed by the scientific discipline under which the problem is being solved.

Values are also inherent in the nature of the problem itself – that is, in the solution which the computer is asked to deliver and the claims that are made as to the use of the solution. One programmer may value cheap, efficient raods and may ask the computer to provide him with specifications for such roads, whereas another may value expensive, beautiful roads. In this second instance, the values are less clearcut and more difficult to measure: what is beautiful? How much should be spent for how much beauty? Not only are such parameters difficult to define and measure; they may in some instances be difficult to admit or recognize. For example, in personnel work, an individual who may not be doing measurably satisfactory work may be recommended by the machine for dismissal. In the event the individual proves to be the aunt of the vice-president or the cornerstone of the morale structure of the office, dismissal may not be the most profitable course. It would be extremely difficult for the administrator to define the office social structure or to admit on the immortal program tape that he valued his job more dearly than his business efficiency. The values at such points become difficult to rank, to relate, and often to bring to the level of conscious realization. And yet such values exert a dynamic influence upon specific and critical choices.

In cases of routine-learning solutions, the values are fixed as classes rather than by specific ranking. That is, it is possible for the machine to rank and order a prescribed set of values on the basis of success in repetitive learning procedures as currently performed in mechanized games of checkers and chess. The existence of the fundamental values, then, must be recognized in the problem structure; the use to which they are put and the effect which they have upon the final result are fixed by the learning process of the computer. Again we see the possibility that certain entries within the decision system structure will be neglected; and, equally important, in cases where the goal of the learning game is poorly defined, the use to which the values and responsibilities are put may yield results which are far from acceptable in actual situations.

Within the category of problems which are inherently special in nature, the parameters, as well as their use within the system structure, are undefined or incompletely defined. In such special solutions one must interpret results of computer simulation as limited in meaning and must impose severe restrictions upon the use of the special solutions in the light of their effect upon the humans for whom they are developed.

Among the special problems, moral values and prejudice belong in the large family of values over which there is no governing discipline which applies to all people. We will discuss problems involving ethical choices, although, in the context of computer operation, we will not differentiate them for the total value problem.

One major problem in working with value parameters in decisions falling outside the routine-direct category is the sensitivity of the balance of the multiplicity of values involved, and of their interrelationships. Even if it were possible to enumerate, rank, and relate the various values in a given special decision process, it is improbable that the parameters would remain fixed for the same decision in a slightly different context.

Where other than rigid values are involved in the decisional setting, another closely related parameter often is relevant. We talk here of responsibility, for the decisional situation is clearly different when the setting is that of an abstract "game" than when it is one in which the decision is to be actually implemented, and where man is directly responsible. Responsibility, as we see it, means that the cause lies with the decision maker and the decision concerns a personally relevant action. Situations of individual responsibility, and the concurrent increased emotional significance, take on the character of uniqueness, and probability guides are not satisfactory. Therefore, although two different situations may require solution by the same decision structure, the parameters fixed by responsibility will greatly modify the processing of the input. The current discussion of bomb-shelter morality provides an example of this type of problem. In such situations it may be that the parameters of the decisional field are not definable until after the choice has been made, after the decisional action has been taken; it is only then, and not before, that the values of decision maker are defined.

RESULTS OF HUMAN LIMITATIONS

The consideration of values in such decisional contexts leads directly to our concern with the frailty of man. Two of the most responsible and respectable of contemporary social-psychological commentators have characterized today's man as increasingly "other directed"[5] and pressed toward "escape from freedom"[6]. Faced with increasing complexity and massive responsibility, man has tended more and more to work in groups, and committee decision is now commonplace. One major consequence is the decrease in individual identify and the loss of individual responsibility. The computer, coming at this time in man's progress, can and does play a special role in enabling man to escape the freedom of responsible choice. After all, who can be held responsible for a decision by a computer? Moreover, the increased complexity of the world man faces makes him more aware of his own limitations. Such awareness leads to feelings of inadequacy and the desire and need for someone or something outside himself that has the qualities he feels lacking in himself – solidity, infallibility, and so on. He looks for the father, the leader, God, scientific truth. The computer has the proper aura. It can be perfect; it can be right; it can be very nearly infallible; it can produce the truth. Already, in its infancy, it can solve problems quickly that would have taken man many lifetimes to solve. It can make systematic sense out of a gigantic mass of apparently disorganized information. In its solid, efficient, light-flashing way it acts without obsessive hesitation – as if it is sure, as if it knows. It acts without emotional involvements, without commitments, in a manner which can be called objective.

Most subject to the hypnotic effect of the computer are those whose direct contact with computer operation and programming is limited. Scientists trained in the design and operation of computing devices frequently must recognize the limitations of mechanization in communication with human systems. Often, however, these men are the very ones who are working within such a rigid discipline that computers are able to solve their problems, and they may read into this ability the ability to solve all problems.

In this setting there exists a considerable danger that complex decision systems involving human parameters will be broken down into routine segments which are more or less independent of human reaction, and that the combination will then be called a credible simulation of the total system. Such a danger has always existed in all categories of problem solution; however, with the advent of increasingly effective computers, the danger is becoming more seductive and more far-reaching in the scope of its influence. Such a process effectively rules out true simulation, but provides the satisfaction of optimum mechanization, with resultant speed and accuracy. Also, there is an attractive but dangerous precedent for restricting value parameters in the interest of simplicity and neatness; the result is a superficial and predictable decision which may be satisfactory in a "game" simulation but is disastrous in application.

Any method which ignores or explains away that part of the subject matter with which it cannot deal is, or can be in the long run, worse than useless. It raises false hopes, and it misleads if it promises what it cannot fulfill. Under the guise of reliability, usually in cases where general reliability cannot be measured or recognized, unimaginative and partial results may be accepted as accurate simulation. One is tempted to accept a completely accurate processing listing of economic factors inherent in a given society as an analysis, instead of treating it as the routine part of a complex decision system whose validity can only be evaluated in the light of its effect on the human environment. One is tempted to talk of the machines as potentially artistically creative. Machines can create; they can and do write music and plays. Speaking of man, Arthur Miller said recently, "I think there is one confusion to be cleared up. While it is true that all of us are creative, not many of us are artists. That is the crucial difference." The "Illiac Suite for String Quartet" is the result of a creative act, but that does not make it artistic, and artistic values are the appropriate ones to use in evaluating musical creations.

One is tempted, too, to evaluate the effects of a nuclear deterrent force in terms of routine decision making as to casualties or economic loss, without an actual study of exactly what human parameters, at a given time, are appropriate to the basic problem of deterrence. Fighting nuclear war on the machines is obliquely related to the question of adequate deterrence; the latter is, however, at least as much a psychological as a military problem and is very "special" indeed. Here again, issues involving values as applied by man may be fed into a computer for analysis and decision, together with values implicitly if not explicitly programmed. The computer is programmed for a particular solution to be put to a specific use. There is no possibility of avoiding consideration of such values; with or without the computer they are a part of the total decisional field. Machine quantification may make it appear that such values are not appropriate, but amorality is at least as serious as immorality, and a problem may be so reduced that its solution bears no real meaning.

The question of how output data are to be used is, to our mind, crucial and a special problem. Certainly, the ability to gather and to use information carries power with it. Once knowledge is openly available, its use by the public is often far removed from the conception of the discoverer. Present-day nuclear physicists know too well the various uses of knowledge. Present-day medical knowledge is being used to produce biological weapons to destroy man's life and to produce techniques to save man's life. The use, then, of output data often involves ethical questions.

Our concern with the parameters of value and responsibility in decision making stems from our view that machines are now making, and will continue to make, decisions in which such issues are significant and in which they are consistently ignored. In too many cases the computers are instrumental in decision making to the extent that they essentially determine the decision output because of their operational mode and man's reaction to them. In many of these cases the routine solutions are theoretically to be used as data to be inserted into a human decision system. Too often, however, the information is presented in such a way as to imply that the human decision is redundant.

It has been stated[4] that

> *the computer systems already operational are impressive enough, but they do not compare with the sophisticated systems that are under study and on order for delivery [to the federal government] in the early 1970s. In some of them the on-line concept is carried so far that if a reconnaissance satellite should send in a report of Russian rocket launchings, it would automatically generate a retaliatory battle plan from one computer that would automatically be put into action by other computers, aiming and firing Atlases, Titans, and Polarises on and under land and sea. The only interruption in the sequence, except for the system's own safety checks and repeats, would be a token one of a few minutes for the President of the United States to exercise freedom of will and say 'fire.' What to do about this choiceless choice, how to extend the time for decision and make the machines as accurate as possible, is*

the subject of serious concern and study by several groups of computer men who address themselves exclusively to command and control problems.

With the increasing efficiency of missile systems the problem of time becomes increasingly important, and the use of computers in such a situation as that described seems both appropriate and necessary. But the crucial factor, as we see it, is the president's choice, which, like any responsible decision, is neither "token" nor "choiceless." While computer men are trying to make the machine as accurate as possible, others, including the president, are concerned with this choice, as they should and must be. A Russian rocket launching may be an accident, or it may be pointed at the Chinese. These are crucial issues on which the survival of our society may depend, and they are part of the decision environment. Thus, while all pertinent decision input should be determined by the most efficient means, we must use extreme caution not to magnify the significance of the computer processing to such a degree that it appears to be the decision itself.

MACHINE-LEARNING SYSTEMS

Man's frailty plays a crucial part, too, in relation to learning machines, even though they have not yet been developed to a point where they are applied in matters critical to human welfare. The successful learning programs have been applied principally to such games as checkers and chess, in which they show remarkable success and promise. Current learning techniques, as applied to computers, demand that the rules of the game be clearly stated, that the goals be exact and easily measurable, and that the game be of such duration that the machine can learn through repetitive playing. While these routine-learning applications are vastly different from applications in the routine-direct category, it should be clear that, although the value and responsibility parameters are defined, fixed, and recognized, the computer essentially orders and ranks the parameters through the process of repetitive learning in such a way as to yield successful completion of the game. The parameters must be defined, then, but the computer is at liberty to weigh the values and to decide which ones should be used in determining the tactics and strategy that will yield success. It should be emphasized that such programs do not at present allow the computer to develop basic parameters – for example, in a game of checkers, to decide that "one square at a time" actually means "two squares at a time." These games are set up within such rigid disciplines that the defining rules of the game and the fixed goal of success fix the decision parameters. But in the realm of learning machines and their operation, one result of the program techniques is removal from the mind of the designer, and of the operator, of an effective understanding of many of the stages by which the machine comes to its conclusions and of the actual long-range intentions of many of the operations. Wiener states[1]: "This is highly relevant to the problem of our being able to foresee undesired consequences outside the frame of the strategy of the game while the machine is still in action and while intervention on our part may prevent the occurrence of these consequences." Because of the time differential – that is, the balance between the speed of the computer and that of man's operations – the communication between man and machine is incomplete.

Machines can be trained to learn, and computers undoubtedly show originality, particularly in game learning, not only in short-term tactics but also in long-range strategy. The machines can transcend their makers and programmers, and the end point may be creative and new, but not necessarily appreciated by the programmer. Because of the time differential and of inadequate knowledge of the learning machine's tactics and strategy, either man must depend on the machine or he must not. This parameter of time balance is quite different from the parameters hitherto discussed. Even when a problem is one of routine decision within the rigid discipline of mathematics, it is common to find the automatized decision made with such speed that it must be used before it can be completely checked. Checking in this case does not imply possible fallibility of the decision-making mechanism but incomplete recognition of the decision and its environment by the programmer. One acceptable mode of checking is simulation – that is, actual trial of the decision-making operation. This is valid where every possible case can be simulated. In the vast majority of significant decisions, however, such an extensive simulation is impossible, or all possible occurrences cannot be recognized.

THE MAN-COMPUTER PROBLEM

Humans within the decision system are more fallible and less apt to operate in a well-organized, accurate manner than computers. However, human solution involves times which are compatible with human review. Humans have sufficient time to use their self-organizing facilities to vary the perspective of the problem in time as it progresses – for example, to vary parameters, and sometimes the actual input. The time balance is not necessarily dangerous; its very existence is one of the benefits of automation. However, it is a parameter of decision processes which should be consistently considered. Particularly in respect to learning systems in which the goals and rules are well defined, the stratagems used to reach the goal, as developed in the solution, may be, as was noted earlier, completely incompatible with the original goals. If the entire operation takes place at such a rapid rate that only success in reaching the goal can be evaluated before the process is put into effect, the time-balance problem is critical. And if, during the course of a machine action, we stop the machine because we do not like or understand a given tactic, we will destroy the total strategy and make the use of a computer pointless. It is doubtful whether man, faced with a problem he has given to the machine can comfortably contradict, stop, or limit the learning program. It is far more likely, in the man-machine relationship, that man will accept the machine's decisions, whether or not he understands them. The fact that we cannot yet include in mechanized processing the complete and necessary value and responsibility parameters is easily overlooked in the light of positive values such as efficiency, speed, accuracy, and objectivity.

Machines can and do simulate events that cannot be studied in actuality. Examples are the action of a petroleum cracking column, a nuclear attack, the burning of a solid fuel inside a rocket, or the flight of a space ship, as well as business procedures. Mechanized business games have become an integral part of training in certain universities and are used by many large industrial concerns in the training and evaluation of personnel. These games, like war games, are not basically in the learning category but operate with pre-determined and fixed values. Only in the reaction of the human players as a part of a total system do the games fall within the special-problem group. The human reaction to the game, however, is often intense. There have been instances at the University of Washington of students dropping from class because of the emotional reaction to the computer game. Although industrialists explain that the game outcomes will not be used in personnel evaluation, the players feel tense and threatened. While in most of the games the players are not playing against the computer but against each other through the computer, the introduction of a mechanized intermediary of such precision and speed increases the threatening aspects of what is called a "game." Because such games cannot possibly include all the variable parameters of actual business, it is sometimes possible for players to "beat" the game by extremely improbable or unethical decisions. Business games of this type are obviously valuable in emphasizing "cause-and-effect" truisms of specific facets of business. Only when it is assumed that a partial simulation of human reaction and economic structure is a complete and accurate simulation does the problem become manifestly dangerous. As in the other examples discussed, as long as the routine solution is admitted to be a routine solution and used only as a partial simulation of systems involving humans, the solution can be used to decided advantage.

The attitudes of the participants when "playing" with a computer are worthy of note. Here again we see a reaction of humans forced to subject their human – and therefore incompletely defined – decision systems for evaluation by the computer, which has been socially accepted as totally objective and accurate.

The human reaction to war games is different. War, like business and like human mental functions such as problem solving, can be simulated, and the simulation may be of great value. The danger lies in believing that in the results of such simulation one has the complete truth. For example, it has been claimed that predictions of victory in war games are becoming increasingly accurate. In this claim the role of man's fantasy is clear. Military leaders need ways of estimating possibilities and probabilities in planning for war; the machine can do more than man, can handle complexities systematically, and can study far more cases and far more variable systems. Yet it is obvious that the input information must be grossly inadequate, especially in planning for the "new" nuclear warfare. There is no way to completely estimate or evaluate what would happen in a "real" war. And yet, how are these results used? Mechanized war games are used, like business games, to rapidly obtain experience and training and to relate cause and effect. In both instances the games are considered useful bases of implementation. Although man knows that the simulation is imperfect, it is all too easy for him to feel that success in the game is indicative of success in the real situation; that tactics and techniques which are effective in game playing are effective in actuality. The mechanized process, because it catches many of the pressures and human reactions of actuality, seems to provide precise objectivity which permits mechanical evaluation of the strategy or tactics involved. Man's frailty makes him wish to shunt off complex decisions to the machine, with its apparent logic, reason, objectivity, and superhuman capacity. Unfortunately, it is in these most important decisions – involving massive responsibility and reaction and concomitant meagerness of information input and criteria of output evaluation – that man most needs help. Unfortunately, it is in these very cases that the machine *must* be used solely as a routine processing device, that it must not be made to take responsibility from man.

While applications involving military operations and national security emphasize the extreme significance of the man-computer interface problem, other, less traumatic, applications and examples should not be neglected. As has been stated, actual machine-learning techniques are not being widely applied at present. Essentially, all computer applications have been handled as routine-direct problems, even those involving war games or business games. The computer is not directly allowed to map original strategy or tactics; it follows specified cues and preestablished values to organize vast quantities of data with its inimitable speed and accuracy. To our mind it is in these contemporary solutions that the essential parameters of values, the time balance, and the human response to the mechanized system must be analyzed with extreme caution. It should be clearly understood that in calling for caution we do not imply that all use of computers is dangerous, or, for that matter, that any computer application must in itself be dangerous. Rather, it is the use of the computer results that concerns us. In the field of mechanical translation of languages, the routine-direct processing is a great complex of decision systems. In the present phase of development, translation errors are usually recognizable, and the problems of the computer-human relationship are more often matters of irritation than of danger. Caution is needed relative to this decision operation of the computer only as regards the use to which the translation process is put. If translation output were to be placed directly into legal or business documents, serious problems could arise. However, as long as the quality of the translation is recognized for what it is, this esoteric process should not be considered to intrude upon the man-computer relationship.

CONCLUSIONS

The levels of human knowledge of the environment and the universe are increasing, and it is obviously necessary that man's ability to cope with this knowledge should increase – necessary for his usefulness and for his very survival. The processes of automation have provided a functional agent for this purpose. Successful mechanized solution of routine problems has directed attention toward the capacity of the computer to arrive at apparent or real solutions of routine-learning and special problems. Increasing use of the computer in such problems is clearly necessary if our body of knowledge and information is to serve its ultimate function. Along with such use of the computer, however, will come restrictions and cautions which have not hitherto been necessary. We find that the computer is being given responsibilities with which it is less able to cope than man is. It is being called on to act for man in areas where man cannot define his own ability to perform and where he feels uneasy about his own performance — where he would like a neat, well-structured solution and feels that in adopting the machine's partial solution he is closer to the "right" than he is in using his own. An aura of respectability surrounds a computer output, and this, together with the time-balance factor, makes unqualified acceptance tempting. The need for caution, then, already exists and will be much greater in the future. It has little to do with the limited ability of the computer per se, much to do with the ability of man to realistically determine when and how he must use the tremendous ability which he has developed in automation.

Let us continue to work with learning machines, with definitions of meaning and "artificial intelligence." Let us examine these processes as "games" with expanding values, aiming toward developing improved computer techniques as well as increasing our knowledge of human functions. Until machines can satisfy the requirements discussed, until we can more perfectly determine the functions we require of the machines, let us not call upon mechanized decision systems to act upon human systems without intervening realistic human processing. As we proceed with the inevitable development of computers and means of using them, let us be sure that careful analysis is made of all automation (either routine direct, routine-learning, or special) that is used in systems of which man is a part — sure that man reflects upon his own reaction to, and use of mechanization. Let us be certain that, in response to Samuel Butler's question[7]: "May not man himself become a sort of parasite upon the machines; an affectionate machine tickling aphid?", we will always be able to answer, "No."

REFERENCE AND NOTES

[1] N. Wiener, *Science* 131, 1355 (1960).

[2] The study of which this article is a part is supported by the Air Force Office of Research.

[3] Many basic decision units may be required by the logical-design engineer to accomplish the operation of the single basic decision.

[4] D. Bergamini, *Reporter* (17 Aug. 1961).

[5] D. Riesman et al., *The Lonely Crowd* (Yale University Press, New Haven, Conn., 1950).

[6] E. Fromm, *Escape from Freedom* (Holt, Rinehart and Winston, New York, 1941).

[7] S. Butler, *Erewhon* (Doubleday, New York, 1872).

. . . There were some forty men involved each with his own particular area of responsibility, so uniformity wasn't always easy to achieve. Kranz later attributed their success to the intensive drilling they had gone through for months ahead of the flight. Although the flight controllers had never handled either a simulation or a theoretical approximation of anything like the present situation, they had been obliged to solve a great many problems in the course of dozens of simulations, and they had met and divided into groups to do so. During these months of simulations, channels of communication had developed between them, so within seemingly random discussions there was a sort of order, the way there is between the cells in a computer or those of the brain.

Henry S.F. Cooper, Jr., "Annals of Exploration, An Accident in Space — II," The New Yorker, *18 November 1972, p. 90.*

33

Henry Mintzberg

MANAGERIAL WORK: ANALYSIS FROM OBSERVATION

What do managers do? Ask this question and you will likely be told that managers plan, organize, coordinate, and control. Since Henri Fayol[9] first proposed these words in 1916, they have dominated the vocabulary of management. (See, for example,[8,12,17]. How valuable are they in describing managerial work? Consider one morning's work of the president of a large organization:

> *As he enters his office at 8:23, the manager's secretary motions for him to pick up the telephone. "Jerry, there was a bad fire in the plant last night, about $30,000 damage. We should be back in operation by Wednesday. Thought you should know."*
>
> *At 8:45, a Mr. Jamison is ushered into the manager's office. They discuss Mr. Jamison's retirement plans and his cottage in New Hampshire. Then the manager presents a plaque to him commemorating his thirty-two years with the organization.*

> *Mail processing follows: An innocent-looking letter, signed by a Detroit lawyer, reads: "A group of us in Detroit has decided not to buy any of your products because you used that anti-flag, anti-American pinko, Bill Lindell, upon your Thursday night TV show." The manager dictates a restrained reply.*
>
> *The 10:00 meeting is scheduled by a professional staffer. He claims that his superior, a high-ranking vice-president of the organization, mistreats his staff, and that if the man is not fired, they will all walk out. As soon as the meeting ends, the manager rearranges his schedule to investigate the claim and to react to this crisis.*

Which of these activities may be called planning, and which may be called organizing, coordinating, and controlling? Indeed, what do words such as "coordinating" and "planning" mean in the context of real acitivty? In fact, these four words do not describe the actual work of managers at all; they describe certain vague objectives of managerial work: ". . . they are just ways of indicating what we need to explain."[1 p. 537]

Other approaches to the study of managerial work have developed, one dealing with managerial decision-making and policy-making processes, another with the manager's interpersonal activities. (See, for example:[2,10].) And some empirical researchers, using the "diary" method, have studied what might be called managerial "media" – by what means, with whom, how long, and where managers spend their time.[a] But in no part of this literature is the actual content of managerial work systematically and meaningfully described.[b] Thus the question posed at the start – what do managers do? – remains essentially unanswered in the literature of management.

This is indeed an odd situation. We claim to teach management in schools of both business and public administration; we undertake major research programs in management; we find a growing segment of the management science community concerned with the problems of senior management. Most of these people – the planners, information and control theorists, systems analysts, etc. – are attempting to analyze and change working habits that they themselves do not understand. Thus, at a conference called at M.I.T. to access the impact of the computer on the manager, and attended by a number of America's foremost management scientists, a participant found it necessary to comment after lengthy discussion[20 p. 198]

> *I'd like to return to an earlier point. It seems to me that until we get into the question of what the top manager does or what the functions are that define the top management job, we're not going to get out of the kind of difficulty that keeps cropping up. What I'm really doing is leading up to my earlier question which no one really answered. And that is: Is it possible to arrive at a specification of what constitutes the job of a top manager?*

His question was not answered.

RESEARCH STUDY ON MANAGERIAL WORK

In late 1966, I began research on this question, seeking to replace Fayol's words by a set that would more accurately describe what managers do. In essence, I sought to develop by the process of induction a statement of managerial work that would have empirical validity. Using a method called "structured observation," I observed for 1-week periods the chief executives of five medium to large organizations (a consulting firm, a school system, a technology firm, a consumer goods manufacturer, and a hospital).

Structured as well as unstructured (i.e., anecdotal) data were collected in three "records." In the *chronology record,* activity patterns throughout the working day were recorded. In the *mail record,* for each of 890 pieces of mail processed during the 5-weeks, were recorded its purpose, format and sender, the attention it received, and the action it elicited. And recorded in the *contact record,* for each of 368 verbal interactions, were the purpose, the medium (telephone call, scheduled or unscheduled meeting, tour), the participants, the form of initiation, and the location. It should be noted that all categorizing was done during and after observation so as to ensure that the categories reflected only the work under observation. Reference 19 contains a fuller description of this methodology and a tabulation of the results of the study.

Two sets of conclusions are presented below. The first deals with certain characteristics of managerial work, as they appeared from analysis of the numerical data (e.g. How much time is spent with peers? What is the average duration of meetings? What proportion of contacts are initiated by the manager?). The second describes the basic content of managerial work in terms of ten roles. This description derives from an analysis of the data on the recorded *purpose* of each contact and piece of mail.

The liberty is taken of referring to these findings as descriptive of managerial, as opposed to chief executive, work. This is done because many of the findings are supported by studies of other types of managers. Specifically, most of the conclusions on work characteristics are to be found in the combined results of a group of studies of foremen[11,16], middle managers[4,5,15,25], and chief executives[6]. And although there is little useful material on managerial roles, three studies do provide some evidence of the applicability of the role set. Most important, Sayles' empirical study of production managers[24] suggests that at least five of the ten roles are performed at the lower end of the managerial hierarchy. And some further evidence is provided by comments in Whyte's study of leadership in a street gang[26] and Neustadt's study of three U.S. Presidents[21]. (Reference is made to these findings where appropriate.) Thus, although most of the illustrations are drawn from my study of chief executives, there is some justification in asking the reader to consider when he sees the terms "manager" and his "organization" not only "presidents" and their "companies," but also "foremen" and their "shops," "directors" and their "branches," "vice-presidents" and their "divisions." The term *manager* shall be used with reference to all those people in charge of formal organizations or their subunits.

SOME CHARACTERISTICS OF MANAGERIAL WORK

Six sets of characteristics of managerial work derive from analysis of the data of this study. Each has a significant bearing on the manager's ability to administer a complex organization.

1. The Manager Performs a Great Quantity of Work at an Unrelenting Pace

Despite a semblance of normal working hours, in truth managerial work appears to be very taxing. The five men in this study processed an average of 36 pieces of mail each day, participated in eight meetings (half of which were scheduled), engaged in five telephone calls, and took one tour. In his study of foremen, Guest[11] found that the number of activities per day averaged 583, with no real break in the pace.

Free time appears to be very rare. If by chance a manager has caught up with the mail, satisfied the callers, dealt with all the disturbances, and avoided scheduled meetings, a subordinate will likely show up to usurp the available time. It seems that the manager cannot expect to have much time for leisurely reflection during office hours. During "off" hours, our chief executives spent much time on work-related reading. High-level managers appear to be able to escape neither from an environment which recognizes the power and status of their positions nor from their own minds, which have been trained to search continually for new information.

2. Managerial Activity is Characterized by Variety, Fragmentation, and Brevity

There seems to be no pattern to managerial activity. Rather, variety and fragmentation appear to be characteristic, as successive activities deal with issues that differ greatly both in type and in content. In effect the manager must be prepared to shift moods quickly and frequently.

A typical chief executive day may begin with a telephone call from a director who asks a favor (a "status request"); then a subordinate calls to tell of a strike at one of the facilities (fast movement of information, termed "instant communication"); this is followed by a relaxed scheduled event at which the manager speaks to a group of visiting dignitaries (ceremony); the manager returns to find a message from a major customer who is demanding the renegotiation of a contract (pressure); and so on. Throughout the day, the managers of our study encountered this great variety of activity. Most surprisingly, the significant activities were interspersed with the trivial in no particular pattern.

Furthermore, these managerial activities were characterized by their brevity. Half of all the activities studied lasted less than 9 minutes and only 10 percent exceeded 1 hour's duration. Guest's foremen averaged 48 seconds per activity, and Carlson[6] stressed that his chief executives were unable to work without frequent interruption.

In my own study of chief executives, I felt that the managers demonstrated a preference for tasks of short duration and encouraged interruption. Perhaps the manager becomes accustomed to variety, or perhaps the flow of "instant communication" cannot be delayed. A more plausible explanation might be that the manager becomes conditioned by his workload. He develops a sensitive appreciation for the opportunity cost of his own time. Also, he is aware of the ever-present assortment of obligations associated with his job – accumulations of mail that cannot be delayed, the callers that must be attended to, the meetings that require his participation. In other words, no matter what he is doing, the manager is plagued by what he must do and what he might do. Thus the manager is forced to treat issues in an abrupt and superficial way.

3. Managers Prefer Issues That Are Current, Specific, and Ad Hoc

Ad hoc operating reports received more attention than did routine ones; current, uncertain information – gossip, speculation, hearsay – which flows quickly was preferred to historical, certain information; "instant communication" received first consideration; few contacts were held on a routine or "clocked" basis; almost all contacts concerned well-defined issues. The managerial environment is clearly one of stimulus-response. It breeds not reflective planners, but adaptable information manipulators who prefer the live, concrete situation, men who demonstrate a marked action orientation.

4. The Manager Sits Between His Organization and a Net work of Contacts

In virtually every empirical study of managerial time allocation, it was reported that managers spent a surprisingly large amount of time in horizontal or lateral (nonline) communication. It is clear from this study and from that of Sayles[24] that that manager is surrounded by a diverse and complex web of contacts which serves as his self-designed external information system. Included in this web can be clients, associates and suppliers, outside staff experts, peers (managers of related or similar organizations), trade organizations, government officials, independents (those with no relevant organizational affiliation), and directors or superiors. (Among these, directors in this study and superiors in other studies did *not* stand out as particularly active individuals.)

The managers in this study received far more information than they emitted, much of it coming from contacts, and more from subordinates who acted as filters. Figuratively, the manager appears as the neck of an hourglass, sifting information into his own organization from its environment.

5. The Manager Demonstrates a Strong Preference for the Verbal Media

The manager has five media at his command – mail (documented), telephone (purely verbal), unscheduled meeting (informal face-to-face), sheduled meeting (formal face-to-face), and tour (observational). Along with all the other empirical studies of work characteristics, I found a strong predominance of verbal forms of communication.

Mail - By all indications, managers dislike the documented form of communication. In this study, they gave cursory attention to such items as operating reports and periodicals. It was estimated that only 13 percent of the input mail was of specific and immediate use to the managers. Much of the rest dealt with formalities and provided general reference data. The managers studied initiated very little mail, only 25 pieces in the 5 weeks. The rest of the outgoing mail was sent in reaction to mail received – a reply to a request, an acknowledgment, some information forwarded to a part of the organization. The managers appeared to dislike this form of communication, perhaps because the mail is a relatively slow and tedious medium to use.

Telephone and Unscheduled Meetings - The less formal means of verbal communication – the telephone, a purely verbal form, and the unscheduled meeting, face-to-face form – were used frequently (two thirds of the contacts in the study) but for brief encounters (average duration of 6 and 12 minutes, respectively). They were used primarily to deliver requests and to transmit pressing information to those outsiders and subordinates who had informal relationships with the manager.

Scheduled Meetings - These tended to be of long duration, averaging 68 minutes in this study, and absorbing over half the managers' time. Such meetings provided the managers with their main opportunities to interact with large groups and to leave the confines of their own offices. Scheduled meetings were used when the participants were unfamiliar to the manager (e.g., students who request that he speak at a university), when a large quantity of information had to be transmitted (e.g., presentation of a report), when ceremony had to take place, and when complex strategy making or negotiation had to be undertaken. An important feature of the scheduled meeting was the incidental, but by no means irrelevant, information that flowed at the start and end of such meetings.

Tours - Although the walking tour would appear to be a powerful tool for gaining information in an informal way, in this study tours accounted for only 3 percent of the managers' time.

In general, it can be concluded that the manager uses each medium for particular purposes. Nevertheless, where possible, he appears to gravitate to verbal media, since these provide greater flexibility, require less effort, and bring faster response. It should be noted here that the manager does not leave the telephone or the meeting to get back to work. Rather, communication is his work, and these media are his tools. The operating work of the organization – producing a product, doing research, purchasing a part – appears to be undertaken infrequently by the senior manager. The manager's productive output must be measured in terms of information, a great part of which is transmitted verbally.

6. Despite the Preponderance of Obligations, the Manager Appears to Be Able to Control His Own Affairs

Carlson suggested in his study of Swedish chief executives that these men were puppets, with little control over their own affairs. A cursory examination of our data indicates that this is true. Our managers were responsible for the initiation of only 32 percent of their verbal contacts and a smaller proportion of their mail. Activities were also classified as to the nature of the managers' participation, and the active ones were outnumbered by the passive ones (e.g., making requests versus receiving requests). On the surface, the manager is indeed a puppet, answering requests in the mail, returning telephone calls, attending meetings initiated by others, yielding to subordinates' requests for time, reacting to crises.

However, such a view is misleading. There is evidence that the senior manager can exert control over his own affairs in two significant ways: (1) it is he who defines many of his own long-term commitments, by developing appropriate information channels which later feed him information, by initiating projects which later demand his time, by joining committees or outside boards which provide contacts in return for his services, and so on. (2) The manager can exploit situations that appear as obligations. He can lobby at ceremonial speeches; he can impose his values on his organization when his authorization is requested; he can motivate his subordinates whenever he interacts with them; he can use the crisis situation as an opportunity to innovate.

Perhaps these are two points that help distinguish successful and unsuccessful managers. All managers appear to be puppets. Some decide who will pull the strings and how, and they then take advantage of each move that they are forced to make. Others, unable to exploit this high-tension environment, are swallowed up by this most demanding of jobs.

THE MANAGER'S WORK ROLES

In describing the essential content of managerial work, one should aim to model managerial activity, that is, to describe it as a set of programs. But an undertaking as complex as this must be preceded by the development of a useful typological description of managerial work. In other words, we must first understand the distinct components of managerial work. At the present time we do not.

In this study, 890 pieces of mail and 368 verbal contacts were categorized as to purpose. The incoming mail was found to carry acknowledgments, requests and solicitations of various kinds, reference data, news, analytical reports, reports on events and on operations, advice on various situations, and statements of problems, pressures, and ideas. In reacting to mail, the managers acknowledged some, replied to the requests (e.g., by sending information), and forwarded much to subordinates (usually for their information). Verbal contacts involved a variety of purposes. In 15 percent of them activities were scheduled, in 6 percent ceremonial events took place, and a few involved external board work. About 34 percent involved requests of various kinds, some insignificant, some for information, some for authorization of proposed actions. Another 36 percent essentially involved the flow of information to and from the manager, while the remainder dealt specifically with issues of strategy and with negotiations. (For details, see:[19].)

In this study, each piece of mail and verbal contact categorized in this way was subjected to one question: why did the manager do this? The answers were collected and grouped and regrouped in various ways (over the course of 3 years) until a typology emerged that was felt to be satisfactory. While an example, presented below, will partially explain this process to the reader, it must be remembered that (in the words of Bronowski[3, p. 62]) "Every induction is a speculation and it guesses at a unity which the facts present but do not strictly imply."

Consider the following sequence of two episodes: a chief executive attends a meeting of an external board on which he sits. Upon his return to his organization, he immediately goes to the office of a subordinate, tells of a conversation he had with a fellow board member, and concludes with the statement, "It looks like we shall get the contract."

The purposes of these two contacts are clear – to attend an external board meeting and to give current information (instant communication) to a subordinate. But why did the manager attend the meeting? Indeed, why does he belong to the board? And why did he give this particular information to his subordinate?

Basing analysis on this incident, one can argue as follows: the manager belongs to the board in part so that he can be exposed to special information which is of use to his organization. The subordinate needs the information but has not the status which would give him access to it. The chief executive does. Board memberships bring chief executives in contact with one another for the purpose of trading information.

Two aspects of managerial work emerge from this brief analysis. The manager serves in a "liaison" capacity because of the status of his office, and what he learns here enables him to act as "disseminator" of information into his organization. We refer to these as *roles* – organized sets of behaviors belonging to identifiable offices or positions[23]. Ten roles were chosen to capture all the activities observed during this study.

All activities were found to involve one or more of three basic behaviors – interpersonal contact, the processing of information, and the making of decisions. As a result, our ten roles are divided into three corresponding groups. Three roles – labeled *figurehead, liaison,* and *leader* – deal with behavior that is essentially interpersonal in nature. Three others – *nerve center, disseminator,* and *spokesman* – deal with information-processing activities performed by the manager. And the remaining four – *entrepreneur, disturbance handler, resource allocator,* and *negotiator* – cover the decision-making activities of the manager. We describe each of these roles in turn, asking the reader to note that they form a *gestalt,* a unified whole whose parts cannot be considered in isolation.

The Interpersonal Roles

Three roles relate to the manager's behavior that focuses on interpersonal contact. These roles derive directly from the authority and status associated with holding managerial office.

Figurehead - As legal authority in his organization, the manager is a symbol, obliged to perform a number of duties. He must preside at ceremonial events, sign legal documents, receive visitors, make himself available to many of those who feel, in the words of one of the men studied, "that the only way to get something done is to get to the top." There is evidence that this role applies at other levels as well. Davis[7 pp. 43-44] cites the case of the field sales manager who must deal with those customers who believe that their accounts deserve his attention.

Leader - Leadership is the most widely recognized of managerial roles. It describes the manager's relationship with his subordinates – his attempts to motivate them and his development of the milieu in which they work. Leadership actions pervade all activity – in contrast to most roles, it is possible to designate only a few activities as dealing exclusively with leadership (these mostly related to staffing duties). Each time a manager encourages a subordinate, or meddles in his affairs, or replies to one of his requests he is playing the *leader* role. Subordinates seek out and react to these leadership clues, and, as a result, they impart significant power to the manager.

Liaison - As noted earlier, the empirical studies have emphasized the importance of lateral or horizontal communication in the work of managers at all levels. It is clear from our study that this is explained largely in terms of the *liaison* role. The manager establishes his network of contacts essentially to bring information and favors to his organization. As Sayles notes in his study of production supervisors[24, p. 258], "The one enduring objective [of the manager] is the effort to build and maintain a predictable, system of relationships. . . ."

Making use of his status, the manager interacts with a variety of peers and other people outside his organization. He provides time, information, and favors in return for the same from others. Foremen deal with staff groups and other foremen; chief executives join boards of directors, and maintain extensive networks of individual relationships. Neustadt notes this behavior in analyzing the work of President Roosevelt[21, p. 150]:

> *His personal sources were the product of a sociability and curiosity that reached back to the other Roosevelt's time. He had an enormous acquaintance in various phases of national life and at various levels of government; he also had his wife and her variety of contacts. He extended his acquaintanceships abroad; in the war years Winston Churchill, among others, became a "personal source." Roosevelt quite deliberately exploited these relationships and mixed them up to widen his own range of information. He changed his sources as his interests changed, but no one who had ever interested him was quite forgotten or immune to sudden use.*

The Informational Roles

A second set of managerial activities relate primarily to the processing of information. Together they suggest three significant managerial roles, one describing the manager as a focal point for a certain kind of organizational information, the other two describing relatively simply transmission of this information.

Nerve Center – There is indication, both from this study and from those by Neustadt and Whyte, that the manager serves as the focal point in his organization for the movement of nonroutine information. Homans, who analyzed Whyte's study, draws the following conclusions[26, p. 187]:

> *Since interaction flowed toward [the leaders], they were better informed about the problems and desires of group members than were any of the followers and therefore better able to decide on an appropriate course of action. Since they were in close touch with other gang leaders, they were also better informed than their followers about conditions in Cornerville at large. Moreover, in their positions at the focus of the chains of interaction, they were better able than any follower to pass on to the group decisions that had been reached.*

The term *nerve center* is chosen to encompass those many activities in which the manager receives information.

Within his own organization, the manager has legal authority that formally connects him – and only him – to *every* member. Hence, the manager emerges as *nerve center* of internal information. He may not know as much about any one function as the subordinate who specializes in it, but he comes to know more about his total organization than any other member. He is the information generalist. Furthermore, because of the manager's status and its manifestation in the *liaison* role, the manager gains unique access to a variety of knowledgeable outsiders, including peers who are themselves *nerve centers* of their own organizations. Hence the manager emerges as his organization's *nerve center* of external information as well.

As noted earlier, the manager's nerve center

information is of a special kind. He appears to find it most important to get his information quickly and informally. As a result, he will not hesitate to bypass formal information channels to get it, and he is prepared to deal with a large amount of gossip, hearsay, and opinion which has not yet become substantiated fact.

Disseminator - Much of the manager's information must be transmitted to subordinates. Some of this is of a *factual* nature, received from outside the organization or from other subordinates. And some is of a *value* nature. Here the manager acts as the mechanism by which organizational influencers (owners, governments, employee groups, the general public, etc., or simply the "boss") make their preferences known to the organization. It is the manager's duty to integrate these value positions, and to express general organizational preferences as a guide to decisions made by subordinates. One of the men studied commented, "One of the principal functions of this position is to integrate the hospital interests with the public interests." Papandreou describes this duty in a paper published in 1952, referring to management as the "peak coordinator"[22].

Spokesman - In his *spokesman* role, the manager is obliged to transmit his information to outsiders. He informs influencers and other interested parties about his organization's performance, its policies, and its plans. Furthermore, he is expected to serve outside his organization as an expert in its industry. Hospital administrators are expected to spend some time serving outside as public experts on health, and corportation presidents, perhaps as chamber or commerce executives.

The Decisional Roles

The manager's legal authority requires that he assume responsibility for all of his organization's important actions. The *nerve center* role suggests that only he can fully understand complex decisions, particularly those involving difficult value tradeoffs. As a result, the manager emerges as the key figure in the making and interrelating of all significant decisions in his organization, a process that can be referred to as *strategy making.* Four roles describe the manager's control over the strategy-making system in his organization.

Entrepreneur - The *entrepreneur* role describes the manager as initiator and designer of much of the controlled change in his organization. The manager looks for opportunities and potential problems which may cause him to initiate action. Action takes the form of *improvement projects* – the marketing of a new product, the strengthening of a weak department, the purchasing of new equipment, the reorganization of formal structure, and so on.

The manager can involve himself in each improvement project in one of three ways: (1) he may *delegate* all responsibility for its design and approval, implicitly retaining the right to replace that subordinate who takes charge of it; (2) he may delegate the design work to a subordinate, but retain the right to *approve* it before implementation; (3) he may actively *supervise* the design work himself.

Improvement projects exhibit a number of interesting characteristics. They appear to involve a number of subdecisions, consciously sequenced over long periods of time and separated by delays of various kinds. Furthermore, the manager appears to supervise a great many of these at any one time – perhaps 50 to 100 in the case of chief executives. In fact, in his handling of improvement projects, the manager may be likened to a juggler. At any one point, he maintains a number of balls in the air. Periodically, one comes down, receives a short burst of energy, and goes up again. Meanwhile, an inventory of new balls waits on the sidelines and, at random intervals, old balls are discarded and new ones added. Both Lindblom[2] and Marples[18] touch on these aspects of strategy making, the former stressing the disjointed and incremental nature of the decisions, and the latter depicting the sequential episodes in terms of a stranded rope made up of fibers of different lengths each of which surfaces periodically.

Disturbance Handler - While the entrepreneur role focuses on voluntary change, the *disturbance handler* role deals with corrections which the manager is forced to make. We may describe this role as follows: the organization consists basically of specialist operating programs. From time to time, it experiences a stimulus that cannot be handled routinely, either because an operating program has broken down or because the stimulus is new and it is not clear which operating program should handle it. These situations constitute disturbances. As generalist, the manager is obliged to assume responsibility for dealing with the stimulus. Thus the handling of disturbances is an essential duty of the manager.

There is clear evidence for this role both in our study of chief executives and in Sayles' study of production supervisors[24, p. 162]:

The achievement of this stability, which is the manager's objective, is a never-to-be-attained ideal. He is like a symphony orchestra conductor, endeavoring to maintain a melodious performance in which contributions of the various instruments are coordinated and sequenced, patterned and paced, while the orchestra members are having various personal difficulties, stage hands are moving music stands, alternating excessive heat and cold are creating audience and instrument problems, and the sponsor of the concert is insisting on irrational changes in the program.

Sayles goes further to point out the very important balance that the manager must maintain between change and stability. To Sayles, the manager seeks "a dynamic type of stability"[24, p. 162]. Most disturbances elicit short-term adjustments which bring back equilibrium; persistent ones require the introduction of long-term structural change.

Resource Allocator - The manager maintains ultimate authority over his organization's strategy-making system by controlling the allocation of its resources. By deciding who will get what (and who will do what), the manager directs the course of his organization. He does this in three ways:

1. *In scheduling his own time,* the manager allocates his most precious resource and thereby determines organizational priorities. Issues that receive low priority do not reach the *nerve center* of the organization and are blocked for want of resources.
2. In designing the organizational structure and in carrying out many improvment projects, the manager *programs the work of his subordinates.* In other words, he allocates their time by deciding what will be done and who will do it.
3. Most significantly, the manager maintains control over resource allocation by the requirement that he *authorize all significant decisions* before they are implemented. By retaining this power, the manager ensures that different decisions are interrelated – that conflicts are avoided, that resource constraints are respected, and that decisions complement one another.

Decisions appear to be authorized in one of two ways. Where the costs and benefits of a proposal can be quantified, where it is competing for specified resources with other known proposals, and where it can wait for a certain time of year, approval for a proposal is sought in the context of a formal *budgeting* procedure. But these conditions are most often not met – timing may be crucial, nonmonetary costs may predominate, and so on. In these cases, approval is sought in terms of an *ad hoc request for authorization.* Subordinate and manager meet (perhaps informally) to discuss one proposal alone.

Authorization choices are enormously complex ones for the manager. A myriad of factors must be considered (resource constraints, influencer preferences, consistency with other decisions, feasibility, payoff, timing, subordinate feelings, etc.). But the fact that the manager is authorizing the decision rather than supervising its design suggests that he has little time to give to it. To alleviate this difficulty, it appears that managers use special kinds of *models* and *plans* in their decision making. These exist only in their minds and are loose, but they serve to guide behavior. Models may answer questions such as, "Does this proposal make sense in terms of the trends that I see in tariff legislation?" or "Will the EDP department be able to get along with marketing on this?" Plans exist in the sense that, on questioning, managers reveal images (in terms of proposed improvement projects) of where they would like their organizations to go: "Well, once I get these foreign operations fully developed, I would like to begin to look into a reorganization," said one subject of this study.

Negotiator - The final role describes the manager as participant in negotiation activity. To some students of the management process[8, p. 343], this is not truly part of the job of managing. But such distinctions are arbitrary. Negotiation is an integral part of managerial work, as this study notes for chief executives and as that of Sayles made very clear for production supervisors[24, p. 131]: "Sophisticated managers place great stress on negotiations as a way of life. They negotiate with groups who are setting standards for their work, who are performing support activity for them, and to whom they wish to 'sell' their services."

The manager must participate in important negotiation sessions because he is his organization's legal authority, its *spokesman* and its *resource allocator.* Negotiation is resource trading in real time. If the resource commitments are to be large, the legal authority must be present.

These ten roles suggest that the manager of an organization bears a great burden of responsibility. He must oversee his organization's status system; he must serve as a crucial informational link between it and its environment; he must interpret and reflect its basic values; he must maintain the stability of its operations; and he must adapt it in a controlled and balanced way to a changing environment.

MANAGEMENT AS A PROFESSION AND AS A SCIENCE

Is management a profession? To the extent that different managers perform one set of basic roles, management satisfies one criterion for becoming a profession. But a profession must require, in the words of the Random House Dictionary, "knowledge of some department of learning or science." Which of the ten roles now requires specialized learning? Indeed, what school of business or public administration teaches its students how to disseminate information, allocate resources, perform as figurehead, make contacts, or handle disturbances? We simply know very little about teaching these things. The reason is that we have never tried to document and describe in a meaningful way the procedures (or programs) that managers use.

The evidence of this research suggests that there is as yet no science in managerial work – that managers do not work according to procedures that have been prescribed by scientific analysis. Indeed, except for his use of the telephone, the airplane, and the dictating machine, it would appear that the manager of today is indistinguishable from his predecessors. He may seek different information, but he gets much of it in the same way – from word-of-mouth. He may make decisions dealing with modern technology but he uses the same intuitive (that is, nonexplicit) procedures in making them. Even the computer, which has had such a great impact on other kinds of organizational work, has apparently done little to alter the working methods of the general manager.

How do we develop a scientific base to understand the work of the manager? The description of roles is a first and necessary step. But tighter forms of research are necessary. Specifically, we must attempt to model managerial work – to describe it as a system of programs. First, it will be necessary to decide what programs managers actually use. Among a great number of programs in the manager's repertoire, we might expect to find a time scheduling program, an information disseminating program, and a disturbance handling program. Then researchers will have to devote a considerable amount of effort to studying and accurately describing the content of each of these programs – the information and heuristics used. Finally, it will be necessary to describe the interrelationships among all these programs so that they may be combined into an integrated descriptive model of managerial work.

When the management scientist begins to understand the programs that managers use, he can begin to design meaningful systems and provide help for the manager. He may ask, which managerial activities can be fully reprogrammed (i.e., automated)? Which cannot be reprogrammed because they require human responses? Which can be partially reprogrammed to operate in a man-machine system? Perhaps scheduling, information collecting, and resource allocating activities lend themselves to varying degrees of reprogramming. Management will emerge as a science to the extent that such efforts are successful.

IMPROVING THE MANAGER'S EFFECTIVENESS

Fayol's 50-year-old description of managerial work is no longer of use to us. And we shall not disentangle the complexity of managerial work if we insist on viewing the manager simply as a decision maker or simply as a motivator of subordinates. In fact, we are unlikely to overestimate the complexity of the manager's work, and we shall make little headway if we take overly simple or narrow points of view in our research.

A major problem faces today's manager. Despite the growing size of modern organizations and the growing complexity of their problems (particularly those in the public sector), the manager can expect little help. He must design his own information system, and he must take full charge of his organization's strategy-making system. Furthermore, the manager faces what might be called the *dilemma of delegation.* He has unique access to much important information but he lacks a formal means of disseminating it. As much of it is verbal, he cannot spread it around in an efficient manner. How can he delegate a task with confidence when he has neither the time nor the means to send the necessary information along with it?

Thus the manager is usually forced to carry a great burden of responsibility in his organization. As organizations become increasingly large and complex, this burden increases. Unfortunately, the man cannot significantly increase his available time or significantly improve his abilities to manage. Hence, in the large, complex bureaucracy, the top manager's time assumes an enormous opportunity cost and he faces the real danger of becoming a major obstruction in the flow of decisions and information.

Because of this, as we have seen, managerial work assumes a number of distinctive characteristics. The quantity of work is great; the pace is unrelenting; there is great variety, fragmentation, and brevity in the work activities; the manager must concentrate on issues that are current, specific, and ad hoc, and, to do so, he finds that he must rely on verbal forms of communications. Yet it is on this man that the burden lies for designing and operating strategy-making and information-processing systems that are to solve his organization's (and society's) problems.

The manager can do something to alleviate these problems. He can learn more about his own roles in his organization, and he can use this information to schedule his time in a more efficient manner. He can recognize that only he has much of the information needed by his organization. Then he can seek to find better means of disseminating it into the organization. Finally, he can turn to the skills of his management scientists to help reduce his workload and to improve his ability to make decisions.

The management scientist can learn to help the manager to the extent he can develop an understanding of the manager's work and the manager's information. To date, strategic planners, operations researchers, and information system designers have provided little help for the senior manager. They simply have had no framework available by which to understand the work of the men who employed them, and they have had poor access to the information which has never been documented. It is folly to believe that a man with poor access to the organization's true *nerve center* can design a formal management information system. Similarly, how can the long-range planner, a man usually uninformed about many of the *current* events that take place in and around his organization, design meaningful strategic plans? For good reason, the literature documents many manager complaints of naïve planning and many planner complaints of disinterested managers. In my view, our lack of understanding of managerial work has been the greatest block to the progress of management science.

The ultimate solution to the problem – to the overburdened manager seeking meaningful help – must derive from research. We must observe, describe, and understand the real work of managing; then and only then shall we significantly improve it.

NOTES

[a] Carlson [6] carried out the classic study just after World War II. He asked nine Swedish managing directors to record on diary pads details of each activity in which they engaged. His method was used by a group of other researchers, many of them working in the United Kingdom. (See [4, 5, 15, 25].)

[b] One major project, involving numerous publications, took place at Ohio State University and spanned three decades. Some of the vocabulary used followed Fayol. The results have generated little interest in this area. (See, for example, [13].)

REFERENCES

[1] Braybrooke, David, "The Mystery of Executives Success Re-examined," *Administrative Science Quarterly,* Vol. 8 (1964), pp. 533-560.

[2] ——— and Lindblom, Charles E., *A Strategy of Decision,* Free Press, New York, 1963.

[3] Bronowski, J., "The Creative Process," *Scientific American,* Vol. 199 (Sept. 1958), pp. 59-65.

[4] Burns, Tom, "The Directions of Activity and Communications in a Departmental Executive Group," *Human Relations,* Vol. 7 (1954), pp. 73-97.

[5] ———, "Management in Action," *Operational Research Quarterly,* Vol. 8 (1957), pp. 45-60.

[6] Carlson, Sune, *Executive Behaviour,* Strömbergs, Stockholm, 1951.

[7] Davis, Robert T., *Performance and Development of Field Sales Managers,* Division of Research, Graduate School of Business Administration, Harvard University, Boston, 1957.

[8] Drucker, Peter F., *The Practice of Management,* Harper & Row, New York, 1954.

[9] Fayol, Henri, *Administration industrielle et gênérale,* Dunod, Paris, 1950 (first published 1916).

[10] Gibb, Cecil A., "Leadership," Chapter 31 in Gardner Lindzey and Elliot A. Aronson (eds.), *The Handbook of Social Psychology,* Vol. 4, 2nd ed., Addison-Wesley, Reading, Mass., 1969.

[11] Guest, Robert H., "Of Time and the Foreman," *Personnel,* Vol. 32 (1955-56) pp. 478-486.

[12] Gulick, Luther H., "Notes on the Theory of Organization," in Luther Gulick and Lyndall Urwick (eds.), *Papers on the Science of Administration,* Columbia University Press, New York, 1937.

[13] Hemphill, J. K., *Dimensions of Executive Positions,* Bureau of Business Research Monograph Number 98, The Ohio State University, Columbus, Ohio, 1960.

[14] Homans, George C., *The Human Group,* Harcourt Brace, New York, 1950.

[15] Horne, J. H., and Lupton, Tom, "The Work Activities of Middle Managers – An Exploratory Study," *The Journal of Management Studies,* Vol. 2 (Feb. 1965), pp. 14-33.

[16] Kelly, Joe, "The Study of Executive Behavior by Activity Sampling," *Human Relations,* Vol. 17 (August 1964), pp. 277-287.

[17] Mackenzie, R. Alex, "The Management Process in 3D," *Harvard Business Review* (November-December 1969), pp. 80-87.

[18] Marples, D. L., "Studies of Managers – A Fresh Start?," *The Journal of Management Studies,* Vol. 4 (October 1967), pp. 282-299.

[19] Mintzberg, Henry "Structured Observation as a Method to Study Managerial Work," *The Journal of Management Studies,* Vol. 7 (February 1970), pp. 87-104.

[20] Myers, Charles A. (editor), *The Impact of Computers on Management,* The M.I.T. Press, Cambridge, Mass., 1967.

[21] Neustadt, Richard E., *Presidential Power: The Politics of Leadership,* The New American Library, New York, 1964.

[22] Papandreou, Andreas G., "Some Basic Problems in the Theory of the Firm," in Bernard F. Haley (editor), *A Survey of Contemporary Economics,* Vol. II, Irwin, Homewood, Illinois, 1952, pp. 183-219.

[23] Sarbin, T. R. and Allen, V. L., "Role Theory," in Gardner Lindzey and Elliot A. Aronson (editors), *The Handbook of Social Psychology,* Vol. I, Second edition, Addison-Wesley, Reading, Mass., 1968, pp. 488-567.

[24] Sayles, Leonard R., *Managerial Behavior: Administration in Complex Enterprises,* McGraw-Hill, New York, 1964.

[25] Stewart, Rosemary, *Managers and Their Jobs,* Macmillan, London, 1967.

[26] Whyte, William F., *Street Corner Society,* 2nd edition, University of Chicago Press, Chicago, 1955.

34

Samuel G. Trull

SOME FACTORS INVOLVED IN DETERMINING TOTAL DECISION SUCCESS

The current vogue surrounding decision making, in the recent literature, most often refers to problem solving under conditions where time as a constraint is either not made explicit or is assumed away. It is assumed that there will be rational behavior on the part of the firm or the individuals who comprise the decision-making centers of the firm. It is also assumed that knowledge of the firm's environment (both internal and external) is either known or knowable.[1]

Organization theorists have pointed out that once a decision is reached then decision rules or procedures are established that act as constraints upon future decisions.[2] Finally, decision making has been extended and formalized with techniques based upon statistical and mathematical evaluations. These often allow a more rigorous approach as well as the generation of useful insights – rather than hindsights – into decision making as a process.

The objective of a manager's decision, undertaken within an economic context, is to maximize a single

---

objective function, (e.g., to obtain a maximum profit for a factory or to operate a department with the lowest possible cost), at the same time being consistent with and subject to other constraints (e.g., service to customers, quality, maintenance of equipment, etc., or other longer-run considerations). From the management frame of reference, the objective is to optimize the reward (however defined) return from all plants within the total system by bringing about maximum gains while sustaining minimum losses.

Furthermore, as mentioned, decision making in the current literature tends to neglect time as an important variable. An exception to this observation is Folsom, who points out in his book *Executive Decision Making* that, at upper management levels where uncertainty prevails in many decisions,

> *... It is often hard to pinpoint the exact stage at which a decision is reached. More often than not, the decision comes about naturally during discussions, when the consensus seems to be reached among those whose judgment and opinion the executive seeks.*[3]

The lack of a measure or index of effectiveness, coupled with a surfeit of alternatives and an absence or reliable probabilities associated with decision variables (even when identified), has led to the labeling of this kind of open or noncomputable decision as heuristic.[4]

Unfortunately, a vast amount of decision making, particularly in organizational administration, falls into the class of heuristic problem solving by virtue of the fact that the overwhelming number of unknowables surrounding the decision dictate the abandonment of a search for a true optimum. Currently there is a trend toward attempting to understand the mechanics involved in such decisions and thereby to increase the chances for effectiveness in decisions of this general type. This is being undertaken by studies directed toward the *process* of decision reaching.

Nevertheless, it would seem that the issue of building improvement into the subjective decision-reaching process encountered most frequently by the executive in a day-to-day situation has not been joined. This lack of attention to an exceedingly important area of decision making most probably stems from (1) the absence of an adequate frame of reference and (2) the somewhat bewildering complexity of the process. However, as far back as the late 1930s Barnard recognized the importance of this problem when he wrote:

> *It seems to me clear that, whatever else may be desirable, it is certainly well to develop the efficiency of the non-logical processes. How can this be done? No direct method seems applicable. This task seems to be one of "conditioning" the mind and to let nature do what it then can. The conditioning will consist of stocking the mind properly and in exercising the non-logical faculties.*[5]

In an attempt to obtain more information about successful decision making, an investigation was made of 100 case examples of decision reaching.[6] The examples studied were drawn from the following general areas: (1) industrial (sales, personnel, production, and financial), (2) military, (3) medical, (4) political, and (5) commercial.

In nearly all the observed situations, the decision process was not explicit. Decisions were reached through an ill-defined procedure that involved a series of interacting events over a period of time.

Examination of the cases showed certain "clustering" of key variables which appeared to be a common feature of the decision-reaching process. These "clusters" of key variables seemed to influence the success of the decision in a causal fashion. This was particularly in evidence when these "clusters" were analyzed in the broader framework of their impact upon the total system and effect upon decision success.

DECISION SUCCESS

Decision success is defined as primarily economic attainment. Also important are the saving of absolute time and survival of the organization.[7]

Successful decisions require more than good decisions. The decision, once made, must be carried out efficiently, so that its effect may be obtained in such a fashion as to satisfy the original problem. A decision of good quality may be offset by poor acceptance.[8]

The Model

Total decision success = *f*(decision quality) + *f*(implementation)

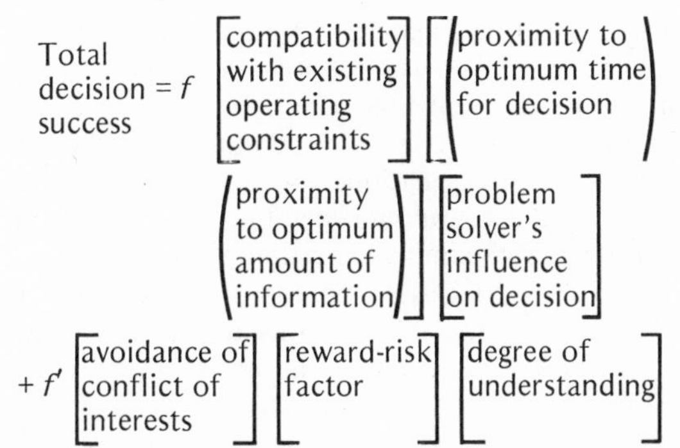

While each of the "clusters" is not mutually exclusive, they offer a frame of reference for the identification and consideration of the most important desiderata in decision reaching. The "clusters" are interacting when integrated into the decision-reaching process. Decomposing the decision-reaching process into block components allows effective analysis of the significant factors which bear upon decision success. In spite of the fact that one is dealing with subjective decisions (lack of information, unspecified value scale, presence of uncertainty, etc.), the more explicit the process of decision reaching, the more feedback may be used in learning to evaluate and improve the system.

Compatability with Existing Operating Constraints - In the cases studied, it was found that the majority of decisions took place within well-defined constraints. When the constraints were known, the executive was actually quite limited in the number of available choices for a successful decision. In addition, it was discovered that prior events and happenings within the organization offered a fund of historical data that defined boundary conditions within which the decision could have been processed. In spite of availability, these historical data were frequently either ignored or discounted. In some instances these data were intentionally passed over as a new innovation or planned changes were instituted. On other occasions they were simply ignored in the expediency to decide and get on with the business at hand.

It was observed that decisions stemming from the same or similar conditions were very often made in the same manner as previously successful decisions. That is, previously successful decisions under similar circumstances were repeated. Occasionally, there were conscious attempts to effect marginal improvements over the previously successful results.

Institutionalized policies of the organizations studied most often presented well-marked channels within which decisions were reached, sometimes resulting in improving a winning combination, when there is no degree of winning, through marginal improvement. In the majority of cases, this previously formulated frame of reference for the decision process was utilized. Two cases were observed where failure to communicate this frame of reference to the operating level resulted in a disasterous decision in one case and a poor decision in the other. The effective preparation, maintenance, and distribution of policy manuals and position guides throughout a firm's organization quite possibly could have simplified the decision-reaching process and reduced the probability of error or inferior results. The distribution procedure acts as a catalyst to bring about the exchange of information and meaning between the individuals involved in the decision.

Proximity of Optimum Time for Decision - In the majority of cases studied, decisions had an optimum time dimension at which the maximum probability for success occurred. The relative success of the decision appeared to be directly related to the point on a time path in which the decision was made.

This observation can be stated graphically where relative decision success varies over a time scale from a premature decision (too early) through a delayed decision (too late). The data surveyed tend to yield values that could be translated into a bell-shaped function with varying kurtosis and range (see Figure 1). A comment on the timing of decision making came from a plant manager who recalled that his general foreman stated, "If you decide now, either yes or no, there is a good probability that either will be correct. If you wait until Monday, either way will probably be wrong!"

As mentioned, the more successful decision was predicated upon the determination of a time path for the decision. Once the time path for the decision is

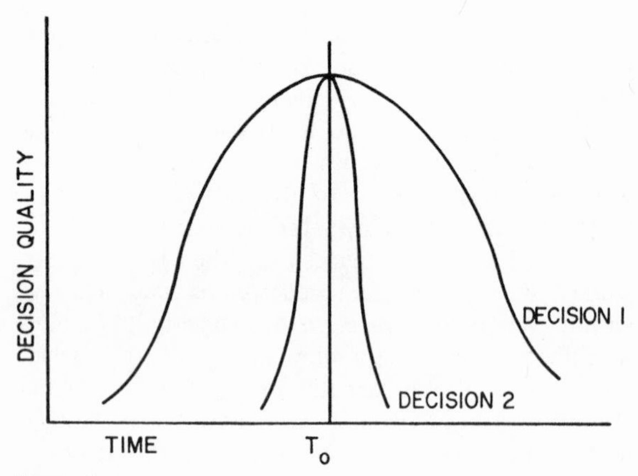

Figure 1

established, the possibility of utilizing time as an ally becomes possible. In most of the cases studied, the ideal point – in time – to make the decision was usually prior to the point where implementation of a decision had to take place. This is the ideal time owing to the fact that a maximum of information pertaining to the decision is theoretically available. The ideal time in practice, however, must be tempered by the fact that the decision maker, acting as an information processing center, is only able to absorb a portion of all the information available.

Several cases in the study offered illustrations of the effect of a premature decision. The decision maker, by failing to map out a time path, predecided or allocated organizational resources for the purpose of obtaining information for a decision that was self-reaching, if given enough time. These observations lend credence to the oft repeated maxim that decisions tend to solve themselves. Recognizing these decisions is one of the skills of management. The criticism of "too much too soon" decision making is applicable only when a framework for the systematic analysis of a time path for decision reaching can be determined.

Converely, on the downhill side of the curve (negative time after optimum time for decision), time is working against the probabilities of any successful outcome.

Exceptions to the bell-shaped time function were observed when decisions shaped a curve which became asymmetrical with a moderate or marked skew to the right or left. Another deviation occurred in an abrupt cutoff condition. An example of this was the execution of a condemned prisoner. In all cases, this abrupt cutoff occurred after the optimum time for the decision (see Figure 2), and no future action by the decision maker was possible. A binodal distribution of the decision time function served to complicate the process. This came about with additional breakthrough information, or from a distinct pronouncement, e.g., announcement of strike settlement, which favored the implementation of the decision.

Generally, the cases pointed up the fact that the human decision maker was either not able or unwilling to frame the decision in terms of the crucial variable of time. Also, it was clear that the decision makers displayed no conscious effort to determine the optimum time for making the judgment, which is one of the most important aspects of any decision. Frequently, management decisions are by crises, as required, or the decision date affects the time when it will be implemented, which is not necessarily optimum time.

Perhaps the insight to be gained from this section of our observations is this: (1) the time path for a decision as related to planning and scheduling should be established, and (2) then the optimum time for the decision might be determined by relating the information processing ability of the decision-making body (the rate at which the decision maker is able to assimilate information) to the previously determined time path for the decision. Once these two factors have been related, it is possible to plan the overall strategy for reaching the decision.

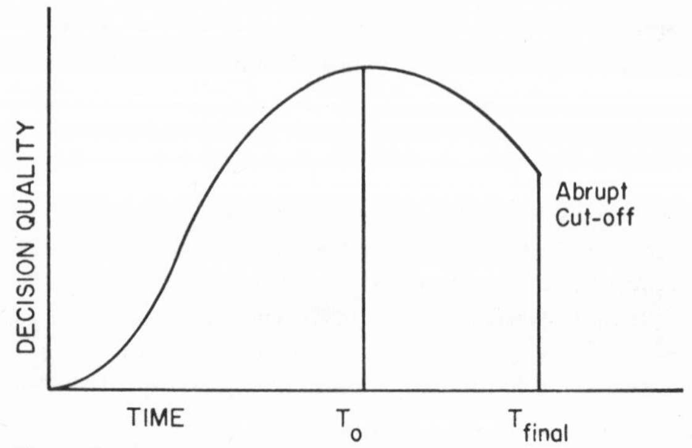

Figure 2

Proximity to Optimum Amount of Information - The determination of the correct amount of data is, or should be, contingent upon the probable total decision reward. Under this rubric, the continuum of the amount of information presents itself. It was observed, in several incidents, that the data acquisition costs exceeded the total decision reward (see Figure 3). In these cases, rather than endure uncertainty, the decision maker had the compulsion to obtain information – the cost of which far exceeded any possible gain. Generally, the decision makers were unaware of the nature of the costs associated with obtaining information. Hence there was very little systematic effort expended toward relating the kind of decision and its probable payoff to data acquisition and the attendant costs.

Figure 4 illustrates two different kinds of decisions.[9] Case A occurs when the system is able to utilize all possible information and there is ample time available to acquire this information. Evaluation time can then continue on until the optimum time for decision. Case B is present when the system is unable to utilize effectively all possible information. This is demonstrated when an aircraft, attempting a landing, discovers just prior to touching down that its relative ground speed is excessive, and it must be waved off to try again. Here any information acquired beyond the capacity for assimilation by the decision-making center brings about dysfunctional consequences. In short, the computational ability of the decision-making center breaks down resulting in a diminished probability of a successful decision.

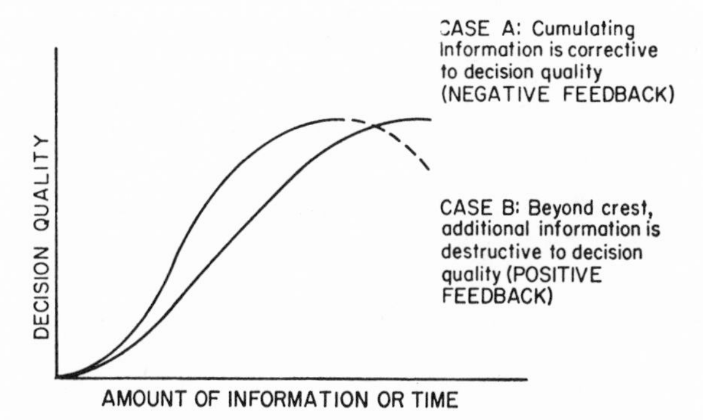

Figure 4

Another factor that tends to complicate the time dimension is implementation time (see Figure 5). Implementation time is most often given by virtue of the quasi-technological nature of decision execution. The communication system and its attendant electromechanical apparatus together with the organizational structure often provide an easily computable constraint. This variable, important in the total decision success function, appeared to behave rather predictably in most cases under observation. It should be pointed out, however, that the decision maker invariably tended to underestimate implementation time.

The Problem Solver's Influence on the Decision - In a totally enclosed decision-making system (one man) there would exist the highest probability for decision success, given a *ceterus parabus* assumption. As more men are introduced into the decision process, the number of communication synapses also

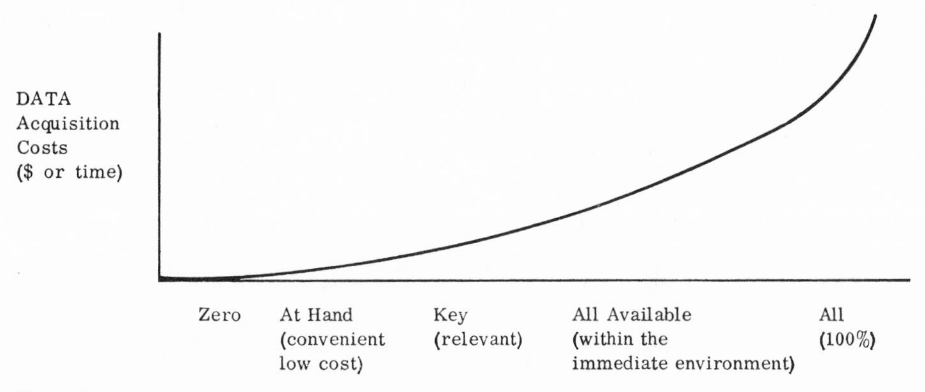

Figure 3

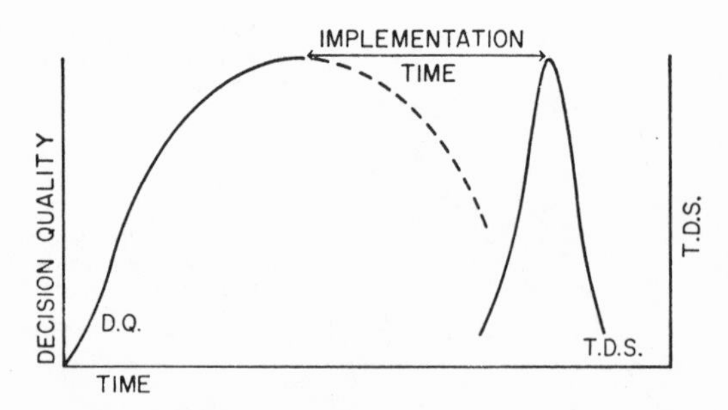

Figure 5

increases and results in a more complex process. Similarly, as the decision process becomes more complex, the decision-making center must exercise proportionately more control or care on the synaptic junctions if the decision is to be successful.

It was observed that if the appropriate level within the organization made the decision, then the result enjoyed the appropriate perspective for success. The adjective *appropriate* applies only within the specific content of a particular decision. It has little analytical value owing to the distinct *ex post* nature that surrounds the term. The significant factor in the process of decision making is how appropriate or relatively appropriate the relationship between the kind of decision and the level within the organization that actually makes the decision is. For example, it was observed that decisions were immediately assigned an unfavorable bias in terms of their ultimate success simply because the "boss" refused to decide – often under his guise of proper delegation – and forced the decision down to a lower executive level.[10]

The perceived authority of the person making the decision tended to have an important influence on the decision-making process. The greater the perceived authority of the decision maker, the greater the extent of informal effort expended by the organization to ensure decision success.[11]

Avoidance of Conflict of Interests - With any decision, there are different individual and organizational goals that either had to be displaced or altered in order to allow the decision process to take place. These might be categorized as follows:

1. Firm (external environmental adjustments).
2. Work group (various pressures occasioned by goal conflicts).
3. Individual (realignment of norms).

The displacement of existing priorities by a new task was one of the more usual and simple illustrations of this category. This required significant changes since it usually involved the reassessment of relative customer worth by the firm, a reassignment of work groups to different tasks, and a readjustment by individuals to allow for their adaptation to an overtime schedule. With any change, there were adjustments. The magnitude and frequency of these adjustments determine in part the acceptability of any particular change by the firm, the work group, and the individual.

Reward-Risk Factor - In discussions of risk and uncertainty it is assumed that the decision maker is rational if the potential gain from his decision increases as the risk of potential loss also increases. In actuality, it appeared that this straightforward theoretical formulation of rationality was ignored to a large extent.

Risk was observed to increase as uncertainty, which stemmed mostly from unreliable knowledge, appeared in the parameters that were utilized to estimate outcomes; however, the potential rewards associated with more risky decisions seldom increased proportionately. It appeared that as uncertainty increased the contingency allowance calculated to offset uncertainty was not commensurate with the possible loss involvement.[12]

Degree of Understanding - In observations of decision reaching, one of the more important variables in the relative success of any decision is the following: the degree to which the individuals involved in implementing the decision understand the basis upon which the decision was made, the means or agencies that are to carry out the decision, and the implications that stem from the decision.

To ideally facilitate these desiderata, the decision environment should be previously conditioned, if optimal results are to be obtained. It is not always possible to obtain ideal conditions; yet a systematic effort directed toward preparation for the decision prior to the actual event was found to be vitally important where marked deviations from accepted procedures were to be instituted.

Eliciting the participation of those concerned with the implementation of any given decision – perhaps the strongest motivational aid – has definite limitations, particularly with respect to the amount of time that is required to achieve. In one case, several individuals who had not taken part in, or become connected with, the formulation of the particular decision were enthusiastically supporting and promoting its implementation. This behavior appeared unusual because, in previous situations with similar conditions, the behavior of individuals not involved in the decision was less than enthusiastic. Upon investigation, it was found that the primary reason for their interest in the decision was their *perceived involvement,* rather than their actual participation. The notable achievement, in this case, was accomplished by bringing the people who would be affected by the decision into discussion groups and by communicating with them by mass media.

Another example occurred in which the decision had been completed, but the implementation procedure was given to a group of subordinates in order that they might design and complete the final stage of the decision. After completing the commission, they voiced the opinion that it was "their" decision as well. Here participation in the implementation phase of the decision process produced a perceived involvement in the decision that substantially enhanced the probability of a successful decision.

This occurs when the leader cites *the* goal and then involves the followers in looking for ways to achieve his goal. For example, the goal may be to paint a room. The leader asks which of three paints they would like best. (Any of the colors is all right with him.) Soon the room is painted, and all feel a successful decision was made. No one even questioned whether the room should be painted, but became involved immediately with implementing the task. Parkinson hints at this type of leadership in *In-laws and Outlaws.*[13]

In some decisions the individuals involved are unable to comprehend the decision or its implications, and consequently the decision suffers. This inability was due to a lack of compatible experience, comprehensive, intellectual capacity, and individual psychological disturbances. Overall, any disruptions or weaknesses in the communication set that surrounds a decision tends to handicap the ultimate success of that decision.

In evaluating the motivational aspect of the decision-reaching process, it was assumed that this dimension involved adequate pursuit and implementation of the decision, and the enlistment of a personal commitment by those individuals involved or affected by the decision. This commitment, although unfeasible in many instances, does represent an aspiration level which is attainable. The ability to establish this as a conditional goal was in many occasions a fault or attribute of the individual decision maker or decision-making center.

CONCLUSION

Within the 100 cases studied, there are several areas that are common to the process in general. By identifying what seem to be the critical stages and elements in decision reaching, the process itself becomes more amenable to a complete and systematic analysis.

A delineated and more explicit model facilitates negative (corrective) feedback. By decomposing the process into components, the procedure of analysis is enhanced through more logical and less cumbersome methods. Information thus gained can be synthesized into the decision-reaching process, with inherent learning and a resultant improvement in the probability of total decision success.

Evaluation time and implementation time are often dynamically integrated into the process in contrast to the absolute time interval used in the illustrations. Here the constant tradeoff for optimum total decision success represents a changing decision-processing program, usually unique to the specific decision.

Decision reaching, varying from traditional decision making, is subjective by nature and includes a high degree of uncertainty. It involves a unique program for the decision process with little or no duplication and with interrelated variables of shifting weight functions. The result is obtained through successive interacting steps which, when completed, allow the ultimate determination. The term – decision reaching – rather than being a neologist's venture, should be useful in the delineation of a specific process. It should likewise set aside a specific area for future research.

REFERENCES AND NOTES

[1] Ward Edwards, "The Theory of Decision Making," *Psychological Bulletin,* Vol. 51, No. 4 (July 1954). This is an excellent review article on theories of decision making.

[2] J. G. March and H. A. Simon, *Organizations,* New York: John Wiley & Sons, 1957.

[3] Marion B. Folsom, *Executive Decision Making,* New York: McGraw-Hill Book Company, 1962, p. 4.

[4] Herbert A. Simon, *The New Science of Management Decision,* New York: Harpers & Row, 1960, pp. 21-34.

[5] Chester I. Barnard, *The Functions of the Executive,* Cambridge, Mass.: Harvard University Press, 1938, p. 321.

[6] Funded, in part, by grants received from Case Collection Program, Harvard Graduate School of Business Administration, Intercollegiate Case Clearing House, Soldiers Field, Boston 63, Mass., and the Schools of Business Administration, University of California, Berkeley, Calif.

[7] A. A. Alchian, *American Economic Review,* 1954.

[8] N. Maier, "Fit Decisions to Your Needs," *Nation's Business,* Vol. 48 (Mar. 1960), pp. 48-50.

[9] It is assumed that in all examples the marginal cost of additional information is equated with the expected marginal product in terms of a better decision.

[10] Perhaps organizations should give thought to delegating authority up to the appropriate level as well as the more conventional – and continually emphasized – formulation.

[11] Often an organization was obliged to attempt to prove a high-status problem solver's answer correct.

[12] Occasionally it was noticed that an organization attempted to account for uncertainty by such means as diversification of supplies and by hedging through subcontracting. Nevertheless, the procedures for estimating decision outcomes under certainty were at least unsystematic.

[13] C. Northcote Parkinson, *In-Laws and Outlaws,* Boston: Houghton Mifflin, 1962.

In some respects a great deal is known about executives in the American political system; in other respects, very little. An extensive and rich literature describes in detail the formal office occupied by the President, governors, and some local executives. In addition, many biographies have been written about and by former incumbents. There are case studies of particular decisions, crises, and brief spans of time during which major new programs were developed. Many analyses of the problems of executive leadership exist. Yet, despite such a substantial body of literature on executives at all levels, it adds up to no meaningful whole. One comes away from it feeling he knows a lot about a given man and his times or about his office, but very little about either the man or his office in relation to other actors who participate in political decisions. An overview or an integrating framework that relates executives to other dynamic elements of the political system seems to be lacking.

John C. Ries, Executives in the American Political System, *Belmont, California (Dickenson), 1969, p. xi.*

PART IX INSTITUTION

. . . does our newly organized [city] state contain any kind of knowledge residing in any section of the citizens, which takes measures, not in behalf of anything in *the state, but in behalf of the state as a whole, devising in what manner its internal and foreign relations may be best regulated?*

The Republic of Plato, *John Llewelyn Davies and David James Vaughan (Translators), London (Macmillan) 1914, p. 128.*

. . . But legislators, and particularly legislative leaders, tend by the very nature of their job to have work attitudes that don't transfer so well to [executive management]. They react rather than initiate, focus on one issue at a time rather than an overall program. They push popular positions rather than unpopular ones, try to work out quiet compromises rather to shape and lead opinion. They like to do themselves, and as a result frequently have mediocre staffs. . . .

The legislative leader rarely focuses on a broad, long-range program. He's always working on the bill coming to the floor tomorrow, getting his troops in line and working out the compromises and deals to pick up missing votes. He's rarely too concerned about the way the bill might or might not fit in with a dozen others in the legislative pipeline.

. . . "The legislator operates in a very short frame of reference. He doesn't plan. The executive must look far down the road, and that's hard to learn."

Alan L. Otten, "Politics & People," The Wall Street Journal, *19 December 1974, p. 12.*

35

James H. Pickford

THE LOCAL PLANNING AGENCY: ORGANIZATION AND STRUCTURE

FUNCTIONS OF A PLANNING AGENCY

. . . Today there is general agreement that the two lines of development – stress on physical development and growing emphasis on the general staff role for local planning agencies – are both necessary for effective operation. Urban planning may be considered, therefore, in part a line activity and in part a staff activity, that is, a mixed line-staff function.

In carrying out its activities, the planning agency is often engaged in acts that are clearly both in the nature of line and staff work. Wherever the agency is integrated into local government, it must perform in this dual capacity. On the one hand, the agency's task is to provide insight into the physical, social, and economic characteristics of the community. It is involved in the work of guiding development, the performance of an operating (line) function. It must

---

also be responsible for the execution of those policies concerning land subdivision and land use by regulating private decisions regarding urban growth. The agency is concerned, too, with numerous public actions that produce growth and development in the urban area through preparation and refinement of plans for transportation, public buildings, public utilities, and recreation areas, urban renewal, capital improvements, etc.

But the agency is also involved in assisting with coordination of the program of public improvements, and is thereby, as a staff unit, frequently drawn into the field of policy making. The agency advises the chief executive on how public improvements conform to the land use plan or to a particular public facilities plan. It is also deeply involved in helping to pinpoint and choose among alternative goals for the community, recommending one course of action as against another, and in advocating specific plans. The staff nature of planning, an indispensable element in the administration of the agency, calls for more than a skilled administrator: substantive knowledge and experience in the area of physical development is essential to the effect performance of the staff role. The major problem, however, is maintaining a requisite balance between line and staff activities.

> *If the planning agency is to make its whole contribution to more effective municipal government and orderly planned urban growth, it must emphasize both its substantive (line) and its administration (staff) responsibility. Preoccupation with the day-to-day pressures of zoning administration or review of proposed subdivision plats leaves little time for advanced planning or for the research that must precede and support long-range planning. Furthermore, neglect of staff advisory and coordinating responsibilities can lead in time to isolation of the planning agency so that its advice is given little heed.*[1]

A local planning program may be classified into seven functions or categories:

1. To establish community development objectives.
2. To conduct research on growth and development of the city.
3. To make development plans and programs.
4. To increase public understanding and acceptance of planning.
5. To provide technical service to other governmental agencies and private groups.
6. To coordinate development activities affecting city growth.
7. To administer land use controls (zoning and subdivision regulations).

The planning program needs a framework of overall, officially approved development policy to serve as both a checklist and directive for the planning agency's activities. Broad development objectives should be established, approved officially, and periodically reviewed (1) to give the government and public a clear sense of the future city, and (2) to give the agency a clear sense of direction for its activities.

The agency must collect and analyze, on a continuing basis, all pertinent data on city growth and development in order to provide a foundation for a rational planning program. The agency must be organized and staffed to analyze and use data as a matter of course in its planning programs. It should be known as the community's central intelligence headquarters on matters affecting community growth and development and should be a depository for and distributor of all relevant data and statistics needed to aid other city departments and private citizens in making informed development decisions.

The central function of a planning agency is to make development plans and programs. These should be based on careful study and analysis and geared to city development objectives, and they should provide a clear guide for governmental decisions in the construction of public facilities and in the regulation of the pace and character of private development through land use controls. Once objectives are defined, the task of the agency is to develop specific plans and programs to help achieve those objectives.

An integral activity of the planning agency is the continuing effort aimed at increasing public acceptance, understanding, and support of the program and the principles of planning. The job of informing residents of the community and public officials about the program will be handled in large part by the staff of the planning agency. This should be recognized as a part of the agency's work program and adequate time allocated to perform the continuing tasks required by public education activities. . . .

Furnishing of technical resources and advice is an important function of the planning agency vis-à-vis other municipal departments, governmental agencies generally, and private citizens. This is a service activity of the agency to which the staff and budget must be adjusted. At the same time, these "short-run" projects should not dominate the operation of the agency. Staff assigned to the development plans and programs functions should not be continually pulled off to work on brush-fire-type projects.

Coordination is a principal objective that is related to many of the activities of the planning agency. Public and private development action must be coordinated within a frame-work of comprehensive development policy, if the planning program is to be effective. The agency should be given responsibility and commensurate authority to seek coherence in the development activities of the municipality and to coordinate decisions made by many private actions and other governmental levels. State legislation often requires that many proposals for public development be submitted to the planning agency for review and comment before action is taken. The agency may be required or directed by the appropriate executive to review development proposals submitted by other municipal departments, special districts, and state and county agencies that are locally based. The principal instrument for coordinating local development activity is the capital improvement program – the tentative list of public improvements to be built within a 5- or 6-year period. The agency may also collaborate with other governments to prepare joint plans or to develop policy on planning matters of mutual concern.

One additional function of a planning agency is the administration of land use controls, fairly and equitably enforced, and based on plans and development policies representative of the public interest.

As previously noted, the major burden of performing these seven functions falls upon the technical staff. The lay citizen commission, however, also has important tasks to perform. In addition to serving as a review and recommending body on current development proposals – zoning amendments, plat approvals, street vacations, and similar actions – the commission is also especially suited to perform four other roles:

1. A representation role on behalf of the public, subjecting planning decisions to citizen examination, by establishing technical advisory committees of informed citizens and officials on specific subjects.
2. A promotional role to stimulate interest in planning.
3. An advisory role to municipal officials on development policies of local government.
4. A coordinative role in working with other public and private agencies to integrate the total governmental planning effort.

Because the typical planning commission spends much of its time on development control administration and other short-run matters, the public tends to see this area as its only responsibility. However, it is important that the public also understand as clearly the total responsibilities of the lay commission. To achieve this goal may require more effort than is currently expended by many commissions. Probably the most effective method for the commission to find the time is by relinquishing to the staff as many routine duties as possible.

MAJOR TYPES OF PLANNING AGENCIES

There are four major patterns of organization of local planning agencies in the United States: (1) the independent planning commission with or without staff, (2) the planning department, (3) the community development department, (4) the administrative planning agency. There are, of course, variations or blends within these general categories.

Independent Planning Commission (Figure 1)

The commission, stemming from voluntary activities of earlier citizens' groups, is intended to foster an objective and neutral attitude in its recommendations. Its advisory role to elected officials is determined by virtue of its independence from the local administration.

Members of the commission are usually appointed as specified by the state planning enabling legislation. Appointing authority usually lies with the mayor and city council or the county governing body. Members are selected on the basis of their demonstrated interest in the community and tend to come largely from its business and professional segments. They serve overlapping terms for a number of years that extend beyond the elected term of office of the appointing authority.

A large number of commissions serve small municipalities, and they may function without a professional staff. Such commissions operate outside the mainstream of government and are mainly concerned with the administration of zoning and subdivision ordinances. Few of these commissions attempt to coordinate development or to initiate and prepare general or functional plans, zoning ordinances, and subdivision regulations. They may rely instead on consultants for such services.

Larger communities utilizing the independent or autonomous planning commission usually have a full-time technical staff. The commission is responsible for the assignment and review of the work of its staff. Formal lines of communication between the planning staff and the chief executive of the municipality usually pass through the commission. The planning staff is theoretically outside the basic structure of the executive branch, and its relationships with personnel in the operating municipal departments may be distant, unless special attention is given to building and maintaining liaisons.

Even though many persons feel that the independent agency is limited in effectiveness, there are communities where the commission still is the most workable device for achieving an effective planning program. These include small communities, unable to hire full-time staffs because of limited financial resources, where good candidates can be persuaded to serve, and also local governments with a weak or dispersed executive and administrative structure where the commission members themselves may be in a position to force changes on behalf of planning.

Planning Department (Figure 2)

The major difference in organization between the independent planning commission which employs a staff and the planning department is that the technical staff of the latter is integrated within a department that is directly responsible to the chief executive of the municipality. The staff is not appointed by, nor subject to, the supervision or control of the planning commission. The commission exists but performs largely an advisory function to the department staff. In some communities, no planning commission exists. The advisory role of the

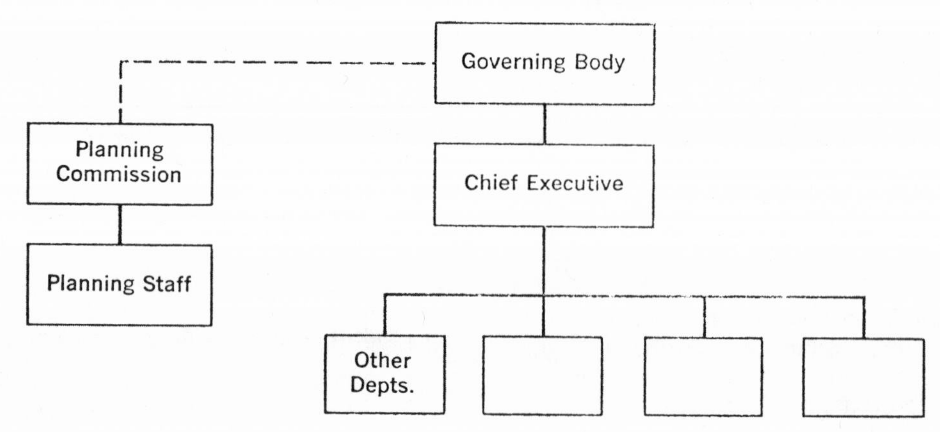

Figure 1
Organization chart, independent planning commission.

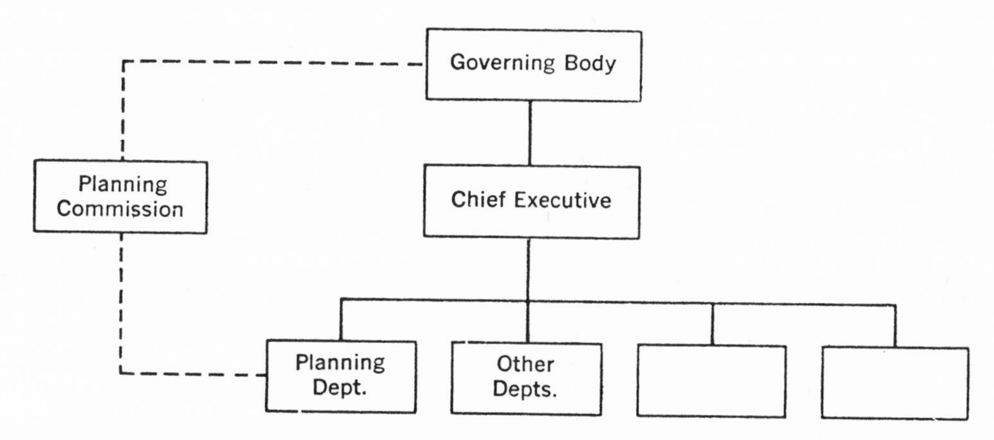

Figure 2
Organization chart, planning department.

commission is performced by special ad hoc committees appointed to review and comment on certain problems under study.

The department form of organization enables the chief executive of the municipality to assume clear authority to coordinate physical planning policy with other kinds of municipal policy covering many aspects of the administrative operations of local government. The planning staff has a direct channel to the chief executive and has a strong and influential sponsor for its recommendations. A departmental form of organization is most effective where a strong executive form of government exists – the strong mayor-council, or the council-manager form of local government.

Several cities, especially those with strong urban renewal programs, have sought better coordination between the planning agency and the executive branch by interposing a "development coordinator" between the chief executive and the planning agency. The cities of Philadelphia, Providence, New Haven, and San Francisco have established this arrangement. In a few council-manager cities, another shift has weakened the independence of the planning commission. In these cities responsibility for directing the planning operation is given to an "assistant city manager for planning." One version of this form of organization may be found in Berkeley, California. Two assistant city managers' positions were created: one in charge of plans and programs (with the planning staff reporting directly to him); and the other assistant in charge of administration with responsibility for seeing that approved plans and projects are completed in accordance with established schedules.

The Community Development Department (Figure 3)

A few cities have shifted the planning function toward the chief executive by combining into a single administrative unit the departments that are broadly responsible for development. A combined planning and development agency – the community development department – has been created to combine planning, urban renewal, and, on occasion, code enforcement activities.

Cities have tended to assign the urban renewal function to a separate department. Coordination of the activities of the planning and urban renewal agencies often proves difficult to achieve. Since both agencies have a common objective of improving the physical fabric of the community, city officials have taken steps to merge the tasks of overall planning, rebuilding, and renewing the city within a single department. This permits renewal to be based on broad planning considerations and also enables the more specific renewal projects to be fed back to the planning staff and to test the plan. This type of consolidation has taken place in at least 15 cities, among them Milwaukee, Tucson, and Evanston,

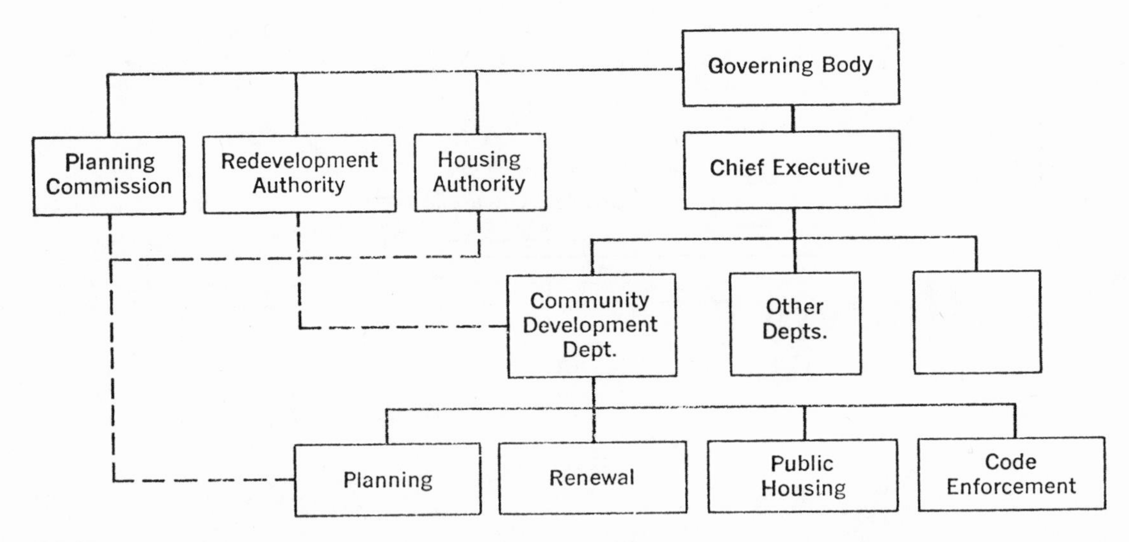

Figure 3
Organization chart, community development department.

Illinois, within the past several years. In most instances the planning commission is still retained as an advisory body but with considerably reduced responsibilities.

Administrative Planning Agency

Nearly 30 years ago, Rexford G. Tugwell proposed that planning be considered as a fourth branch of government because of its importance in the administrative structure of government. Planning would thus be raised to the level of the legislative, judicial, and executive arms of government.

Two major planning agencies have been organized somewhat along the pattern proposed by Tugwell: the New York City Plan Commission and the Puerto Rico Planning Board. Tugwell played a major role in both agencies, serving as the first chairman for the New York agency and later, as governor of Puerto Rico, establishing the territory's first planning board. While neither agency has assumed an importance equal to the traditional three branches of government, their organization patterns are quite different from those of planning agencies in other parts of the United States. The statutory framework in Puerto Rico, as well as the procedures followed by planning agencies, is considerably different from practices on the mainland, since the commonwealth is free from many of the specifications imposed on governmental jurisdictions here.

The administrative planning agency, as its name implies, is essentially an independent regulatory agency. The Puerto Rico planning board has most of the characteristics of an administrative agency – it makes, adopts, and enforces its plans and regulatory ordinances. It also has quasi-legislative powers. The New York City Plan Commission, however, more nearly resembles planning agencies in other cities.

One distinguishing characteristic of the administrative planning agency is that the chairman of the agency is a full-time, salaried employee, who acts as the chief administrator of the technical staff as well as chairman of the policy-making commission. This type of planning agency offers some advantages in dealing with ordinances pertaining to regulation of land use, where the burden of administering such regulations is fast becoming too great for a nonpaid citizen group to handle.

REFERENCE

[1] *Local Planning Administration* (Chicago: International City Managers' Association, 3rd ed., 1959), p. 52.

36

Robert A. Walker

THE IMPLEMENTATION OF PLANNING MEASURES

It is a common error of technicians to confuse what is technically possible with what is politically possible. In city planning, as in other fields of human relations, our specialized skill has far outstripped our ability to translate it into governmental action. Thus it is essential, in talking about the implementation of planning measures, that we distinguish clearly between the determination of what is technically possible, which is the art of the planner, and what it is now possible to do, which is the art of the politician and the administrator. Much of the confusion that has attended the discussion of this problem in the United States stems, I believe, from the failure to make clear this distinction.

Organization for city planning in this country has confused the role of the planner, on the one hand, with that of the politician and administrator, on the other. Rather than subordinating the expert to those whose function it is to determine public policy, the effort has been to make the planning agencies independent of these officials. We have given them distinct legal powers to make it difficult for their recommendations to be overruled. They have been given authority to go to the public with whatever proposals they desired, whether or not these might conflict with the views of responsible public officials. Members of planning commissions have been given overlapping terms, so that a newly elected administration could not quickly change the composition and outlook of the commission. We have, in a word, tried to insulate the planning function from the other aspects of governmental administration. It has not worked well. I think the reason is fairly clear. It has not worked because official planning agencies are dependent upon public funds. These must come from city councils and chief executives, and if planners embarrass these officials, they can remove the source of irritation simply by eliminating their budgets. . . .

I mention these things not to be critical of American planning, which I think has many notable accomplishments behind it, but to indicate that I think we have misunderstood the true nature of the problem of implementing planning measures. Planning can be carried on, as an official governmental activity, only within the normal limits of political and administrative action. It is not enough to say, as is commonly heard, that we must "sell" planning to the public. We have sold it as a general idea. The creation of some 1,000 city planning commissions and passage of over 1,300 zoning ordinances testifies to its public appeal. The real difficulty is that "selling" the idea of planning and getting specific planning measures into effect are two different matters. A few proposals, like zoning, can be dramatized and made the subject of widespread public discussion. Most measures, however, will be carried out only if they influence the daily decisions of administrators and politicians. It is they who must be sold in the first instance.

Our whole technique of making plans and trying to put them into effect has tended to confuse a citizens' movement for civic improvement and the function of government. As a result, we have neglected some of the basic facts about the nature of political and administrative action. The tendency, stemming from the early success of the Plan of Chicago prepared by Daniel Burnham, has been to view a plan as something done once, adopted officially, and then "carried out" over the years. Unfortunately, the Chicago experience has misled other cities. The Burnham plan was carried as far as it was because of the unusual leadership and energy of a few distinguished citizens over a period of years. It called for extensive bond issues and public expenditure which only an energetic and costly propaganda campaign could persuade the public to accept, and it was rooted in a social and political condition which is rapidly waning – i.e., the availability of wealthy men with the leisure and sense of public service which leads them to give generously of their time to civil affairs. Whatever the reason, the later efforts of American cities to imitate the nature and methods of Chicago planning have seldom succeeded.

Today the planned development of a municipality can be realized only if there is a paid staff to do the planning and all the many programs and activities of the city government are directed toward the desired goal. This means, first, that those who make the important decisions must be convinced that the plans are good and that they are politically feasible. It means also that the specific day-to-day decisions which administrators and city councils make must be considered in the light of the effect which they will have on the plan. This will not happen unless planning is accepted as important and as an integral part of governmental administration.

As anyone knows who has had experience with administration at the level of top policy determination, the question of whether or not a proposal or decision conforms with a predetermined plan is only one of many questions which must be answered before action can be taken. Decisions at this level, if they are of any importance, always involve the weighing of many influences – the numerous pressure groups which may be interested, the recommendations of technical departments of the government, the reaction of the press, the possible loss of a valued subordinate if he is overruled. These are but a few of the factors which enter into an important governmental decision. It should be clear, therefore, that unless there is someone constantly available to defend a long-range plan, or to modify it to meet changing conditions, the plan will be gradually submerged. The influence immediately at hand, particularly the aspects which are vigorously presented by individual department heads or supported by interested civic groups, will overwhelm the more remote objectives of the planning scheme.

The importance of the administrative process just outlined for planning have been obscured by the fact that much of the planning in American cities has been done by firms of professional city planning consultants or landscape architects. Without attempting to evaluate the technical quality of these plans, it is quite clear that the implementation of such plans has been weak. The reason, I believe, stems from what has just been said. After the plan prepared in this manner is presented to the city, and the firm has withdrawn, there is usually no high-ranking municipal officer retained to defend it or modify it to meet changing conditions. If any staff at all is retained, it is often a younger technician with neither the experience nor the status to make his counsel felt in the top levels of municipal administration. His attachment to a semi-independent planning commission, rather than to the hierarchy of municipal administration, has further weakened his position. Planning cannot hope to be truly effective in modern city government unless a competent staff, under-ranking official with direct access to the top policy-making officials is employed. The "planning point of view" must be presented to the mayor or city manager as vigorously as are the recommendations of the director of public works or the viewpoints of the downtown merchants' association.

To be fully effective, also, city planning must take into account all significant aspects of city growth and governmental services. This fact has been gradually realized in the United States, and the scope of planning has broadened as municipalities have widened the range of public services. Early plans included little more than proposals for public buildings, streets, and parks. Later plans concerned themselves with transit problems, zoning, subdivision control, water supply, and occasionally schools and libraries. In the decade following 1930, the depression and the availability of workers from relief roles led planning agencies into the field of housing, principally through a nationwide program of real-property inventories. World War II, with its violent shifts of population to congested centers of war industry, again focused attention on housing and transit problems. The tendency has been, therefore, for planning to give more attention to the specifically social and economic problems of urban living rather than to confine itself to designing the physical layout of the city.

Postwar developments are continuing this trend. Far-reaching urban redevelopment and public housing legislation have heightened the interest of planners in slum clearance and housing. The industrial and population changes brought about by the war have made clear to an increasing number of communities that they can plan for the future only on the basis of adequate information about the economic and industrial foundation upon which they rest. . . .

Despite the progress which has been made toward a truly comprehensive concept of planning, there has not been widespread acceptance of the idea that planning commissions would take the leadership in such fields as public housing and privately financed urban redevelopment projects.

On the other hand, it is perfectly clear that projects of this kind are major factors in modern city planning. This fact is given some formal recognition in the vague language which one finds in both state and federal housing and redevelopment legislation to the effect that projects "should conform to the plan for the community." But such language means little in practice unless there is in fact a plan and a planning agency which is taken seriously by those who make the vital decisions about housing or slum clearance. It is seldom indeed that these conditions obtain. On the contrary, both public housing and urban redevelopment authorities are being added to the already confused structure of municipal government in many of the major cities. The result is that the public, city officials, and the members of these agencies are all in varying degrees confused as to just what the relationship between these agencies, on the one hand, and the planning commission, on the other, ought properly to be. In Chicago, as one example, the mayor found it necessary to appoint a coordinator in his office to help develop a consistent program among them.

The situation in housing and slum clearance has been mentioned because it illustrates the fact that the implementation of planning is handicapped by lack of acceptance of the idea that planning should provide the framework for community development, and hence that other municipal services should adapt their activities to it. There are a number of reasons why planning is not accepted in this sense and they cannot all be discussed in this brief paper. Three difficulties should, however, be mentioned.

One limitation is the narrow view which many planners and planning commission members take of their own field. This is likely to be particularly true of the citizens who serve on the commissions themselves, for as I have indicated below the professional group of planning technicians has been broadening its horizons in recent years. Older concepts of planning, concentrating on zoning and problems in architectural or engineering design, remain widespread, however, and are a major element in severely limiting the leadership which planners are prepared to take in community development as a whole. Zoning is particularly responsible for this, because of the heavy burden it commonly places on planning agencies to consider applications for exceptions or "variances" in the uses permitted by the zoning ordinance. Only a few cities have, like Los Angeles, California, provided separate bodies for appeals of this kind. In most cases they are heard by the planning commission. As a result there is all too little time left for planning. Unfortunately, too, many planning technicians seem to accept this state of affairs or to become preoccupied with specialized design problems. . . .

The second handicap to leadership under which planning agencies are laboring is the widespread custom of appointing successful businessmen, prominent lawyers, representatives of the construction industry, engineers, and architects to planning commissions. A decade ago these groups almost completely dominated the commissions. While no recent detailed study is available, my general observations lead me to believe that labor leaders, social workers, and persons of similar interests are still scarce among commission members. Thus the political and social outlook of the planning commissions is often definitely conservative. It follows that the more liberal groups in the community, particularly those concerned with promoting public housing and the newer social service programs, lack confidence in the leadership of the planning commissions. This situation may change in time, but the present fact is clear.

A third factor which limits the effectiveness of planning is that relatively few public officials as yet think of the planning agency as something essential to day-to-day administration. The organization and content of American planning have militated against such a view. As has been suggested, only in recent years have planners begun to interest themselves in the role of planning in municipal administration. . . .

Mayors and city managers, however, are only gradually coming to think of the planning staff as an aid to them in integrating and coordinating all municipal activities. Many of them feel the need for more systematic planning, better coordination, and closer integration of complementary programs, but they have not yet come to think of the planning staff as serving them in this capacity. The reason has already been indicated – the history of American planning has, until recently, been one of stressing independence of action rather than assistance to incumbent officials. The changing attitude within the planning profession will take time to make itself felt.

I believe the most important step ahead in making planning effective is the education of top administrators to think of planning as essentially their responsibility, and to regard the planning staff as a group of specialized assistants to aid them in carrying out this responsibility. The function of the planning staff is to keep before the administrator a total picture of the community. It should help him to see the city as a place where human beings live, earn a livelihood, seek recreation, and raise families. It should keep before him the economic and social facts which shape the city today, and are likely to guide its destinies tomorrow. Above all, it should provide a pattern of desirable future development with which the myriad proposals and decisions arising from day to day can be compared.

Thus the planning staff must be able to help the chief executive modify the enthusiasm of particular municipal departments for their own projects, if such projects warp a pattern of sound development. It must help him meet the demands of pressure groups, by helping him present in the strongest terms the interests of the whole community as against those of any particular group. As Chester I. Barnard has said, in his excellent book *The Functions of the Executive*, "seeing the whole" is one of the most difficult executive functions. Our need is to get administrators to recognize this fact, and hence to make use of assistance in grasping the whole. Planning cannot hope to rise much above the ability and foresight of those who must make the most important decisions affecting the future of the community.

It must be stressed, in thinking about planning in this light, that the function of the planning agency is to advise and not to propagandize. If a planning staff takes its proposals to the press and the public, in its own name, it will become embroiled in the political turmoil which these are likely to create. An official planning agency, dependent upon annual appropriations, must keep out of politics if it is to survive. As I said at the outset, if it embarrasses elected officials by going beyond policies they are prepared to support, appropriations will disappear. On the other hand, it is essential that interested citizen groups maintain an active interest in planning and call upon elected public officials for an integrated program of future community development.

My point here is that the basic responsibility for planning must be with those having political responsibility to the community. The present situation in the United States, whereby city planning is carried on by citizen boards without political responsibility, is what makes it so frequently ineffective. Elected officials do not feel that planning proposals are part of their program to defend; they do not feel an obligation to make their official actions consistent with such proposals. Thus the gap between planning and its implementation is a tremendous one. It will remain so until planning, like other municipal services, is viewed as primarily the concern of politically responsible officials.

. . . [It] has been suggested that planning operates against an unfavorable psychological background. I believe this is true in several respects. One is that the whole idea of governmental planning is viewed with some apprehension by citizens of democratic countries, because they think it will mean arbitrary direction of their individual activities. The idea that "planning" should be insulated from "politics" lends credence to this fear, for the political process is essential to maintaining democratic control over governmental action. The acceptability of planning proposals must be fought out in the political arena along with all other proposals for government action. If they cannot stand the test of politics, they have no place in democratic government. The quality and far-sightedness of the plans which the public will accept is, like all other political decisions, a measure of the level of public intelligence. It follows that educational efforts on the part of civic groups interested in better cities are a vital necessity. Public understanding of the purposes of planning is essential, if a scheme of planned development is to withstand conflicting pressures and arguments.

Another aspect of the unfavorable psychological background of planning, at least as it has developed in the United States, is the tradition that a "plan" is something prepared once, accepted by the city fathers at a ceremony, and then filed for future reference. This leads the public to assume that the city has a plan, and that developments will follow it. The psychological effect is one of lack of concern, or a vague feeling that all is well and that the average citizen has no further concern with the problem of planning. It is only after the "plan" has long gathered dust and someone discovers that it is hopelessly out of date that a new burst of energy is likely to revive general interest in planning. This may be called the cyclical approach to planning. It is seldom effective.

A final aspect of the psychological problem is that "practical" administrators and planners seldom understand one another – they do not "talk the same language," as it is commonly expressed. The tendency is for planners, full of vision for the future, to feel that other departments of government are peopled with narrow-minded technicians myopically concentration on their individual problems. Administrators and technicians, for their part, are likely to conclude that the planners live in a never-never land where technical problems, budget limitations, and practical politics do not exist. As I have indicated, part of the planning profession in America is striving to overcome this barrier to cooperation and the road to better understanding lies fairly well defined.

The principal administrative obstacles to implementing specific planning proposals have, I believe, been indicated. These obstacles can be removed only as the planning staff enters into the whole process of day-to-day administration, advising on all important decisions affecting the future of the community and aiding the chief executive to keep a picture of the whole before him when beleaguered by advocates of partial measures. When this concept of the planning function gains acceptance, administrative obstacles as such will largely lose their present meaning. The remaining obstacles will be those with which all public officials must deal – different views to be reconciled, personality conflicts to be considered, jock-eying for influence to be counterbalanced. But if the planning staff operates as part of the chief executive's organization, the mechanism for overcoming these obstacles is fairly clear. If the planning agency is detached, the obstacles can be practically insurmountable.

. . . It has been suggested that failure to appoint radicals and others of more liberal political views to planning groups has been a serious limitation. As I indicated earlier, I agree that the tendency to appoint more conservative elements of the community has interfered seriously with the acceptance of the planning agency's leadership in housing and related fields. The political attitudes of those who get appointed to planning commissions is, however, simply a matter of the political climate of the community. Planning must be adjusted to that climate, or the resulting proposals will not be accepted. Thus the problem is misconceived if viewed as one of simply appointing "radicals." It is better conceived as one of keeping planning "in tune with the times," of permitting elected officials to determine the composition of planning agencies to suit their needs and to appoint directors of planning with whom they can work cooperatively. Our custom of overlapping terms for commission members, and a semi-independent status for the planning agency itself, has militated against this type of adjustment and shifts in political tides are reflected but slowly in the composition of planning commissions. This is exactly what those who designed this pattern desired, but they overlooked the fact that chief executives are not likely to find much in the budget for agencies with which they are markedly out of sympathy. And it may be added that since political fortunes measure public sentiment, the composition of planning agencies must either reflect changes in political outlook or they will find their foundations in public support eroding from under them.

Legal obstacles to carrying out planning measures are real, but I think generally overrated. Use of the police power, through zoning, to guide main lines of urban development has been firmly established in the courts. As a matter of fact, I believe it might be argued that the courts have been at least as willing as most planning agencies to venture into such difficult fields as retroactive zoning and the removal of nonconforming uses. Zoning of this kind was upheld as far back as 1915 in the Supreme Court of the United States.

The opposition to this kind of control over the uses of private property lies more in the political than the legal field. The power of eminent domain remains expensive to invoke, and excess condemnation is still not widely accepted in the United States. . . . These things are important, but the restrictions thus imposed are relatively minor as compared with the great possibilities for accomplishment which existing governmental resources and authority present. The more urgent problem is whether existing programs and authority are to be used to achieve well-planned or poorly planned objectives for the community. At present, we suffer more from poor planning and coordination than from inadequate legal authority.

The economic obstacles to planning are substantial from two points of view. First, entrenched economic interests will oppose planning schemes which threaten those interests. And, second, those with taxable incomes are likely to oppose increased expenditures for government services. Thus holders of slum property, hoping for high prices in some commercial development, may oppose slum clearance projects which involve sale or condemnation of their land at prices based on present use. The higher-income group may oppose the expenditures entailed in the construction of public housing on the site. Once again, it is essential to look upon planning as an aspect of municipal government, and as such subject to all the vicissitudes of political influence. The problem is one of marshaling popular support for deserving projects. Politicians and top administrators should be held responsible for the decisions made. This means active civic groups behind planning, on the one hand, and good planning advice for the public official, on the other. It does not mean that the planning staff, as public servants, should be expected to enter the political arena and joust with either Right ot Left.

This can lead only to annihilation. In the long run, economic interests will be compromised with innumerable other interests in the community.

In conclusion, then, I hold that the implementation of planning proposals must be sought in the actions of those public officials who carry major responsibility for municipal services. Planning, to be effective, must influence the decisions and actions which they undertake from day to day, month to month, year to year. This means that the planning staff must occupy a place in municipal organization close to the top officials of the city, and its function must be seen as one of advising. Much such advice will be confidential. Officials must be free to take it or leave it. If the planning staff goes to the public to bring pressure to bear on responsible officials, it is inevitable that the latter will reduce it to impotence in time. The measures that originate with planning agencies should, like all public policy, run the course of the political and administrative process. Where free men share political power, competing ideas and compromise will leave their marks on the final product. This is inevitable and, until planners are exempt from human frailties, it is eminently desirable.

We need new kinds of institutions. Existing organizations can obviously do much of value. R&D firms, commercial management consultants, university based urban studies groups, client-funded think tanks, consulting engineering firms — all have a legitimate and useful role to play . . . if urban services are to be made more effective and urban policies more rational to a more than trivial degree, then two conditions must be met: (1) analysts must provide advice that springs from a much richer body of theory and of experience; and (2) analysts and researchers must be placed in a novel kind of relation with officials. The analysts must be independent enough to be critical of city policies, insulated enough from daily operational concerns to work persistently on underlying problems, but close enough in their working relations with city agencies to produce recommendations that are timely, realistic, and usable. This combination of conditions is not well met within traditional organizational forms.

Peter L. Szanton (The New York City-Rand Institute), "Systems Problems in the City," Operations Research, *Vol. 20, No. 3, May-June 1972, p. 471.*

37

Rexford G. Tugwell

THE FOURTH POWER

When historians look back, after several decades, they may be able to see how a directive power offered to range itself alongside the executive, the legislative, and the judicial. If, by then, it has developed into a fourth division within our governmental system, there need not have been at any time the theatrical recognition which came to the executive out of the administrative futility inherent in parliamentary government during the eighteenth century. The process can be evolutionary and adaptive; it can be, that is, unless it is deliberately so delayed that opposing physical and social forces reduce the American state to relative ineffectiveness. If this last should happen it would be sufficiently dramatic and obvious; but it would not result in the development of a fourth power. For the whole system would either be subjected to a foreign executive or submerged in a chaos out of which anything might emerge — anything, that is, except institutions with fundamental provision for the participation of every citizen after his sort, which is, after all, the democratic *sine qua non*. . . .

Selected from Planning and Civic Comment, *April-June 1939, Part II, 1-2,4,6,11,16-17,26,30-31.*

The duties to be undertaken and the problems to be solved, even with the restricted American view of what is properly governmental, are more weighty and difficult than ever before. The necessities imposed by this circumstance, it must be insisted, make simple planning, at least, inevitable. Regard, for instance, the growth of the federal budget or of municipal budgets in recent years. This is some sort of index to responsibility. And if the percentage of those budgets which is devoted to duties thrust on government (directly or indirectly) by technical change is measured, it is apparent that the whole growth – and perhaps more – is of this sort. And government has hardly begun its extension into industry. It is not that government has "gone into business," as we say, extensively. On the contrary, one reason for the recurrent fiscal troubles of government is the prevalent unwillingness to have anything done publicly for which an adequate charge can thinkably be made. There are wanted, even by most tax payers, only such extensions of public service as are unprofitable. Revenue has, therefore, to be got by taxation, a kind of price which is universally disliked; it is so unpopular, indeed, and the demand for expansion of non-paying activities is often so great that administrators are forever tempted to unbalance their budgets far beyond the amounts put aside for capital-investment. . . .

. . . If one set of men is always making problems and another set always having to face their consequences, and if they are responsible to antagonistic principals, the situation may well get out of hand. Indeed it has. The harassed executive is right who finds that his problems increase more rapidly than the instruments for their solution. His solutions are really only to be found in a diminution of his problems – particularly those deliberately created for him as an incident to irrelevant private conflicts or in the evangelical disciplining of dissenters from either 100 percent socialism or perfected individualism. . . .

The competitive system, as a *system* – an automatic regulator – has failed. The years since the Great War have seen the intensification of strain, the perfection of instruments for communication, for transport, for measurement, the final victory of scientific management, the making available of marvellous new materials in profusion. And the national income is less at the end than at the beginning. It may be that it cannot be sustained even at that level except by a system of deficit financing which will contribute continually to class antagonism. The truth is that the system of individualistic and uncoordinated businesses is one which cannot operate successfully in an advanced technical system. It is suited only to an age of horse locomotion, of communication by post, of heavy materials, clumsy design, and an ignorant personnel. . . .

A plan for an industry, a city, a nation, is not something which can be experimented with in the old sense. Much more is involved – more people, more property in a wider space and over a longer time. Damage is done by mistakes which may be irreparable. But there is another consideration. The plan or policy cannot be built up from constituent units. It has to grow out of a concept of a functioning whole. An industry cannot place its plants, warehouses, outlets, sources of materials without relation to each other, and it cannot place them without relation to all other related activities: finance, insurance, communication, substitute goods, tariffs, and the like. A city cannot provide for schools, fire protection, police, sewers, water and light, and all its other services except through what has come to be called a "Master Plan" implemented by control of the capital budget.

The planner faced with problems of this sort in industry or in government is forced to think from the center out, to use a concept of the whole which will comprehend the parts, to have in mind a vast complex of meshing arrangements each of which has relation to all others. None, of course, can undergo experimentation without affecting all. Change becomes a serious matter, one for reference to a board of directors or to a planning commission, and safeguards are thrown about the process to ensure deliberation and the exercise of a judgment which includes the whole. . . .

. . . Everyone knows that efficiency in industry has progressed infinitely further than it has in government in spite of strenuous attempts to prevent or to break up integration. And everyone knows that city government has progressed much further, in spite of frequent corruption, than has the federal government. Indeed, our central government, faced with the most gigantic of planning tasks and with the immediate necessity of preventing the disintegration of society, possesses only the most rudimentary mechanisms for the purpose. Is it because of a written constitution which has often been too literally interpreted; is it because the natural divisiveness of a legislature allowed wholly inappropriate powers has prevented change; or because industrial interests, intent on their own profitable evolution, have deliberately kept government weak in their own interest; or, again, because the federal government has been kept more closely under the scrutiny of moralists, educators, and others who were insulated from the evolution of institutions and who lingered in a half-imaginary past from which they sought to prevent departure? Whatever the reason, it is the supreme political tragedy of our time that the central government should have suffered an arrested development. The instruments of wholeness are not ones which can be invented and perfected overnight. They require long preparation and maturation in a period when time is the one thing lacking. . . .

. . . The suggestion that a strengthened executive would be sufficient can be looked at more clearly. If he had the full powers which belong to his office and are necessary to its satisfactory operation, other defects would appear. They can be seen now in some cities and states. There is no denying the fact that democracy frequently turns up irresponsible demagogues with regularity as elected executives; and even that corrupt and venal candidates sometimes have a temporary success. Not all American presidents would have seemed as adequate as they did seem if their duties had been more exacting. A power is needed which is longer-run, wider-minded, differently allied, than a reformed executive would be. This new agency would need to be severely hedged about with limitations on qualification, the persons chosen would need to be given longer-term appointments than any other except judicial officials, but with the canons of selection carefully worked out, a body useful to democracy and not farther removed from its rewards and penalties than would serve to resolve its worst paradoxes and to protect it from itself ought to be feasible. But it would have to be beyond and independent of the executive almost as certainly as the legislative. . . .

[Such a directive power] is subject to much more rigorous limitations than might be gathered from what is said about planning by representatives of the other powers of government who recognize so few limitations that they find difficulty in appreciating the situation of a power which by its nature is subject to the control of existent fact and circumstance. If the directive is examined in a detached way, it is seen at once that it cannot become an *arbitrary* regimenting power, but must always be ruled by the necessity for deliberately gathering up wisdom from wherever it may come, and for applying it under the most strictly given conditions. This gathering-up process can only be accomplished by a rigorously fixed procedure of expert preparation, public hearings, agreed findings, and careful translation into law – which are in turn subject to legislative ratification. The directive has an advantage over the executive from not having to operate any organization, over the legislative from not representing any faction or region, and over the judicial from dealing with a volume of fact rather than a volume of precedent.

The margin of safety which the community possesses in entrusting power to the directive is widened by its persistent orientation to the future, a future discovered by charting the trends of the past through the present. And this projection is not subject to opinion or to change as a result of pressure from special interests. In this forecasting of the shape of things to come, it can succeed, aside from maintaining the most honorable relation with facts, only by possessing and using the most modern techniques for discovering them. It thus has an interest in progress and in modernization which is quite different from the traditional interests of the other powers. The discipline of fact is a more impressive one than the discipline of legal ethics or even of a watchful constituency.

All this is of the nature of theory at present, since there are few instances in which governments of any sort have admitted the directive to effective status. It seems clear, however, that if the directive is permitted to evolve these will be features of its operation. It may thus establish a genuinely social policy, as contrasted with private policies, dictated by contemporary resources, techniques, and circumstances rather than by political expediency; tuned to the universe, the continent, the region, and the times, rather than to an imaginary environment in some past Utopia for speculators in private advantage. It will not be pursued because it suits a whim, a prejudice, an economic interest, or a political gain. It will be distilled with modern devices from the then controlling conditions for the success of society. It will take account of all there is to work with and allow itself to be guided only by the interests of all there are to work for. It appears to be the best way, in a modern society, of carrying out the brave commitment made in the preamble to the American Constitution.

THE TUGWELL PROPOSED CONSTITUTION

Tugwell contends that the division of authority built into the present Constitution stifles creative action and precludes accountability, and that the sectional interests it established in the eighteenth century block the national interests and needs of the twentieth century. Tugwell's constitution is designed to alleviate what has been felt by some to be the inability of the 1787 Constitution to respond to twentieth century needs and institutions.

Planners will readily embrace Tugwell's recommendation of a more regional concept of boundaries. Twenty republics based on population – each having 5 percent of the total – would replace the existing 50 states, whose boundaries are historically and politically accidental more than regionally and demographically rational.

Planners will also be interested in the proposed document because it legalizes the planning function, which has heretofore existed only in extralegal fashion. Tugwell simply recommends that planning and regulation, which have assumed large roles in twentieth-century government activity, be acknowledged as constitutionally part of the governmental process.

Selected from Planning, *ASPO Newsletter, American Society of Planning Officials, Chicago, Illinois, December 1970, 143-144.*

In addition to the executive, judicial, and legislative, the proposed constitution has three new branches: planning, regulatory, and electoral.

The electoral branch would provide guidelines for conducting federal elections and would establish an Overseer of Elections who is directed, among other things, to "arrange for discussion, in annual district meeting, of the President's views, of the findings of the Planning Branch. . . ."

A National Regulatory Board would make general rules for the regulation of those groups, organizations, or industries which affect the public interest. Proponents of civil rights and ecology take note: the Regulator is given the power to restrain enterprises which discriminate, or which do ecological damage.

In the Tugwell legislative branch, the Senate – composed of past chairmen of the federal planning board, former high government officials, appointees of the President, the House, and the Judiciary – would be responsible together with the President for consideration and approval of the development plans as submitted by the planning branch. The House of 400 members – 100 elected at large and 300 elected from districts much in the same manner as is now current practice – would receive and approve the budget, also originating in the planning branch.

Provisions for the planning branch are contained in Article IV (reprinted here). These planning axioms – 6- and 12-year development plans, separation of the planning function from the executive, custody of the official maps – will have a familiar and traditional ring to all planners.

A planning board of 15 members is to consider 6- and 12-year development plans which are annually submitted to the President and then to the Senate for debate and approval. As established by the constitution, the Republics are invited to submit development proposals for inclusion in the federal plans. Direct involvement on the part of the Republics in the preparation of the plans is anticipated by Mr. Tugwell when he allows for delegation of research and administration to the Republics when and where convenient. The President is admonished by the proposed constitution to seriously "pay attention" to the recommendations of the planning board in formulating policies.

Karen E. Hapgood, ASPO

ARTICLE IV
The Planning Branch

Section 1. There shall be a Planning Branch to formulate and administer plans and to prepare budgets for the uses of expected income in pursuit of policies formulated by the processes provided herein.

Section 2. There shall be a National Planning Board of fifteen members appointed by the President; the first members shall have terms designated by the President of one to fifteen years, thereafter one shall be appointed each year; the President shall appoint a Chairman who shall serve for fifteen years unless removed by him.

Section 3. The Chairman shall appoint, and shall supervise, a planning administrator, together with such deputies as may be agreed to by the Board.

Section 4. The Chairman shall present to the Board six- and twelve-year development plans prepared by the planning staff. They shall be revised each year after public hearings, and finally in the year before they are to take effect. They shall be submitted to the President on the fourth Tuesday in July for transmission to the Senate on September 1 with is comments.

If members of the Board fail to approve the budget proposals by the forwarding date, the Chairman shall nevertheless make submission to the President with notations of reservation by such members. The President shall transmit this proposal, with his comments, to the House of Representatives on September 1.

Section 5. It shall be recognized that the six- and twelve-year development plans represent national intentions tempered by the appraisal of possibilities. The twelve-year plan shall be a general estimate of probable progress, both governmental and private; the six-year plan shall be more specific as to estimated income and expenditure and shall take account of necessary revisions.

The purpose shall be to advance, through every agency of government, the excellence of national life. It shall be the further purpose to anticipate innovations, to estimate their impact, to assimilate them into existing institutions, and to moderate deleterious effects on the environment and on society.

The six- and twelve-year plans shall be disseminated for discussion and the opinions expressed shall be considered in the formulation of plans for each succeeding year with special attention to detail in proposing the budget.

Section 6. For both plans an extension of one year into the future shall be made each year and the estimates for all other years shall be revised accordingly. For nongovernmental activities the estimate of developments shall be calculated to indicate the need for enlargement or restriction.

Section 7. If there be objection by the President or the Senate to the six- or twelve-year plans, they shall be returned for restudy and resubmission. If there still be differences, and if the President and the Senate agree, they shall prevail. If they do not agree, the Senate shall prevail and the plan shall be revised accordingly.

Section 8. The Newstates, on June 1, shall submit proposals for development to be considered for inclusion in those for the Newstates of America. Researches and administration shall be delegated, when convenient, to planning agencies of the Newstates.

Section 9. There shall be submissions from private individuals or from organized associations affected with a public interest, as defined by the Board. They shall report intentions to expand or contract, estimates of production and demand, probable uses of resources, numbers expected to be employed, and other essential information.

Section 10. The Planning Branch shall make and have custody of official maps, and these shall be documents of reference for future developments both public and private; on them the location of facilities, with extension indicated, and the intended use of all areas shall be marked out.

Official maps shall also be maintained by the planning agencies of the Newstates, and in matters not exclusively national the National Planning Board may rely on these.

Undertakings in violation of official designation shall be at the risk of the venturer, and there shall be no recourse; but losses from designations after acquisition shall be recoverable in actions before the Court of Claims.

Section 11. The Planning Branch shall have available to its funds equal to one-half of one percent of the approved national budget (not including debt services or payments from trust funds). They shall be held by the Chancellor of Financial Affairs and expended according to rules approved by the Board; but funds not expended within six years shall be available for other uses.

Section 12. Allocations may be made for the planning agencies of the Newstates; but only the maps and plans of the national Board, or those approved by them, shall have status at law.

Section 13. In making plans, there shall be due regard to the interests of other nations and such cooperation with their intentions as may be approved by the Board.

Section 14. There may also be cooperation with international agencies and such contributions to their work as are not disapproved by the President.

Perhaps the most remarkable administrative change, however, has come about through the establishment in 1942 of the Planning Board and the Bureau of the Budget. . . . As it is now you have before you at all times, not just during the session, a program, adjusted to a careful appraisal of income, of projects which are most needed in all Departments and for all purposes. These have not only been seriously proposed and defended, they have been subjected to analysis, to the most meticulous examination and comparison, and to public hearing. Only after all this have they been included in the six-year plan. Moreover they have gone through the appraisal of the Governor and his staff before they have come to you. . . . You are able to answer local critics in terms of the total income available, of the need for other projects, and of the relation of the proposal to a carefully constructed scheme of improvements. Nothing is more discouraging to the lobbyists; nothing is of more assistance to the conscientious legislator.

R.G. Tugwell, Governor of Puerto Rico, Message to the Sixteenth Legislature at Its Second Regular Session, *San Juan, Puerto Rico, 12 February 1946, pp. 18 & 19.*

38

Charles Abrams

THE CITY PLANNER AND THE PUBLIC INTEREST

Almost nowhere on the American landscape are the problems that beset our current society so conspicuous as in the city. Here the effects of racial segregation are most visible; slum life and poverty are most concentrated. Here the physical environment is the least suited to children, and housing and school problems seem insuperable.

The profession that is closest to dealing with these problems as an entity is city planning. While planning as a means to an end is not new, city planning as a profession is new and has become increasingly important as urbanization has advanced. To deal with the problems of urban environment, a vast armory of political power has become available to the city planner and the planning official, but it is this very increase in power that is confronting them with serious ethical conflicts. Prospects of resolving these conflicts are being thwarted by two peculiarities in the American scene: first, the increasingly complex

---

Negro question, and second, our continued reverence for states rights and local autonomy.

Until the 1930s, city planning in America was mainly a local regulatory process, sustained now and then by the beautification urge and the desire to maintain real estate values or preserve neighborhood prestige. Zoning, park creation, and street design were its principal devices. While city planners professed occasional concern for salvaging the central city, they also manifested an anti-city, pro-suburban bias that derived its logic from an image of new towns immunized against the slums and slovenliness of older settlements. The federal government remained free of any planning involvement because it lacked regulatory powers at the local level; and the state, except for some park programs, simply delegated to the cities and suburbs whatever planning responsibilities it was thought to have.

The New Deal enlarged the planner's horizons, his functions, and his responsibilities. The tools he already worked with under the local regulatory process gained greater power with health, housing, and works programs; simultaneously, new powers – more extensive eminent domain, greater tax and spending powers – were added to the planner's tool kit. As planning power expanded, however, so did the potential for its pervasions.

The susceptibility of the planning power to perversion by the majority as a means of oppressing minorities has a history going back to its earliest exercise as a restrictive power. The licensing power to regulate substandard buildings in California, for example, was employed to oppress Chinese. A San Francisco ordinance made it all but impossible for Orientals to operate their laundries and another local law forced them to move from the sections in which they had established their homes.

Similarly, no sooner was zoning introduced into American cities than private interests and municipal officials forged it into an instrument for restricting the movements of unwelcome ethnic groups. When land was needed for expanding park programs, the victims of public acquisition were often minorities. The introduction of the *cul-de-sac* saw it turned into a device for fending off dark-skinned neighbors; the dead-end street became a method for keeping out dead-end kids; the greenbelt became the medium for separating the black belt from the white belt; the "neighborhood unit" became the means of safeguarding the elite against infiltration by the unwanted. Rejection of subdivision plans and oppressive interpretations of building plans became part of the administrative perversions.

After racial zoning was struck down by the Supreme Court, the restrictive covenant designed to preserve beauty was forged into a device to preserve the all-white neighborhood against challenge by minorities. And although the avowed intent of the more recent slum clearance and urban renewal programs has been to upgrade living conditions, their effect has often been to uproot Negro settlements and institutions.

Painful experience has taught us that there are no easy formulas for replacing slums with something better or for breaking up the ghettoes in which Negroes are living. If, therefore, the slum and the ghetto are to be with us for a while – and it will be difficult to eliminate them entirely in the foreseeable future – there are many better ways to improve the houses of the poor than by obliterating them by "renewal." In any case, while slums and Negro settlements are often one and the same, the fact that a race composing only 11 percent of the nation's population accounts for 70 percent of renewal's displaced persons illustrates how oppression can be injected into the very planning programs we regard as curative.

As things stand today, city planning is a recognized function of our emerging welfare society. Effective planning, however, is being frustrated at each level of government – federal, state, and local – by the continued inability of their several constituencies to reach any sort of practical consensus with regard to the problems of minority groups. In the main, this fundamental difficulty is concealed behind the reluctance at each level of government to trespass on the other's domain. The main impetus given to planning by the federal government, for example, is in housing, with honorable mention being given to the enhancement of beauty and the reduction of poverty. But federal programs can be implemented only by consent of the states, and if the planning power is abused at the local level, i.e., the suburb, it is usually overlooked in deference to local autonomy and states rights. Whatever restrictions against oppression the federal government puts on the use of its funds amount to no more than a gesture.

Nor are matters better at the state level of government, where the plenary police power is vested. Few states take any interest in city planning. Most delegate the planning function to the localities, even in situations where urban problems are clearly regional in nature. State planning laws today are mostly dead letters – promising in preamble, but palsied in power and poor in purse. Fifteen states in 1961 had no planning laws at all; most others have laws that, at best, are advisory or exhortatory. Like the federal government, the states may sometimes persuade, plead, and press, but they will never compel. When urged to act for regional cooperation, state governors readily preach the gospel of *inter*state cooperation – the achievement of which is difficult and therefore politically palatable – while ignoring almost completely the job of *intra*state municipal cooperation, which is legally enforceable and therefore politically embarrassing. It is embarrassing because the suburbs have little inclination to become involved in the social, racial, and financial problems of the city. The states, still dominated by suburban and rural interests, continue to invoke principles of local autonomy as a screen for justifying state withdrawal from responsibility for growing intercommunity concerns.

I might note in passing that this resistance to regionalization runs contrary to the trend in other parts of the democratic world, where artificial barriers that obstruct regional rationalization of the epicentric city and the satellite suburbs are being removed. The well-publicized Toronto example of regional controls is significant, but it is far less significant than the unpublicized consolidation in 1965 of 13 communities on Ile-Jesu into a single municipality by the Province of Quebec. The province acted because it felt that the fragmented autonomous localities in this area adjoining Montreal made no sense as isolated operating units of government. Another example is the voluntary consolidation of five cities into a single major municipality in the Osaka region of Japan.

Although these types of consolidations were common in the United States up to the turn of the century, they are a rarity today, and despite much hue and cry about the need for regional approaches to urban problems, there is no evidence that any of our own states are prepared to move in this direction. In a speech on March 2, 1965, President Johnson defined a city as "the whole urban area – the central city and its suburbs" but if this definition were to be translated into policy, the whole structure of local autonomy and states rights would require radical reassessment, and at this point in the Great Society the prospect is extremely remote.

The movement of Negroes into northern and western communities and the rise of the Negro as a force in political life . . . spurred a movement to outlaw discriminatory practices in housing [resulting in certain provisions in the 1968 Civil Rights Act]. But enforcement is generally mild. Nor can such laws ease the minority's shelter problem unless shelter is simultaneously made available at costs that the poor can afford. And truly low income shelter has been achieved only in slums and in the public housing program – which in the entire 27 years of its existence has produced fewer dwelling units than private enterprise produces in 5 months of a single year.

When one regards the lowest level of government, the municipality, the prospect is dimmest of all. The central city, once a dominant force in American life, is at bay. The poverty of people and the poverty of cities have become part of the same problem. The plight of the city dwellers cannot be dealt with if the cities are financially crippled. But the cities have been given only the limping public housing program, the groping urban renewal program and the war on poverty, which, while well intentioned, is less a "war" than a series of skirmishes, and which, unlike the Peace Corps, is not designed to supplement, finance, expand, and improve existing programs, but to innovate demonstration projects and pilot efforts. The cities are performing their historical function as havens for the poor and the oppressed, but what they need are *not* a few pilot efforts but more funds to improve their existing school systems and to meet their policing, relief, and other commitments.

Although they are the cores of widening urban areas, 41 out of 62 northeastern central cities lost population between 1950 and 1960, with 14 of them losing more than 10 percent. There has been an outflow of the middle class and an inflow of the underprivileged minorities and the elderly. In 1960, less than one-third of the urban-suburban white population lived in cities, while 78 percent of all nonwhites lived in cities. Since the end of World War II, moreover, local governments have increased their debts more than fivefold, while the federal debt per capital has actually declined. Although population migration and racial problems, poverty, ignorance, and slums are the concerns of a federal government with a welfare power and the funds to implement it, federal assistance to cities, which have been bearing the main weight of these problems, has remained minuscule.

Where the executive and legislative branches of government cannot lead, it is futile to expect the judiciary to offer much effective help. The judicial power, although valiantly trying to enforce school desegregation, is able to protect the individual against the local majority in a diminishing number of instances. While courts occasionally strike down racial zoning ordinances or compel a reapportionment based upon population shifts, the judicial power can no longer review every intricate question or check each infringement of constitutional rights in our increasingly complex society. The proliferation of administrative agencies, such as local zoning commisions, school boards, urban renewal, housing, and city planning departments, has been accompanied by an increase in the effective power of these agencies, and our courts are simply unequipped to review their findings or discover abuses. Only where there is a glaring perversion of power will the courts now intervene, and most perversions are becoming increasingly undiscoverable.

Underlying lack of innovation in government, however, is often lack of consensus in the body politic, and one of the most difficult issues on which to attain a consensus is the race issue. One reason is that minority groups as well as liberal thinkers are caught in conflicts between one right and another. The right to dwell where one chooses is pitted against the right to choose one's neighbors. Equality under law is confronted by the claim that the long subordination of the Negro's rights demands that they be given preferential treatment, which in turn is attacked as "discrimination in reverse." The right of a Negro child to an integrated school is confronted by the right of a white child to a school in his own neighborhood. Meanwhile, the racial issue has become entangled in a jungle of verbal abstractions such as "discrimination," "segration," "integration," "open occupancy," "deliberate speed," "ghettoization," "quota system," and "color blindness," none of which have been clearly defined. While there is much froth in the debate, specific mechanisms for realizing individual rights and individual protection through well-considered programs are yet to be devised.

In this political drama the city planner is clearly a central character. But what is to be his role? Should he be simply the agent of his public or private employer, conforming to the employer's whims while suppressing his own scruples? Or is he supposed to go out on every limb at once, spanning the gaps between beauty and freedom, environmental and moral decency, the demands of the suburb and the needs of the urb, the pressure of a vested interest to exclude and the press of his conscience to assure free movement to people? The issues involve more than city planning — they raise ethical, political and philosophical questions that have not been faced since the ascent of the planning power and the rise of the welfare state. These are not easy questions for the planner: like many another professional, he is caught in the maelstrom of change the new welfare society is bringing.

Still, despite the difficulty of such questions, the planner has the responsibility to grapple with them, and to do so he must play a dual role. He must answer to his client and to his soul, for planning is not only design, politics, beauty, housing, urban renewal, zoning, land use, abuse, misuse, disuse, nonuse; it is also ethics. The planner is citizen as well as servant, an individual as well as a contractor, and my main complaint about him, much as I appreciate his quandary, is that I find him rarely on the hustings. Most of the programs which give him his bread have come from the public housing lobbyists. The planner himself has had little to do with the great wave of congressional interest in urban aesthetics or with urban renewal, new towns, and housing programs; this is because he tends to regard himself solely in his professional, rather than his ethical, capacity.

Now it is only fair to say that other professional groups face the same dilemma of identification. Many professional men think they should keep their expertise above the level of political battle. Others feel they should speak only through their professional organizations, which usually engage in research rather than politics. This withdrawal from political activity has been encouraged by the federal tax policy that forbids charitable deductions of gifts to organizations engaged in promoting or opposing legislation. Thus, for example, the bulk of foundation money goes to organizations that conduct research, and financial aid for organizations that take public positions on legislation, however worthy, is hard to come by. As a result, lobbies representing private interests operate freely and effectively, while those groups that might represent the public interest stay out of politics. City planning, although we have entrusted it with the power to manipulate environment, is one of the professions that has chosen to remain silent.

Other professions have resolved the dilemma in various ways. In some, the professional association and its legislative pressure group are divided into separate, although closely allied, organizations. I think we need something of this sort in city planning, for if there were an organization for political action within the planning profession, the planner could more easily express his social principles and better fulfill his social functions.

Our generation is privileged to live in a time of trouble and excitement and challenge. It is a period in which disciplinary overspecialization calls upon the planner to integrate, to combine in himself talents of the Renaissance man and the *chef de cuisine;* to be at once the savant, the oracle, and the Admirable Crichton. There are few such men, of course, in city planning or in the related professions, and it is questionable that in an age of specialization their numbers will increase. Still, the city planner is blessed in the challenge and there might be a few who could yet speak to the occasion. But unless the planner makes himself heard — as expert and as citizen alike — he can scarcely disclaim responsibility for the continuing blight of our cities or for the continuing perversion of the planning power.

Whether justified or not, there has long been a tendency in the United States for those requiring local governmental decisions, and the public at large, to regard municipal agencies with skepticism. This tendency is naturally strengthened when information, descriptive of procedure and practice, is not readily available except to participants or those closely concerned with government. Furthermore, without this information, constructive discussion and reaction by interested groups and individual citizens in the community is unlikely, if not impossible.

. . . in California, all elected and appointed local governmental bodies operate in open meetings in accordance with the Ralph M. Brown Act (California Government Code, Sections 54950-54960). This state legislation not only prohibits closed meetings or executive sessions, but any unofficial gathering of a majority of the members of a local governmental agency which leads to the resolution of public business in private. Its intent is clearly to require discussion and resolution of local governmental affairs in the open. Only personnel and administrative matters are exempt when immediate disclosure or public discussion would be self-defeating or unjustifiably harmful to persons involved. In essence and effect, the Brown Act confirms the public's right of informational access to local governmental decision-making, its "right-to-know" what transpires during this process.

M.C. Branch, "Establishing and Disclosing the Decision-Making Process in Local Planning," Tomorrow — Today, *Los Angeles, California (City Planning Department) December 1967, p. 1.*

39

Edmund M. Burke

CITIZEN PARTICIPATION STRATEGIES

The participation of citizens in community planning, public as well as private, has increased rapidly in the past few years to the point where it is now a fairly common and frequently praised practice. Federal legislation and, more telling, the demands of citizens themselves have combined to make citizen participation an essential requirement in any urban project. Yet, nothing in community planning to date has caused more contention. In city after city, program after program, citizen participation is the principal source of confusion and conflict.

Urban renewal agencies are a case in point. They have been subject to constant criticism for their citizen participation practices. Even the few which have conscientiously attempted to involve citizens at the grass-roots level find their motives suspect.[1] The poverty program committed to citizen participation from the start, is another and more publicized example. Its difficulties over citizen participation have involved congressmen, Office of Economic

Opportunity (OEO) officials, majors, community organization workers, citizens, and even the Bureau of the Budget. The resulting imbroglio now is being blamed for the recent congressional antagonism to the Community Action Program (CAP). Nor, surprisingly, are private agencies, based on the principle of voluntarism, immune from attack. Mobilization for Youth in New York City, as well as other Ford Foundation-financed "gray area" projects, have stumbled over the application of citizen participation and have been forced to curtail program goals. Why has this happened?

DILEMMAS OF CITIZEN PARTICIPATION

Part of the difficulty stems from society's idealized value premise concerning citizen participation, coupled with an inability to make it work in policy making. Citizen participation is part of our democratic heritage, often proclaimed as a means to perfect the democratic process. Stated most simply, it views the citizen as the ultimate voice in community decision making. Citizens *should* share in decisions affecting their destinies. Anything less is a betrayal of our democratic tradition.

Yet, even its most ardent supporters admit that citizens cannot participate in all decision-making functions. Questions of national security are the most extreme examples. Decisions requiring technical competency may be others. On what basis, though, are decisions defined as technically outside the purview of the citizen?

The answer is not easily found. Indeed, there may not be one. For example, when neighborhood groups insist on participating in local school issues, including selection of the school principal – is that citizen participation? Or as Fred Hechinger, education editor of the *New York Times,* describes it, "citizen control"?[2] The question is perplexing because many of those currently sympathetic to the New York City citizens abhorred attempts of other citizen groups who, during the McCarthy period, demanded their kind of participation in school affairs.

Moreover, arrayed against the objectives of citizen participation are those of experts. Jealous of their own prerogatives, they may be unwilling to admit nonprofessionals into the decision-making arena. In New York, for example, a $3.5 million legal aid project funded by OEO was put in jeopardy because the legal profession refused to allow the poor to have a direct voice in policy making.[3]

But this, of course, is the basis of the dilemma – the demand both for participatory democracy and expertise in decision making. Certainly, it is not possible to maximize both value preferences. Accommodations have to be made. Generally, value conflicts – the conflict between freedom and control is one example – tend to be resolved pragmatically. Mechanisms are developed to minimize differences and conditions are used as criteria for determining when and how to maximize one value over another.

Probably the most troublesome area of all is the choice of *strategy* objectives for citizen participation. Commonly advocated as serving fairly specific objectives, citizen participation is often predicated less upon value premises than upon practical considerations. In many cases, this is what makes it acceptable. It can, according to some claims, rebuild deteriorating neighborhoods, devise realistic and better plans, pave the way for the initiation of the poor and the powerless into the mainstream of American life, achieve support and sanction for an organization's objectives, end the drift toward alienation in cities, halt the rise in juvenile deliquency, and recreate small town democracy in a complex urban society.

This suggests that citizens can be used as instruments for the attainment of specific ends. Citizen participation, in other words, is a *strategy.* But the ends are sometimes conflicting. In one case, citizen participation is advocated as an administrative technique to protect the stability or even the existence of an organization; in another, it is viewed as an educational or therapeutic tool for changing attitudes; in still another case, it is proposed as a means for assisting an organization to define its goals and objectives.

To imply that citizen participation is a single, undifferentiated, and overriding strategy is misleading. It is more accurate to speak of several strategies of citizen participation, defined in terms of given objectives. These objectives will be limited by available resources, as well as the organizational character of community activities, particularly community planning. Because planning operates through formal organizations, any strategy will be influenced by organizational demands – the necessity for coordinated efforts, the orientation toward purposeful (ideally, rational) action, and the demands of the environment, which, for public agencies, are often the requirements of extragovernmental jurisdictions. Thus the relevancy of a strategy depends both upon an organization's abilities to fulfill the requirements necessary for the strategy's effectiveness, and upon the adaptability of the strategy to an organizational environment.

The intent of this paper is to analyze citizen participation not as a value, but as the basis for various strategies. The more common uses of citizen participation will be reviewed, indicating the assumptions, conditions, and organizational requirements of each. Five strategies will be identified: *education therapy, behavioral change, staff supplement, cooptation,* and *community power.*

EDUCATION THERAPY STRATEGY

A frequently proclaimed but rarely viable strategy of citizen participation focuses upon the presumed need for improvement of the individual participants. Accomplishing a specific task is irrelevant; rather, the participants become clients who are the objects of treatment. Consequently, this strategy has often been defined as an end in itself.

One focus is education. In this context the act of participation is held to be a form of citizenship training, in which citizens working together to solve community problems not only learn how democracy works but also learn to value and appreciate cooperation as a problem-solving method. This would strengthen local government, spur community development, and create a sense of community or community identification.

Utilizing participation in community affairs as an educational device has had a profound and controversial impact on the practice of community organization employed by social workers. Early writers advocated participation not as the means but as the goal of community organization. At this stage there was a strong social reform orientation attached to community organization, and one pioneer in the field, Eduard Lindeman, termed it the "community movement."[4] Later writers continued this emphasis, but referred to participation as the "process goal" of community organization. Murray Ross, one of the principal spokesmen of this school of thought, explains that the aim of community organization is to help communities develop their own capacities to solve problems. Achievement of planning goals is secondary.[5]

More recently this view has come under criticism. Some maintain that process is secondary – in fact, not a goal at all but only a means. Cooperative attitudes, learned through the medium of participation, are preliminary to problem solving, and therefore are the means by which tasks are successfully undertaken.[6] One writer has suggested that in practice this takes place automatically – the goal of integration or process is abandoned as the demands of achieving specific tasks arise.

> *Broadly based decision-making must be replaced by decision-making by a few, who then "sell" the task objective to others. The process of encouraging people to make their own decisions as to what is good for them thus gives way to the process of convincing them of what a change agent or a small group of "leaders" thinks is good for them.*[7]

Another way to focus this strategy is to use participation therapeutically as a means for developing self-confidence, and, indeed, self-reliance – an underlying theme, incidentally, of the citizen participation objectives of both urban renewal and poverty programs. Individuals, according to this logic, will discover that by cooperating with their neighbors they can bring about changes affecting their community. More significantly, they will inspire each other, communicating an elan of hope and self-confidence. The participants will learn that they can reform their own lives; or, according to the hopes of OEO, turn away from the self-defeating and despairing culture of poverty; or, according to the Department of Housing and Urban Development (HUD), increase their sense of responsibility for their dwelling unit.

However meritorious the aim, there appears to be considerable difficulty in implementing this strategy. Admittedly, social group workers use participation as a device to achieve therapeutic or educational objectives. Then, too, those working with citizen groups report that positive changes do occur among individuals participating in community projects. Oscar Lewis, the anthropologist, has suggested that organizing the poor and giving them a sense of power and leadership through participating has been one method of abolishing the subculture of poverty in certain countries, notably Cuba.[8] Black Power advocates, as well, adopt as one of their premises that the organization of the black community will bring about the self-confidence and hope that American society has consistently denied its Negro citizens. But the formal and deliberate organization of citizens for this purpose has rarely been tried, and, if so, seldom for any appreciable time.

One such attempt was in Cincinnati more than 50 years ago. Called the Cincinnati Social Unit Plan, it organized a number of neighborhood districts and involved citizens in major health planning programs. The program engendered criticism from professional groups – chiefly medical – and local government officials. Some held the view that the citizens became *too* self-reliant. "There are still those who do not trust the voice of the people," commented Eduard Lindeman at the time.[9]

Mobilization for Youth in New York, a prototype of the Community Action Agencies, is another attempt, and similar in many respects to the Cincinnati program. It too defines participation as a means for increasing the independence of individuals in deprived areas. "Participation by adults," according to the Mobilization for Youth project, "in decision-making about matters that affect their interests increases their sense of identification with the community and the larger social order. People who identify with their neighborhood and share common values are more likely to control juvenile behavior."[10] But, like the Cincinnati program, it has run into difficulty with public officials and many of its goals have been emasculated.

What frustrates the use of this strategy in community planning is the inability to accommodate it to organizational demands. The focus is upon the means; participation is the overriding objective, not the accomplishment of goals or group tasks. The participants, therefore, must be determiners of decisions and policies, even to the point of allowing them to make unwise decisions or to create conflict and controversy. If, for example, the aim is to build self-reliance into the poor, any attempt to deter or inhibit their role in decision making will only reinforce their alienation and their belief that they are incapable of making decisions. Public officials in Cincinnati or New York would not take such chances. Similarly the Bureau of the Budget, governed by norms of efficiency and performance, has discouraged the use of this strategy by local antipoverty agencies.[11]

BEHAVIORAL CHANGE STRATEGY

Group participation has been found to be a major force for changing individual behavior. Individuals tend to be influenced by the groups to which they belong and will more readily accept group-made decisions than lectures or individual exhortations to change. This has led to a strategy of participation which, although somewhat similar to the *education therapy* strategy, is sufficiently different to require a separate classification. The strategy is deliberately change oriented and is aimed at influencing individual behavior through group membership. It is a strategy commonly associated with community organization practice and more recently with increasing importance in certain schools of management science.[12] Moreover, it is a strategy reflected in much of the urban renewal literature on citizen participation, and, in fact, in even enunciated in a President's Housing Message to Congress.[13]

Briefly, the objective is to induce change in a system or subsystem by changing the behavior of either the system's members or influential representatives of the system. The group is seen as a source of influence over its members. Therefore, by focusing upon group standards – its style of leadership, or its emotional atmosphere – it is considered possible to change the behavior of the individual members. The group itself becomes a target of change even though the goal may be to change individual behavior.[14] This particular emphasis distinguishes this strategy from the *education therapy* strategy, for though many of the techniques may be similar the objective is different. Whether an individual personally benefits from participating in the process is not necessarily relevant. The focus is upon the task and upon helping the group accomplish the task goal.

Two major premises underlie the *behavioral change* strategy. First, it has been found that it is easier to change the behavior of individuals when they are members of a group than to change any one of them separately. Second, individuals and groups resist decisions which are imposed upon them. They are more likely to support a decision and, equally important, more likely to assist in carrying it out if they have had a part in discovering the need for change and if they share in the decision-making process. Participation in the decision-making process, in other words, can create commitment to new objectives.

The effectiveness of this strategy, however, depends upon the existence of certain conditions. In the first place, the participants must have a strong sense of identification with the group, and feel assured that their contributions and activities are meaningful both to themselves and to the group. There must, too, be some satisfactions or gains from participation, either through personal and group accomplishments or from the mere fact of the association with other members. The awareness of the need for change, and consequent pressure for change, must come from within the group as a shared perception. Facts, data, and persuasion are not enough.

There is a necessity, too, for participants to be actively involved in the decision-making process. The making of the decisions, the working through of the problem, so to speak, are the dynamic factors that change behavior. Communication channels, consequently, need to be open and undistorted. "Information relating to the need for change, plans for change, and consequences of change must be shared by all relevant people in the group."[15]

Planning agencies, particularly publicly supported ones, find it difficult to fulfill these conditions. Even though committed to the strategy, as many often are, intra- and extraorganizational demands often dictate a change in strategy. The complexity of many planning projects and more important, the commitment of planners themselves, obstruct the citizen from becoming actively involved in decision making. Citizens frequently complain that they are unable to understand the planners and consequently unable to become committed to a policy or goal they do not understand.

Extraorganizational demands have the effect also of closing off communication channels to the citizens. Organizations faced with adhering to performance norms, such as budget deadlines, discover that they are unable to apply the strategy. The demands for submission of program proposals (for example, the initial planning period in the poverty program is 6 months) or the priority demands emanating from a national agency, such as HUD or OEO, precludes the possibility of involving citizens for the purpose of changing their behavior. Local poverty agencies' staff complain that their time is spent in selling proposals to citizens to gain their support. They have neither the time nor the sanction to effectively foster group deliberation and initiative, however much they would like to.

A further difficulty is relating the participant group to another influential or decision-making centers of the community. It is rarely possible to include all members of a system in a community planning project. Frequently, then, the planning organization is dealing with system representatives. The group becomes not merely a medium of change, but also an agent of change – an action group designed to influence much larger systems. One example would be a representative neighborhood renewal committee attempting to influence other residents and city officials to improve its neighborhood area. But it is not always possible to assume that those involved are in a position to carry out the group's intentions. For the strategy to be effective in community planning, therefore, the participant must not only commit himself to a course of action, but also be in a position to commit others. This has been a vexing problem in community planning. It is not uncommon to involve someone who has little or no influence in the group be represents, or who may not be truly representative of his group.

If, on the other hand, the system representatives can influence change in their own reference groups, the strategy is a highly effective model for planned change. Experiments in industry with this strategy have been quite persuasive.[16] Moreover, a group highly committed to a change objective has proved to be a more effective change agent than an equivalent number of individuals.

STAFF SUPPLEMENT STRATEGY

Probably one of the oldest and certainly one of the most prevalent reasons for citizen participation is the simple principle of voluntarism – the recruitment of citizens to carry out tasks for an organization which does not have the staff resources to carry them out itself. This is a strategy basic to voluntary associations. Hospitals, family casework agencies, recreation services such as the YMCA and the Scouts, and fund-raising agencies rely upon citizen volunteers to perform many essential agency functions. In some instances, agencies depend entirely upon citizens to achieve their objectives. The clearest example is the voluntary fund-raising agency.

In community planning this strategy has been used to supplement the expertise of the planning agency's staff with the expertise of particular citizens. Basically this is what Nash and Durden proposed in their suggestion to replace the planning commission with citizen task forces.[17] Moreover, it is a strategy widely used by Welfare Councils and a premise which underlies the Welfare Councils' reliance (overreliance, some suggest[18]) on the committee approach to social planning. The assumption of the Welfare Council is that its own staff need not be experts on substantive planning issues, rather they should be experts in knowing how to involve and work with citizens who are the presumed experts.

The objective of the strategy is to exploit the abilities, free time, and/or the expertise of individuals to achieve a desired goal. Ideally, it is a procedure whereby the citizen volunteer is matched with the specifications of the task. Interestingly, some agencies actually write up detailed job descriptions for volunteer roles. Much attention, therefore, has to be given to perfecting techniques for recruiting and holding volunteers. Incentives to stimulate willingness to participate become crucial because of the desire to recruit specific individuals.

The use of skilled volunteers as supplementary staff is easily compatible with the requirements of many organizations. It is assumed that the volunteer is in agreement with the organization's objectives and is recruited to assist in carrying out those objectives. Few citizen participants are actually involved in policy-making roles. Scout leaders, case aides, and fund-raising solicitors, for example, are recruited to carry out the policies and directions of the organization. Incidentally, it is this auxiliary role which the Bureau of the Budget prefers for the poor in the poverty program.[19]

There are opportunities in community planning, nevertheless, for the participant to play a significant role in policy making, and where, in fact, this is the assumption upon which he is recruited.[20] The particular expertise of the citizen participant – a juvenile court judge in a study of delinquency, a public welfare recipient in an analysis of poverty, or a public health doctor in an air pollution study – is supposed to assist in determining policy. There is the possibility, however, that the citizen's expertise can become merely a sanctioning element, that is, a symbol through which in actuality the staff's voice becomes policy. But that is another strategy – *cooptation* – and is discussed later.

It is difficult to assess the usefulness of the participation of skilled volunteers in community planning. The overall strategy depends, of course, upon the classical notions of rationality in planning, about which there is now considerable doubt.[21] On many issues, particularly in welfare, urban renewal, and city planning, there are few "correct" decisions. The absence of any valid data and, more important, the ambiguity of assigning values creates a situation in which decision making arises out of bargaining, negotiation, and compromise. The advice of an expert, whether he be citizen or professional, often becomes merely another opinion. And this is a decided limitation in relying upon this strategy exclusively in community planning. Additional strategies need to be employed which take into accout the politics of decision making or are aimed at overcoming value differences.

COOPTATION

Another citizen participation practice is to involve citizens in an organization in order to prevent anticipated obstructionism. In this sense citizens are not seen as a means to achieve better planning goals nor are they seen as partners in assisting an organization in achieving its goal; rather, they are viewed as potential elements of obstruction or frustration whose cooperation and sanction are found necessary. This strategy, *cooptation*, has been defined as "the process of absorbing new elements into the leadership or policy-determining structure of an organization as a means of averting threats to its stability and existence."[22]

Cooptation is neither a new technique nor does it apply only to voluntary or welfare organizations. Corporations, for example, elect representatives of banking institutions to their boards of directors to provide access to financial resources. Politicians have been notably imaginative in this art. For instance, to ward off predictions that his administration would be fiscally irresponsible, President Kennedy appointed a highly respected Republican as Secretary of the Treasury.

Cooptation can take two forms, both of which are applicable to organizations involving citizens. One is employed in response to specific power forces. Certain individuals are considered to have sufficient resources or influence – financial, decision making, legislative – to vitally affect the operation of the organization. To capture this influence or at least neutralize it, not only are they brought into the organization, but, more significantly, they are included at the policy-making level because their influence is crucial to the continuation of current organizational policy. This has been termed "informal" cooptation, and its key characteristic is that it is a technique "of meeting the pressure of specific individuals or interest-groups which are in a position to enforce demands."[23]

Although informal cooptation has obvious advantages, it also exerts its own toll. Choice becomes constrained. Those coopted will want to share in influencing policy and thus become one more definer of organization policy. Stability and security may be gained by cooptation, but frequently at a price. An organization will thus have to weigh the benefits against the costs.

A more prevalent practice of welfare and planning organizations is to rely upon what has been termed "formal" cooptation. It is a device for winning consent and legitimacy from the citizenry at large. The underlying belief is that the need the organization purports to serve is not in itself sufficiently persuasive to gain community support. Thus groups who reflect the sentiments of the community are absorbed into the organization in order to gain legitimacy. Clergymen, for example, are inevitably involved in community projects because they bestow credibility upon the projects. Other groups reflecting community sentiments, who consequently are invariably involved, are representatives of labor, business, the professions, and women's organizations.

Formal cooptation also describes the practice of setting up and maintaining communication networks in a community. Any organization needs to establish reliable and readily accessible channels of communication through which information and requests may be transmitted to all relevant segments and participants. An organization depending upon community support and sanction is obliged to relate itself to the community as a particpant. A common method is to tap into already existing citizen groups – neighborhood organizations or block clubs, for instance. In this way the local citizens, through their voluntary associations or committees, become identified and committed to the program and, ideally, the apparatus of the operating agency.[24]

The participants' ability to affect policy, according to Philip Selznick, is the basis for the distinction between informal and formal cooptation. Informal cooptation implies a sharing of power in response to specific or potential pressures. Formal cooptation, on the other hand, merely seeks public acknowledgment of the agency-constituency relationship, since it is not anticipated that organizational policies will be put in jeopardy. What is shared "is the *responsibility* for power," explains Selznick, "not the power itself."[25]

It is not possible to assume, however, that voluntary groups formally coopted by an organization will be willing to remain passive with respect to policy. Where citizen groups are in general agreement with the goals of the host agency, as may have been the case in Selznick's analysis of the TVA, the observation may be applicable. But with changing conditions and possible disagreement on goals, the citizen group may endeavor to capture or at least influence the policy-making centers to ensure that policies are made in their interest.

Urban renewal is an example. On the whole,

renewal agencies have tended to adopt the formal cooptation strategy. Relationships are established with neighborhood groups or block committees, which serve both as a means for sanctioning renewal objectives and as a network of communication, especially in project areas marked for rehabilitation. Citizen groups, however, soon resist the role of sanctioning agents and information carriers, and push for more of a voice in planning decisions. Consequently, local urban renewal agencies have been hard pressed to establish procedures for citizen involvement, turning to trial-and-error applications of different practices.[26]

Yet, despite the usual disparaging connotation attached to cooptation, it does provide a means for achieving social goals. Certain groups not normally included in community policy making are given an entrance into the decision-making arena. Moreover, because it provides overlapping memberships, it is also a device that increases the opportunity for organizations to relate to one another and, thus, find compatible goals. From the organization's viewpoint, it provides a means for giving "outsiders" an awareness and understanding of the problem it faces.[27] At the same time, the strategy is an administrative device. Facilitating the achievement of social goals is incidental. The aim is to permit the limited participation of citizens as a means of achieving organization goals, but not to the extent that these goals are impeded.

COMMUNITY POWER STRATEGIES

Power may be defined as the ability to exercise one's will even over the opposition of others. Individuals are capable of obtaining power and influence through the control of wealth or institutions. Whether such power can be exercised in all instances, or whether a small group can control all community decisions is a matter of dispute. Not disputed, however, is the fact that centers of power do exist outside the formal political structure of a community and such centers are influential in shaping community decisions.[28]

Most community organizations are interested in exerting influence. Frequently, organizations come into being exclusively for the purpose of bringing their will to bear on community decisions. There are two strategies of citizen participation based on theories of community power, both designed to exploit community power. The first is to capture influentials by involving them as participants in the organization in order to achieve organizational objectives. This is the informal cooptation previously explained.

Another significantly different strategy accepts the premises of community power theories, but not the conclusions. Change, it is suggested, can be caused by confronting existing power centers with the power of numbers – an organized and committed mass of citizenry. In effect, a new center of power is created, based not upon control of wealth and institutions but upon size and dedication.[29] This type of organization has the ability to obtain accommodation from existing power centers, both from its inherent strength and its choice of tactics.

Demonstrations, boycotts, and picketing are the common weapons of such mass organizations. Negotiation on issues is inevitable, but negotiation from strength is a prerequisite. The power structure must first be put into a position of willingness to negotiate and this occurs only after they have been pushed to do so. "When those prominent in the status quo turn and label you an agitator," says Saul Alinsky, the chief ideologist of the conflict-oriented strategists, to his organizers, "they are completely correct, for that is, in one word, your function – to agitate to the point of conflict."[30]

The conflict strategy works best for organizations committed to a cause rather than to specific issues or services. In securing the involvement of individuals identified with the basic cause, the organization serves as the unifying vehicle for achievement of individual aims. There is, then, little necessity to include the participants in the goal-defining process. Agreement is assumed. But on specific means to achieve the goal, disagreement may arise. Because the participants are emotionally involved in the ends, detached, pragmatic analysis of alternatives is difficult. Concerns are immediate and give rise to impatience, which, coupled with emotional involvement, can often lead to internal squabbling and dissension. Such conflict over means can immobilize an organization and lead to schisms. Certain race relations agencies have exhibited this difficulty.

Moreover, the effectiveness of the strategy appears

limited in duration. Maintaining citizen interest appears to be the chief difficulty. The organization has only its goal, its idealized purpose, to sustain interest and create satisfactions. It is difficult to maintain interest in idealized goals over long periods of time. The emotional commitment required is too personally enervating. Often the leader of the organization is forced to depend upon exhortations or the manufacturing of crises to recharge interest. Membership dwindles, or frequently the organization changes, tending to rely less upon conflict tactics and more upon cooperation. New classes of participants, reflecting community sentiments or power forces, are invited to join. Goals are modified and the organization becomes undistinguishable from other service-oriented organizations.

CONCLUSIONS

It is apparent that the effectiveness of a particular strategy of citizen participation depends upon certain conditions and assumptions peculiar to itself; likewise, each strategy has its own advantages and limitations. The principal difficulty is in adapting a strategy or strategies to the demands of the particular type of organization and the environment within which it functions. Not all strategies are appropriate for all organizations. Conflict-oriented strategies, as many local antipoverty agencies have demonstrated, are inappropriate in governmentally sponsored programs which demand coordination and cooperation.

A strategy of conflict appears best suited to social reform organizations which are privately supported, or, even more advantageous, self-supporting. Most disadvantaged groups seeking social change have had to depend upon either their own resources or the resources of groups highly sympathetic to their cause. The civil rights struggle is one good example; organized labor is another.

The *behavioral change* strategy and the *staff supplement* strategy appear to be the most appropriate for community planning. The latter permits the planning agency to employ on a voluntary basis the expertise of community individuals. Citizens are recruited for their particular talents – knowledge of the problem (and this can include people who are affected by the problem itself, such as clients of social welfare agencies), skill in publicity and promotion techniques, influence with community decision centers, and representation of community sentiment groups. Such people are recruited into the organization and encouraged to contribute their specialized knowledge to the solution of problems, functioning as full-fledged organizational participants.

The *behavioral change* strategy would appear to be useful in overcoming what is commonly referred to as the "politics" of the planning process. Given the debatable preference characteristics of planning goals and the free-market concept of competing community organizations, it would seem advisable to employ a strategy of participation aimed at accommodating various interests. The *behavioral change* strategy has the advantage of subjecting value preferences to a dialogue, allowing them to be aired within the context of the planning process. Other involved organizations are also encouraged to participate in order to allay their fears, gain their advice, and seek their cooperation.

Obviously, this implies a more purposeful approach to citizen participation than is commonly assumed by planning agencies. One issue, of course, is the ability of the staff to work with citizen groups. The appropriateness of any strategy of citizen participation will depend in large measure upon the capabilities and knowledge of the staff to implement it. A strategy of *cooptation,* for instance, requires skill primarily in administration – relating citizen participants to the organization in such a way that they will not interfere with organization goal achievement. Power-conflict strategies appear to demand leadership of a particular type; often a charismatic leader is needed. He has to be skillful in exhorting his followers, giving them a sense of purpose, and helping them to identify with the goals of the organization.

The *behavioral change* and *staff supplement* strategies, on the other hand, require knowledge and skill in handling the dynamics of individual and group behavior.[31] While constantly seeking to maximize rationality, the staff needs to be sensitive to the individual differences of participants, enabling them to contribute to the planning process. The staff also must be able to analyze community systems in order to locate decision centers, identify representatives of community sentiment groups, and suggest individuals who can contribute knowledge and information to the solution of a problem. Moreover, the staff role is the direct antithesis of the executive leadership role. Although direction often is warranted, the aim is to give the citizen a sense of participation and an opportunity for leadership. The intention is to work *with* citizens in a collaborative process in much the same way that David Godschalk and William Mills suggest.[32]

Finally, there is the issue of organizational commitment to, but limited grasp of, citizen participation. The objective in this paper has been to provide an analytical understanding of citizen participation, in its various forms and functions. Clearly, understanding the particular conditions requisite for the success of a particular strategy frequently is a source of difficulty, which contributes to the confusion and contention over the efficacy of citizen participation in general. Not clear about strategy implications of citizen participation, many planning organizations find a gap between what they purport to do and what they actually can do. Federally sponsored programs, such as urban renewal, the poverty program, and model cities, are a case in point.

Federal agencies at the national level, constrained by congressional critics and bureaucratic practices, are forced to specific priorities and program guidelines, inhibiting participation by citizens on the local level. In turn, local staffs are often reduced to grinding out programs for Washington's approval. Many have been disillusioned and demoralized.[33]

Whether or not this is an inherent conflict is difficult to say. At this time it seems so. Yet it is likely that within the constraints imposed on organizations at both national and local levels a new strategy of citizen participation may evolve. Too many federal agencies are too committed to the general principle of citizen participation not to find a solution.

To do so, the premise that citizen participation is self-evident has to be discarded. Planning agencies must be more precise about what they mean by citizen participation, how they intend to implement it, what agency resources will be used to organize and involve citizens, and what voice citizens will have in planning decisions. This may mean a redefinition of planning agencies' goals toward a new focus where a citizen group assumes the responsibility for defining the goals and aims of the planning agency. But it also may mean less contentious citizen participation.

REFERENCES

[1] Conflict and controversy is also found among staff members within urban renewal agencies. See "Citizen Participation in Urban Renewal," *Columbia Law Review, 66* (Mar. 1966), 500-505.

[2] Fred M. Hechinger, "I. S. 201 Teaches Lesson on Race," *The New York Times.* Sept. 25, 1966, p. E9.

[3] "Lawyers for the Poor" (editorial), *The New York Times,* Dec. 12, 1966.

[4] Eduard C. Lindeman, *The Community* (New York: Association Press, 1921), pp. 58-76.

[5] Murray G. Ross, *Case Histories in Community Organization* (New York: Harper & Row, 1958), pp. 10-11; and Murray G. Ross, *Community Organization* (New York: Harper & Row, 1955), pp. 13, 21-22, 48-53.

[6] Bernard Coughlin, "Community Planning: A Challenge to Social Work," *Social Work* (Oct. 1961), 37-42.

[7] Roland L. Warren, *The Community in America* (Chicago: Rand McNally, 1963), pp. 329-330.

[8] Oscar Lewis, *La Vida* (New York: Random House, 1966), pp. xlii-lii.

[9] Eduard Lindeman, "New Patterns of Community Organization," *Proceedings of the National Conference of Social Work, 1937* (Chicago: University of Chicago Press, 1937), p. 321. See also, Roy Lubove, *The Professional Altruist* (Cambridge, Mass.: Harvard University Press, 1965), pp. 175-178.

[10] Mobilization for Youth, Inc., *A Proposal for the Prevention and Control of Delinquency by Expanding Opportunities* (New York: Mobilization for Youth, Inc., 1961), p. 126.

[11] *The New York Times,* Nov. 5, 1965, p. 1.

[12] See, for example, Douglas McGregor, *The Human Side of Enterprise* (New York: McGraw-Hill, 1960).

[13] John F. Kennedy, *Housing Message to Congress,* March 1961.

[14] Dorwin Cartwright, "Achieving Change in People: Some Applications of Group Dynamics Theory," *Human Relations,* IV (1951), 387.

[15] *Ibid.,* p. 390.

[16] See L. Coch and J. R. P. French, Jr., "Overcoming Resistance to Change," *Human Relations,* 1:4 (1948), 512-532.

[17]Peter Nash and Dennis Durden, "A Task Force Approach to Replace the Planning Board," *Journal of the American Institute of Planners,* XXX (Feb. 1964), 10-22.

[18]Robert Morris, "Social Work Preparation for Effectiveness in Planned Change," *Proceedings of the Council on Social Work Education* (New York: Council on Social Work Education, 1963), pp. 166-180.

[19]*The New York Times, loc. cit.*

[20]This also holds true to some extent if the citizen is recruited to serve on a board of directors.

[21]Richard S. Bolan, "Emerging Views of Planning in an Emerging Urban Society," *Journal of the American Institute of Planners,* XXXIII (July 1967), 233-245.

[22]Philip Selznick, "Foundations of the Theory of Organization," *American Sociological Review,* 13 (Feb. 1948), 34.

[23]*Ibid.,* p. 35.

[24]Philip Selznick, *TVA and the Grassroots* (Berkeley, Calif.: University of California Press, 1953), pp. 224-225.

[25]*Ibid.,* pp. 34-35 (his emphasis).

[26]Edmund M. Burke, "Citizen Participation in Renewal," *Journal of Housing* (Jan. 1966), 18-21.

[27]James D. Thompson and William J. McEwen, "Organizational Goals and Environment: Goal Setting as an Interaction Process," *American Sociological Review,* 23 (Feb. 1958), 28.

[28]For an excellent summation of power and influence, see Dorwin Cartwright, "Influence, Leadership, Control," in James March, ed., *Handbook of Organizations* (Chicago: Rand McNally, 1965), pp. 1-47.

[29]Advocacy planning appears also to stress the concept of community power as a strategy of change. The power the advocate planner is stressing, however, is the power of knowledge – the technical apparatus that he can offer local interest groups which thus enables them to gain concessions from city hall. See Lisa R. Peattie, "Reflections on Advocacy Planning," *Journal of the American Institute of Planners,* XXXIV (Mar. 1968), 80-88.

[30]Quoted in Charles E. Silberman, *Crisis in Black and White* (New York: Vintage Books, 1965), p. 335.

[31]Staff requirements for implementing the client-oriented strategy are difficult to define. In fact, the advocates of using participation as an educational or therapeutic device have not been too clear on the requirements of the strategy itself. More emphasis has been placed on its merits than on its utility and consequences in community planning. Conceivably this is why it tends to be vitiated in practice.

[32]David R. Godschalk and William E. Mills, "A Collaborative Approach to Planning Through Urban Activities," *Journal of the American Institute of Planners,* XXXII (Mar. 1966), 86-95.

[33]For an unintentional indictment of citizen participation in the poverty program, see Memorandum to Participants in ABCD (Boston) Staff-Community Conference, held on Jan. 7, 1967, entitled "Evaluation of Conference" (mimeographed, Feb. 10, 1967), p. 4.

PART X ENVIRONMENT

When the population was small or scattered, the resource base of water, soil, wildlife, air, open space, and other essentials for human existence was not taxed beyond its physical capacity to deliver an expected product or service. In addition, the technology for severe and massive resource depletion remained at a miniscule level compared with the massive engineering potential for resource exploitation that is available today. It was not until the mid-twentieth century that the sheer numbers of people and the consumption rate jointed together to form environmental crises of worldwide dimensions. As urbanization *and* suburbanization *increased, the burdens of urban wastes, along with the rapidly rising social costs that were imposed upon the metropolitan environment, exceeded the traditional delivery systems. That is, the* metabolic by-products *of immense human settlements were difficult to manage once the urban organism grew to metropolitan size without appropriate funds and technology to undertake the ponderous jobs of metropolitan solid waste disposal, air and water pollution control, and transportation and energy supply.*

Spenser W. Havlick, The Urban Organism, *The City's Natural Resources from an Environmental Perspective, New York (Macmillan) 1974, pp. 7 & 10.*

40

Ian L. McHarg

THE PLIGHT: A STEP FORWARD

. . . The country is not a remedy for the industrial city, but it does offer surcease and some balm to the spirit. Indeed, during the Depression there were many young men who would not submit to the indignity of the dole or its queues and who chose to live off the land, selling their strength where they could for food and poaching when they could not, sleeping in the bracken or a shepherd's bothy in good weather, living in hostels and public libraries in winter. They found independence, came to know the land and live from it, and sustained their spirit.

So, when first I encountered the problem of the place of nature in man's world it was not a beleagured nature, but merely the local deprivation that was the industrial city. Scotland was wild enough, protected by those great conservators, poverty and inaccessibility. . . . But this has changed dramatically in the intervening decades, so that today in Europe and the United States a great erosion has been accomplished which has diminished nature – not only in the countryside at large, but within the enlarging cities and, not least, in man as a natural being.

There are large numbers of urban poor for whom the countryside is known only as the backdrop to westerns or television advertisements. Paul Goodman speaks of poor children who would not eat carrots pulled from the ground because they were dirty, terror-stricken at the sight of a cow, who screamed in fear during a thunderstorm. The Army regularly absorbs young men who have not the faintest conception of living off the land, who know nothing of nature and its processes. In classical times the barbarians in fields and forest could only say "bar bar" like sheep; today their barbaric, sheepish descendants are asphalt men.

Clearly, the problem of man and nature is not one of providing a decorative background for the human play, or even ameliorating the grim city: it is the necessity of sustaining nature as source of life, milieu, teacher, sanctum, challenge and, most of all, of rediscovering nature's corollary of the unknown in the self, the source of meaning.

There are still great realms of empty ocean, deserts reaching to the curvature of the earth, silent, ancient forests and rocky coasts, glaciers and volcanoes, but what will we do with them? There are rich contented farms and idyllic villages, strong barns and white-steepled churches, tree-lined streets and covered bridges, but these are residues of another time. There are, too, the silhouettes of all the Manhattans, great and small, the gleaming golden windows of corporate images – expressionless prisms suddenly menaced by another of our creations, the supersonic transport whose sonic boom may reduce this image to a sea of shattered glass.

But what do we say now, with our acts in city and countryside? While I first addressed this question to Scotland in my youth; today the world directs the same question to the United States. What is our performance and example? What are the visible testaments to the American mercantile creed – the hamburger stand, gas station, diner, the ubiquitous billboards, sagging wires, the parking lot, car cemetery, and that most complete conjunction of land rapacity and human disillusion, the subdivision. It is all but impossible to avoid the highway out of town, for here, arrayed in all its glory, is the quintessence of vulgarity, bedecked to give the maximum visibility to the least of our accomplishments.

And what of the cities? Think of the imprisoning gray areas that encircle the center. From here the said suburb is an unrealizable dream. Call them no-place although they have many names. Race and hate, disease, poverty, rancor and despair, urine and spit live here in the shadows. United in poverty and ugliness, their symbol is the abandoned carcasses of automobiles, broken glass, alleys of rubbish and garbage. Crime consorts with disease, group fights group, the only emancipation is the parked car.

What of the heart of the city, where the gleaming towers rise from the dirty skirts of poverty? Is it like midtown Manhattan where 20 percent of the population was found to be indistinguishable from the patients in mental hospitals? Both stimulus and stress live here with the bitch goddess success. As you look at the faceless prisms do you recognize the home of *anomie.*

Can you find the river that first made the city? Look behind the unkempt industry, cross the grassy railroad tracks and you will find the rotting piers, and there is the great river, scummy and brown, wastes and sewage bobbing easily up and down with the tide, endlessly renewed.

If you fly to the city by day, you will see it first as a smudge of smoke on the horizon. As you approach, the outlines of its towers will be revealed as soft silhouettes in the hazardous haze. Nearer you will perceive conspicuous plumes which, you learn, belong to the proudest names in industry. Our products are households words but it is clear that our industries are not yet housebroken. Drive from the airport through the banks of gas storage tanks and the interminable refineries. Consider how dangerous they are, see their cynical spume, observe their ugliness. Refine they may, but refined they are not.

You will drive on an expressway, a clumsy concrete form, untouched by either humanity or art, testament to the sad illusion that there can be a solution for the unbridled automobile. It is ironic that this greatest public investment in cities has also financed their conquest. See the scars of the battle in the remorseless carving, the dismembered neighborhoods, the despoiled parks. Manufacturers are producing automobiles faster than babies are being born. Think of the depredations yet to be accomplished by myopic highway builders to accommodate these toxic vehicles. You have plenty of time to consider in the long, peak-hour pauses of spasmodic driving in the blue gas corridors.

You leave the city and turn toward the countryside. But can you find it? To do so you will follow the paths of those who tried before you. Many stayed to build. But those who did so first are now deeply embedded in the fabric of the city. So as you go you transect the rings of the thwarted and disillusioned who are encapsulated in the city as nature endlessly eludes pursuit.

You can tell when you have reached the edge of the countryside for there are many emblems – the cadavers of old trees piled in untidy heaps at the edge of the razed deserts, the magnificent machines for land despoliation, for felling forests, filling marches, culverting streams, and sterilizing farmland, making thick brown sediments of the creeks.

Is this the countryside, the green belt – or rather the greed belt, where the farmer sells land rather than crops, where the developer takes the public resource of the city's hinterland and subdivides to create a private profit and a public cost? Certainly, here is the area where public powers are weakest – either absent or elastic – where the future costs of streets, sidewalks and sewers, schools, police and fire protection are unspoken. Here are the meek mulcted, and refugees thwarted.

Rural land persists around the metropolis, not because we have managed the land more wisely but because it is larger, more resistant to man's smear, more resilient. Nature regenerates faster in the country than in the city where the marks of men are well nigh irreversible. But it still wears the imprint of man's toil. DDT is in the arctic ice, in the ocean deeps, in the rivers and on the land, atomic wastes rest on the continental shelf, many creatures are forever extinguished, the primeval forests have all but gone, and only the uninitiated imagine that these third- and fourth-growth stands are more than shadows of their forebears. Although we can still see great fat farms, their once deep soils, a geological resource, are thinner now, and we might well know that farming is another kind of mining, dissipating the substance of eons of summers and multitudes of life. The Mississippi is engorged with 5 cubic miles of soil each year, a mammoth prodigality in a starving world. Lake Erie is on the verge of becoming septic, New York City suffers from water shortages while the Hudson flows foully past, salt water encroaches in the Delaware, floods alternate with drought, the fruits of two centuries of land mismanagement. Forest fires, mudslides, and smog become a way of life in Los Angeles, and the San Andreas Fault rises in temperature to menace San Franciscans.

The maps all show the continent to be green wild landscapes save for the sepia cities huddled on lakes and seaboards, but look from a plane as it crosses the continent and makes an idiocy of distance, see the wild green sectioned as rigorously as the city. In the great plains nature persists only in the meandering stream and the flood plain forest, a meaningful geometry in the Mondrian patterns of unknowning men.

It matters not if you choose to proceed to the next city or return to the first. You can confirm an urban destination from the increased shrillness of the neon shills, the diminished horizon, the loss of nature's companions until you are alone, with men, in the heart of the city, God's Junkyard – or should it be called Bedlam, for cacophony lives here. It is the expression of the inalienable right to create ugliness and disorder for private greed, the maximum expression of man's inhumanity to man. And so our cities grow, coalescing into a continental necklace of megalopoles, dead gray tissue encircling the nation.

Surely the indictment is too severe – there must be redeeming buildings, spaces, places, landscapes. Of course there are – random chance alone would have ensured some successful accidents. But there are also positive affirmations; yet it is important to recognize that many of these are bequests from earlier times. Independence, Carpenter and Faneuil Halls symbolize the small but precious heritage of the eighteenth century: the great state houses, city halls, museums, concert halls, city universities and churches, the great urban park systems were products of the last century. Here in these older areas you will find humane, generous suburbs where spacious men built their concern into houses and spaces so that dignity and peace, safety and quiet live there, shaded by old trees, warmed by neighborliness.

You may also see hints of a new vitality and new forms in the cities, promising resurgence. You may even have found, although I have not, an expressway that gives structure to a city, or, as I have, a parkway that both reveals and enhances the landscape. There are farmlands in good heart; there are landowners – few it is true – who have decided that growth is inevitable, but that it need not lead to despoliation but to enlargement. New towns are being constructed and concepts of regional planning are beginning to emerge. There is an increased awareness for the need to manage resources and even a title for this concern – the New Conservation. There is a widening certainty that the gross national product does not measure health or happiness, dignity, compassion, beauty or delight, and that these are, if not all inalienable rights, at least most worthy aspirations.

But these are rare among the countless city slums and scabrous towns, pathetic subdivisions, derelict industries, raped land, befouled rivers, and filthy air.

At the time of the founding of the republic – and for millenia before – the city had been considered the inevitable residence for the urbane, civilized, and polite. Indeed, all these names *say* city. It was as widely believed that rich countries and empires were inevitably built upon the wealth of the land. The original cities and towns of the American eighteenth century were admirable – Charleston and Savannah, Williamsburg, Boston, Philadelphia, New Orleans. The land was rich and beautiful, canons of taste espoused the eighteenth-century forms of architecture and town building, a wonder of humanity and elegance.

How then did our plight come to be and what can be done about it? It is a long story which must be told briefly and, for that reason, it is necessary to use a broad brush and paint with coarse strokes. This method inevitably offends for it omits qualifying statements, employs broad generalities, and often extrapolates from too slender evidence. Yet the basic question is so broad that one need not be concerned with niceties. The United States is the stage on which great populations have achieved emancipation from oppression, slavery, peonage and serfdom, where a heterogeneity of peoples has become one and where an unparalleled wealth has been widely distributed. These are the jewels of the American diadem. But the setting, the environment of this most successful social revolution, is a major indictment against the United States and a threat to her success and continued evolution.

Our failure is that of the Western world and lies in prevailing values. Show me a man-oriented society in which it is believed that reality exists only because man can perceive it, that the cosmos is a structure erected to support man on its pinnacle, that man exclusively is divine and given dominion over all things, indeed that God is made in the image of man, and I will predict the nature of its cities and their landscapes. I need not look far for we have seen them – the hot-dog stands, the neon shill, the ticky-tacky houses, dysgenic city, and mined landscapes. This is the image of the anthropomorphic, anthropocentric man; he seeks not unity with nature but conquest. Yet unity he finally finds, but only when his arrogance and ignorance are stilled and he lies dead under the greensward. We need this unity to survive.

Among us it is widely believed that the world consists solely of a dialogue between men, or men and God, while nature is a faintly decorative backdrop to the human play. If nature receives attention, then it is only for the purpose of conquest or, even better, exploitation – for the latter not only accomplishes the first objective, but provides a financial reward for the conqueror.

We have but one explicit model of the world and that is built upon economics. The present face of the land of the free is its clearest testimony, even as the gross national product is the proof of its success. Money is our measure, convenience is its cohort, the short term is its span, and the devil may take the hindmost is the morality.

Perhaps there is a time and place for everything; and with wars and revolutions, with the opening and development of continents, the major purposes of exploration and settlement override all lesser concerns, and one concludes in favor of the enterprises while regretting the wastages and losses which are incurred in these extreme events. But if this was once acceptable as the inevitable way, that time has passed.

The pioneers, the builders of railroads and canals, the great industrialists who built the foundations for future growth were hard-driven, single-minded men. Like soldiers and revolutionaries, they destroyed much in disdain and in ignorance, but there are fruits from their energies and we share them today. Their successors, the merchants, are a different breed, more obsequious and insidious. The shock of the assassination of a president stilled for only one day their wheedling and coercive blandishments for our money. It is their ethos, with our consent, that sustains the slumlord and the land rapist, the polluters of rivers and atmosphere. In the name of profit they preempt the seashore and sterilize the landscape, fell the great forests, fill the protective marshes, build cynically in the flood plain. It is the claim of convenience for commerce – or its illusion – that drives the expressway through neighborhoods, homes, and priceless parks, a taximeter of indifferent greed. Only the merchant's creed can justify the slum as a sound investment or offer tomato stakes as the highest utility for the priceless and irreplaceable redwoods.

The economists, with a few exceptions, are the merchants' minions and together they ask with the most barefaced effrontery that we accommodate our value system to theirs. Neither love nor compassion, health nor beauty, dignity nor freedom, grace nor delight are important unless they can be priced. If they are nonprice benefits or costs, they are relegated to inconsequence. The economic model proceeds inexorably towards its self-fulfillment of more and more despoliation, uglification, and inhibition to life, all in the name of progress – yet, paradoxically, the components which the model excludes are the most important human ambitions and accomplishments and the requirements for survival. . . .

For me the indictment of city, suburb, and countryside becomes comprehensible in terms of the attitudes to nature that society has and does espouse. These environmental degradations are the inevitable consequence of such views. It is not incongruous but inevitable that the most beautiful landscapes and the richest farmlands should be less highly valued than the most scabrous slum and loathsome roadside stand. Inevitably, an anthropocentric society will choose tomato stakes as a higher utility than the priceless and irreplaceable redwoods they have supplanted.

Where you find a people who believe that man and nature are indivisible, and that survival and health are contingent upon an understanding of nature and her processes, these societies will be very different from ours, as will be their towns, cities, and landscapes. The hydraulic civilizations, the good farmer through time, the vernacular city builders have all displayed this acuity. But it is in the traditional society of Japan that the full integration of this view is revealed. That people, as we know, has absorbed a little of the best of the West and much of the worst while relinquishing accomplishments that we have not yet attained and can only envy.

In that culture there was sustained an agriculture at once incredibly productive and beautiful, testimony to an astonishing acuity to nature. This perception is reflected in a language rich in descriptive power in which the nuances of natural processes, the tilth of the soil, the dryness of wind, the burgeoning seed, are all precisely describable. The poetry of this culture is rich and succinct, the graphic arts reveal the landscape as the icon. Architecture, village, and town building use natural materials directly with stirring power, but it is garden making that is the unequaled art form of this society. The garden is the metaphysical symbol of society in Tao, Shinto, and Zen – man in nature.

Yet this view is not enough: man has fared less well than nature here. The jewel of the western tradition is the insistence upon the uniqueness of the individual and the preoccupation with justice and compassion. The Japanese medieval feudal view has been casual to the individual human life and rights. The western assumption of superiority has been achieved at the expense of nature. The oriental harmony of man-nature has been achieved at the expense of the individuality of man. Surely a united duality can be achieved by accounting for man as a unique individual rather than as a species, man in nature. . . .

Two widely divergent views have been discussed, the raucous anthropocentrism which insists upon the exclusive divinity of man, his role of dominion and subjugation on one hand, and the oriental view of man submerged in nature on the other. Each view has distinct advantages, both have adaptive value. Are the benefits of each mutually exclusive? I think not; but in order to achieve the best of both worlds it is necessary to retreat from polar extremes. There is indisputable evidence that man exists in nature; but it is important to recognize the uniqueness of the individual and thus his especial opportunities and responsibilities. . . .

Surely the minimum requirement today for any attitude to man-nature is that it approximate reality. One could reasonably expect that if such a view prevailed, not only would it affect the value system, but also the expressions accomplished by society.

Where else can we turn for an accurate model of the world and ourselves but to science? We can accept that scientific knowledge is incomplete and will forever be so, but it is the best we have and it has that great merit, which religions lack, of being self-correcting. Moreover, if we wish to understand the phenomenal world, then we will reasonably direct our questions to those scientists who are concerned with this realm – the natural scientists. More precisely, when our preoccupation is with the interaction of organisms and environment – and I can think of no better description for our concern – then we must turn to ecologists, for that is their competence.

We will agree that science is not the only mode of perception – that the poet, painter, playwright, and author can often reveal in metaphor that which science is unable to demonstrate. But, if we seek a workman's creed which approximates reality and can be used as a model of the world and ourselves, then science does provide the best evidence.

From the ecological view one can see that, since life is only transmitted by life, then, by living, each one of us is physically linked to the origins of life and thus – literally, not metaphorically – to all life. Moreover, since life originated from matter, then, by living, man is physically united back through the evolution of matter to the primeval hydrogen. The planet Earth has been the one home for all its processes and all its myriad inhabitants since the beginning of time, from hydrogen to men. Only the bathing sunlight changes. Our phenomenal world contains our origins, our history, our milieu; it is our home. It is in this sense that ecology (derived from *oikos*) is the science of the home. . . .

The ecological view requires that we look upon the world, listen, and learn. The place, creatures, and men were, have been, are now, and are in the process of becoming. We and they are here now, co-tenants of the phenomenal world, united in its origins and destiny.

As we contemplate the squalid city and the pathetic subdivision, suitcase agriculture and the cynical industrialist, the insidious merchant, and the product of all these in the necklace of megalopoles around the continent, their entrails coalescing, we fervently hope that there is another way. There is. The ecological view is the essential component in the search for the face of the land of the free and the home of the brave. This work seeks to persuade to that effect. It consists of borrowings from the thoughts and dreams of other men, forged into a workman's code – an ecological manual for the good steward who aspires to art. . . .

With highways: the engineer is most competent when considering the automobile as a projectile that responds to the laws of dynamics and statics. He understands structures and pavements very well indeed and his services are indispensable. But the matter of the man in the automobile as a creature with senses is outside his ken; the nature of the land as interacting biophysical processes is unknown to him. His competence is not the design of highways, merely of the structures that compose them – but only after they have been designed by persons more knowing of man and the land.

The method that has been used traditionally by the Bureau of Public Roads and state highway departments involves calculating the savings and costs derived from a proposed highway facility. Savings include savings in time, operating costs, and reduction in accidents. Costs are those of construction and maintenance. It is necessary to obtain a minimum ratio of savings to costs of 1.2:1.0. Any qualitative factors are considered after the conclusion of the cost-benefit analysis, and then only descriptively.

The objective of an improved method should be to incorporate resource values, social values and aesthetic values in addition to the normal criteria of physiographic, traffic, and engineering considerations. In short, the method should reveal the highway alignment having the maximum social benefit and the minimum social cost. This poses difficult problems. It is clear that new considerations must be interjected into the cost-benefit equation and that many of these are considered nonprice factors. Yet the present method of highway cost-benefit analysis merely allocates approximate money values to convenience, a commodity as difficult to quantify as either health or beauty.

Interstate highways should maximize public and private benefits:

1. By increasing the facility, convenience, pleasure, and safety of traffic movement.
2. By safeguarding and enhancing land, water, air, and biotic resources.
3. By contributing to public and private objectives of urban renewal, metropolitan and regional development, industry, commerce, residence, recreation, public health, conservation, and beautification.
4. By generating new productive land uses and by sustaining or enhancing existing ones.

Such criteria include the orthodoxies of route selection, but place them in a larger context of social responsibility. The highway is no longer considered only in terms of automotive movement within its right-of-way, but in context of the physical, biological and social processes within its area of influence.

The highway is thus considered as a major public investment, which will affect the economy, the way of life, health, and visual experience of the entire population within its sphere of influence. It is in relation to this expanded role that it should be located and designed.

It is clear that the highway route should be considered a multipurpose rather than a single-purpose facility. It is also clear that, when a highway route is so considered, there may be conflicting objectives. As in other multipurpose planning, the objective should be to maximize all potential complementary social benefits at the least social cost.

This means that the shortest distance between two points, meeting predetermined geometric standards, is not the best route. Nor is the shortest distance over the cheapest land. *The best route is the one that provides the maximum social benefit at the least social cost.*

The present method of cost-benefit analysis, as employed for route selection, has two major components: (1) the savings in time, operating costs, and safety provided by the proposed facility, and (2) the sum of engineering, land, and building purchase, financing, administrative, construction, operation, and maintenance costs.

On the credit side it seems reasonable to allocate all economic benefits derived from the highway. These benefits accrue from the upgrading of land use, frequently from agricultural to industrial, commercial, or residential uses. Great indeed are these values. In certain favored locations they may be multiples of the cost of the highway. But highways do reduce economic values; they do constitute a health hazard, a nuisance, and danger; they can destroy the integrity of communities, institutions, residential quality, scenic, historic, and recreational value.

This being so, it appears necessary to represent the sum of effects attributable to a proposed highway alignment and to distinguish these as benefits, savings, and costs. In certain cases these can be priced and can be designated price benefits, price savings, or price costs. In other cases, where valuation is difficult, certain factors can be identified as nonprice benefits, savings, or costs.

A balance sheet in which most of the components of benefit and cost are shown should reveal the alignments of maximum social utility.

Considerations of traffic benefits as calculated by the Bureau of Public Roads can be computed for alternative alignments. The cost of alternative routes can be calculated. Areas in which increased land and building values may result can be located, if only tentatively, in relation to the highway and prospective intersections. Prospective depreciation of land and building value can also be approximately located. Increased convenience, safety, and pleasure will presumably be provided within the highway right-of-way; inconvenience, danger, and displeasure will parallel its path on both sides. The degree to which the highway sustains certain community values can be described, as can the offense to health, community, scenery, and other important resources.

The method proposed here is an attempt to remedy deficiencies in route-selection method. It consists, in essence, of identifying both social and natural processes as social values. We will agree that

SUGGESTED CRITERIA FOR INTERSTATE HIGHWAY ROUTE SELECTION

Benefits and Savings

Price Benefits
Reduced time distance
Reduced gasoline costs
Reduced oil costs
Reduced tire costs
Reduced vehicle depreciation
Increased traffic volume
Increase in value (Land and Buildings)
  Industrial values
  Commercial values
  Residential values
  Recreational values
  Institutional values
  Agricultural land values

Nonprice Benefits
Increased convenience
Increased safety
Increased pleasure

Price Savings
Nonlimiting topography
Adequate foundation conditions present
Adequate drainage conditions present
Available sands, gravels, etc.
Minimum bridge crossings, culverts, and other structures required

Nonprice Savings
Community values maintained
Institutional values maintained
Residential quality maintained
Scenic quality maintained
Historic values maintained
Recreational values maintained
Surface water system unimpaired
Groundwater resources unimpaired
Forest sources maintained
Wildlife resources maintained

Costs

Price Costs
Survey
Engineering
Land and building acquisition
Construction costs
Financing costs
Administrative costs, operation and maintenance costs
Reduction in value (Land and Buildings)
  Industrial values
  Commercial values
  Residential values
  Recreational values
  Institutional values
  Agricultural land values

Nonprice Costs
Reduced convenience to adjacent properties
Reduced safety to adjacent populations
Reduced pleasure to adjacent populations
Health hazard and nuisance toxic fumes, noise, glare, dust

Price Costs
Difficult topography
Poor foundations
Poor drainage
Absence of construction materials
Abundant structures required

Nonprice Costs
Community values lost
Institutional values lost
Residential values lost
Scenic values lost
Historic values lost
Recreational values lost
Surface Water resources impaired
Groundwater resources impaired
Forest resources impaired
Wildlife resources impaired

land and building values do reflect a price value system; we can also agree that for institutions that have no market value there is still a hierarchy in values — the Capitol is more valuable than an undifferentiated house in Washington, Independence Hall more precious than a house in Philadelphia's Society Hill or Central Park more valuable than any other in New York. So too with natural processes. It is not difficult to agree that different rocks have a variety of compressive strengths and thus offer both values and penalties for building; that some areas are subject to inundation during hurricanes and other areas are immune; that certain soils are more susceptible to erosion than others. Additionally, there are comparative measures of water quantity and quality, soil drainage characteristics. It is possible to rank forest or marsh quality, in terms of species, numbers, age, and health in order of value. Wildlife habitats, scenic quality, the importance of historic buildings, recreational facilities can all be ranked.

If we can evaluate and rank aesthetic, natural-resource and social values, we can then proceed. Thus, if destruction or despoliation of existing social values were to be caused by proposed highway alignment, that alignment value would be decreased by the amount of the social costs. The physical costs of construction are social costs too. Therefore, we can conclude that any alignment that transects areas of high social values and also incurs penalties in heightened construction costs will represent a maximum-social-cost solution. The alternative is always to be sought – an alignment that avoids areas of high social costs and incurs the least penalties in construction costs and creates new values. The basis of the method is constant for all case studies – that nature is interacting process, a seamless web, that it is responsive to laws, that it constitutes a value system with intrinsic opportunities and constraints to human use.

If we can accept the initial proposition, we can advance to a second. That is, if physical, biological and social processes can be represented as values, then *any* proposals will affect these. One would ask that such changes be beneficial, that they add value. But changes to land use often incur costs. The best of all possible worlds would be a proposal that provided new values and incurred no costs. In the absence of this unlikely circumstance, we might be satisfied if new values exceeded the costs incurred. Preferably, these costs should not involve irreversible losses. The solution of maximum social benefit at least social cost might be the optimum. This could be called the solution of maximum social utility.

In essense, the method consists of identifying the area of concern as consisting of certain processes in land, water, and air – which represent values. These can be ranked – the most valuable land and the least, the most valuable water resources and the least, the most and least productive agricultural land, the richest wildlife habitats and those of no value, the areas of great and little scenic beauty, historic buildings and their absence, and so on. The interjection of a highway will transect this area; it will destroy certain values. Where will it destroy the least? Positively, the highway requires certain conditions – propitious slopes, good foundation materials, rock, sand, and gravel for its construction, and other factors. Propitious circumstances represent savings, adverse factors are costs. Moreover, the highway can be consciously located to produce new values – more intense and productive land uses adjacent to intersections, a delightful experience for the motorist, an added convenience to the traveler. The method requires that we obtain the most benefit for the least cost, but that we include as values social process, natural resources, and beauty.

We can identify the critical factors affecting the physical construction of a highway and rank these from least to greatest cost. We can identify social values and rank them from high to low. Physiographic obstructions – the need for structures, poor foundations, etc. – will incur high social costs. We can represent these identically. For instance, let us map physiographic factors so that the darker the tone, the greater the cost. Let us similarly map social values so that the darker the tone, the higher the value. Let us make the maps transparent. When these are superimposed, the least-social-cost areas are revealed by the lighest tone.

However, there is one important qualification that must be recognized. While in every case there should be little doubt as to the ranking within a category, there is no possibility of ranking the categories themselves. For example, it is quite impossible to compare a unit of wildlife value with a unit of land value or to compare a unit of recreational value with one of hurricane danger. All that can be done is to identify natural and social processes and superimpose these. By so doing we can observe the maximum concurrence of either high or low social values and seek that corridor which transects the areas of least social value in all categories. Exact resolution of this problem seems unrealizable. Economists have developed price values for many commodities, but there seems no prospect that institutions, scenic quality, historic buildings, and those other social values considered can be given exact price values.

It is immediately conceded that the parameters are not co-equal. In a given area, considered by itself, existing urbanization and residential quality are likely to be more important than scenic value or wildlife. Yet is is reasonable to presume that, where there is an overwhelming concentration of physiographic obstruction and social value, such areas should be excluded from consideration; where these factors are absent, there is a presumption that such areas justify consideration.

This is not yet a precise method for highway route selection; yet it has the merit of incorporating the parameters currently employed and adding new and important social considerations, revealing their locational characteristics, permitting comparison, disclosing aggregates of social values and costs. Whatever limitations of imprecision it may have, it does enlarge and improve existing method.

Within limits set by the points of origin and destination, responsive to physiographic obstructions and the pressure of social values, *the highway can be used as conscious public policy to create new and productive land uses at appropriate locations.* In any such analysis cost-benefit calculations would require that any depreciation of values would be discounted from value added. In addition, scenic value should be considered as possible added value. It is, of course, possible that a route could be physiographically satisfactory, avoid social costs, create new economic values at appropriate locations, and also provide a satisfactory scenic experience.

The highway is likely to create new values whether or not this is an act of conscious policy. Without planning, new values may displace existing ones, but even if a net gain results, there may well be considerable losses.

Some years ago I gave an address at Princeton on "The Ecological View." I extolled the diagnostic and prescriptive powers of this integrative science. The following day I was asked to employ ecology in the selection of a 30-mile route for I-95 between the Delaware and Raritan rivers. The inhabitants of this bucolic region were threatened by an alignment that appeared to select almost all that was precious and beautiful – the maximum destruction to be accomplished with the least benefit at the greatest cost. The enraged citizenry constituted themselves into the Delaware-Raritan Committee on I-95. Faced with the problem, little time, and less money, the method we have just outlined was developed and applied. Through the transparencies – like light shining through a stained glass window – was visible that alignment of least social cost. Its influence was felt and, one after another, through 34 alternative alignments, the proposed highway moved nearer and nearer to that ultimately proposed by the author.

To claim this as an ecological method is to flatter it. It is enough to say that it did use data reflecting social, resource, and aesthetic values, but the data were hurridly assembled and gross. Residential value was derived from land and building values that gave high social value to the wealthy and too little to the poor; urbanization was classed into a few gross categories, excluding the enormous variety of conditions within this description. Nonetheless, it offered a large measure of success. It provided a method whereby the values employed were explicit, where the selection method was explicit – where any man, assembling the same evidence, would come to the same conclusion. It introduced the least-social-cost/maximum-social-benefit solution, a relative-value system that could consider many nonprice benefits, savings, and costs, and, not least, the measure of scenic experience as a potential value.

Subsequently, the method was employed in the Borough of Richmond in New York where, as is now commonplace, a treasured open space was threatened by highway destruction. Here the subject of traffic was not in dispute; no intersections were proposed for the controversial 5-mile section of the Richmond Parkway, and social benefit was thus limited to the convenience of the trip and the scenic experience of the motorist. In this example the matter of reducing social costs to maintain social values was preponderant – but increasingly this is the overwhelming problem.

The issue was a simple one. Should the highway select the greenbelt for its route in order to reveal it to the public or should it serve the greenbelt, but avoid the destruction of transection? The character of the highway is not changed by entitling it a parkway, but this title has been used to describe highways in areas of great natural beauty – the Blue Ridge and Palisades Parkways, for example. Here, where beautiful landscapes are abundant, there is little social loss and great social benefit. Where resources are as precious as the greenbelt in Staten Island, this conception is not appropriate. Better, follow the example of the Bronx River Parkway and create new values while avoiding destruction of the few oases that remain for 12 million New Yorkers.

We can now apply the method to the Richmond Parkway. The first group of factors included some of those orthodox criteria normally employed by engineers – slope, bedrock geology, soil foundation conditions, soil drainage, and susceptibility to erosion. The degree of opportunity or limitation they afford is reflected directly in the cost of highway construction. The next category concerns danger to life and property and includes areas vulnerable to flood inundation from hurricanes. The remaining categories are evaluations of natural and social processes, including historic values, water values, forest values, wildlife values, scenic values, recreation values, residential values, institutional values, and land values. Each factor, with its three grades of values, is photographed as a transparent print. The transparencies of the first group are superimposed upon one another, and from this a summary map is produced that reveals the sum of physiographic factors influencing highway route alignment. Each subsequent parameter is then superimposed upon the preceding until all parameters are overlaid. The darkest tone then represents the sum of social values and physiographic obstructions to a highway corridor; the lightest tone reveals the areas of least social value representing the least direct cost for highway construction. The highway should be located in that corridor of least social value and cost, connecting points of origin and destination. Moreover, it should provide new values – not only of convenience, but also of scenic experience – as a product of public investment.

It is important to observe that the reader parallels the experience of the author at the beginning of the study. The method was known but the evidence was not. It was necessary to await its compilation, make the transparent maps, superimpose them over a light table, and scrutinize them for their conclusion. One after another they were laid down, layer after layer of social values, an elaborate representation of the Island, like a complex X-ray photograph with dark and light tones. Yet in the increasing opacity there were always lighter areas and we can see their conclusion.

REFERENCE

[1] *Srole, Leo,* et al., Mental Health in the Metropolis: The Midtown Manhattan Study. *New York, McGraw-Hill, 1962.*

. . . The purposes of this Act are: To declare a national policy which will encourage productive and enjoyable harmony between man and his environment; to promote efforts which will prevent or eliminate damage to the environment and biosphere and stimulate the health and welfare of man; to enrich the understanding of the ecological systems and natural resources important to the nation . . .

. . . all agencies of the Federal Government shall – (A) utilize a systematic, interdisciplinary approach which ensures the integrated use of the natural and social sciences and the environmental design arts in planning and in decision-making which may have an impact on man's environment; (B) identify and develop methods and procedures . . . which will insure that presently unquantified environmental amenities and values may be given appropriate consideration in decision-making along with economic and technical considerations. . . .

The National Environmental Policy Act of 1969.

41

George Macinko

SATURATION: A PROBLEM EVADED IN PLANNING LAND USE

. . . The primary goal of planning appears to be the promotion and maintenance of an environment which will allow for "optimum human living." Planners generally believe this goal is most likely to be realized in an environment in which provision is made for solitude, for public open space, and for the aesthetic pleasures provided by a landscape which embodies some aspects of a "natural" or at least a semirural flavor. Much of the impetus of the planning movement throughout the nation, as in the Brandywine Valley, arises from an aversion to a completely built-up landscape.

Perhaps the most important operating assumption of contemporary planning is that the conditions necessary for "optimum human living" can be attained by means of various technical planning measures. This belief is typified by a statement made recently by the managing editor of *Architectural Forum* to the effect that the "foolish" idea that every family should have its own house on its own plot of

---

land is the basis of our present land chaos, and by his advocacy of cluster housing and variable density zoning as the solution to this chaos[1]. These and related measures are all based on the assumption that increasing demands for space can be met by exercising ingenuity in the allocation of space.

One of the most pervasive of contemporary philosophical beliefs is that progress, which one might presume to be judged in terms of human welfare, is intimately and inexorably linked with growth – growth both in numbers of people and in their institutions, especially industry. Planners did not originate this belief, but their planning philosophy reveals its widespread acceptance. Briefly, this philosophy maintains that growth is good, for progress depends on it; or at least that growth is inevitable. Progress in this context is defined in strongly economic terms[2]. It is reasoned that economic considerations are the key to human welfare and that economic advance depends primarily on a steadily increasing demand for the fruits of production. Achievement and maintenance of this increasing demand are thought to depend largely on increasing the number of consumers. Growth in population is therefore held to be a condition of progress[3].

In analyzing these objectives, assumptions, and philosophy, I will make use of a specific example which I believe to be typical of the present state of thinking in the land planning movement. In so doing I intend to demonstrate that the assumptions and philosophy that guide current planning efforts are inappropriate to the goals professed.

THE PLANNING PROCESS

In November 1963, the Greater Wilmington Development Council, Inc., sponsored a forum at which the future use of land in Delaware was forecast. The forecast, "Delaware's Tomorrow? – 1982 Impact Visualized"[4] was prepared by the Delaware Chapter of the American Institute of Architects and dealt with that portion of Delaware north of the Chesapeake and Delaware canal.

The forecast was based on the facts that there were at the time 75,000 acres of open land in northern Delaware and that this land was being used for residential and industrial purposes at the rate of 4,000 acres annually. From this the architects concluded that the continuation of present trends, with their accent on the single, detached dwelling on a uniform plot of land, would result in the disappearance of open land by 1982. This was viewed as undesirable, and the architects proposed that Delaware adopt more flexible zoning regulations which would provide for cluster housing and variable-density zoning.

Only a continuation of present trends and policies would make the architects' forecast come true. The disappearance of open land could be prevented by a change in trend or policy. Evidently the architects regarded their proposal as the change in policy which would allow deflection of the forecast. In a very limited sense they were correct, for, given the projected level of demand, under their program some open land would remain in 1982. However, even if the architects' proposal had been adopted in its entirety, some open land would have been used up between 1963 and 1982. Suppose we arbitrarily allow that the rate of consumption of open space after adoption of the proposal would have been reduced by 50 percent. In 1982 there would then be about 37,500 acres of open land. But what about 1983 and thereafter? Presumably the demand for land will continue beyond 1982. If development is inevitable[5], even the most intelligently guided development will eventually lead to the destruction of open space just as surely as would random development in a shorter period of time.

The operating assumption that a continuing demand for space can be met by ingenuity in allocation of space is untenable for a limited space subject to a continuing demand. Such space allocation is a delaying or rearguard action that slows down the ultimate confrontation. It does not "solve the problem," and may in the long run have adverse effects. By appearing to be a solution, it temporarily hides one of the most pressing reasons for public concern – the fact that open land is in danger of becoming exceedingly short in supply.

The analysis is not intended to deny the usefulness of recent land planning proposals but, instead, to delimit more closely their capabilities and limitations. Measures such as cluster housing can provide certain real advantages in the economics of street and utility layout and in the arrangement of buildings to fit the physical characteristics of their sites, to name but a few. But to hold that, in the absence of some measure of population control, cluster housing creates "permanent open space"[6] is to practice self-delusion. Consider, for example, what happens to a countrywide area when all its land is under cluster development. How do you keep land between clusters open unless you stop all further growth? And if you are willing and able to limit growth, then land planning takes on an entirely different character and many new opportunities present themselves.

Because land planning as currently practiced appears to have serious weaknesses which would prevent the attainment of its announced goals, I have questioned a number of people associated with planning at either the academic or the practical level, not only in Delaware, but also in Michigan, Idaho, California, Maryland, and Pennsylvania. The general purpose of my questions was to determine how planners proposed to handle the problem of using a finite amount of space without seriously considering the implications of sustained population growth. More specifically, I wished to determine how planners proposed to handle open-space requirements after their present plans were fully realized. That is, what are the prospects for land use after 1982?

The replies I have received to these questions have shown a remarkable degree of uniformity, and they make it difficult to give planners credit for having fully thought out the implications of their position. A question of the type "Why can't you keep 75,000 acres of land open?" is greeted with incredulity. After considerable ambiguity, the most common standby is that "one can't stop progress." But if progress will not defer to the need for 75,000 acres of open land, what miracle can be expected to restrain progress when but 37,500 acres remain open? Or 18,750? The question "If you do not plan to keep 75,000 acres open, then what amount of open space is planned for?" is generally replied to in this fashion: "Planning is not a document or a blueprint of the future, but is, instead, a process." One is left to wonder how provisions for open space will be met by this "process" which presumes an indefinite continuation of growth with its concomitant space requirements. If the situation which we will have in 1982 with no planning is undesirable, it will be no less undesirable at some later date under the sanction of planning.

Several things appear moderately clear at this time. It seems that planning requires a plan, and that if the preservation of open land is regarded as a valuable objective, then in this particular the plan must be relatively inflexible – one cannot, in a finite area, plan both to preserve open space and to use it up. Inflexibility in this context need not be interpreted to mean that a given segment of open land can never be used for some other purpose, but, instead, signifies that if the goal is to have a determinate amount of open land in the planned area, then any use of the original open land must be compensated for from within the planned area. Failure to compensate for used land must, of course, lead to the loss of open land[7].

Furthermore, though planners are ostensibly committed to the preservation of a wide array of environmental features, the mandate for preservation suffers in actual practice. Leopold[8] points out that planners pay more attention to encouraging development than to protecting the valuable attributes of the environment which will be destroyed or diminished in the process of development. When pushed on this issue, planners contend that preservation of land must ultimately bow to the inevitability of ever-continuing growth. Proponents of this view of growth ignore the ecological doctrine which sets limits on all forms of organic growth. A fundamental tenet of ecology is that any species has the biotic potential to occupy any given finite space, and that under favorable conditions the species in question will increase until the population density is such that growth must cease[9]. In a recent report on the growth of world population, the Committee on Science and Public Policy of the National Academy of Sciences also emphasizes limits on growth: "There can be no doubt concerning the long-term prognosis: Either the birth rate of the world must come down or the death rate must go back up"[10]. An even more recent Academy report indicates that the United States is not exempted from these limiting conditions: ". . . continued growth of the United States population would first become intolerable and then physically impossible"[11].

Thus it appears that never-ending growth is not

only not inevitable, but in fact is impossible, for the mathematics of biology and space set constraints if man does not choose to do so. This insight finds no cognizance in planning theory, which looks upon the suggestion that growth may have limits as being too political or too farfetched for frank discussion. It is my belief that, when faced with the space situation that a long-term perspective on growth discloses, planners all too readily subscribe to the popular supposition that the ecological law of space saturation under favorable conditions is inapplicable to man. However, the fantastic growth of world population over the past half-century[12] indicates that this law does have relevance to man, and, in fact, is more relevant to man than to any other species, for man has developed and is in the process of developing powers that will enable him to extend conditions favorable to his increase throughout the entire planet.

Fremlin[13], in a chilling essay, reminds us that progress in technology (allowing for vast increases in human numbers) does not negate the fact that population growth has limits, but, instead, merely emphasizes that mankind faces the collective choice of determining at what population density it wishes to call a halt – or, in the absence of deliberate choice, of having limiting conditions imposed on it. It is thus seen that the law will be inapplicable to man *only if* man chooses to make it so by exercising his power of foreseeing the consequences of his actions and by then taking appropriate measures to avert those consequences he deems undesirable.

But if one accepts the conclusion that growth does have limits, it is important to attempt to determine what takes place as these limits are approached. While one can debate the extent to which rapid population growth in the United States – our population is increasing at least twice as fast as is the population of any other major industrial country of the Western World[14] – contributes to social, economic, and political problems, the effect of our rapid population growth on land utilization is far less debatable. It is quite clear that growth forces planners to follow a policy of accommodation.

Saying that planning must be flexible, they must alter plans to accommodate more industrial and residential growth than was planned for. That the alterations in planning which are required to accommodate growth invariably cause reductions in the space originally reserved for public functions, playgrounds, parks, and nature areas in order to make way for parking lots, expressways, and residential and industrial sites is dismissed as unfortunate but beyond human control. The further indulgence of this presumably never-ending spiral of growth can be expected to result in the progressive deterioration of many environmental features which are now judged desirable.

RECONSTRUCTION OF LAND PLANNING

My criticism of the way planning is now being done should not be construed as an attack on the very idea of planning, for the future will require more rather than less planning. Furthermore it would be a serious mistake to hold that planners are more responsible for increased densities of population, or that they have any more control over development, than the real estate agent, the highway engineer, and many other public and private agencies. Nor can the planner be expected to modify social, economic, and political conditions through a "Master Plan."

What I am here concerned with is to point out what the planner can reasonably be expected to contribute to the solution of a major problem, and then to suggest a means by which this contribution might be effected.

Reconstruction of land planning must begin with recognition that any land-use policy that completely evades the issue of population control can be no more than a temporary luxury which can lead only to an increasingly painful reckoning in the not-too-distant future. The problems posed by population growth will not disappear if they are ignored; their solution in a democratic society must come by way of common consent, and this will require time and understanding; therefore, the sooner these problems are confronted honestly and directly, the more likely it is that measures designed to alleviate problem situations will be successful. In the area of land planning such a confrontation would, it is hoped, reveal the true nature of the land problem by showing that the chaotic land situation cannot be attributed solely to sprawl resulting from development of large, single-family lots, but, instead, would show that the amount of available open land at any time depends on both (1) the size of the individual bites taken from a stock fund and (2) the number of biters.

Planners have worked exclusively on measures designed to affect rates of usage of open space without giving any serious thought to reversing trends, while at the same time they have given the public the impression that trends are being taken care of. However, I have yet to encounter a plan which makes explicit the fact that only by reversing the trend of ever-continuing use of land for construction purposes can future open space be assured. Instead, the public is enjoined to make more efficient use of the land, with no apparent recognition that use, if continued, uses up. Demographers hold as a truism the statement that in a finite space any rate of human increase, no matter how small, if maintained, will lead to saturation conditions[15]. Planners of land use should realize that, similarly, any rate of open land usage must, if maintained, lead to saturation conditions – that is, no more open land. This is merely to paraphrase the ecologists, who insist it is not the rate that is of ultimate importance but the trend.

OTHER ALTERNATIVES

Because planners have thus far failed to face the logical implications of their position, the public has not had laid before it the widest range of possible planning alternatives. The choices actually presented to the public today are severely circumscribed. Most often, choice is limited to one of two alternatives: one depicting future environmental conditions (for example in 1982) in the absence of planning, the other presenting conditions that might be realized at that future date if planning is implemented. In either instance the forces of growth are accepted uncritically, the main distinction being between growth taking place in a completely unregulated fashion, and growth taking place with its areal aspects subjected to some degree of regulation. Nowhere can one find a plan which portrays the type of environment that could be developed if growth were deliberately curbed or restrained. This is surprising, for planners freely admit that growth presents them with their most vexing problems, many of which definitely lead to a decline in the quality of the human habitat. For example, the population of the greater Wilmington area is expected to grow from the present 213,000 to 583,000 by 1980[16]. Almost everyone involved in planning for this area agrees that a much more desirable environment could be achieved for 1980 if the population was less than the 583,000 projected. In other words, by almost any index chosen – education, housing, transportation, recreation, water, or wildlife – the habitat designed for 300,000 or fewer apparently would be superior to that which must accommodate nearly double that number. But, despite this private admission, the general public remains largely uninformed on the matter.

Thus the public may in fact be dissatisfied with the limited choices now made available, but nowhere can it find any details of other alternatives. To argue, a priori, that the public would not choose any alternative that involved a conscious effort to restrain growth is spurious, for human motivation is complex. As Caldwell[17] points out,

> *One might as convincingly argue that one presumably likes the environment in which circumstances place him if he makes no effort to escape his surroundings or to change them. To the extent that... they [the public] consider efforts to change it hopeless or unwise, they may endure an environment that they consider far from ideal. Dissatisfied with what they have, they have no clear vision of what the ideal might be or have no notion of how a better environment might be attained at a price they would be willing to pay.*

The land planner is in a position to play a significant role in broadening the basis for public choice.

But before the land planner can be expected to provide the vision Caldwell seeks and, therefore, before planning can be expected to yield the cultural and aesthetic harvest of which it is capable, it is necessary that an attack be mounted against the assumption that the population explosion is inevitable. Here the land planner can be immensely useful, for, though the American public remains apathetic to statistical predictions of population growth, it evidences a genuine and growing concern with the landscape this population is producing. In large part, the land planning movement owes its existence to the public's aversion to a completely man-dominated landscape, and this fact can be capitazlied on.

Mumford[18] tells us that statistics can provide us with essential information if we treat them for what they are worth, and would have us use "statistical predictions as road guides that indicate what will happen if we go further, at the same pace, on the same route, not as commands to continue on this road if we find by consulting the map that we are headed in the wrong direction." When confronted with the statistic predicting a national population topping the billion mark in less than a century, the planner is in a position to inform the public of the increasingly undesirable environmental effects this route entails, and thereby to dramatize the fact that the map reveals us to be heading in the wrong direction.

Surely we can expect that, if encouraged to do so, the planner will provide widespread dissemination of the insight set forth by Stewart L. Udall[19], who, in discussing the irresistible pressure that continued population increase places on even the most dedicated of public lands (national parks and wilderness areas), suggests that, "We might formulate a law governing population and open space: *The amount of open space available per person will tend to decrease at a faster rate than the population increases.*" When messages such as Udall's are combined with the many other predictable environmental consequences of continued population growth – increasing problems of environmental pollution, the threat to outdoor recreation, the decline and then demise of wildlife – when these are made abundantly clear to the public, then perhaps the land planner will be able to work within a demographic situation that offers a reasonable promise of success. Cook states that there is conclusive evidence that in the United States the birth rate is largely under voluntary control and that, as a consequence, if we "give the people an accurate picture of what lies ahead populationwise . . . they can be expected to cut their fertility to fit . . . the realities of the modern world"[14, pp. 67-68]. While Cook's optimism may prove to be unfounded, we can surely do no less than to give the public the reasonably accurate depiction of the environmental consequences of sustained population growth which has been so notably lacking in the past.

I believe the planner's reluctance to deal directly with the problems posed by population growth results from the facts that (1) most planners have not yet realized the truly profound implications population growth presents to their practice, and (2) other planners, noting the potentially serious consequences of population growth, believe these matters lie outside their domain[20]. I hope that my effort here will serve in some measure to overcome this reluctance by convincing planners of the first persuasion that a new perspective on growth is called for, and, by bringing this problem to the attention of a wide segment of the American scientific community, I hope that planners of the second persuasion will be encouraged to enlarge their conception of what planning should be. If we as a society are to create and maintain a suitable human environment, we must ask more of our planners, and we must also be prepared to give them the understanding and support they will need.

Man's recent and phenomenal increase in numbers, coupled with his tendency to spread construction activities over ever-wider areas, has led to a growing concern for the quality of his future environment. Land planners, though they acknowledge the relevance of both of these factors, have concentrated exclusively on measures designed to modify and guide construction activities and have ignored the problems posed by unlimited population growth in a limited space. The mathematics of biology and space indicate that this oversight can be, at best, a short-term luxury.

Because land planners have not yet chosen to face squarely the implications of sustained population growth, contemporary planning exhibits serious weaknesses and poses a dilemma. The opinion that optimum human living is to be found under certain environmental conditions clashes head on with the principle of unlimited growth which precludes developing and sustaining the type of environment judged most desirable.

John Dewey[21], in discussing the basic needs of modern society, stated:

> *What is needed is intelligent examination of the consequences that are actually effected by inherited institutions and customs, in order that*

> *there may be intelligent consideration of the ways in which they are to be intentionally modified in behalf of the generation of different consequences.*

I suggest that an examination of the environmental consequences of our inherited belief that a perpetual increase in the number of men and, perforce, in their space-using proclivities is good will show us an environment that becomes increasingly undesirable with the passage of time. Therefore, to bring about more desirable consequences, it is well past time for the serious reexamination and intentional modification of these uncritical beliefs. It is a gross understatement to say that a major revision of land planning would be warranted if the idea of growth were more fully explored. Only the willfully irrational can ignore the implications of such an exploration.

REFERENCES AND NOTES

[1] P. Blake, *Saturday Evening Post,* No. 34, 14 (5 Oct. 1963). Later these ideas received fuller expression in P. Blake, *God's Own Junkyard: The Planned Deterioration of America's Landscape* (Holt, Rinehart and Winston, New York, 1964). A more technical and detailed elaboration of basically the same approach is found in W.H. Whyte, *Cluster Development* (Woodhaven Press Association, for the American Conservation Association, New York, 1964).

[2] For a different point of view on the relation of economics to progress see: J.K. Galbraith, "What the Future Holds for America," *U.S. News & World Report,* 50, No. 25, 58 (1964). In *Science,* **145**, 117 (1964), Galbraith presents a more complete statement on this subject. Central to his thesis is the notion that in a poor society nothing is more important than the poverty which characterizes it. Therefore, in such a society economic considerations dominate social attitudes and rigidly specify the problems that will be accorded priority. However, Galbraith argues that to assume that economic considerations must be an equally dominant influence on social thought and action in a rich society is to set up a barrier to rational thought and needed action when social problems have ceased to be primarily economic.

[3] The argument that economic health is best stimulated by population growth is not accorded universal agreement, and those who take exception to the argument grow in number and influence. See statements by A.W. Schmidt, L. duP. Copeland, and M.S. Eccles in *The Economic Consequences of the Population Explosion,* report on a conference sponsored by the Planned Parenthood Federation of America-World Population Emergency Campaign (New York, 1963). These exceptions have not yet made any significant impression on land planning philosophy.

[4] See Wilmington *Morning News,* 19 Nov. 1963. Approximately 400 people attended the forum to see the architect's slide lecture. The program generated enough interest to warrant a repeat performance before more than 900 people on 28 Jan. 1964. Both showings took place in Wilmington, a city of 90,000, and they provide another example of the public's growing concern for the future of its landscape.

[5] Virtually every contemporary land planning organization accepts this premise uncritically, and in fact justifies its existence largely on the basis that the burgeoning of real estate activities is an inexorable process destined to go on forever. The process is regarded as being subject to partial control in that it might be channelled or guided, but, at the same time, it is held to be outside of human control in that it cannot be curtailed. The question to be resolved, however, is whether such a conclusion may not be more dogmatic than axiomatic.

[6] A.L. Strong, *Open Space in the Penjerdel Region Now or Never* (Pennsylvania-New Jersey-Delaware Metropolitan Project, Inc., Philadelphia, 1963), p. 40.

[7] I am familiar with the argument that various renewal programs may convert built-up lands to open lands. Inasmuch as the nation's annual loss of open land for various construction purposes – residential, industrial, and transportation – has exceeded one million acres for more than a decade, and the amount of land which reverts to open status is but a very small percentage of this figure, the argument has little significance.

[8] L.B. Leopold, *U.S. Geol. Surv. Circ. 414-A* (1960). pp. 3-4.

[9] A.N. Woodbury, in *Principles of General Ecology* (McGraw-Hill, New York, 1954), states the principle thus: "It is easy to show mathematically that if each of the young ones of any species which are started into life in each generation could grow, develop, and reproduce, a relatively small number of generations would produce standing room only for that species." Quoted by W.P. Taylor, in *Natural Resources,* M.R. Huberty and W.L. Flock, eds. (McGraw-Hill, New York, 1959), p. 241.

[10] National Academy of Sciences-National Research Council, *The Growth of World Population.* A report prepared by the Committee on Science and Public Policy, National Academy of Sciences, (NAS-NRC Publ. No. 1091, Washington, D.C., 1963), p. 9. For a pithy, philosophical synopsis of the ecological and demographic dimensions of our burgeoning space problems, see P.B. Sears, *Science 127,* 9 (1958).

[11] National Academy of Sciences-National Research Council, *The Growth of U.S. Population.* A report prepared by the Committee on Population, National Academy of Sciences. Quoted by E. Langer in *Science 148,* 1205 (1965).

[12] National Academy of Sciences, *The Growth of World Population.* A report prepared by the Committee on Science and Public Policy, National Academy of Sciences (NAS-NRC Publ. No. 1091, Washington, D.C., 1963), pp. 8-9.

[13] J.H. Fremlin, *New Scientist 415,* 287 (1964).

[14] R.C. Cook, *Social Educ. 24,* No. 2, 65 (1965).

[15] F.N. Notestein, in *World Population and Future Resources,* P.K. Hatt, ed. (American Book Co., New York, 1952), p. 58.

[16] Wilmington *Morning News,* 13 Oct. 1964.

[17] L.K. Caldwell, reprint from the *Transactions of the Twenty-Sixth North American Wildlife and Natural Resources Conference,* 6-8 March 1961 (Wildlife Management Institute, Washington, D.C., 1961), pp. 41-42.

[18] L. Mumford, from a lecture at Princeton University, Nov. 1964. Quoted in the Philadelphia *Evening Bulletin,* 16 May 1965.

[19] S.L. Udall, *Population Bull. 20,* No. 4, 99 (1964).

[20] In a personal communication, dated 12 Apr. 1965, a Delaware planner states, "The planners' public service is essentially finite and statistical work. Casting future balances is our social contribution. We can't preach and sound the alarm, except as a quiet sideline. It is the responsibility of others to change attitudes, I fear. However, noting the grave portents of over-population, planners are naturally grateful when the alarms are sounded."

[21] J. Dewey, in *Readings in Ethics,* G.H. Clark and T.V. Smith, eds. (Appleton-Century-Crofts, New York, ed. 2, 1935), p. 418.

[22] This study was supported by a University of Delaware Summer Faculty Fellowship.

42

Irving Rosow

THE SOCIAL EFFECTS OF THE PHYSICAL ENVIRONMENT

Much of the force behind the movement for housing reform is epitomized by one of its most articulate exponents: "The tenants' entire social life may hang on the smallest whim of the greenest draftsman or rent collector"[3]. Although it is an extreme statement – and one which, if taken too literally, imposes a severe responsibility – it sums up a basic operational assumption of idealistic housing practitioners. As a movement and ideology, housing and planning rest on the premise that, by the manipulation of the physical environment, we can control social patterns. If housing exerts an independent influence on how people live, then the creation of certain housing conditions can change social relationships. We can affect the choice of friends, family adjustments, and generally how people spend their time. All this is subject, of course, to given cost limitations. But apparently, within these restrictions, different housing decisions may have different social consequences. The problem of the housers is to learn more about how and in what way factors of design do indeed affect patterns of social life.

In this paper, we should like to assess this premise against the findings of a growing body of housing research. This is by no means a systematic, exhaustive coverage of the literature or of housers' working "hypotheses." It is more of an interim clarification, now that "some of the early research returns are in," of those assumptions which tend to be more effective than others in realizing social policies which are at once the housers' goals and guides.

The assumptions may be classified into several general categories to which they refer: (1) social pathology and social efficiency; (2) "livability" of the dwelling unit; (3) neighborhood structure and integration; and (4) aesthetics. Our concern here is mainly, though not exclusively, with low- and middle-income public housing or similar planned neighborhoods.

SOCIAL PATHOLOGY

The housers have effectively won their point that slum clearance pays dividends in terms of social welfare and hard cash. The correlation between poor housing and the incidence of crime, disease, juvenile delinquency, mortality, etc., have been established beyond doubt. Planners no longer have to deal seriously with the objection that correlations do not prove causal relationships. Enough work has already shown rehousing to be a sufficient condition to produce a sharp, significant decline in these morbidities[5,6,40]. Slum clearance and the elimination of obviously substandard housing have reduced some virulent social problems and contributed to health and welfare among underprivileged groups.

Furthermore, Rumney, among others, has demonstrated impressively that the dividends of rehousing are not only to be reaped in social values like health, but also in dollars and cents[38,39]. Slums have been highly profitable rental and speculative properties to their owners, but largely at public expense. Municipalities have subsidized slums indirectly. For 13 major cities, the public expenditures on blighted areas have exceeded revenues from them by ratios of 2.2 to 9.9[39]. In other words, the direct costs of public services (relief, police and fire, welfare, etc.) were between two and ten times as high as the taxes which these areas contributed to the public coffers. These ratios are reversed for high-rental residential districts, which tend to yield considerably more revenue than expense, thereby providing the funds for slum services. Despite certain difficulties with such indices[14], the overall outline is clear. Depressed housing areas represent social and financial liabilities which are greatly eased by clearance and rehousing.

LIVABILITY

The factor of livability is the most instrumental aspect of design and is most commonly related to the individual dwelling unit[33,34,35]. It refers to the utilitarian organization of space and facilities which best accommodates the needs of the occupants and minimizes frictions and frustrations from factors of layout and design[17]. According to one sociologist of housing:

> *Modern architecture does its best to accommodate in the most utilitarian manner the informal aspects of private family living. . . . Room arrangements [are favored] that serve the everyday life of the family and reduce household chores to a minimum. . . . Relaxation and informality in the relations between different family members are promoted*[32].

Livability thus becomes an expression of the "functional" goals of modern architecture and design, and its norms are efficiency. It implies a careful adaptation of design to use, and frequently specific features or space are designed for specific purposes.

But there is evidence of "nonconforming" usage. Many features are neglected, used for purposes other than those intended, the activity takes place elsewhere, or does not take place at all in the confines of the plan. This applies both to the dwelling unit and to the neighborhood.

In the private dwellings, for example, study space is commonly provided in the "children's bedroom" where, at desks or built-in desk shelves, the younger generation can do its lessons in privacy. This nominally removes them from the distractions of family intercourse without imposing undue restraint on the rest of the family. Yet the picture of the teenager with homework scattered over the living room floor and the radio or TV set blaring in his ear has become almost a stereotype, although to the writer's knowledge it has not been examined in research.

Other features have had a varying fate in nonconforming usage. The short-lived experiment with the tiny "Pullman" kitchen is a case in point. Another, in private houses, is the basement recreation room which matures into a conventional storage space. Or in residential suburbs, the facade of house after house may have enormous "picture windows" which are covered by venetian blinds or drapes (often made-to-measure) to give some privacy to the occupants.

Similar nonconforming usage is found in community facilities, with the community center a conspicuously neglected amenity[16]. In one interracial development, for example, a professionally staffed community center was created to promote interracial activities and contact. But only 15 percent of the women reported meeting women of the other race in the center compared with almost 60 percent who named the community laundry[25].

These few illustrations are typical. They can be multiplied, although these suffice to make the point here. Nonconforming usage is important because it almost invariably represents fixed installations which become an economic liability. The space and cost might have been otherwise invested. The occupants are thereby penalized in some sense for incorrect predictions about use or the plasticity of habits.

In mass housing, livability of the domicile necessarily takes on a more restricted meaning than the ideals represented by modern architecture. Cost limitations and family patterns, which vary in time, in stage of the family cycle, and from one group to another, forbid detailed attention to individual preferences and force standardization of design. In effect, this reduces livability largely to considerations of housekeeping and mechanics – choices of kitchen layout; easy-to-clean wall, floor, and window surfaces; convenient storage; etc. Not only are these decisions sharply restricted by cost factors (e.g., consolidating fixtures about plumbing cores), but they frequently involve choices between space and appliances. In this respect, public housing may represent rather few differences from *any* new housing. This is especially true to the extent that, aside from *gross blunders* of design, housekeeping ease may be increased more by appliances than by design decisions. The availability, for example, of a clothes dryer may save more time and exasperation than a brilliant subtle detail of planning.

Beyond housekeeping mechanics, livability problems can be essentially reduced to factors of space which afford the room for group activities *and* for privacy. In the arrangement of space the designers can make ingenious decisions which minimize frictions. But this is most true in relatively expensive residential housing. In middle- and low-income dwellings, the cost limitations severely restrict the amount of space available. And only so many alternatives exist as realistic choices. The different livability consequences among them may be even more limited.

The effect of the space variable on livability is not simply in crowding (whether persons per room or use crowding) or space for social activities. Usually, there is enough room for those activities in which the family engages together, although larger social affairs, such as parties, may suffer in small dwellings. The basic problem boils down to privacy. To some extent it is possible to isolate part of the dwelling by clever design: careful solutions of circulation, sound insulation from closet placement, etc. But real privacy requires room for comfortable retirement; and unless this space is "manufactured," the lack of privacy may become a source of friction and frustration. The difficulty of creating space without sacrificing other indispensables is a problem of which designers are only too well aware.

The problem of livability may be viewed in a somewhat different perspective. We may properly ask under what conditions livability factors are positive causes of frustration, friction, and tension, and the extent to which they are significant in the social adjustment of the family. In Westgate, a prefabricated housing project for married students, there were extremely serious livability difficulties[20]. Notwithstanding, Festinger reported:

> *This general satisfaction [with living in the community] existed in spite of, and seemed to compensate for, many physical inadequacies of*

the houses. At the time of our study there were many physical nuisances in the houses. Some were incompletely equipped, the grounds were muddy and had not yet been landscaped, they were difficult to heat in the winter, and the like. One example of the reaction to such physical inadequacies will suffice, however, to illustrate the point. At the time of the investigation many of the houses had trouble with the roofs. The houses were prefabricated, and many of the roofs had not been assembled properly . . . in the interviews about one-third of the residents reported that the roofs leaked. Any rain accompanied by a moderately strong wind would apparently raise the roof slightly, and water would pour down the walls. One family reported that in a particular strong rain the roof had started to blow off; the husband had to go outside and hold the roof down until the wind subsided.

It is remarkable, however, that even such serious physical inconveniences did not create a strong *impression on the residents. Typically the reaction was, "Oh yes, there are many things wrong with these houses, but we love it here and wouldn't want to move."*

The adequate and satisfying social life was sufficient to override many inconveniences. The result was a rather happy social and psychological existence[19] *(italics inserted).*

This is not to recommend such Spartan trials as built-in housing features. But housing attitudes are far too complex to ascribe them specifically to livability frustrations which may be much less relevant than housers suppose. The adjustment of family members to one another is a function of social and personal factors to which the dwelling may contribute relatively little. In another study, for example, of 33 families who built homes, five had absolutely no dissatisfaction with their previous dwelling and the complaints of the others centered about highly discrete details which were annoying[36].

The almost standard response of housers to such evidence is, "Aha, but these people don't know what it can be like to live in a well-designed dwelling." This is largely true. But on the other hand, there is little evidence that satisfaction with new housing is directly related to livability resulting *from design per se* except when there is a significant improvement in housing, especially where people came from substandard housing, or occupants are particularly conscious of housing in highly literate, sophisticated terms.

This brings us to a second point regarding space and privacy. There is evidence of important class differences in the meaning and valuation of privacy. In a study of 50 families in New York state, Cutler found that exactly one half of the people in lower-class families complained about the lack of privacy in comparison with only 10 percent of the middle class and none of the upper class[9]. Furthermore, in defining the elements of privacy, lower-class respondents mentioned having a room of one's own twice as frequently as upper-class people (70 percent versus 34 percent). Conversely, 44 percent of the upper class compared with 8 percent of the lower class mentioned such factors as outdoor privacy, rooms that could be closed off, extra baths, extra guest rooms, and the maid living away from the family. In other words, higher social groups take for granted amenities which the lower classes would like. The lower-class groups basically want more space than they have available. Although satisfaction with housing is clearly related to size of dwelling[8,9,13] — which is in turn related to many other features as well — crowding, privacy, and space limitations may not be so important to working-class groups as to other segments of the population. Dean found in a Steubenville sample that, although 21 percent of the semiskilled and unskilled workers were doubled up with relatives, only 6 percent specifically complained about this in terms of overcrowding, whereas in the white collar group 6 percent were living with relatives, but 29 percent complained of overcrowding[13]. This is consistent with other findings in which space was subordinate to other features in housing complaints of working-class groups.

Thus conceptions of privacy and adequate space have different class meanings; and there is little evidence that these assume drastic importance in family adjustment *provided* that some adequate space standards are met and that the class culture does not demand private space for highly individual personal activities. Chapin, for example, observes:

> *Thus privacy becomes a value. One may question the validity of imputing to others the desires, needs, and wants that are characteristic in this respect of nervously high strung, sophisticated, and responsive intellectual persons. Perhaps the common run of home occupants is not as sensitive to deprivation of privacy as some, but it is safer to assume that some individuals born to the common run of humanity will be sensitive. . . . Privacy is needed for thinking, reflection, reading and study, and for aesthetic enjoyment and contemplation. Intrusions on the fulfillment of personal desires need to be shut off. . . .*[7]

This is a statement of highly personal goals pursued in the home. As Chapin indicates, it tends to be highly class selective in its relevance – or, indeed, to characterize particular social types within classes, particularly middle- and upper-class intellectuals and aesthetes. Their needs can ultimately only be satisfied with space, which again involves cost more than design factors.

In general there tends to be an incompatibility between highly individuated housing goals and standardization imposed by mass housing. Virtually the only manipulable variable in large-scale projects is the diversification of dwelling unit sizes within the overall budget. This is perhaps the most opportune way to juggle space to satisfy the needs of specialized groups – whether by providing larger units for larger families and those with high privacy activities or by smaller units for smaller families (young and old couples).

The provision of space poses additional problems, since large increments of floor space may have to be provided to realize small increments of actual free space. In studies of the Pierce Foundation, for example, middle-class families tended to fill free space with furnishings so that, from one family to another, similar amounts of open space were found despite different sizes of comparable rooms. Some of these furnishings were for storage and others for decoration. It is important to note, however, that when the designer assumes that he is squeezing out several more cubic feet of open space in reality he may be creating a "vacuum" which the occupant will "abhor" and fill at the earliest possible opportunity.

Further, although surveys and the like may reveal considerable agreement about the *categories* of housing complaints or desired housing features, the research on livability has not "weighted" these factors, especially by class and social typology variables, to reveal how important housing values actually are to different groups. Who is willing to sacrifice how much of what (including money) to get the kind of housing he wants? For example, a New York realtor not long ago expressed amazement at his middle-class tenants who preferred to give up a bedroom and sleep on a studio couch in the living room in order to own a car. Or as one houser put it:

> *We are inclined to look down our noses at the family who lives in a shack so they can own an automobile, or the six persons who live in one room and yet pool their resources to buy a television set; but are we sure we are right?*[18]

Nowhere does the problem of relative class values become so acute as the question of livability. The planners, designers, architects, and housers must operate with assumptions about how people do live, how they want to live, how they would live if given a chance[37]. Most of them conventionally assume that the people for whom they are designing must be "educated" to appreciate and exploit the housing advantages being placed at their disposal. When the tenants fail to respond, this is often written off as "no fair test" because the housing itself was too restricted and not enough design influence was brought to bear. On the contrary, there is ample reason to believe that design factors generally represent only conditions of the physical environment which do not significantly alter human outlooks apart from the significant social experience which housing may not provide. And it remains to be seen how the average new project or even the better ones which have been built with strong cost restrictions can affect the relations of family members to one another – for those people who did not come from substandard housing. Or, more precisely, it is an open question whether livability factors can be significantly separated from the sheer fact of rehousing lower-class groups or custom building for upper-class families. The intermediate range may be slightly but not substantially affected by livability provisions.

In summary, the factors of livability which can be

influenced by design tend to apply selectively to class groups and to those with highly specialized housing needs. It is mainly the lower-class group, moving into new housing from conspicuously poor dwellings, who benefit most from factors of livability. The more specialized needs of people with strong privacy desires can scarcely be met in new dwellings except at high cost.

COMMUNITY INTEGRATION

Intimately bound up with the planning movement is a reaction to the fragmentation and segmentalization of urban life. There is an effort to recapture an "organic" environment in which people will be integrated into communities on a residential basis. This is epitomized in the controversy over the neighborhood plan in its various names and guises[1,10,11,15,24,31,42].

A series of studies has amply demonstrated the effect of residential placement on group formation and the selection of friends[4,16,20,21,23,28,29,30 et al]. The evidence is well known and so clear that it warrants little elaboration. In planned communities friendship groups are determined by two variables: proximity of neighbors and orientation of dwellings. People select their friends primarily from those who live nearby and those whom their home faces.

The full significance of these patterns, however, is less clear. Housers interpret this to mean that people are indeed being integrated into community or neighborhood structures. But these patterns are adumbrated by several factors: deviant cases in which planned communities did *not* result in this spontaneous cohesiveness[12,19]: that they characterized the earliest period in the life of the community, but then became instable and gave way to the extension of friendships further afield, outside the neighborhood[16,22]; that the neighborhood community developed a system of social stratification which was not solidary in its effects[12,16,21,23,26,29]; that community integration was most directly related to homogeneous social composition and inversely related to length of residence[22]. In other words, when people of a similar type (e.g., students, war plant workers) are brought together into a new community for a relatively impermanent period, considerable social solidarity springs up. But the longer people stay and the more diversified the group, the more is solidarity affected by status differentiation and the establishment of friendships elsewhere.

The deeper issues presented here center about the significance of homogeneous social composition. Some critics of the neighborhood concept have contended that homogeneous communities tend to formalize social segregation and defeat the very democratic objectives which neighborhood planners seek[24]. According to others, the contrary desire to establish socially heterogeneous communities will not necessarily achieve democratic aims. Some argue that there is no reason to assume that people will spontaneously or willingly enter heterogeneous planned neighborhoods any more than similar unplanned neighborhoods[22]. If they do, there is no assurance that they will interact harmoniously, but may, on the contrary, perpetuate existing differences[12,26,43].

Some changes are perceptible. For example, in interracial housing, attitudes toward Negroes became more favorable on closer contact[25]. In two projects with internal segregation of Negroes, between two thirds and three fourths of the white occupants favored segregation while in two others with no internal segregation, about 40 percent of the whites favored segregation. These attitudes were accompanied by extremely sharp differences in association between the races. In the segregated projects, less than 5 percent of the white women knew any Negro woman by first name or engaged in such cooperative activity as babysitting, shopping together, helping in illness, etc. In contrast to this, the proportion of white women who had such associations with Negroes in the nonsegregated projects varied from one fourth to three fourths. The occupancy patterns evidently served to reinforce negative attitudes in the segregated projects.

Jahoda and West indicate the strong influence exerted by previous interracial experience on attitudes[25]. Of those whites who had previously lived in racially mixed neighborhoods, 19 percent expected trouble and 17 percent found relations better than they had expected; but of those who had no prior biracial experience, 56 percent expected trouble and 43 percent were pleasantly surprised by the absence of conflict. But under the best conditions, among the whites who had *both* previous residential experience *and* worked with Negroes, only 45 percent favored interracial housing. Although whites' desire to move from an interracial project was directly related with their expectations of a Negro invasion, among those who expected no change in the proportion of Negro residents, 50 percent nonetheless planned to move – not necessarily, of course, as a reflection of their racial attitudes. In this project, the management's policy of a "quota" system maintained a stable proportion of the races through time. But in the same town, a comparable project without a "quota" showed the proportion of Negroes gradually reaching between 80 to 90 percent.

Unquestionably, the influence of greater contact on racial images had an impact on stereotypes, but the overall picture is one of accommodation rather than community integration. Changes in attitude ultimately depended on more than sheer contact. Jahoda and West indicate, "But in part, also, the more favorable expectations [of whites] are the result of that kind of *sustained* interracial contact which displaces racial stereotypes"[25] (italics inserted).

The factor of residential mobility has bearing both on the necessary condition of sustained contact in changing racial stereotypes and on the friendship patterns of homogeneous communities. Apparently, about three fourths of the American people changed their residence during the 1940-1950 decade[2]. Furthermore, among the working classes most eligible for planned projects, there is the highest residential mobility and the lowest integration into the local neighborhood. Higher-class groups who are less mobile and have longer residential tenure have more friends within their immediate neighborhood. Nonetheless, in terms of *sources* of friendships, the lower-class groups draw upon immediate neighbors more than twice as frequently as middle- and upper-class groups (33 percent versus 14 percent)[41]. The reasons for the mobility may vary – economic opportunity, desire to own a home, middle-class aspirations, changing stages in the family cycle, etc.[14,36]. The motives are less important than the implications for the integration of the community. High mobility is not a condition favoring *sustained* contacts necessary to change racial attitudes. Nor is either high mobility or long tenure in planned neighborhoods conducive to the sustained solidarity and friendship patterns observed in the newly formed communities.

One may properly inquire what the alternative friendship patterns may have been in homogeneous *unplanned* neighborhoods. Apparently, they were of a similar character to those of the planned neighborhood. Lower-class groups draw heavily upon the local area for their friendships, but because they move frequently, their friends are spread about; upper-class people have more diverse sources of friendships, but with more stable residence they gradually extend their local contacts and become integrated into the community[11]. Planned communities at a given density and layout provide X number of conveniently located neighbors with whom friendships are established. It remains to be shown that unplanned neighborhoods of equal density and equally homogeneous composition do not provide the identical patterns of friendship formation and group structure.

We may here be the innocent victims of a research bias. Housing research has concentrated heavily on the planned community and findings have been interpreted as changed social patterns, although no base point for change was established. In fact, we may have inadvertently discovered basic patterns which have been operating in urban environments but which were not intensively investigated prior to the modern housing developments.

Under these circumstances, one is forced to ask anew, "What are the *social patterns* which housing and design have changed?" If anything, one is impressed perhaps less by the changes than by the continuities and the persistence of previous social patterns – with the exception of the easing of social pathology by the movement from substandard housing areas. There is little conclusive evidence of more than ephemeral changes in social patterns through the medium of planned communities. Particularly, the integration of the community does not seem to be significantly greater than is found in homogeneous, unplanned neighborhoods. Stratification and racial divisions remain effective forces.

Thus, to all intents and purposes, it remains to be established how planning does significantly more than shift or regroup active – not latent – social relations into new settings.

AESTHETICS

The final factor which concerns us is the commitment of planners and housers to an aesthetic way of life. In this, their assumptions of psychological effects of aesthetic atmospheres may be on firmer ground. Light, air, greenery, variety of color, materials, forms – within the dwelling unit and the neighborhood – may create interest which affects people's moods. In extreme cases this is clear. One need only allude to women's customary responses to blue fluorescent light.

There is reason to suspect class differentials on the importance of the variable of aesthetics. But planners have here a more intangible factor which, despite its subjectivity, may have subtle effects on mood and thereby potentially affect tolerance thresholds and the texture of social interaction. Very little research has been done on the psychological impact of different aesthetic environments (beyond some preliminary research into effects of colors), so there is little ground on which to evaluate the importance of design from this standpoint. Certainly we know its importance in merchandising – although housers may not be identified with the aesthetics in some proven packaging. In housing, however, we would expect the aesthetic element to involve much more of a Gestalt perception process rather than segmental responses. Aesthetic judgments will have to await further research.

REFERENCES

[1] American Public Health Association, Committee on the Hygiene of Housing, *Planning the Neighborhood,* Chicago: Public Administration Service, 1948.

[2] Bauer, Catherine, "Social Questions in Housing and Community Planning," *Journal of Social Issues, 7,* Nos. 1 and 2 (1951), pp. 1-34, Special Housing Issue.

[3] Bauer, Catherine, "Good Neighborhoods," *Annals,* 242 (1945), pp. 104-15.

[4] Caplow, Theodore, and R. Forman, "Neighborhood Interaction in a Homogeneous Community," *American Sociological Review,* 15 (June 1950), pp. 357-66.

[5] Chapin, F. Stuart, "The Effects of Slum Clearance and Rehousing on Family and Community Relationships in Minneapolis," *American Journal of Sociology,* 43 (1938), pp. 744-63.

[6] Chapin, F. Stuart, "An Experiment on the Social Effects of Good Housing," *American Sociological Review,* 5 (Dec. 1940), pp. 868-79.

[7] Chapin, F. Stuart, "Some Factors Related to Mental Hygiene," *Journal of Social Issues,* 7, Nos. 1 and 2 (1951), Special Housing Issue.

[8] Cottam, Howard, *Housing and Attitudes Toward Housing in Rural Pennsylvania,* State College, Pa.: Pennsylvania State College School of Agriculture, 1942.

[9] Cutler, Virginia, *Personal and Family Values in the Choice of a Home,* Ithaca, N.Y.: Cornell University Agricultural Experimental Station, 1947.

[10] Dahir, James, *Communities for Better Living,* New York: Harter & Row, 1950.

[11] Dahir, James, *The Neighborhood Unit Plan,* New York: Russell Sage Foundation, 1947.

[12] Danhof, R., "The Accommodation and Integration of Conflicting Cultures in a Newly Established Community," *American Journal of Sociology,* 48 (1943), pp. 14-43.

[13] Dean, John, "The Ghosts of Home Ownership," *Journal of Social Issues,* 7, Nos. 1 and 2 (1951), Special Housing Issue.

[14] Dean, John, "The Myths of Housing Reform," *American Journal of Sociology,* 54 (1949), pp. 271-88.

[15] Dewey, Richard, "The Neighborhood, Urban Ecology and City Planners." *American Sociological Review,* 15 (Aug. 1950), pp. 502-07.

[16] Durant, Ruth, *Watling,* London: P.S. King & Son, 1939.

[17] Federal Public Housing Authority, *The Livability Problems of 1000 Families,* Washington: National Housing Agency, 1945.

[18] Ferrier, Clarence, "Frontiers of Housing Research," *Land Economics* 25 (1949), supplement.

[19] Festinger, Leon, "Architecture and Group Membership," *Journal of Social Issues,* 7, Nos. 1 and 2 (1951), Special Housing Issue.

[20] Festinger, Leon, Stanley Schachter, and Kurt Back, *Social Pressures in Informal Groups,* New York: Harper & Row, 1950.

[21] Form, William, "Status Stratification in a Planned Community," *American Sociological Review,* (10 (Oct. 1945), pp. 605-13.

[22] Form, William, "Stratification in Low and Middle Income Housing Areas," *Journal of Social Issues,* 7, Nos. 1 and 2 (1951), pp. 109-31, Special Housing Issue.

[23] Infield, H., "A Veterans' Cooperative Land Settlement and Its Sociometric Structure," *Sociometry,* 6 (1947), pp. 50-70.

[24] Isaacs, Reginald, "The Neighborhood Theory," *Journal of the American Institute of Planners,* 14 (1948), pp. 15-23.

[25] Jahoda, Marie, and Patricia Salter West, "Race Relations in Public Housing," *Journal of Social Issues,* 7, Nos. 1 and 2 (1951), Special Housing Issue.

[26] Jevons, R., and J. Madge, *Housing Estates,* Bristol, England: University of Bristol, 1946.

[27] Kilbourn, Charlotte, and Margaret Lantis, "Elements of Tenant Instability in a War Housing Project," *American Sociological Review,* 11 (Feb. 1946), pp. 57-64.

[28] Kuper, Leo, *Living in Towns,* London: Cresset Press, 1953.

[29] Merton, Robert, "Patterns of Interpersonal Influence and Communications Behavior in a Local Community," in Paul Lazarsfeld and Frank Stanton (eds.), *Communications Research, 1948-49,* New York: Harper & Row, 1949.

[30] Merton, Robert, "The Social Psychology of Housing," in Wayne Dennis (ed.), *Current Trends in Social Psychology,* Pittsburgh: University of Pittsburgh Press, 1948, pp. 163-217.

[31] Perry, Clarence, *Housing for the Machine Age,* New York: Russell Sage Foundation, 1939.

[32] Riemer, Svend, "Architecture for Family Living," *Journal of Social Issues,* 7, Nos. 1 and 2 (1951), Special Housing Issue.

[33] Riemer, Svend, "Designing the Family Home," in H. Becker and R. Hill (eds.), *Family, Marriage and Parenthood,* Boston: D.C. Heath, 1948.

[34] Riemer, Svend, "Maladjustment to the Family Home," *American Sociological Review,* 10 (Oct. 1945), pp. 442-48.

[35] Riemer, Svend, "Sociological Perspective in Home Planning," *American Sociological Review,* 12 (Apr. 1947), pp. 155-59.

[36] Rosow, Irving, "Home Ownership Motives," *American Sociological Review,* 13 (1948), pp. 751-55.

[37] Rosow, Irving, "Housing Research and Administrative Decisions," *Journal of Housing,* 8 (1951), pp. 285-87.

[38] Rumney, Jay, and S. Shuman, *The Cost of Slums in Newark,* Newark, N.J.: 1946.

[39] Rumney, Jay, "The Social Costs of Slums," *Journal of Social Issues,* 7, Nos. 1 and 2 (1951), Special Housing Issue.

[40] Rumney, Jay, and S. Shuman, *The Social Effects of Public Housing,* Newark: Newark Housing Authority, 1944.

[41] Smith, Joel, William Form, and Gregory Stone, "Local Intimacy in a Middle-Sized City," *American Journal of Sociology,* 60 (Nov. 1954), pp. 276-85.

[42] Sweetser, Jr., Frank, "A New Emphasis for Neighborhood Research," *American Sociological Review,* 7 (Aug. 1942), pp. 525-33.

[43] Wright, H., *Rehousing Urban America,* New York: Columbia University Press, 1935.

43

Bernard Asbell

THE DANGER SIGNALS OF CROWDING

One of the engaging superstitions of our time, a three-part myth, is that overpopulation is just around the corner, that a shortage of food will do us in by the millions, and that only mass reduction of births (especially among the proliferating poor) will prevent disaster.

This is not an essay to encourage complacency about our rapidly growing population. The dangers facing us are real, perhaps more imminent than most of us think. But to deal with them properly, we need to catch up on some newly emerging scientific research that contradicts widely held assumptions.

First, while there must be some level of population that would constitute *over*population of our finite-sized planet, none of us has any idea what that level is. What will threaten us in advance of overpopulation is *crowding,* a wholly different idea. We are starting to learn something about crowding, thanks to a handful of scholars who are pooling an unlikely mixture of insights ranging from anthropology to

---

biochemistry. One thing they are finding is that under some circumstances millions, perhaps scores of millions, may live in densely packed harmony (say, in stacked dwellings of well-designed apartment houses in a Boston-to-Washington megalopolis). But under other circumstances the old saw, "two's company but three's a crowd," may be a sound scientific warning.

Second, food supply is, at best, only indirectly linked with our possible doom. If we were to die of crowding by the millions, that would happen long before the food supply ran out. We would die not of hunger, but of shock, lowered resistance, rampaging disease, nervous breakdowns, and, possibly, mass mayhem and widespread murder. Of the last, we are already witnessing early warning signals in our cities — collections of humanity that are not overpopulated but, in certain neighborhoods, dangerously crowded.

Finally, campaigns for birth control may do little to lessen the oppression of crowding, at least in the short run. In fact, there is evidence that a runaway birthrate among the poor is not so much a cause of crowding as a result of it.

Crowding is a specific happening, clinically observable and definable. In simplified terms, crowding occurs when organisms are brought together in such manner and numbers as to produce *physical* reactions of stress. Important among these reactions is stepped-up activity of the adrenal glands. When these reactions to stress are widespread and sustained, they are followed by physical weakening, sometimes rage and violence or extreme passivity, a rise in sexual aberrations, and a breakdown of orderly group behavior. What may follow is a tidal wave of deaths, ending when the population is no longer crowded.

Those things have happened time and again, in various combinations, to all kinds of animals, from lemmings in Scandinavia to deer trapped on an island in Maryland. Ethologists — those who study group behavior of animals — have been cautious about projecting their findings on man. But Dr. Edward T. Hall, prominent Northwestern University anthropologist, and some colleagues in psychiatry are stitching together evidence that the animal in man may be governed by somewhat the same system of stress reactions. Hall discusses this in *The Hidden Dimension*, his definitive book on the subject of crowding.

"If man does pay attention to animal studies," says Hall, "he can detect the gradually emerging outlines of an endocrine servomechanism not unlike the thermostat in his house. The only difference is that instead of regulating heat the endocrine control system regulates the population."

John J. Christian, an ethologist also trained in medical pathology, was a pioneer in discovering that population buildup, leading to stress, brings on an endocrine reaction and, finally, population collapse — which he calls a "die-off."

WHAT KILLED THE SIKA DEER?

About a mile out in Chesapeake Bay lies a small patch of land, half a square mile, called James Island. It is uninhabited, or at least was until 1916 when someone adorned the island by releasing four or five Sika deer. The deer were fruitful and multiplied until, by 1955, they had procreated a herd of almost 300 — about one per acre, extremely dense for deer.

In that year, Christian visited the island, bringing a hypothesis and a gun. He shot five animals and made detailed examinations of their adrenal glands, thyroid, heart, lungs, gonads, and other tissues. Their organs appeared normal in every way except one. The adrenal glands were immensely oversized, bulging like overused muscles. When animals are under frequent or sustained stress, their adrenals — which are important to regulation of growth, reproduction, and defenses against disease — become overactive and enlarged. If this abnormality was related to crowdedness — and if the herd population was still growing — clearly James Island was in for an interesting time. Christian waited and watched.

For the next 2 years, herd size stayed about the same. Then in the third year, 1958, more than half the herd inexplicably dropped dead. The island was strewn with 190 carcasses in two years, chiefly females and young, leaving 80 survivors.

What had killed so many? It was not malnutrition, for food was abundant. The coats of the dead deer shone healthily, their muscles were well developed, plenty of body fat. For that matter, if the epidemic of whatever-it-was was so severe, how come 80 survived it — and now appeared robust?

After the die-off, Christian revisited the island in 1960, shot a few more animals, and examined them. For one thing, they were substantially — more than 30 percent — larger in body size than those shot at the climax of the crowding. But the more striking thing was that their adrenals were *half* the size of those examined earlier – back to normal. In young deer, they were one fifth the size of their overstressed counterparts.

"Mortality evidently resulted," Christian later reported to a symposium on crowding, "from shock following severe metabolic disturbance, probably as a result of prolonged adrenocortical hyperactivity. . . . There was no evidence of infection, starvation, or other obvious cause to explain the mass mortality." Subsequently, he says, it was found that the hyperactivity had in all probability resulted in potassium deficiency.

A landmark study it was, finding out how those deer died. But it doesn't tell us how they *lived*, what their behavior was like just before the agonies of emotional and physical stress killed them. A search for clues brings us to another study, which by chance was taking place at the same time in the same state.

WHY THE RATS WENT BERSERK

In a stone barn at the outskirts of Rockville, Maryland, John B. Calhoun began breeding populations of Norway rats, a deliberate creation of crowding. In each of several rooms, Calhoun set up four pens, connecting them in a row by ramps arching over their separating walls. In the wild, these animals normally organize in sexually balanced groups of about 12. The penned rats soon multiplied to an adult population of 80, almost twice the comfortable number of 12 to a pen.

Rats are busybodies, and so got in the habit – at least at first – of scurrying over the ramps from pen to pen. Also they were conditioned to eat in the presence of others, cheek to cheek. So the two central pens, where they were most likely to find companions, became popular "eating clubs." Calhoun at times observed as many as 60 eaters crowding into a single inner pen.

This crowding soon led to what Calhoun calls a "behavioral sink." (A striking term, Webster's *New World Dictionary* defines the noun *sink* as 1. a cesspool or sewer; 2. any place or thing considered morally filthy or corrupted.)

A single dominant male took charge of each of the less-populated end pens, preventing the entrance of other males, but freely permitting the comings and goings of his females. *His* females. The lord rat of each end pen established a harem of a half-dozen or more females.

Because of these end-pen harems, females were distributed among the four pens fairly evenly. Males, however, were overwhelmingly crowded into the middle pens. And their natural manners, under the stress of crowding and shortage of ladies, gave way to havoc. The more dominant males took to violence. They would suddenly go berserk, attacking females, juveniles, and passive males, by biting their tails, sometimes severing them entirely. The floor was almost always bloody from these carryings-on, which Calhoun had never before seen among the species. Then there emerged a group of males that made sexual advances on unreceptive females, often those not in heat, and later on other males, and finally juveniles. Their ability to perceive appropriate sex partners seemed to have vanished.

Two other types of male emerged, almost opposite in levels of activity. "The first," Calhoun reports, "were completely passive and moved through the community like somnambulists. They ignored all the other rats of both sexes, and all the other rats ignored them. . . . To the casual observer the passive animals would have appeared to be the healthiest and most attractive members of the community. . . . But their social disorientation was nearly complete."

Perhaps the strangest type was what Calhoun called the "probers," who moved in packs of three or four. They would confound a female by courting her as a group, harass lactating females, and upset nests of pups.

END OF MOTHERHOOD

Under these strains, motherhood in the crowded pens began going to pieces. Mothers grew sloppy about nestbuilding, often losing interest and leaving it incomplete. Litters got all mixed up, so no mother seemed to know whose babies were whose – nor seemed to care. Frequently, abandoned young were cannibalized by groups of male probers.

Of 558 born at the height of the sink, only one in four survived weaning. Miscarriages were common. Autopsies of females revealed tumors of the uterus, ovaries, fallopian tubes, and mammary glands. And, hardly surprising, adrenals were conspicuously enlarged. As on James Island, the stress took its greatest death toll on the young and the female, contributing heavily to halting the population growth.

There is reason to believe that the behavioral sink could have been prevented without increasing available space – if Calhoun had divided the same space into a greater number of smaller pens and closed them off from one another. Thus a small group of rats, say, the instinctive group of about 12, although pressed for space, would have its own inviolate territory. An English ethologist, H. Shoemaker, tried this with canaries. First he placed a large number in a single large cage. A hierarchy developed in which the dominant birds interfered with the nesting of low-ranking families. Then he transferred them to small cages so that each adult male, including the low-ranking, was master over his family's territory. Brooding then proceeded more normally.

The canary experiment, simple as it is, may have vast significance in considering the development of – and prevention of – behavioral sinks in cities of human beings. Edward Hall, who as an anthropologist is chiefly concerned with human behavior, emphasizes the critical importance of architecture in avoiding the stress of urban crowding. He also emphasizes that architectural needs vary greatly from culture to culture; that one man's company may be another man's crowd. And the alarming thing is that, while some human populations are already showing clear signs of crowding stress, we are doing next to nothing about learning to design territories – proper homes and neighborhoods – to prevent behavioral sinks among the crowded poor.

One of the few studies linking human dwelling space with stress was made by a French couple, Paul and Marie Chombart de Lauwe, among working-class families of France. First they tried to correlate behavior with the number of residents per dwelling unit. This revealed little. Then they got the idea of considering the number of *square meters per person* in the home, regardless of the number who lived in it. They found that, when each person had less than 8 to 10 square meters, instances of physical illness and criminal behavior were double those in less crowded homes. Thus human crowding was clearly linked with illness and violence.

Space and violence were linked in a different kind of study recently completed by a Columbia University psychiatrist, Augustus F. Kinzel. Small both in size and subject scope, the study was limited to 14 men in a federal prison. Eight had histories of violent behavior; six were considered nonviolent. Standing his subject in the center of a bare room, Dr. Kinzel would say, "I'm going to step toward you. Tell me to stop when you feel I'm too close." He would try this from several directions. It turned out that all the men seemed encircled by an invisible "buffer zone" which, when intruded upon, made them feel intensely uncomfortable. The violent men felt "crowded" at an average distance of 3 feet, the nonviolent at half that distance. Perhaps the most important finding of the study was that the violent men were more sensitive to approach from the rear. Their reactions were often clearly physical, particularly among the violent types. Some reported tingling or "goose pimples" across their shoulders and backs. Some literally stepped away with clenched fists as Kinzel entered the buffer zone, even though he was hardly within touching distance. They accused Kinzel of "rushing" them. One commented, "If I didn't know you, I might be ready for anything."

Just as the violent and nonviolent prisoners felt crowded at different distances, the reactions of French workers establish no rule to measure what constitutes crowding in other ethnic groups, even other classes. Crowding differs from people to people, class to class. Hall describes customs of various national groups – particularly the Japanese, Germans, and Arabs – to show the great variations in their sense of proper space between persons.

Most Japanese, for example, are happiest when family and friends are huddled together in the center of a room, or all making body contact under a huge quilt before a fireplace. They feel it is congenial for whole families to sleep close together on the floor. Their dwelling spaces are small but as variegated in purpose as the many rooms of a large American house. The Japanese change the size, moods, and uses of their rooms by rearranging screens and sliding their doors open and shut. Their concepts of "togetherness" and "aloneness" are so different from ours that the Japanese language contains no word that translates into our word "privacy." Yet a Japanese has strong feelings against two houses having a common wall. If his house is not separated from his neighbor's by a strip of land, no matter how narrow, he feels crowded. It is an important sign of his territorial integrity. When strangers are out of sight, Japanese are entirely untroubled by their sounds. In a Japanese inn, where a Westerner would toss and turn angrily at the sounds of a party in the adjoining room, a Japanese would sleep unmindful of it.

In contrast, Germans are especially sensitive to intrusion by sounds of strangers, one reason their hotels often have double doors and thick walls. A German has a strong sense of his own space – *lebensraum,* another word not readily translatable – and is disturbed if that space is not respected. In an office, he keeps his door closed. American visitors often misread this trait as something unfriendly. On the other hand, Germans regard the American habit of leaving doors open as unbusinesslike. Hall tells of an American camp for German prisoners of war in which men were bunked four to a small hut. The men went to great lengths to find materials for building partitions to separate themselves. German families, too, require clear definitions of territory. During the postwar housing shortage, American occupiers blithely ordered Berliners to share kitchens and baths, having no idea of the extreme stress – and violence – their order invited. New arrangements had to be made, Hall reports, "when the already overstressed Germans started killing each other over the shared facilities."

Arabs are happiest among crowds of people, a high noise level of conversation. They require great human involvement, closeness. Conversing, they look at each other piercingly, with much touching of hand to hand, hand to body. In his home, however, an Arab prefers spaciousness – large, high rooms with a commanding view – or he feels crowded. For all his love of involvement, the Arab needs privacy too. The way he gets it is by falling silent, retreating into himself. To talk to an Arab who appears conversationally withdrawn is to exercise bad manners in the extreme – an act of aggression certain to induce stress.

These examples provide the sketchiest of hints of how complex, delicate, and explosive the matter of human crowding can be. Little is formally known about the elements of crowding that are at work in the impoverished ghettoes of American cities – except that they *are* at work, rapidly creating behavioral sinks.

"It is fairly obvious," says Hall, "that American Negroes and people of Spanish culture who are flocking to our cities are being very severely stressed. Not only are they in a setting that does not fit them, but they have passed the limits of their tolerance to stress. The United States is faced with the fact that two of its creative and sensitive peoples are in the process of being destroyed and like Samson could bring down the structure that houses us all."

FACTORIES OF STRESS

When the stress of newly urbanized Negroes is discussed, the solutions proposed are almost always limited to ending discrimination, improving education, providing jobs – and housing that is seldom described beyond being low cost. Without these social improvements, clearly stress will not be eliminated. But Hall's thesis is that these alone cannot halt the growth of behavioral sinks. Space – and the architecture of that space – must be designed for the specific cultural needs of these urban newcomers. A great deal is now known about how *not* to design this space, but little about ways it should be designed.

For example, highrise apartments, no matter how low cost, for people recently of an agrarian tradition are factories of stress. "It's no place to raise a family," a typical tenant complains. "A mother can't look out for her kids if they are 15 floors down on a playground. When I want to go up or down, I think twice because it may take me half an hour to get the elevator."

For a starting point in planning proper spaces for urban newcomers, Hall urges planners to consider that "Puerto Ricans and Negroes have a much higher involvement ratio than New Englanders and Americans of German or Scandinavian stock." As an example of architecture for "involvement," Hall recommends a look at the Spanish plaza and the Italian piazza, "whereas the strung-out Main Street so characteristic of the United States reflects . . . our lack of involvement in others."

One enterprising planner, Neal Mitchell, a professor of design at Harvard, has worked out a novel way of finding out what impoverished Negroes want in their housing. He consults with impoverished Negroes. Mitchell bought $80 worth of doll houses and furniture and invited poor people to arrange it according to their preferences. He found – contradicting the assumptions of almost all low-cost housing architects – that nobody wanted a dining room. They wanted a kitchen large enough to eat in. They also complained that public housing apartments they lived in were much too small. Yet, Mitchell reports, "every single person who played with our game wound up with a smaller square footage than the one they thought insufficient. It was just a question of design."

Next, Mitchell brought out blocks marked "house," "school," "church," "store," and so forth, and let people arrange their communities. Most people in the ghetto, he found, reject the suburban single-family house. One welfare mother told him, "That green front yard is useless. I want to sit out on my front steps and see all those neighbors. I want to be close enough to holler at them."

"I THINK WE ARE GOING TO PULL THROUGH . . ."

Thus that welfare mother confirms Hall's suggestion of "architecture for involvement."

Mitchell is hopeful: "I think we are going to pull through because, you know, there is one thing about this country: it is flexible and it is willing to learn."

If Edward Hall and his "crowding" colleagues are more apprehensive, it is because they fear that our willingness to learn may lose in a race against the onrushing development of the urban sink.

We do not recognize any danger in crowding as long as we can produce enough food for physical growth. Yet overpopulation can destroy the quality of human life through many mechanisms such as traffic jams, water shortages, and environmental pollution; spreading urban and suburban blight; deterioration in professional and social services; destruction of beaches, parks, and other recreational facilities; restrictions on personal freedom owing to the increased need for central controls; the narrowing of horizons as classes and ethnic groups become more segregated, with the attendant deepening of racial tensions.

René Jules Dubos, "Man Adapting: His Limitations and Potentialities," in William R. Ewald, Jr. (Editor), Environment for Man, *The Next Fifty Years, Bloomington, Illinois (Indiana University Press) 1967, p. 21.*

PART XI EMERGING CONCEPTS

. . . In significant areas of the world and within a period as short as fifty years, the majority of the world's population will be accommodated in vast metropolitan complexes, each on a scale of twenty million people or more. Staggering as this figure may seem, it is by no means unreasonable, or even if exaggerated, it sets the stage for an evaluation of this new phenomenon in the history of society. [It is important] to consider what might be lost in such a future world, and even more importantly to define the inherent values of metropolitan life and to suggest how they could be enhanced. If the metropolis is inevitable, we might also speculate on the form it should take and the facilities it should have, if this new way of living is to be the best man has enjoyed. [We should] explore the possibilities of action along two lines: the measures within our present reach and also those on the horizon that require an increase in our understanding or acceptance, or that demand technical, economic, or administrative means not yet in existence. . . .

. . . When applied to cities, plans and dreams have usually been aimed at solving problems of the present or an inducing a return to some image of the past . . . Only rarely do we find a contemporary plan that anticipates the future with pleasure. Men are attracted to the metropolis by real values — choice, freedom, privacy, opportunity, culture, entertainment. How can we ensure the realization of these ends? More importantly, what are the possibilities for metropolitan life that are as yet undreamed of? And what kind of power, knowledge, or guidance must be applied to achieve them? The spirit of hopeful intervention should prove at least as effective as the desire to escape present discomfort.

Lloyd Rodwin (Editor), The Future Metropolis, *New York (Brazilia) 1961, pp. 10 & 16.*

44

Richard S. Bolan

EMERGING VIEWS OF PLANNING

City planners view the world of policy making and public decision making in a very particular way. Some critics, in fact, suggest that one reason planning is not as effective as it might be can be traced to this view. Until recently, city planners have felt that because the policy maker could not, or would not, adopt the planners' view, community decision making represented something less than optimum rationality. There are signs, however, that city planners are changing their view, and there is some hope that in so doing they might become more effective.

A number of recent probes into the nature of the planning process suggest a new understanding of social movements, governmental processes, alternative "styles" of planning, and, indeed, the very nature of cognitive processes and rationality. Robert Hoover's study of historical social movement and change,[1] John Friedmann's analysis of varying planning styles associated with comparative national political environments,[2] the host of community decision

---

making studies beginning with Floyd Hunter's study of Atlanta and including the classic study of public housing in Chicago by Meyerson and Banfield,[3] the "Symposium on Programming and the New Urban Planning" in the recent issue of the *AIP Journal*,[4] and other writings indicate a convergence of thought dominated by the theme which breaks down the classical, simplistic ideals which have been characteristic of city planning for half a century.

This essay is an effort to describe that convergence, with the basic purpose of indicating possible directions for a broader, more detailed conceptual framework for the planning process. It is hoped that the discussion will provoke and stimulate new thought and new research in a vein which can aid planners of all persuasions to come to grips with the complex and crucial process of managing urban change.

Two central questions seem to be paramount as the new viewpoints evolve. The first recognizes the wide disparity that exists between the planner's traditional notions of rationality and the actual social (or "political") processes by which policies are actually chosen, and asks what adaptations must be made in the method, strategy, or content of the planning process in order to yield more rational public policies within a democratic framework. The second question asks how the planner could deal more effectively with the elusive problems of goals and values. Unlike engineers or other professionals, the planner has no clearly circumscribed area of technical expertise which is universally recognized, and he finds himself constantly attempting to deal with value choices which clearly transcend technical judgment alone.

THE CLASSICAL MODEL UNDER ATTACK

The salient features of the planner's view of the world of policymaking are the following:

1. The planning commission (with its professional staff) is an advisory body which assists government in formulating policy. Its view is comprehensive in that no aspect of community development is assumed to be beyond its responsibility. It is also comprehensive in the sense that the planning commission is the guardian of the *whole* public interest rather than any particular special interest.

2. From this, it is assumed that the planning commission is both capable and responsible for establishing long-term development goals which provide a broad perspective and give substance to short-term particularistic community decisions. Planning is construed to be the antithesis of nihilism and is thus responsible for developing the broadest and highest aspirations to give meaning and purpose to the community's day-to-day activities.

3. These long-term goals are expressed by a long-range comprehensive or master plan whose salient features include a map of what the pattern of land use development will be at some distant point in the future and some general policies as to how the community should be guided as it attempts to strive for that end state.

4. With this, it is assumed that short-term, small-scale development decisions are to be measured against the yardstick of the master plan. The master plan would essentially eliminate debate on goals and on general means so that debate could focus on relatively narrow grounds of particular means.

This view has had its difficulties, and an added characteristic has been introduced. Planning is now viewed as a *process* (still largely undefined) and the master plan is a *flexible* guide to public policy. This came about because it was sometimes found desirable to change goals; predictions of the future did not always turn out to be accurate; and new values, new opportunities, and unforeseen side effects kept cropping up. These difficulties have never shaken the planners' faith that this is the ideal model (on the contrary, they provide a justification for doing a new master plan).

Reality, of course, has never measured up to the ideal. In small, homogeneous communities it sometimes comes close; but in large urban centers, master plans and the accompanying political features of the classical model have seldom served as the fountainhead of guidance for particular community decisions. Even worse, the ideal itself is beginning to be questioned in principle.

Banfield cites four reasons why he considers comprehensive master planning an impossible ideal[5]: an inability to predict the future much beyond 5 years at a time (if that), an inability to discover the goals of the community on which all can agree, the decentralized character of our political system, and a lack of knowledge of effective means to achieve ends. Friedmann[6] adds another: the notion of comprehensive planning serving the total public interest is incompatible with the narrow interests involved in the competition for power and influence which is characteristic of any political system. Mann[7] suggests that, given the character and composition of the decision-making process, the best that the planner can hope for is "problem solving" or "opportunity seizing." Altshuler[8] adds that, in reality, decision makers prefer operating at levels where comprehension and prediction are more certain "even if this means fragmenting policy choices rather than integrating them." Wingo[9] suggests the added difficulty that the comprehensive ideal overlooks the substantial human and social costs of achieving the desired goal. He adds that "such a logic implies that the payoff to the community is realized only as the sought-after environment emerges. . . . The critical policy question is not only how much the community is prepared to give up to realize the goals implicit in the master plan, but *who* gives up *how* much so that the fruits of the plan can be realized – *quite frequently by others."* (Emphasis mine.)

Finally, the most "comprehensive" attack on comprehensive planning is made by Braybrooke and Lindblom,[10] largely on the grounds of the practical limits of rationality. As Banfield has done, Braybrooke and Lindblom question one's ability to determine goals or values, which, in any event, are always in a state of flux. They then go on to list the following as practical limits to comprehensive (or "synoptic") policy making:

1. Man's limited problem-solving capabilities.
2. The lack of truly comprehensive information.
3. The costliness of comprehensive analysis.
4. The inability to construct a satisfactory method for evaluating values or goals.
5. The closeness of observed relationships between fact and value.
6. The openness of systems of variables.
7. The analyst's need for strategic sequences of analytical moves.
8. The diverse forms in which policy problems arise.

Braybrooke and Lindblom further hold that decision making is (1) incremental or tending toward relatively small changes, (2) remedial, in that decisions are made to move away from ills rather than toward goals, (3) serial, in that problems are not solved at one stroke but rather successively attacked, (4) exploratory, in that goals are continually being redefined or newly discovered, (5) fragmented or limited, in that problems are attacked by considering a limited number of alternatives rather than all possible alternatives, and (6) disjointed, in that there are many dispersed "decision points" (Reading No. 45).

Most critics do not suggest that planning be abandoned or that we relegate public issues to a fatalistic reliance on some "unseen hand," which will somehow help us "muddle through." The real question is, can, or should, the process of planning adapt to the difficulties that the critics have raised? If comprehensive planning is difficult either in principle or in practical terms, are there reasonable alternative approaches which can be effective? Both the proponents of comprehensive planning and its critics have polarized themselves into discussing the extremes. Granted that the world can never be as the comprehensive planner dreams it to be, neither is it so totally incremental as Lindblom suggests. What is suggested is that there are many possible positions between these extremes, and that planning needs to respond in a manner carefully calculated to be appropriate to circumstances.

THE ENVIRONMENT FOR DECISION

As a part of examining these possibilities in more systematic fashion, it is important to review briefly the public environment of American community life in which the planning process is attempting to function.

Historical Environmental Factors

American communities possess attributes which are a function of place, historical development, economic rationale, and social composition. The fact that the Irish settled in Boston and the Germans in

Milwaukee implies two quite different social and political environments in these two cities, and the kind of planning process which may work well in one may not work well in the other. Similarly, the fact that Miami finds its economy based on tourism and Cleveland on manufacturing has its implications for their planning processes. These different environments give rise to variations in the kinds of interest groups which evolve and the differences in social and political strength found among similar interest groups.

Size and stage of development similarly have their impact on the planning process. Generally speaking, the larger and more mature urban communities possess greater diversity in social and economic structure, leading to greater diversity in public and private goals and a broader spectrum of political interest. The mature urban area undergoing a period of stabilization and decline is also likely to be more conservative and more tolerant of built-in diseconomies and inefficiencies. On the other hand, the young, rapidly growing area is more receptive to change and risk taking.

Stability and uncertainty themselves have an impact on the planning process. Rapid turnover of political leadership, frequent crises, boom-and-bust economic conditions, and racial and ethnic conflict product conditions which are most difficult for classical long-range planning, or at least drastically different for it than those in communities that experience stable leadership and near social and economic equilibrium.

These statements are almost trivial. Most responsible planners have intuitively recognized these differences in communities, and most of them instinctively realize that these differences have ramifications for the planning process. The important point is that most planners, in accounting for these differences, have not thought through alternative planning strategies or "styles." Instead they have tried to develop alternative ways of selling comprehensive or "master" planning.[11] Planning method has not adjusted and adapted to circumstances; instead the ways of imposing the classical ideal have been altered. Accompanying this is the usual wishful thinking that some day, given enough time and education, circumstances will change and the classical ideal will come into its own — even though more careful examination could reveal that the classical ideal might never be appropriate to present or future circumstances.

Factors in the Decision-Making Environment

The decision-making environment has been the focus of considerable attention during the last few years. A host of case studies has been completed, which has essentially sought answers to the questions how are public decisions made, who makes them, who influences them, what goals (if any) are being sought, what kinds of information inputs are most critical, and what political, social, and economic obstacles have to be overcome? The authors of each case study characteristically attempt to generalize and develop some kind of conceptual framework which might yield greater insight and understanding of public decision making. Not unexpectedly, this yields as many theoretical constructs as there are case studies. Some general characteristics are evident, however.

The most consistent notion running through the case studies is the "structure of influence" concept. This ranges from Floyd Hunter's[12] initial idea of a "power elite" to Robert Dahl's[13] more sophisticated notion of many dispersed, multicentered "elites." Perhaps more useful, and more fundamental, is Banfield's[14] description of the structure of influence, which is comprised of the *formal* structure (the collection of agencies and institutions of any community, their expressed goals, and the legal framework which circumscribes their activities and responsibilities) and the *informal* structure of influence (or the behavior of individuals within the formal structure and the actions they take in order that the elaborate checks and balances of formal decentralization can be overcome by informal centralization). He suggests that there "are hundreds, perhaps thousands, of bodies, each of which has a measure of legal authority and none of which has enough of it to carry out a course of action which other bodies oppose . . . therefore every opponent's terms must always be met if there is to be action."

Banfield's Chicago studies lead him to suggest the most important mechanism for doing this is the political party or machine. The Syracuse studies also tend to support this conclusion.[15] Many other communities, however, lack a machine and, in fact, proudly proclaim nonpartisan local politics. Thus what might be accomplished through party machinery in Chicago would have to be accomplished in quite different ways in a nonpartisan city. Moreover, there are hundreds of variations in between. New York and Cincinnati, which have more than two local parties, suggest even more complex processes (although the latter is somewhat simplified by virtue of commitment to a strong city manager).[16] In cities like Boston and Los Angeles, each elected official tends to have his own private "machine," so that in these cities Banfield's "most important mechanism" is as fully dispersed as the formal structure.[17]

Thus from city to city the structure of influence is highly varied. Yet, as all the case studies show, this structure is extremely important to planning proposals, whether such proposals involve areawide plans (such as the Chicago Plan Commission's CBD Plan or the St. Paul land use plan) or particularized proposals (such as the sewage disposal plant for Syracuse). To think a single mold of planning method or planning idealism can be imposed in such diverse circumstances is suggested as naive, at best.

Other factors in the decision-making environment have their effects on planning. Friedmann[18] has suggested that measures of bureaucratic status and professional competency are of crucial importance. The anti-bureaucratic habits of the New England town can be contrasted with the pro-bureaucratic tendencies found in Washington, D.C., or the strong commitment to professional managers in the Midwest. Clearly, these present quite different environments for planning. In addition, Friedmann suggests that variations in the efficiency of relevant information systems will affect the style or character of planning activities. Finally, all the case studies exhibit the importance of private interest groups influencing public decisions. Business, labor, neighborhood associations, minority groups, civic associations all leave their mark in greater or lesser degree. As suggested earlier, such groups are strongly influenced by historical development of the community and prevailing social and economic character. Their behavior is also influenced by the skill and character of their leadership, the degree to which any given proposal threatens or enriches them, and the internal character of decision making within an individual organization of institution.[19]

Factors in the Dynamics of Decision Making

Rounding out the description of the environment for planning are factors which deal with the dynamics involved in reaching decisions. Broadly speaking, these include the characteristics of the people involved in making decisions, the techniques of debate and negotiation that they use, and the methods of transmitting influence.

In speaking of the characteristics of the people involved in making decisions, what is implied is their zeal, energy, and skill, their relative bargaining power based on past participation in public debate, and their self-interest and status perceptions. All these not only have a bearing on how decisions will be made but also on who will make them. These factors can be seen to have been quite strong in the various case studies where actors playing key roles in some public decisions were almost totally absent in others, and actors capable of exerting considerable leverage in some decisions are largely ineffectual in their participation in others.

Techniques of debate and negotiation are also important, not only in terms of the skill and substantive content involved but also in terms of the manner in which nonobjective criteria are handled. In virtually every public issue there are value choices which transcend objective analysis, and, indeed, objective measurement. These value choices are often more crucial to the debate than formal logic, and the degree to which any actor can highlight or obscure these value choices (depending on his purposes) will tend to be extremely instrumental in the final outcomes of decisions.[20] Similarly, skill in bargaining is a very important dynamic ingredient in the process: that is, the ability to create and effectively use exchange processes, compromise, and shared interests necessary in settling an issue.

THE NATURE OF THE PUBLIC AGENDA

The Attributes of Decisions to Be Made

Often overlooked by both planners and the case study investigators is the character of decisions facing a community. This is crucial to the planner and more than once frustrations have arisen through failure to recognize this.

Of major importance in any proposal is the degree

to which it can be argued on ideological grounds. A decision to construct a playground, for example, is a relatively simple one because at this point in the history of most American cities there is little ground for argument in fundamental values. Recreation is a traditional public responsibility and playgrounds are a generally valued public good, which threaten no major structural alterations in the community. Public housing, on the other hand, is still being fought in many communities on the battleground of government intervention in a traditional private domain. Much of the recent protest against urban renewal relies on similar argument. Under these conditions, simulation models, cost-benefit analyses, and the planner's traditional land use studies are introduced to no avail. Settlements of such issues are more often based on peremptory rules based on long-standing value concepts.

Also of major importance is the scope of any given proposal. This relates to its distributive properties or, in short, how many people benefit and how many are threatened. Clearly, urban renewal for an entire neighborhood is a quite different public debate than a proposed "pilot" project on a single block; a proposed metropolitan open space system is argued quite differently than a proposed tot-lot. Generally speaking, the broader the scope of any given proposal, in terms of how many are affected, the more difficult it is to reach a decision and the greater likelihood that substantial compromise, or even stalemate, exists.

In similar fashion, the time horizon and flexibility of any given proposal has an impact on the character of decision making. Proposals which promise to be permanent fixtures of long life allowing little or no flexibility over time are likely to stir more heated debate than improvements or programs which are not expected to last long or can be easily altered or modified over time. Construction of an expressway or a rail rapid transit line comes under the first category. The introduction of an innovation in an education program is an example of the latter.

Other factors in the character of decisions to be made lie in the complexity of organization necessary to carry out proposals and the confidence one might have in the anticipated consequences actually coming to pass. It is extremely difficult to reach agreement on proposals which require cooperation and coordination among many agencies. A latent goal of any agency is its own maintenance – a goal which may be seriously threatened in the political environment by cooperation with other agencies. Any such arrangement has built-in questions of who leads and who follows and whose interests are dominant or subsidiary. Similarly, proposals whose consequences are highly uncertain are extremely difficult to resolve, and the intrinsic conservative tendencies of decision makers will likely whittle down such proposals so that the best one might hope for would be small-scale "pilot" efforts where greater confidence in outcomes might be felt.

Generating Forces Behind Public Issues

The decision system of urban communities does not function of its own accord. Particular issues have to be fed in, and the way in which this occurs is extremely important. Generally, two kinds of forces set the system in motion, both largely initiated by the maintenance and enhancement needs of organizations, institutions, groups, and individuals.[21]

The first of these sets of forces is viewed as those emerging from the general community (or beyond). Such forces arise from notions of general social movements as people strive for a better life, more even distribution of resources, or for individuality in an increasingly technology oriented society. Postwar suburbanization has been just such a movement. Today civil rights is a key social force in the public decisions of most U.S. urban areas. Another movement evident in contemporary society is based on general antiestablishment notions, represented by Levin's "alienated voter."[22] Many of today's successful politicians are winning office by appealing to this movement with promises of "little city halls" in local neighborhoods.[23]

These broad social forces are an essential part of determining the value system of the public at large and they force themselves into the formalized day-to-day decision-making and planning systems whether those who regularly participate in those systems wish them to or not. Many years of frustration and apparent apathy may precede this, but once such movements gain headway they are extremely difficult to stop, and may sweep over a community or a nation. For example, the intellectual urbanist's vain exhortations about the evils of suburbanization have never slowed that movement down for a moment.

The second set of forces are those characteristic of regular or frequent participants in the decision making process: elected officials, appointed officials, professional staff, organized interest groups, institutions, and various "elite" groups. The driving forces here are the general individual motivations of competition, success, recognition, influence, and "rising to the top." Perhaps one of the most useful concepts in considering these forces is Norton Long's "ecology of games":

> *Within each game there is a well established set of goals whose achievement indicates success or failure for the participants, a set of socialized roles making participant behavior highly predictable, a set of strategies and tactics handed down through experience and occasionally subject to improvement and change, an elite public whose approbation is appreciated, and, finally, a general public which has some appreciation for the standing of the players. Within the game the players can be rational in the varying degrees that the structure permits. At the very least they know how to behave, and they know the score.*[24]

Participants in Long's various games are in a continual state of warding off threats to their position or attempting to better their position. In each instance there are varying degrees of concern for abstract notions of an overall public interest. Each is primarily motivated, however, by narrow or special interests either held individually or shared by members of a group or organization.

In general these two forces provide the origins of public issues. In part, they stem from perceived problems and, as suggested by Braybrooke and Lindblom, seek remedial solutions for these problems. But contrary to what Braybrooke and Lindblom suggest, they may also stem from desires for enhancement of position, status, or wealth, and consequently may have little to do with ills or problems.[25] Finally, of course, many public issues combine seeking remedies with achieving enhancement goals.

These distinctions in the origins of public issues are important because they can have a significant influence on the decision making environment. A social movement seeking redress (such as the civil rights movement) enters into the system and is handled within it in quite different ways from the enhancement symbol of a special interest group (such as the promotion of a convention center by a chamber of commerce).

ALTERNATIVE VARIATIONS IN A PLANNING SYSTEM

In raising the question as to how the planning process functions in the decision making environment previously observed, Banfield states the dilemma incisively:

> *Chicagoans, like other Americans, want their city's policies to be comprehensive and consistent. But they also want to exercise influence in making and carrying out these policies; they want to be able to force the government to bargain with them when its policy threatens particular interests of theirs. It will be a long time, probably, before they will be willing to sacrifice as much of the second end as would be necessary to achieve the first.* The tension between the nature of the system and the requirements of planning is, for all practical purposes, ineradicable. *(Italics mine.)*[26]

Banfield goes on to make a fundamental distinction between what he calls "central decision" and "social choice." The former may be likened to the kinds of decisions which might be made within the staff of a planning agency or by a mayor's "cabinet." They attempt to follow logical, controlled processes of analysis, development of alternatives, and evaluation of consequences. The latter "is the accidental by-product of the action of two or more actors . . . who have no common intention and who make their selections competitively or without regard to each other . . . but, it is an outcome which no one has planned as a 'solution' to a 'problem.' It is a 'resultant' rather than a 'solution'."[27]

It is the wide gap between "central decisions" and "social choice" which frustrates and thwarts the efforts of the professional planner. The "tension" between the two may be "ineradicable," but substantial evidence suggests that planners are making a wide variety of efforts to accommodate it. And these efforts seem to lie in the direction of developing alternative "styles" of planning which attempt to take more explicit account of the processes of social choice than does the classical ideal. Four components of planning "style" are suggested here as salient in describing this search for valid linkages: (1) variations in planning strategy, (2) variations in planning method, (3) variations in planning content, and (4) variations in planning organization.

Variations in Planning Strategy

Strategy is viewed here as the means by which the planner attempts to persuade government officials and others that governmental policies and plans ought to be influenced by the information, criteria, and values which he (the planner) is specifically able to bring to bear.

Two extremes might be considered. The first would be a strategy aimed at compulsion or regulation wherein the planner would seek to get laws passed which require that government officials and others must take the planners' data, criteria, and standards into account in the course of their decision making (such as mandatory referral laws and compulsory capital budgeting). The other extreme would be that of attempting to inject a planning system into the policy-making process by voluntary, "incentive" means. In short, the planner would seek to make it appear more profitable for government officials, community leaders, and voters to voluntarily submit to or be influenced by what the planner offers.

The first extreme is unlikely to be found to any effective degree in American cities (not for want of trying, however). Planners and planning agencies have seldom been able to develop sufficient political strength to create a truly compulsory situation (except for nominal and innocuous laws such as mandatory referrals which only give the planner the right to see and comment on proposals but not initiate, veto, or alter them). Thus the planner is usually forced to seek the means to persuade government officials that it is to their benefit to listen to him and follow his advice. Some of the ways which he might do this are suggested here.

Probabilistic Programming (or the "Politics of Information") - Melvin Webber, in a recent *Journal* article,[28] provided full and explicit recognition of the dilemma facing planners. As a result he postulated what he termed as an "interim" strategy "until some resolutions of the inconsistencies within planning theory can be found." He describes this strategy as follows:

> *It is the tactic of programming, as alternative to program-making or plan-making.* It calls for the installation of decision-aiding processes *that might never yield a formal program for even middle-run actions. Rather it would support the incremental, multicentered processes of deciding and acting;* but it would expand. the probabilities *that these decisions and actions* would be taken more rationally. *(Italics mine.)*[29]

His strategy is intended to take explicit recognition of the limitations on man's cognitive capacities cited by Braybrooke and Lindblom earlier in this paper. His fundamental tactic in this strategy is the installation of new urban information systems designed to be injected into the processes of social choice so that those participating are armed with better knowledge, more facts, and hopefully slightly better predictions. If they still act irrationally (as in the planners' eyes they doubtless will), they are at least better aware of their irrationality and its consequences.

This strategy has not been without some success. As Webber points out: "To inform a shopping center investor about consumer travel behavior and . . . market potential is to shape his decisions about shopping center locations and tenant mix."[30] Experience suggests that such information does indeed influence those who take the trouble to acquaint themselves with the facts which planners have assembled. Wholly unperceived opportunities often become exposed in this process.

However, while Webber expresses faith that the strategy would expand the *probabilities* of more rational decisions, it really only expands the *possibilities.* Again there is the inherent difficulty in assuming that more facts, more knowledge, and more certain predictions lead to a more rational process of social choice. Webber also points out that the installation of such decision-aiding processes is difficult and costly. One wonders immediately, then, if the benefits will be commensurate with the cost, especially in view of the fact that the benefits are recognized as uncertain at the outset.

Webber's strategy might be viewed as an up-to-date version of Meyerson's "middle-range bridge."[31] The computer has given the strategy tremendous new power. But in the final analysis the effectiveness of "middle-range" functions is entirely dependent on whether the planner is a meaningful participant in the political environment involved in patching together "social choices." The door to this participation is not open to him simply because he is armed with exhaustive and elaborate data files. The "politics of information" arises not from any intrinsic character or value of information, as such, but rather in *who* uses it and *how* it is used.

The Informal Coordinator-Catalyst - Another strategy for the planning process is for the planner to place himself squarely in the cross-fire of conflict in public issues. Armed with his particular type of methodological skills, he would attempt to be so positioned as to be able to scan the entire range of private interests and, with rational and objective criteria, directly involve himself with working out the compromises and other bases for settlement of issues. From the planner's point of view this would reinsure that the results of the social choice process have a higher degree of technical validity.

This strategy is frequently recommended for a metropolitan planning agency which usually has no formal governmental base of operation but rather sits in a "limbo" between different levels of government – financially supported by everyone but responsible to no one. The main difficulty with the strategy is obtaining the acquiescence of the other participants in the process to permit the planner to play this role. It is most likely to occur on a metropolitan issue which cuts across both functional and territorial jurisdictions and where a real leadership vacuum exists (or, more specifically, in those issues in which stalemate is publicly intolerable but where the political environment is nonetheless likely to produce stalemate, such as the sewage disposal system in Syracuse).

Unfortunately, the planner aspiring to this strategy is on delicate ground. For this role is exactly what the American voter expects of his *elected* officials. If his elected officials claim that the checks and balances of the system are an impediment, the American voter has shown some signs in the recent past of a willingness to remove such impediments by permitting increasing degrees of centralized authority and responsibility. Given a choice between a party machine or a strong executive, invariably the latter is chosen, as witnessed by the professional manager movement in the earlier part of this century and the trend toward strong mayor governments in more recent years. Clearly, the planner employed by a strong mayor can follow this strategy only insofar as his employer permits and is benefited by it. Even for the metropolitan planning agency, choice of this strategy depends very much on the structure of influence, the relationships between municipal, county, and state governments, the degree of dominance of the central city, and the mechanics by which metropolitan issues have been traditionally handled.

The Disjointed Incrementalist - This is the strategy of Braybrooke and Lindblom.[32] It is based on what is purported to be the empirical character of policy making and asserts that strategy which conforms to this character is more likely to be effective than traditional notions of "synoptic" (or comprehensive) idealism, which the authors maintain is impossible to attain anyhow. These so-called empirical characteristics were listed earlier in the paper. Their implications when thought of as a strategy for planning would follow somewhat along these lines.

The planner would not attempt comprehensiveness (since he is unable to achieve it), but would rather work with segmental and incremental policy problems as these problems arise. His analysis would always be partial, focusing on the particular aspects of the problem at hand without placing too much stock in vague, hard-to-measure externalities or spillover effects. He would recognize that any actions or programs stemming from his analysis would be experimental and the problem at issue would be successively attacked over time so that his current concern should be less toward ultimate solutions and more toward immediate (albeit partial) remedies. He would not attempt to define or articulate goals which his community should be moving toward, but would rather describe the ills currently present in the community and how they might be remedied. As a corollary, he would not attempt to devise means of achieving particular ends, but would instead select ends appropriate to available (or possibly available) means. He would not attempt to analyze or even identify all possible alternative solutions to a problem (since he probably could not do this in any event) nor would he be much concerned with "system" effects which might arise from particular decisions. Finally, he would recognize that many other decision centers are concerned with the same issue and will be preparing their own analyses. The planner would, therefore, focus on the particular areas in which he has particular expertise.

There is a certain pragmatic attractiveness to this strategy (and, if Braybrooke and Lindblom are right, we are already behaving this way anyhow). It is true that certain purists would argue that Braybrooke and Lindblom make the same mistake as Kinsey; the fact that people do not behave according to an ideal does not necessarily mean that the ideal is wrong. Empiricists also argue that very little actual evidence was presented by Braybrooke and Lindblom, and that counterevidence does exist to suggest that governments and institutions have indeed made decisions which are not incremental, remedial, or serial. As Webber points out: "The success of the British town planners in changing the rules of city building in England, and the success of the Puerto Rican planners in helping to transform an agrarian society to industrial status in less than a generation, suggest that even though planned parametric change may be uncommon, it is nonetheless possible."[33]

The strategy under some circumstances may be as effective as its authors claim. There are serious risks to be taken into account, however. It is very often the potential externalities, spillover, and side effects of policy decisions which are most crucial, and if the analyst does not anticipate them, Banfield's process of "social choice" will surely expose them for they are the very "stuff" of political controversy. If only the primary benefits and costs of a given proposal were to result, there frequently would be very little to argue about. Moreover, as was pointed out earlier, while unwilling to pay the full price, the American voter at least wants his public officials to *appear* to be comprehensive and consistent. The Braybrooke-Lindblom strategy thus carries with it high political risks.

Advocacy and Plural Planning - In a recent paper, Davidoff spelled out the case for what he called advocacy planning.[34] His approach to this strategy stems from the perpetual tension within the professional between personal convictions and the supposedly value-free objectivity of professional practice. Davidoff concludes (probably correctly) that these can never be separated and agrees with Banfield that very few, if any, public issues are wholly free of value choice to be settled on objective technical grounds alone. He complains that alternative plans have always been prepared on technical grounds so that they were never really valid alternatives at all. He holds "that there is or should be a Republican or Democratic way of viewing city development; that there should be conservative and liberal plans; plans to support the private market and plans to support greater government control."[35] He further suggests that the executive and the legislative branches of government should each have their own planners. He goes on to suggest that special interest groups should have their own plans and planners. Frieden, in a companion article, cites the evidence that this is already occurring in neighborhood associations and in poverty programs: "Professionals in several different fields are currently attempting a new approach for assisting groups of people, usually on a neighborhood basis, in their dealings with local officials. A handful of urban planners, social workers and lawyers have been working directly for neighborhood groups rather than the citywide agencies."[36]

Two considerations seem to be dominant in Frieden's comment. The first is an effort to restore the integrity of suboptimization to those who in the past have shouldered the burdens or costs of citywide or systemwide optimization. The second is to give this same group an effective and skillful voice in public decision making, where they had no voice before.

The implications of this as a planning strategy are not entirely clear. The planner distinctly abandons the pretense that he serves the whole public interest. To some degree the environment for decision making is altered, but only in that new participants with new skill in protecting their interests are added. However, Banfield's notion that public decision making is a "resultant" rather than a "planned solution" is not substantially affected in spite of the fact that many more planners are putting their oar into the process. There is no guarantee that a neighborhood association will get just the kind of program and public policies they want simply because they have a planner in their camp. Moreover, at the citywide level, it is not certain if there were Republican and Democratic teams of planners (or executive and legislative teams) that this would in any way develop different, "better," or more rational decision making. One might anticipate, in fact, that it would add to the number of incommensurables to be pondered and make the decision process even more difficult.

On a practical note, how one mobilizes the professional and financial resources to implement this strategy can be a serious question, especially if individual planners find better salaries and safer havens inside city hall.

Adaptive and Contingency Planning - John Friedmann, in a recent *AIP Journal* article, makes a distinction between "developmental" and "adaptive" planning.[37] In his words, "The former is concerned with achieving a high rate of cumulative-investment for a given area by activating unused resource capabilities; the latter is interested chiefly in qualitative adaptations to the changing interplay of economic forces within the area. To put it another way, adaptive planning generally takes place in response to externally induced development." He goes on to say, "Adaptive planning attempts merely to relieve temporary crises in housing, education, local transportation, municipal water supply, or outdoor recreation which may be caused by . . . exogenous changes." He argues that given contemporary characteristics of regional development, the character of current economic decision making, and the high degree of interregional interdependency, the planner of a city, municipality, and perhaps even a metropolis, can only, with any realism, perform adaptive planning. He can only respond to exogenous forces, take advantage of exogenously developed resource capabilities (such as a new federal program), and attempt to create the proper response to pressures created by those forces and opportunities.

Such planning may be less than comprehensive since the planner is principally concerned with foreseeable "temporary crises." Also it need not be purposeful planning — it is simply responding to external stimuli. It would not be wholly reflex action, however, since the planner would presumably be guided by some standards and criteria as to how the community *ought* to react. Thus adaptive planning does not fully answer the dilemma posed by the gap between the political system and the planning system since it, too, eventually has to come to grips with normative values.

Adaptive planning has many advantages, however, and can be a useful approach to the planner's dilemma. Planning is immediately focused on the critical predictions and their likely consequences, and it pays attention primarily to those public actions specifically responsive to those predictions. It is, in effect, "problem-solving" or remedial planning in the sense of Braybrooke and Lindblom except that it is focused on ills (or "temporary crises") which may exist in the future as well as those already existing. It can be a quite simplistic approach understandable to all, since the likelihood of future crises is strongly based on past experience and fairly simple cause-and-effect chains of reasoning. It is not, for example, difficult for anyone to envision that, if a community's population grows, its need for school buildings will grow in some related proportion.

A corollary strategy might be termed "contingency planning," of which the most universal example is the purchase of life insurance. It is, in effect, that approach which is aimed at creating conditions whereby the effects of contingencies or unforeseen crises can be deflected or absorbed at minimum cost of inconvenience. Much of civil engineering has this focus wherein large safety factors are introduced into designs to reduce to a tolerable minimum the likelihood of failure. Occasionally, the engineer even offers a cost-reliability tradeoff, such as in a drainage system which can be designed to accommodate a storm whose likely frequency is once every 10 years or, alternatively, can be designed to accommodate a storm whose probability is only once every 100 years.

Where contingency planning deals with *probabilities* based on previously observed experience, it seems to work quite well. The tradeoffs are explicit, and costs and benefits can be reasonably calculated. Optimization techniques are becoming well known.[38] On the other hand, contingency planning which deals only in *possibilities* does not seem to fare so well. The outstanding example of this is civil defense planning.

More fundamentally, however, contingency planning shares with adaptive planning a basic problem – it seeks to avoid the worst with no guidance as to how to achieve the best. Again, however, this may be all that is possible – not because (as Friedmann implies) the local puppet must dance to the national or regional piper (even if this were true), but rather because at any level no two people fully agree as to what "the best" is that should be sought. These two closely related forms of planning, then, are very much akin to "disjointed incrementalism," with the main distinction being that they focus on reasonably likely future crises rather than only present ills.

Variations in Planning Method

The planning method, or scheme of logic, is also subject to varying degrees of choice. It might range from various forms of partial or limited analysis to the broadest efforts at comprehensive vision where all interdependencies might be accounted for and all disparate elements integrated into a meaningful whole. It might be highly mathematical and quantitative, seeking to express order in precise and dimensionally consistent terms; or it may be Utopian and qualitative, seeking order through the subjective qualities of beauty or social harmony. It might be bold and far-seeing or hesitant and experimental. It might focus on processes or end states, achievement goals or performance standards.

The choice of planning method is independent of planning strategy yet it is affected by it. Certain strategies could involve the use of almost any method, while others might be appropriate only in conjunction with a specific method. For example, the "coordinator-catalyst" strategy could involve virtually any planning method, and the selection would be guided by other factors in the planning or decision system. On the other hand, the "disjointed incrementalist," by nature, would shun comprehensive methods, favoring partial or fragmentary schemes of logic. Similarly, the advocate planner, oriented to the ends of a particular interest group, would select a planning method consistent with those limited ends.

Four alternative methods are described here to illustrate their variability. They range from efforts at global optimization of a single welfare function to suboptimization among disparate value systems. They traverse the scale from comprehensiveness to reasoned selectivity to sheer opportunism. Other schemes of logic can be described, but those listed are typical, in current use, and well serve to illustrate the variability possible.

Systems Analysis and Simulation - The planning profession currently finds it fashionable to engage in mathematical model building and systems analysis.[39] The focus is on "planning" models as opposed to purely predictive devices, since what is sought is an instrument for more thoroughly exploring the consequences of alternative public and private actions. Such models attempt to build in "systems" effects in order that an action taken regarding, say, the transportation system can be analyzed in terms of its effects on the land system and social or economic distributions. Beginning to be of important concern in the development of these models are techniques which can also analyze alternative time-space sequences of public and private action. In addition, efforts to build in the costs and benefits of public actions are underway.

These efforts clearly reflect a desire for more refined and more varied comprehensive techniques. Because these models are dependent on the use of the computer, they also are highly flexible and can theoretically evaluate many more alternatives and can provide information about short-run, as well as long-range, effects. Thus the desires of interest groups, both individually and collectively, can presumably be plugged into the model, and very rapid predictions of their immediate and future consequences and their impacts on other systems can be explicitly determined. In short, the models should enable the decision makers to be much more explicit about the varying effects of the social choice processes (with no guarantee the resultant decisions will be any "better" or "worse" than they have been in the past).

In the final analysis, however, this is clearly comprehensive planning dressed up in fancier garb. It has an important place in the diverse professional interests of planners if only to add to the still meager stock of knowledge about how cities function. It does little about overcoming the tension described by Banfield between the political system and the requirements of comprehensive planning. This alternative, in fact, makes far more exacting the requirements of planning and continues in the blind faith that more elegant and more elaborate analytical techniques are the answer to bringing together the political system and the planning system.

Cost Effectiveness and Program Planning - Management techniques recently enunciated by the RAND Corporation and applied with outstanding success in the Department of Defense[40] have inspired many planners to consider their application to urban problems. In part, the efforts at developing mathematical models and "systems" analysis described above stem from this example. Some of the Community Renewal Program efforts currently underway attempt to carry it even further. Essentially all public activities in the urban area are broken into individual program "elements." From this a 5- or 6-year program, force structure and budget are developed after explicit objectives of each program element have been identified. The program and budget for a given element evolve from a systematic analysis of alternatives and their likely consequences. Individual programs thus optimized are then analyzed collectively and optimization takes place at successively higher levels, with continual analysis of alternative cost-benefit relationships, and effective tradeoffs between program elements are developed. Feedback is presumed to be built into the system wherein expectation and actual performance are evaluated.

This approach to planning differs from comprehensive planning only slightly and many of the same conceptual problems persist. It does require a more intensive focus on the relationships between means, ends, and resources. It puts a heavy stress on suboptimization of individual functional systems as an essential part of achieving broader optimization; a stress which is quite different from that of traditional comprehensive planning, which purportedly strives primarily for overall territorial optimization. It does not, however, materially contribute to bridging the gap between central decision making and social choice. On the contrary, it relies heavily on centralization of authority and its success in the Department of Defense goes hand in hand with the Department Secretary's parallel success of creating one of the most centralized government agencies in the nation's history.

Quasi-Keynesian Planning - The present crop of national economic planners now lay claim to having developed the techniques to "manage" the economy. Such management is based on Keynesian notions of strategic and selective control of certain key levers in a basically private economy, and the results of such control have yielded an unprecedented period of national prosperity. Not without envy, some planners suggest that, in order to "manage" our cities better, comprehensive planning should give way to more selective effort to regulate and control certain key features in the environment. Doebele[41] has suggested the planner strive to establish "the control positions or leverage points from which . . . [the planner] might exercise an influence over the development of the metropolitan region far greater than . . . resources would seem to make possible." His list of key levers includes control over accessibility (highways and public transit), basic utilities (water and sewer), and basic public institutions and facilities (education, health, open space, and so forth). By careful control of these levers (and an improved regulation of private development than that found in current zoning and subdivision control) private development for the most part would be subtly channeled in accordance with the desires of public policy. Such planning would be in keeping with traditional notions of the relations between laissez-faire private activity and strategic public intervention to correct inequities or to curb the abuse or wastage of natural resources.

There are many questions one might raise about this approach. There are, for example, no a priori notions as to the most critical or strategic leverage points. There is no Keynes among urban scholars who has adequately theorized how the city functions and, therefore, how it might be managed. However, this may not be a fatal flaw, and the approach suggests a very easy answer to Banfield's tension. Certain strategic policy areas would be agreed by all to be relegated to "central" decision making, and everything else would be "social choice" constrained only by certain basic, but minimal, "rules of the game."

Ad Hoc Opportunism - One finds little discussion of this method in the planning literature. It has had some formal recognition and many examples of how it operates can be found. One such example is in a current demonstration program of housing rehabilitation being carried on by the New York City Rent and Rehabilitation Administration.[42] Essentially, the method is as follows: operating within certain predetermined rules of the game, opportunities to move toward some highly generalized goal are seized as circumstances permit. No particular program is articulated nor is any definite schedule set. No preconceived notions or detailed goals are set forth and, consequently, particularized goals may vary considerably over time.

The New York case will serve to illustrate. The general objective is to improve housing through rehabilitation. The ground rules are that families are not permanently displaced and that rehabilitation will take place within the framework of certain technical standards: that is, those imposed by the city building code and by applicable FHA programs. The rehabilitation work will be carried out by private owners in return for which the owners will be entitled to charge higher rents and/or take advantage of the tax exemption and tax abatement provisions in local law. Rent supplements are used to prevent displacement. In working with the people affected, the program is not announced as a project (urban renewal or otherwise) but rather as an effort of the city to improve their housing. As a first step, vacant buildings, wherever found, are taken over and rehabilitated, creating both a tangible symbol of what is to be done and a permanent or temporary relocation resource. From this opportunities are seized as they occur – a vacant lot is converted into a playground; other vacant lots or dilapidated buildings are acquired for pedestrian paths or expansions to community facilities; additional buildings are rehabilitated. None of the actions are taken as part of a predetermined plan, but are decided with the people of the neighborhood as opportunities present themselves. This method of "planning" is unlikely to produce Utopian neighborhoods, but both the housing and the neighborhood are visibly improved with little of the political controversy normally found in urban renewal.

This method has many advantages, but also many obvious faults. Clearly, it only works so long as it does not run afoul of effects outside the area. One must work within a citywide framework of transportation, utility, and community facility systems which, sooner or later, have to be accounted for.

Alternatives in Planning Content

The previous discussion suggests that planning could be flexible in its basic strategy and method. Many alternatives are available, none of which is philosophically or methodologically right by the familiar criteria. Similarly, variations in elements of content are also appropriate. Choice of content is not totally independent of choice of method, although there are no impediments to discussing them separately.

The selection of issues and problems to be concerned about is extremely difficult; a determination of priorities can seldom be made on wholly objective grounds and frequently incommensurables are competing equally for attention. Yet every planning agency must decide its agenda. A survey of planning agencies throughout the United States would likely yield a substantial amount of variation on the kinds of problems receiving emphasis. This might largely be the result of external pressures on the planning agency. Many times, however, issues are also generated by the planning agency itself, and how and why the agency selects these issues and not others are key questions.

In addition to the scope of planning, variations are possible in the time horizons of planning. Planners have long recognized varying time horizons, ranging from immediate problems to policies intended to guide a community for as much as 40 (or more) years in the future. Where the emphasis lies may again be (and likely will be) a matter of conscious and deliberate choice on the part of the planning agency. The planning agency in the nation's capital found it possible and appropriate to its environment to expend considerable time and effort on a 40-year plan. For other communities, however, the 6-year capital improvements program may have more relevance and therefore more emphasis in the planning program. In still others, much of the planning effort may emphasize work on immediately pending urban renewal projects.

Similarly, there may be varying emphases in ends-means relationships. The planner may try to stimulate considerable public discussion and concern for community goals and strive for public decisions on broad objectives. Or, alternatively, he may pay considerably less attention to goals and focus more on implementing actions and programs. Means and ends are, as suggested by Braybrooke and Lindblom, intimately related, and one might expect that undue emphasis on one at the expense of the other may lead to serious shortcomings. Thus planners need to create some meaningful balance in emphasis between the two – that balance again being related to the environment in which the planning process takes place.

Finally, there is the opportunity for selectivity in the kinds of information the planner chooses to insert into the decision-making process. At one extreme he may provide only the information which supports or justifies his recommendation. At the other extreme, he may flood the scene with any information which has the faintest relevance to the issue in order to indicate objectively all sides of an issue or all possible alternatives. This latter extreme is, of course, impossible to achieve even if one were able to pay the cost. The former is easily achievable but raises some uncomfortable ethical questions. Some choice between these extremes is inherent in the planner's behavior for virtually every public issue.

Alternatives in Positioning the Planning System

It is frequently said that organization charts have little meaning, and that what is really crucial to effective organizational behavior are the lines of communication and the interactions which take place as decisions are made. Nevertheless, the formal organization chart does tend to position people and groups of people in a way which implies roles and lines of communication. Thus the position of the planning function with respect to the planning process *is* important, and various discussions about whether the planner ought to serve in the mayor's office, the city council chamber, or the isolated sanctuary of an independent commission are quite relevant.

For the purposes of this paper, four different positions for planning within the decision system are listed, although it is recognized that there are many other conceivable variations. The first is the Walker position that the planner be employed by the chief executive.[43] The second is that suggested by Kent, that the planner be employed by the legislative body.[44] The third is the traditional Bassett-Bettman concept of the independent, nonpolitical advisory commission. A fourth position might be that in which planning is dispersed; perhaps the extreme example is found in the New England town where planning is largely dispersed among a number of special-purpose ad hoc committees created as needed by the town meeting.[45]

A point of interest about the debate over which position is best is that invariably the decision environment is assumed to be uniform from community to community. Most of the followers of the Walker position insist that a strong mayor aided directly by a well-conceived comprehensive planning program will accomplish far more than a similarly endowed mayor who has to deal indirectly through an advisory commission. Whether this is true or not is of little consequence to a community which does not have a strong mayor. Altshuler gives a convincing argument in support of the independent commission in the Twin Cities.[46] Here, however, the planning function existed in an extremely conservative community with no clear leadership apparent among either public or private groups. Consequently, Altshuler's arguments perhaps lose their impact if applied to present-day Boston, Detroit, or St. Louis. Kent advocates that the planner be employed by the city council. This may be appropriate in council-manager cities where it is sometimes difficult to draw a line between executive and legislative functions (or in those few cities in which the voters elect professional planners to the council). Applying Kent's arguments to Banfield's Chicago cases, one imagines that only a planner with an extreme case of schizophrenia could work for such a city council – it would be a superhuman task to attempt to reconcile the fundamental differences between the inner and suburban ward councillors. Thus the argument over the position of the planning process has meaning only insofar as it bears some relevance to the decision environment in which the planning process is to function . . .

REFERENCES

[1] Robert C. Hoover, "Innovation and Historic Social Change," *Proceedings of the 48th Annual Conference of the American Institute of Planners*, Washington, D.C., 1965, pp. 50-55.

[2] John Friedmann, "The Social Context of National Planning Decisions: A Comparative Approach," paper delivered at the 1964 Annual Meeting of the American Political Science Association, Chicago, Ill., Sept. 1964.

[3] Martin Meyerson and Edward C. Banfield, *Politics, Planning, and the Public Interest* (New York: The Free Press, 1955); and Floyd Hunter, *Community Power Structure* (Chapel Hill, N.C.: University of North Carolina Press, 1953).

[4] *Journal of the American Institute of Planners,* Vol. XXXI (Nov. 1965).

[5] Edward C. Banfield, "The Use and Limitations of Metropolitan Planning in Massachusetts," paper presented at the Fifth Working Conference on Metropolitan Planning and Regional Development, Joint Center for Urban Studies, Metropolitan Area Planning Council, June 1965, pp. 12-14. See also Edward C. Banfield, "Ends and Means in Planning," *UNESCO International Social Science Journal,* XI (1959), 365-368.

[6] John Friedmann, "Introduction: The Study and Practice of Planning," *UNESCO International Social Science Journal,* XI, (1959), 336.

[7] Lawrence D. Mann, "Studies in Community Decision-Making," *Journal of the American Institute of Planners,* XXX (Feb. 1964), 64.

[8] Alan Altshuler, "The Goals of Comprehensive Planning," *Journal of the American Institute of Planners,* XXXI (Aug. 1965), 190.

[9] Lowdon Wingo, *Cities and Space: The Future Use of Urban Land* (Baltimore: Johns Hopkins Press, 1963), p. 5.

[10] David Braybrooke and Charles Lindblom, *A Strategy of Decision* (New York: The Free Press, 1963), Chaps. 2 and 3.

[11] See Alan A. Altshuler, *The City Planning Process: A Political Analysis* (Ithaca, N.Y.: Cornell University Press, 1965), Chap. II.

[12] Hunter, *loc. cit.*

[13] Robert Dahl, *Who Governs?* (New Haven, Conn.: Yale University Press, 1961).

[14] Edward C. Banfield, *Political Influence* (New York: The Free Press, 1961), pp. 235-236.

[15] Roscoe Martin, Frank Munger, et al., *Decisions in Syracuse* (Bloomington, Ind.: University of Indiana Press, 1961), p. 46.

[16] Kenneth Gray, *A Report on Politics in Cincinnati,* City Politics Reports (Cambridge, Mass.: Harvard and MIT Joint Center for Urban Studies, 1959).

[17] Edward C. Banfield and Martha Derthick, *A Report on the Politics of Boston,* City Politics Report (Cambridge, Mass.: Harvard and MIT Joint Center for Urban Studies, 1960); and James Q. Wilson, *A Report on the Politics of Los Angeles,* City Politics Report (Cambridge, Mass.: Harvard and MIT Joint Center for Urban Studies, 1960).

[18] John Friedmann, "The Social Context of National Planning Decisions," *op. cit.,* pp. 13, 19-21.

[19] Edward C. Banfield and James Q. Wilson, *City Politics* (Cambridge, Mass.: Harvard and MIT Presses, Joint Center for Urban Studies, 1963).

[20] David Braybrooke and Charles Lindblom, *op. cit.,* Chap. 7.

[21] Banfield, *Political Influence, op. cit.,* p. 263.

[22] Murray B. Levin, *The Alienated Voter* (New York: Holt, Rinehart and Winston, 1960).

[23] The 1965 mayoralty campaign in New York City is one outstanding case of this.

[24] Norton E. Long, "The Local Community as an Ecology of Games," *American Journal of Sociology* (Nov. 1958), pp. 251-261.

[25] For example, Boston's Government Center Project was not conceived to solve a particular problem but rather to serve a deliberate symbolic goal. Technically speaking, the need for a new city hall and other government offices might be considered a "problem," but one which could have been solved more easily in other ways.

[26] Banfield, *Political Influence, op. cit.,* pp. 325-326.

[27] *Ibid.,* pp. 326-327.

[28] Melvin M. Webber, "The Roles of Intelligence Systems in Urban-Systems Planning," *Journal of the American Institute of Planners,* XXXI (Nov. 1965).

[29] *Ibid.,* p. 293.

[30] *Ibid.,* p. 295.

[31] Martin Meyerson, "Building the Middle Range Bridge for Comprehensive Planning," *Journal of the American Institute of Planners,* XXII (Spring 1956).

[32] Braybrooke and Lindblom, *op. cit.*

[33] Webber, *op cit.,* p. 293.

[34] Paul Davidoff, "Advocacy and Pluralism in Plan-

ning," *Journal of the American Institute of Planners,* XXXI (Nov. 1965).

[35] *Ibid.,* p. 335.

[36] Bernard J. Frieden, "Toward Equality of Urban Opportunity," *Journal of the American Institute of Planners,* XXXI (Nov. 1965).

[37] John Friedmann, "Regional Development in Post-Industrial Society," *Journal of the American Institute of Planners,* XXX (May 1964).

[38] Arthur Maas, et al., *Design of Water-Resource Systems* (Cambridge, Mass.: Harvard University Press, 1962).

[39] See Special Issue, "Urban Development Models: New Tools for Planning," *Journal of the American Institute of Planners,* XXXI (May 1965).

[40] Charles Hitch, *Decision-Making for Defense* (Berkeley, Calif.: University of California Press, 1965).

[41] William A. Doebele, "Techniques for Stimulating and Controlling Physical Development," paper delivered at the Fourth Working Conference on Metropolitan Planning and Regional Development, Joint Center for Urban Studies, Metropolitan Area Planning Council, Boston, June 1965.

[42] For a full description of this program see Hope Mandarin, "Combined Rent Supplement, Rehabilitation Demonstrations Under Both Public and Private Sponsorship Highlight Potentials and Problems," *Journal of Housing,* No. 5 (May 1966), p. 255.

[43] Robert A. Walker, *The Planning Function in Urban Government* (Social Science Studies No. 39), (Chicago: University of Chicago Press, Second Edition, 1950).

[44] T.J. Kent, Jr., *The Urban General Plan* (San Francisco: Chandler Publishing Company, 1964).

[45] Richard S. Bolan, "Local Planning in the Boston Area," presented at the Fourth Working Conference on Metropolitan Planning and Regional Development, Joint Center for Urban Studies, Metropolitan Area Planning Council, Boston, June 1965.

[46] Alan A. Altshuler, *The City Planning Process, op. cit.,* 384-391.

. . . The process is itself the actuality, and requires no antecedent static cabinet. Also the processes of the past, in their perishing, are themselves, energizing as the complex origin of each novel occasion. The past is the reality at the base of each new actuality. The process is its absorption into a new unity with ideals and with anticipation,. by the operation of the creative Eros.

I now pass to the second metaphysical principle. It is the doctrine that every occasion of actuality is in its own nature finite. There is no totality which is the harmony of all perfections. Whatever is realized in any one occasion of experience necessarily excludes the unbounded welter of contrary possibilities. There are always 'others,' which might have been and are not. . . .

Alfred North Whitehead, Adventures of Ideas, *New York (Macmillan) 1933, p. 356.*

45

Charles E. Lindblom

THE SCIENCE OF "MUDDLING THROUGH"

Suppose that an administrator is given responsibility for formulating policy with respect to inflation. He might start by trying to list all related values in order of importance, e.g., full employment, reasonable business profit, protection of small savings, prevention of a stock market crash. Then all possible policy outcomes could be rated as more or less efficient in attaining a maximum of these values. This would of course require a prodigious inquiry into values held by members of society and an equally prodigious set of calculations on how much of each value is equal to how much of each other value. He could then proceed to outline all possible policy alternatives. In a third step, he would undertake systematic comparison of his multitude of alternatives to determine which attains the greatest amount of values.

In comparing policies, he would take advantage of any theory available that generalized about classes of policies. In considering inflation, for example, he

---

would compare all policies in the light of the theory of prices. Since no alternatives are beyond his investigation, he would consider strict central control and the abolition of all prices and markets, on the one hand, and elimination of all public controls with reliance completely on the free market, on the other, both in the light of whatever theoretical generalizations he could find on such hypothetical economies.

Finally, he would try to make the choice that would in fact maximize his values.

An alternative line of attack would be to set as his principal objective, either explicitly or without conscious thought, the relatively simple goal of keeping prices level. This objective might be compromised or complicated by only a few other goals, such as full employment. He would in fact disregard most other social values as beyond his present interest, and he would for the moment not even attempt to rank the few values that he regarded as immediately relevant. Were he pressed, he would quickly admit that he was ignoring many related values and many possible important consequences of his policies.

As a second step, he would outline those relatively few policy alternatives that occurred to him. He would then compare them. In comparing his limited number of alternatives, most of them familiar from past controversies, he would not ordinarily find a body of theory precise enough to carry him through a comparison of their respective consequences. Instead, he would rely heavily on the record of past experience with small policy steps to predict the consequences of similar steps extended into the future.

Moreover, he would find that the policy alternatives combined objectives or values in different ways. For example, one policy might offer price level stability at the cost of some risk of unemployment; another might offer less price stability but also less risk of unemployment. Hence the next step in his approach – the final selection – would combine into one the choice among values and the choice among instruments for reaching values. It would not, as in the first method of policy making, approximate a more mechanical process of choosing the means that best satisfied goals that were previously clarified and ranked. Because practitioners of the second approach expect to achieve their goals only partially, they would expect to repeat endlessly the sequence just described, as conditions and aspirations changed and as accuracy of prediction improved.

BY ROOT OR BY BRANCH

For complex problems, the first of these two approaches is of course impossible. Although such an approach can be described, it cannot be practiced except for relatively simple problems and even then only in a somewhat modified form. It assumes intellectual capacities and sources of information that men simply do not possess, and it is even more absurd as an approach to policy when the time and money that can be allocated to a policy problem are limited, as is always the case. Of particular importance to public administrators is the fact that public agencies are in effect usually instructed not to practice the first method. That is, their prescribed functions and constraints – the politically or legally possible – restrict their attention to relatively few values and relatively few alternative policies among the countless alternatives that might be imagined. It is the second method that is practiced.

Curiously, however, the literatures of decision making, policy formulation, planning, and public administration formalize the first approach rather than the second, leaving public administrators who handle complex decisions in the position of practicing what few preach. For emphasis I run some risk of overstatement. True enough, the literature is well aware of limits on man's capacities and of the inevitability that policies will be approached in some such style as the second. But attempts to formalize rational policy formulation – to lay out explicitly the necessary steps in the process – usually describe the first approach and not the second.[1]

The common tendency to describe policy formulation even for complex problems as though it followed the first approach has been strengthened by the attention given to, and successes enjoyed by, operations research, statistical decision theory, and systems analysis. The hallmarks of these procedures, typical of the first approach, are clarity of objective, explicitness of evaluation, a high degree of comprehensiveness of overview, and, wherever possible, quantification of values for mathematical analysis. But these advanced procedures remain largely the appropriate techniques of relatively small scale problem solving, where the total number of variables to be considered is small and value problems restricted. Charles Hitch, head of the Economics Division of RAND Corporation, one of the leading centers for application of these techniques, has written:

> *I would make the empirical generalization from my experience at RAND and elsewhere that operations research is the art of sub-optimizing, i.e., of solving some lower-level problems, and that difficulties increase and our special competence diminishes by an order of magnitude with every level of decision making we attempt to ascend. The sort of simple explicit model which operations researchers are so proficient in using can certainly reflect most of the significant factors influencing traffic control on the George Washington Bridge, but the proportion of the relevant reality which we can represent by any such model or models in studying, say, a major foreign-policy decision, appears to be almost trivial.*[2]

Accordingly, I propose in this paper to clarify and formalize the second method, much neglected in the literature. This might be described as the method of *successive limited comparisons.* I will contrast it with the first approach, which might be called the rational comprehensive method.[3] More impressionistically and briefly – and therefore generally used in this article – they could be characterized as the branch method and root method, the former continually building out from the current situation, step by step and by small degrees; the latter starting from fundamentals anew each time, building on the past only as experience is embodied in a theory, and always prepared to start completely from the ground up.

Let us put the characteristics of the two methods side by side in simplest terms.

Assuming that the root method is familiar and understandable, we proceed directly to clarification of its alternative by contrast. In explaining the second, we shall be describing how most administrators do in fact approach complex questions; for the root method, the "best" way as a blueprint or model, is in fact not workable for complex policy questions, and administrators are forced to use the method of successive limited comparisons.

INTERTWINING EVALUATION AND EMPIRICAL ANALYSIS (1b, Table below)

The quickest way to understand how values are handled in the method of successive limited comparisons is to see how the root method often breaks down in *its* handling of values or objectives. The idea that values should be clarified, and in advance of the examination of alternative policies, is appealing. But what happens when we attempt it for complex social

Rational Comprehensive (Root)	*Successive Limited Comparisons (Branch)*
1a. Clarification of values or objectives distinct from and usually prerequisite to empirical analysis of alternative policies.	1b. Selection of value goals and empirical analysis of the needed action are not distinct from one another but are closely intertwined.
2a. Policy formulation is therefore approached through means-end analysis: First the ends are isolated, then the means to achieve them are sought.	2b. Since means and ends are not distinct, means-end analysis is often inappropriate or limited.
3a. The test of a "good" policy is that it can be shown to be the most appropriate means to desired ends.	3b. The test of a "good" policy is typically that various analysts find themselves directly agreeing on a policy (without their agreeing that it is the most appropriate means to an agreed objective).
4a. Analysis is comprehensive; every important relevant factor is taken into account.	4b. Analysis is drastically limited: (1) Important possible outcomes are neglected. (2) Important alternative potential policies are neglected. (3) Important affected values are neglected.
5a. Theory is often heavily relied upon.	5b. A succession of comparisons greatly reduces or eliminates reliance on theory.

problems? The first difficulty is that, on many critical values or objectives, citizens disagree, congressmen disagree, and public administrators disagree. Even where a fairly specific objective is prescribed for the administrator, there remains considerable room for disagreement on subobjectives. Consider, for example, the conflict with respect to locating public housing, described in Meyerson and Banfield's study of the Chicago Housing Authority[4] – disagreement which occurred despite the clear objective of providing a certain number of public housing units in the city. Similarly conflicting are objectives in highway location, traffic control, minimum wage administration, development of tourist facilities in national parks, or insect control.

Administrators cannot escape these conflicts by ascertaining the majority's preference, for preferences have not been registered on most issues; indeed, there often *are* no preferences in the absence of public discussion sufficient to bring an issue to the attention of the electorate. Furthermore, there is a question of whether intensity of feeling should be considered as well as the number of persons preferring each alternative. By the impossibility of doing otherwise, administrators often are reduced to deciding policy without clarifying objectives first.

Even when an administrator resolves to follow his own values as a criterion for decisions, he often will not know how to rank them when they conflict with one another, as they usually do. Suppose, for example, that an administrator must relocate tenants living in tenements scheduled for destruction. One objective is to empty the buildings fairly promptly, another is to find suitable accommodation for persons displaced, another is to avoid friction with residents in other areas in which a large influx would be unwelcome, another is to deal with all concerned through persuasion if possible, and so on.

How does one state even to himself the relative importance of these partially conflicting values? A simple ranking of them is not enough; one needs ideally to know how much of one value is worth sacrificing for some of another value. The answer is that typically the administrator chooses – and must choose – directly among policies in which these values are combined in different ways. He cannot first clarify his values and then choose among policies.

A more subtle third point underlies both the first two. Social objectives do not always have the same relative values. One objective may be highly prized in one circumstance, another in another circumstance. If, for example, an administrator values highly both the dispatch with which his agency can carry through its projects *and* good public relations, it matters little which of the two possibly conflicting values he favors in some abstract or general sense. Policy questions arise in forms which put to administrators such a question as, given the degree to which we are or are not already achieving the values of dispatch and the values of good public relations, is it worth sacrificing a little speed for a happier clientele, or is it better to risk offending the clientele so that we can get on with our work? The answer to such a question varies with circumstances.

The value problem is, as the example shows, always a problem of adjustments at a margin. But there is no practicable way to state marginal objectives or values except in terms of particular policies. That one value is preferred to another in one decision situation does not mean that it will be preferred in another decision situation in which it can be had only at great sacrifice of another value. Attempts to rank or order values in general and abstract terms so that they do not shift from decision to decision end up by ignoring the relevant marginal preferences. The significance of this third point thus goes very far. Even if all administrators had at hand an agreed set of values, objectives, and constraints, and an agreed ranking of these values, objectives, and constraints, their marginal values in actual choice situations would be impossible to formulate.

Unable consequently to formulate the relevant values first and then choose among policies to achieve them, administrators must choose directly among alternative policies that offer different marginal combinations of values. Somewhat paradoxically, the only practicable way to disclose one's relevant marginal values even to oneself is to describe the policy one chooses to achieve them. Except roughly and vaguely, I know of no way to describe – or even to understand – what my relative evaluations are for, say, freedom and security, speed and accuracy in governmental decisions, or low taxes and better schools than to describe my preferences among specific policy choices that might be made between the alternatives in each of the pairs.

In summary, two aspects of the process by which values are actually handled can be distinguished. The first is clear: evaluation and empirical analysis are intertwined; that is, one chooses among values and among policies at one and the same time. Put a little more elaborately, one simultaneously chooses a policy to attain certain objectives and chooses the objectives themselves. The second aspect is related but distinct: the administrator focuses his attention on marginal or incremental values. Whether he is aware of it or not, he does not find general formulations of objectives very helpful and in fact make specific marginal or incremental comparisons. Two policies, *X* and *Y*, confront him. Both promise the same degree of attainment of objectives *a,b,c,d,* and *e.* But *X* promises him somewhat more of *f* than does *Y*, while *Y* promises him somewhat more of *g* than does *X*. In choosing between them, he is in fact offered the alternative of a marginal or incremental amount of *f* at the expense of a marginal or incremental amount of *g*. The only values that are relevant to his choice are these increments by which the two policies differ; and, when he finally chooses between the two marginal values, he does so by making a choice between policies.[5]

As to whether the attempt to clarify objectives in advance of policy selection is more or less rational than the close intertwining of marginal evaluation and empirical analysis, the principal difference established is that for complex problems the first is impossible and irrelevant, and the second is both possible and relevant. The second is possible because the administrator need not try to analyze any values except the values by which alternative policies differ and need not be concerned with them except as they differ marginally. His need for information on values or objectives is drastically reduced as compared with the root method; and his capacity for grasping, comprehending, and relating values to one another is not strained beyond the breaking point.

RELATIONS BETWEEN MEANS AND ENDS (2b)

Decision making is ordinarily formalized as a means-end relationship: means are conceived to be evaluated and chosen in the light of ends finally selected independently of and prior to the choice of means. This is the means-end relationship of the root method. But it follows from all that has just been said that such a means-end relationship is possible only to the extent that values are agreed upon, are reconcilable, and are stable at the margin. Typically, therefore, such a means-end relationship is absent from the branch method, where means and ends are simultaneously chosen.

Yet any departure from the means-end relationship of the root method will strike some readers as inconceivable. For it will appear to them that only in such a relationship is it possible to determine whether one policy choice is better or worse than another. How can an administrator know whether he has made a wise or foolish decision if he is without prior values or objectives by which to judge his decisions? The answer to this question calls up the third distinctive difference between root and branch methods: how to decide the best policy.

THE TEST OF "GOOD" POLICY (3b)

In the root method, a decision is "correct," "good," or "rational" if it can be shown to attain some specified objective, where the objective can be specified without simply describing the decision itself. Where objectives are defined only through the marginal or incremental approach to values described above, it is still sometimes possible to test whether a policy does in fact attain the desired objectives; but a precise statement of the objectives takes the form of a description of the policy chosen or some alternative to it. To show that a policy is mistaken, one cannot offer an abstract argument that important objectives are not achieved; one must instead argue that another policy is more to be preferred.

So far, the departure from customary ways of looking at problem solving is not troublesome, for many administrators will be quick to agree that the most effective discussion of the correctness of policy does take the form of comparison with other policies that might have been chosen. But what of the situation in which administrators cannot agree on values or objectives, either abstractly or in marginal terms? What then is the test of "good" policy? For the root method, there is no test. Agreement on objectives failing, there is no standard of "correctness." For the method of successive limited comparisons, the test is agreement on policy itself, which remains possible even when agreement on values is not.

It has been suggested that continuing agreement in Congress on the desirability of extending old-age insurance stems from liberal desires to strengthen the welfare programs of the federal government and from conservative desires to reduce union demands for private pension plans. If so, this is an excellent demonstration of the ease with which individuals of different ideologies often can agree on concrete policy. Labor mediators report a similar phenomenon: the contestants cannot agree on criteria for settling their disputes but can agree on specific proposals. Similarly, when one administrator's objective turns out to be another's means, they often can agree on policy.

Agreement on policy thus becomes the only practicable test of the policy's correctness. And for one administrator to seek to win the other over to agreement on ends as well would accomplish nothing and create quite unnecessary controversy.

If agreement directly on policy as a test for "best" policy seems a poor substitute for testing the policy against its objectives, it ought to be remembered that objectives themselves have no ultimate validity other than they are agreed upon. Hence agreement is the test of "best" policy in both methods. But where the root method requires agreement on what elements in the decision constitute objectives and on which of these objectives should be sought, the branch method falls back on agreement wherever it can be found.

In an important sense, therefore, it is not irrational for an administrator to defend a policy as good without being able to specify what it is good for.

NONCOMPREHENSIVE ANALYSIS (4b)

Ideally, rational comprehensive analysis leaves out nothing important. But it is impossible to take everything important into consideration unless "important" is so narrowly defined that analysis is in fact quite limited. Limits on human intellectual capacities and on available information set definite limits to man's capacity to be comprehensive. In actual fact, therefore, no one can practice the rational comprehensive method for really complex problems, and every administrator faced with a sufficiently complex problem must find ways drastically to simplify.

An administrator assisting in the formulation of agricultural economic policy cannot in the first place be competent on all possible policies. He cannot even comprehend one policy entirely. In planning a soil bank program, he cannot successfully anticipate the impact of higher or lower farm income on, say, urbanization – the possible consequent loosening of family ties, possible consequent eventual need for revisions in social security, and further implications for tax problems arising out of new federal responsibilities for social security and municipal responsibilities for urban services. Nor, to follow another line of repercussions, can he work through the soil bank program's effects on prices for agricultural products in foreign markets and consequent implications for foreign relations, including those arising out of economic rivalry between the United States and the USSR.

In the method of successive limited comparisons, simplification is systematically achieved in two principal ways. First, it is achieved through limitation of policy comparisons to those policies that differ in relatively small degree from policies presently in effect. Such a limitation immediately reduces the number of alternatives to be investigated and also drastically simplifies the character of the investigation of each. For it is not necessary to undertake fundamental inquiry into an alternative and its consequences; it is necessary only to study those respects in which the proposed alternative and its consequences differ from the status quo. The empirical comparison of marginal differences among alternative policies that differ only marginally is, of course, a counterpart to the incremental or marginal comparison of values discussed above.[6]

Relevance as Well as Realism

It is a matter of common observation that in Western democracies public administrators and policy analysts in general do largely limit their analyses to incremental or marginal differences in policies that are chosen to differ only incrementally. They do not do so, however, solely because they desperately need some way to simplify their problems; they also do so in order to be relevant. Democracies change their policies almost entirely through incremental adjustments. Policy does not move in leaps and bounds.

The incremental character of political change in the United States has often been remarked. The two major political parties agree on fundamentals; they offer alternative policies to the voters only on relatively small points of difference. Both parties favor full employment, but they define it somewhat differently; both favor the development of water power resources, but in slightly different ways; and both favor unemployment compensation, but not the same level of benefits. Similarly, shifts of policy within a party take place largely through a series of relatively small changes, as can be seen in their only gradual acceptance of the idea of governmental responsibility for support of the unemployed, a change in party positions beginning in the early 1930s and culminating in a sense in the Employment Act of 1946.

Party behavior is in turn rooted in public attitudes, and political theorists cannot conceive of democracy's surviving in the United States in the absence of fundamental agreement on potentially disruptive issues, with consequent limitation of policy debates to relatively small differences in policy.

Since the policies ignored by the administrator are politically impossible and so irrelevant, the simplification of analysis achieved by concentrating on policies that differ only incrementally is not a capricious kind of simplification. In addition, it can be argued that, given the limits on knowledge within which policy makers are confined, simplifying by limiting the focus to small variations from present policy makes the most of available knowledge. Because policies being considered are like present and past policies, the administrator can obtain information and claim some insight. Nonincremental policy proposals are therefore typically not only politically irrelevant but also unpredictable in their consequences.

The second method of simplification of analysis is the practice of ignoring important possible consequences of possible policies, as well as the values attached to the neglected consequences. If this appears to disclose a shocking shortcoming of successive limited comparisons, it can be replied that, even if the exclusions are random, policies may nevertheless be more intelligently formulated than through futile attempts to achieve a comprehensiveness beyond human capacity. Actually, however, the exclusions, seeming arbitrary or random from one point of view, need be neither.

Achieving a Degree of Comprehensiveness

Suppose that each value neglected by one policy-making agency were a major concern of at least one other agency. In that case, a helpful division of labor would be achieved, and no agency need find its task beyond its capacities. The shortcomings of such a system would be that one agency might destroy a value either before another agency could be activated to safeguard it or in spite of another agency's efforts. But the possibility that important values may be lost is present in any form of organization, even where agencies attempt to comprehend in planning more than is humanly possible.

The virtue of such a hypothetical division of labor is that every important interest or value has its watchdog. And these watchdogs can protect the interests in their jurisdiction in two quite different ways: first, by redressing damages done by other agencies; and, second, by anticipating and heading off injury before it occurs.

In a society like that of the United States in which individuals are free to combine to pursue almost any possible common interest they might have and in which government agencies are sensitive to the pressures of these groups, the system described is approximated. Almost every interest has its watchdog. Without claiming that every interest has a sufficiently powerful watchdog, it can be argued that our system often can assure a more comprehensive regard for the values of the whole society than any attempt at intellectual comprehensiveness.

In the United States, for example, no part of government attempts a comprehensive overview of policy on income distribution. A policy nevertheless evolves, and one responding to a wide variety of interests. A process of mutual adjustment among farm groups, labor unions, municipalities and school boards, tax authorities, and government agencies with responsibilities in the fields of housing, health, highways, national parks, fire, and police accomplishes a distribution of income in which particular income problems neglected at one point in the decision processes become central at another point.

Mutual admustment is more pervasive than the explicit forms it takes in negotiation between groups; it persists through the mutual impacts of groups upon each other even where they are not in communication. For all the imperfections and latent dangers in this ubiquitous process of mutual adjustment, it will often accomplish an adaptation of policies to a wider range of interests than could be done by one group centrally.

Note too, how the incremental pattern of policy making fits with the multiple pressure pattern. For when decisions are only incremental – closely related to known policies, it is easier for one group to anticipate the kind of moves another might make and easier too for it to make correction for injury already accomplished.[7]

Even partisanship and narrowness, to use pejorative terms, will sometimes be assets to rational decision making, for they can doubly reinsure that what one agency neglects, another will not; they specialize personnel to distinct points of view. The claim is valid that effective rational coordination of the federal administration, if possible to achieve at all, would require an agreed set of values[8] – if "rational" is defined as the practice of the root method of decision making. But a high degree of administrative coordination occurs as each agency adjusts its policies to the concerns of the other agencies in the process of fragmented decision making I have just described.

For all the apparent shortcomings of the incremental approach to policy alternatives with its arbitrary exclusion coupled with fragmentation, when compared to the root method, the branch method often looks far superior. In the root method, the inevitable exclusion of factors is accidental, unsystematic, and not defensible by any argument so far developed, while in the branch method the exclusions are deliberate, systematic, and defensible. Ideally, of course, the root method does not exclude; in practice it must.

Nor does the branch method necessarily neglect long-run considerations and objectives. It is clear that important values must be omitted in considering policy, and sometimes the only way long-run objectives can be given adequate attention is through the neglect of short-run considerations. But the values omitted can be either long run or short run.

SUCCESSION OF COMPARISONS (5b)

The final distinctive element in the branch method is that the comparisons, together with the policy choice, proceed in a chronological series. Policy is not made once and for all; it is made and remade endlessly. Policy making is a process of successive approximation to some desired objectives in which what is desired itself continues to change under reconsideration.

Making policy is at best a very rough process. Neither social scientists, nor politicians, nor public administrators yet know enough about the social world to avoid repeated error in predicting the consequences of policy moves. A wise policy maker consequently expects that his policies will achieve only part of what he hopes and at the same time will produce unanticipated consequences he would have preferred to avoid. If he proceeds through a *succession* of incremental changes, he avoids serious lasting mistakes in several ways.

In the first place, past sequences of policy steps have given him knowledge about the probable consequences of further similar steps. Second, he need not attempt big jumps toward his goals that would require predictions beyond his or anyone else's knowledge, because he never expects his policy to be a final resolution of a problem. His decision is only one step, one that if successful can quickly be followed by another. Third, he is in effect able to test his previous predictions as he moves on to each further step. Fourth, he often can remedy a past error fairly quickly – more quickly than if policy proceeded through more distinct steps widely spaced in time.

Compare this comparative analysis of incremental changes with the aspiration to employ theory in the root method. Man cannot think without classifying, without subsuming one experience under a more general category of experiences. The attempt to push categorization as far as possible and to find general propositions which can be applied to specific situations is what I refer to with the word "theory." Where root analysis often leans heavily on theory in this sense, the branch method does not.

The assumption of root analysts is that theory is the most systematic and economical way to bring relevant knowledge to bear on a specific problem. Granting the assumption, an unhappy fact is that we do not have adequate theory to apply to problems in any policy area, although theory is more adequate in some areas – monetary policy, for example – than in others. Comparative analysis, as in the branch method, is sometimes a systematic alternative to theory.

Suppose that an administrator must choose among a small group of policies that differ only incrementally from each other and from present policy. He might aspire to "understand" each of the alternatives – for example, to know all the consequences of each aspect of each policy. If so, he would indeed require theory. In fact, however, he would usually decide that, *for policy-making purposes,* he need know, as explained above, only the consequences of each of those aspects of the policies in which they differed from one another. For this much more modest aspiration, he requires no theory (although it might be helpful, if available), for he can proceed to isolate probable differences by examining the differences in consequences associated with past differences in policies, a feasible program because he can take his observations from a long sequence of incremental changes.

For example, without a more comprehensive social theory about juvenile delinquency than scholars have yet produced, one cannot possibly understand the ways in which a variety of public policies – say on education, housing, recreation, employment, race relations, and policing – might encourage or discourage delinquency. And one needs such an understanding if he undertakes the comprehensive overview of the problem prescribed in the models of the root method. If, however, one merely wants to mobilize knowledge sufficient to assist in a choice among a small group of similar policies – alternative policies on juvenile court procedures, for example – he can do so by comparative analysis of the results of similar past policy moves.

THEORISTS AND PRACTITIONERS

This difference explains – in some cases at least – why the administrator often feels that the outside expert or academic problem solver is sometimes not helpful and why they, in turn, often urge more theory on him. And it explains why an administrator often feels more confident when "flying by the seat of his pants" than when following the advice of theorists. Theorists often ask the administrator to go the long way round to the solution of his problems, in effect, to follow the best canons of the scientific method, when the administrator knows that the best available theory will work less well than more modest incremental comparisons. Theorists do not realize that the administrator is often in fact practicing a systematic method. It would be foolish to push this explanation too far, for sometimes practical decision makers are pursuing neither a theoretical approach nor successive comparisons, nor any other systematic method.

It may be worth emphasizing that theory is sometimes of extremely limited helpfulness in policy making for at least two rather different reasons. It is greedy for facts; it can be constructed only through a great collection of observations. And it is typically insufficiently precise for application to a policy process that moves through small changes. In contrast, the comparative method both economizes on the need for facts and directs the analyst's attention to just those facts that are relevant to the fine choices faced by the decision maker.

With respect to precision of theory, economic theory serves as an example. It predicts that an economy without money or prices would in certain specified ways misallocate resources, but this finding pertains to an alternative far removed from the kind of policies on which administrators need help. On the other hand, it is not precise enough to predict the consequences of policies restricting business mergers, and this is the kind of issue on which the administrators need help. Only in relatively restricted areas does economic theory achieve sufficient precision to go far in resolving policy questions; its helpfulness in policy making is always so limited that it requires supplementation through comparative analysis.

SUCCESSIVE COMPARISON AS A SYSTEM

Successive limited comparisons is, then, indeed a method or system; it is not a failure of method for which administrators ought to apologize. Nonetheless, its imperfections, which have not been explored in this paper, are many. For example, the method is without a built-in safeguard for all relevant values, and it also may lead the decision maker to overlook excellent policies for no other reason than that they are not suggested by the chain of successive policy steps leading up to the present. Hence it ought to be said that under this method, as well as under some of the most sophisticated variants of the root method – operations research, for example – policies will continue to be as foolish as they are wise.

Why then bother to describe the method in all the above detail? Because it is in fact a common method of policy formulation, and is, for complex problems, the principal reliance of administrators as well as of other policy analysts.[9] And because it will be superior to any other decision-making method available for complex problems in many circumstances, certainly superior to a futile attempt at superhuman comprehensiveness. The reaction of the public administrator to the exposition of method doubtless will be less a discovery of a new method than a better acquaintance with an old. But by becoming more conscious of their practice of this method, administrators might practice it with more skill and know when to extend or constrict its use. (That they sometimes practice it effectively and sometimes not may explain the extremes of opinion on "muddling through," which is both praised as a highly sophisticated form of problem solving and denounced as no method at all. For I suspect that insofar as there is a system in what is known as "muddling through," this method is it.)

One of the noteworthy incidental consequences of clarification of the method is the light it throws on the suspicion an administrator sometimes entertains that a consultant or adviser is not speaking relevantly and responsibly when in fact by all ordinary objective evidence he is. The trouble lies in the fact that most of us approach policy problems within a framework given by our view of a chain of successive policy choices made up to the present. One's thinking about appropriate policies with respect, say, to urban traffic control is greatly influenced by one's knowledge of the incremental steps taken up to the present. An administrator enjoys an intimate knowledge of his past sequences that "outsiders" do not share, and his thinking and that of the "outsider" will consequently be different in ways that may puzzle both. Both may appear to be talking intelligently, yet each may find the other unsatisfactory. The relevance of the policy chain of succession is even more clear when an American tries to discuss, say, antitrust policy with a Swiss, for the chains of policy in the two countries are strikingly different and the two individuals consequently have organized their knowledge in quite different ways.

If this phenomenon is a barrier to communication, an understanding of it promises an enrichment of intellectual interaction in policy formulation. Once the source of difference is understood, it will sometimes be stimulating for an administrator to seek out a policy analyst whose recent experience is with a policy chain different from his own.

This raises again a question only briefly discussed above on the merits of like-mindedness among government administrators. While much of organization theory argues the virtues of common values and agreed organizational objectives, for complex problems in which the root method is inapplicable, agencies will want among their own personnel two types of diversification; administrators whose thinking is organized by reference to policy chains other than those familiar to most members of the organization and, even more commonly, administrators whose professional or personal values or interests create diversity of view (perhaps coming from different specialties, social classes, geographical areas) so that, even within a single agency, decision making can be fragmented and parts of the agency can serve as watchdogs for other parts.

REFERENCES

[1] James G. March and Herbert A. Simon similarly characterize the literature. They also take some important steps, as have Simon's recent articles, to

describe a less heroic model of policy making. See *Organizations* (New York: John Wiley & Sons, 1958), p. 137.

[2] "Operations Research and National Planning – A Dissent," 5 *Operations Research* 718 (Oct. 1957). Hitch's dissent is from particular points made in the article to which his paper is a reply; his claim that operations research is for low-level problems is widely accepted.

For examples of the kind of problems to which operations research is applied, see C.W. Churchman, R.L. Ackoff, and E.L. Arnoff, *Introduction to Operations Research* (New York: John Wiley & Sons, 1957); and J.F. McCloskey and J.M. Coppinger (eds.), *Operations Research for Management*, Vol. II, (Baltimore: The Johns Hopkins Press, 1956).

[3] I am assuming that administrators often make policy and advise in the making of policy and am treating decision making and policy making as synonymous for purposes of this paper.

[4] Martin Meyerson and Edward C. Banfield, *Politics, Planning and the Public Interest* (New York: The Free Press, 1955).

[5] The line of argument is, of course, an extension of the theory of market choice, especially the theory of consumer choice, to public policy choices.

[6] A more precise definition of incremental policies and a discussion of whether a change that appears "small" to one observer might be seen differently by another is to be found in my "Policy Analysis," 48 *American Economic Review* 298 (June 1958).

[7] The link between the practice of the method of successive limited comparisons and mutual adjustment of interests in a highly fragmented decision-making process adds a new facet to pluralist theories of government and administration.

[8] Herbert Simon, Donald W. Smithburg, and Victor A. Thompson, *Public Administration* (New York: Alfred A. Knopf, 1950), p. 434.

[9] Elsewhere I have explored this same method of policy formulation as practiced by academic analysts of policy ("Policy Analysis," 48 *American Economic Review* 298 [June 1958]). Although it has been here presented as a method for public administrators, it is no less necessary to analysts more removed from immediate policy questions, despite their tendencies to describe their own analytical efforts as though they were the rational comprehensive method with an especially heavy use of theory. Similarly, this same method is inevitably resorted to in personal problem solving, where means and ends are sometimes impossible to separate, where aspirations or objectives undergo constant development, and where drastic simplification of the complexity of the real world is urgent if problems are to be solved in the time that can be given to them. To an economist accustomed to dealing with the marginal or incremental concept in market processes, the central idea in the method is that both evaluation and empirical analysis are incremental. Accordingly, I have referred to the method elsewhere as "the incremental method."

46

Dennis A. Rondinelli

URBAN PLANNING AS POLICY ANALYSIS: MANAGEMENT OF URBAN CHANGE

CHARACTERISTICS OF URBAN POLICY MAKING

Proposition 1

Policy making is an inherently political rather than a deliberative process. Policy is made through socio-political processes — resolution of conflict among groups with divergent interests — rather than by intellectual and deliberative choice (Banfield, 1961; Bauer, Pool, and Dexter, 1963; Bauer and Gergen, 1968; Braybrooke and Lindblom, 1963). As a process of political interaction, policy evolves from a process of interorganizational conflict over a wide variety of values, criteria, ends, means, and interpretations of rationality. It is a generator as well as a product of conflict, evolving through a process of "social weighting" that rarely can be comprehensively planned or centrally guided. Groups seeking particular goals or allocations of resources induce response from other groups that stand to gain or lose from enactment of

---

the policy proposals. Through political interaction and social adjustment, the decisions and priorities of the participants in policy making are ratified, altered, compromised, or rejected (Lindblom, 1959). In some cases, as Bachrach and Baratz (1970) note in their study of antipoverty programs in Baltimore, policies evolve from a history of actions that restrict the choice of alternatives and the margin of acceptable change. Policy can be made through "nondecisions" by the unchallenged drift of events or by deliberate attempts to repress conflict. It may evolve indirectly, generated by unanticipated results of previous decisions. Indeed, policy making often transcends deliberative problem solving; as a process of political interaction, it is more complex and distinctly different from individual decision making.

Proposition 2

Policy is formulated and implemented through highly fragmented and multinucleated structures of semi-independent groups and organizations in both the public and private sectors, and through a complex system of formal and informal delegation of responsibility and control. If Great Society policy making taught one lesson, it is the difficulty of controlling either the evolution of policy proposals through legislative enactment or the implementation of policy through administrative management. Power resources are fragmented and widely dispersed. Points of leverage are multiple and decentralized. Policy is formulated and implemented by a multitude of organizations with highly specialized personnel, information, technical expertise, analytical skills, and influence resources. Each group pursues its own perceptions of its interests and its own conception of the public interest. A potentially large number of them gain veto or delaying power over enactment of urban policy proposals and carve out domains or spheres of influence over program implementation. The history of the Model Cities program, antipoverty legislation (Donovan, 1967), the Area Redevelopment Act (Levitan, 1964), the Public Works and Economic Development Act (Rondinelli, 1969), the Appalachian Regional Development program, and federal highway assistance (Morehouse, 1969; Levin and Abend, 1971) document the complex interaction of groups at all stages of policy formulation.

Once enacted, policies must be implemented through a highly decentralized governmental structure (Grodzins, 1966). Discretionary authority, regulatory control, allocational responsibility, and approval powers are fragmented through systems of interagency, intergovernmental, and intersectoral delegation. At the federal level, for example, when the Economic Development Administration provides business loans to private firms in economically depressed areas, it must submit the applications to the Small Business Administration for clearance. SBA reviews the proposals and performs credit investigations and market feasibility studies. EDA has little control over the time SBA takes to review the applications, the criteria it uses in the review process, or the people making decisions. EDA supplementary technical assistance programs, moreover, are related to programs administered by other federal agencies. Congress decreed in EDA's enabling legislation that grant proposals dealing with specialized aspects of area development must be reviewed by the Department of Labor's Manpower Training Development Office, the Community Facilities Administration of the Department of Housing and Urban Development, the Farmer's Home Administration in the Department of Agriculture, health facilities program agencies within the Department of Health, Education, and Welfare, and the Bureau of the Budget.

Regional development policy is implemented through a quagmire of intergovernmental hybrids. In the Appalachian Regional Commission, for instance, administrative power is shared among a federal cochairman appointed by the president, the governors of 12 states, and an executive staff responsible to the commission as a corporate body. Substantial influence over decisions is delegated to state government agencies and semi-independent local development district corporations. In Pennsylvania, responsibility for implementation of Appalachian Regional Development policy is delegated by the governor to the commonwealth's Department of Commerce, which redelegates planning, clearance, allocation, and review powers to more than a dozen specialized agencies, departments, and commissions over which the Department of Commerce exercises little direct control (Commonwealth of Pennsylvania, 1967).

The reticulated pattern of delegation and fragmentation of power extends beyond government into the private sector. The policy boards of the Model Cities, antipoverty, and Economic Development District agencies are composed of local special interest groups, neighborhood target groups, business and labor representatives, civic and service organizations, as well as local, state, and other government officials. In the traditional sense of administrative responsibility, federal departments providing urban development assistance cannot be held accountable for the outcomes of policies they are assigned to implement. They must rely increasingly on state and local government officials to define local problems, formulate appropriate policy responses, and interpret and implement federal guidelines. Administrative accountability in the Model Cities program, for instance, is shifted almost entirely by formal contracts from HUD to local government agencies. "We are spelling out in clear terms," notes Secretary Romney, "that local government officials must exercise final control and responsibility for the content and administration of a local Model Cities program" (U.S. Congress, 1969). Lines of power and responsibility are intertwined by interdepartmental agreements, delegate agency mandates, and intergovernmental contracts, most of which are nearly impossible to enforce formally and have little legal standing. Enforcement comes through informal pressure and manipulation. The ability to guide, let alone comprehensively plan, national development policies is highly complicated and narrowly constrained by delegation.

Proposition 3

Policy problems are complex, amorphous, and difficult to define concisely. Urban policy planning is limited, moreover, by political parameters on defining the problem. Problems become the focus of policy making to the extent that specialized groups and coalitions can bring public attention to them. Few issues are defined in the same way by all who participate in the policy making process. Each interested organization places a different emphasis on a different component of the problem or defines the whole problem in terms of a part. Interest groups proposing legislation to assist urban depressed areas in the 1960s, for example, saw the problem as one of high unemployment, declining physical plant, changing technological and economic advantages, and obsolete infrastructure in industrial communities. A strong coalition of southern congressmen, who significantly amended the original policy proposals, viewed the problem as underemployment, inability of rural areas to mobilize resources, outmigration of unskilled labor from rural areas to urban centers, and failure to exploit natural resources in agricultural areas. The Appalachian governors saw it as multistate competitive disadvantage. The Department of Labor defined it in terms of the need for massive manpower retraining. Social welfare groups were sure that the crux of the problem was racial injustice, discrimination, lack of educational opportunities, and the need for both "black capitalism" and economic development of center city ghettos. Each group mobilized support for its own definition of the problem (Rondinelli, 1969).

Proposition 4

Problem perception, policy response, and program implementation are characterized by long lead and lag times. Comprehensive analysis and coordinated control of policy implementation are constrained, further, by the long lag and lead times inherent in political interaction. Lags develop between emergence of a problem and public recognition. Acknowledgement of a problem's existence does not assure allocations of public resources for its solution. A lag exists until proponents can mobilize a coalition of support, resolve conflicts with opponents, and gain consensus on appropriate policy responses. The ARA, Public Works and Economic Development and Appalachian Regional Development acts of the 1960s were policies designed to ameliorate problems of urban and regional economic decline that first arose prior to the 1930s depression. Lags exist, moreover, between the proposal of policies and their legislative enactment. The initial proposal for creation of ARA was made in late 1954; the bill creating the agency was not signed into law until 1961. Although proposals for an Appalachian Regional assistance program first appeared in the mid-1950s, the Appalachian Regional Commission was not created until a decade later. Many components of the Economic Opportunity Act of 1964 were initially introduced during the New Deal. Furthermore, long leads occur between organization of the programs and identification and evaluation of their effects. Thirty years were required to recognize publicly the failure of New Deal social welfare policies. Conditions under which programs were formulated change during both lag and lead times. Perceptions and definitions of the problem, personalities, and motivations of participants change; the strength of demands and support of sponsoring and opposing interest groups shift. The problem itself may be partially or totally displaced from public attention.

Proposition 5

Systematic analysis and evaluation are complicated by the difficulty of determining real policy output. Dror's (1968) distinction between the nominal output of a program (reports, projects, rules, trained manpower, and so forth) and the real output (substantive effects of policies on conditions they were designed to correct) has significance for policy planning. The experience of ARA, EDA, OEO, and Model Cities is one of extreme difficulty in identifying and measuring real policy outputs. By the end of its first year in operation, EDA found it had no way of proving that its activities were responsible for bettering conditions in areas where unemployment rates fell below 6 percent, the termination level for EDA assistance (Rauner, 1967). In order to justify its program to Congress and the Bureau of the Budget, EDA was forced to adopt a "worst first" strategy. Investments were concentrated in areas with the highest rates of unemployment and lowest family incomes – those least likely to be affected by national economic growth. Only in this way could EDA isolate the influence of its program on regional recovery. But the "worst first" strategy raised claims by other federal agencies that EDA's policies obstructed their own plans. The cochairman of the Appalachian Regional Commission testified to Congress of the difficulties encountered by EDA's plans for development of the most hopeless regions, while the Appalachian Commission was attempting to concentrate resources in the areas of highest growth potential (U.S. Congress, 1967b). Systematic analysis and quantitative evaluation yielded to political and social subjectivity: "Ultimately a value judgment is required to decide whether one unemployed person in Lowville, New York, is equivalent to one low-income family in Wolf, Kentucky," argues a former EDA assistant administrator. "In the same view, determining how much of EDA's program appropriation should be assigned to each of its seven program sets is properly a matter of administrative judgment. No mathematical computations or maximizing formula can solve this problem" (Rauner, 1967).

Proposition 6

Facts, information, and statistics used to analyze policy alternatives are subjectively interpreted through preconceived specialized interests. Even if "objective" indicators of "optimal" courses of action could be determined, the data would not be treated objectively. Not only the substance of policy but facts and statistics also become the subject of debate and conflict. Quantitative data are rarely interpreted by participants independnently of their role perceptions, subjective expectations, preconceived interests, and ideological predispositions. Congressional hearings on regional economic development legislation reveal that neither supporting nor opposing policy analysts allowed facts to complicate the preconceived logic of their arguments. Experts provided congressional committees with analysis yielding diametrically opposed conclusions. Some used the same set of data to support different arguments before different committees. Provisions to prohibit industrial piracy – subsidies that would induce industries to move from one urban area to another – were written into ARA, EDA, and Appalachian bills largely because of the "evidence" presented by national business lobbies that opposed passage of the legislation. "Is it not ironic that the same witnesses who tell us that this legislation is bad because it will be so effective that it will lead to pirating are invariably the same witnesses who tell us that this legislation is bad because it will not help the depressed areas at all?" asked Pennsylvania Representative William W. Scranton at a congressional hearing on the EDA bill. "These arguments, frequently offered by the same witnesses, are not even consistent" *(Congressional Record,* 1961). A favorite tactic of congressional committee members themselves is to invite policy analysts and experts who will present evidence favorable to their own predisposition toward a policy proposal.

Proposition 7

The number of possible alternatives for ameliorating policy problems is indeterminate. Alternatives evolve through processes of political interaction. Traditional planning theory requires systematic evaluation of alternatives in order to make optimal choices. Gans (1970: 223) defines planning as a "method and process of decision making that proposes or identifies goals (or ends) and determines effective policies (or means) – those which can be shown analytically to achieve the goals while minimizing undesirable financial, social, and other consequences." Yet, in reality, the choice of alternative means is dictated by the possibilities evolving from political interaction rather than from deliberative, a priori design and analysis. Alternatives are gradually invented out of compromises among participants with different perceptions of the problem, interests, and criteria. Acceptance of one alternative – optimal to one set of interests – need not result in the rejection of others. Mutual adjustments result in creation of new courses of action from combinations of existing alternatives (Diesing, 1955). A priori delineation and evaluation of alternatives is complicated, moreover, by the fact that groups participating in policy making rarely perceive their goals clearly or define their objectives explicitly. Goal formation is often situational, that is, dependent on expectations of what can actually be achieved under given political conditions at a particular point in time. As expectations change, goals are altered. In most cases, the ends-means chain of which Gans and others speak is not a chain at all. Goals may be instrumental rather than terminal. Ends become means: attainment of one set of goals may merely pave the way to pursue another set. Thus the number of possible permutations and combinations of feasible or potentially feasible alternatives can be enormous. The alternatives given priority depend in part on the groups drawn into policy-making conflicts and on the strength of their influence.

Evaluation and choice are twice confounded by substantive and political spillovers. Initial policy conflicts often expand into intricate extended networks of secondary conflicts over values, ideology, and socioeconomic and political costs and benefits. Spillovers occur from and to related policy problems. Participants in policy making often come into conflict over questions that have little to do with the substantive content of the problem. They become enmeshed in arguments involving personal political ambitions, personal and organizational prestige, control over funds and other resources, and philosophical doctrine.

Finally, political parameters may make consideration of a wide range of alternatives impossible, limiting evaluation to a restricted set or to only one. The running conflict over delegation of functions and authority in the ARA, EDA, and OEO programs strongly reflected the power of the Budget Bureau to restrict consideration of other forms of organization. Senator Paul Douglas and his associates on the Banking and Currency Committee, for instance, favored creation of independent regional development assistance agencies capable of performing their own planning and operational functions. But the Budget Bureau and some specialized federal departments insisted on delegation of ARA and EDA powers. The intensity with which the Bureau of the Budget fought for delegation foreclosed the possibility of dispassionate, objective analysis and choice of the optimal alternative. "I originally favored a single agency operation," Douglas recalls, "but I was overruled by the bureaucracy downtown in the Budget Bureau, and my head was so bloody after that encounter that I gave up" (U.S. Congress, 1965: 107).

Proposition 8

Each participant in policy formulation and implementation has limited evaluation capacity. Even when the number of alternatives is large, the ability of any participant in policy making to evaluate them comprehensively is limited. In reality, as Simon (1958) notes, decision making often reduces to a choice between two alternatives: "doing X" or "not doing X." "Not doing X" may represent the whole set of possible alternatives that decision makers lack the resources, interest, information, or power to evaluate. These courses of action may be considered vaguely in terms of the opportunity costs of rejecting "doing X," or considered serially and incrementally only if alternative "doing X" is rejected, or if it is accepted and later proves to be ineffective (Braybrooke and Lindblom, 1963). But if "doing X" is considered satisfactory to the participating interest groups, alternatives may never be explicated.

The limited evaluation capacity of one congressional committee working on economic development legisilation is not untypical. When a bill to amend the Appalachian Regional Development Act and Title V of EDA legislation was introduced in 1967, minority party members of the House Public Works Committee strongly opposed further expansion of these programs. They were at a loss, however, to offer positive alternatives to the amendments. They noted:

> *When faced with the monumental task of searching and collecting information and data, visiting representative areas of the country to determine their problems and needs, conferring with state and local officials and business leaders in an effort to develop a really workable and effective program, we soon came to the conclusion that our limited, though capable, staff could not even make a dent in this workload in the time available to us. Reluctantly, we were forced to abandon, at this time, the development of a constructive alternative (U.S. Congress, 1967a:90).*

Proposition 9

Policy planning is done under conditions of uncertainty, risk, incomplete information, and partial ignorance of the situation in which problems evolve, the resources of interested groups, and the effectiveness of proposed solutions. Professional planners and public administrators have done little better than legislators in comprehensive policy analysis. Studies of the Federal Aid Highway Act of 1962 – a law requiring that assisted highway projects be the result of a "cooperative, comprehensive and continuing planning process" – indicate that the Bureau of Public Roads lacks the political power to impose comprehensive analysis requirements (Morehouse, 1969). State, local, and metropolitan planning agencies lack the information, political resources, and analytical ability to comply with areawide planning provisions of later transportation programs (Levin and Abend, 1971). Greer (1965) suggests that local planners often found themselves in the same situation with the Federal Workable Program for 701 Assistance.

The Area Redevelopment Administration's attempts to implement congressional requirements for submission of overall economic development programs (OEDP's) by depressed areas as a condition for financial support were obstructed by uncertainty over congressional standards, lack of competent analysts at the local level, and by political pressures from congressmen themselves to speed up the process of aid distribution. ARA was not able to specify the requirements for comprehensive planning or even to evaluate the OEDP's that were submitted. Most local planning groups, therefore, simply filed superficial reports filled with masses of badly analyzed data to satisfy minimum standards. ARA could not disqualify localities for not performing a task that it could neither define nor evaluate. "The agency resolved this dilemma," reports Levitan (1964: 200), "by accepting each OEDP submitted by communities as a token of good faith and an indication that the community desired to plan its economic future on a sound basis." Attempts to formulate Model Cities guidelines to allow maximum freedom for analysis and planning by localities failed miserably. "They did no good," one former Model Cities Administration deputy director complained. "Most of the cities didn't understand the process but were willing to play our silly little game for money. What was meant as a challenge, a prod, was interpreted as a regulation, a cage. Regulations you can relate to; freedom is something else" (Jorden, 1971: 46).

IMPLICATIONS FOR PLANNING

Policy analysis requires drastic modifications in the concept of and approaches to planning, and fundamental changes in planning education. Research, analytical techniques, and skills required for traditional plan making are not adequate for policy planning. Planners trained as policy analysts must develop a view of the planning process that is substantially different from that of comprehensive planning. While existing planning curricula focus heavily on substantive urban problems, few provide the knowledge and skills required for effective intervention in the policy-making process. Conflict resolution and the management of social change are intrinsic to policy planning.

A Political Interaction View of Policy Planning

Comprehensive planning was prescriptive – seeking to design an ideal end state for urban development – rather than interventional. The plan-making approach to education stressed objective, synoptic analysis, the search for endless numbers of alternatives and a combination of best choices into a long-range master plan for urban growth. Details of "routine" decisions were relegated to politicians and administrators. But grand schemes, long-range comprehensive plans, and systematic policy scenarios were ignored conveniently in a political system that renders rational, comprehensive evaluation of urban policies highly improbable and synoptic policy changes nearly impossible.

Analysis, to be of value to policy makers, must isolate components of urban problems and reduce them to calculable proportions. Policy planners must indicate how resources can be mobilized and focused on remediable aspects of problems in such a way that urban areas can be moved marginally, through successive approximation, away from unsatisfactory social and economic conditions. Policy planners must delineate those alternatives upon which a variety of interest can act jointly and seek ways of binding together some of the disparate participants in policy making to promote mutual cooperation along lines of specialization and common interest. An integral part of policy analysis is the search for ways of reconciling differences among specialized interests, where possible, and evaluating compromise positions, bases for mutual exchange, incentives, and instruments of manipulation and persuasion.

A political interaction view of policy planning defines one of the planner's roles as that of identifying "strategic factors." Strategic factors, Barnard (1938: 203) notes, are those "whose control, in the right form, at the right place and time, will establish a new system or set of conditions which meets the purpose." Policy planning must search out limiting factors inhibiting desired social change and delineate the types of complementary factors needed to enact and implement appropriate programs or controls. Given the complexity of the pluralistic political system in which he must operate, the policy planner may focus on calculating the opportunity costs of pursuing alternative courses of action or of taking no deliberate action. By explicating the losses incurred by urban interests from the lag between socio-economic change and the public response to that change, strong incentives might be provided for the formation of effective coalitions to reduce their losses from inaction, delayed action, or inappropriate action. Policy planning is adjunctive – a process of facilitating adjustment among competing interests within a multinucleated governmental structure, to encourage policy outputs of marginally better quality measured against the status quo (Rondinelli, 1971).

Research Needs and Planning Skills

The characteristics of public policy making have been explored by the social sciences in recent years, but little thought has been given to manipulating processes of political interaction in order to plan urban policy more effectively. Nor have the skills and knowledge needed by policy planners been identified. Research is scarce on the relationships among the policy-making structures, the characteristics of the policy-making process, and techniques of interaction and knowledge needed to manage urban change and to design strategies of intervention. The attempt here is to raise research questions and to suggest some categories of skills and knowledge needed by policy planners to intervene effectively in urban policy making (see Figure 1).

Adaptive adjustments among groups seeking to influence policy through tacit interaction (Schelling, 1960; Lindblom, 1965) strongly characterized the evolution of urban development legislation during the 1960s (Rondinelli, 1969; Cleaveland, 1969). Indirect adjustments often take place without direct communication among policy-making participants, either because they cannot or do not want to communicate with each other. Each participant, instead, takes an action that he believes will avoid or resolve conflict based on expectations of what other participants will do or what they have done in the past. To some degree, it is based on intuitive rapport, "second

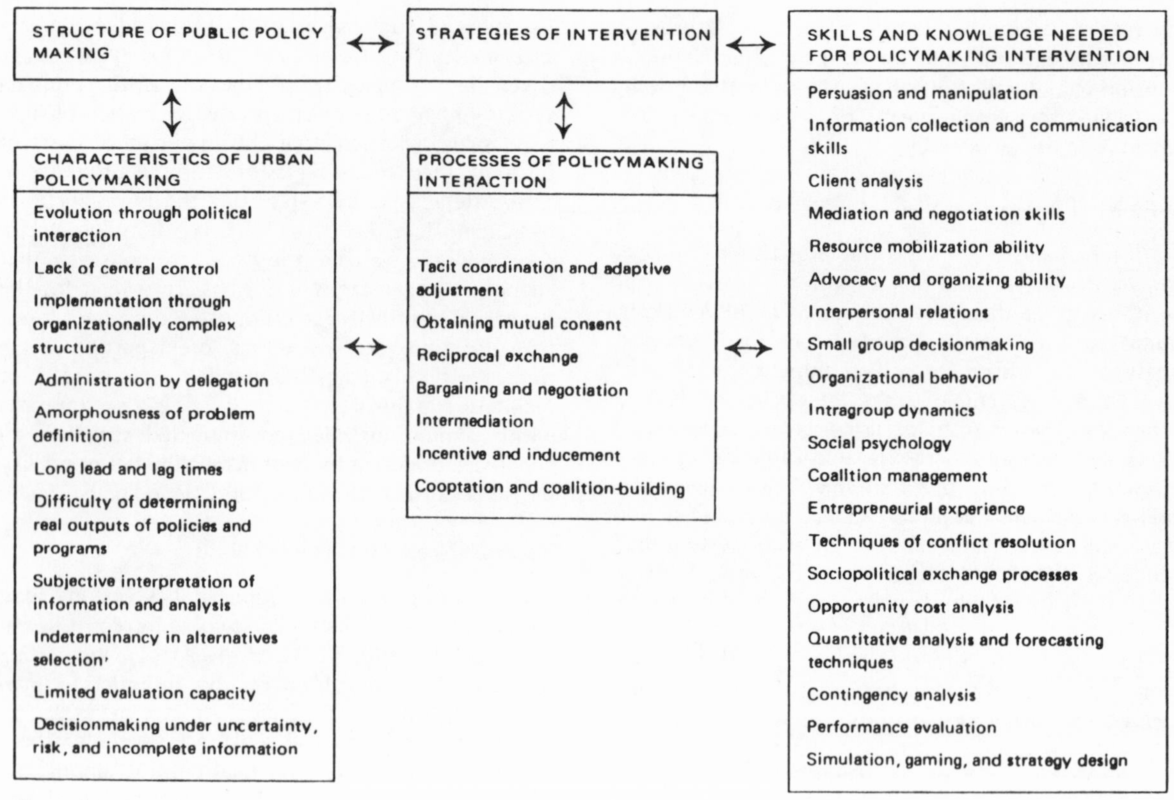

Figure 1
Skills and Knowledge Needed in Urban Policy Making.

guessing," or mutual recognition of a desirable goal. In other cases, it arises from uncoordinated reactions to the same basic conditions or perceptions of the same problem.

How do these techniques influence the content of policy proposals? How do they affect the structure and dynamics of conflict resolution? Are adaptive, noncentrally coordinated processes of interaction more successful in implementing policy proposals than techniques of direct coordination? In what types of issues are they least effective? Can groups be induced to tacit agreement by third parties? A planner attempting to use adaptive adjustment techniques to guide urban policies through formulation and implementation must understand processes of small-group decision making, organizational behavior, intragroup dynamics, and interorganizational interaction.

The pluralistic, multinucleated structure of policy making involves vast networks of specialized groups linked together in intertwining "decision chains" (Wheaton, 1964). Through decision chains, coalitions create and participate in spheres of influence over specific types of issues and programs. To enact a policy or reform a program often requires that multiple consent be obtained from the myriad of interests composing an organized sphere of influence. Indeed, failure to seek approval from the clientele of an established program, as proponents of a Federal Department of Urban Affairs discovered in the early 1960s (Parris, 1969), may activate a coalition of opposition to plans for policy change. What are the techniques of obtaining mutual consent? How do the techniques of obtaining formal, legal consent ("review and approval," "sign-off consent") differ from those of obtaining informal, political consent (log-rolling, reciprocity, vote swapping). What tradeoffs, exchanges, and spillovers arise in the act of obtaining mutual consent? Can some links in the decision chain be avoided or neglected in certain types of urban issues without jeopardizing the policy outcome? Can the process be standardized for particular issues to reduce lag and lead times in policy ratification and implementation? To deal effectively with decision chains in urban policy making, planners must possess persuasion and manipulation skills, experience with client analysis, and information and communication skills. Knowledge of general principles of social psychology can assist in designing tactics of influence and manipulation (Mehrabian, 1970).

Policy conflicts are settled through processes of reciprocal exchange, negotiation, intermediation, and bargaining. The complex delegate agency arrangements characterizing urban policy implementation mandate the use of exchange-bargaining techniques among proponents and opponents of urban assistance policies. Failure to negotiate settlements among disparate groups – federal departments, state and local government agencies, clientele and target groups, and political factions – has led to serious complications in implementing Model Cities programs. What are the channels of exchange and bargaining among participants in urban policy making? What groups and organizations serve as intermediaries in urban policy conflicts? What functions can the planning agency play in facilitating process of exchange and negotiation? Which inputs into the bargaining processes influence policy decisions? Do quantitative analysis, evaluation of data, and trend projection play an important role in influencing the environment for negotiation and bargaining? What are the terms of and parameters on bargaining and exchange among opposing groups in urban policy formulation? Mediation and negotiation skills become important for planners involved in bargaining – exchange relationships in urban policy making. Knowledge of organizational behavior, processes of conflict resolution, sociopolitical exchange processes, and strategy design are imperative. The ability to simulate and design games of strategy involving urban issues could provide a means of assisting political and administrative decision makers to test alternative proposals and political tactics.

Coalition building is the essence of conflict management. Ultimately, urban policies evolve from compromises among groups with sufficient power and resources to persuade other participants of the desirability of a particular course of action. Incentives and inducements change both the parameters of decision making and the costs and benefits of policy alternatives to interested groups. What are the techniques of coalition building? How are interested groups brought together? How can initial policy proposals be designed so as to control the scope of conflict and attract allies into a coalition strong enough to reinsure enactment and effective implementation? How can conflict that is being repressed by groups attempting to prevent an issue from being discussed be socialized to ensure that the issues become the focus of public attention? Alternatively, how can issues that are being socialized to the point that effective coalition formation becomes impossible be repressed? What are the costs of building urban policy coalitions? What are the dynamics of interaction among coalition participants? Which techniques are necessary to maintain a coalition of support for urban policies? The use of incentives requires entrepreneurial experience. Skills in organizational leadership, advocacy, resource mobilization, and coalition management are essential to this aspect of policy planning.

Intervention in policy making is a continuous process of strategic analysis. Forecasting, quantitative measurement, contingency planning, and identification of opportunity costs are integral parts of strategy design. Monitoring and performance evaluation skills are as necessary for effective policy planning as substantive expertise in the urban problem issues. If planners are serious about redirecting the profession's energies toward policy planning, curricula must be redesigned to provide the skills, knowledge, and experience necessary for effective management of urban change.

REFERENCES

Altshuler, A. (1970) "Decision-making and the Trend Toward Pluralistic Planning," in E. Erber, ed., *Urban Planning in Transition.* New York: Grossman, pp. 183-186.

Bachrach, P., and M.S. Baratz (1970) *Power and Poverty.* New York: Oxford University Press.

Banfield, E.C. (1961) *Political Influence.* New York: The Free Press.
___(1970) *The Unheavenly City.* Boston: Little, Brown.
Barnard, C.I. (1938) *The Functions of the Executive.* Cambridge, Mass.: Harvard University Press.
Bauer, R.I., de Sola Pool, and L. Dexter (1963) *American Business and Public Policy.* New York: Atherton.
Bauer, R., and K. Gergen (1968) *The Study of Policy Formation.* New York: The Free Press.
Bolan, R.S. (1967) "Emerging Views of Planning," *Journal of the American Institute of Planners* 33 (July): 233-245 (Reading No. 44).
___(1969) "Community Decision Behavior: The Culture of Planning," *Journal of the American Institute of Planners* 35 (Sept.): 301-310.
Braybrooke, D., and C.E. Lindblom (1963) *A Strategy of Decision: Policy Evaluation as a Social Process.* New York: The Free Press.
Cleaveland, F. (1969) *Congress and Urban Problems.* Washington, D.C.: The Brookings Institution.
Commonwealth of Pennsylvania, Governor's Office of Administration (1967) Executive Directive No. 18, "The Administration of the Appalachian Program in Pennsylvania," mimeographed, April 13.
Congressional Record (1961) 107, no. 4 (March 29): 521.
Diesing, P. (1955) "Noneconomic Decision-Making," *Ethics* 46 (Oct.): 18-35.
Donovan, J.C. (1967) *The Politics of Poverty.* New York: Western Publishing Company.
Dror, Y. (1968) *Public Policy Making Re-examined.* San Francisco: Chandler.
Friedmann, J. (1969) "Notes on Societal Action," *Journal of the American Institute of Planners* 35 (Sept.): 311-318.
Gans, H.J. (1970) "From Urbanism to Policy Planning," *Journal of the American Institute of Planners* 36 (July): 223-226.
Greer, S. (1965) *Urban Renewal and American Cities.* New York: Bobbs-Merrill.
Grodzins, M. (1966) *The American System.* Chicago: Rand McNally.
Hyman, A.A. (1971) "The Management of Planned Change," in S.E. Seashore and R.J. McNeill, eds., *Management of the Urban Crisis.* New York: The Free Press.
Jordan, Fred (1971) "The Confessions of a Former Grantsman," *City* (Summer): 45-47.
Levin, M., and N. Abend (1971) *Bureaucrats in Collision: Case Studies in Area Transportation.* Cambridge, Mass.: The MIT Press.
Levitan, S. (1964) *Federal Aid to Depressed Areas.* Baltimore, Md.: Johns Hopkins Press.
Lindblom, C.E. (1959) "The Handling of Policy Norms in Analysis," in M. Abromovitz et al., *Allocation of Economic Resources.* Stanford, Calif.: Stanford University Press.
---- (1965) *The Intelligence of Democracy.* New York: The Free Press.
Mehrabian, A. (1970) *Tactics of Social Influence.* Englewood Cliffs, N.J.: Prentice-Hall.
Morehouse, T.A. (1969) "The 1962 Highway Act: A Study in Artful Interpretation," *Journal of the American Institute of Planners* 35 (May): 160-168.
Parris, J.H. (1969) "Congress Rejects the President's Urban Department," in F. Cleaveland, ed., *Congress and Urban Problems.* Washington, D.C.: The Brookings Institution.
Perin, C. (1967) "The Noiseless Secession from the Comprehensive Plan," *Journal of the American Institute of Planners* 33 (Sept.): 336-346.
Rauner, R.M. (1967) "Regional and Area Planning: The EDA Experience." Washington, D.C.: U.S. Department of Commerce, Economic Development Administration, mimeographed.
Rondinelli, D.A. (1969) "Policy Analysis and Planning Administration: Toward Adjunctive Planning for Regional Development," unpublished Ph.D. dissertation, Cornell University, Ithaca, N.Y.
___(1971) "Adjunctive Planning and Urban Development Policy," *Urban Affairs Quarterly* 7 (Sept.): 13-39.
Schelling, T.C. (1960) *The Strategy of Conflict.* New York: Oxford University Press.
Simon, H.A. (1958) "The Role of Expectations in an Adaptive or Behavioristic Model," in M.J. Bowman, ed., *Expectations, Uncertainty and Business Behavior.* New York: Social Science Research Council.
U.S. Congress (1965) Senate Committee on Banking and Currency, "Public Works and Economic Development," Hearings on S. 1648, 89th Congress, 1st Session, (May).
U.S. Congress (1967a) House Committee on Public

Works, "Appalachian Regional Development Act Amendments of 1967 and Amendments to the Public Works and Economic Development Act of 1965," Hearings, 90th Congress, 1st Session.
U.S. Congress (1967b) Senate Committee on Government Operations, "Creative Federalism," Hearings, 89th Congress, 2nd Session.
U.S. Congress (1969) Testimony of George Romney, Senate Committee on Banking and Currency, "Progress in the Model Cities Program," Hearings before the Subcommittee on Housing and Urban Affairs, 91st Congress, 1st Session.
Wheaton, W.L.C. (1964) "Public and Private Agents of Change in Urban Expansion," in M. Webber et al., *Explorations into Urban Structure.* Philadelphia: University of Pennsylvania Press.

If systems analysis were indeed a useful approach to social problems, this would be too bad for the behavioral sciences, but would not matter from a broader social point of view. The trouble is that systems analysis in its present state-of-the-art – while being the most useful available approach – is still quite helpless in facing complex social issues . . . (1) Systems analysis focuses on proposing preferable policies, neglecting the institutional contexts . . . (2) . . . does not take into account political needs . . . (3) . . . has difficulties in dealing with "irrational" phenomena . . . (4) . . . is unable to deal with basic value issues . . . (5) . . . deals with identifying preferable alternatives among available or easily synthesized areas . . . (6) . . . requires some predictability in respect to alternatives . . . (7) . . . requires significant quantification of main relevant variables. (8) Basic strategic choices – such as attitudes to risk and time – are not explicitly faced by systems analysis. . . .

Yehezkel Dror, Policy Analysis: A Theoretical Framework and Some Basic Concepts, *P-4156, Santa Monica, California (RAND Corporation) July 1969, pp. 3,4.*

47

Beryl L. Crowe

THE TRAGEDY OF THE COMMONS REVISITED

There has developed in the contemporary natural sciences a recognition that there is a subset of problems, such as population, atomic war, and environmental corruption, for which there are no *technical* solutions[1,2]. There is also an increasing recognition among contemporary social scientists that there is a subset of problems, such as population, atomic war, environmental corruption, and the recovery of a livable urban environment for which there are no current *political* solutions[3]. The thesis of this article is that the common area shared by these two subsets contains most of the critical problems that threaten the very existence of contemporary man.

The importance of this area has not been raised previously because of the very structure of modern society. This society, with its emphasis on differentiation and specialization, has led to the development of two insular scientific communities – the natural and the social – between which there is very little

---

communication and a great deal of envy, suspicion, disdain, and competition for scarce resources. Indeed, these two communities more closely resemble tribes living in close geographic proximity on university campuses than they resemble the "scientific culture" that C.P. Snow placed in contrast to and opposition to the "humanistic culture"[4].

Perhaps the major problems of modern society have, in large part, been allowed to develop and intensify through this structure of insularity and specialization because it serves both psychological and professional functions for both scientific communities. Under such conditions, the natural sciences can recognize that some problems are not technically soluble and relegate them to the nether land of politics, while the social sciences recognize that some problems have no current political solutions and then postpone a search for solutions while they wait for new technologies with which to attack the problem. Both sciences can thus avoid responsibility and protect their respective myths of competence and relevance, while they avoid having to face the awesome possibility that each has independently isolated the same subset of problems and given them different names. Thus both never have to face the consequences of their respective findings. Meanwhile, due to the specialization and insularity of modern society, man's most critical problems lie in limbo, while the specialists in problem solving go on to less critical problems for which they can find technical or political solutions.

In this circumstance, one psychologically brave, but professional foolhardy soul, Garrett Hardin, has dared to cross the tribal boundaries in his article "The Tragedy of the Commons"[1]. In it, he gives vivid proof of the insularity of the two scientific tribes in at least two respects: first, his "rediscovery" of the tragedy was in part wasted effort, for the knowledge of this tragedy is so common in the social sciences that it has generated some fairly sophisticated mathematical models[5]; second, the recognition of the existence of a subset of problems for which science neither offers nor aspires to offer technical solutions is not likely, under the contemporary conditions of insularity, to gain wide currency in the social sciences. Like Hardin, I will attempt to avoid the psychological and professional benefits of this insularity by tracing some of the political and social implications of his proposed solution to the tragedy of the commons.

The commons is a fundamental social institution that has a history going back through our own colonial experience to a body of English common law which antidates the Roman conquest. That law recognized that in societies there are some environmental objects which have never been, and should never be, exclusively appropriated to any individual or group of individuals. In England the classic example of the commons is the pasturage set aside for public use, and the "tragedy of the commons" to which Hardin refers was a tragedy of overgrazing and lack of care and fertilization which resulted in erosion and underproduction so destructive that there developed in the late nineteenth century an enclosure movement. Hardin applies this social institution to other environmental objects such as water, atmosphere, and living space.

The cause of this tragedy is exposed by a very simple mathematical model, utilizing the concept of utility drawn from economics. Allowing the utilities to range between a positive value of 1 and a negative value of 1, we may ask, as did the individual English herdsman, what is the utility to me of adding one more animal to my herd that grazes on the commons? His answer is that the positive utility is near 1 and the negative utility is only a fraction of minus 1. Adding together the component partial utilities, the herdsman concludes that it is rational for him to add another animal to his herd; then another, and so on. The tragedy to which Hardin refers develops because the same rational conclusion is reached by each and every herdsman sharing the commons.

ASSUMPTIONS NECESSARY TO AVOID THE TRAGEDY

In passing the technically insoluble problems over to the political and social realm for solution, Hardin has made three critical assumptions: (1) that there exists, or can be developed, a "criterion of judgment and a system of weighting . . ." that will "render the incommensurables . . . commensurable . . ." in real life; (2) that, possessing this criterion of judgment, "coercion can be mutually agreed upon," and that the application of coercion to effect a solution to problems will be effective in modern society; and (3) that the administrative system, supported by the criterion of judgment and access to coercion, can and will protect the commons from further desecration.

If all three of these assumptions were correct, the tragedy which Hardin has recognized would dissolve into a rather facile melodrama of setting up administrative agencies. I believe these three assumptions are so questionable in contemporary society that a tragedy remains in the full sense in which Hardin used the term. Under contemporary conditions, the subset of technically insoluble problems is also politically insoluble, and thus we witness a full-blown tragedy wherein "the essence of dramatic tragedy is not unhappiness. It resides in the remorseless working of things."

The remorseless working of things in modern society is the erosion of three social myths which form the basis for Hardin's assumptions, and this erosion is proceeding at such a swift rate that perhaps the myths can neither revitalize nor reformulate in time to prevent the "population bomb" from going off, or before an accelerating "pollution immersion," or perhaps even an "atomic fallout."

ERODING MYTH OF THE COMMON VALUE SYSTEM

Hardin is theoretically correct, from the point of view of the behavioral sciences, in his argument that "in real life incommensurables *are* commensurable." He is, moreover, on firm ground in his assertion that to fulfill this condition in real life one needs only "a criterion of judgment and a system of weighting." In real life, however, values are the criteria of judgment, and the system of weighting is dependent upon the ranging of a number of conflicting values in a hierarchy. That such a system of values exists beyond the confines of the nation state is hardly tenable. At this point in time one is more likely to find such a system of values within the boundaries of the nation state. Moreover, the nation state is the only political unit of sufficient dimension to find and enforce political solutions to Hardin's subset of "technically insoluble problems." It is on this political unit that we will fix our attention.

In America there existed, until very recently, a set of conditions which perhaps made the solution to Hardin's problem subset possible: we lived with the myth that we were "one people, indivisible. . . ." This myth postulated that we were the great "melting pot" of the world wherein the diverse cultural ores of Europe were poured into the crucible of the frontier experience to produce a new alloy – an American civilization. This new civilization was presumably united by a common value system that was democratic, equalitarian, and existing under universally enforceable rules contained in the Constitution and the Bill of Rights.

In the United States today, however, there is emerging a new set of behavior patterns which suggest that the myth is either dead or dying. Instead of believing and behaving in accordance with the myth, large sectors of the population are developing life styles and value hierarchies that give contemporary Americans an appearance more closely analogous to the particularistic, primitive forms of "tribal" organizations living in geographic proximity than to that shining new alloy, the American civilization.

With respect to American politics, for example, it is increasingly evident that the 1960 election was the last election in the United States to be played out according to the rules of pluralistic politics in a two-party system. Certainly 1964 was, even in terms of voting behavior, a contest between the larger tribe that was still committed to the pluralistic model of compromise and accommodation within a winning coalition, and an emerging tribe that is best seen as a millennial revitalization movement directed against mass society – a movement so committed to the revitalization of old values that it would rather lose the election than compromise its values. Under such circumstances former real-life commensurables within the Republican Party suddenly became incommensurable.

In 1968 it was the Democratic Party's turn to suffer the degeneration of commensurables into incommensurables as both the Wallace tribe and the McCarthy tribe refused to play by the old rules of compromise, accommodation, and exchange of interests. Indeed, as one looks back on the 1968 election, there seems to be a common theme in both these camps – a theme of return to more simple and direct participation in decision making that is only possible in the tribal setting. Yet, despite this similarity, both the Wallaceites and the McCarthyites responded with a value perspective that ruled out compromise, and they both demanded a drastic change in the dimension in which politics is played. So firm were the value commitments in both of these tribes that neither (as was the case with the Goldwater forces in 1964) was willing to settle for a modicum of power that could accrue through the processes of compromise with the national party leadership.

Still another dimension of this radical change in behavior is to be seen in the black community, where the main trend of the argument seems to be, not in the direction of accommodation, compromise, and integration, but rather in the direction of fragmentation from the larger community, intransigence in the areas where black values and black culture are concerned, and the structuring of a new community of like-minded and like-colored people. But to all appearances even the concept of color is not enough to sustain commensurables in their emerging community as it fragments into religious nationalism, secular nationalism, integrationists, separationists, and so forth. Thus those problems which were commensurable, both interracial and intraracial, in the era of integration became incommensurable in the era of Black Nationalism.

Nor can the growth of commensurable views be seen in the contemporary youth movements. On most of the American campuses today there are at least 10 tribes involved in "tribal wars" among themselves and against the "imperialistic" powers of those "over 30." Just to tick them off, without any attempt to be comprehensive, there are the up-tight protectors of the status quo who are looking for middle-class union cards, the revitalization movements of the Young Americans for Freedom, the reformists of pluralism represented by the Young Democrats and the Young Republicans, those committed to New Politics, the Students for a Democratic Society, the Yippies, the Flower Children, the Black Students Union, and the Third World Liberation Front. The critical change in this instance is not the rise of new groups; this is expected within the pluralistic model of politics. What is new are value positions assumed by these groups which lead them to make demands, not as points for bargaining and compromise with the opposition, but rather as points which are "not negotiable." Hence they consciously set the stage for either confrontation or surrender, but not for rendering incommensurables commensurable.

Moving out of formalized politics and off the campus, we see the remnants of the "hippie" movement which show clear-cut tribal overtones in their commune movements. This movement has, moreover, already fragmented into an urban tribe which can talk of guerrilla warfare against the city fathers, while another tribe finds accommodation to urban life untenable without sacrificing its values and therefore moves out to the "Hog Farm," "Morning Star," or "Big Sur." Both hippie tribes have reduced the commensurables with the dominant WASP tribe to the point at which one of the cities on the Monterey Peninsula felt sufficiently threatened to pass a city ordinance against sleeping in trees, and the city of San Francisco passed a law against sitting on sidewalks.

Even among those who still adhere to the pluralistic middle-class American image, we can observe an increasing demand for a change in the dimension of life and politics that has disrupted the elementary social processes: the demand for neighborhood (tribal?) schools, control over redevelopment projects, and autonomy in the setting and payment of rents to slumlords. All these trends are more suggestive of tribalism than of the growth of the range of commensurables with respect to the commons.

We are, moreover, rediscovering other kinds of tribes in some very odd ways. For example, in the educational process, we have found that one of our first and best empirical measures in terms both of validity and reproducibility – the IQ test – is a much better measure of the existence of different linguistic tribes than it is a measure of "native intellect"[6]. In the elementary school, the different languages and different values of these diverse tribal children have even rendered the commensurables that obtained in the educational system suddenly incommensurable.

Nor are the empirical contradictions of the common value myth as new as one might suspect. For example, with respect to the urban environment, at least 7 years ago Scott Greer was arguing that the core city was sick and would remain sick until a basic sociological movement took place in our urban environment that would move all the middle classes to the suburbs and surrender the core city to the "...segregated, the insulted, and the injured"[7]. This argument by Greer came at a time when most of us were still talking about compromise and accommodation of interests, and was based upon a perception that the life styles, values, and needs of these two groups were so disparate that a healthy, creative restructuring of life in the core city could not take place until pluralism had been replaced by what amounted to geographic or territorial tribalism; only when this occurred would urban incommensurables become commensurable.

Looking at a more recent analysis of the sickness of the core city, Wallace F. Smith has argued that the productive model of the city is no longer viable for the purposes of economic analysis[8]. Instead, he develops a model of the city as a site for leisure consumption, and then seems to suggest that the nature of this model is such that the city cannot regain its health because it cannot make decisions, and that it cannot make decisions because the leisure demands are value-based and, hence, do not admit of compromise and accommodation; consequently, there is no way of deciding among these various value-oriented demands that are being made on the core city.

In looking for the cause of the erosion of the myth of a common value system, it seems to me that as long as our perceptions and knowledge of other groups were formed largely through the written media of communication, the American myth that we were a giant melting pot of equalitarians could be sustained. In such a perceptual field it is tenable, if not obvious, that men are motivated by interests. Interests can always be compromised and accommodated without undermining our very being by sacrificing values. Under the impact of the electronic media, however, this psychological distance has broken down, and we now discover that these people with whom we could formerly compromise on interests are not, after all, really motivated by interests but by values. Their behavior in our very living room betrays a set of values, moreover, that are incompatible with our own, and consequently the compromises that we make are not those of contract but of culture. While the former are acceptable, any form of compromise on the latter is not a form of rational behavior but is rather a clear case of either apostasy or heresy. Thus we have arrived not at an age of accommodation but one of confrontation. In such an age "incommensurables" remain "incommensurable" in real life.

EROSION OF THE MYTH OF THE MONOPOLY OF COERCIVE FORCE

In the past, those who no longer subscribed to the values of the dominant culture were held in check by the myth that the state possessed a monopoly on coercive force. This myth has undergone continual erosion since the end of World War II, owing to the success of the strategy of guerilla warfare, as first revealed to the French in Indochina, and later conclusively demonstrated in Algeria. Suffering as we do from what Senator Fulbright has called "the arrogance of power," we have been extremely slow to learn the lesson in Vietnam, although we now realize that war is political and cannot be won by military means. It is apparent that the myth of the monopoly of coercive force as it was first qualified in the civil rights conflict in the South, then in our urban ghettos, next on the streets of Chicago, and now on our college campuses has lost its hold over the minds of Americans. The technology of guerrilla warfare has made it evident that, while the state can win battles, it cannot win wars of values. Coercive force which is centered in the modern state cannot be sustained in the face of the active resistance of some 10 percent of its population unless the state is willing to embark on a deliberate policy of genocide directed against the value dissident groups. The factor that sustained the myth of coercive force in the past was the acceptance of a common value system. Whether the latter exists is questionable in the modern nation state. But, even if most members of the nation state remain united around a common value system which makes incommensurables for the majority commensurable, that majority is incapable of enforcing its decisions upon the minority in the face of the diminished coercive power of the governing body of the nation state.

EROSION OF THE MYTH OF ADMINISTRATORS OF THE COMMONS

Hardin's thesis that the administrative arm of the state is capable of legislating temperance accords with current administrative theory in political science and touches on one of the concerns of that body of theory when he suggests that the ". . . great challenge facing us now is to invent the corrective feedbacks that are needed to keep the custodians honest."

Our best empirical answers to the question – *Quis custodiet ipsos custodes?* – "Who shall watch the watchers themselves?" – have shown fairly conclusively[9] that the decisions, orders, hearings, and press releases of the custodians of the commons, such as the Federal Communications Commission, the Interstate Commerce Commission, the Federal Trade Commission, and even the Bureau of Internal Revenue, give the large but unorganized groups in American society symbolic satisfaction and assurances. Yet, the actual day-to-day decisions and operations of these administrative agencies contribute, foster, aid, and indeed legitimate the special claims of small but highly organized groups of differential access to tangible resources which are extracted from the commons. This has been so well documented in the social sciences that the best answer to the question of who watches over the custodians of the common is the regulated interests that make incursions on the commons.

Indeed, the process has been so widely commented upon that one writer has postulated a common life cycle for all the attempts to develop regulatory policies[10]. This life cycle is launched by an outcry so widespread and demanding that it generates enough political force to bring about the establishment of a regulatory agency to reinsure the equitable, just, and rational distribution of the advantages among all holders of interest in the commons. This phase is followed by the symbolic reassurance of the offended as the agency goes into operation, developing a period of political quiescence among the great majority of those who hold a general but unorganized interest in the commons. Once this political quiescence has developed, the highly organized and specifically interested groups who wish to make incursions into the commons bring sufficient pressure to bear through other political processes to convert the agency to the protection and furthering of their interests. In the last phase even staffing of the regulating agency is accomplished by drawing the agency administrators from the ranks of the regulated.

Thus it would seem that, even with the existence of a common value system accompanied by a viable myth of the monopoly of coercive force, the prospects are very dim for saving the commons from differential exploitation or spoliation by the administrative devices in which Hardin places his hope. This being the case, the natural sciences may absolve themselves of responsibility for meeting the environmental challenges of the contemporary world by relegating those problems for which there are no technical solutions to the political or social realm. This action will, however, make little contribution to the solution of the problem.

ARE THE CRITICAL PROBLEMS OF MODERN SOCIETY INSOLUBLE?

Earlier in this article I agreed that perhaps until very recently there existed a set of conditions which made the solution of Hardin's problem subset possible; now I suggest that the concession is questionable. There is evidence of structural as well as value problems which make comprehensive solutions impossible and these conditions have been present for some time.

For example, Aaron Wildavsky, in a comprehensive study of the budgetary process, has found that, in the absence of a calculus for resolving "intrapersonal comparison of utilities," the governmental budgetary process proceeds by a calculus that is sequential and incremental rather than comprehensive. This being the case, ". . . if one looks at politics as a process by which the government mobilizes resources to meet pressing problems"[11] the budget is the focus of these problem responses and the responses to problems in contemporary America are not the sort of comprehensive responses required to bring order to a disordered environment. Another example of the operation of this type of rationality is the American involvement in Vietnam; for what is the policy of escalation but the policy of sequential incrementalism given a new Madison Avenue euphemism? The question facing us all is the question of whether incremental rationality is sufficient to deal with twentieth-century problems.

The operational requirements of modern institutions makes incremental rationality the only viable form of decision making, but this only raises the prior question of whethere there are solutions to any of the major problems raised in modern society. It may well be that the emerging forms of tribal behavior noted in this article are the last hope of reducing political and social institutions to a level where incommensurables become commensurable in terms of values *and* in terms of comprehensive responses to problems. After all, in the history of man on earth we might well assume that the departure from the tribal experience is a short-run deviant experiment that failed. As we stand "on the eve of destruction," it may well be that the return to the face-to-face life in the small community unmediated by the electronic media is a very functional response in terms of the perpetuation of the species.

There is, I believe, a significant sense in which the human environment is directly in conflict with the source of man's ascendancy among the other species of the earth. Man's evolutionary position hinges not on specialization, but rather on generalized adaptability. Modern social and political institutions, however, hinge on specialized, sequential, incremental decision making and not on generalized adaptability. This being the case, life in the nation state will continue to require a singleness of purpose for success, but in a very critical sense this singleness of purpose becomes a straightjacket that makes generalized adaptation impossible. Nowhere is this conflict more evident than in our urban centers where there has been a decline in the livability of the total environment that is almost directly proportionate to the rise of special-purpose districts. Nowhere is this conflict between institutional singleness of purpose and the human dimension of the modern environment more evident than in the recent warning of S. Goran Lofroth, chairman of a committee studying pesticides for the Swedish National Research Council, that many breast-fed children ingest from their mother's milk "more than the recommended daily intake of DDT"[12] and should perhaps be switched to cow's milk because cows secrete only 2 to 10 percent of the DDT they ingest.

HOW CAN SCIENCE CONTRIBUTE TO THE SAVING OF THE COMMONS?

It would seem that, despite the nearly remorseless working of things, science has some interim contributions to make to the alleviation of those problems of the commons which Hardin has pointed out.

These contributions can come at two levels:

1. Science can concentrate more of its attention on the development of technological responses which at once alleviate those problems and reward those people who no longer desecrate the commons. This approach would seem more likely to be successful than the "... fundamental extension in morality ..." by administrative law; the engagement of interest seems to be a more reliable and consistent motivator of advantage-seeking groups than does administrative wrist slapping or constituency pressure from the general public.
2. Science can perhaps, by using the widely proposed environmental monitoring systems, use them in such a way as to sustain a high level of "symbolic dis-assurance" among the holders of generalized interests in the commons – thus sustaining their political interest to a point where they would provide a constituency for the administrator other than those bent on denuding the commons. This latter approach would seem to be a first step toward the "... invention of the corrective feedbacks that are needed to keep custodians honest." This would require a major change in the behavior of science, however, for it could no longer rest content with development of the technology of monitoring and with turning the technology over to some new agency. Past administrative experience suggests that the use of technology to sustain a high level of "dis-assurance" among the general population would also require science to take up the role and the responsibility for maintaining, controlling, and disseminating the information.

Neither of these contributions to maintaining a habitable environment will be made by science unless there is a significant break in the insularity of the two scientific tribes. For, if science must, in its own insularity, embark on the independent discovery of "the tragedy of the commons," along with the parameters that produce the tragedy, it may be too slow a process to save us from the total destruction of the planet. Just as important, however, science will, by pursuing such a course, divert its attention from the production of technical tools, information, and solutions which will contribute to the political and social solutions for the problems of the commons.

Because I remain very suspicious of the success of either demands or pleas for fundamental extensions in morality, I would suggest that such a conscious turning by both the social and the natural sciences is, at this time, in their immediate self-interest. As Michael Polanyi has pointed out, ". . . encircled today between the crude utilitarianism of the philistine and the ideological utilitarianism of the modern revolutionary movement, the love of pure science may falter and die"[13]. The sciences, both social and natural, can function only in a very special intellectual environment that is neither universal or unchanging, and that environment is in jeopardy. The questions of humanistic relevance raised by the students at MIT, Stanford Research Institute, Berkeley, and wherever the headlines may carry us tomorrow, pose serious threats to the maintenance of that intellectual environment. However ill-founded *some* of the questions raised by the new generation may be, it behooves us to be ready with at least some collective, tentative answers – if only to maintain an environment in which both sciences will be allowed and fostered. This will not be accomplished so long as the social sciences continue to defer the most critical problems that face mankind to future technical advances, while the natural sciences continue to defer those same problems which are about to overwhelm all mankind to false expectations in the political realm.

REFERENCES

[1] G. Hardin, *Science* **162,** 1243 (1968).

[2] J.B. Wiesner and H.F. York, *Sci. Amer.* **211** (No. 4), 27 (1964).

[3] C. Woodbury, *Amer. J. Public Health* **45,** 1 (1955); S. Marquis, *Amer. Behav. Sci.* **11,** 11 (1968); W.H. Ferry, *Center Mag.* **2,** 2 (1969).

[4] C.P. Snow, *The Two Cultures and the Scientific Revolution* (Cambridge University Press, New York, 1959).

[5] M. Olson, Jr., *The Logic of Collective Action* (Harverd University Press, Cambridge, Mass., 1965.

[6] G.A. Harrison, et al., *Human Biology* (Oxford University Press, New York, 1964), p. 292; W.W. Charters, Jr. in *School Children in the Urban Slum* (Free Press, New York, 1967).

[7] S. Greer, *Governing the Metropolis* (Wiley, New York, 1962), p. 148.

[8] W.F. Smith, "The Class Struggle and the Disquieted City," a paper presented at the 1969 annual meeting of the Western Economic Association, Oregon State University, Corvallis, Ore.

[9] M. Bernstein, *Regulating Business by Independent Commissions* (Princeton University Press, Princeton, N.J., 1955); E.P. Herring, *Public Administration and the Public Interest* (McGraw-Hill, New York, 1936); E.M. Redford, *Administration of National Economic Control* (Macmillan, New York, 1952).

[10] M. Edelman, *The Symbolic Uses of Politics* (University of Illinois Press, Urbana, Ill., 1964).

[11] A Wildavsky, *The Politics of the Budgetary Process* (Little Brown, Boston, 1964).

[12] Corvallis *Gazette-Times,* 6 May 1969, p. 6.

[13] M. Polanyi, *Personal Knowledge* (Harper & Row, New York, 1964), p. 182.

48

Walton J. Francis

A REPORT OF THE MEASUREMENT AND THE QUALITY OF LIFE

SOCIAL INDICATORS

Lowering our vision from a grand scheme which aggregates social welfare in common units, we can seek to devise "social indicators" which measure progress or problems in particular areas of social concern. In *Toward a Social Report* a social indicator was defined as

> *a statistic of direct normative interest which facilitates . . . judgments about the condition of major aspects of a society. It is . . . a direct measure of welfare and . . . if it changes in the right direction, while other things remain equal . . . people are "better off." Thus statistics on the number of doctors or policemen could not be social indicators, whereas figures on health or crime rates could be.*[1]

We doubt that any social indicator will ever meet the criterion of "a direct measure of welfare" in the strictest sense. Even ignoring fundamental episte-

---

Selected from A Report on Measurement and Quality of Life and the Implications for Government Action of "The Limits to Growth," *by Walton J. Francis, Department of Health, Education, and Welfare, Washington, D.C. (1973), 6-16.*

mological issues, we face serious conceptual and practical difficulties.

For example, what is health? A positive feeling of mental and physical well-being, normal functioning in society, or absence of pain and disease? Depending on our definition, we might develop considerably different indicators (e.g., self-assessment of well-being on subjective rating scale versus "objective" counts of disability-free days). Similarly, if we accept disability-free days, how do we measure them – by counting days outside of hospital beds[2], or as subsequent researchers have proposed, by counting days in which no impairment of normal functioning occurs? The difference is substantial, since Americans average about 2 years of institutional bed disability and 5 years of impaired functioning.

Thus social indicators are often limited proxies for what we really want to measure and, in the narrowest sense, are "mere" social statistics such as "the number of doctors per 1,000 population." Of course, such data can be extraordinarily useful, as in the use of physician visits to measure improved access to health care by the poor, and coliform counts to serve as a measure of potentially harmful microorganisms in water. But the data are not direct measures of welfare.

A major cause of this reliance on proxies is, of course, not only conceptual. Some things are just hard to measure, however they are defined. Edward R. Murrow used to say that a cash register doesn't ring when a man changes his mind for freedom. Nor do we have good ways to measure privacy or government responsiveness or a host of other attributes of life quality.

These difficulties mean that lists of social indicators simply do not speak for themselves, independent of context. The very labeling of data as "social indicators," with the normative implications which labeling carries, may be misleading in its implications. Indeed, the omission of items which cannot be measured is itself a major bias – often the most important facts about social conditions are qualitative, derivative, or interactive.

CONCEPTUAL AND PRACTICAL DIFFICULTIES: SPECIFIC PROBLEMS

In what follows we discuss problems specifically in the context of social indicators. However, all these difficulties apply to any approach to measuring the quality of life.

Which Direction Is Better?

Related to the surrogate nature of most social indicators is the difficulty of interpreting their message simply and unambiguously. Is an increase in the divorce rate a bad thing, because it indicates increasing decay in the family structure, is it a good thing because it means that fewer unhappy couples are forced to stay together, or do we evaluate it at an even higher level and question the very notion of a husband-wife family structure? Is an increase in years of educational attainment a measure of increased skill development, of changing tastes or income availability for consumption, of future discontent as unrealistic aspirations are dashed, or of all three?

The divorce example presents primarily a problem of valuation. People disagree about goals, and social indicators are not value neutral. A Marxist and a businessman might both want to see unemployment go up, because the former believes it hastens the day of revolution, and the latter that it reduces the rate of inflation and disciplines labor unions. Similarly, a little unemployment might be a good thing in most people's view, but a lot intolerable. So which direction is better depends both on where we are and on our fundamental values. The measure itself is not an "objective" test independent of our beliefs.

The education example presents another problem. The phenomena that indicators measure have many facets, and a single indicator thus represents multiple outputs, both good and bad. Some of these outputs are subtle or indirect and often not well known. Perhaps only a few economists and sociologists fully appreciate the dangers of continued increases in educational attainment for income distribution and social turmoil. And perhaps they are wrong. But the indicator cannot be accepted at face value simply because it is widely perceived to represent an important benefit.

Many examples of indicators ambiguously representing multiple outputs are found in a "White Paper on Living Conditions," published by the Economic Planning Agency of Japan. One indicator used by this document is a measure of "social security" representing the ratio of government income maintenance programs to the GNP. The U.S. ranks low, implicitly because we provide a less adequate income maintenance system than the developed countries to which we are compared. In fact, this measure primarily reflects the American reliance on a progressive income tax, which reduces the need for direct transfers, and the relatively large role of private health insurance in this country. Surely we would not be improving our income maintenance performance if welfare and unemployment payments increased during a recession, or if we raised taxes on lower incomes and had to increase Social Security and other benefits to leave beneficiaries as well off as before – but our ratio would rise.

How Do We Measure It?

Even if we agree on values, and multiple outputs are not a problem, development and interpretation of indicators present substantial conceptual and practical difficulties. Some countries measure unemployment by the number of people visiting state unemployment services, a procedure roughly as valid as measuring crime by reports from the police to the FBI (which we do) or poverty by number of welfare cases (which we do not). In this context, a health index based on deductions for hospital bed days would look *worse* as we increased access to health services for the poor.

The avoidance of such "institutional" indicators does not prevent foundering. The unemployment rate counts rich teenagers looking for summer jobs in resorts as well as fathers who cannot feed their families. But it omits those so discouraged they have quit searching for work. And the employment and unemployment data omit entirely the dimension of job satisfaction, except as it is reflected (along with other things such as increased opportunities) in high turnover rates. Of course, where we are sensitive to the limits of possible indicators, these deficiencies can be reduced by improving measures and/or careful use for any particular indicator. So one approach is to develop improved indicators.

Sometimes such improvement is difficult. Less widely known but more important than the deficiencies in the unemployment rate are some major biases in our income distribution statistics. For example, independent evidence (such as total unemployment benefits paid) shows that the cash income of the lowest (and highest) income groups is underreported substantially in the Census and Current Population Surveys. Moreover, certain kinds of income – such as the cash value of Food Stamps and Medicaid – are not counted at all in these surveys. This means that the total number of poor people and the "poverty gap" are exaggerated substantially in the official poverty data. This bias might not matter per se – on the grounds that poverty is relative – except that the most underreported items have increased especially rapidly in the last decade. Therefore, our trend data are substantially biased over time, and policy decisions concerning such issues as the relative balance between more money versus improved incentives and equities may be distorted. Unfortunately, the conceptual (where should we draw the line on what we count as income?) and practical (how can we correct the data?) problems in improving the data are severe.

But even perfect meausrement has its limitations. Our data on annual income are broken out in great demographic detail. These data show that the gap between median black and white incomes decreased substantially on a percentage basis in the 1960s, but increased in terms of absolute dollars. Depending on their political proclivities, and on the issue at hand, various social commentators use one measure of progress or the other. (The most honest use both if they have space for explanation.) There is simply no unambiguous measure of progress in this area, or probably in any other.

MEASURING IT HIGHER AND ENJOYING IT LESS

One approach to improving measures of the quality of life is to bypass the "objective" social indicators and test our progress against subjective perceptions. In some ultimate sense we cannot measure psychic satisfaction directly or satisfactorily. (What does it really mean if two people both tell us they are "very happy" – perhaps one of them has been stunted in aspiration and the other in potential, and does "very" mean the same to both?) But we can elicit states, attitudes, and values from public opinion surveys and avoid some of the interpretive problems of the objective counts (though surveys present other dangers – as evidenced by the underreporting of income). Divorce is good for some people and bad for others – why not just ask?

One serious danger lurking behind the subjective approach was, as usual, described first and best by de Tocqueville:

> *The evil which was suffered patiently as inevitable seems unendurable as soon as the idea of escaping from it crosses men's minds. All the abuses then removed call attention to those that remain, and they now appear more galling. The evil, it is true, has become less, but sensibility to it has become more acute.*[3]

Such a phenomenon may account for much of the malaise of our times (Crane Brinton has argued plausibly that most revolutions occur at times when conditions are improving rapidly, and Schumpeter has argued that the very successes of capitalism will create a class of alientated intellectuals who will destroy capitalism). The problem this phenomenon presents is that all our social indicators might be rising but public disaffection increasing. Is the quality of life, then, rising or falling?

Perhaps the dilemma will not occur. Cantrill's *Hope's and Fears of the American People*[4] shows that in 1971 virtually all groups in society, including blacks and youth, believed that their personal lot had improved in the 1960s and would improve in the 1970s (the exception is the poor, but since there is substantial turnover in the poverty pool, this probably reflects the problems of those most recently displaced). Yet, seemingly inconsistently, but perhaps reflecting "intellectual" opinion, every single group except blacks believed that the country as a whole had lost ground in the recent past.

CAUSAL IGNORANCE AND SOCIAL INDICATORS

Much of this problem of ambiguity of meaning for social indicators, even those which have the most surface plausibility, lies in our lack of understanding of cause and effect in human behavior, particularly over long periods of time and in complicated situations. Even if we had the best imaginable indicator for positive health, how would it be affected by, and relate to, indicators of environmental quality, unemployment, and alienation? Since the underlying phenomena are linked in various more or less direct ways, so too are the indicators linked. But what if we cannot measure, let alone understand, the links?

Even in the "simplest" cases we become trapped in causal quagmires. If we plumb black-white income statistics, for example, we become faced with the need to control for age, education, region, family size, and labor force participation to attempt to get a meaningful measure of the residual gap related "solely" to race. This residual itself might be due to discrimination, motivational differences, quality of education, or a variety of complicating interactions (e.g., if blacks face discrimination they may rationally restrict their job search, which lowers their exposure to good on-the-job experience, which lowers their future earnings and reinforces black decisions to invest less in education and white decisions to discriminate). Is Moynihan right in calling attention to the income parity that young, married black couples in the north have recently reached with their white counterparts, a statistic which implicitly controls many variables, or would a critic be right who emphasized that smaller proportions of young, northern blacks are married couples, and that more of the black wives work?

It is no wonder, then, that as ambitious an attempt at exploring causal linkages as *The Limits to Growth*[5] has been judged by many critics to be so oversimplified as to be fundamentally misleading. Indeed, the very force of *Limits'* conclusions derives from its use of supposedly fundamental physical properties, and the model completely ignores the price system, the political system, evolution in values and customs, and other processes of social and economic adjustment. It is as incomplete as models which assume that, because the HEW budget goes up, educational attainment of children goes up, ignoring the complicated real-world forces which mediate the process. *Limits* could be viewed as a strictly hypothetical projection rather than a prediction (we doubt, though, that the authors are so modest in intention), but on either interpretation its causal modeling is primitive.

We mention these deficiencies in *Limits* to illustrate how difficult causal analysis is. Nobody else has a well-developed countermodel capable of producing reliable predictions covering the same variables as *Limits,* let alone a model which deals with broader dimensions of the quality of life.

One example of an area in which even simple feedback loops of the kinds used by *Limits* could aid our understanding substantially lies in manpower training, rehabilitation, and job placement. Virtually without exception, no analyst of the benefits and costs of such investments has made a serious attempt to deal with the "displacement" effect – the filling of a job by a newly trained worker may mean that another worker continues unemployed. That this effect occurs to some substantial degree, especially during periods of high unemployment, can scarcely be doubted. And to some extent we may want displacement. The problem is how much occurs, under what circumstances, and for whom? Unfortunately, standard procedures for evaluating training result in attributing greater net benefits the greater the displacement and the larger the program, since the comparison is with persons who did not receive training or other services – part of the group made worse off. Similarly, the economic benefits of higher education may be due substantially to displacement from better paying jobs of those with only a high school education, regardless of real differences in skills (for an extended argument making this and related points, see Lester Thurow, "Education and Economic Equality," in *The Public Interest,* Summer 1972). Of course these problems in no way vitiate the real benefits of training and education – but no one knows even approximately where the balance lies.

As a final example, the Coleman Report's[6] main contributions were to document the lack of measurable inequality of inputs in our schools, to shake naive faith in the benefits of increased educational resources, and to stimulate rethinking of the meaning of "inequality." But even so sophisticated an analysis has misled (primarily through reviewers) innumerable people into believing that its findings imply that socioeconomic integration will improve disadvantaged student performance. Perhaps integration will, but statistical correlation does not imply causation (even had the Coleman statistical procedures been fully adequate) and the Coleman Report, like social science research in general, is a very thin reed to rely on.

Lacking such causal knowledge, the validity and usefulness of any set of social indicators (or social statistics) for policy diagnosis becomes suspect. Policy analysis becomes an art rather than a science. Straightforward quantitative measures of progress are inadequate not only because they fail to disclose the underlying processes, but also because they may even obscure them.

MATHEMATICAL MANIPULATION AND SOCIAL INDICATORS

Even if we had the most perfect indicators imaginable, how do we present the numbers? What criteria should guide us? This problem has at least three dimensions – the selection of the indicator, its quantitative presentation, and quantitative pro-

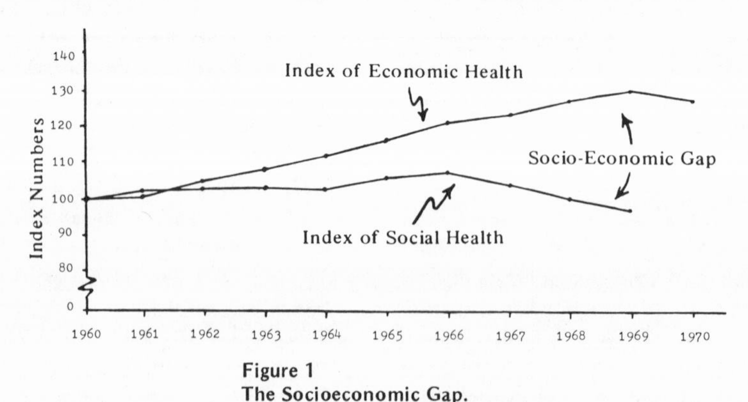

Figure 1
The Socioeconomic Gap.

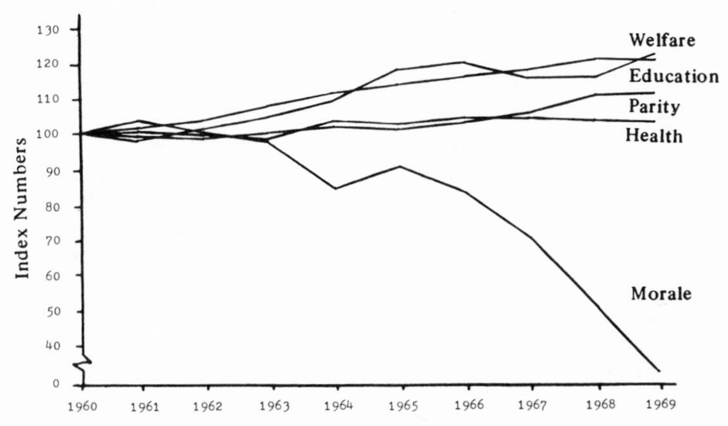

Figure 2
Area Indicators. (Note: Each of these five area indicators is the arithmetic mean of several components. In turn, the arithmetic mean of the five area indicators constitutes the overall index of social health, which is shown in Figure 1.

cedures used to link it to other indicators. Each presents substantial problems.

One recent effort at measuring the quality of American life illustrates well all these problems (Michael Spautz, "The Socio-Economic Gap," in *Social Science Research*, June 1972). The summary indexes of this article appear in the figures reproduced here (with permission of the publishers). They look reasonable. Yet Spautz commits every single fallacy we discuss and his results are capricious (he makes clear in his article that he is aware of the dangers and that his results are illustrative – in which spirit the following criticisms are made).

Spautz prepares each of his "area" indicators on the basis of a large number of items selected from standard statistical series. It should not be surprising, therefore, that many of his individual indicators stand out for ambiguity or error. Is a big increase in draftees medically disqualified a sign of decreasing health (Spautz), of increasing skill among cooperating lawyers, doctors, and draftees, or of decreasing morale? Is the increasing size of social welfare expenditures (the HEW budget) a measure of welfare (Spautz), of increasing problems, of Parkinson's law, or of anything important at all? Is the rise in the medical care CPI[7] a measure of decreasing health (Spautz), of our inability to adjust this and other service price indexes for subtle quality increases, or of increasing access to health by the poor? Does anyone seriously believe that changes in the number of addicts reported to the Bureau of Narcotics and Dangerous Drugs by police and hospitals are an approximately accurate measure of changes in addiction? While these are among the most glaring questions, every single indicator he picks could be similarly challenged.

But even supposing each indicator were "perfect," problems remain. Spautz has no explicit basis other than expediency (and if he had one, no supportable basis) for the particular indicators he picked versus a larger or smaller number of entirely different indicators. Why not use Cantrill's ladder index of personal satisfaction to measure morale? Why exclude measures of freedom and political participation and thus give these dimensions of life quality a zero weight? Changes in the ratio of social to defense expenditures have been dramatic in recent years, in spite of the war, but surely this duplicates in large part separate indexes of defense and social expenditures.

Given an indicator, and a set of raw data for it, how should it be quantified for presentation? Spautz presents indexes which show changes each year from a base value of 100, to provide a common format for the data (a procedure which is unobjectionable in itself). But by what mathematical procedure should the data be transformed? Linearly or logarithmically, using the raw data or the data expressed as percentages, or what? To take an extreme case, his work stoppage rate index goes from 100 to 0 because the percentage of days lost from strikes doubled in 9 years. Yet the actual change, from 0.14 to 0.28 percent of worker time lost, could have been expressed as a decrease from 99.86 to 99.72 percent of worker time *not* lost, and the index would have shown a drop of 100 to only 99.9 on this basis.

Finally, in aggregating individual indexes to his area and gap indexes, Spautz weights each indicator equally. This is a natural and "obvious" procedure, but logically indefensible. Would anyone defend the notion that the percentage of the population in the military is exactly as important a measure of "welfare" as the poverty rate, or that economic consumption is as important, alone, as all other aspects of society?

Some of the absurdities could be eliminated. Better measures could be developed. But the inherent arbitrariness would remain. Using different weighting procedures alone, and accepting every other element of Spautz's paper, every single aggregated index could have been shown to decrease substantially. If any single component of the morale index has risen, different weighting could have resulted in the overall morale index rising. The danger is that the better the surface plausibility of the various indicators (i.e., the better each appears to meet the *Social Report* criterion of "a direct measure of welfare"), the less obvious would be the arbitrariness of the overall index. In constructing overall indexes of social welfare, we lack a theoretical basis (such as we might have for black-white income differences) which even allows a point of departure.

THE DANGERS OF BELIEF

Because a set of social indicators, or even a single indicator, is inherently arbitrary, it presents a special political danger. We may delude ourselves into believing that the index really measures what it purports to measure. This danger increases in proportion to the claims made or inferred for the indicator – as in the case of the GNP.

This is a danger over and above the normal use or abuse of statistics in the political process, increasing with the honesty, competence, and good will of the authors. For example, the Council on Environmental Quality annual reports question the wisdom of current environmental policies only in the most oblique ways. Special analyses printed as part of the U.S. budget never address the benefits of various social programs, even to the point of admitting our ignorance. *Consumer Reports* has never told its readers that its automobile ranking system is completely arbitrary nor that very minor changes in weights or test results would change the rankings substantially.

When we consider the need for political expediency together with the probability that those who prepare social indicators will feel impelled to defend their effort, the possibilities for misleading results multiply. One can imagine how a Council of Social Advisors would have wrestled 5 years ago among more or less apocalyptic versions of social collapse and more or less strident calls for the Great Society. One need hardly guess just how much scholarly investigation or even speculation such a council would have devoted either then or today to such questions as the effects of the Vietnam War on American morale. One can wonder whether *Toward a Social Report* could have preserved its objective, analytic tone had it tried to examine critically the effects of HEW programs on various social indicators or attempted an overall assessment of the quality of life. Indeed, *Toward a Social Report* was originally intended to be a social report without the "Toward." Because of difficulties of the kinds we have been discussing, this ambitious goal was dropped and an explicitly exploratory approach was taken.

The greatest danger comes from committing the easiest sins of all – those of omission. Economists footnote in their texts the dangers of interpreting the GNP as a welfare measure (and sometimes don't even bother since this is taken for granted), but this is much like selling prescriptions without wrap-around labels giving directions for use.

A related danger lies in ignoring things which have not, or cannot, be measured. Virtually every benefit-cost study states in passing that many important outputs are not taken into account because they are nonquantifiable, and defends its approach on the impeccable ground that any reduction in ignorance is useful. What other posture can be taken in a world in which at best we can gain partial understanding? So we ignore displacement, self-respect and alienation, and crime reduction, in a long list of the unmeasured consequences of training programs. The problem, of course, is that a little bit of the truth is sometimes taken for the whole – half-right analysis can be worse than none.

These propensities are compounded because truth is contextual – who is to say where to draw the line, and must the universe be surveyed in every statement? Few would argue that the failure of the Census to publish prominently "corrected" estimates of poverty (and of income at the upper end) or at least to suggest the magnitude of error is reasonable, and most would agree that Food Stamps should be counted as if they were cash. But where do we stop – do we insist on counting the cash value of all government services, correcting for tax incidence, and adding in the "value" of leisure? The official definition of poverty is itself arbitrary, as is the decision to use annual average rather than monthly or lifetime income data. The problems are paralyzing, and were we to insist on accuracy, let alone theoretical nicety or contextual "exposure," we would still be waiting for poverty data.

It could be argued that in spite of the inherent pitfalls and inescapable political overtones, social indicator efforts would hardly be likely to leave us worse off than we are now. Surely the Council of Economic Advisors and Council on Environmental Quality have raised the level of economic and environmental debate. The dangers of actively but unintentionally misleading attempts can be coped with – there is rarely a shortage of critics as the authors of *The Limits to Growth* have discovered. And the need for better assessment of where we are going is patent. But truth does not necessarily win, and the prospect of major advance is small.

REFERENCES

[1] U.S. Department of Health, Education, and Welfare, *Towards a Social Report,* Washington, D.C., 1969.

[2] *ibid.,* p. 3.

[3] *ibid.,* p. xii.

[4] Hadley Cantrill, *Hopes and Fears of the American People,* New York (Universe), 1971, 93 pp.

[5] Donella H. Meadows, Dennis L. Meadows, Jorgen Randers, William W. Behrens, III, *The Limits to Growth,* New York (Universe), 1972.

[6] James Samuel Coleman, *Equality of Educational Opportunity,* Washington, D.C. (U.S. Government Printing Office, 1966) Vol. 1, 737 pp.; Vol. 2, 548 pp.

[7] Consumer Price Index.

49

Robert M. Griffin, Jr.

ETHOLOGICAL CONCEPTS FOR PLANNING

In the last few years an increasing number of books and articles have attempted to apply concepts of animal behavior to man. Among these are Desmond Morris's *The Naked Ape* (1), in which man's behavior is pictured as essentially governed by primitive instincts, and Robert Ardrey's *The Territorial Imperative* (2), in which the possession of territory is viewed as a fundamental need of man. These popular writers have been severely criticized for oversimplifying both human and animal behavior.[1] Despite these criticisms, it would be a mistake to conclude that the new science of ethology – the biological study of behavior[2] – has little serious significance for human behavior. At the very least, ethology has challenged pivotal assumptions in the behavioral sciences and has stimulated new research and theoretical concepts in such diverse fields as anthropology, child development, psychoanalysis, epidemiology, and social psychology.[3] Some ethologists of solid reputation have even begun to take an interest in scientific analysis of urban problems.[4] It may therefore be useful for planners and urban scholars to become generally familiar with this new field of knowledge.

Analysis of biological origins and controls over human behavior may suggest new limitations and possibilities for planned arrangements of the environment. This may result from demonstrations of some invariable patterns which are common to all forms of human behavior. The plasticity of human behavior suggests, however, that ethology may prove more enlightening in explaining how discrepancies may arise between overt behavior and biological responses to the environment. Conflicts of this kind may affect states of human health and well-being.

CONCEPTS OF ETHOLOGY

Charles Darwin's *Origin of the Species* (1859) and Herbert Spencer's *Principles of Psychology* (1855) laid the intellectual foundation for ethology. The theory of evolution rested on the assumption that there was competition among organisms for limited resources. This competition usually proceeded indirectly rather than through combat. Those animals with more efficient organs set a standard that the less talented could not meet, and the latter would disappear through less frequent reproduction. This process of *natural selection* and genetic *mutation* (which Darwin postulated, but never explained) resulted in an evolution of biological characteristics. Darwin (6) later studied expressions of emotion in man and animals. Most versions of animal behavior were, however, inaccurate, either tending to picture competition among animals as bloodthirsty, in keeping with the assumptions of Social Darwinism, or sentimentally emphasizing human qualities in unscientific, subjective anthropomorphism.

In reaction to anthropomorphism, a school of behavioral studies, which became known as the Mechanists, emerged around 1900 and flourished primarily in the United States and Russia. In terms of Darwinian theory, these progenitors of American experimental psychologists focused on capacity of the brain and influence of environmental factors. In contrast to the Mechanists, William McDougall (along with other biologists) was impressed with the "purposiveness" or "goal-directedness" of much behavior which could not be reconciled with the hypothesis that all behavior patterns are acquired from environmental interaction. To some extent psychoanalysis and Gestalt psychology may be considered to be of this Vitalist school of thought. The assumption of "purpose" instead of "directiveness" in the behavior of an organism was unacceptable teleology to the Mechanists (13).

The early ethologists were European naturalists or zoomen who were personally acquainted with the overwhelming variety of puzzling behavior among animals which was not satisfactorily explained by either the Mechanists or Vitalists. Scattered research from about 1890 through 1920 influenced the Austrian, Konrad Lorenz, who is generally considered to be primarily responsible for formulating the principal concepts of ethology.

Theory and Method

Encouraged by scientific advances which reestablished the genetic basis for the influence of mutation in evolution, ethologists applied neo-Darwinian logic to the analysis of species behavior *characteristics,* which were conceived as organs in the same way that morphological characteristics have been viewed. It was noted that behavior patterns enabled the species to survive. As Niko Tinbergen, another early ethologist, has put it, Lorenz was asking about behavior: "What is it good for?" and "How does it work'?" (31, p. 417)

The study of the survival value of behavior was not guesswork. Observable behavior in such studies is considered the cause and survival is the effect. Ethologists assumed that behavior must be sufficiently adaptive to maintain the species. If survival value of a behavior pattern was to be examined, then it was the behavior of the species rather than the individual which was of interest. To discover how the species behaved, it was necessary to observe animals under natural conditions and to prepare descriptive dossiers of their life histories. These were known as ethograms. The ultimate aim was an accurate picture of muscle action patterns, but for some purposes either characterizations or selected peculiarities of animals' responses to one another were considered sufficient.

After the ethogram was prepared, behavior was classified by presumed function and experimentally studied. For example, simple cardboard models of the parent gull might be held in front of chicks to find what visual stimuli elicited the latter's begging; or dummies with different characteristics would be used to see what controlled the movement of fish. Animals were raised in isolation from their parents to separate individually learned behavior from species-specific behavior. Controlled laboratory experiments have been used with increasing frequency in recent years.

Mechanisms of Behavior

Ethologists developed a series of ideas to conceptualize the functions of physiological processes in behavior.

Appetitive and Consummatory Behavior - A distinction was made between "appetitive" and "consummatory" behavior.

> *Appetitive behavior is the more variable, searching phase of behavioral sequences (e.g., looking for food); consummatory behavior is the more stereotyped phase which often leads to the termination of the sequence (e.g., eating food) (16, pp. 17-18).*

This distinction provided hypotheses to explain behaviors observed in animals:

> *The "goal" of appetitive behavior is the expression of the consummatory act and not, as we are tempted to think, the satisfaction of a physiological need such as hunger, nesting, etc.... In other words, what the hungry hawk ... is "seeking" is the opportunity to pounce upon a prey, not the food as such. The activity of pouncing depletes the instinct center and thus quiets the restless seeking behavior (7, pp. 175-176).*

Later, it was recognized that reafferent feedback as well as depletion of energy may account for the termination of appetitive behavior.

Stimuli and Releasers - To explain the releasing of energy, ethologists devised the concept that discharge of the consummatory act was blocked by an internal mechanism until released by a sign stimuli. These signals were considered to be distinctive combinations of characters, or even single characters, which identify appropriate objects and rarely occur in inappropriate objects. Response was to these characters, not to the situation as a whole. Behavior caused the animal to do the "right" thing at the "right" moment because the consummatory act was withheld until it fitted the situation. Thus "directiveness" of behavior was a consequence of selective sensitivity to specific situations. All could be explained by the pressure of natural selection.

Instinct Center - Coordination of motor activities was carried out by the "instinct center" – an organized part of the central nervous system. Its parts were linked together to provide for the appropriate sequence of muscular and glandular activities which constituted behavior. This concept of behavior has sometimes been explained as a hydraulic model (or so-called "flush toilet") in which endogenous drives build up action-specific energy in the appetitive phase of behavior. This raises the energy level in the "container," which was released when the appropriate sign stimuli appeared and acted like a weight which pulls the "valve" of the "innate releasing mechanism," allowing energy to be channeled into specific pathways, or consummatory behavior.

Displacement and Vacuum Activities - When the releaser for an intention movement was present, but action was in conflict with another drive, the accumulated energy was viewed by ethologists as overflowing the hydraulic tank into displacement activities. This behavior would often appear irrelevant to the situation. Vacuum activities were seemingly aimless movements which occurred when the accumulation of energy was so great that it overflowed into activities when a releaser was not present.

Instincts or Innate Characteristics of Behavior - Species-specific behavior, which appeared in animals raised in isolation, was considered to be based on instinct. These instincts operated in an animal's normal life as means of accomplishing its survival activities. Ethologists viewed instincts as directed toward the goal of the consummatory act, originating from information stored in genes rather than individually acquired. This innate behavior was not, however, purposive, as McDougall's concept of instincts. The cause of this behavior was natural selection. Innate character was originally considered a property of stereotyped patterns of behavior; later, however, this view was modified. It was recognized that the degree of stereotypy was not a reliable criterion of innateness, and that liable patterns of behavior may be innate and stereotyped patterns of behavior may be learned.

Social Relations - The mechanisms of behavior also explained animal social relations: "The basic concept here is that animals carry in their own structure and behavior the releasers that evoke appropriate social responses in their fellow group members" (7, p. 92). These were simple morphological sign stimuli usually supplemented by peculiarities of behavior. Stimuli released "behavior adequate for a particular social interaction" through presentation by a member of the species without regard to the learned recognition of the individual (7, p. 183).

> *The Herring Gull chick aims its pecking response at the parent's bill tip from the first, without having to learn it. The male Stickleback raised in isolation reacts to other males by fighting, to females by courting. It could not have learned this. In other words, it is not only the capacity to perform these motor patterns which is innate, but their sensitivity to special releasing and directing stimuli as well (32).*

The early ethologists neglected some learned patterns of social relations. They concentrated on aspects of social organization which developed from the interaction of innate sign stimuli and releasers and were adapted to survival. Learning is now considered to be of great importance in modifying and directing innate tendencies of behavior and in elaborating behavior repertories, especially in primate animals.

Ritualization and Displays - Releasers' effectiveness depended upon their distinctiveness and conspicuousness in communication with other members of the species. They tended to be clearly differentiated by bodily structure, color, or type of instinctive action. The actions which functioned as releasers were considered to evolve from displacement activities into conspicuous sign stimuli by *ritualization.* Under the pressure of natural selection, movements acquired formality, stiffness, slowness, or exaggeration which served no purpose except to make them stand out sharply from ordinary functional movements. These stylized movements were called *displays.*

Imprinting - A special phenomenon observed by Lorenz was an animal's capacity, during a limited, specific period of its early life, to develop attachments for people and things. Lorenz maintained that impressions which occurred at these times were different from learning because they involved no rewards and lasted a lifetime (13, p. 49; 7, pp. 182-183).

Characteristics of Behavior

The great variety in behavior among different species limits generalizations about animals. There are, however, a few behavior patterns which are fairly common among numerous species.

Spontaneous Rhythms and Movements - Even in the lowest forms of animal life, it is evident that behavior occurs in the absence of immediate environmental stimuli. Karl S. Ashley demonstrated that flatworms and sea anemones have spontaneous and constant rhythms of movement which are independent of external stimuli. Bees, starlings, and fish display consistent orientation in determining direction. Recently, cellular behaviors of endogenous origins have been demonstrated. The evidence is sufficient to suggest that animal life is characterized by behavior which cannot be explained by external stimuli and therefore is of endogenous origins.

Social Aggregation - Even the most primitive forms of animal life tend to aggregate in groups. This may result from local environmental conditions or from directional responses to specific environmental conditions (tropistic aggregation). Aggregation may benefit the species by providing buffers against environmental pressures, such as wind, dilution of toxins, protection from predation, possibilities for polymorphisms, and stimulation and synchronization of reproduction.

Aggression - One common form of species-specific behavior is aggression. Contrary to popular use of the word, aggression is not seen by ethologists as naked warfare. Aggressive action may be an apparently placid grazing cow's subtle change of direction. In response, the subordinate cow gradually veers away.

The term "aggressive behavior" usually excludes predatory behavior and defensive action against predators. Different muscular actions are involved in each. Aggression between members of a species usually consists of threatening or bluff. It consumes enormous amounts of time in many species.

In view of some of the survival advantages of social aggregation, the question arises as to the survival advantages of behavior which tends to repel members of the species from one another. At one time, ethologists thought that aggression divided certain objects, which are indispensable for reproduction, among as many males as possible. The objects differed by species. It is now generally held that the function of aggression is to spread out animals as evenly as possible in order to conserve the food supply.

Defended Territory - Frequent physical combat among species members would not have survival value. To effect a regular spacing out of the species without resort to actual combat, many species have evolved relatively simple, but conspicuous and specific (or "improbable"), morphological features and behaviors which stimulate the withdrawal of other animals. They also mark their territories in different ways: certain tropical fish, by their own vivid coloring; birds, by song; and mammals, by scent. Within these boundaries, the resident is dominant, and space functions – in Julian Huxley's analogy – as a compressible rubber ball.

Many animals spend hours bickering and confronting one another along the boundaries of their territories; but near the center, the owner is confident and energetic while the invader is timid and slow. Observers note that the energy with which an animal will attack an invading member of his species is inversely proportional to the distance from his territory's center to the place of attack, and that the defending animal, regardless of the size of the aggressor, will almost always win. At some point in the compression of space, an animal will defend territory to his death.

Social Cooperation - Species could not survive, in most cases, if they did not overcome spacing for social contacts such as breeding and raising young. Submissive behavior and appeasement gestures are two methods for reducing combat and aggressive displays. Frequently antithetical gestures to aggressive ones, such as looking elsewhere, are employed. Mutual grooming and sexual presentations may also be used to curb aggression.

Physical combat at a territory's boundary would not be adaptive. Species frequently inhibit their aggression by displacement activities. One of the best known examples is the behavior of the three-spined stickleback fish when confronted by a neighbor at his boundary. The stickleback signals to his neighbor to stay away by standing on his head in the water, turning broadside with erect ventral spines in an aggressive display, and then vents his aggression by biting sand furiously in a spurious nesting act. Interaction by signaling devices also achieves cooperation in tasks which the individual cannot accomplish alone, such as mating, spacing out, or raising offspring.

Social Organization - Creation and maintenance of community life depend on social organization. While there are many varieties of social organization in the animal kingdom, the type of social organization used appears to be adapted to behavior characteristic of the species. Social organization may be based on (1) individually defended territory; (2) pairs, either expedient and temporary alliances for reproduction or permanent "marriages" of animals tied to one another by affectionate bonds; (3) dominance hierarchy – lineal, as in the case of chickens' peck order; triangular, with one dominant at the apex; or oligarchial, with a group of dominants cooperating in the tasks of governing, as in the case of apes; (4) leadership – where the leader without coercion determines direction, rate of movement, or mood – for example, schools of fish and some flocks of birds.

In species such as rats and wolves, society is organized in dominance hierarchies. Dominance is not severe; fights rarely occur; and the young are given preferential treatment. There is, however, a highly developed mutual recognition of in-group members and antagonism toward out groups. Rat packs are continually at war with one another, apparently as a result of intraspecific selection.[5] If a strange rat is introduced into an established community, he is attacked and destroyed by the pack. Lorenz considers this type of community to be characteristic of man (20, pp. 151-158). Others seem to agree with his basic premise, citing the wolf pack as their example (7, p. 279).

"Home Range" Territoriality - As animals achieve integration into communities, territory serves as an important means of preserving group stability and cohesion by continuous association and reinforcement. This form of territory is sometimes considered to be the "home range."

Some ethologists look at communal territory in terms of its function as an external reference to dominance; however, it is clear that territory serves to coordinate physiological and behavior processes beyond mere dominance. Hediger explains the survival value of such territoriality by its function in (1) reinsuring propagation by regulating densities, (2) providing a frame in which things are done, (3) keeping animals within communication distance so that presence of food or enemy can be signaled, and (4) developing an inventory of reflex responses to terrain features so that animals can act without thinking when danger strikes.[6]

Human Behavior - Ethologists have not systematically investigated human behavior. Under the theory of convergent adaptation, however, very similar behavior patterns in unlike species are regarded as having the same function because of the extreme mathematical improbability of their occurring by chance. This theory is the basis for Lorenz's (20) use of animal behavior to illuminate the nature of the problem of human aggression.[7]

Criticisms and Modifications of Concepts

Some of the foregoing concepts of mechanisms and patterns of behavior have been criticized and subsequently modified. The concept of innate behavior, for example, has been criticized on several grounds (17,9,12, and 15). Many ethologists now refer to innate characteristics of behavior as "species-specific" (invariable in that species) or "unlearnt" to avoid the implication that isolation from conspecifics deprives an animal of all opportunities to be influenced by its environment (31).[8] Earlier simplified notions of the relation between learned behavior and innate capacity to learn have been replaced by more complex ones. The simple models of physiological processes are giving way to concrete descriptions of neurophysiological responses to specific stimuli (13, pp. 42-43; 15, pp. 324-325; 18, pp. 8-20). It is apparent that there is great variety and complexity in behavior among animal species. The implication that explanations for all behavior have been reduced to a few simple principles, so eagerly seized upon by Ardrey and Morris, seems to be a philosophical faith rather than a scientific finding. But the fact remains that concepts of ethology have brought about a major reexamination of problems of human behavior.[9]

SOME IMPLICATIONS OF ETHOLOGY

Probably the most profound effect of ethological research has been its challenge to the dogma that all significant human behavior may be explained by environmental influences.[10] American psychologists, in the Mechanist tradition, tend to ignore behavior's genetic origins in their preoccupation with learning theory (2, p. 28; 16, p. 22; 33, pp. 1-6; 18, pp. 79-82). Cultural anthropologists treat the evolution of human culture as a separate process from biological evolution. The result has been the assumption in the social sciences that man is a completely unique species, distinguished by his capacity to transmit and learn culture.

Studies of the development of animal behavior have, however, provided concepts which seem to explain problems of human development. For example, imprinting seems to explain formation of a child's early attachments to his mother. Studies of maternal and peer group deprivations among monkeys and dogs seem to explain the effect of similar deprivations on the development of social attachments, sexual behavior, and nurturance in humans. These and other studies suggest that human emotional development and early learning may not be very different from some animal species.[11]

Neo-Darwinism has also stimulated development of behavioral genetics.[12] Physical anthropologist Alexander Alland draws upon neo-Darwinian theories and studies of animal behavior to reconceptualize cultural evolution as part of systemic biocultural adaptation to environment rather than a self-contained process.[13] Alland's emphasis upon interaction of genetic and cultural factors in environment adaptation reflects a general trend in evolutionary thought.[14]

This trend toward consideration of the genetic basis for patterns of behavior has raised the further question of the extent to which man is genetically adapted to urban life. Some ethologists maintain that man is phylogenetically adapted to a pattern of small-group life in which interaction is less frequent and less extensive than in urban life (3, 20).[15] There is a growing body of research which indicates that group membership and position may produce physiological changes which greatly alter susceptibility to disease and may even, at high densities, become predominant determinants despite lack of any emotional disturbance. It has been suggested that this may be the most important explanation of current rising incidences of mental illness and chronic disease.[16] The implications of some of this research for spatial relations have been reexamined by Edward Hall in his study of proxemetics.[17] Konrad Lorenz has perhaps developed the most specific theory of the importance of small groups in human behavior. In common with Sigmund Freud, he contends that man's behavior arises spontaneously from instincts and is not necessarily subject to rational control. Cultural rites, customs, manners, and habits displayed by individuals with which man is united in bonds of affection control innate releasing mechanisms so that drives are expressed in ways appropriate for the culture. Lorenz warns that danger of human aggression is increased by the assumption that it is not innate and that it may be eradicated by appropriate conditioning (28).

The basis for these theories of behavior is that *Homo sapiens* has lived in small hunting packs and agricultural communities for all but a minute period of its existence. Members of the species who most frequently survived to breed would therefore seem to have genetic characteristics appropriate for small-group living. In the transition to urban industrial conditions of life, the most important change may not be increased density, but specialization of functions which leads to far more complex, frequent, and unsynchronized rhythms of human interaction than occurred in simpler societies, and which results in loss of orientation and the psychological support of continuous, modulated small-group life (4).[18]

Perhaps the most fundamental issue raised by ethology, however, is whether behavioral models in which man consciously maximizes certain values are appropriate in urban planning.[19] It is not necessary to accept fully the various hypotheses concerning man's physiological responses to crowding, spontaneity of aggression, and instincts for territory to suspect that subtle, delicate, infinitely complex, and unconsciously patterned interrelationships between man and his environment may vitally affect human well-being. Studies of animal species suggest that survival is often a consequence of highly interrelated, partly genetic, and partly learned individual responses to the group, and group adjustment to its ecological niche. Certainly, man is able to use social organization and technology to protect himself from deleterious biological effects of the environment, but this fact only emphasizes the burden placed on planning. How should planning be used to control these adaptive mechanisms? Past behavior and present preferences are not sufficient criteria. The survival value of a particular behavior has no relationship to the tenacity with which organisms persist in that behavior. Conscious preferences do not necessarily take into account biological stresses. Is it possible that the planning approach of maximizing or optimizing a narrow selection of explicit societal values may tend to perpetuate behaviors which would otherwise be checked by unconscious constraints and to implement destructive side effects which would otherwise be minimized? It may be that the rational planning model must be bounded by a biological or biocultural model of man that takes into account the full range of human transactions with environment. This seems to be no more than the wisdom of Olmstead, Howard, Geddes, Mumford, and Stein.[20]

Author's Note: The writer was supported by a Public Health Service Fellowship (I-F3-CH-36,544-01) from the Division of Medical Care Administration while preparing this article. I am grateful to F. Stuart Chapin, Jr., for encouragement in the pursuit of these studies and to Helmut Mueller for an introduction to the literature.

NOTES

[1] See Morton Fried on Morris in "Tales of the Unhairy," *Saturday Review of Literature*, Feb. 17, 1968, pp. 34-35; and Peter H. Klopfer, "From

Ardrey to Altruism," *Behavioral Science* (in press). But Morris also has his defenders: Peter M. Driver, "Book Review," *Behavioral Science,* XIII, No. 3 (May 1968), 240-44.

[2] One of the foremost authorities in the field has noted that ethologists ". . . differ widely in their opinions of what the science is about" (31). However, the definition of ethology as the biological study of behavior seems to be supported by several recent writers, possibly because it is sufficiently inclusive and vague to embrace differences of opinion.

[3] See, for a few selected examples, F.T. Cloak, Sr., "Is a Cultural Ethology Possible?" *Research Previews,* XV, No. 1 (April 1968), cultural anthropology and evolution; Edward T. Hall, *The Hidden Dimension* (Garden City, N.Y.: Doubleday, 1966), anthropology and urban space; Clare Russell and H.M.S. Russell, *Human Behavior* (Boston: Little Brown, 1961), child development; John Bowlby, "The Nature of the Child's Tie to the Mother," *International Journal of Psychoanalysis,* XXXIX (1958), 35-73, child development and psychoanalytic theory; John Cassel, "Physical Illness in Response to Stress" (School of Public Health, University of North Carolina, Chapel Hill, N.C.) (mimeographed), epidemiology; Mortimer J. Adler, *The Difference of Man and the Difference It Makes* (New York: Holt, Rinehart and Winston, 1968), social psychology; and Robert Sommer, "Man's Proximate Environment," *The Journal of Social Issues,* XXII, No. 4, 59-75, human space.

[4] Books by Peter H. Klopfer and Robert Sommer are expected to appear soon. They should provide a conceptual background for considering problems of cities.

[5] This is selection of a trait as a result of competition *within* the species. The resulting characteristic may be irrelevant or even unfavorable to the species' survival.

[6] As reported in Edward T. Hall, *The Hidden Dimension,* p. 8.

[7] Man's use of symbolic language has been considered by some scholars to distinguish his behavior from similar behavior in animals. However, Dobzhansky, a geneticist and evolutionary theorist, found communication among honey bees that indicates the distance and direction of food is a use of symbolic language. *Mankind Evolving* (New Haven, Conn.: Yale University Press, 1962), p. 210. Cloudsley-Thompson (5, p. 64) found that ". . . chimpanzees, dogs and other animals can solve problems of such complexity that some process analogous to visualization must be involved. Indeed, the only characteristic of human language which appears to be unique is the fact that it is learned largely independently of heredity; yet even there the evolution and spread of vowel sounds has been traced across Europe."

There are, of course, differences between man and animals in the degree to which language is exploited. The point here is only that there can be some doubt that the element of symbolic language in communication absolutely distinguishes human from animal behavior.

[8] Lorenz in *Evolution and Modification of Behavior* (18) accepts neither criticisms nor revisions of the concept of the innate – his defense is probably the best argument for that concept. Konishi (15) presents the most objective review of the controversy.

[9] It is quite possible that there are some problems and deficiencies in ethological theory – as there are in any scientific theory – but nothing suggests that it is generally invalid. Much of the disagreement among ethologists seems to reduce the different priorities which are attached to the development of theory and experimental research at this time. Lorenz (18, p. 3) explains such a view in his statement that the surrender of the concept of the innate by "most modern English-speaking ethologists" is "partly from overcaution, partly because they work to compromise with the behaviorist critique, but mostly in consequence of a rebound phenomenon in detecting some errors in 'naive' attitudes. . . ." Qualified specialists, such as Konishi (15) and Etkin (7, pp. 168-70), find many criticized concepts are still valid. My review of the controversies suggests that they are frequently marred by ignorance and misunderstanding of the evolutionary context which gives a specific meaning to terms employed by ethologists.

[10] In a perhaps extravagant appraisal – yet a well-informed one – Julian Huxley (11, p. 409) states

that Lorenz has disposed of the claim that all useful behavior is learned behavior and that all learning is a matter of conditioning. He has, according to Huxley, changed the study of behavior to one of adaptive value and paved the way for transcending the "mind-body conflict in an integral and truly monistic approach . . ." in which emotion is taken into account.

[11]See, for example, Harry F. Harlow, "Love in Infant Monkeys," *Scientific American* (June 1959); John Bowlby, "The Nature of the Child's Tie to His Mother," *International Journal of Psychoanalysis,* XXXIX (1958), 350-73; Harry F. Harlow and Margaret K. Harlow, "Social Deprivation in Monkeys," *Scientific American* (Nov. 1962), pp. 137-46; Harry F. Harlow, "The Heterosexual Affectional System in Monkeys," *American Psychologist,* XVII (1962), 1-9; and J.P. Scott in Etkin (7) and Scott (24 and 25).

[12]For discussion of some of the implications of genetics for human behavior, see Jerry Hirsch, "Behavior Genetics and Individuality Understood," *Science* CXXXXII (Dec. 1963), 1436-42; and Bruce K. Eckland, "Genetics and Sociology: A Reconsideration," *American Sociological Review,* XXXII (April 1967), 173-94.

[13]Alexander Alland, Jr., *Evolution and Human Behavior,* (Garden City, N.Y.: Doubleday, 1967).

[14]For examples, T. Dobzhanski, *Mankind Evolving;* René Jules Dubos, *Man Adapting* (New Haven, Conn.: Yale University Press, 1967); Anne Roe and George Gaylord Simpson (eds.), *Behavior and Evolution* (New Haven, Conn.: Yale University Press, 1958).

[15]René Jules Dubos, "Man Adapting: His Limitations and Potential," in William R. Ewald (ed.), *Environment for Man* (Bloomington, Ind.: Indiana University Press, 1967).

[16]See John Cassel, "Physical Illness in Response to Stress," (School of Public Health, University of North Carolina at Chapel Hill). (Mimeographed.)

[17]*The Hidden Dimension* (Garden City, N.Y.: Doubleday, 1966).

[18]For this reason, this writer believes that the debate over the implications of pathological behavior by rats in overcrowded conditions has largely missed the point. The problem is not overcrowding per se, but the effect of overcrowding on social relations. The outstanding fact about the Chinese in Hong Kong has been that apparently high densities have not affected traditional forms of Chinese family and group life. In mainland China, however, the Communists have found it necessary to break down these traditions in the pursuit of forced industrialization. Whether Chinese refugees in Hong Kong will be content to live under British paternalism as nonparticipating marginal beneficiaries of western industrialization remains doubtful in the light of history of paternalism in the United States. Full participation in modern industrial life has everywhere destroyed traditional forms of extended family solidarity similar to the Chinese. Only when the Chinese in Hong Kong feel the effects of adaptation to urban industrial technology and social organization will evaluation of the effects of density be possible. For background on this debate, see Michael Hugo Brunt, "Hong Kong Housing," in H. Wentworth Eldredge (ed.), *Taming Megalopolis,* I (New York: Praeger, 1967); and Robert C. Schmitt, "Implications of Density in Hong Kong," *Journal of the American Institute of Planners,* XXIX, No. 3 (May 1963), 210-217. Descriptions of Chinese behavior might be compared with Wilbert Moore, *Man, Time and Society* (New York: Wiley, 1963).

[19]The existence of a wide range of variables affecting spatial behavior is not in dispute, but the strategy chosen by most analysts is to sacrifice scope for simplicity. For example, "An individual who arrives in a city and wishes to buy some land to live upon will be faced with the double decision of how large a lot he should purchase and how close to the center of the city he should settle. In reality he would also consider the apparent character and social composition of the neighborhood, the quality of the schools in the vicinity, how far away he would be from any relatives he might have in the city, and a thousand other factors. However, the individual in question is an 'economic man,' defined and simplified in a way such that we can handle the analysis of his decision-making. He merely wishes to maximize his satisfaction by owning and consuming the goods he likes and avoiding those he dislikes. . . . We are not concerned with how tastes are formed, but simply what they are." William Alonso, *Location and*

Land Use (Cambridge, Mass.: Harvard University Press, 1964), p. 18.

"In addition to the journey-to-work, the choice (of location of households) is influenced by the distribution of the stock and quality of housing, prestige and cultural group associations, varieties in the quality of highly valued local services, such as schools, and many other considerations. Complex substitution effects are certain to take place among these if rational behavior asserts itself at all. . . . The importance of such considerations in the locational decisions of individual households is not to be depreciated, but this study has chosen to focus rather on the manner in which some critical technical variables influence the spatial organization of the urban community through their effect on the individual decision in the market. . . ." Lowdon Wingo, *Transportation and Urban Land* (Washington, D.C.: Resources for the Future, 1961), p. 92.

[20] The difficulties of this task are not underrated. It may well be that it is simply impossible to model biocultural factors in the mechanistic manner to which urban planners have become accustomed. Perhaps the fact that we only attribute authority to mechanistic versions of science has been a major obstacle to the development of policy sciences, in Harold Lasswell's phrase, as a basis for planning.

REFERENCES

1. Ardrey, Robert, *African Genesis* (New York: Dell Publishing Company, 1967).
2. Ardrey, Robert, *The Territorial Imperative* (New York: Atheneum Publishers, 1967).
3. Calhoun, John B., "The Role of Space in Animal Sociology," *The Journal of Social Issues,* XXII, No. 4, (Oct. 1966).
4. Calhoun, John B., "Population Density of Social Pathology," in Leonard Duhl (ed.), *The Urban Condition,* (New York: Basic Books, Inc., 1963).
5. Cloudsley-Thompson, J.L., *Animal Behavior* (New York: Macmillan Publishing Co., 1961).
6. Darwin, Charles, *The Expressions of Emotions in Man and Animals* (London: John Murry, 1872).
7. Etkin, William (ed.), *Social Behavior and Organization Among Vertebrates* (Chicago: University of Chicago Press, 1964).
8. Etkin, William, *Social Behavior from Fish to Man* (Chicago: University of Chicago Press, 1967).
9. Hebb, D.C., "Heredity and Environment in Mammalian Behavior," *British Journal of Animal Behavior,* I (1953), 43-47.
10. Hind, R.A., and Tinbergen, N., "The Comparative Study of Species-Specific Behavior," in Ann Roe and George Gaylord Simpson (eds.), *Behavior and Evolution* (New Haven, Conn.: Yale University Press, 1958).
11. Huxley, Julian, "Lorenzian Ethology," *Zeitschrift für Turpsychologie,* XX (1963), 402-409.
12. Jensen, D.C., "Operationism and the Question 'Is This Behavior Learned or Innate?' " *Behavior,* XVII (1967), 1-8.
13. Klopfer, Peter H., and Hailman, Jack P., *An Introduction to Animal Behavior: Ethology's First Century* (Englewood Cliffs, N.J.: Prentice-Hall, Inc., 1967).
14. Klopfer, Peter H., *Behavioral Aspects of Ecology* (Englewood Cliffs, N.J.: Prentice-Hall, Inc., 1962).
15. Konishi, Masakazu, "The Attributes of Instinct," *Behavior,* XXVII (1966), 316-28.
16. Marler, Peter, and Hamilton, William J., III, *Mechanisms of Animal Behavior* (New York: John Wiley & Sons, Inc., 1966).
17. Lehman, Daniel S., "A Critique of Konrad Lorenz's Theory of Instinctive Behavior," The *Quarterly Review of Biology,* XXVIII, No. 4, 337-59.
18. Lorenz, Konrad, *Evolution and Modification of Behavior* (Chicago: University of Chicago Press, 1965).
19. Lorenz, Konrad, *King Solomon's Ring* (New York: Thomas Y. Crowell Company, 1952).
20. Lorenz, Konrad, *On Aggression* (New York: Bantam Books, Inc., 1967).
21. Morris, Robert, *The Naked Ape: A Zoologist's Study of the Human Animal* (London: Cope, 1967).
22. Nielson, E.T., "The Method of Ethology," *Proceedings of the 10th International Congress of Entomology,* II (1958), 263-65.
23. Scott, John Paul, *Animal Behavior* (Chicago: University of Chicago Press, 1958).
24. Scott, John Paul, and Fuller, John L., *Genetics and the Social Behavior of the Dog* (Chicago: University of Chicago Press, 1965).

25. Southwick, Charles H. (ed.), *Primate Social Behavior* (New York: Van Nostrand Reinhold Company, 1963).
26. Sommers, Robert, "Man's Proximate Environment," *The Journal of Social Issues,* XXII, No. 4 (Oct. 1966), 59-70.
27. Thorpe, W.H., *Learning and Instincts in Animals* (London: Muethuen and Company, 1956).
28. Thorpe, W.H., *Biology and the Nature of Man* (New York: Oxford University Press, 1962).
29. Thorpe, W.H., and Zangwill, O.L., *Current Problems in Animal Behavior* (Cambridge: The University Press, 1966).
30. Tinbergen, N., "Behavior, Systematics and Natural Selection," *Ibis,* CI (1959), 318-30.
31. Tinbergen, W., "On Aims and Methods of Ethology," *Zeitschrift fur Turpsychologie,* XX (1963), 410-33.
32. Tinbergen, N., *Social Behavior in Animals* (London: Methuen and Company, 1953).
33. Tinbergen, N., *The Study of Instincts* (New York: Oxford University Press, 1951).

. . . we cannot escape the question why reasonable beings do behave so unreasonably. Undeniably, there must be superlatively strong factors which are able to overcome the commands of individual reason so completely and which are so obviously impervious to experience and learning. As Hegel said, "What experience and history teach us is this — that people and governments never have learned anything from history, or acted on principles deduced from it."

All these amazing paradoxes, however, find an unconstrained explanation, falling into place like the pieces of a jigsaw puzzle, if one assumes that human behavior, and particularly human social behavior, far from being determined by reason and cultural tradition alone, is still subject to all laws prevailing in all phylogenetically adapted instinctive behavior. Of these laws we possess a fair amount of knowledge from studying the instincts of animals . . .

Konrad Lorenz, On Aggression, *New York (Harcourt, Brace & World) 1963, p. 237.*

50

Stanley Milgram

THE EXPERIENCE OF LIVING IN CITIES

When I first came to New York it seemed like a nightmare. As soon as I got off the train at Grand Central I was caught up in pushing, shoving crowds on 42nd Street. Sometimes people bumped into me without apology; what really frightened me was to see two people literally engaged in combat for possession of a cab. Why were they so rushed? Even drunks on the street were bypassed without a glance. People didn't seem to care about each other at all.

This statement represents a common reaction to a great city, but it does not tell the whole story. Obviously, cities have great appeal because of their variety, eventfulness, possibility of choice, and the stimulation of an intense atmosphere that many individuals find a desirable background to their lives. Where face-to-face contacts are important, the city offers unparalleled possibilities. It has been calculated

---

by the Regional Plan Association[1] that in Nassau County, a suburb of New York City, an individual can meet 11,000 others within a 10-minute radius of his office by foot or car. In Newark, a moderate-sized city, he can meet more than 20,000 persons within this radius. But in midtown Manhattan he can meet fully 220,000. So there is an order-of-magnitude increment in the communication possibilities offered by a great city. That is one of the bases of its appeal and, indeed, of its functional necessity. The city provides options that no other social arrangement permits. But there is a negative side also, as we shall see.

Granted that cities are indispensable in complex society, we may still ask what contribution psychology can make to understanding the experience of living in them. What theories are relevant? How can we extend our knowledge of the psychological aspects of life in cities through empirical inquiry? If empirical inquiry is possible, along what lines should it proceed? In short, where do we start in constructing urban theory and in laying out lines of research?

Observation is the indispensable starting point. Any observer in the streets of midtown Manhattan will see (1) large numbers of people, (2) a high population density, and (3) heterogeneity of population. These three factors need to be at the root of any sociopsychological theory of city life, for they condition all aspects of our experience in the metropolis. Louis Wirth[2], if not the first to point to these factors, is nonetheless the sociologist who relied most heavily on them in his analysis of the city. Yet, for a psychologist, there is something unsatisfactory about Wirth's theoretical variables. Numbers, density, and heterogeneity are demographic facts but they are not yet psychological facts. They are external to the individual. Psychology needs an idea that links the individual's *experience* to the demographic circumstances of urban life.

One link is provided by the concept of overload. This term, drawn from systems analysis, refers to a system's inability to process inputs from the environment because they are too many inputs for the system to cope with, or because successive inputs come so fast that input A cannot be processed when input B is presented. When overload is present, adaptations occur. The system must set priorities and make choices. A may be processed first while B is kept in abeyance, or one input may be sacrificed altogether. City life, as we experience it, constitutes a continuous set of encounters with overload and of resultant adaptations. Overload characteristically deforms daily life on several levels, impinging on role performance, the evolution of social norms, cognitive functioning, and the use of facilities.

The concept has been implicit in several theories of urban experience. In 1903 George Simmel[3] pointed out that, since urban dwellers come into contact with vast numbers of people each day, they conserve psychic energy by becoming acquainted with a far smaller proportion of people than their rural counterparts do, and by maintaining more superficial relationships even with these acquaintances. Wirth[2] points specifically to "the superficiality, the anonymity, and the transitory character of urban social relations."

One adaptive response to overload, therefore, is the allocation of less time to each input. A second adaptive mechanism is disregard of low-priority inputs. Principles of selectivity are formulated such that investment of time and energy are reserved for carefully defined inputs (the urbanite disregards the drunk sick on the street as he purposefully navigates through the crowd). Third, boundaries are redrawn in certain social transactions so that the overloaded system can shift the burden to the other party in the exchange; thus harried New York bus drivers once made change for customers, but now this responsibility has been shifted to the client, who must have the exact fare ready. Fourth, reception is blocked off prior to entrance into a system; city dwellers increasingly use unlisted telephone numbers to prevent individuals from calling them, and a small but growing number resort to keeping the telephone off the hook to prevent incoming calls. More subtly, a city dweller blocks inputs by assuming an unfriendly countenance, which discourages others from initiating contact. Additionally, social screening devices are interposed between the individual and environmental inputs (in a town of 5,000 anyone can drop in to chat with the mayor, but in the metropolis organizational screening devices deflect inputs to other destinations). Fifth, the intensity of inputs is diminished by filtering devices, so that only weak and relatively superficial forms of involvement with others are allowed. Sixth, specialized institutions are created to absorb inputs that would otherwise swamp the individual (welfare departments handle the financial needs of a million individuals in New York City, who would otherwise create an army of mendicants continuously importuning the pedestrian). The interposition of institutions between the individual and the social world, a characteristic of all modern society, and most notably of the large metropolis, has its negative side. It deprives the individual of a sense of direct contact and spontaneous integration in the life around him. It simultaneously protects and estranges the individual from his social environment.

Many of these adaptive mechanisms apply not only to individuals but to institutional systems as well, as Meier[4] has so brilliantly shown in connection with the library and the stock exchange.

In sum, the observed behavior of the urbanite in a wide range of situations appears to be determined largely by a variety of adaptations to overload. I now deal with several specific consequences of responses to overload, which make for differences in the tone of city and town.

SOCIAL RESPONSIBILITY

The principal point of interest for a social psychology of the city is that moral and social involvement with individuals

The principal point of interest for a social psychology of the city is that moral and social involvement with individuals is necessarily restricted. This is a direct and necessary function of excess of input over capacity to process. Such restriction of involvement runs a broad spectrum from refusal to become involved in the needs of another person, even when the person desperately needs assistance, through refusal to do favors, to the simple withdrawal of courtesies (such as offering a lady a seat, or saying "sorry" when a pedestrian collision occurs). In any transaction more and more details need to be dropped as the total number of units to be processed increases and assaults an instrument of limited processing capacity.

The ultimate adaptation to an overloaded social environment is to totally disregard the needs, interests, and demands of those whom one does not define as relevant to the satisfaction of personal needs, and to develop highly efficient perceptual means of determining whether an individual falls into the category of friend or stranger. The disparity in the treatment of friends and strangers ought to be greater in cities than in towns; the time allotment and willingness to become involved with those who have no personal claim on one's time is likely to be less in cities than in towns.

Bystander Intervention in Crises - The most striking deficiencies in social responsibility in cities occur in crisis situations, such as the Genovese murder in Queens. In 1964, Catherine Genovese, coming home from a night job in the early hours of an April morning, was stabbed repeatedly, over an extended period of time. Thirty-eight residents of a respectable New York City neighborhood admit to having witnessed at least a part of the attack, but none went to her aid or called the police until after she was dead. Milgram and Hollander, writing in *The Nation*[5], analyzed the event in these terms:

> *Urban friendships and associations are not primarily formed on the basis of physical proximity. A person with numerous close friends in different parts of the city may not know the occupant of an adjacent apartment. This does not mean that a city dweller has fewer friends than does a villager, or knows fewer persons who will come to his aid; however, it does mean that his allies are not constantly at hand. Miss Genovese required immediate aid from those physically present. There is no evidence that the city had deprived Miss Genovese of human associations, but the friends who might have rushed to her side were miles from the scene of her tragedy.*
>
> *Further, it is known that her cries for help were not directed to a specific person; they were general. But only individuals can act, and as the cries were not specifically directed, no particular person felt a special responsibility. The crime and the failure of community response seem absurd to us. At the time, it may well have seemed equally absurd to the Kew Gardens residents that not one of the neighbors*

would have called the police. A collective paralysis may have developed from the belief of each of the witnesses that someone else must surely have taken that obvious step.

Latane and Darley[6] have reported laboratory approaches to the study of bystander intervention and have established experimentally the following principle: the larger the number of bystanders, the less the likelihood that any one of them will intervene in an emergency. Gaertner and Bickman[7] of The City University of New York have extended the bystander studies to an examination of help across ethnic lines. Blacks and whites, with clearly identifiable accents, called strangers (through what the caller represented as an error in telephone dialing), gave them a plausible story of being stranded on an outlying highway without more dimes, and asked the stranger to call a garage. The experimenters found that the white callers had a significantly better chance of obtaining assistance than the black callers. This suggests that ethnic allegiance may well be another means of coping with overload: the city dweller can reduce excessive demands and screen out urban heterogeneity by responding along ethnic lines; overload is made more manageable by limiting the "span of sympathy."

In any quantitative characterization of the social texture of city life, a necessary first step is the application of such experimental methods as these to field situations in large cities and small towns. Theorists argue that the indifference shown in the Genovese case would not be found in a small town, but in the absence of solid experimental evidence the question remains an open one.

More than just callousness prevents bystanders from participating in altercations between people. A rule of urban life is respect for other people's emotional and social privacy, perhaps because physical privacy is so hard to achieve. And in situations for which the standards are heterogeneous, it is much harder to know whether taking an active role is unwarranted meddling or an appropriate response to a critical situation. If a husband and wife are quarreling in public, at what point should a bystander step in? On the one hand, the heterogeneity of the city produces substantially greater tolerance about behavior, dress, and codes of ethics than is generally found in the small town, but this diversity also encourages people to withhold aid for fear of antagonizing the participants or crossing an inappropriate and difficult to define line.

Moreover, the frequency of demands present in the city gives rise to norms of noninvolvement. There are practical limitations to the Samaritan impulse in a major city. If a citizen attended to every needy person, if he were sensitive to and acted on every altruistic impulse that was evoked in the city, he could scarcely keep his own affairs in order.

Willingness to Trust and Assist Strangers - We now move away from crisis situations to less urgent examples of social responsibility. For it is not only in situations of dramatic need but in the ordinary, everyday willingness to lend a hand that the city dweller is said to be deficient relative to his small-town cousin. The comparative method must be used in any empirical examination of this question. A commonplace social situation is staged in an urban setting and in a small town – a situation to which a subject can respond by either extending help or withholding it. The responses in town and city are compared.

One factor in the purported unwillingness of urbanites to be helpful to strangers may well be their heightened sense of physical (and emotional) vulnerability – a feeling that is supported by urban crime statistics. A key test for distinguishing between city and town behavior, therefore, is determining how city dwellers compare with town dwellers in offering aid that increases their personal vulnerability and requires some trust of strangers. Altman, Levine, Nadien, and Villena[8] of The City University of New York devised a study to compare the behaviors of city and town dwellers in this respect. The criterion used in this study was the willingness of householders to allow strangers to enter their home to use the telephone. The student investigators individually range doorbells, explained that they had misplaced the address of a friend nearby, and asked to use the phone. The investigators (two males and two females) made 100 requests for entry into homes in the city and 60 requests in the small towns. The results for middle-income housing developments in Manhattan were compared with data for several small towns

(Stony Point, Spring Valley, Ramapo, Nyack, New City, and West Clarkstown) in Rockland County, outside of New York City. As Table 1 shows, in all cases there was a sharp increase in the proportion of entries achieved by an experimenter when he moved from the city to a small town. In the most extreme case the experimenter was five times as likely to gain admission to homes in a small town as to homes in Manhattan. Although the female experimenters had notably greater success both in cities and in towns than the male experimenters had, each of the four students did at least twice as well in towns as in cities. This suggests that the city-town distinction overrides even the predictably greater fear of male strangers than of female ones.

Table 1
Percentage of Entries Achieved by Investigators for City and Town Dwellings (see text)

Experimenter	*Entries Achieved (%)*	
	*City**	*Small Town†*
Male		
No. 1	16	40
No. 2	12	60
Female		
No. 3	40	87
No. 4	40	100

*Number of requests for entry, 100.
†Number of requests for entry, 60.

The lower level of helpfulness by city dwellers seems due in part to recognition of the dangers of living in Manhattan, rather than to mere indifference or coldness. It is significant that 75 percent of all the city respondents received and answered messages by shouting through closed doors and by peering out through peepholes; in the towns, by contrast, about 75 percent of the respondents opened the door.

Supporting the experimenters' quantitative results was their general observation that the town dwellers were noticeably more friendly and less suspicious than the city dwellers. In seeking to explain the reasons for the greater sense of psychological vulnerability city dwellers feel, above and beyond the differences in crime statistics, Villena[8] points out that, if a crime is committed in a village, a resident of a neighboring village may not perceive the crime as personally relevant, though the geographic distance may be small, whereas a criminal act committed anywhere in the city, though miles from the city-dweller's home, is still verbally located within the city; thus, Villena says, "the inhabitant of the city possesses a larger vulnerable space."

Civilities - Even at the most superficial level of involvement — the exercise of everyday civilities — urbanites are reputedly deficient. People bump into each other and often do not apologize. They knock over another person's packages and, as often as not, proceed on their way with a grumpy explanation instead of an offer of assistance. Such behavior, which may visitors to great cities find distasteful, is less common, we are told, in smaller communities, where traditional courtesies are more likely to be observed.

In some instances it is not simply that, in the city, traditional courtesies are violated; rather, the cities develop new norms of noninvolvement. These are so well defined and so deeply a part of city life that *they* constitute the norms people are reluctant to violate. Men are actually embarrassed to give up a seat on the subway to an old woman; they mumble "I was getting off anyway," instead of making the gesture in a straightforward and gracious way. These norms develop because everyone realizes that, in situations of high population density, people cannot implicate themselves in each others' affairs, for to do so would create conditions of continual distraction which would frustrate purposeful action.

In discussing the effects of overload I do not imply that at every instant the city dweller is bombarded with an unmanageable number of inputs, and that his responses are determined by the excess of input at any given instant. Rather, adaptation occurs in the form of gradual evolution of norms of behavior. Norms are evolved in response to frequent discrete experiences of overload; they persist and become generalized modes of responding.

Overload on Cognitive Capacities: Anonymity - That we respond differently toward those whom we know and those who are strangers to us is a truism. An eager patron aggressively cuts in front of someone in a long movie line to save time only to confront a friend; he then behaves sheepishly. A man is involved in an automobile accident caused by another driver, emerges from his car shouting in rage, then moderates his behavior on discovering a friend driving the other car. The city dweller, when walking through midtown streets, is in a state of continual anonymity vis-a-vis the other pedestrians.

Anonymity is part of a continuous spectrum ranging from total anonymity to full acquaintance, and it may well be that measurement of the precise degrees of anonymity in cities and towns would help to explain important distinctions between the quality of life in each. Conditions of full acquaintance, for example, offer security and familiarity, but they may also be stifling, because the individual is caught in a web of established relationships. Conditions of complete anonymity, by contrast, provide freedom from routinized social ties, but they may also create feelings of alienation and detachment.

Empirically, one could investigate the proportion of activities in which the city dweller or the town dweller is known by others at given times in his daily life, and the proportion of activities in the course of which he interacts with individuals who know him. At his job, for instance, the city dweller may be known to as many people as his rural counterpart. However, when he is not fulfilling his occupational role – say, when merely traveling about the city – the urbanite is doubtless more anonymous than his rural counterpart.

Limited empirical work on anonymity has begun. Zimbardo[9] has tested whether the social anonymity and impersonality of the big city encourage greater vandalism than do small towns. Zimbardo arranged for one automobile to be left for 64 hours near the Bronx campus of New York University and for a counterpart to be left for the same number of hours near Stanford University in Palo Alto. The license plates on the two cars were removed and the hoods were opened, to provide "releaser cues" for potential vandals. The New York car was stripped of all movable parts within the first 24 hours, and by the end of 3 days was only a hunk of metal rubble. Unexpectedly, however, most of the destruction occurred during daylight hours, usually under the scrutiny of observers, and the leaders in the vandalism were well-dressed, white adults. The Palo Alto car was left untouched.

Zimbardo attributes the difference in the treatment accorded the two cars to the "acquired feelings of social anonymity provided by life in a city like New York," and he supports his conclusions with several other anecdotes illustrating casual, wanton vandalism in the city. In any comparative study of the effects of anonymity in city and town, however, there must be satisfactory control for other confounding factors: the large number of drug addicts in a city like New York; the higher proportion of slum dwellers in the city; and so on.

Another direction for empirical study is investigation of the beneficial effects of anonymity. The impersonality of city life breeds its own tolerance for the private lives of the inhabitants. Individuality and even eccentricity, we may assume, can flourish more readily in the metropolis than in the small town. Stigmatized persons may find it easier to lead comfortable lives in the city, free of the constant scrutiny of neighbors. To what degree can this assumed difference between city and town be shown empirically? Judith Waters[10], at The City University of New York, hypothesized that avowed homosexuals would be more likely to be accepted as tenants in a large city than in small towns, and she dispatched letters from homosexuals and from normal individuals to real estate agents in cities and towns across the country. The results of her study were inconclusive. But the general idea of examining the protective benefits of city life to the stigmatized ought to be pursued.

Role Behavior in Cities and Towns - Another product of urban overload is the adjustment in roles made by urbanites in daily interactions. As Wirth has said[2], "Urbanites meet one another in highly segmental roles. . . . They are less dependent upon particular persons, and their dependence upon others is confined to a highly fractionalized aspect of the other's round of activity." This tendency is particularly noticeable in transactions between customers and individuals offering professional or sales services. The owner of a country store has time to become well acquainted with his dozen-or-so daily customers, but the girl at the checkout counter of a busy A&P, serving hundreds of customers a day, barely has time to toss the green stamps into one customer's shopping bag before the next customer confronts her with his pile of groceries.

Meier, in his stimulating analysis of the city[4], discusses several adaptations a system may make when confronted by inputs that exceed its capacity to process them. Meier argues that, according to the principle of competition for scarce resources, the scope and time of the transaction shrink as customer volume and daily turnover rise. This, in fact, is what is meant by the "brusque" quality of city life. New standards have developed in cities concerning what levels of services are appropriate in business transactions (see Figure 1).

McKenna and Morgenthau[11], in a seminar at The City University of New York, devised a study (1) to compare the willingness of city dwellers and small-town dwellers to do favors for strangers that entailed expenditure of a small amount of time and slight inconvenience but no personal vulnerability, and (2) to determine whether the more compartmentalized, transitory relationships of the city would make urban salesgirls less likely than small-town salesgirls to carry out, for strangers, tasks not related to their customary roles.

To test for differences between city dwellers and small-town dwellers, a simple experiment was devised in which persons from both settings were asked (by telephone) to perform increasingly onerous favors for anonymous strangers.

Within the cities (Chicago, New York, and Philadelphia), half the calls were to housewives and the other half to salesgirls in women's apparel shops; the division was the same for the 37 small towns of the study, which were in the same states as the cities. Each experimenter represented herself as a long-distance caller who had, through error, been connected with the respondent by the operator. The experimenter began by asking for simple information about the weather for purposes of travel. Next the experimenter excused herself on some pretext (asking the respondent to "please hold on"), put the phone down for almost a full minute, and then picked it up again and asked the respondent to provide the phone number of a hotel or motel in her vicinity at which the experimenter might stay during a forthcoming visit. Scores were assigned the subjects on the basis of how helpful they had been. McKenna summarizes her results in this manner:

> *People in the city, whether they are engaged in a specific job or not, are less helpful and informative than people in small towns; . . . People at home, regardless of where they live, are less helpful and informative than people working in shops.*

However, the absolute level of cooperativeness for urban subjects was found to be quite high, and does not accord with the stereotype of the urbanite as aloof, self-centered, and unwilling to help strangers. The quantitative differences obtained by McKenna and Morgenthau are less great than one might have expected. This again points up the need for extensive empirical research in rural-urban differences, research that goes far beyond that provided in the few illustrative pilot studies presented here. At this point we have very limited objective evidence on differences in the quality of social encounters in city and small town.

But the research needs to be guided by unifying theoretical concepts. As I have tried to demonstrate, the concept of overload helps to explain a wide variety of contrasts between city behavior and town behavior: (1) the differences in role enactment (the tendency of urban dwellers to deal with one another in highly segmented, functional terms, and of urban sales personnel to devote limited time and attention to their customers); (2) the evolution of urban norms quite different from traditional town values (such as the acceptance of noninvolvement, impersonality, and aloofness in urban life; (3) the adaptation of the

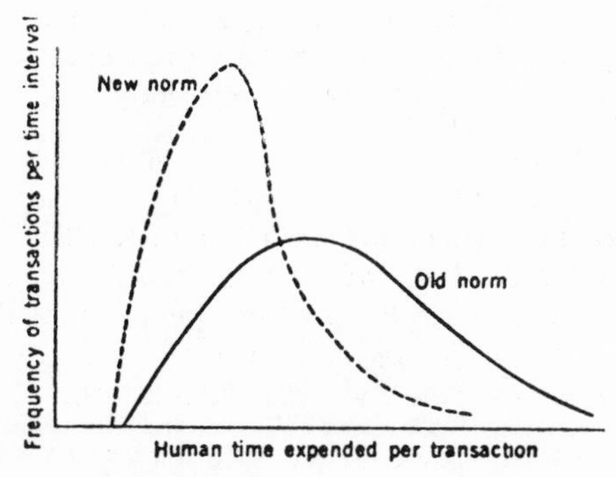

Figure 1
Changes in the demand for time for a given task when the overall transaction frequency increases in a social system. (Reprinted with permission from R.L. Meier, "A Communications Theory of Urban Growth," 1962. Copyrighted by MIT Press, 1962.)

urban dweller's cognitive processes (his inability to identify most of the people he sees daily, his screening of sensory stimuli, his development of blasé attitudes toward deviant or bizarre behavior, and his selectivity in responding to human demands); and (4) the competition for scarce facilities in the city (the subway rush; the fight for taxis; traffic jams; standing in line to await services). I suggest that contrasts between city and rural behavior probably reflect the responses of similar people to very different situations, rather than intrinsic differences in the personalities of rural and city dwellers. The city is a situation to which individuals respond adaptively.

FURTHER ASPECTS OF URBAN EXPERIENCE

Some features of urban experience do not fit neatly into the system of analysis presented thus far. They are no less important for that reason. The issues raised next are difficult to treat in quantitative fashion. Yet I prefer discussing them in a loose way to excluding them because appropriate language and data have not yet been developed. My aim is to suggest how phenomena such as "urban atmosphere" can be pinned down through techniques of measurement.

The "Atmosphere" of Great Cities - The contrast in the behavior of city and town dwellers has been a natural starting point for urban social scientists. But even among great cities there are marked differences in "atmosphere." The tone, pacing, and texture of social encounters are different in London and New York, and many persons willingly make financial sacrifices for the privilege of living within a specific urban atmosphere which they find pleasing or stimulating. A second perspective in the study of cities, therefore, is to define exactly what is meant by the atmosphere of a city and to pinpoint the factors that give rise to it. It may seem that urban atmosphere is too evanescent a quality to be reduced to a set of measurable variables, but I do not believe the matter can be judged before substantial effort has been made in this direction. It is obvious that any such approach must be comparative. It makes no sense at all to say that New York is "vibrant" and "frenetic" unless one has some specific city in mind as a basis of comparison.

In an undergraduate tutorial that I conducted at Harvard University some years ago, New York, London, and Paris were selected as reference points for attempts to measure urban atmosphere. We began with a simple question: does any consensus exist about the qualities that typify given cities? To answer this question one could undertake a content analysis of travel-book, literary, and journalistic accounts of cities. A second approach, which we adopted, is to ask people to characterize (with descriptive terms and accounts of typical experiences) cities they have lived in or visited. In advertisements placed in the *New York Times* and the *Harvard Crimson* we asked people to give us accounts of specific incidents in London, Paris, or New York that best illuminated the character of that particular city. Questionnaires were then developed and administered to persons who were familiar with at least two of the three cities.

Some distinctive patterns emerged[12]. The distinguishing themes concerning New York, for example, dealt with its diversity, its great size, its pace and level of activity, its cultural and entertainment opportunities, and the heterogeneity and segmentation ("ghetoization") of its population. New York elicited more descriptions in terms of physical qualities, pace, and emotional impact than Paris or London did, a fact which suggests that these are particularly important aspects of New York's ambiance.

A contrasting profile emerges for London; in this case respondents placed far greater emphasis on their interactions with the inhabitants than on physical surroundings. There was near unanimity on certain themes: those dealing with the tolerance and courtesy of London's inhabitants. One respondent said:

> *When I was 12, my grandfather took me to the British Museum . . . one day by tube and recited the* Aeneid *in Latin for my benefit. . . . He is rather deaf, speaks very loudly and it embarrassed the hell out of me, until I realized that nobody was paying any attention. Londoners are extremely worldly and tolerant.*

In contrast, respondents who described New Yorkers as aloof, cold, and rude referred to such incidents in the following:

> *I saw a boy of 19 passing out anti-war leaflets to passersby. When he stopped at a corner, a*

> *man dressed in a business suit walked by him at a brisk pace, hit the boy's arm, and scattered the leaflets all over the street. The man kept walking at the same pace down the block.*

We need to obtain many more such descriptions of incidents, using careful methods of sampling. By the application of factor-analytic techniques, relevant dimensions for each city can be discerned.

The responses for Paris were about equally divided between responses concerning its inhabitants and those regarding its physical and sensory attributes. Cafés and parks were often mentioned as contributing to the sense that Paris is a city of amenities, but many respondents complained that Parisians were inhospitable, nasty, and cold.

We cannot be certain, of course, to what degree these statements reflect actual characteristics of the cities in question and to what degree they simply tap the respondents' knowledge of widely held preconceptions. Indeed, one may point to three factors, apart from the actual atmospheres of the cities, that determine the subjects' responses.

1. A person's impression of a given city depends on his implicit standard of comparison. A New Yorker who visits Paris may well describe that city as "leisurely," whereas a compatriot from Richmond, Virginia, may consider Paris too "hectic." Obtaining reciprocal judgment, in which New Yorkers judge Londoners and Londoners judge New Yorkers, seems a useful way to take into account not only the city being judged but also the home city that serves as the visitor's base line.

2. Perceptions of a city are also affected by whether the observer is a tourist, a newcomer, or a longer-term resident. First, a tourist will be exposed to features of the city different from those familiar to a long-time resident. Second, a prerequisite for adapting to continuing life in a given city seems to be the filtering out of many observations about the city that the newcomer or tourist finds particularly arresting; this selective process seems to be part of the long-term resident's mechanism for coping with overload. In the interest of psychic economy, the resident simply learns to tune out many aspects of daily life. One method for studying the specific impact of adaptation on perception of the city is to ask several pairs of newcomers and old-timers (one newcomer and one old-timer to a pair) to walk down certain city blocks and then report separately what each has observed.

Additionally, many persons have noted that when travelers return to New York from an extended sojourn abroad they often feel themselves confronted with "brutal ugliness"[13] and a distinctive, frenetic atmosphere whose contributing details are, for a few hours or days, remarkably sharp and clear. This period of fresh perception should receive special attention in the study of city atmosphere. For, in a few days, details which are initially arresting become less easy to specify. They are assimilated into an increasingly familiar background atmosphere which, though important in setting the tone of things, is difficult to analyze. There is no better point at which to begin the study of city atmosphere than at the moment when a traveler returns from abroad.

3. The popular myths and expectations each visitor brings to the city will also affect the way in which he perceives it[14]. Sometimes a person's preconceptions about a city are relatively accurate distillations of its character, but preconceptions may also reinforce myths by filtering the visitor's perceptions to conform with his expectations. Preconceptions affect not only a person's perceptions of a city but what he reports about it.

The influence of a person's urban base line on his perceptions of a given city, the differences between the observations of the long-time inhabitant and those of the newcomer, and the filtering effect of personal expectations and stereotypes raise serious questions about the validity of travelers' reports. Moreover, no social psychologist wants to rely exclusively on verbal accounts if he is attempting to obtain an accurate and objective description of the cities' social texture, pace, and general atmosphere. What he needs to do is to devise means of embedding objective experimental measures in the daily flux of city life, measures that can accurately index the qualities of a given urban atmosphere.

EXPERIMENTAL COMPARISONS OF BEHAVIOR

Roy Feldman[15] incorporated these principles in a comparative study of behavior toward compatriots and foreigners in Paris, Athens, and Boston. Feldman wanted to see (1) whether absolute levels and patterns of helpfulness varied significantly from city to city, and (2) whether inhabitants in each city tended to treat compatriots differently from foreigners. He examined five concrete behavioral episodes, each carried out by a team of native experimenters and a team of American experimenters in the three cities. The episodes involved (1) asking natives of the city for street directions; (2) asking natives to mail a letter for the experimenter; (3) asking natives if they had just dropped a dollar bill (or the Greek or French equivalent) when the money actually belonged to the experimenter himself; (4) deliberately overpaying for goods in a store to see if the cashier would correct the mistake and return the excess money; and (5) determining whether taxicab drivers overcharged strangers and whether they took the most direct route available.

Feldman's results suggest some interesting contrasts in the profiles of the three cities. In Paris, for instance, certain stereotypes were born out. Parisian cab drivers overcharged foreigners significantly more often than they overcharged compatriots. But other aspects of the Parisians' behavior were not in accord with American preconceptions: in mailing a letter for a stranger, Parisians treated foreigners significantly better than Athenians or Bostonians did, and, when asked to mail letters that were already stamped, Parisians actually treated foreigners better than they treated compatriots. Similarly, Parisians were significantly more honest than Athenians or Bostonians in resisting to temptation to claim money that was not theirs, and Parisians were the only citizens who were more honest with foreigners than with compatriots in this experiment.

Feldman's studies not only begin to quantify some of the variables that give a city its distinctive texture, but they also provide a methodological model for other comparative research. His most important contribution is his successful application of objective, experimental measures to everyday situations, a mode of study which provides conclusions about urban life that are more pertinent than those achieved through laboratory experiments.

TEMPO AND PACE

Another important component of a city's atmosphere is its tempo or pace, an attribute frequently remarked on but less often studied. Does a city have a frenetic, hectic quality, or is it easygoing and leisurely? In any empirical treatment of this question, it is best to start in a very simple way. Walking speeds of pedestrians in different cities and in cities and towns should be measured and compared. William Berkowitz[16] of Lafayette College has undertaken an extensive series of studies of walking speeds in Philadelphia, New York, and Boston, as well as in small and moderate-sized towns. Berkowitz writes that "there does appear to be a significant linear relation between walking speed and size of municipality, but the absolute size of the difference varies by less than ten percent."

Perhaps the feeling of rapid tempo is due not so much to absolute pedestrian speeds as to the constant need to dodge others in a large city to avoid collisions with other pedestrians. (One basis for computing the adjustments needed to avoid collisions is to hypothesize a set of mechanical manikins sent walking along a city street and to calculate the number of collisions when no adjustments are made. Clearly, the higher the density of manikins the greater the number of collisions per unit of time, or, conversely, the greater the frequency of adjustments needed in higher population densities to avoid collisions.)

Patterns of automobile traffic contribute to a city's tempo. Driving an automobile provides a direct means of translating feelings about tempo into measurable acceleration, and a city's pace should be particularly evident in vehicular velocities, patterns of acceleration, and latency of response to traffic signals. The inexorable tempo of New York is expressed, further, in the manner in which pedestrians stand at busy intersections, impatiently awaiting a change in traffic light, making tentative excursions into the intersection, and frequently surging into the street even before the green light appears.

VISUAL COMPONENTS

Hall has remarked[17] that the physical layout of the city also affects its atmosphere. A gridiron pattern of streets gives the visitor a feeling of rationality, orderliness, and predictability but is sometimes monotonous. Winding lanes or streets branching off at strange angles, with many forks (as in Paris or Greenwich Village), create feelings of surprise and aesthetic pleasure, while forcing greater decision making in plotting one's course. Some would argue that the visual component is all-important — that the "look" of Paris or New York can almost be equated with its atmosphere. To investigate this hypothesis, we might conduct studies in which only blind, or at least blindfolded, respondents were used. We would no doubt discover that each city has a distinctive texture even when the visual component is eliminated.

SOURCES OF AMBIANCE

Thus far we have tried to pinpoint and measure some of the factors that contribute to the distinctive atmosphere of a great city. But we may also ask, why do differences in urban atmosphere exist? How did they come about, and are they in any way related to the factors of density, large numbers, and heterogeneity discussed above?

First, there is the obvious factor that, even among great cities, populations and densities differ. The metropolitan areas of New York, London, and Paris, for example, contain 15 million, 12 million, and 8 million persons, respectively. London has average densities of 43 persons per acre, while Paris is more congested, with average densities of 114 persons per acre[18]. Whatever characteristics are specifically attributable to density are more likely to be pronounced in Paris than in London.

A second factor affecting the atmosphere of cities is the source from which the populations are drawn[19]. It is a characteristic of great cities that they do not reproduce their own populations, but that their numbers are constantly maintained and augmented by the influx of residents from other parts of the country. This can have a determining effect on the city's atmosphere. For example, Oslo is a city in which almost all the residents are only one or two generations removed from a purely rural existence, and this contributes to its almost agricultural norms.

A third source of atmosphere is the general national culture. Paris combines adaptations to the demography of cities *and* certain values specific to French culture. New York is an admixture of American values and values that arise as a result of extraordinarily high density and large population.

Finally, one could speculate that the atmosphere of a great city is traceable to the specific historical conditions under which adaptations to urban overload occurred. For example, a city which acquired its mass and density during a period of commercial expansion will respond to new demographic conditions by adaptations designed to serve purely commercial needs. Thus Chicago, which grew and became a great city under a purely commercial stimulus, adapted in a manner that emphasizes business needs. European capitals, on the other hand, incorporate many of the adaptations which were appropriate to the period of their increasing numbers and density. Because aristocratic values were prevalent at the time of the growth of these cities, the mechanisms developed for coping with overload were based on considerations other than pure efficiency. Thus the manners, norms, and facilities of Paris and Vienna continue to reflect aesthetic values and the idealization of leisure.

COGNITIVE MAPS OF CITIES

When we speak of "behavioral comparisons" among cities, we must specify which parts of the city are most relevant for sampling purposes. In a sampling of "New Yorkers," should we include residents of Bay Ridge or Flatbush as well as inhabitants of Manhattan? And, if so, how should we weight our sample distribution? One approach to defining relevant boundaries in sampling is to determine which areas form the psychological or cognitive core of the city. We weight our samples most heavily in the areas considered by most people to represent the "essence" of the city.

The psychologist is less interested in the geographic layout of a city or in its political boundaries than in the cognitive representation of the city. Hans Blumenfeld[20] points out that the perceptual structure of a modern city can be expressed by the "silhouette" of the group of skyscrapers at its center and that of smaller groups of office buildings at its "subcenters," but that urban areas can no longer, because of their vast extent, be experienced as fully articulated sets of streets, squares, and space.

In *The Image of the City*[21], Kevin Lynch created a cognitive map of Boston by interviewing Bostonians. Perhaps his most significant finding was that, while certain landmarks, such as Paul Revere's house and the Boston Common, as well as the paths linking them, are known to almost all Bostonians, vast areas of the city are simply unknown to its inhabitants.

Using Lynch's technique, Donald Hooper[22] created a psychological map of New York from the answers to the study questionnaire on Paris, London, and New York. Hooper's results were similar to those of Lynch: New York appears to have a dense core of well-known landmarks in midtown Manhattan, surrounded by the vast unknown reaches of Queens, Brooklyn, and the Bronx. Times Square, Rockefeller Center, and the Fifth Avenue department stores alone comprise half the places specifically cited by respondents as the haunts in which they spent most of their time. However, outside the midtown area, only scattered landmarks were recognized. Another interesting pattern is evident: even the best-known symbols of New York are relatively self-contained, and the pathways joining them appear to be insignificant on the map.

The psychological map can be used for more than just sampling techniques. Lynch[21] argues, for instance, that a good city is highly "imageable," having many known symbols joined by widely known pathways, whereas dull cities are gray and nondescript. We might test the relative "imagibility" of several cities by determining the proportion of residents who recognize sampled geographic points and their accompanying pathways.

If we wanted to be even more precise, we could construct a cognitive map that would not only show the symbols of the city, but would measure the precise degree of cognitive significance of any given point in the city relative to any other. By applying a pattern of points to a map of New York City, for example, and taking photographs from each point, we could determine what proportion of a sample of the city's inhabitants could identify the locale specified by each point (see Figure 2). We might even take the subjects blindfolded to a point represented on the map, then remove the blindfold and ask them to identify their location from the view around them.

One might also use psychological maps to gain insight into the differing perceptions of a given city that are held by members of its cultural subgroups, and into the manner in which their perceptions may change. In the earlier stages of life, whites and Negroes alike probably have only a limited view of the city, centering on the immediate neighborhood in which they are raised. In adolescence, however, the field of knowledge of the white teen-ager probably undergoes rapid enlargement; he learns of opportunities in midtown and outlying sections and comes to see himself as functioning in a larger urban field. But the process of ghettoization, to which the black teen-ager is subjected, may well hamper the expansion of his sense of the city. These are speculative notions, but they are readily subject to precise test.

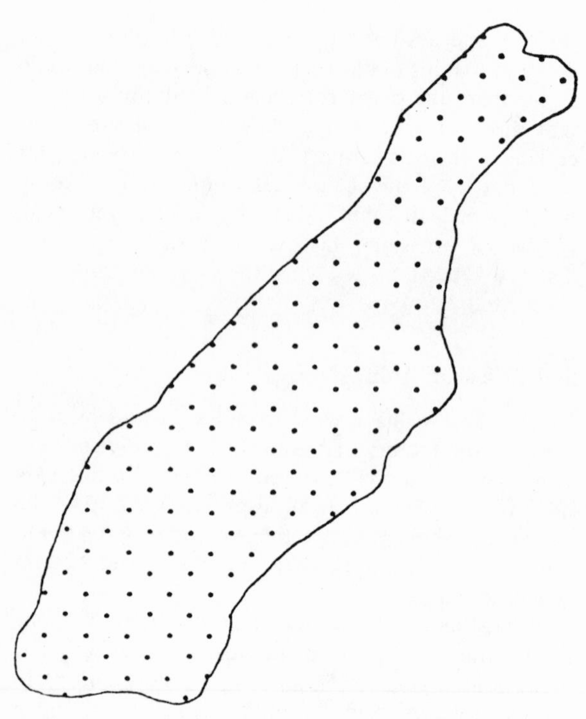

Figure 2
To create a psychological map of Manhattan, geographic points are sampled, and, from photographs, the subjects attempt to identify the location of each point. To each point a numerical index is assigned indicating the proportion of persons able to identify its location.

CONCLUSION

I have tried to indicate some organizing theory that starts with the basic facts of city life: large numbers, density, and heterogeneity. These are external to the individual. He experiences these factors as overloads at the level of roles, norms, cognitive functions, and facilities. These overloads lead to adaptive mechanisms, which create the distinctive tone and behaviors of city life. These notions, of course, need to be examined by objective comparative studies of cities and towns.

A second perspective concerns the differing atmospheres of great cities, such as Paris, London, and New York. Each has a distinctive flavor, offering a differentiable quality of experience. More precise knowledge of urban atmosphere seems attainable through application of the tools of experimental inquiry.

REFERENCES

A 51-minute film depicting the experiments described in this reading is available to educational groups. It is entitled The City and the Self *and is distributed by Time-Life Films, Inc., Time-Life Building, Rockefeller center, New York, N.Y. 10020.*

[1] *New York Times* (15 June 1969).

[2] L. Wirth, *Amer. J. Soc.* **44**, 1 (1938). Wirth's ideas have come under heavy criticism by contemporary city planners, who point out that the city is broken down into neighborhoods, which fulfill many of the functions of small towns. See, for example, H.J. Gans, *People and Plans: Essays on Urban Problems and Solutions* (Basic Books, New York, 1968); J. Jacobs, *The Death and Life of Great American Cities* (Random House, New York, 1961); G.D. Suttles, *The Social Order of the Slum* (University of Chicago Press, Chicago, 1968).

[3] G. Simmel, *The Sociology of Georg Simmel,* K.H. Wolff, ed. (Macmillan, New York, 1950) [English translation of G. Simmel, *Die Grosssstadte und das Geistesleben Die Grossstadt* (Jansch, Dresden, 1903].

[4] R.L. Meier, *A Communications Theory of Urban Growth* (MIT Press, Cambridge, Mass., 1962).

[5] S. Milgram and P. Hollander, *Nation* **25**, 602 (1964).

[6] B. Latané and J. Darley, *Amer. Sci.* **57**, 244 (1969).

[7] S. Gaertner and L. Bickman (Graduate Center, The City University of New York), unpublished research.

[8] D. Altman, M. Levine, M. Nadien, J. Villena (Graduate Center, The City University of New York), unpublished research.

[9] P.G. Zimbardo, paper presented at the Nebraska Symposium on Motivation (1969).

[10] J. Waters (Graduate Center, The City University of New York), unpublished research.

[11] W. McKenna and S. Morgenthau (Graduate Center, The City University of New York), unpublished research.

[12] N. Abuza (Harvard University), "The Paris-London-New York Questionnaires," unpublished.

[13] P. Abelson, *Science* **165**, 853 (1969).

[14] A. L. Strauss, ed., *The American City: A Sourcebook of Urban Imagery* (Aldine, Chicago, 1968).

[15] R.E. Feldman, *J. Personality Soc. Psychol.* **10**, 202 (1968).

[16] W. Berkowitz, personal communication.

[17] E.T. Hall, *The Hidden Dimension* (Doubleday, Garden City, N.Y., 1966).

[18] P. Hall, *The World Cities* (McGraw-Hill, New York, 1966).

[19] R.E. Park, E.W. Burgess, R.D. McKenzie, *The City* (University of Chicago Press, Chicago, 1967), pp. 1-45.

[20] H. Blumenfeld, in *The Quality of Urban Life* (Sage, Beverly Hills, Calif., 1969).

[21] K. Lynch, *The Image of the City* (MIT Press and Harvard University Press, Cambridge, Mass., 1960).

[22] D. Hooper (Harvard University), unpublished.

[23] Barbara Bengen worked closely with me in preparing the present version of this article. I thank Dr. Gary Winkel, editor of *Environment and Behavior,* for useful suggestions and advice.

51

Dennis L. Meadows

TOWARD A SCIENCE OF SOCIAL FORECASTING

Today the United States is searching for improved policies on energy production, population growth, rural development, income distribution, foreign relations, and other important social areas. It will take decades to work out such policies and to implement them, and decades more to assess completely their consequences. Should the policies prove ineffective or undesirable, the process of revision will require still more years.

In spite of these delays, the political and economic institutions that make most social choices are structured to give little weight to the consequences of their actions more than a few years into the future. Politicians are mainly concerned with those outcomes of their decisions that may appear before the next election. The normative and descriptive models of our economy generally disregard the delayed, nonlinear, and irreversible nature of the consequences that may derive from a policy. As a result, most of our society's actions are based implicitly upon a concern for only the next 5 to 10 years.

Selected from Proceedings of the National Academy of Sciences, Washington, D.C., *69(12), 3828-3831 (December 1972).*

This myopia is reflected in the lack of any systematic effort to develop a comprehensive view of our society's long-term evolution. The Council on Environmental Quality recently conducted an informal survey of forecasting efforts by government agencies[1]. The survey found that some agencies, such as the Council of Economic Advisors, make no attempt at comprehensive projections of social and economic changes more than 5 years into the future. Interestingly, the Soviets have assigned an economist to make comprehensive forecasts of the United States' society in the year 2000. However, this economist complained recently that he is the only specialist in his institute who has been unable to find a counterpart in the United States.

If an ocean liner takes 5 miles or more to change course, one cannot successfully steer it on the basis of information only about obstacles a few hundred yards ahead of it. Instead, radar or some other mechanism must be used to project the course and the speed of the ship well ahead of its minimum maneuvering distance. Because it may take 30 years or more to alter the course of economic, social, and political institutions, society also needs some form of projective process which will indicate necessary changes well in advance of the time they must actually be effective. We need to develop a social radar function that encompasses the full set of important interactions and whose time horizons are commensurate with the inertia of our institutions.

During the past 2 years I have directed a group of scientists and students at M.I.T. in a preliminary forecast of the long-term consequences of global population growth and economic development. Drawing on that experience, I want to:

1. Describe the cause for concern over the lack of adequate long-range social forecasting methods.
2. Indicate several minimum requirements for the new forecasting techniques we need.
3. Point to some of the unanswered questions about the mechanics and the ethics of the process through which formal models of social systems are used to influence the development of social policy.

I hope I will leave you with the impression that the development of improved methods for long-range social forecasting is a legitimate and urgent area of scientific investigation.

THREE CHARACTERISTICS OF THE GLOBAL SYSTEM

The results of our research on current growth patterns are summarized elsewhere[2]. The basis for our conclusions can be found in three dynamic characteristics of the human socioeconomic system. These characteristics emerge most clearly when the globe is viewed as a whole, but they are also true for each individual nation.

First, most physical attributes of the global system are characterized by exponential growth. Population, mineral resource consumption, pollution generation, and food production are all examples of major global elements that are currently growing exponentially at rates unprecedented in human history. Current rates of growth would lead the world's population and food production to double in about 30 years; annual rates of resource consumption and pollution generation would double in 17 and 13 years, respectively[3]. Of course, current growth may not continue at such rates. Nevertheless, it remains true that numerous social institutions promote and profit by physical growth, and it is unlikely that global growth rates will change very rapidly.

Second, the earth's finite stock of resources and the finite capacity of the ecosystem to absorb material emissions place some limits to material growth. Should the unabated growth of any physical quantity press on one of these limits, it could impose unacceptable costs on the global society. The concept of a limit to material growth is imprecise, for the nature of any limit depends in a complex fashion on the available technology, and on the magnitude, composition, and geographical distribution of the existing material flows. Nevertheless, most people accept as axiomatic the notion that no material quantity can continue to grow indefinitely on a finite planet. Any sustained material growth trend will eventually deplete a finite resource stock or precipitate the collapse of some important natural ecosystem.

To state as an axiom that material growth must stop is not, of course, to suggest that human progress must stop or even to imply that global economic production must eventually stagnate. Human activities include many functions that are not material intensive. Education, basic research, athletics, social development, and cultural activities of all kinds can continue to expand more or less indefinitely, even after the use of materials comes into balance with the finite environment.

Third, there is typically a very long delay in the effective response of society to any problem associated with material and population growth. The delay arises in several ways:

1. It may take many years for the growing quantity to cross a threshold above which its costs begin to outweigh its benefits.
2. Because our information about the functioning of complex systems is incomplete, several years may elapse before the cause of the increasing costs is perceived.
3. Since there are both costs and benefits associated with most activities, and since different individuals, institutions, or nations generally receive the costs and benefits unequally, it will often take many years to obtain agreement on the need to respond to some problem.
4. Once action is agreed upon it may take years to develop alternative technologies or to make the economic investments and institutional changes needed to reduce the magnitude of a material flow.
5. Finally, the physical and biological processes of the globe have a certain inertia. The response of the environment to a change in man's materials use is not immediate.

The global response to DDT usage illustrates the nature and the magnitude of these delays. The global use of DDT as an insecticide was initiated in 1940. It took several years for the level of use to rise to the point where observable biological damage becan to occur to species other than the target pests. Not until about 1960 was significant public attention focused on possible harmful consequences of widespread DDT use. Even in the United States, where economic and technical factors favor the use of alternative pest control methods, it has taken until 1972 to ban DDT from most uses. Poorer countries are still very far from adopting similar bans. Finally, even when DDT usage begins to decline, its levels in the marine environment will continue to rise for several years because of the delays in transmission and degradation of the chemical. One study suggests that if we were to begin decreasing the use of DDT today so that global application of the chemical declined linearly to zero by the year 2000, levels of DDT in marine fish would continue to rise for about 10 more years, would stay above current levels until 1995, and would still be present in significant amounts in the year 2020[4].

Similar delays exist in most other sectors of our global system. For example, after more than 20 years of strenuous development, fission power sources still provide only a fraction of 1 percent of the total United States power needs. Even after the average family size declines to replacement level, about two children per family, it will take the population 70 years to stabilize.

These three dynamic characteristics of social systems, a rapid rate of physical growth, limits to physical growth, and long delays in social response to changing conditions, have serious implications. Any engineer would recognize these three conditions as sufficient to introduce a pronounced tendency toward system instability. A growing system characterized by these three conditions will tend to expand beyond its ultimate limits, in a behavior mode we call "overshoot," and eventually fall back to a sustainable level. If, during the period of overshoot, the overloaded resource base is consumed, eroded, or otherwise degraded, the final sustainable level may be greatly decreased.

POSSIBLE WAYS TO REDUCE INSTABILITY

The above analysis indicates three possible ways of decreasing the instability of our growing socioeconomic system, one directed toward each of the three dynamic characteristics that produce the potential for instability.

One approach is to raise the effective limits to population and economic growth through technologies that allow more efficient use of resources or create less harmful impact on the environment. Such technologies can easily be envisioned and some have been developed to meet other goals and have tended to be environmentally destructive. Raising effective limits by technological advance does not make the system inherently more stable; in fact, it may ultimately make the magnitude and the consequences of an overshoot more severe. However, technologies that conserve materials would provide time to make more permanent system adjustments; therefore, such technologies are to be encouraged.

A second approach to system stability would be to decrease the rate of physical growth, through deliberate social and economic changes. The goal of such a process would be a stable population and an economy based on a constant flow of energy and materials. Ultimately, ending growth is the only viable policy on a finite planet, but it is an extremely long-term policy, which can only be planned and implemented on a time scale of 50 to 100 years. Thus, while this approach should be adopted, it must be augmented.

A third way to decrease the probability and magnitude of a physical overshoot is to reduce the length of system delays. This approach has only a limited range of application, since many physical and biological delays are fixed – for example, the time it takes for populations to age, pollutants to be degraded, or radioactive materials to decay are essentially outside of our control.

However, many *social* delays could be circumvented if policies could be based on anticipation of social needs, rather than on responses to them. Today, we typically evaluate a policy by examining the costs and benefits it has yielded in the past. Were we to shift from historical analysis to projective planning, we would instead ask what costs and benefits could be expected 20 or more years from now if a new policy were implemented. With the availability of suitable projective techniques for evaluating the future costs and benefits of a decision, the projective approach would serve to decrease many of the social response delays in the system.

Of course tentative efforts are already being made to develop improved forecasting methods. There is a significant difference between the "try it and see" attitude implicit in the decision to use DDT back in 1940 and environmental impact statements we now require before certifying new pesticides. Unfortunately, in spite of recent efforts, most areas of social decision making still use no formal long-range assessment techniques in choosing among alternative policies. This is true, at least in part, because we have no generally accepted techniques for projecting the social consequences of our policies. Therefore, given the fact that our policies do have important, long-term implications for our social, economic, and ecological system, we should assign a high priority to developing the appropriate forecasting methodologies.

REQUIREMENTS FOR A NEW FORECASTING METHODOLOGY

To be useful, new social forecasting methods must have several features. First, they must be able to integrate into one conceptual framework information that ranges in precision from intuitive perceptions to controlled measurements of physical systems. There already exists a great amount of information about the determinants of long-term societal evolution. However, our confidence in the various pieces of that information varies widely, perhaps by several orders of magnitude. Many of our current analytical methods require data that are more numerous or more precise than those typically available. These methods are thus unable to deal with many of the more important long-term problems.

If we wish to understand the behavior of total systems, we cannot ignore several relevant areas simply because the data are in a form that cannot be handled by our particular methods of analysis. Social changes come through the interaction of demographic, economic, technical, cultural, and other factors. When we ignore elements in one or more of these areas, we may overlook the fundamental cause of the problem. Instead, we should incorporate into our studies the best information available, whatever its form or precision. Of course, care must also be taken to test the potential impact of errors in the data on the conclusions derived from the analysis.

Second, the analytical frameworks we need should provide a neutral vocabulary that permits professionals from many different fields to cooperate directly in pooling their knowledge. No demographer, no economist, no political scientist, no engineer can make by himself the forecasts we need. The behavior of social systems comes from the interactions among variables that are included within the boundaries of many different traditional disciplines. To study short-run phenomena, professionals in one discipline often can usefully consider most of the influences from factors outside their discipline to be exogenous or constant, and then they can restrict their study only to the factors within their area of specialty. Over the longer run, however, the interactions of real systems fail to confine themselves to man's artificial boundaries. Interdisciplinary factors are not constant and cannot be excluded from explicit analysis. Unfortunately, the vocabulary, paradigms, and analytical methods of one profession are generally not shared by others. Interdisciplinary research has often taken the form of one person acquiring information from many others, then attempting to distill out everything relevant to a particular problem, and finally incorporating the accumulated information into an analytical methodology from his own field. The inescapable preconceptions and values inherent in any one professional's outlook on the world make that approach far less than an optimal procedure. We need general methodologies that can be understood and used by people from many different professions.

Third, we require a philosophy of system structure that acknowledges the complexity, the nonlinearities, the delays, and the tenuous causal relationships that determine the behavior of real-world systems. Associated with that philosophy should be analytical techniques that can accommodate such mathematically difficult relationships.

Finally, we need a new theory of inference and a set of formal techniques that will permit analysts to assess the confidence that may be assigned to conclusions derived from complex, nonlinear, underidentified, simulation models. At the moment, we have well-developed procedures for assessing confidence intervals only for the results of a limited set of models based on some rigid mathematical prerequisites. These procedures cannot be applied to the nonlinear and complex models needed to represent the total behavior of large social systems.

In the absence of formal validation techniques, we rely on the qualitative and subjective interpretation of sensitivity analyses to determine whether the results of a model study are insensitive to error. When it is found that no member of a reasonable set of changes in underlying assumptions leads to different conclusions, then it is declared that more confidence can be placed in the model. This is not a satisfying mode of validation for someone who has come from a background in the physical sciences, and it is a poor basis for social choice. The validation techniques we need are probably some combination of formal control engineering methods with the statistical formalisms that have been developed in the field of econometrics.

It is important to recognize that methods which will meet the above requirements are not simple extensions of techniques currently used in the natural sciences. The natural sciences have advanced through reductionism by isolation of individual elements of a system, control of exogenous influences on their behavior, and then systematic variation of a few "independent" variables to measure their influence. There are very few truly "independent" variables in social systems, and the problems of interest often come from the simultaneous interaction of all the parts. The techniques we require must be holistic – based on the recognition that social change does not depend on the attributes of a single factor, but on the interaction of many.

MODELS AND THE PROCESS OF SOCIAL CHOICE

Even when we have developed modeling methods with the above characteristics, we will not have provided society with an effective social radar. The nature of human and social decision making will prevent any form of formal model from directly making decisions. Policies are set and decisions are made through the interaction of many individuals as they assess the possible consequences of alternative actions and compare the expected costs and benefits of each possible outcome. Formal models can only assist in the first half of that decision process, establishing what *could* be. Deciding what *should* be – i.e., exercising social values – will remain the prerogative of individuals and the institutions that represent them. Thus we must view the nature of social decision making, even with vastly better models than are available today, as a process through which the formal model of the system analyst complements the value system and verbal models of the decision maker. Using verbal models, social policy makers may identify appropriate problems. With formal models, system analysts may determine the possible consequences of alternative responses to the problem. Finally, the subjective judgment of those involved in the outcome can be used in picking from the possible outcomes those that most satisfy social goals. This cooperative process between verbal and formal models involves many stages that are still poorly understood. Thus an important focus of research on improved models would be the process of model development and use. . . .

Until recently it has been possible for most individuals, particularly natural scientists, to regard social systems as outside the realm of predictive science. However, the availability of computers that can handle complex models and the urgent need to put current actions in the context of their future consequences suggest we should change that attitude. Relatively little effort is now invested in the development of long-term forecasting methodologies, yet the effectiveness of many current decisions will be impaired without a longer-term view.

REFERENCES

[1] Council on Environmental Quality (ed.) (1972) *Summary of Long-Range Forecasting Activities Performed by the Federal Agencies,* Aug 11, 1972, Interagency memorandum.

[2] Meadows, D.H. et al. (1972) *The Limits to Growth* (Universe Books, New York); Meadows, D.L., Meadows, D.H. (eds.) (1972) *Toward Global Equilibrium – Collected Papers* (Wright-Allen Press, Boston, Mass.), in press; Meadows, D.L., et al. (1972) *The Dynamics of Growth in a Finite World,* forthcoming.

[3] The U.S. Bureau of Mines projects that the world's primary demand for minerals may increase annually by 3.6 to 5.5 percent. This forecast is equivalent to a doubling time of from 20 to 13 years. U.S. Bureau of Mines (1970) *Mineral Facts and Problems* (Government Printing Office, Washington, D.C.), p. 3; [Time series on pollution emissions are poor or nonexistent in most cases. However, energy production and, therefore, thermal emissions are crude but useful indices of overall pollution. Thermal wastes were projected by the SCEP study to increase globally by 5.7 percent per year. (1970) *Man's Impact on the Global Environment* (MIT Press), p. 64.]

[4] Randers, J. (1972) "System Simulation to Test Environmental Policy: A Sample Study of DDT Movement in the Environment," *International Journal of Environmental Studies,* London, England, Nov. 1972.

AUTHOR IDENTIFICATION

1. *Leonard Reissman:* Professor of Sociology and Human Relations, Tulane University, New Orleans, Louisiana.
2. *Sir Ebenezer Howard (d. 1928):* Author, founder of the British new towns movement and the towns of Letchworth and Welwyn, England.
3. *Clarence A. Perry (d. 1944):* Social Scientist, New York City.
4. *Robert B. Mitchell:* Emeritus Professor of Urban and Regional Planning, University of Pennsylvania, Philadelphia, Pennsylvania.
5. *Richard Babcock:* Attorney at Law, Chicago, Illinois.
6. *Paul D. Spreiregan, A.I.A.:* Washington, D.C.
7. *Kevin Lynch:* Professor of Planning and Urban Studies, Massachusetts Institute of Technology, Cambridge, Massachusetts.
8. *Eliel Saarinen (d. 1950):* Architect and Town Planner, Bloomfield Hills, Michigan.
9. *Albert Z. Guttenberg:* Professor of Urban and Regional Planning, University of Illinois, Urbana-Champaigne, Illinois.
10. *F. Stuart Chapin, Jr.:* Professor of Planning, University of North Carolina, Chapel Hill, North Carolina.
11. *Edwin von Böventer:* Professor of Economics, University of Munich, Munich, Federal Republic of Germany.
12. *Ida R. Hoos:* Research Sociologist, University of California, Berkeley, California.
13. *Andrew Vazsonyi:* Professor of Management, University of Rochester, Rochester, New York.
14. *John Dearden:* Professor of Business Administration, Harvard Business School, Boston, Massachusetts.
15. *Alan Black:* Associate, Creighton Hamburg Inc., Delmar, New York.
16. *Rexford G. Tugwell:* Fellow, Center for the Study of Democratic Institutions, Santa Barbara, California.
17. *E.S. Savas:* Professor of Public Systems Management, Columbia University, New York City.
18. *Melville C. Branch:* Professor of Planning and Urban Studies, University of Southern California, Los Angeles, California.
19. *Charles J. Hitch:* President, Resources for the Future Inc., Washington, D.C.
20. *Peter Wood:* Merseyside Structure Plan, Liverpool, England.
21. *Melville C. Branch* (18 above).
22. *James Hughes:* South Bend, Indiana.
    Lawrence D. Mann: Professor of City and Regional Planning, Harvard University, Cambridge, Massachusetts.
23. *Richard A. Johnson:* Professor of Management and Organization, University of Washington, Seattle, Washington.
    Fremont E. Kast: Professor of Management and Organization, University of Washington, Seattle, Washington.
    James E. Rosenzweig: Professor of Management and Organization, University of Washington, Seattle, Washington.
24. *Jay W. Forrester:* Professor of Management, Massachusetts Institute of Technology, Cambridge, Massachusetts.
25. *Jay W. Forrester* (24 above).
26. *Harper Q. North:* Associate Director of Research for Electronics, Naval Research Laboratory, Washington, D.C.
    Donald L. Pyke: Coordinator, Academic Planning, University of Southern California, Los Angeles, California.
27. *Alan J. Rowe:* Professor of Management, University of Southern California, Los Angeles, California.
28. *Richard L. Van Horn:* Professor of Industrial Administration, Carnegie-Mellon University, Pittsburgh, Pennsylvania.
29. *Ira S. Lowry:* Management Sciences Department, Rand Corporation, Santa Monica, California.
30. *Martin Shubik:* Professor of the Economics of Organization, Yale University, New Haven, Connecticut.

31. *Melville C. Branch* (18 above).
32. *David L. Johnson:* Professor of Electrical Engineering and Computer Science, University of Washington, Seattle, Washington.
    Arthur L. Kobler: Clinical Professor of Psychology, University of Washington, Seattle, Washington.
33. *Henry Mintzberg:* Faculty of Management, McGill University, Montreal, Quebec, Canada.
34. *Samuel G. Trull:* Professor of Management, California State University, San Francisco, California.
35. *James H. Pickford:* Director of Program Development, Appalachian Regional Commission, Washington, D.C.
36. *Robert A. Walker:* Professor of Political Science, Stanford University, Stanford, California.
37. *Rexford G. Tugwell* (16 above).
38. *Charles Abrams (d. 1970):* Professor, Division of Urban Planning, School of Architecture, Columbia University, New York City.
39. *Edmund M. Burke:* Professor of Social Work, Boston College, Chestnut Hill, Massachusetts.
40. *Ian L. McHarg:* Professor of Landscape Architecture and Regional Planning, University of Pennsylvania, Philadelphia, Pennsylvania.
41. *George Macinko:* Professor of Geography, Central Washington State College, Ellensburg, Washington.
42. *Irving Rosow:* Professor, Langley Porter Neuropsychiatric Institute, University of California, San Francisco, California.
43. *Bernard Asbell:* Author and Consultant, Guilford, Connecticut.
44. *Richard S. Bolan:* Associate Professor of Social Work, Boston College, Chestnut Hill, Massachusetts.
45. *Charles E. Lindblom:* Professors of Economics, Yale University, New Haven, Connecticut.
46. *Dennis A. Rondinelli:* Assistant Professor of Business Management, Vanderbilt University, Nashville, Tennessee.
47. *Beryl L. Crowe:* Member of the Faculty, The Evergreen State College, Olympia, Washington.
48. *Walter J. Francis:* U.S. Department of Health, Education, and Welfare, Washington, D.C.
49. *Robert M. Griffin, Jr.:* Associate Professor of Environmental Planning, Pennsylvania State University, University Park, Pennsylvania.
50. *Stanley Milgram:* Professor of Psychology, City University of New York, New York City.
51. *Dennis L. Meadows:* Assistant Professor of Engineering, Dartmouth College, Hanover, New Hampshire.

SUBJECT INDEX

AUTHORS AND INDIVIDUALS REFERENCED